TODAY'S TECHNICIAN™

CLASSROOM MANUAL
FOR AUTOMOTIVE SUSPENSION & STEERING SYSTEMS

TODAY'S TECHNICIAN™

CLASSROOM MANUAL FOR AUTOMOTIVE SUSPENSION & STEERING SYSTEMS

SIXTH EDITION

Mark Schnubel

CENGAGE Learning®

Australia • Brazil • Mexico • Singapore • United Kingdom • United States

Today's Technician™: Automotive Suspension & Steering Systems, 6th Edition

Mark Schnubel

VP, General Manager, Skills and Planning: Dawn Gerrain

Director, Development, Global Product Management, Skills: Marah Bellegarde

Product Manager: Erin Brennan

Senior Product Development Manager: Larry Main

Senior Content Developer: Meaghan Tomaso

Product Assistant: Scott Royael

Marketing Manager: Linda Kuper

Market Development Manager: Jonathon Sheehan

Senior Production Director: Wendy Troeger

Production Manager: Mark Bernard

Content Project Manager: Christopher Chien

Art Director: GEX

Media Developer: Debbie Bordeaux

Cover Image(s): © cla78/Shutterstock

Library of Congress Control Number: 2014930382

ISBN-13: 978-1-285-43812-2

ISBN-10: 1-285-43812-4

Cengage Learning
200 First Stamford Place, 4th Floor
Stamford, CT 06902
USA

Printed in the United States of America
2 3 4 5 6 7 18 17 16 15

CONTENTS

CONTENTS

CONTENTS

CONTENTS

PREFACE

Thanks to the support the *Today's Technician* series has received from those who teach automotive technology, Cengage Learning, the leader in automotive-related textbooks, is able to live up to its promise to regularly provide new editions of texts of this series. We have listened and responded to our critics and our fans and present this new updated and revised sixth edition. By revising this series on a regular basis, we can respond to changes in the industry, changes in technology, changes in the certification process, and to the ever-changing needs of those who teach automotive technology.

We also listened to instructors who said something was missing or incomplete in the last edition. We responded to those and the results are included in this sixth edition.

The *Today's Technician* series features textbooks that cover all mechanical and electrical systems of automobiles and light trucks. Principally, the individual titles correspond to the certification areas for 2013 in areas of National Institute for Automotive Service Excellence (ASE) certification.

Additional titles include remedial skills and theories common to all of the certification areas and advanced or specific subject areas that reflect the latest technological trends.

This new edition, like the last, was designed to give students a chance to develop the same skills and gain the same knowledge that today's successful technician has. This edition also reflects the changes in the guidelines established by the National Automotive Technicians Education Foundation (NATEF) in 2013.

The purpose of NATEF is to evaluate technician training programs against standards developed by the automotive industry and recommend qualifying programs for certification (accreditation) by ASE. Programs can earn ASE certification upon the recommendation of NATEF. NATEF's national standards reflect the skills that students must master. ASE certification through NATEF evaluation ensures that certified training programs meet or exceed industry-recognized, uniform standards of excellence.

The technician of today and for the future must know the underlying theory of all automotive systems and be able to service and maintain those systems. Dividing the material into two volumes, a Classroom Manual and a Shop Manual, provides the reader with the information needed to begin a successful career as an automotive technician without interrupting the learning process by mixing cognitive and performance learning objectives into one volume.

The design of Cengage Learning's *Today's Technician* series was based on features that are known to promote improved student learning. The design was further enhanced by a careful study of survey results, in which the respondents were asked to value particular features. Some of these features can be found in other textbooks, while others are unique to this series.

Each Classroom Manual contains the principles of operation for each system and subsystem. The Classroom Manual also contains discussions on design variations of key components used by the different vehicle manufacturers. It also looks into emerging technologies that will be standard or optional features in the near future. This volume is organized to build upon basic facts and theories. The primary objective of this volume is to allow the reader to gain an understanding of how each system and subsystem operates. This understanding is necessary to diagnose the complex automobiles of today and tomorrow. Although the basics contained in the Classroom Manual provide the knowledge needed for diagnostics, diagnostic procedures appear only in the Shop Manual. An understanding of the underlying theories is also a requirement for competence in the skill areas covered in the Shop Manual.

A spiral-bound Shop Manual covers the "how-tos." This volume includes step-by-step instructions for diagnostic and repair procedures. Photo Sequences are used to illustrate some of the common service procedures. Other common procedures are listed and are accompanied with fine line drawings and photos that allow the reader to visualize and conceptualize the finest details of the procedure. This volume also contains the reasons for performing the procedures, as well as when that particular service is appropriate.

The two volumes are designed to be used together and are arranged in corresponding chapters. Not only are the chapters in the volumes linked together, the contents of the chapters are also linked. This linking of content is evidenced by marginal callouts that refer the reader to the chapter and page that the same topic is addressed in the other volume. This feature is valuable to instructors. Without this feature, users of other two-volume textbooks must search the index or table of contents to locate supporting information in the other volume. This is not only cumbersome but also creates additional work for an instructor when planning the presentation of material and when making reading assignments. It is also valuable to the students; with the page references, they also know exactly where to look for supportive information.

Both volumes contain clear and thoughtfully selected illustrations. Many of which are original drawings or photos specially prepared for inclusion in this series. This means that the art is a vital part of each textbook and not merely inserted to increase the number of illustrations.

The page layout, used in the series, is designed to include information that would otherwise break up the flow of information presented to the reader. The main body of the text includes all of the "need-to-know" information and illustrations. In the wide side margins of each page are many of the special features of the series. Items that are truly "nice-to-know" information such as: simple examples of concepts just introduced in the text, explanations or definitions of terms that are not defined in the text, examples of common trade jargon used to describe a part or operation, and exceptions to the norm explained in the text. This type of information is placed in the margin, out of the normal flow of information. Many textbooks attempt to include this type of information and insert it in the main body of text; this tends to interrupt the thought process and cannot be pedagogically justified. By placing this information off to the side of the main text, the reader can select when to refer to it.

Jack Erjavec
Series Editor

Highlights of this Edition—Classroom Manual

The text was updated to include the latest technology in suspension and steering systems. Updated and expanded coverage of basic electrical theory and hybrid electric vehicle safety, hybrid vehicle steering systems, active steering systems, rear active steering (RAS), CAN bus networking, computer-controlled suspension systems, and adaptive cruise control systems. Expanded coverage of tires and wheels including tire plus sizing and rim offset considerations as well as expanded alignment theory coverage and sequencing of information. The text also includes the latest technology in vehicle stability control systems, traction control systems, active roll control, lane departure warning (LDW) systems, collision mitigation systems, telematics, and tire pressure monitoring systems (TPMS).

PREFACE

The first chapter explains the design and purpose of basic suspension and steering systems. This chapter provides students with the necessary basic understanding of suspension and steering systems. The other chapters in the book allow the student to build upon his or her understanding of these basic systems.

The second chapter explains all the basic theories required to understand the latest suspension and steering systems described in the other chapters. Students must understand these basic theories to comprehend the complex systems explained later in the text.

The other chapters in the book are designed to be stand alone to allow reordering of topics covered to fit individual program needs and explain all the current model systems and components such as wheel bearings, tires and wheels, shock absorbers and struts, front and rear suspension systems, computer-controlled suspension systems, steering columns and linkages, power steering pumps, steering gears and systems, four-wheel steering systems, frames, and four-wheel alignment. All of the art pieces have been replaced with color photos and color diagrams throughout the text to improve visual concepts of suspension and steering systems and components.

Highlights of this Edition—Shop Manual

The chapters in the Shop Manual have been updated to explain the diagnostic and service procedures for the latest systems and components described in the Classroom Manual. Diagnostics is a very important part of an automotive technician's job. Therefore, proper diagnostic procedures are emphasized in the Shop Manual.

All Photo Sequences are now in color. These Photo Sequences illustrate the correct diagnostic or service procedure for a specific system or component. These Photo Sequences allow the students to visualize the diagnostic or service procedure. Visualization of these diagnostic and service procedures helps students to remember the procedures, and perform them more accurately and efficiently. The text covers the information required to pass an ASE test in Suspension and Steering Systems. New and updated Job Sheets have been created to meet current NATEF tasks.

Chapter 1 explains the necessary safety precautions and procedures in an automotive repair shop. General shop safety and the required shop safety equipment are explained in the text. The text describes safety procedures when operating vehicles and various types of automotive service equipment. Correct procedures for handling hazardous waste materials are detailed in the text.

Chapter 2 describes suspension and steering diagnostic and service equipment and the use of service manuals. This chapter also explains employer and employee obligations and ASE certification requirements.

The other chapters in the text have been updated to explain the diagnostic and service procedures for the latest suspension and steering systems explained in the Classroom Manual. New job sheets related to the new systems and components have been added in the text. All of the art pieces have been replaced with color photos and color diagrams throughout the text to improve the student's visualization of diagnostic and service procedures.

Mark Schnubel

Features of the Classroom Manual include the following:

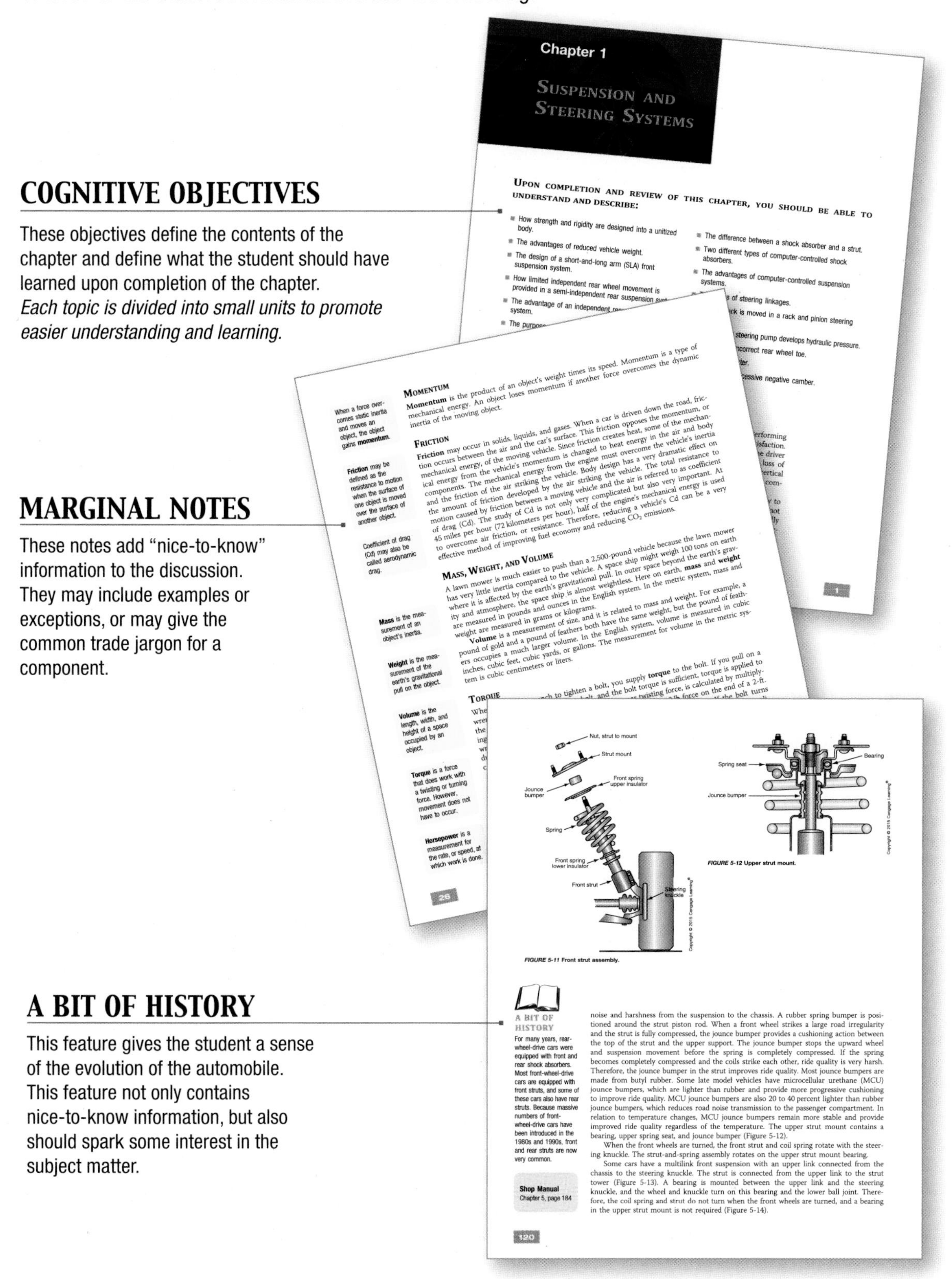

COGNITIVE OBJECTIVES

These objectives define the contents of the chapter and define what the student should have learned upon completion of the chapter. *Each topic is divided into small units to promote easier understanding and learning.*

MARGINAL NOTES

These notes add "nice-to-know" information to the discussion. They may include examples or exceptions, or may give the common trade jargon for a component.

A BIT OF HISTORY

This feature gives the student a sense of the evolution of the automobile. This feature not only contains nice-to-know information, but also should spark some interest in the subject matter.

AUTHOR'S NOTES

This feature includes simple explanations, stories, or examples of complex topics. These are included to help students understand difficult concepts.

CAUTIONS AND WARNINGS

Throughout the text, warnings are given to alert the reader to potentially hazardous materials or unsafe conditions. Cautions are given to advise the student of things that can go wrong if instructions are not followed or if a nonacceptable part or tool is used.

CROSS-REFERENCES TO THE SHOP MANUAL

Reference to the appropriate page in the Shop Manual is given whenever necessary. Although the chapters of the two manuals are synchronized, material covered in other chapters of the Shop Manual may be fundamental to the topic discussed in the Classroom Manual.

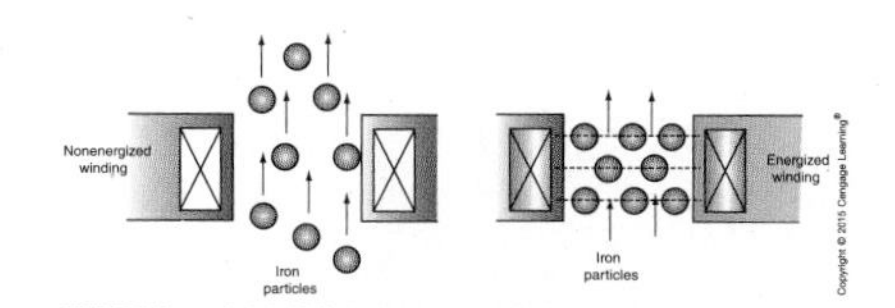

FIGURE 5-24 Magneto-rheological fluid action in a strut or shock absorber.

jelly-like consistency for a firm ride (Figure 5-24). The computer can change the shock absorber damping characteristics almost instantaneously, in 1 millisecond, and can also supply a varying amount of current through the shock absorber windings to provide a wide variety of shock absorber damping characteristics. Depending on the amount of current supplied, the fluid's viscosity or resistance to flow can be varied from thinner than water to a near plastic or solid state or any consistency in-between. This makes the fluid infinitely adjustable in viscosity and achieves continuously variable real-time damping to suit almost every driving condition instantly.

The advantages of magneto-rheological fluid-controlled shock absorber are:

- Ability to smooth out the action of each tire.
- Reduced noise, bounce, and vibration giving a flatter ride by controlling body motions.
- Exceptional roll control during evasive steering maneuvers.
- Excellent handling by controlling weight transfer during lateral and longitudinal maneuvers.
- Enhanced road isolation by reducing high-frequency road noise through the dampers.
- When integrated with ABS and traction control, ensures maximum stability and balance on slippery surfaces and gravel.

SUMMARIES

Each chapter concludes with a summary of key points from the chapter. These are designed to help the reader review the chapter contents.

SUMMARY

- Shock absorbers or struts play a very important role in ride quality, steering control, and tire life.
- Tire and wheel jounce travel occurs when a tire strikes a bump in the road surface and the tire and wheel move upward.
- Rebound tire and wheel travel occurs when the tire and wheel move downward after jounce travel.
- When a spring is deflected upward during jounce travel, it stores energy. The spring then expands downward in rebound travel with all the energy it stored during the jounce travel. If the spring action is not controlled, the energy in the spring during rebound travel drives the tire against the road surface with excessive force. This action drives the tire and wheel back upward in jounce travel and the wheel continues oscillating up and down.
- The shock absorb... control spring action and prevent excessive tire and wheel oscillations.

TERMS TO KNOW
Active suspension systems
Adjustable struts
Gas-filled shock absorber
Heavy-duty shock absorbers
Jounce travel
Load-leveling shock absorbers
Magneto-rheological fluid

TERMS TO KNOW LIST

A list of new terms appears next to the Summary.

TERMS TO KNOW (continued)
Rebound travel
Shock absorbers
Shock absorber ratios
Spring insulator
Struts
Travel-sensitive struts
Upper strut mount

- During jounce travel, the piston moves downward in the lower shock absorber chamber; during rebound travel, this piston moves upward. Because the lower shock absorber chamber is sealed and filled with oil, this oil must flow past the piston during any piston movement.
- Valves and openings in the shock absorber piston provide precision control of the oil flow past the piston to control spring action. Shock absorber valves are matched to the amount of energy that may be stored in the spring.
- A nitrogen gas charge is located in the oil reservoir of many shock absorbers and struts to prevent oil cavitation or foaming, which provides more positive shock absorber action.
- Shock absorber ratio refers to the difference between the shock absorber control on the compression and extension cycle. Many shock absorbers provide more control on the extension cycle.
- Internal design is similar in shock absorbers and struts, but struts also support the coil spring.
- Most front struts are connected between the steering knuckle and the upper strut mount.
- Many rear struts are connected between the spindle and the upper support.
- A travel-sensitive shock absorber provides increased resistance to piston movement as the shock absorber is extended.
- Adjustable shock absorbers and struts have a manual adjustment that allows the technician or owner to adjust the strut orifice opening.
- Load-leveling shock absorbers have air pumped into the shock absorbers from an onboard compressor to maintain a specific rear suspension height regardless of the rear suspension load.
- Most front struts rotate on the upper strut mount bearing as the front wheels are turned.

REVIEW QUESTIONS

Short answer essay, fill-in-the-blank, and multiple-choice questions are found at the end of each chapter. These questions are designed to accurately assess the student's competence in the stated objectives at the beginning of the chapter.

REVIEW QUESTIONS

Short Answer Essays

1. Describe spring action during jounce and rebound travel.
2. Describe uncontrolled spring action without a shock absorber.
3. Explain shock absorber operation.
4. Describe the vehicle safety hazards created by worn-out shock absorbers.
5. Explain shock absorber ratio.
6. Explain the purpose of the nitrogen gas charge in shock absorbers and struts.
7. Explain the differences between heavy-duty mono-tube and conventional twin tube shock absorbers.
8. Identify the purposes of an upper strut mount.
9. Explain the purpose of strut-and-spring insulators.
10. Describe the purpose of the rubber spring bumper on a strut piston rod.

Fill-in-the-Blanks

1. Shock absorbers control spring action and prevent excessive spring ________.
2. Shock absorber design is matched to the ________ rate of the spring.
3. Modern shock absorbers are ________ damping devices that increase resistance the faster the suspension moves.
4. The nitrogen gas charge in a shock absorber prevents oil ________ and ________.
5. The single tube design in a heavy-duty shock absorber prevents excessive ________.
6. A shock absorber with a 70/30 ratio provides more control on the ________ cycle.
7. The lower end of many front struts is bolted to the steering ________.

128

SHOP MANUAL

To stress the importance of safe work habits, the Shop Manual also dedicates one full chapter to safety. Other important features of this manual include:

PERFORMANCE-BASED OBJECTIVES

These objectives define the contents of the chapter and define what the student should have learned upon completion of the chapter. These objectives also correspond with the list of required tasks for NATEF certification. *Each NATEF task is addressed.*

Although this textbook is not designed to simply prepare someone for the certification exams, it is organized around the NATEF task list. These tasks are defined generically when the procedure is commonly followed and specifically when the procedure is unique for specific vehicle models. Imported and domestic model automobiles and light trucks are included in the procedures.

BASIC TOOLS LISTS

Each chapter begins with a list of the basic tools needed to perform the tasks included in the chapter.

PHOTO SEQUENCES

Many procedures are illustrated in detailed Photo Sequences. These detailed photographs show the students what to expect when they perform particular procedures. They also can provide the student a familiarity with a system or type of equipment, which the school may not have.

SERVICE TIPS

Whenever a short-cut or special procedure is appropriate, it is described in the text. These tips are generally those things commonly done by experienced technicians.

CUSTOMER CARE

This feature highlights those little things a technician can do or say to enhance customer relations.

SPECIAL TOOLS LISTS

Whenever a special tool is required to complete a task, it is listed in the margin next to the procedure.

CROSS-REFERENCES TO THE CLASSROOM MANUAL

Reference to the appropriate page in the Classroom Manual is given whenever necessary. Although the chapters of the two manuals are synchronized, material covered in other chapters of the Classroom Manual may be fundamental to the topic discussed in the Shop Manual.

MARGINAL NOTES

These notes add "nice-to-know" information to the discussion. They may include examples or exceptions, or may give the common trade jargon for a component.

CAUTIONS AND WARNINGS

Throughout the text, warnings are given to alert the reader to potentially hazardous materials or unsafe conditions. Cautions are given to advise the student of things that can go wrong if instructions are not followed or if a nonacceptable part or tool is used.

TERMS TO KNOW LIST

Terms in this list can be found in the Glossary at the end of the manual.

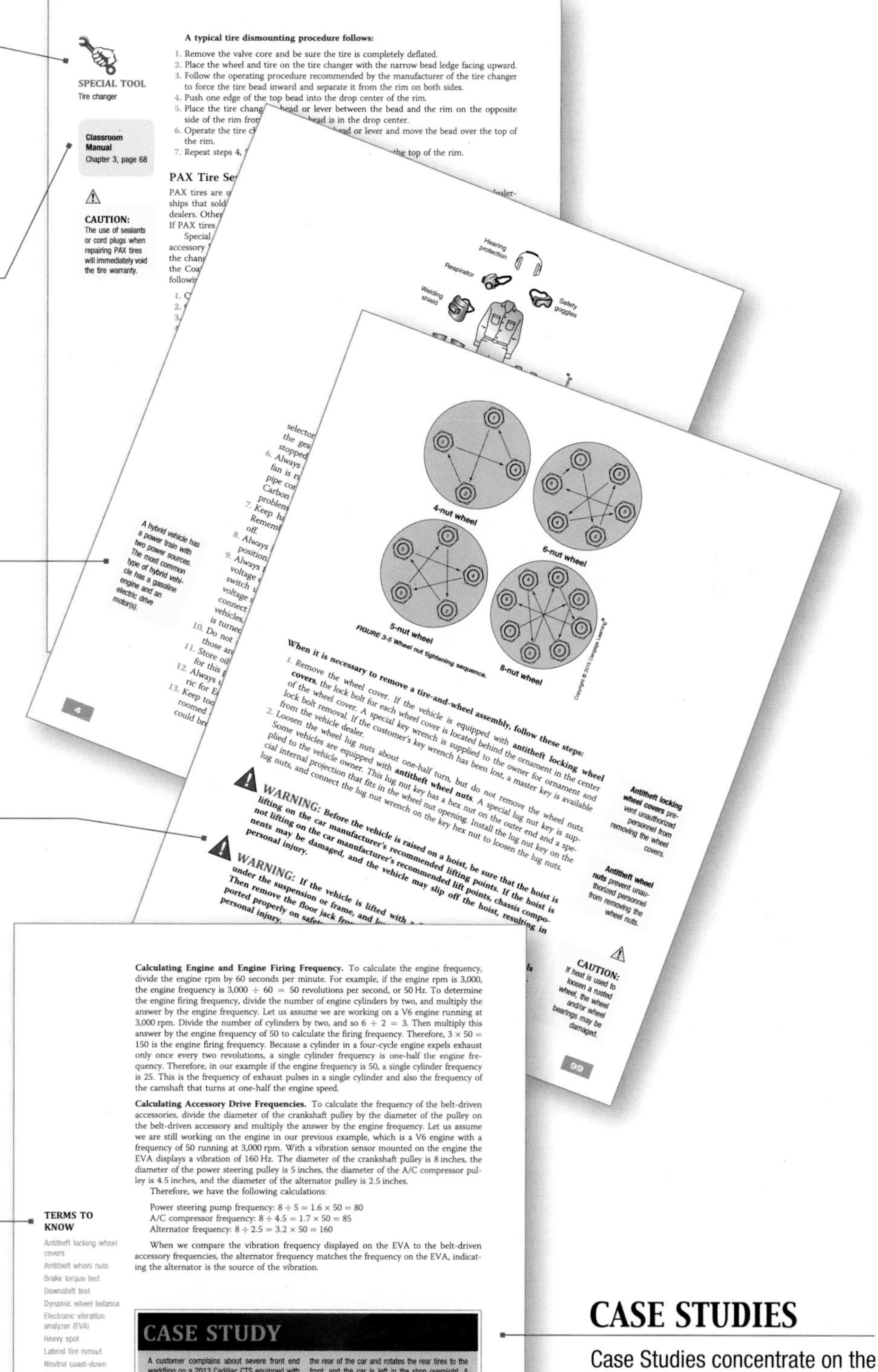

Calculating Engine and Engine Firing Frequency. To calculate the engine frequency, divide the engine rpm by 60 seconds per minute. For example, if the engine rpm is 3,000, the engine frequency is 3,000 ÷ 60 = 50 revolutions per second, or 50 Hz. To determine the engine firing frequency, divide the number of engine cylinders by two, and multiply the answer by the engine frequency. Let us assume we are working on a V6 engine running at 3,000 rpm. Divide the number of cylinders by two, and so 6 ÷ 2 = 3. Then multiply this answer by the engine frequency of 50 to calculate the firing frequency. Therefore, 3 × 50 = 150 is the engine firing frequency. Because a cylinder in a four-cycle engine expels exhaust only once every two revolutions, a single cylinder frequency is one-half the engine frequency. Therefore, in our example if the engine frequency is 50, a single cylinder frequency is 25. This is the frequency of exhaust pulses in a single cylinder and also the frequency of the camshaft that turns at one-half the engine speed.

Calculating Accessory Drive Frequencies. To calculate the frequency of the belt-driven accessories, divide the diameter of the crankshaft pulley by the diameter of the pulley on the belt-driven accessory and multiply the answer by the engine frequency. Let us assume we are still working on the engine in our previous example, which is a V6 engine with a frequency of 50 running at 3,000 rpm. With a vibration sensor mounted on the engine the EVA displays a vibration of 160 Hz. The diameter of the crankshaft pulley is 8 inches, the diameter of the power steering pulley is 5 inches, the diameter of the A/C compressor pulley is 4.5 inches, and the diameter of the alternator pulley is 2.5 inches.

Therefore, we have the following calculations:

Power steering pump frequency: 8 ÷ 5 = 1.6 × 50 = 80
A/C compressor frequency: 8 ÷ 4.5 = 1.7 × 50 = 85
Alternator frequency: 8 ÷ 2.5 = 3.2 × 50 = 160

When we compare the vibration frequency displayed on the EVA to the belt-driven accessory frequencies, the alternator frequency matches the frequency on the EVA, indicating the alternator is the source of the vibration.

TERMS TO KNOW

Antitheft locking wheel covers
Antitheft wheel nuts
Brake torque test
Downshift test
Dynamic wheel balance
Electronic vibration analyzer (EVA)
Heavy spot
Lateral tire runout
Neutral coast-down
Neutral run-up test
Ply separation
Radial runout
Slow acceleration test
Standing acceleration test
Static balance
Steering input test
Steering pull
Tire conicity
Tire thump
Tire vibration
Tread wear indicators

CASE STUDY

A customer complains about severe front end waddling on a 2013 Cadillac CTS equipped with steel-belted radial tires, which have 11/32" of tread depth. The technician questions the customer further regarding this problem and learns that the problem occurs only after the car has sat overnight, and it lasts only for approximately six blocks when the customer starts driving the car.

The technician test drove the vehicle but was unable to duplicate the customer complaint. The technician carefully checks all steering and suspension components, wheel bearing adjustment, and tire runout. No problems are found during these checks, and the technician asks the customer to leave the car in the shop overnight. The technician installs the front tires on the rear of the car and rotates the rear tires to the front, and the car is left in the shop overnight. A road test the next morning indicates rear end waddling, which proves the problem must be caused by a tire problem such as a radial cord ply imperfection, since moving the tires original position was the only thing changed on the car. The technician removed the two rear tires and checked them on a Road Force tire balancer once they cooled down again. The tire that had been moved to the left rear position was found to have excessive road force variation, which was beyond the limits of match mounting. The technician replaced the defective tire and checks the balance of the other three tires. The customer reports later that the problem has been completely corrected.

132

CASE STUDIES

Case Studies concentrate on the ability to properly diagnose the systems. Beginning with Chapter 3, each chapter ends with a case study in which a vehicle has a problem, and the logic used by a technician to solve the problem is explained.

ASE-STYLE REVIEW QUESTIONS

Each chapter contains ASE-style review questions that reflect the performance-based objectives listed at the beginning of the chapter. These questions can be used to review the chapter as well as to prepare for the ASE certification exam.

ASE CHALLENGE QUESTIONS

Each technical chapter ends with five ASE challenge questions. These are not more review questions; rather they test the students' ability to apply general knowledge to the contents of the chapter.

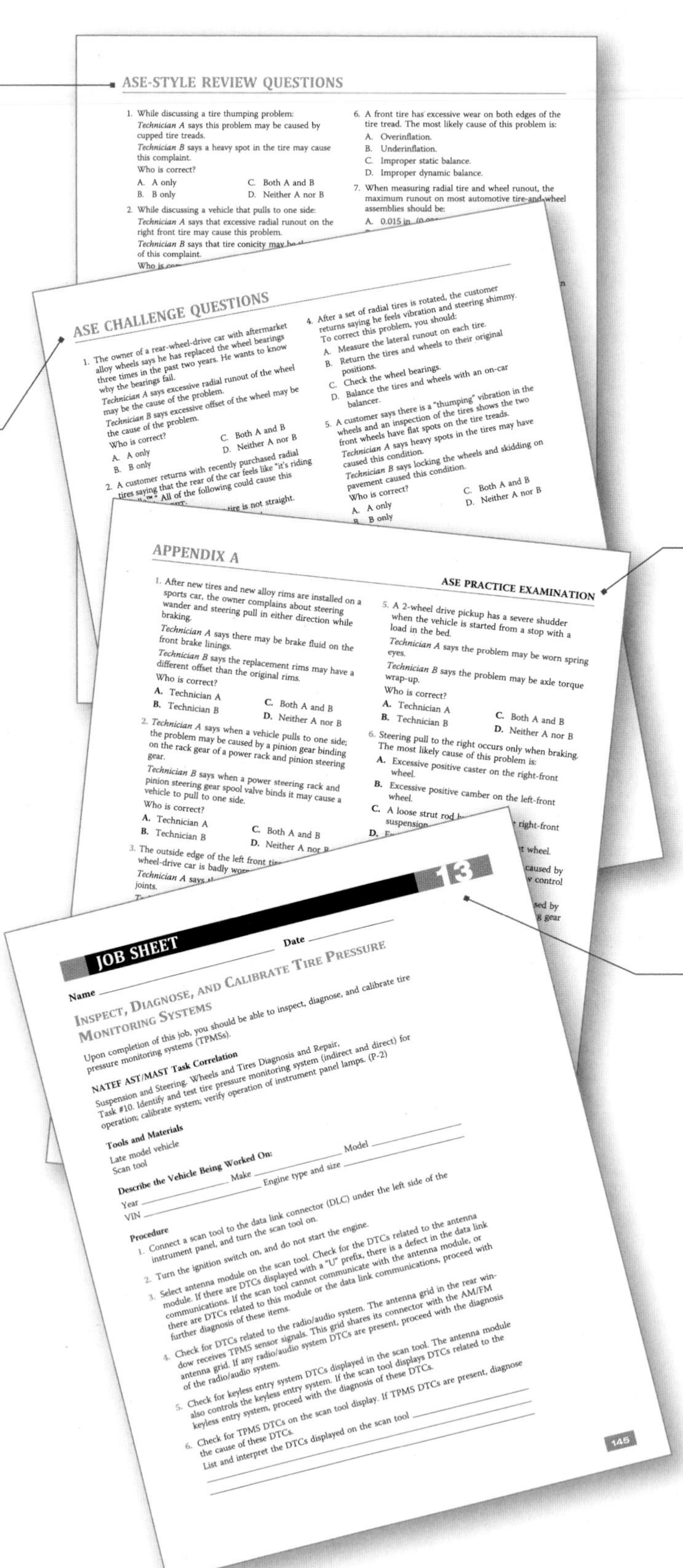

ASE PRACTICE EXAMINATION

A 50 question ASE practice exam, located in the appendix, is included to test students on the contents of the Shop Manual.

JOB SHEETS

Located at the end of each chapter, the Job Sheets provide a format for students to perform procedures covered in the chapter. A reference to the NATEF Task addressed by the procedure is referenced on the Job Sheet.

SUPPLEMENTS

INSTRUCTOR RESOURCES

The Instructor Resources, now available online, are a robust ancillary product that contains all preparation tools to meet any instructor's classroom needs. It includes chapter outlines in PowerPoint with images, video clips, and animations that coincide with each chapter's content coverage, chapter tests powered by Cognero with hundreds of test questions, a searchable Image Library with all photos and illustrations from the text, theory-based Worksheets in Word that provide homework or in-class assignments, the Job Sheets from the Shop Manual in Word, a NATEF correlation chart, and an Instructor's Guide in electronic format.

To access these Instructor Resources, go to **login.cengagebrain.com**, and create an account or log into your existing account.

COURSEMATE

The all new CourseMate for *Today's Technician: Automotive Suspension & Steering Systems*, 6e offers students and instructors access to important tools and resources, all in an online environment. The CourseMate includes an Interactive eBook of the core text, interactive quizzes, flashcards, as well as an Engagement Tracker tool for monitoring students' progress in the CourseMate product.

To learn more about this resource and access free demo CourseMate resources, go to **www.cengagebrain.com**, and search by this book's ISBN (9781285438108). To access CourseMate materials that you have purchased, go to **login.cengagebrain.com**, enter your access code, and create an account or log into your existing account.

REVIEWERS

The author and publisher would like to extend a special thanks to the instructors who reviewed this text and offered invaluable feedback:

Clay Brown
McHenry County College
Crystal Lake, IL

Steve Calvert
University of Northwestern Ohio
Lima, OH

Christopher J. Marker
University of Northwestern Ohio
Lima, OH

Chapter 1

Suspension and Steering Systems

Upon completion and review of this chapter, you should be able to understand and describe:

- How strength and rigidity are designed into a unitized body.
- The advantages of reduced vehicle weight.
- The design of a short-and-long arm (SLA) front suspension system.
- How limited independent rear wheel movement is provided in a semi-independent rear suspension system.
- The advantage of an independent rear suspension system.
- The purposes of vehicle tires.
- The terms *positive* and *negative offset* as they relate to vehicle wheel rims.
- Three different loads that are applied to wheel bearings.
- The purposes of shock absorbers.
- The difference between a shock absorber and a strut.
- Two different types of computer-controlled shock absorbers.
- The advantages of computer-controlled suspension systems.
- Two types of steering linkages.
- How the rack is moved in a rack and pinion steering gear.
- How a power steering pump develops hydraulic pressure.
- The result of incorrect rear wheel toe.
- Front wheel caster.
- The results of excessive negative camber.

Introduction

The suspension system must provide proper steering control and ride quality. Performing these functions is extremely important to maintain vehicle safety and customer satisfaction. For example, if the suspension system allows excessive vertical wheel oscillations, the driver may lose control of the steering when driving on an irregular road surface. This loss of steering control can result in a vehicle collision and personal injury. Excessive vertical wheel oscillations transfer undesirable vibrations from the wheel(s) to the passenger compartment, which causes customer dissatisfaction with the ride quality.

The suspension system and frame must also position the wheels and tires properly to provide normal tire life and proper steering control. If the suspension system does not position each wheel and tire properly, wheel alignment angles are incorrect and usually cause excessive tire tread wear. Improper wheel and tire position can also cause the steering to pull to one side. When the suspension system positions the wheels and tires properly, the steering should remain in the straight-ahead position if the car is driven straight ahead on a reasonably straight, smooth road surface. However, if the wheels and tires are not properly positioned, the steering can be erratic, and excessive steering effort is required to maintain the steering in the straight-ahead position.

The steering system is also extremely important to maintain vehicle safety and reduce driver fatigue. For example, if a steering system component is suddenly disconnected, the driver may experience a complete loss of steering control, resulting in a vehicle collision and personal injury. Loose steering system components can cause erratic steering, which causes the driver to continually turn the steering wheel in either direction to try and keep the vehicle moving straight ahead. This condition results in premature driver fatigue.

Frames and Unitized Bodies

Some vehicles, such as rear-wheel-drive cars, sport utility vehicles (SUVs), and trucks, have a frame that is separate from the body (Figure 1-1). Other vehicles have a unitized body that combines the frame and body in one unit, eliminating the external frame (Figure 1-2). In a unitized body, the body design, rather than a heavy steel frame, provides strength and rigidity. All parts of a unitized body are load-carrying members, and these body parts are welded together to form a strong assembly.

The frame or unitized body serves the following purposes:

1. Allows the vehicle to support its total weight, including the weight of the vehicle and cargo.
2. Allows the vehicle to absorb stress when driving on rough road surfaces.
3. Enables the vehicle to absorb torque from the engine and drive train.
4. Provides attachment points for suspension and other components.

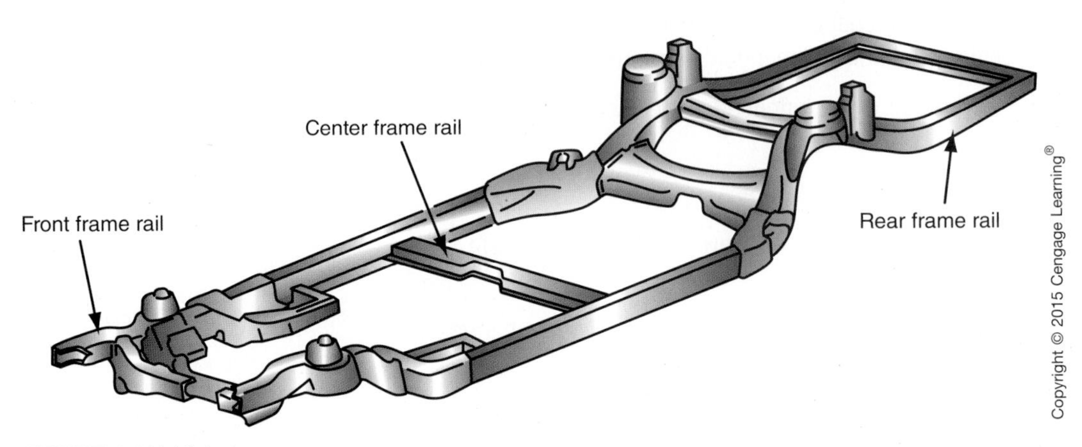

FIGURE 1-1 **Vehicle frame.**

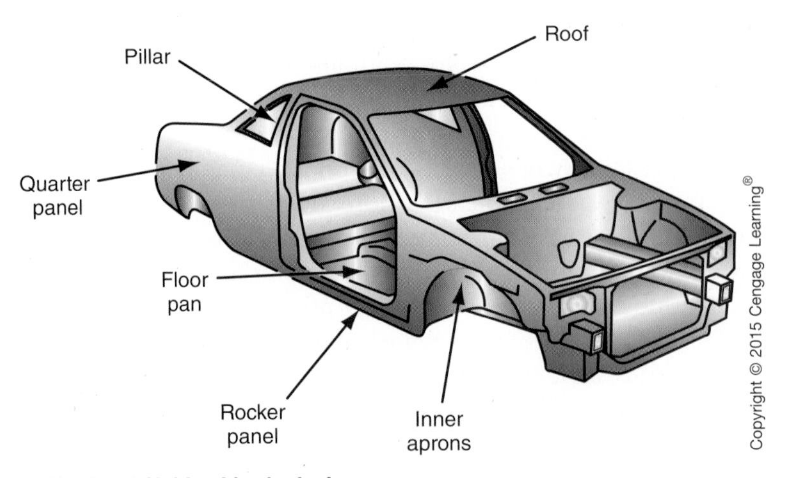

FIGURE 1-2 **Unitized body design.**

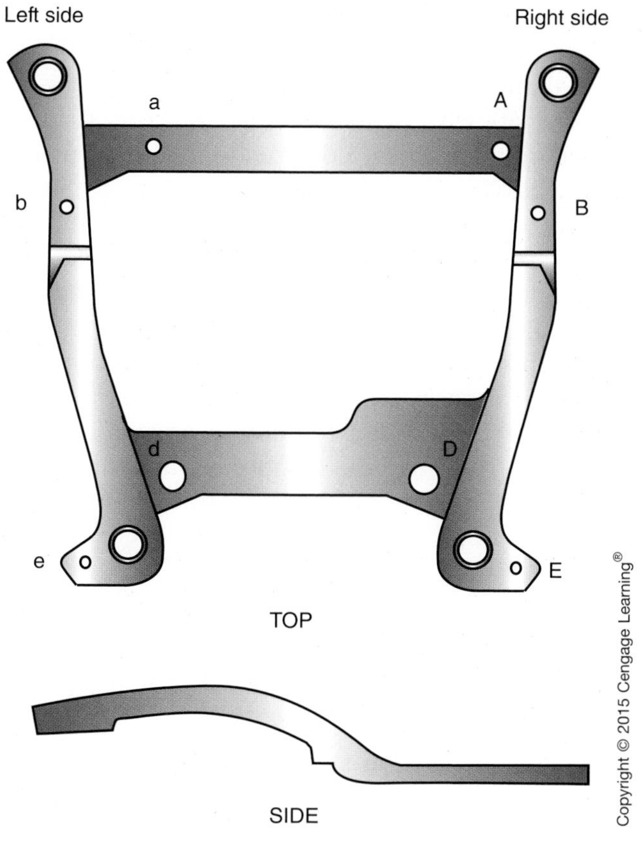

FIGURE 1-3 **Engine cradle.**

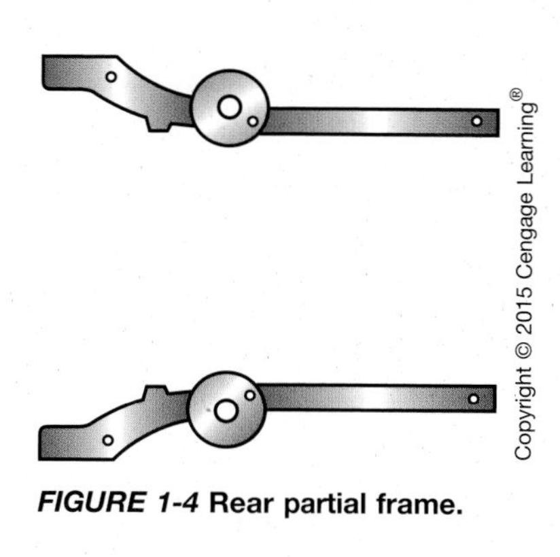

FIGURE 1-4 **Rear partial frame.**

Shop Manual
Chapter 10, page 402

Carbon dioxide is a by-product of the gasoline, diesel fuel, or ethanol combustion process.

A **greenhouse gas** is one that collects high in the earth's atmosphere and covers the earth like a blanket, resulting in more heat retained on the planet.

In the United States, it is estimated that automobiles contribute 1.5 billion tons of CO_2 to the atmosphere each year. Coal-burning power plants produce 2.5 billion tons of CO_2 each year.

The unitized body provides a steel box around the passenger compartment to provide passenger protection in a collision. In most unitized bodies, special steel panels are inserted in the doors to protect the vehicle occupants in a side collision. Some unitized body components are manufactured from high-strength or ultra high-strength steels. The unitized body design is the most popular design today. A steel cradle is mounted under the front of the unitized body to support the engine and transaxle (Figure 1-3). Rubber and steel mounts support the engine and transaxle on the cradle. Large rubber bushings are mounted between the cradle and the unitized body to help prevent engine vibration from reaching the passenger compartment. Some unitized bodies have a partial frame mounted under the rear of the vehicle to provide additional strength and facilitate the attachment of rear suspension components (Figure 1-4).

Vehicle weight plays a significant role in fuel consumption. One automotive design engineer states, "Fuel economy improvements are almost linear with weight reduction. A 30 percent reduction in vehicle weight provides approximately a 30 percent improvement in fuel economy." If a Toyota Prius weighs 3,300 lb (1,497 kg) and provides 50 miles per gallon (mpg), the same Prius would provide 55 mpg if it weighed 3,000 lb (1,360 kg).

Carbon dioxide (CO_2) emissions are a major concern for automotive manufacturers, because CO_2 is a **greenhouse gas** that contributes to global warming. Vehicle manufacturers are facing increasingly stringent CO_2 emission standards. CO_2 emissions are proportional to fuel consumption. Reduced fuel consumption results in lower CO_2 emissions. Therefore, reducing vehicle weight results in less fuel consumption and lower CO_2 emissions. Reduced weight also contributes to improved vehicle performance.

When vehicle bodies, front and rear suspension systems, and steering systems are built from lighter-weight components, these items can make an important contribution to improved fuel economy and reduced CO_2 emissions. The components in these systems

It is estimated that for every gallon of gasoline burned, 28 lb (12.7 kg) of CO_2 is released into the atmosphere. Approximately, 21 lb (9.53 kg) comes directly from the tailpipe of a vehicle, and 7 lb (3.18 kg) comes from the extraction, transportation, and refining of the gasoline.

may be manufactured from high-strength steels, aluminum, magnesium, titanium, or carbon composites to reduce vehicle weight. The Corvette Z06 has a hydro-formed aluminum frame to reduce vehicle weight. Lightweight aluminum construction throughout the Jaguar XK results in a vehicle weight of 3,671 lb (1,665 kg), which is 450 lb (204 kg) lighter than its Mercedes-Benz SL500 competitor that primarily uses steel construction. A new model Ford Fiesta is available in Europe since October 2008 and later on in the United States and other countries. In this model, 55 percent of the unitized body structure is made from high-strength steels, making the body 10 percent stiffer torsionally than its predecessor and providing a very rigid safety cell surrounding the occupants. Ford says this model is 88 lb (44 kg) lighter than the previous model and more fuel and CO_2 efficient.

High material and production costs have prevented the use of carbon composites in production vehicles. Carbon composites have been used in some exotic ultra high-performance or race cars, where cost was not a factor. Carbon composites provide reduced weight and improved crash energy absorption. New carbon composites and improved manufacturing processes may soon make the use of some carbon composite components a reality in high-production vehicles.

When coil-type suspension springs are manufactured from titanium rather than steel, the weight saving per vehicle is 44 to 55 lbs (20–25 kg).

Front Suspension Systems

The front and rear suspension systems are extremely important to provide proper wheel position, steering control, ride quality, and tire life. The impact of the tires striking road irregularities must be absorbed by the suspension systems. The suspension systems must supply proper ride quality to maintain customer satisfaction and reduce driver fatigue, as well as provide proper wheel and tire position to maintain directional stability when driving. Proper wheel position also ensures normal tire tread life.

Typical components in a short-and-long arm (SLA) front suspension system are illustrated in Figure 1-5. This type of front suspension system has a long lower control arm and a shorter upper control arm. The main front suspension components serve the following purposes:

The average car manufactured in the United States contains 347 lb (157 kg) of aluminum. A decade ago, the average car contained 242 lb (110 kg) of aluminum, and in 1973, the average car contained 81 lb (37 kg) of aluminum.

1. Upper and lower control arms—control lateral (side-to-side) wheel movement.
2. Upper and lower control arm bushings—allow upward and downward control arm movement and absorb wheel impacts and vibrations.
3. Coil springs—allow proper suspension ride height and control suspension travel during driving maneuvers.
4. Ball joints—allow the knuckle and wheels to turn to the right or left.

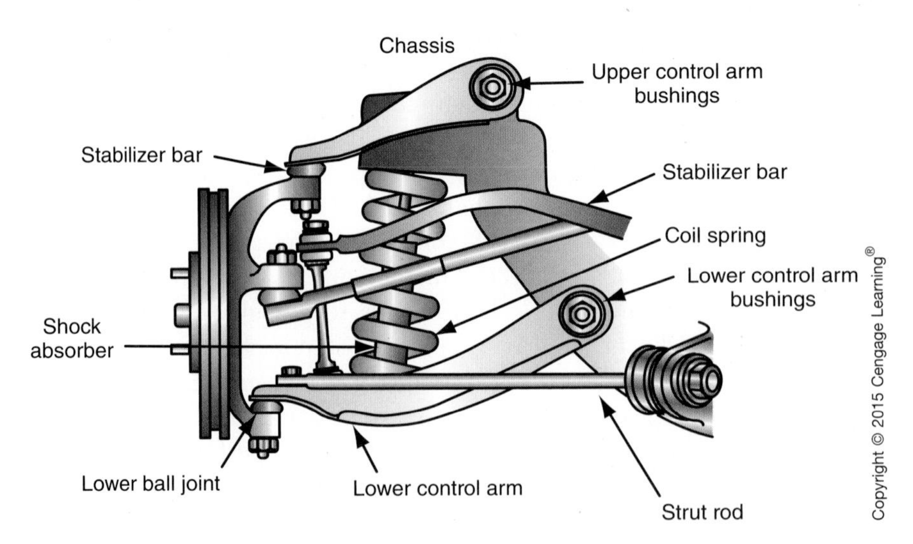

FIGURE 1-5 **Typical short-and-long arm (SLA) front suspension system.**

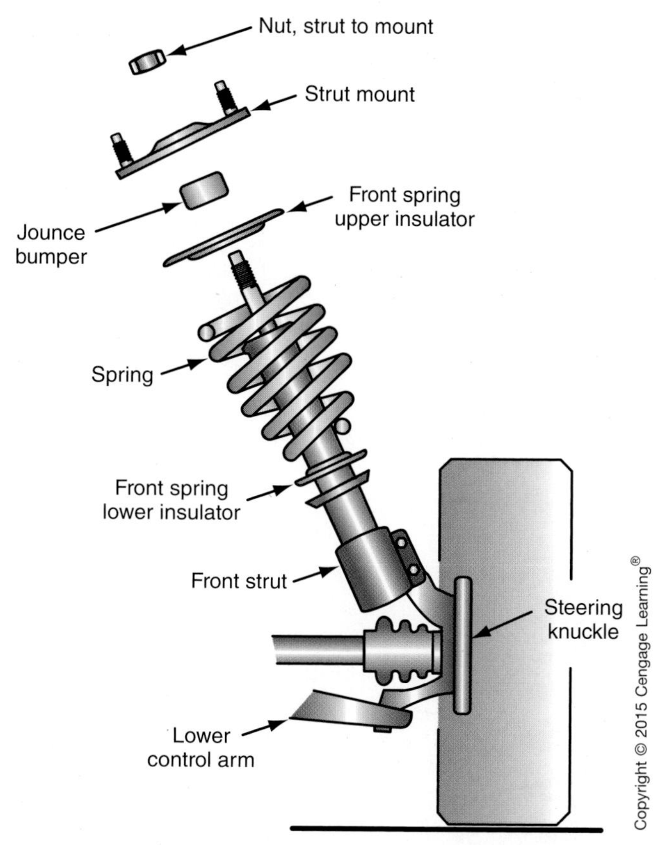

FIGURE 1-6 **Typical MacPherson strut front suspension system.**

5. Steering knuckles—provide mounting surfaces for the wheel bearings and hubs.
6. Shock absorbers—control spring action when driving on irregular road surfaces.
7. Strut rod—controls fore-and-aft wheel movement.
8. Stabilizer bar—reduces body sway when a front wheel strikes a road irregularity.

Shop Manual
Chapter 6, page 215

A MacPherson strut front suspension system has no upper control arm and ball joint; instead, a strut is connected from the top of the knuckle to an upper strut mount bolted to the reinforced strut tower in the unitized body (Figure 1-6). The strut supports the top of the knuckle and also performs the same function as the shock absorber in an SLA suspension system. The coil spring is mounted between a lower support on the strut and the upper strut mount. Insulators are mounted between the ends of the coil spring and the mounting locations. A bearing in the upper strut mount allows the strut and coil spring to rotate with the spindle when the front wheels are turned.

Rear Suspension Systems

A typical live axle rear suspension system has a one-piece rear axle housing. Trailing arms are connected from the rear axle housing to the chassis through rubber bushings. The coil springs are mounted between the trailing arms and the chassis (Figure 1-7). Because the rear axle housing is a one-piece assembly, vertical movement of one rear wheel causes the opposite rear wheel to be tipped outward at the top. This action increases tire tread wear and reduces ride quality and traction between the tire tread and road surface.

Many front-wheel-drive cars have a semi-independent rear suspension system with an inverted steel U-section connected between the rear spindles (Figure 1-8). The inverted U-section usually contains a tubular stabilizer bar. When one rear wheel strikes a road irregularity, the inverted U-section and stabilizer bar twist, allowing some independent

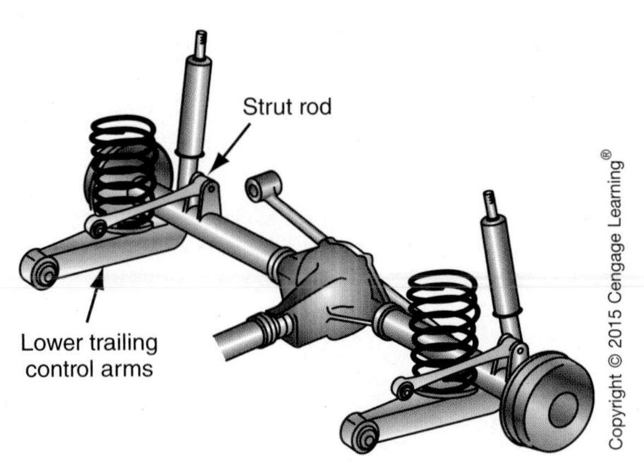

FIGURE 1-7 **Live axle rear suspension system with coil springs.**

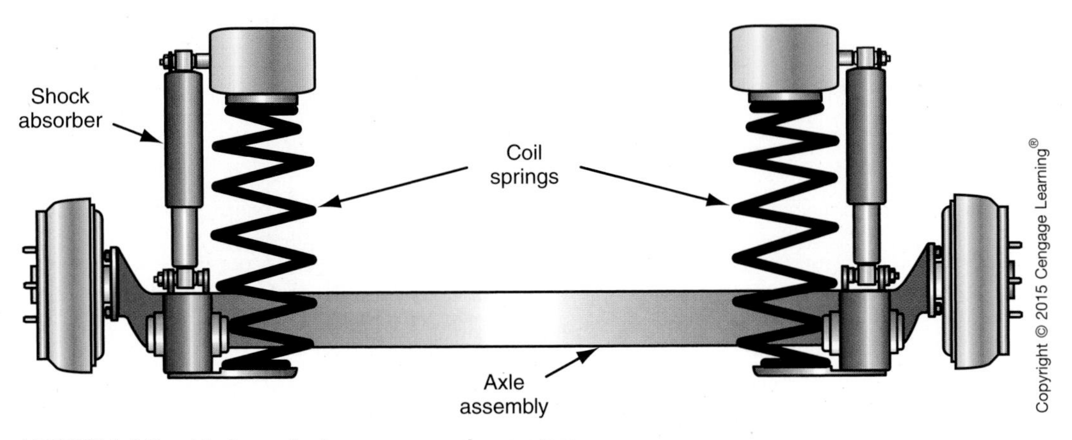

FIGURE 1-8 **Semi-independent rear suspension system.**

rear-wheel movement before the wheel movement affects the opposite rear wheel. Some semi-independent rear suspension systems have a track bar and brace connected from the inverted U-section to the chassis to reduce lateral rear-axle movement (Figure 1-9).

Many vehicles have an independent rear suspension system, wherein each rear wheel can move independently without affecting the position of the opposite rear wheel. This type of suspension system reduces rear tire wear and provides improved steering control. Independent rear suspension systems have a number of different configurations. A MacPherson strut independent rear suspension system has a strut and coil spring assembly connected from the top of the spindle through an upper strut mount to the chassis (Figure 1-10). No provision for strut rotation is required, because the rear wheels are not steered. Some independent rear suspension systems have a multilink design, wherein an adjustment link connected from the rear spindle to the chassis allows rear wheel position adjustment (Figure 1-11).

Tires, Wheels, and Hubs

Tire Purpose

Tires are extremely important because they play a large part in providing vehicle safety and ride quality! Tires are the only point of contact between the vehicle and the road surface. Vehicle tires provide these functions:

1. Tires must support the vehicle weight safely and firmly.
2. Tires must provide a comfortable ride.

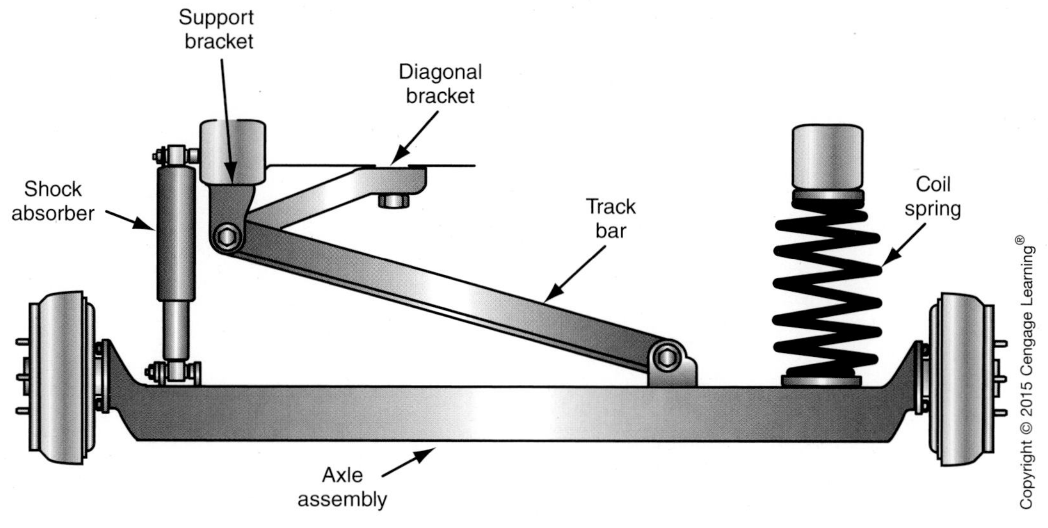

FIGURE 1-9 **Semi-independent rear suspension system with track bar and brace.**

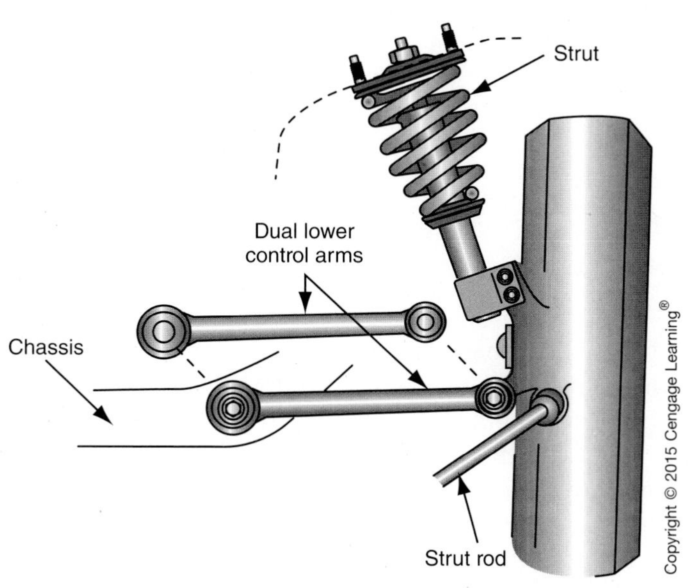

FIGURE 1-10 **MacPherson strut independent rear suspension system.**

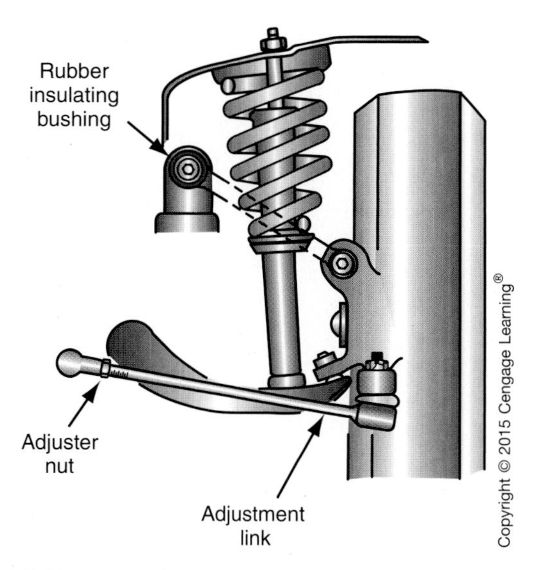

FIGURE 1-11 **Short-and-long arm independent rear suspension system.**

3. Tires must supply adequate traction on various road surfaces to drive and steer the vehicle.
4. Tires must contribute to proper steering control and directional stability of the vehicle.
5. Tires must absorb high stresses when cornering, accelerating, and braking.
6. Tire treads must be designed to propel water off the tread and away from the tire when driving on wet highways. This action prevents water from lifting the tires off the road surface, which decreases tire traction.

Wheel Rim Purpose

Shop Manual
Chapter 3, page 98

Wheel rims can be manufactured from steel, cast aluminum, forged aluminum, pressure-cast chrome-plated aluminum, or magnesium alloy. Wheel rims must retain the tires safely under all operating conditions without distortion. Tires and wheels must form airtight containers at all temperatures so air does not leak from the assembly. Wheel rims must position the tires at the proper distance inward or outward from the vertical mounting surface

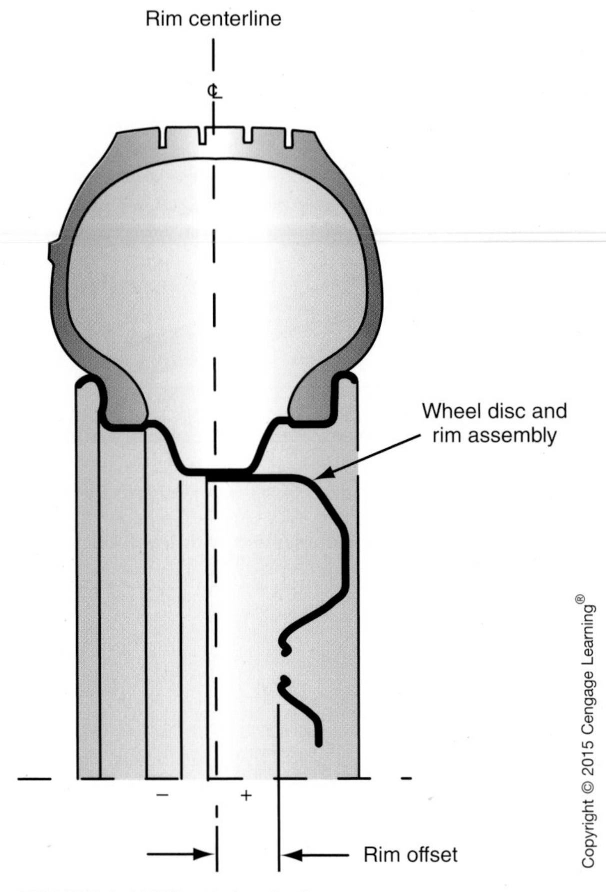

FIGURE 1-12 **Wheel rim design.**

of the wheel. The distance between the vertical wheel rim mounting surface and the centerline of the wheel rim is called **wheel offset** (Figure 1-12). If the wheel centerline is located inboard from the vertical wheel mounting surface, the wheel has **positive offset**. Conversely, if the wheel centerline is located outboard from the vertical wheel mounting surface, the wheel has **negative offset**. Wheel rims typically have four to six mounting openings that fit over studs in the wheel hub. When a wheel rim is installed on the hub studs, tapered nuts are then tightened to the specified torque to retain the wheel and tire assembly on the hub. On many wheel rims, the openings in the wheel rim are tapered to match the tapers on the retaining nuts. These tapered openings and matching tapered nuts center the wheel rim on the hub.

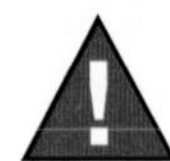

WARNING: Wheel nuts must be tightened in the proper sequence to the specified torque. Failure to follow the proper wheel nut tightening procedure and torque may cause a wheel to come off a car while driving. This action usually results in serious personal injury and extensive vehicle and/or property damage.

Wheel Hubs

Wheel hubs must provide a secure mounting surface for the wheel rim and tire assembly. Wheel hubs also contain the wheel bearings that provide smooth wheel rotation with reduced friction. Wheel bearings must have a minimum amount of end play to greatly reduce wheel lateral movement. The wheel hub and bearing assemblies must carry the load

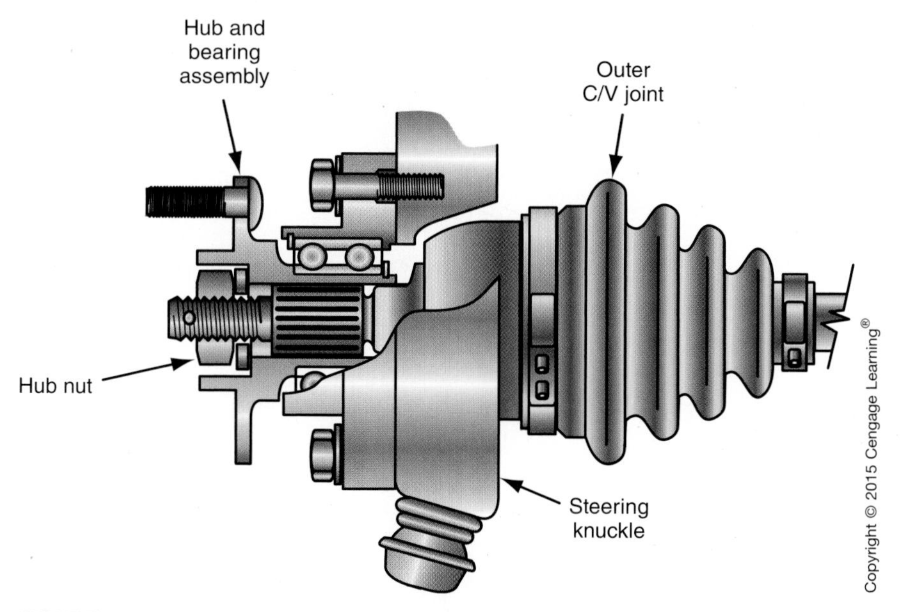

FIGURE 1-13 **Wheel bearing and hub assembly.**

supplied by the vehicle weight, and these assemblies must also guide the wheel and tire assembly (Figure 1-13). The vehicle weight is supplied to the wheel hub and bearing assembly in a vertical direction. This type of bearing load is called a **radial bearing load**. When the vehicle turns a corner, the wheel hubs and bearings must carry **thrust bearing loads** supplied in a horizontal direction and **angular bearing loads** supplied in a direction between the horizontal and the vertical.

Shock Absorbers and Struts

Each corner of the vehicle has a shock absorber or strut connected from the suspension system to the chassis. Shock absorbers control spring action and wheel oscillations to provide a comfortable ride. Controlling spring action and wheel oscillations also improve vehicle safety because the struts help to keep each tire tread in contact with the road surface. If the struts are worn out, excessive wheel oscillations when driving on irregular road surfaces can cause the driver to lose control of the vehicle. Struts also reduce body sway and lean while turning a corner. Struts reduce the tendency of the tire tread to lift off the road surface. This action improves tire tread life, traction, steering control, and directional stability.

Struts contain a sealed lower chamber filled with a special oil. Many shock absorbers have a nitrogen gas charge on top of the oil. This gas charge helps to prevent the shock absorber oil from foaming. A circular steel mount containing a rubber bushing is attached to the bottom end of the lower chamber, and this lower mounting is bolted to the suspension system. The upper strut housing is connected to a piston rod that extends into the lower chamber. A piston valve assembly is attached to the lower end of the piston rod (Figure 1-14). The upper strut mount is similar to the lower mounting, and the upper mount is bolted to the chassis.

Shop Manual
Chapter 5, page 180

When a wheel strikes a road irregularity, the wheel and suspension move upward, and the spring in the suspension system is compressed. This action forces the lower shock absorber chamber to move upward, and the oil must flow from below the shock absorber piston and valve to the area above the valve. Upward wheel movement is called **jounce travel**. The strut valves are designed to provide precise oil flow control, and thus control the speed of upward wheel movement.

When a spring is compressed, it stores energy and then immediately expands with an equal amount of energy. When the spring expands, the tire and wheel assembly is forced

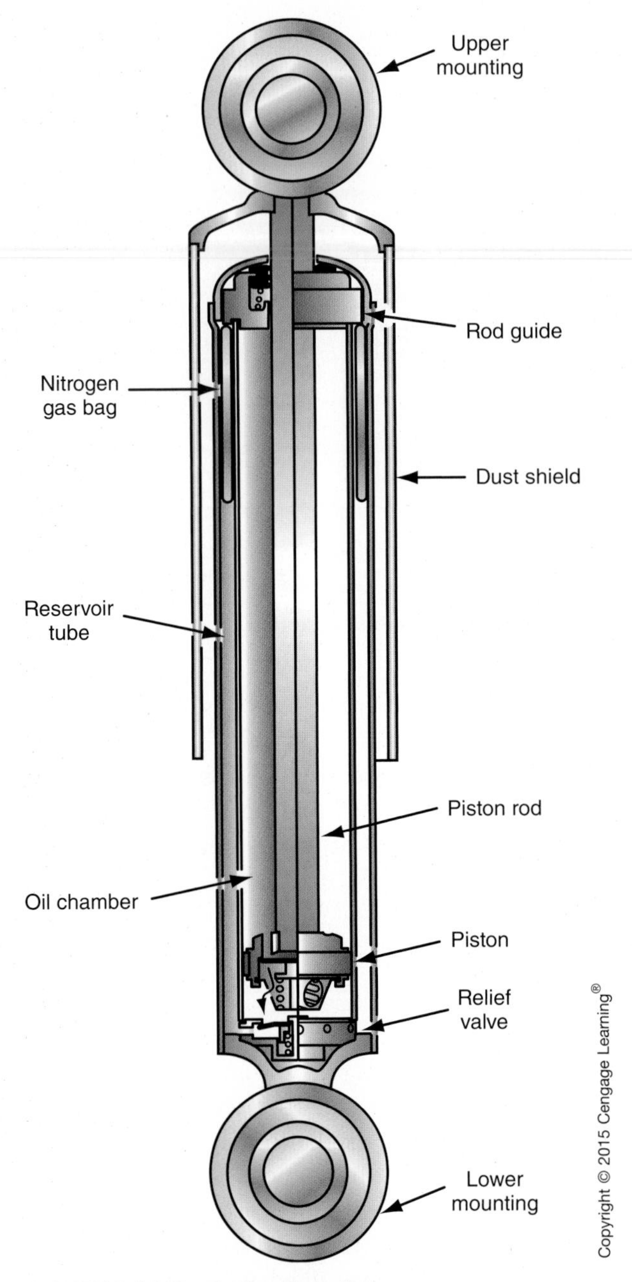

FIGURE 1-14 **Shock absorber design.**

downward. Under this condition, the lower strut chamber is forced downward, and oil must flow from above the shock absorber piston and valve to the area below the valve (Figure 1-15). Downward wheel movement is called **rebound travel**. The strut valves provide precise control of the oil flow, and this action controls spring action and wheel oscillations. Shock absorbers and valves are usually designed to provide more control during the rebound travel compared to the jounce travel.

Internal strut design is similar to shock absorber design, but struts also support the top of the steering knuckle. In most suspension systems, the lower end of the strut is attached to the top of the steering knuckle, and a special mount is connected between the upper end of the strut and the chassis (Figure 1-16). On front suspension systems, the upper strut mount must allow strut and spring rotation when the front wheels are turned to the right or the left. The upper strut mount isolates wheel and suspension vibrations from the chassis.

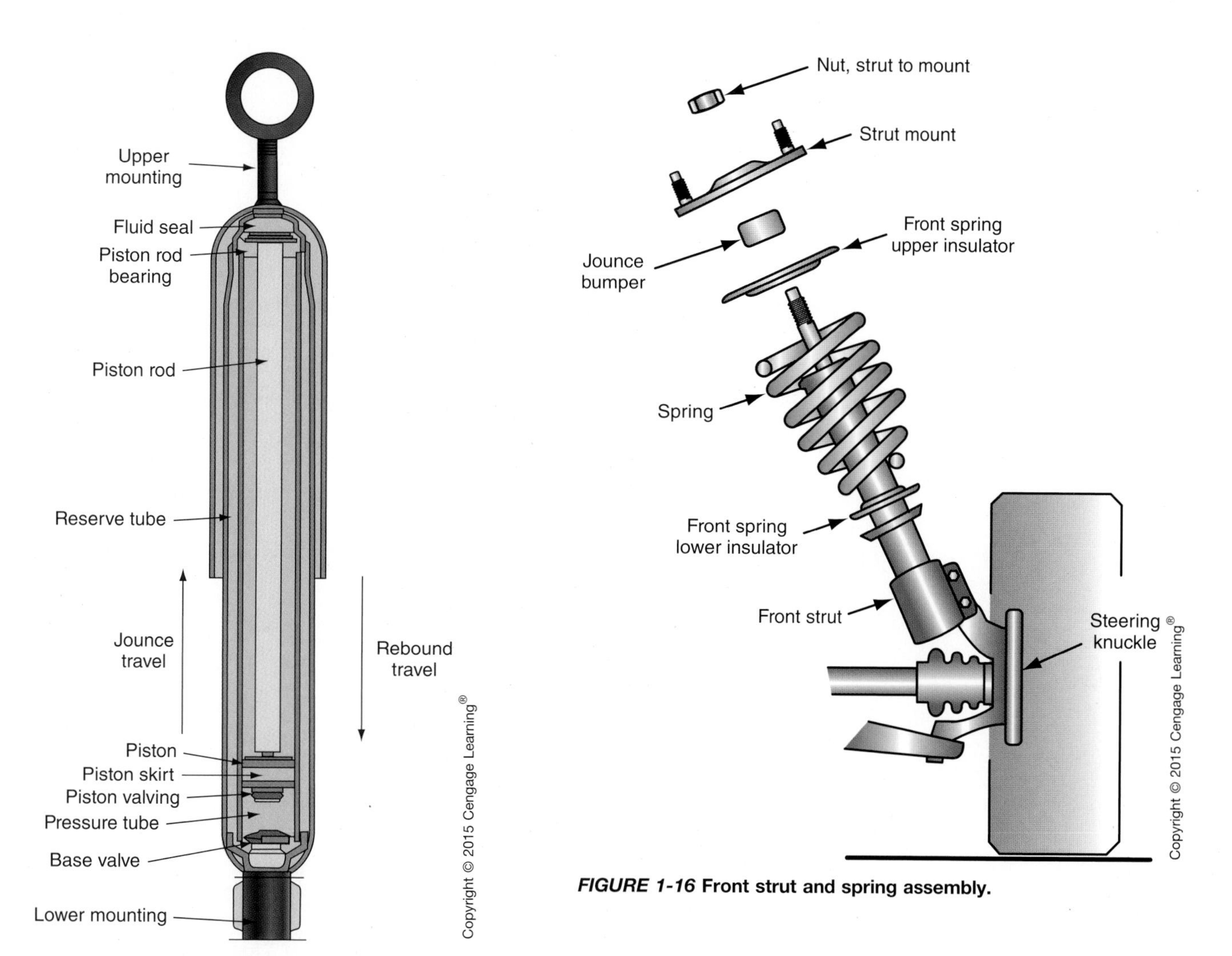

FIGURE 1-15 **Shock absorber action.**

FIGURE 1-16 **Front strut and spring assembly.**

Computer-Controlled Suspension Systems and Shock Absorbers

Many vehicles are equipped with computer-controlled suspension systems that provide a soft, comfortable ride for normal highway driving, and then automatically and very quickly switch to a firm ride for hard cornering, braking, or fast acceleration. Computer-controlled suspension systems reduce body sway during hard cornering, and thus contribute to improved ride quality and vehicle safety. Some computer-controlled suspension systems are driver-adjustable with up to four suspension modes to allow the driver to tailor the ride quality to the driving style.

Some computer-controlled suspensions systems have electronically actuated solenoids in each shock absorber or strut. These solenoids rotate the shock absorber or strut valves to adjust the valve openings and shock absorber control (Figure 1-17). Other shock absorbers or struts contain a magneto-rheological fluid that is a synthetic oil containing suspended iron particles. A computer-controlled electric winding is designed into the shock absorber housing. When there is no current flow through the winding, the iron particles are randomly dispersed in the oil. Under this condition, the oil consistency is thinner and the oil flows easily through the shock absorber valves to provide a softer ride. If the suspension computer supplies current flow to the shock absorber windings, the iron particles are aligned so the oil has a jelly-like consistency (Figure 1-18). This action instantly provides a

Shop Manual
Chapter 11, page 421

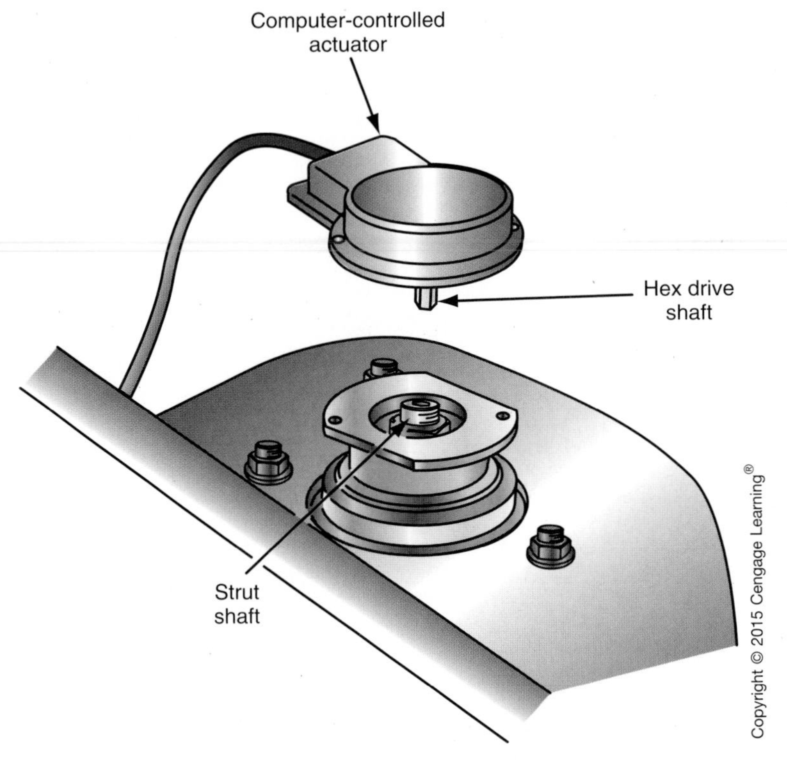

FIGURE 1-17 **Strut actuator.**

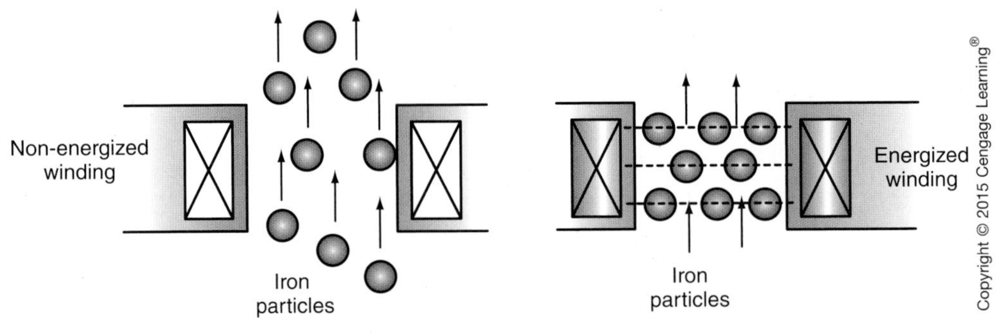

FIGURE 1-18 **Magneto-rheological fluid action in a strut or shock absorber.**

much firmer ride. The computer can provide a large variation in current flow through the shock absorber windings and a wide range of ride control. Input sensors at each corner of the vehicle inform the suspension computer the velocity of the wheel jounce and rebound, and the computer uses these input signals to operate the shock absorber windings or actuators.

AUTHOR'S NOTE: One of the advantages of computer-controlled shock absorbers and suspension systems is the speed at which modern computers can perform output functions. For example, a suspension computer can change the thickness of the magneto-rheological fluid in a shock absorber in about 1 millisecond (ms). When a wheel and tire strike a road irregularity and move upward, this fast computer action adjusts the thickness of the shock absorber fluid in relation to the wheel jounce velocity before the wheel moves downward in the rebound stroke and strikes the road surface.

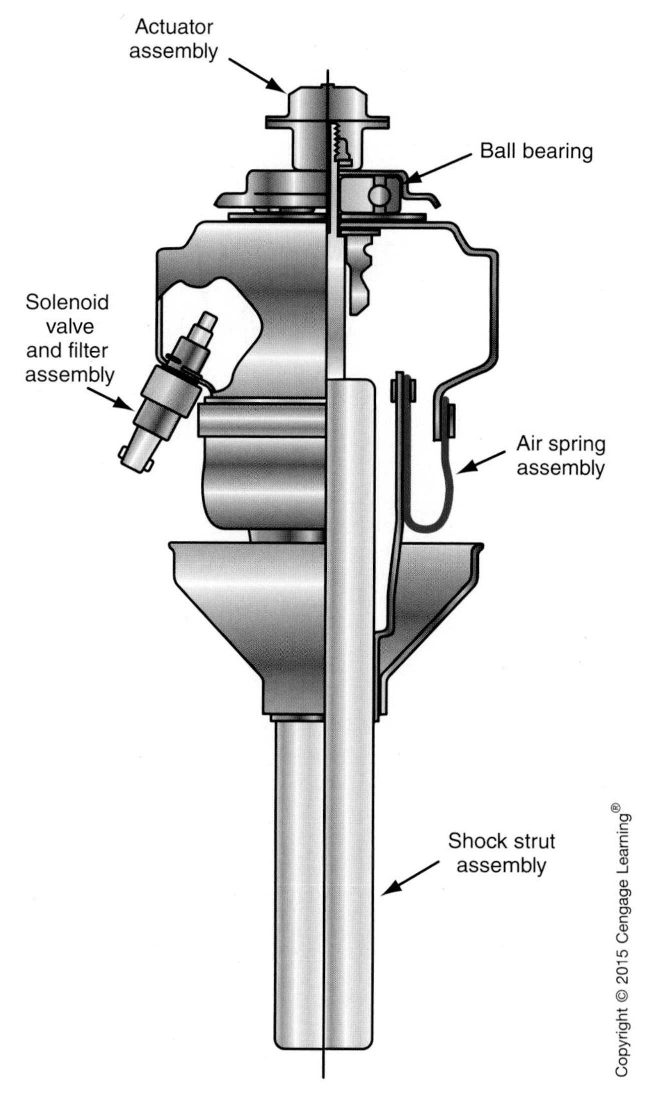

FIGURE 1-19 **Air spring and strut assembly.**

Some computer-controlled suspension systems have air springs in place of coil springs (Figure 1-19). Front and rear height sensors inform the suspension computer regarding the suspension height, and the computer operates an air compressor and air spring control valves to control the amount of air in the air springs, and thus control suspension height. Some air suspension systems also have computer-controlled shock absorbers or struts.

Steering Systems

Steering systems are essential to provide vehicle safety, steering quality, and steering control! Steering system problems can cause the steering to pull to one side when driving straight ahead, excessive steering effort, **wheel shimmy**, or excessive steering wheel free play. These problems all reduce vehicle safety and increase driver fatigue. Therefore, steering systems must be properly maintained.

Wheel shimmy may be defined as rapid inward and outward wheel and tire oscillations.

Steering Columns and Steering Linkage Mechanisms

The steering column connects the steering wheel to the steering gear. The steering wheel is connected to the steering shaft, and this shaft extends through the center of the steering column. The lower end of the steering shaft is connected through a universal joint or flexible coupling to the shaft from the steering gear. The steering shaft is supported on bearings

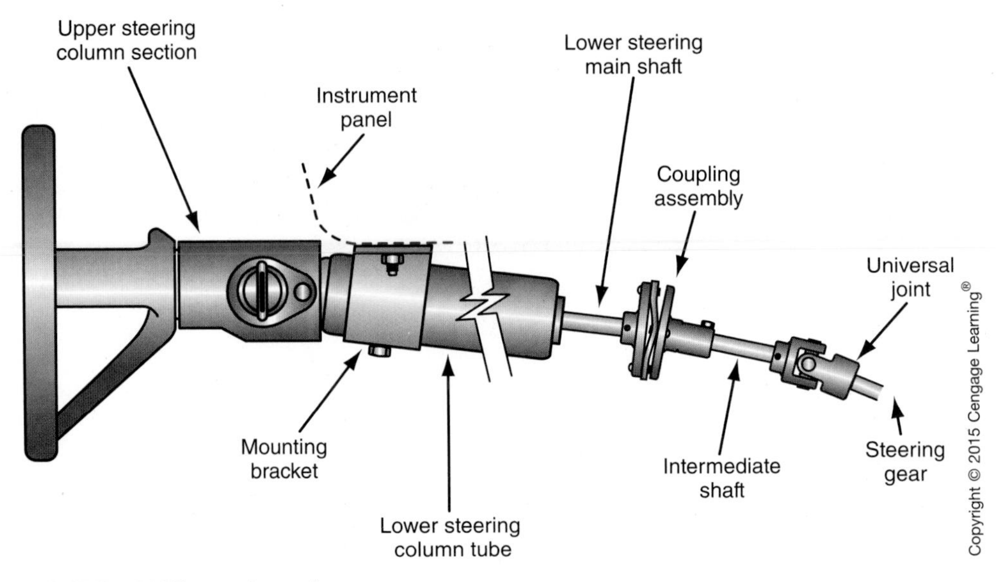

FIGURE 1-20 **Tilt steering column.**

in the steering column. Some steering columns are designed to collapse or move away from the driver, if the driver is thrown against the steering wheel in a collision. Some steering columns are designed so the driver can tilt the steering wheel downward or upward to provide increased driver comfort and facilitate entering and exiting the driver's seat (Figure 1-20). Some steering columns also provide a telescoping action so the steering wheel can be moved closer to, or farther away from, the driver. Other steering columns do not have any tilt or telescoping action. A mounting bracket retains the steering column to the instrument panel.

On most vehicles, the ignition lock cylinder and ignition switch are mounted in the steering column. Removing the key from the ignition switch locks the steering column and the gear shift on many vehicles. The steering column usually contains a combination signal light, wipe/wash, dimmer, and cruise control switch. This switch may be called a smart switch. The switch for the hazard warning lights is also mounted in the steering column. An air bag inflator module is mounted in the top of the steering wheel, and a clock-spring electrical connector under the steering wheel maintains electrical contact between the inflator module and the air bag electrical system.

Steering linkages connect the steering gear to the steering arms on the front wheels. In a parallelogram steering linkage, a pitman arm is connected from the steering gear to a center link (Figure 1-21). A pivoted idler arm bolted to the chassis supports the other end of the center link. Tie rods are connected from the center link to the steering arms attached to the front wheels. Pivoted ball studs are mounted in the inner ends of the tie rods, and outer tie rod ends are threaded into the tie rod adjusting sleeves. The outer tie rod ends contain pivoted ball studs, and these tapered studs fit into matching tapered openings in the outer ends of the steering arms. The pitman arm and idler arm position the center link so the tie rods are parallel to the lower control arms. This tie rod position is very important to maintain proper steering operation.

CAUTION:
Regular inspection and maintenance of steering linkage components is very important to provide normal component life and maintain vehicle safety.

Many vehicles have a rack and pinion steering linkage. In these linkages, the tie rods are connected through inner tie rod ends to the rack in the rack and pinion steering gear. Outer tie rod ends are connected from the tie rods to the steering arms (Figure 1-22). In this type of steering linkage, the steering gear mounting must position the tie rods so they are parallel to the lower control arms. The rack and pinion steering gear can be mounted on the cowl or the front cross member.

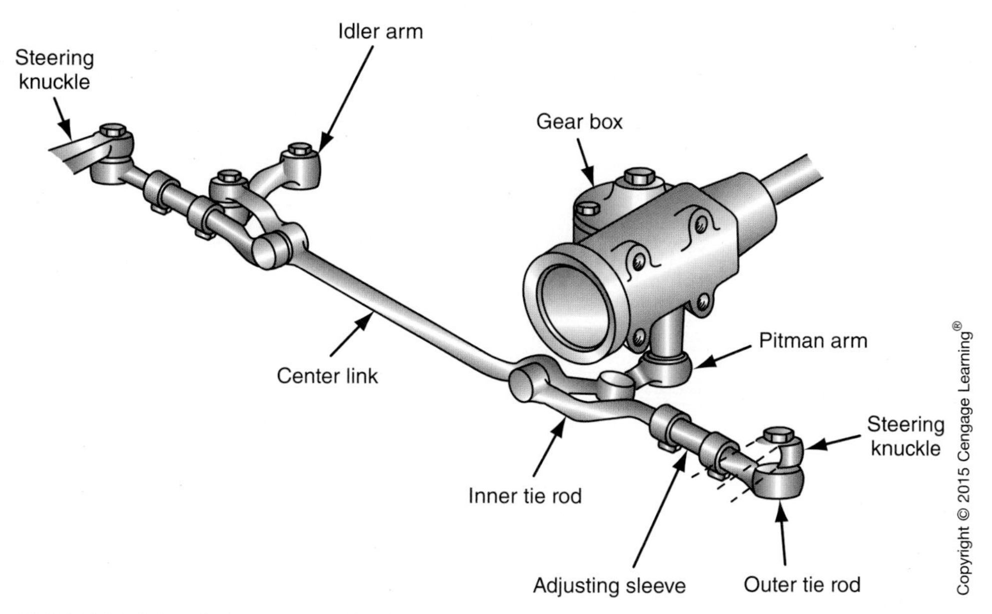

FIGURE 1-21 **Parallelogram steering linkage.**

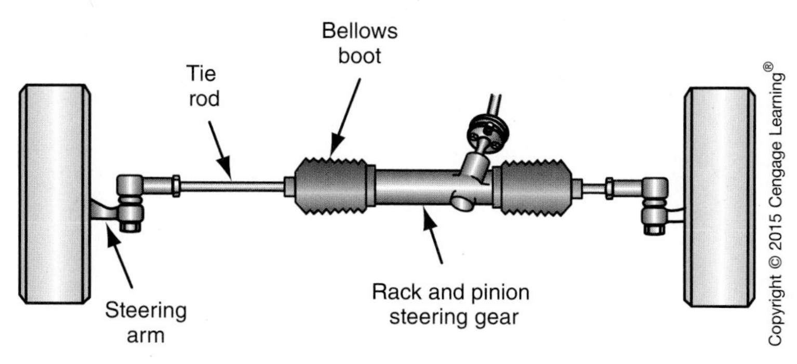

FIGURE 1-22 **Rack and pinion steering gear and linkage.**

Recirculating Ball Steering Gears

Some vehicles are equipped with a recirculating ball steering gear, wherein the steering shaft is attached to a worm gear in the steering gear. A ball nut with internal grooves is mounted over the worm gear. Ball bearings are mounted between the worm gear and ball nut grooves to reduce friction and provide reduced steering effort. Outer grooves on the ball nut are meshed with matching teeth on the sector shaft (Figure 1-23). The lower end of the sector shaft is splined to the pitman arm. When the steering wheel is turned, the ball nut moves upward or downward on the worm gear, which rotates the sector shaft to provide the desired steering action. Recirculating ball steering gears can be manual-type with no hydraulic assist, or power-type with hydraulic assist from the power steering pump.

Shop Manual
Chapter 14,
page 551

Rack and Pinion Steering Gears

Rack and pinion steering gears and linkages are more compact than recirculating ball steering gears and parallelogram steering linkages. Therefore, rack and pinion steering gears are usually installed on smaller, front-wheel-drive vehicles. Rack and pinion steering gears transfer more road shock from the front wheels to the steering gear and steering wheel, because the tie rods are connected directly to the rack in the steering gear.

In a rack and pinion steering gear, a toothed rack is mounted on bushings in the rack housing. The rack teeth are meshed with teeth on a pinion gear mounted near one end of the gear. The pinion gear is mounted on bearings in the gear housing. The steering shaft from the steering column is attached to the upper end of the pinion gear (Figure 1-24).

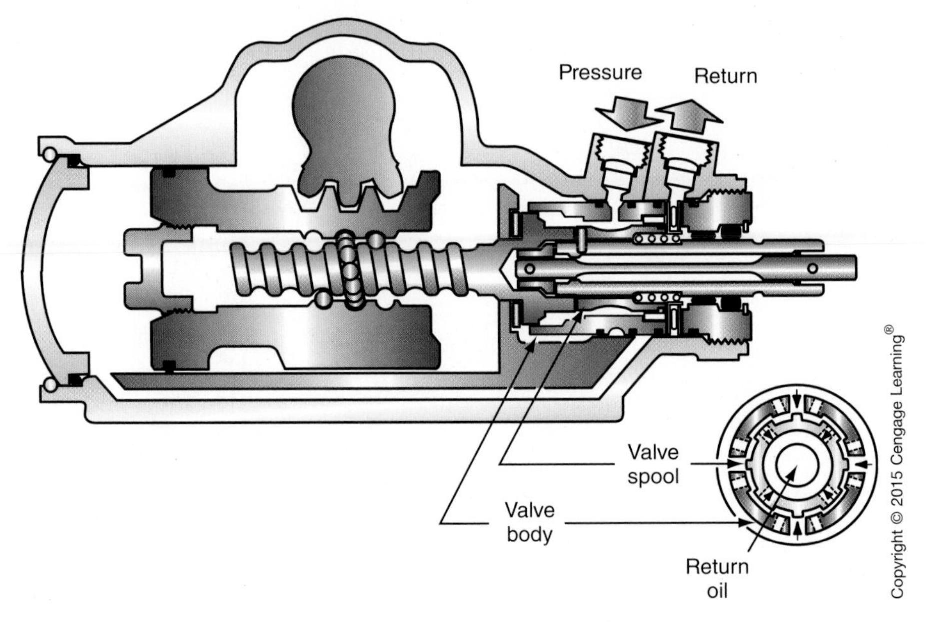

FIGURE 1-23 **Power recirculating ball steering gear.**

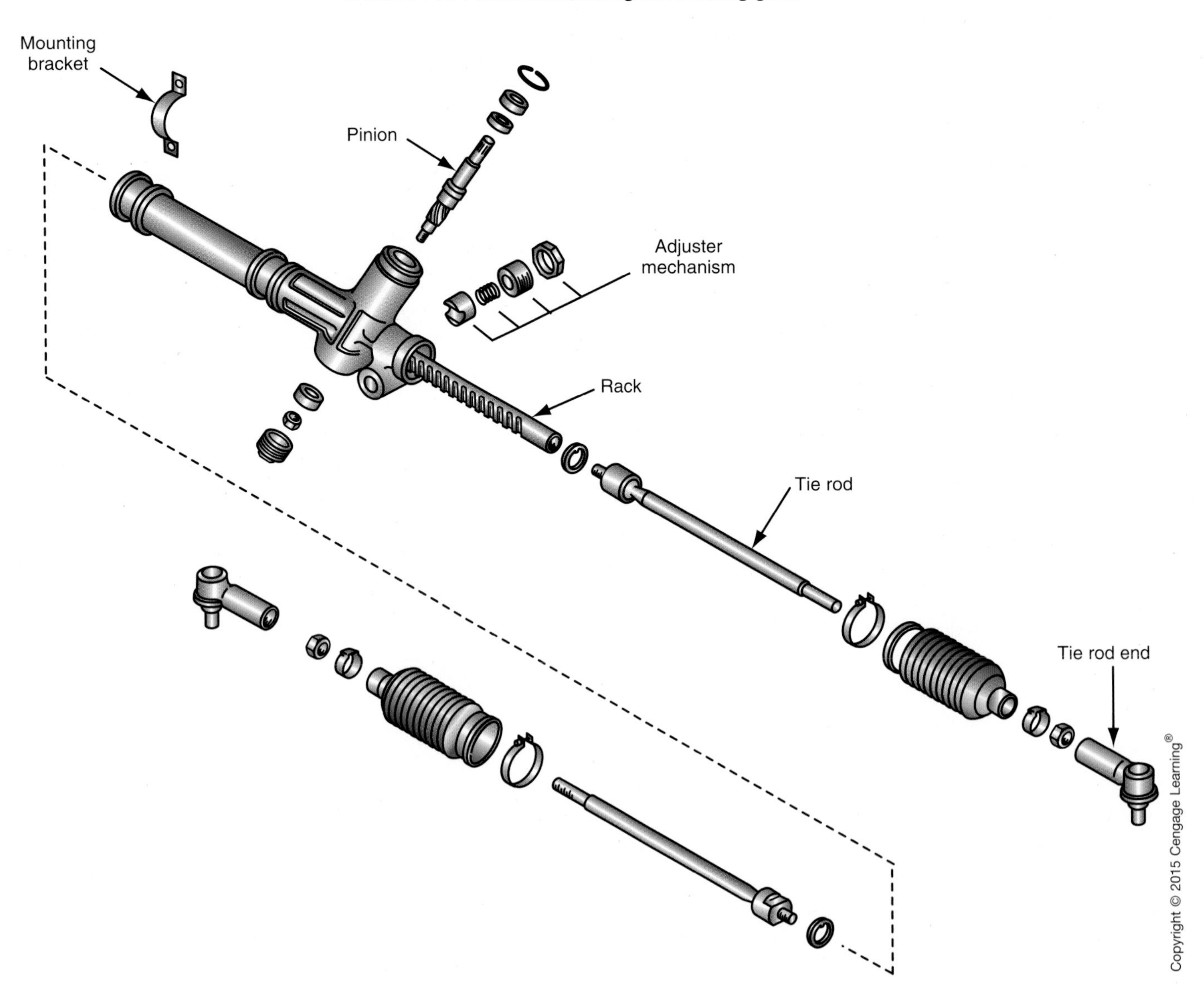

FIGURE 1-24 **Manual rack and pinion steering gear.**

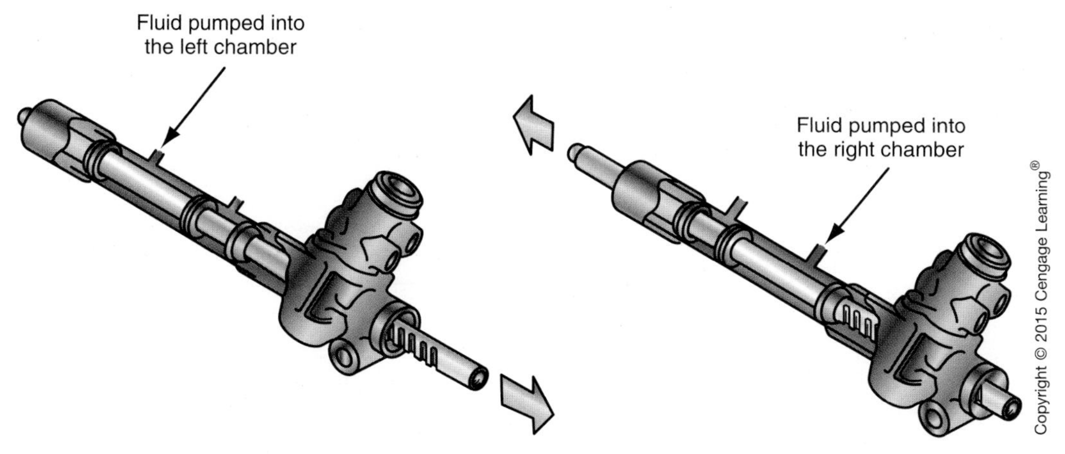

FIGURE 1-25 **Power rack and pinion steering gear with sealed rack piston chambers.**

When the steering wheel is turned, the rotation of the pinion gear moves the rack inward or outward to provide the desired steering action. Rack and pinion steering gears can be manual-type or power assisted by fluid pressure from the power steering pump. Power rack and pinion steering gears have a piston near the center of the rack, and fluid pressure is supplied from the power steering pump to sealed chambers on either side of the rack piston to provide steering assistance (Figure 1-25).

Wheel Alignment

Proper **wheel alignment** is extremely important to provide steering control, ride quality, and normal tire tread life. Improper wheel alignment may cause steering wander, steering pull to the right or left, or improper steering wheel return after turning a corner. Incorrect wheel alignment may contribute to harsh ride quality. Wheel alignment angles that are not within specifications may cause rapid tire tread wear.

Wheel alignment is the proper positioning of the tire and wheel assemblies on the vehicle to provide normal tire tread life, precise steering control, and satisfactory ride quality.

Rear Wheel Tracking

The rear wheels must track directly behind the front wheels to provide proper steering control. The front and rear wheels must be parallel to the vehicle centerline to provide proper rear wheel tracking. If the rear wheels are tracking directly behind the front wheels, the **thrust line** is positioned at the geometric centerline of the vehicle (Figure 1-26). If the left rear wheel has excessive **toe-out**, the thrust line is moved to the left of the geometric centerline (Figure 1-27). This improperly positioned thrust line causes the steering to pull to the right and also results in rapid tread wear on the left rear tire.

The **thrust line** is an imaginary line positioned at a 90 degree angle to the center of the rear axle and extending forward.

Toe-out is a condition that occurs if the distance between the front edges of the front or rear tires is greater than the distance between the rear edges of the front or rear tires.

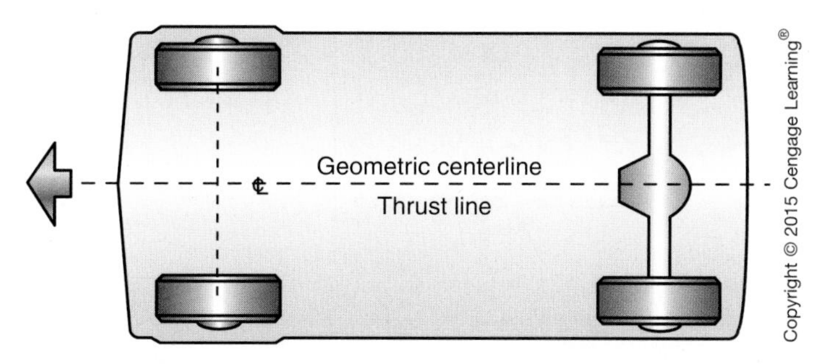

FIGURE 1-26 **Front and rear wheels are parallel to the vehicle centerline.**

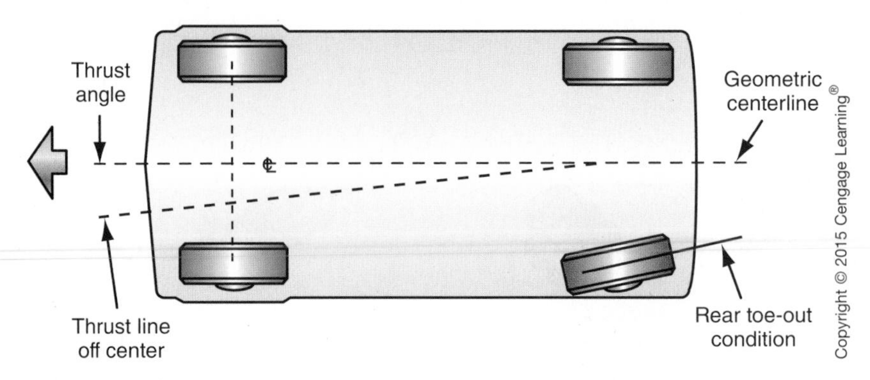

FIGURE 1-27 **Excessive toe-out on the left rear wheel moves the thrust line to the left of the geometric centerline.**

Shop Manual
Chapter 9,
page 340

Rear Wheel Toe

A toe-out condition occurs when the distance between the front edges of the rear tires is greater than the distance between the rear edges of the rear tires. Toe-in occurs if the distance between the rear edges of the rear tires is greater than the distance between the front edges of the rear tires (Figure 1-28). On front-wheel-drive vehicles, driving forces tend to push the rear spindles backward. Therefore, these vehicles are usually designed with a zero toe-in or a slight toe-in. Improper rear wheel toe causes rapid tire tread wear because the wheel and tire assembly is being pushed sideways to a certain extent as the vehicle is driven. Steering pull to one side may also be a result of improper rear wheel toe.

Positive camber is present when the vertical centerline of the tire is tilted outward in relation to the true vertical centerline of the tire.

Negative camber occurs if the vertical centerline of the tire is tilted inward in relation to the true vertical centerline of the tire.

Rear Wheel Camber

Camber is the angle between the centerline of the wheel and tire in relation to the true vertical centerline of the wheel and tire viewed from the front. **Positive camber** occurs when the vertical centerline of the wheel and tire is tilted outward in relation to the true vertical centerline of the wheel and tire. **Negative camber** occurs if the vertical centerline of the wheel and tire is tilted inward in relation to the true vertical centerline of the wheel and tire (Figure 1-29).

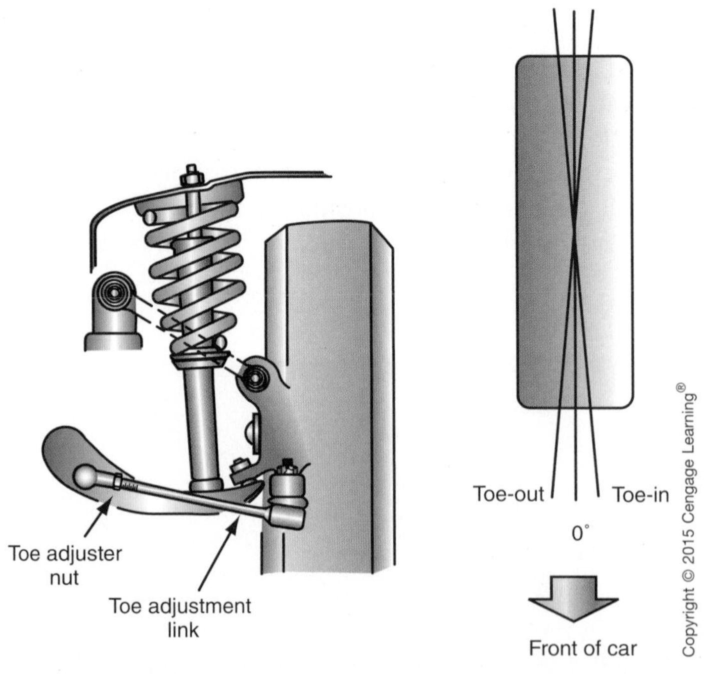

FIGURE 1-28 **Rear wheel toe.**

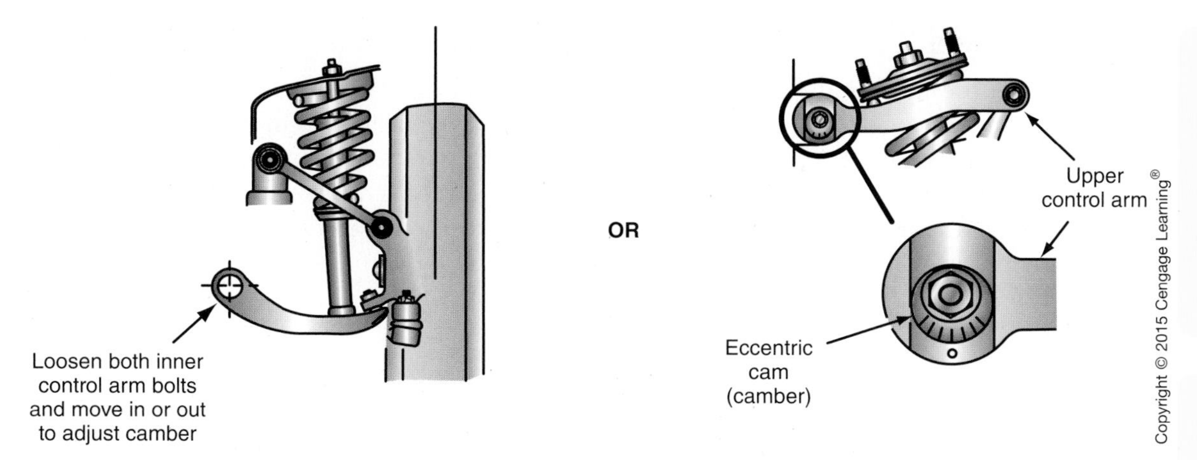

FIGURE 1-29 **Rear wheel camber.**

Positive caster is the angle of a line through the center of the tire that is tilted rearward in relation to the true vertical centerline of the tire viewed from the side.

Negative caster is the angle of a line through the center of the tire that is tilted forward in relation to the true vertical centerline of the tire viewed from the side.

Excessive positive camber causes rapid wear on the outside edge of the tire tread, whereas excessive negative camber results in rapid wear on the inside edge of the tire tread. Because a tilted wheel and tire assembly always rolls in the direction of the tilt, improper camber angle on the front wheels may cause steering pull. Many front-wheel-drive vehicles have a slightly negative rear-wheel camber that improves cornering stability.

Front Wheel Camber

Front wheel camber is the same as rear wheel camber except for the camber setting. Many front suspension systems are designed with a slightly positive camber.

Front Wheel Caster

Caster is the tilt of a line through the tire centerline (steering axis) in relation to the true vertical tire centerline viewed from the side. **Positive caster** occurs when the centerline of the tire is tilted rearward in relation to the true vertical centerline of the tire. **Negative caster** occurs when the centerline of the tire is tilted forward in relation to the centerline of the tire viewed from the side (Figure 1-30). Excessive negative caster causes the steering to wander when the vehicle is driven straight ahead. Excessive positive caster causes increased steering effort and rapid steering wheel return after turning a corner. Harsh ride quality is also a result of excessive positive caster because this condition causes the caster line to be aimed at road irregularities.

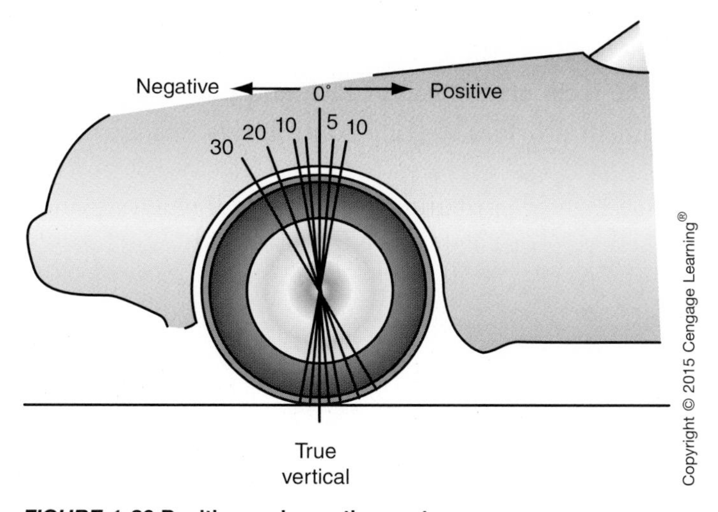

FIGURE 1-30 **Positive and negative caster.**

A BIT OF HISTORY

China is one of the emerging automotive markets in the world. In 2002, Chinese car sales totaled 1,126,000. This was the first year that car sales in China exceeded 1,000,000, and car sales from 2001 to 2002 increased approximately 50 percent. Total sales of cars, trucks, and buses totaled 3,500,00 in 2002. In the first 3 months of 2003, car sales increased 40 percent to 1,360,000. Experts predict that car sales in China may never achieve this rate of growth again, but predict a significant, steady growth rate for the Chinese automotive industry. Some North American vehicle and parts manufacturers are forming partnerships with Chinese automotive

A BIT OF HISTORY *(continued)*

manufacturers and building automotive manufacturing facilities in China.

Front Wheel Toe

Front wheel toe may be defined the same as rear wheel toe. However, the front wheel toe specification usually differs from the rear wheel toe specification. On a rear-wheel-drive car, driving forces tend to move the steering knuckles to a toe-out position. Therefore, these vehicles usually have a slight toe-in. Drive axle forces on a front-wheel-drive vehicle tend to move the front steering knuckles to a toe-in position, and this condition requires a slight toe-out specification.

SUMMARY

TERMS TO KNOW

Angular bearing loads
Carbon dioxide (CO_2)
Greenhouse gas
Jounce travel
Negative camber
Negative caster
Negative offset
Positive camber
Positive caster
Positive offset
Radial bearing load
Rebound travel
Thrust bearing loads
Thrust line
Toe-out
Wheel alignment
Wheel offset
Wheel shimmy

- The suspension system must provide proper steering control and ride quality.
- The steering system must maintain vehicle safety and reduce driver fatigue.
- All parts of a unitized body are load-carrying members, and these parts are welded together to form a strong assembly.
- Carbon dioxide (CO_2) is a by-product of the gasoline, diesel fuel, or ethanol combustion process.
- Greenhouse gasses collect in the earth's upper atmosphere and form a blanket around the earth, which traps heat nearer the earth's surface.
- Front and rear suspension systems must provide proper wheel position, steering control, ride quality, and tire life.
- A short-and-long arm (SLA) front suspension system has a lower control arm that is longer than the upper control arm.
- In a MacPherson strut front suspension system, the top of the steering knuckle is supported by the lower end of the strut.
- Rear suspension systems can be live axle, semi-independent, or independent.
- Wheel rims are manufactured from steel, cast aluminum, forged aluminum, pressure-cast chrome-plated aluminum, or magnesium alloy.
- Wheel hubs contain the wheel bearings and support the load supplied by the vehicle weight.
- Bearing loads can be radial, thrust, or angular.
- Shock absorbers control spring action and wheel oscillations.
- The steering column connects the steering wheel to the steering gear.
- The steering linkage connects the steering gear to the front wheels.
- Steering gears can be recirculating ball or rack-and-pinion type.
- Proper wheel alignment provides steering control, ride quality, and normal tire tread life.
- Improper wheel alignment contributes to steering pull when driving straight ahead, improper steering wheel return, harsh ride quality, rapid tire tread wear, and steering pull while braking.

REVIEW QUESTIONS

Short Answer Essays

1. Explain how the engine and transaxle are supported in a front-wheel-drive vehicle with a unitized body.
2. Explain the purpose of coil springs in a short-and-long arm front suspension system.
3. Describe how the top of the steering knuckle is supported in a MacPherson strut front suspension system.
4. Explain the disadvantages of a live axle rear suspension system.
5. Explain the sources of CO_2 related to gasoline production and vehicle operation.
6. Describe the design of a wheel rim with positive offset.
7. Explain a radial bearing load.
8. Describe jounce wheel travel.
9. Explain the operation of a computer-controlled shock absorber that contains magneto-rheological fluid.
10. Explain why CO_2 is a harmful gas.

Fill-in-the-Blanks

1. In a rack and pinion steering gear, the inner ends of the tie rods are attached to the ends of the ________________.
2. The ________________ provides a steel box around the passenger compartment to provide passenger protection in a collision.
3. Tires are extremely important because they play a large part in providing ________________ and ________________.
4. Proper wheel alignment provides normal ________________ ________________ ________________, precise ________________ ________________, and satisfactory ride quality.
5. Excessive toe-out on a rear wheel may cause ________________ and rapid ________________.
6. The vehicle thrust line should be positioned at the ________________ of the vehicle.
7. If a wheel has negative camber, the tire centerline is tilted ________________ in relation to the true vertical tire centerline.
8. On a wheel rim the tapered openings match the tapers on the ________________.
9. Each corner of the vehicle has a ________________ or ________________ connected from the suspension system to the chassis.
10. Steering column connects the steering wheel to the ________________.

Multiple Choice

1. Unitized vehicle bodies have these special features and applications:
   A. Unitized bodies are typically used in large rear-wheel drive vehicles.
   B. Unitized bodies have special steel panels in the hood and trunk lid to protect the vehicle occupants in a collision.
   C. Some members of a unitized body are load-carrying members.
   D. Unitized bodies have a steel box around the passenger compartment.
2. The purpose of the upper and lower control arms in a short-and-long arm front suspension system is to:
   A. Allow the steering knuckle to turn to the right or left.
   B. Control spring action on irregular road surfaces.
   C. Reduce body sway.
   D. Control lateral (side-to-side) wheel movement.
3. A semi-independent rear suspension system has:
   A. A rear axle with an inverted steel U-section containing a tubular stabilizer bar.
   B. Upper and lower control arms.
   C. A one-piece rear axle housing.
   D. A lower ball joint.
4. All of these statements about wheel rims are true EXCEPT:
   A. Wheel rims must position the tires at the proper distance inward or outward from the vertical wheel mounting surface.
   B. A wheel rim with a positive offset has the center of the tire positioned inboard from the vertical wheel mounting surface.
   C. A wheel rim with a neutral offset has the center of the tire positioned at the vertical center of the brake rotor.
   D. A wheel rim with a negative offset has the center of the tire positioned outboard from the vertical wheel mounting surface.
5. A thrust-type bearing load is applied in a:
   A. Horizontal direction.
   B. Vertical direction.
   C. Angular direction.
   D. Radial direction.

6. Shock absorbers control spring action and help to prevent:
   A. Improper wheel and tire position.
   B. Wheel oscillations.
   C. Excessive steering effort.
   D. Slow steering wheel return after a turn.
7. In a parallelogram steering linkage, the tie rods are parallel to the:
   A. Steering arms.
   B. Center link.
   C. Upper control arms.
   D. Lower control arms.
8. All of these statements about recirculating ball steering gears are true EXCEPT:
   A. Ball bearings are mounted between the worm gear and the ball nut.
   B. The lower end of the sector shaft is splined to the idler arm.
   C. As the steering wheel is turned, the ball nut moves upward or downward on the worm gear.
   D. The steering shaft is attached to the worm gear.
9. Excessive rear wheel toe-out on the left rear wheel causes:
   A. Steering pull to the right.
   B. Harsh ride quality.
   C. Reduced steering effort.
   D. Excessive wheel oscillations.
10. Proper wheel alignment is extremely important to provide:
    A. Steering control.
    B. Ride quality.
    C. Reduce body sway.
    D. Normal tire tread life.

Chapter 2

Basic Theories

Upon completion and review of this chapter, you should be able to understand and describe:

- Newton's Laws of Motion.
- Work and force.
- Power.
- The most common types of energy and energy conversions.
- Inertia and momentum.
- Friction.
- Mass, weight, and volume.
- Static unbalance.
- Dynamic unbalance.
- The compressibility of gases and the noncompressibility of liquids.
- Atmospheric pressure and vacuum.
- Venturi operation.
- Voltage
- Current
- Resistance
- Ohm's Law

Introduction

An understanding of the basics is essential before you attempt a study of complex systems and components. Basic theories such as static balance, dynamic balance, and compressibility must be understood prior to a study of the components and systems in this book. If you have studied basic theories previously, the information in this chapter may be used as a review. A thorough study of this chapter will provide all the necessary background information you need before you study the suspension and steering systems in this book.

A BIT OF HISTORY

The automotive industry in the United States is a very large, dynamic industry. Total production of passenger cars, trucks, buses, and commercial vehicles has increased from 4,192 in 1900 to 12,770,714 in 2000. Since 1920, the lowest vehicle production was 725,215 in 1945.

Newton's Laws of Motion

First Law

A body in motion remains in motion, and a body at rest remains at rest, unless some outside force acts on it. When a car is parked on a level street, it remains stationary unless it is driven or pushed. If the gas pedal is depressed with the engine running and the transmission in drive, the engine delivers power to the drive wheels and this force moves the car.

Second Law

A body's acceleration is directly proportional to the force applied to it, and the body moves in a straight line away from the force. For example, if the engine power supplied to the drive wheels increases, the vehicle accelerates faster.

Third Law

For every action, there is an equal and opposite reaction. A practical application of this law occurs when the wheel on a vehicle strikes a bump in the road surface. This action drives the wheel and suspension upward with a certain force, and a specific amount of energy is stored in the spring. After this action occurs, the spring forces the wheel and suspension downward with a force equal to the initial upward force.

WORK AND FORCE

Force is defined as energy applied to an object.

Work is defined as the result of applying a force.

When a **force** moves a certain mass a specific distance, **work** is produced. When work is accomplished, the mass may be lifted or slid on a surface (Figure 2-1). Because force is measured in pounds and distance is measured in feet, the measurement for work is foot-pounds (ft.-lb). In the metric system, work is measured in Newton meters (Nm). If a force moves a 3,000-pound vehicle for 50 feet, 150,000 foot-pounds of work are produced. Mechanical force acts on an object to start, stop, or change the direction of the object. It is possible to apply force to an object and not move the object. For example, you may push with all your strength on a car stuck in a ditch and not move the car. Under this condition, no work is done. Work is only accomplished when an object is started, stopped, or redirected by mechanical force.

ENERGY

Energy may be defined as the ability to do work.

When **energy** is released to do work, it is called kinetic energy. This type of energy may also be referred to as energy in motion. Stored energy may be called potential energy. Energy is available in one of six forms:

1. *Chemical energy* is contained in the molecules of different atoms. In the automobile, chemical energy is contained in the molecules of gasoline and also in the molecules of electrolyte in the battery.
2. *Electrical energy* is required to move electrons through an electric circuit. In the automobile, the battery is capable of producing electrical energy to start the vehicle, and the alternator produces electrical energy to power the electrical accessories and recharge the battery.
3. *Mechanical energy* is the ability to move objects. In the automobile, the battery supplies electrical energy to the starting motor, and this motor converts the electrical energy to mechanical energy to crank the engine. Because this energy is in motion, it may be called kinetic energy.
4. *Thermal energy* is energy produced by heat. When gasoline burns, thermal energy is released.
5. *Radiant energy* is light energy. In the automobile, radiant energy is produced by the lights.

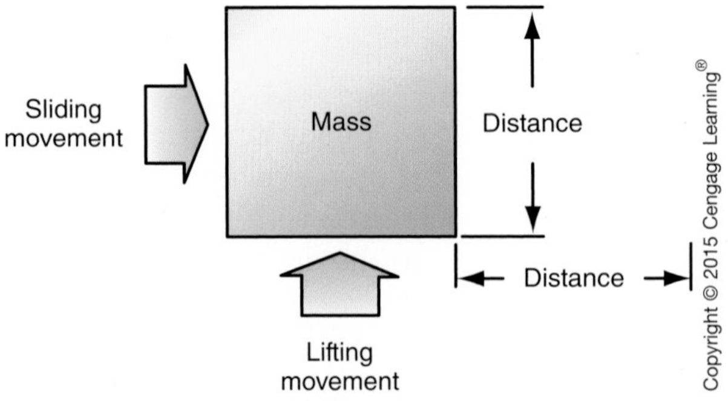

FIGURE 2-1 Work is accomplished when a mass is lifted or slid on a surface.

6. *Nuclear energy* is the energy within atoms when they are split apart or combined. Nuclear power plants generate electricity with this principle. This type of energy is not used in the automobile.

ENERGY CONVERSION

Energy conversion occurs when one form of energy is changed to another form. Since energy is not always in the desired form, it must be converted to a form we can use. Some of the most common automotive energy conversions are discussed in the following sections.

Chemical to Thermal Energy Conversion

Chemical energy in gasoline or diesel fuel is converted to thermal energy when the fuel burns in the engine cylinders.

Thermal to Mechanical Energy Conversion

Mechanical energy is required to rotate the drive wheels and move the vehicle. The piston and crankshaft in the engine and the drive train are designed to convert the thermal energy produced by the burning fuel into mechanical energy (Figure 2-2).

Electrical to Mechanical Energy Conversion

The windshield wiper motor converts electrical energy from the battery or alternator to mechanical energy to drive the windshield wipers.

Mechanical to Electrical Energy Conversion

The alternator is driven by mechanical energy from the engine. The alternator converts this energy to electrical energy, which powers the electrical accessories on the vehicle and recharges the battery.

INERTIA

The **inertia** of an object at rest is called static inertia, whereas dynamic inertia refers to the inertia of an object in motion. Inertia exists in liquids, solids, and gases. When you push and move a parked vehicle, you overcome the static inertia of the vehicle. If you catch a ball in motion, you overcome the dynamic inertia of the ball.

Inertia is defined as the tendency of an object at rest to remain at rest or the tendency of an object in motion to stay in motion.

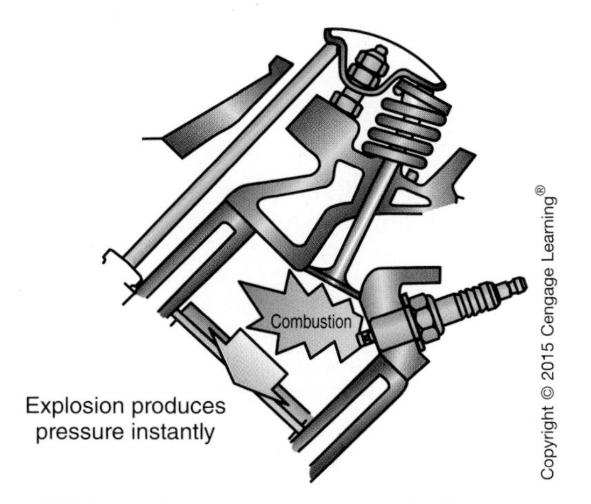

FIGURE 2-2 **Thermal energy in the fuel is converted to mechanical energy in the engine cylinders. The piston, crankshaft, and drive train deliver this mechanical energy to the drive wheels.**

When a force overcomes static inertia and moves an object, the object gains **momentum**.

Momentum

Momentum is the product of an object's weight times its speed. Momentum is a type of mechanical energy. An object loses momentum if another force overcomes the dynamic inertia of the moving object.

Friction may be defined as the resistance to motion when the surface of one object is moved over the surface of another object.

Friction

Friction may occur in solids, liquids, and gases. When a car is driven down the road, friction occurs between the air and the car's surface. This friction opposes the momentum, or mechanical energy, of the moving vehicle. Since friction creates heat, some of the mechanical energy from the vehicle's momentum is changed to heat energy in the air and body components. The mechanical energy from the engine must overcome the vehicle's inertia and the friction of the air striking the vehicle. Body design has a very dramatic effect on the amount of friction developed by the air striking the vehicle. The total resistance to motion caused by friction between a moving vehicle and the air is referred to as coefficient of drag (Cd). The study of Cd is not only very complicated but also very important. At 45 miles per hour (72 kilometers per hour), half of the engine's mechanical energy is used to overcome air friction, or resistance. Therefore, reducing a vehicle's Cd can be a very effective method of improving fuel economy and reducing CO_2 emissions.

Coefficient of drag (Cd) may also be called aerodynamic drag.

Mass, Weight, and Volume

A lawn mower is much easier to push than a 2,500-pound vehicle because the lawn mower has very little inertia compared to the vehicle. A space ship might weigh 100 tons on earth where it is affected by the earth's gravitational pull. In outer space beyond the earth's gravity and atmosphere, the space ship is almost weightless. Here on earth, **mass** and **weight** are measured in pounds and ounces in the English system. In the metric system, mass and weight are measured in grams or kilograms.

Mass is the measurement of an object's inertia.

Volume is a measurement of size, and it is related to mass and weight. For example, a pound of gold and a pound of feathers both have the same weight, but the pound of feathers occupies a much larger volume. In the English system, volume is measured in cubic inches, cubic feet, cubic yards, or gallons. The measurement for volume in the metric system is cubic centimeters or liters.

Weight is the measurement of the earth's gravitational pull on the object.

Volume is the length, width, and height of a space occupied by an object.

Torque

When you pull a wrench to tighten a bolt, you supply **torque** to the bolt. If you pull on a wrench to check the torque on a bolt, and the bolt torque is sufficient, torque is applied to the bolt, but no movement occurs. This torque, or twisting force, is calculated by multiplying the force and the radius. For example, if you supply a 10 lb force on the end of a 2-ft. wrench to tighten a bolt, the torque is $10 \times 2 = 20$ ft.-lb (Figure 2-3). If the bolt turns during torque application, work is done. When a bolt does not rotate during torque application, no work is accomplished.

Torque is a force that does work with a twisting or turning force. However, movement does not have to occur.

Power

James Watt calculated that a horse could move 330 lbs for 100 feet in 1 minute (Figure 2-4). If you multiply 330 lbs by 100 feet, the answer is 33,000 foot-pounds of work. Watt determined that one horse could do 33,000 foot-pounds of work in 1 minute. Thus, 1 **horsepower** (HP) is equal to 33,000 foot-pounds per minute, or 550 foot-pounds per second. Two horsepower could do this same amount of work in one-half minute, or 4 horsepower could complete this work in one-quarter minute. If you push a 3,000-pound (1,360-kilogram) car for 11 feet (3.3 meters) in one-quarter minute, you produce 4 horsepower. From this brief

Horsepower is a measurement for the rate, or speed, at which work is done.

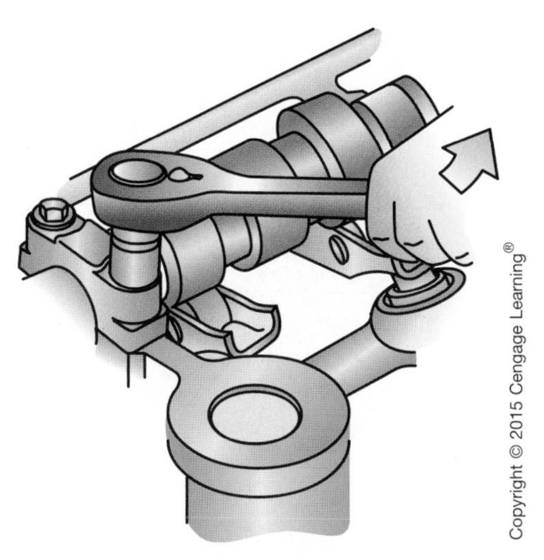

FIGURE 2-3 **Force applied to a wrench produces torque. If the bolt turns, work is accomplished.**

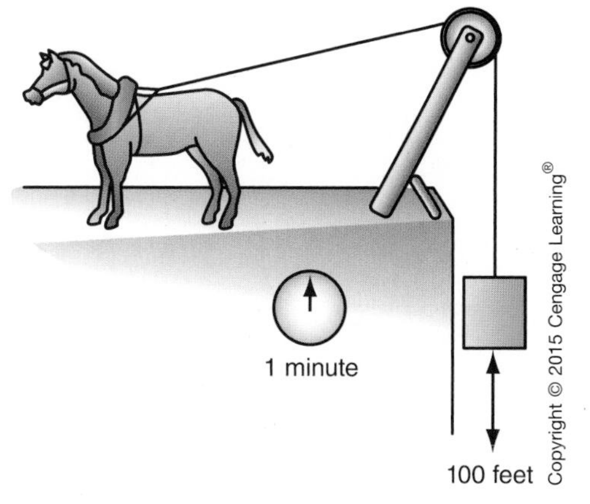

FIGURE 2-4 **One horsepower is produced when 330 lb are moved 100 ft. in 1 minute.**

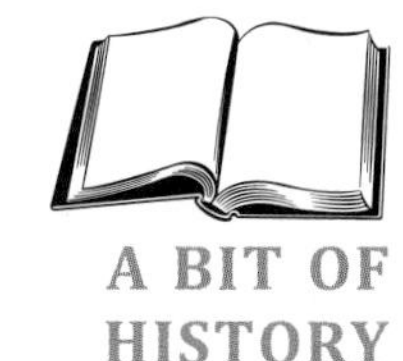

A BIT OF HISTORY

James Watt, a Scotsman, is credited with being the first person to calculate power. He measured the amount of work that a horse could do in a specific time.

discussion about horsepower, we can understand that as power increases, speed also increases, or the time to do work decreases.

Principles Involving Tires and Wheels in Motion

Shop Manual
Chapter 2, page 60

If you roll a cone-shaped piece of metal on a smooth surface, the cone does not move straight ahead. Instead, the cone moves toward the direction of the tilt on the cone. When you are riding a bicycle and want to make a left turn, if you tilt the bicycle to the left, it is much easier to complete the turn. The reason for this action is that a tilted, rolling wheel tends to move in the direction of the tilt. Similarly, if a tire and wheel on a vehicle are tilted, the tire and wheel tend to move in the direction of the tilt (Figure 2-5). This principle is used in front wheel alignment.

The casters on a piece of furniture are angled so the center of the caster wheel is some distance from the pivot center (Figure 2-6). When the furniture is moved, the casters turn on their pivots to bring the caster wheels into line with the pushing force on the furniture (Figure 2-7). This caster action causes the furniture to move easily in a straight line.

The weight of a bicycle rider is projected through the bicycle front fork to the road surface, and the tire pivots on the vertical centerline of the wheel when the handlebars are turned. Notice the centerline of the front fork is tilted rearward in relation to the

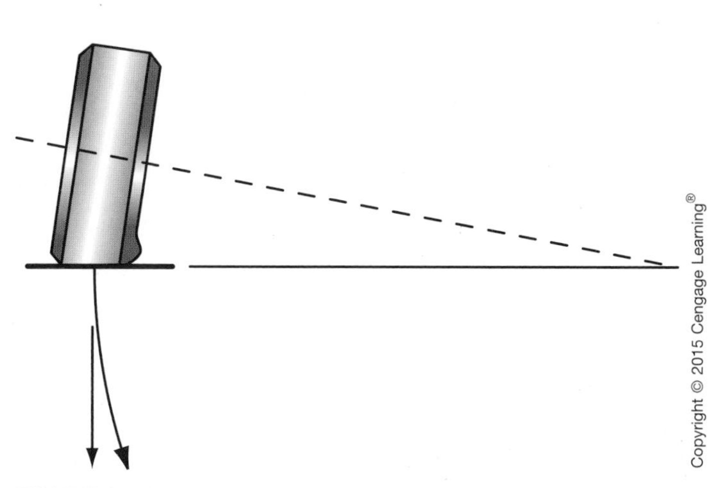

FIGURE 2-5 **A tilted, rolling wheel tends to move in the direction of the tilt.**

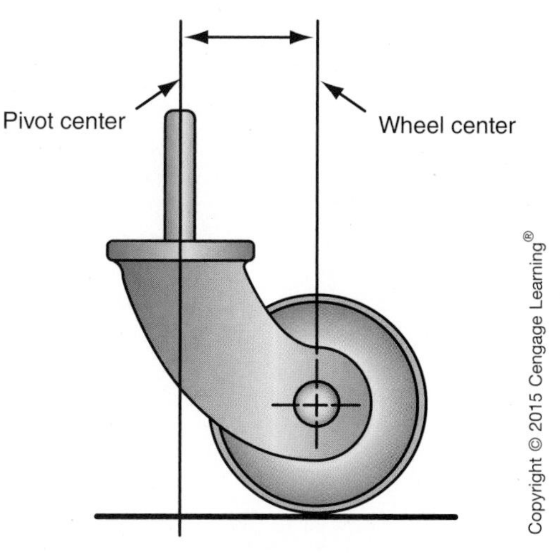

FIGURE 2-6 **Distance between the wheel center and the pivot center on a caster.**

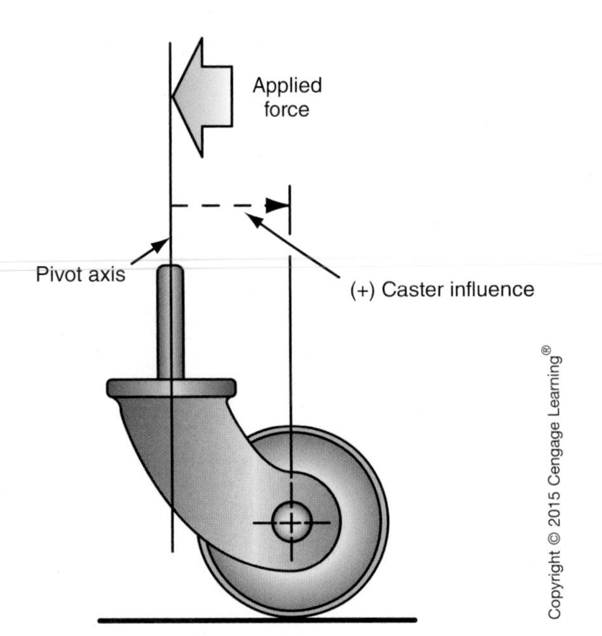

FIGURE 2-7 **Caster aligned with the pushing force provides straight-ahead movement when furniture is pushed.**

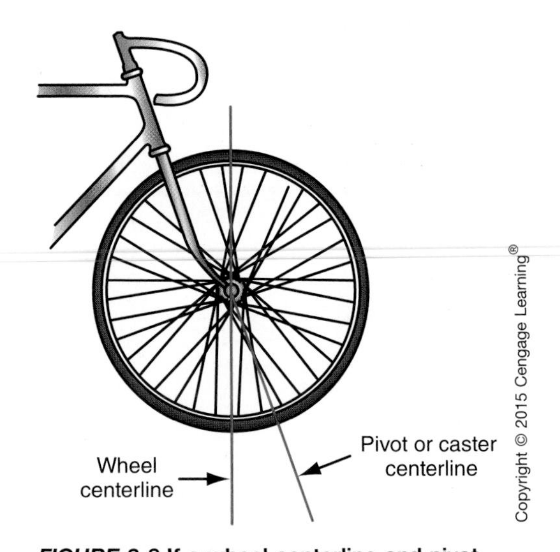

FIGURE 2-8 **If a wheel centerline and pivot point are behind the front fork centerline pivot point, the wheel tends to return to the straight-ahead position after a turn. The wheel also tends to remain in the straight-ahead position as the bicycle is driven.**

Rolling tires and wheels that are tilted always move in the direction of the tilt.

vertical centerline of the wheel (Figure 2-8). Since the tire pivot point is behind the front fork centerline where the weight is projected against the road surface, the front wheel tends to return to the straight-ahead position after a turn. The wheel also tends to remain in the straight-ahead position as the bicycle is driven. This principle is applied in automotive front wheel alignment.

Principles Involving the Balance of Wheels in Motion

Static balance refers to the equal distribution of weight around the center of a tire-and-wheel assembly viewed from the side.

When the weight of a tire and wheel assembly is distributed equally around the center of wheel rotation viewed from the side, the wheel and tire have proper **static balance**. Under this condition, the tire and wheel assembly has no tendency to rotate by itself, regardless of the wheel position. If the weight is not distributed equally around the center of wheel rotation, the wheel and tire are statically unbalanced (Figure 2-9). As the wheel and tire rotate, centrifugal force acts on this static unbalance and causes the wheel to "tramp" or hop. This wheel and tire action can often be seen while looking at tires on vehicles that may pass you on a highway.

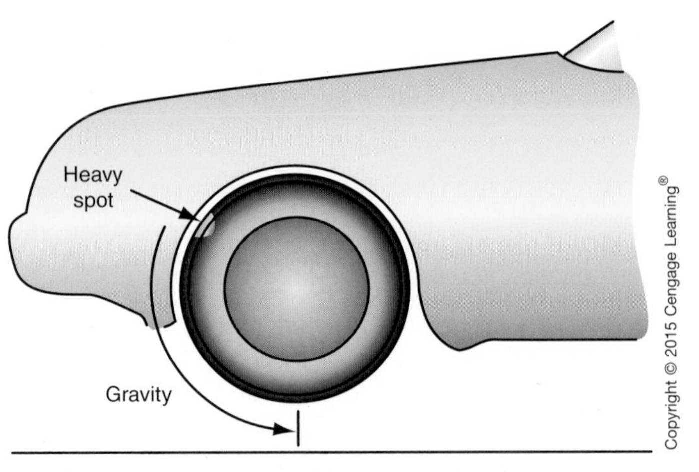

FIGURE 2-9 **Static wheel unbalance caused by unequal weight distribution around the center of the wheel rotation.**

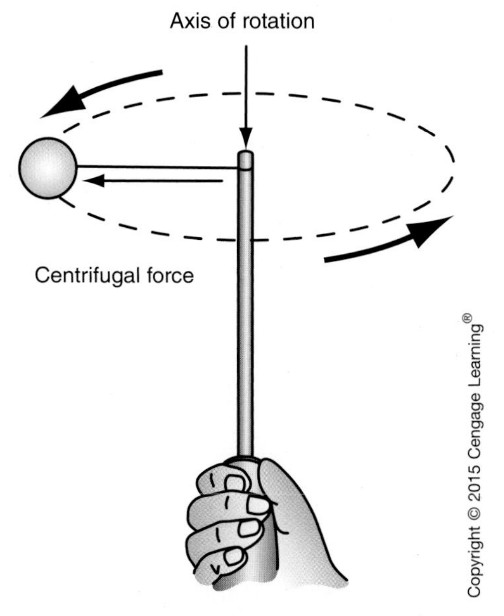

FIGURE 2-10 **A weight tends to rotate at a 90 degree angle in relation to the axis of rotation.**

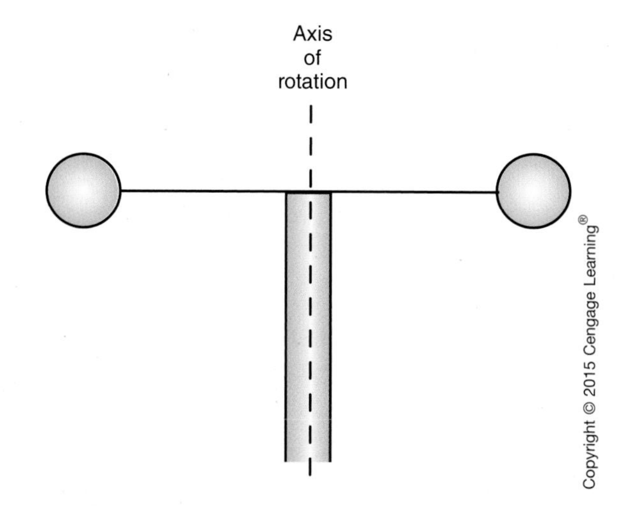

FIGURE 2-11 **When weight on a metal bar is distributed equally on the centerline of the axis of rotation, the bar and balls remain in dynamic balance during rotation.**

When a ball is rotated on the end of a string, the ball and string form an angle with the axis of rotation. If the rotational speed is increased, the ball and string form a 90 degree angle with the axis of rotation (Figure 2-10). Any weight will always tend to rotate at a 90 degree angle to the axis of rotation.

If two balls are positioned on a metal bar so their weight is equally distributed on the centerline of the rotational axis of rotation, the path of rotation remains at a 90 degree angle to the centerline of the axis when the bar is rotated. Under this condition, the metal bar and the balls are in **dynamic balance** (Figure 2-11).

Dynamic balance refers to the equal distribution of weight on each side of a tire and wheel centerline viewed from the front or rear.

If weights are placed on a metal bar so their weights are not equally distributed in relation to the centerline of the rotational axis of rotation, the weights still tend to rotate at a 90 degree angle in relation to the rotational axis. This action forces the pivot out of its vertical axis (Figure 2-12). When the bar is rotated 180 degree, the bar is forced out of its vertical axis in the opposite direction. If this condition is present, the bar has a wobbling action as it rotates. Under this condition, the bar is said to have dynamic unbalance, but static balance is maintained.

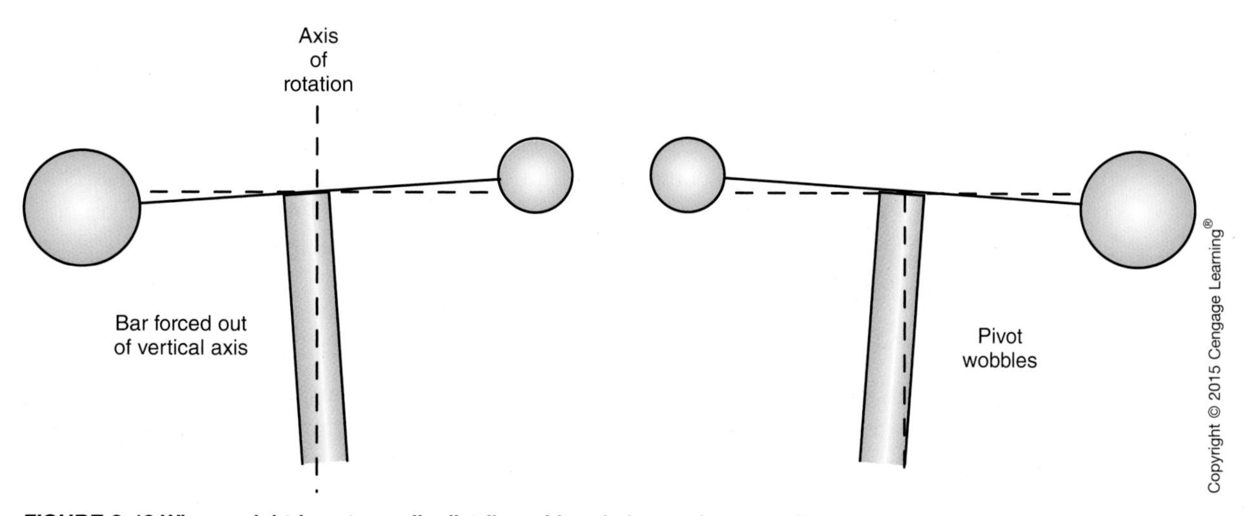

FIGURE 2-12 **When weight is not equally distributed in relation to the centerline of rotational axis, dynamic unbalance causes a wobbling action during rotation.**

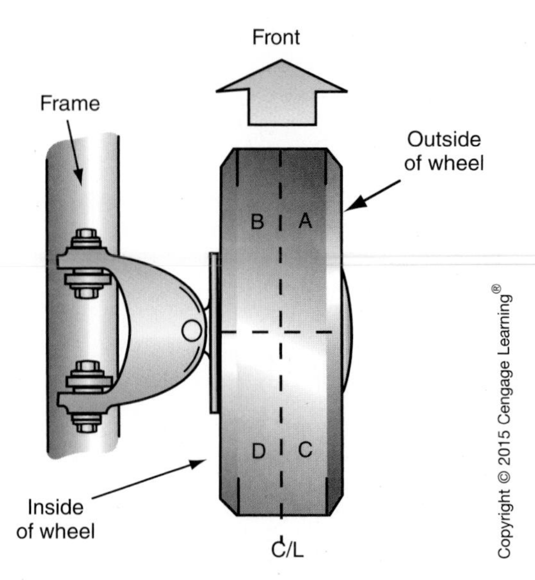

FIGURE 2-13 **Weight must be distributed equally in relation to the tire and wheel centerline.**

Molecular energy may be defined as the kinetic energy available in atoms and molecules because of the constant electron movement within the atoms and molecules.

Similarly, when the weight on a tire and wheel is not distributed equally on both sides of the tire centerline viewed from the front, the tire and wheel are dynamically unbalanced. This condition produces a wobbling action as the tire and wheel rotate. The weight must be distributed equally in relation to the tire centerline to provide proper dynamic balance (Figure 2-13). These principles are used in automotive wheel balancing.

> **AUTHOR'S NOTE: Improper static or dynamic wheel balance causes excessive tire tread wear, increased wear on suspension components, and driver fatigue.**

Pressure may be defined as a force exerted on a given surface area.

PRINCIPLES INVOLVING LIQUIDS AND GASES

Molecular Energy

Remember that kinetic energy refers to energy in motion. Since electrons are constantly in motion around the nucleus in atoms or molecules, kinetic energy is present in all matter. Kinetic energy in atoms and molecules increases as the temperature increases. A decrease in temperature reduces the kinetic energy. Molecules in solids move slowly compared to those in liquids or gases. Gas molecules move quickly compared to liquid molecules. Since gas molecules are in constant motion, they spread out to fill all the space available. At higher temperatures, gas molecules spread out more, whereas at lower temperatures, gas molecules move closer together.

Pressure and temperature are directly related. If you increase one, you also increase the other.

Pressure and volume are inversely proportional. If volume is decreased, pressure is increased.

Temperature

Temperature affects all liquids, solids, and gases. The volume of any matter increases as the temperature increases. Conversely, the volume decreases in relation to a reduction in temperature. When the gases in an engine cylinder are burned, the sudden temperature increase causes rapid gas expansion, which pushes the piston downward and causes engine rotation (Figure 2-14).

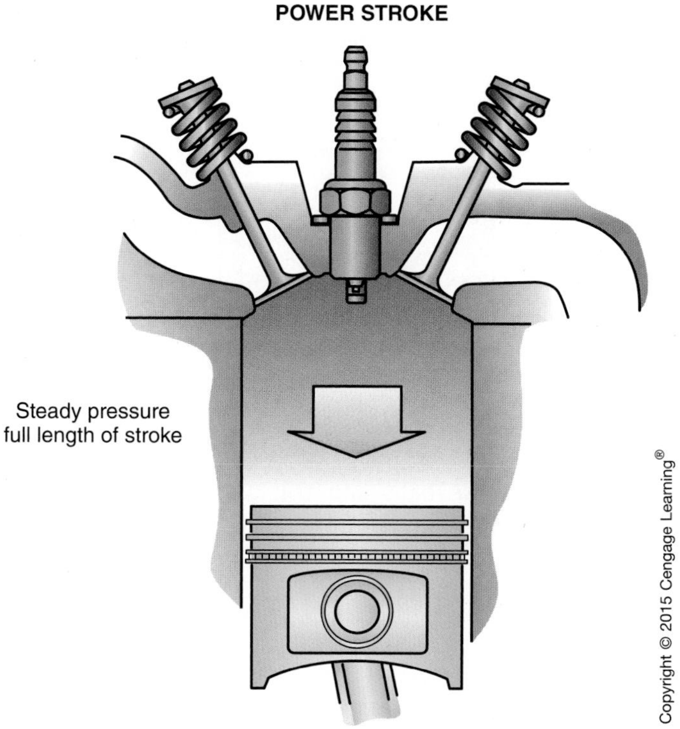

POWER STROKE

Steady pressure
full length of stroke

FIGURE 2-14 **Hot, expanding gases push the piston downward and rotate the crankshaft.**

Pressure and Compressibility

Liquids are not compressible.

Since liquids and gases are both substances that flow, they may be classified as fluids. If a nail punctures an automotive tire, the air escapes until the pressure in the tire is equal to atmospheric pressure outside the tire. When the tire is repaired and inflated, air pressure is forced into the tire. If the tire is inflated to 32 pounds per square inch (psi), or 220 kilopascals (kPa), this pressure is applied to every square inch on the inner tire surface. Pressure is always supplied equally to the entire surface of a container. Since air is a gas, the molecules have plenty of space between them. When the tire is inflated, the pressure in the tire increases, and the air molecules are squeezed closer together, or compressed. Under this condition, the air molecules cannot move as freely, but extra molecules of air can still be forced into the tire. Therefore, gases such as air are said to be **compressible**.

If a substance is **compressible**, its volume decreases as pressure on the substance increases.

The air in the tire may be compared to a few balls on a billiard table without pockets. If a few more balls are placed on the table, the balls are closer together, but they can still move freely (Figure 2-15).

If the vehicle is driven at high speed, friction between the road surface and the tires heats the tires and the air in the tires. When air temperature increases, the pressure in the tire also increases. Conversely, a temperature decrease reduces pressure.

If 100 cubic feet (2.8 cubic meters) of air is forced into a large truck tire, and the same amount of air is forced into a much smaller car tire, the pressure in the car tire is much greater.

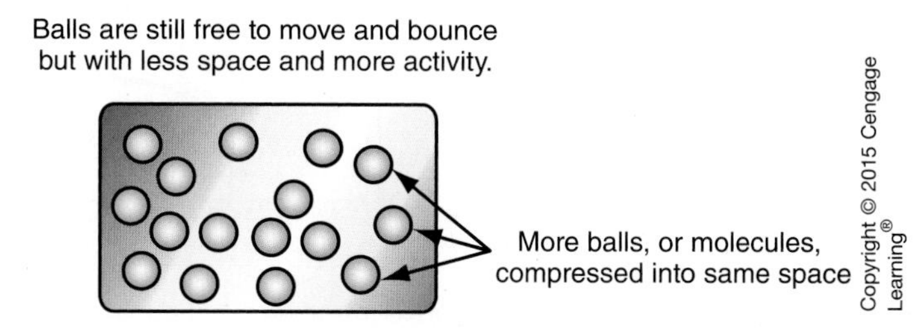
Balls are still free to move and bounce
but with less space and more activity.
More balls, or molecules,
compressed into same space

FIGURE 2-15 **Gases can be compressed much like more balls can be placed on a billiard table containing a few balls.**

Similar to a liquid, molecules occupy all space. In a container filled with liquid, molecules occupy all the space, and more molecules cannot be added to the container.

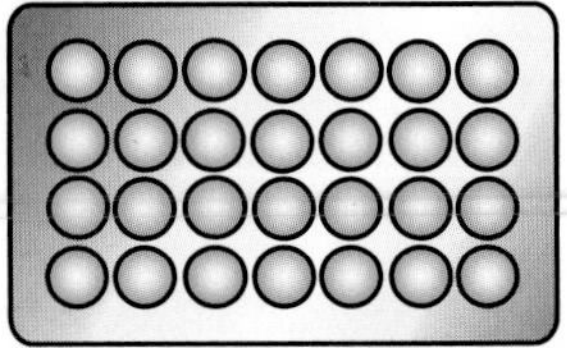
Balls could roll and move
Balls, or molecules, fill all available space
Copyright © 2015 Cengage Learning®

FIGURE 2-16 **Liquids are noncompressible, just as more balls cannot be added to a billiard table with no pockets that is completely filled with balls.**

Molecules in a liquid may be compared to a billiard table without pockets that is completely filled with balls. These balls can roll around, but no additional balls can be placed on the table because the balls cannot be compressed. Similarly, liquid molecules cannot be compressed (Figure 2-16).

Liquid Flow

If a tube is filled with billiard balls and the outlet is open, more balls may be added to the inlet. When each ball is moved into the inlet, a ball is forced from the outlet. If the outlet is closed, no more balls can be forced into the inlet (Figure 2-17).

The billiard balls in the tube may be compared to molecules of power steering fluid in the line between the power steering pump and steering gear. Since noncompressible fluid fills the line and gear chamber, the force developed by the pump pressure is transferred through the line to the gear chamber (Figure 2-18).

Hydraulic Fluid as a Flexible Machine

Hydraulic pressure has the same effect as a mechanical lever, because both of these items can multiply the input force to do more work. If a mechanical lever has a fulcrum at the center point and a 10-pound (lb), or 4.5-kilogram (kg), weight on one end of the lever,

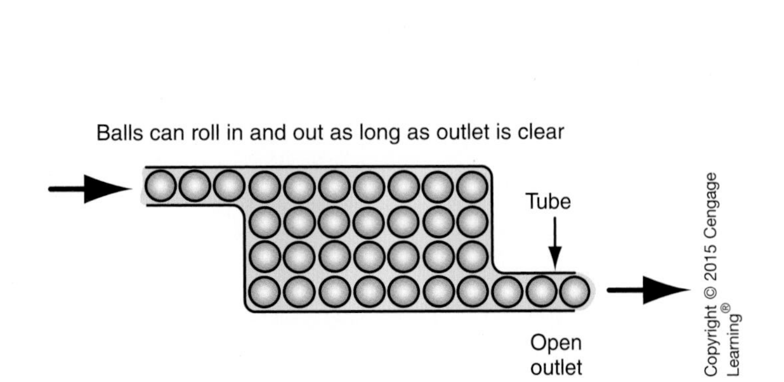
Balls can roll in and out as long as outlet is clear
Tube
Open outlet
Copyright © 2015 Cengage Learning®

FIGURE 2-17 **Billiard ball movement in a tube filled with balls compared to liquid flow.**

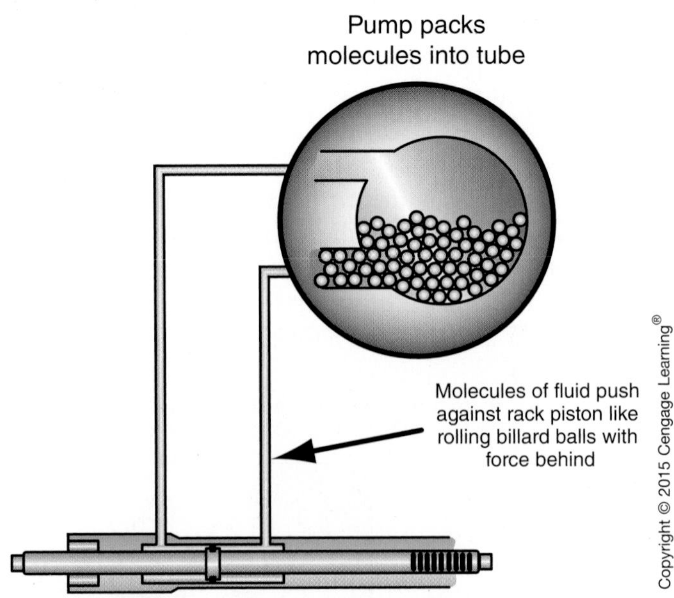
Pump packs molecules into tube
Molecules of fluid push against rack piston like rolling billard balls with force behind
Copyright © 2015 Cengage Learning®

FIGURE 2-18 **Power steering pump pressure supplied to the steering gear chamber.**

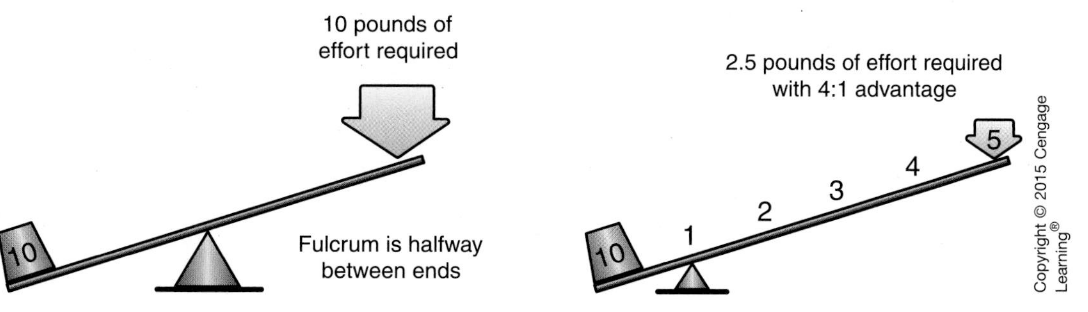

FIGURE 2-19 **A mechanical lever multiplies input force and makes it easier to do work.**

10 lb (4.5 kg) of weight are required on the other end of the lever to force the lever downward and raise the weight on the opposite end. If the lever is 5 ft. long and the fulcrum is placed 1 ft. from the weight to be lifted, the lever has a 4 to 1 mechanical advantage. Under this condition, 2.5 lb (1.1 kg) of weight are required on the other end of the lever to lift the 10-lb (4.5 kg) weight (Figure 2-19). Therefore, a mechanical lever multiplies input force and makes it easier to do work.

When the power steering pump pressure supplied to the steering gear chamber is 1,000 psi (7,000 kPa), this pressure is applied to every square inch in the steering gear chamber (Figure 2-20). This pressure applied to the rack piston in the gear chamber acts like a mechanical lever and helps move the rack piston. Since the rack is connected through steering linkages and arms to the front wheels, the force on the rack piston helps the driver move the front wheels to the left or right during a turn (Figure 2-21).

CAUTION:
Never completely block a pump outlet unless instructed to do so in the vehicle manufacturer's service manual. This action can result in extremely high pump pressure, damaged components, and personal injury.

Atmospheric pressure may be defined as the total weight of the earth's atmosphere.

Atmospheric Pressure

Since air is gaseous matter with mass and weight, it exerts pressure on the earth's surface. A one-square-inch column of air extending from the earth's surface to the outer edge of the atmosphere weighs 14.7 psi at sea level. Therefore, **atmospheric pressure** is 14.7 psi at sea level (Figure 2-22).

An equal volume of hot air weighs less and exerts less pressure on the earth's surface than cold air.

Atmospheric Pressure and Temperature

When air becomes hotter it expands, and this hotter air is lighter compared to an equal volume of cooler air. This hotter, lighter air exerts less pressure on the earth's surface compared to cooler air.

If the temperature decreases, air contracts and becomes heavier. Therefore, an equal volume of cooler air exerts more pressure on the earth's surface compared to hotter air.

Atmospheric pressure decreases as altitude increases.

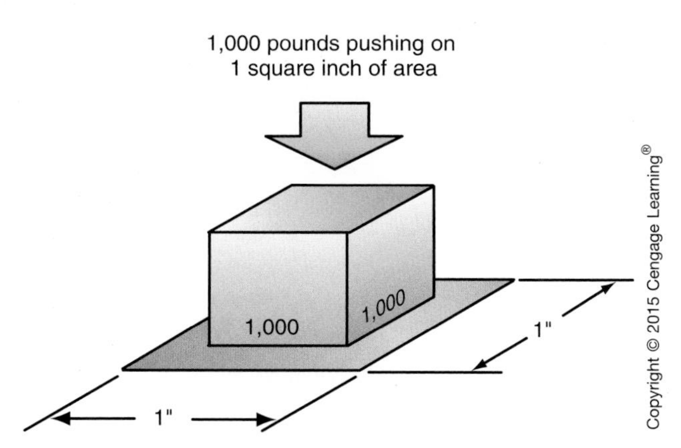

FIGURE 2-20 **Hydraulic pressure is applied equally to every square inch in a container.**

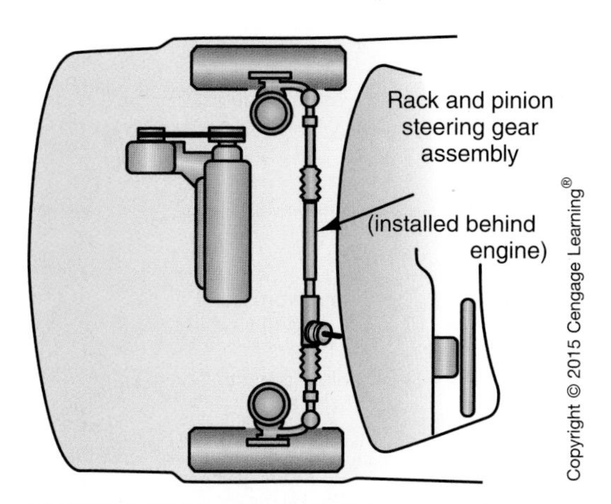

FIGURE 2-21 **Steering gear, linkages, and arms connected to the front wheels.**

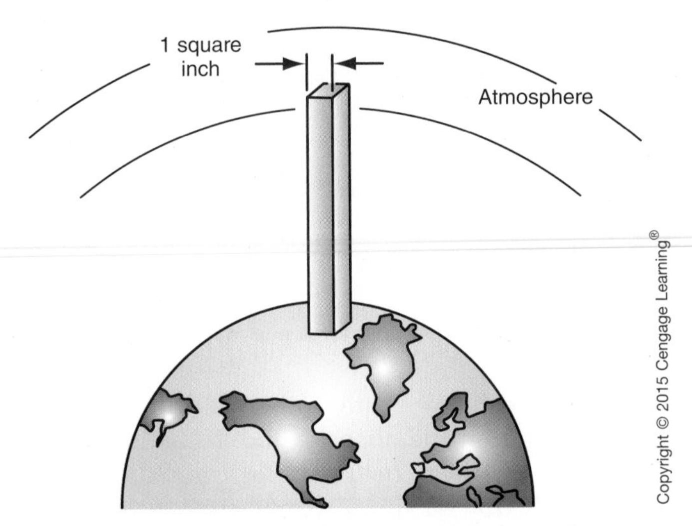

FIGURE 2-22 **A column of air one inch square extending from the earth's surface at sea level to the outer edge of the atmosphere weighs 14.7 lb.**

Atmospheric Pressure and Altitude

When you climb above sea level, atmospheric pressure decreases. The weight of a column of air is less at 5,000 feet (1,524 meters) elevation compared to sea level. As altitude continues to increase, atmospheric pressure and weight continue to decrease. At an altitude of several hundred miles above sea level, the earth's atmosphere ends, and there is only vacuum beyond that point.

Vacuum

Vacuum may be defined as the absence of atmospheric pressure. A vacuum may be called a low pressure because it is a pressure less than atmospheric pressure.

Atmospheric pressure is generally considered to be 14.7 pounds per square inch (psi) at sea level. Pressures greater than atmospheric pressure may be measured in psi, whereas pressures below atmospheric pressure are measured in **vacuum**, or psi absolute (psia) (Figure 2-23).

A conventional pressure gauge is used to measure pressures greater than atmospheric pressure. This type of pressure gauge indicates 0 psi at atmospheric pressure, and as the pressure increases it can read up to 15 psi. A conventional vacuum gauge indicates 0 inches of mercury (in. Hg) when atmospheric pressure is applied, and as the vacuum increases this gauge reads from 0 in. Hg to 29.91 in. Hg, and this may be considered a perfect vacuum.

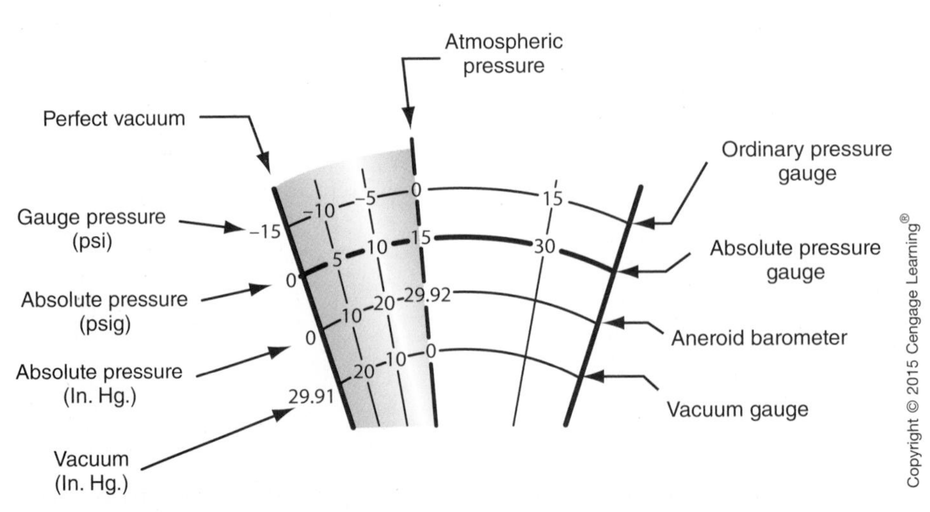

FIGURE 2-23 **Pressure and vacuum scales.**

An absolute pressure gauge indicates absolute pressure in pounds per square inch gauge (psig). An absolute pressure gauge indicates 0 at a perfect vacuum, 15 psig at atmospheric pressure, and 30 psig at 15 psi on a conventional pressure gauge. An aneroid barometer reads in. Hg absolute pressure, and thus it reads 0 in. Hg absolute at a perfect vacuum and 29.92 in. Hg absolute when atmospheric pressure is present.

Pressures above atmospheric pressure measured in psi are found in these automotive systems:

1. Fuel
2. Oil
3. Cooling
4. Power steering
5. Air conditioning
6. Turbocharger boost
7. Air springs
8. Gas-filled shock absorbers
9. Brake system during brake application

Pressures below atmospheric pressure measured in inches of vacuum are found in these automotive systems:

1. Manifold vacuum signal
2. Ported vacuum signal above the throttle
3. Carburetor venturi
4. Air-conditioning evacuation

Vacuum could be measured in psi, but inches of mercury (in. Hg) are most commonly used for this measurement. Let us assume that a manometer is partially filled with mercury, and atmospheric pressure is allowed to enter one end of the tube. If vacuum is supplied to the other end of the manometer, the mercury is forced downward by the atmospheric pressure. When this movement occurs, the mercury also moves upward on the side where the vacuum is supplied. If the mercury moves downward 10 in., or 25.4 centimeters (cm), where the atmospheric pressure is supplied, and upward 10 in. (25.4 cm) where the vacuum is supplied, 20 in. Hg is supplied to the manometer (Figure 2-24). The highest possible, or perfect, vacuum is approximately 29.9 in. Hg.

Liquids, solids, and gases tend to move from an area of high pressure to a low-pressure area.

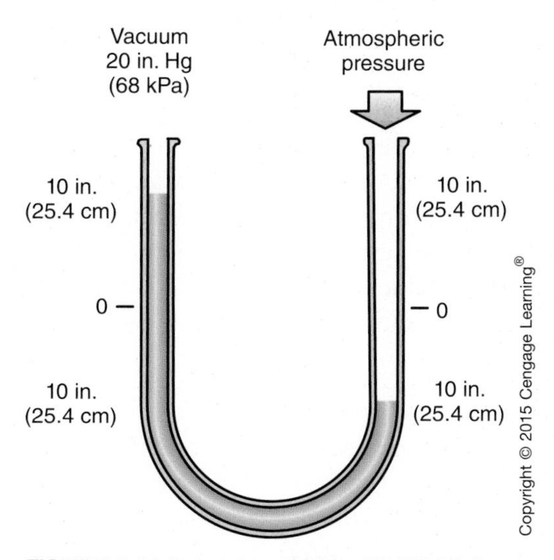

FIGURE 2-24 **A vacuum of 20 in. Hg (68 kPa) supplied to mercury in a manometer.**

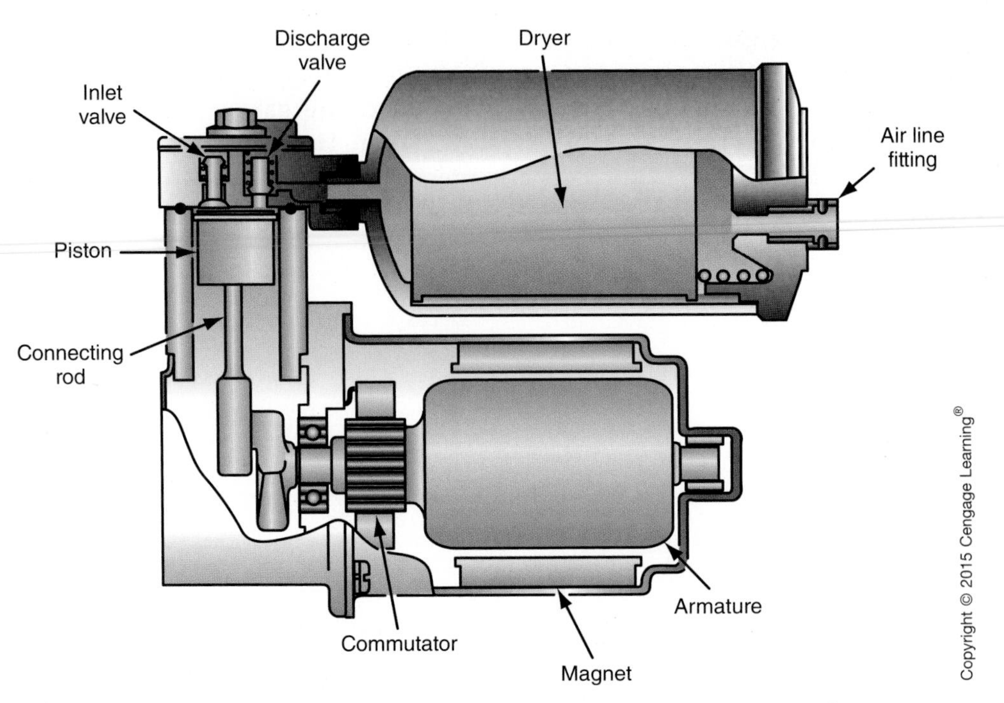

FIGURE 2-25 **An air suspension compressor creates vacuum in the cylinder, and atmospheric pressure at the inlet forces air into the cylinder. The compressor provides high pressure at the discharge valve, which moves air into the air springs.**

Vacuum and atmospheric pressure are used in several automotive systems. For example, atmospheric pressure is present outside the compressor inlet on an electronic air suspension system. When the compressor is running, it creates a vacuum at the inlet and in the compressor cylinder. This pressure difference causes air to move from the atmosphere surrounding the inlet into the cylinder. The compressor develops high pressure at the discharge valve, and this pressure forces air into the air springs when the pressure is lower than at the compressor outlet (Figure 2-25).

Pumps use high and low pressure to move liquids or gases. For example, as a power steering pump rotates, it creates a high pressure at the pump outlet by pumping against a restriction, and a low pressure at the inlet by moving fluid through a restriction. This pressure difference causes fluid to flow through the power steering system.

Venturi Principle

A venturi may be defined as a narrow area in a pipe through which a liquid or a gas is flowing.

If a gas or a liquid is flowing through a pipe and the pipe diameter is narrow in one place, the flow speeds up in the narrow area. This increase in speed causes a lower pressure in the narrow area, which may be defined as a venturi (Figure 2-26). Power steering pumps use a venturi principle to assist in the control of pump pressure.

Technicians must understand basic principles to comprehend the complex systems on modern vehicles. When basic principles are mastered, then technicians understand both the reason for, and the result of, these specific service procedures.

Electrical Principles

We often think of suspension and steering components and systems as mechanical, but many subsystems are controlled or powered by electricity. Examples of these are the active steering system, the computer controlled suspension system. Therefore, a basic understanding of some of the electrical principles, including voltage, amperage, and resistance is required. The following section is meant to be a review or overview of these concepts. It is

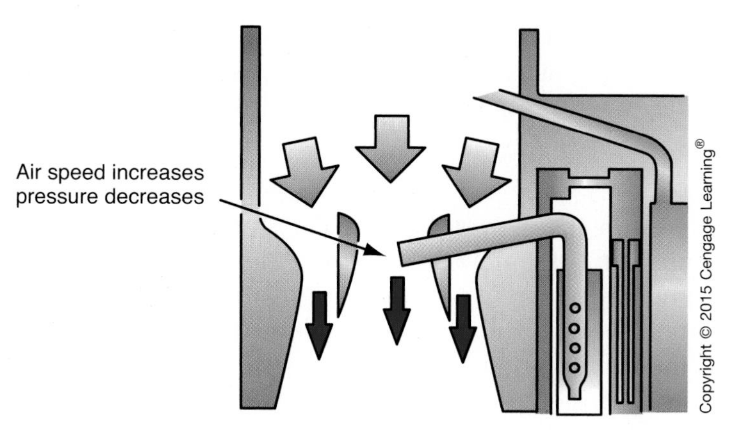

FIGURE 2-26 **A venturi increases air flow speed and produces a vacuum.**

recommended that a technician receive in-depth electrical and electronic training as part of their overall education. Electronics is part of every major system on the automobile and electrical failures have become routine complaints, though not always routine repairs.

The Basics: Voltage, Amperage, and Resistance

For many it is often easier to think of electricity in terms of water flowing in your home, hydraulic system principles, as there is a visible cause and effect. The flow of electricity is similar to the flow of water through your household plumbing. Where the water pressure is similar to electrical pressure or voltage, the water flow is similar to current flow in a conductor or amperage, and a restriction such as a kink in a hose is similar to the resistance in an electrical system. However, unlike water, electricity does not flow out the end of the wire and pour onto the ground if left open.

Voltage (V) or electromotive force (EMF) is the electrical pressure and is measured in volts (V); it may be either direct current (DC) or alternating current (AC). In the automotive industry, especially on hybrid electric vehicle (HEV) and electric vehicle (EV) platforms, we deal with both. Current is the flow or rate of flow of electrons under pressure in a conductor between two points having a difference in potential and is measured in **amperes (A)** or amperage. A complete circuit is required for current to flow. A DC circuit requires a complete circuit or loop between positive (+) and negative (−) for current to flow. Resistance is the friction in an electrical circuit, which restricts the flow of electrons under pressure and is measured in **ohms (Ω)**. Electrical resistance is a load on the moving current that must be present to do any useful work. Resistance controls the amount of current flow in an electrical circuit. Electrical devices that use electricity to operate have a greater amount of resistance than a conductor (wire) and are considered loads in a circuit. A motor, light bulb, or solenoids are examples of electrical loads in a circuit; a load in a circuit is the electrical device that consumes electricity. Poor connections and corrosion are examples of unwanted electrical loads in a circuit. Resistance in an electrical circuit is measured in ohms (Ω). If there is resistance (Ω) in the conductor, electrons will not flow as readily.

Voltage is a unit of measurement of electrical pressure.

Ampere is a unit of measurement of electrical current flow.

Ohm is a unit of measurement of electrical resistance.

For all practical automotive purposes electricity only flows through a good conductor; in the majority of automotive wiring this is copper. It takes a high voltage (V) to flow current (A) through a poor conductor. Air for example is a poor conductor, but with a high enough voltage even air can flow electricity. Lightning with millions of volts is capable of flowing current through air! High voltage electrical systems on HEV and EV platforms do not contain voltage as high as lightning, but it is not the 12V system either. To jump an inch of air, it takes approximately 10,000 volts. Have you ever seen a faulty secondary ignition wire (sparkplug wire) arcing to the engine? The voltage on the HEV and EV orange wiring harness is generally between 200 and 300 volts, which is too low to jump through air, but the capacitors and condensers in the system may contain much higher voltage.

To summarize:

- Voltage is the pressure that moves electrons and is measured in either AC or DC volts (V).
- Current or amperage is the flow or volume of electrons flowing in a conductor and is measured in amperes (A).
- Electrical resistance is a load or opposition on the moving electrons (current) in a circuit and is measured in ohms (Ω).

Ohm's Law

Ohm's law defines the relationship between voltage, resistance, and current.

The amount of current flow is determined by the amount of resistance in the loop of the circuit. In a fixed voltage circuit, if the amount of resistance in the loop of the circuit is high the current flow will be low and if the amount of resistance in the loop of the circuit is low the current flow will be high. This inverse relationship is known as **Ohm's law** and is summarized by the mathematical equation in Figure 2-27. Ohm's law states that it requires 1V of electrical pressure to move 1 A through 1Ω of resistance. Mathematically,

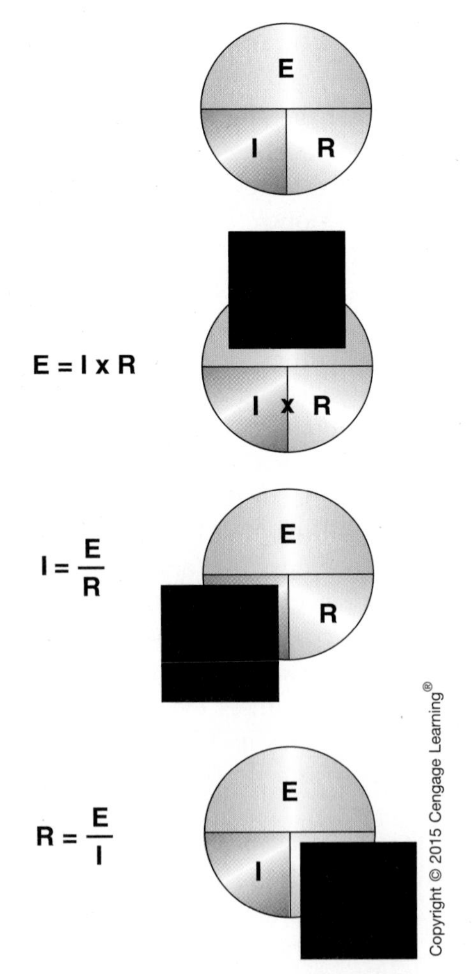

FIGURE 2-27 **With Ohm's law, if you know two of the three electrical factors you can calculate the third.**

Ohm's law is expressed by the following equations, using the symbols E for voltage, I for current, and R for resistance:

To find voltage $E = I \times R$
To find current $I = E \div R$
To find resistance $R = E \div I$

For example, if a 12.6 volt circuit has 2 ohm's of resistance you can use Ohm's law to determine the current flow in the circuit as follows:

$$I = E \div R$$
$$I = 12.6 \div 2$$
$$I = 6.3 \text{ amps}$$

The current flow in the circuit is 6.3 amps.

These equations can be used to calculate the voltage, current, or resistance of a circuit. It is the understanding of this relationship that is most important to you as a technician. You are not generally crunching numbers but instead you are making measurements on a circuit with a digital multimeter (DMM) and trying to understand what this information means. Ohm's law can help to clear up what the meter is telling you for a given reading.

Types of Circuits and Using Digital Multimeters

There are three basic types of circuits that we will be dealing with and a few ways to test these circuits for voltage, amperage, and resistance with a digital multimeter (DMM). It is imperative that you understand these basic concepts and how to test a circuit with a DMM to avoid damage to the electrical circuit and components and/or damage to your expensive DMM. Remember Snap-On does not warrantee misuse of equipment and service managers do not understand expensive mistakes! Take your time and think before you test. Remember the best place to begin your diagnosis is with the electrical diagram and system operation description contained in the vehicle service information database. As stated earlier this section is meant to be a review and not an in-depth training section, please refer to *Today's Technician Automotive Electricity & Electronics* for more information and training.

Series circuit is a circuit that provides a single path for current flow from the electrical source through all the circuit's components and back to the source (Figure 2-28).

Series Circuit Laws:

- Current flow is the same at any point in the circuit (Figure 2-29).
- The sum of the individual voltage drops equals the source voltage (Figure 2-30).
- Total circuit resistance is the sum of the individual circuit resistances (Figure 2-31).

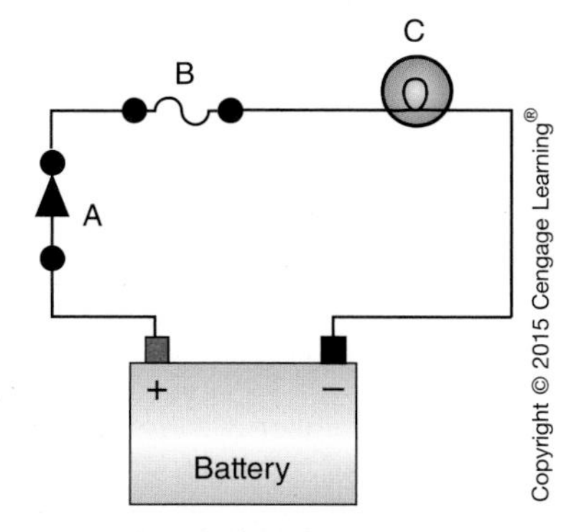

FIGURE 2-28 **A simple series circuit including a switch (A), a fuse (B), and a lamp (C). For all article purposes the only load in the circuit is the lamp.**

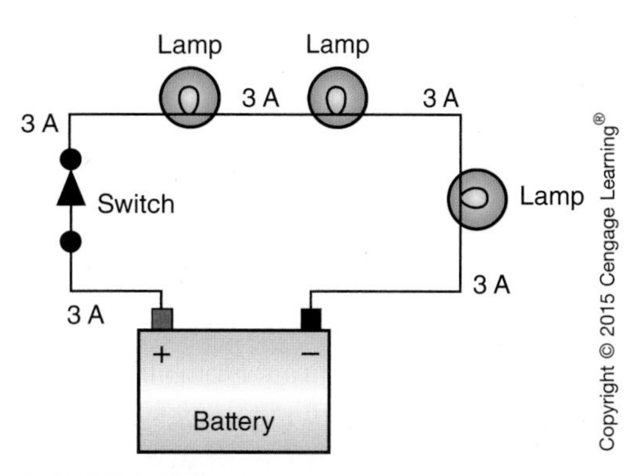

FIGURE 2-29 **Current flow is the same at any point in the circuit.**

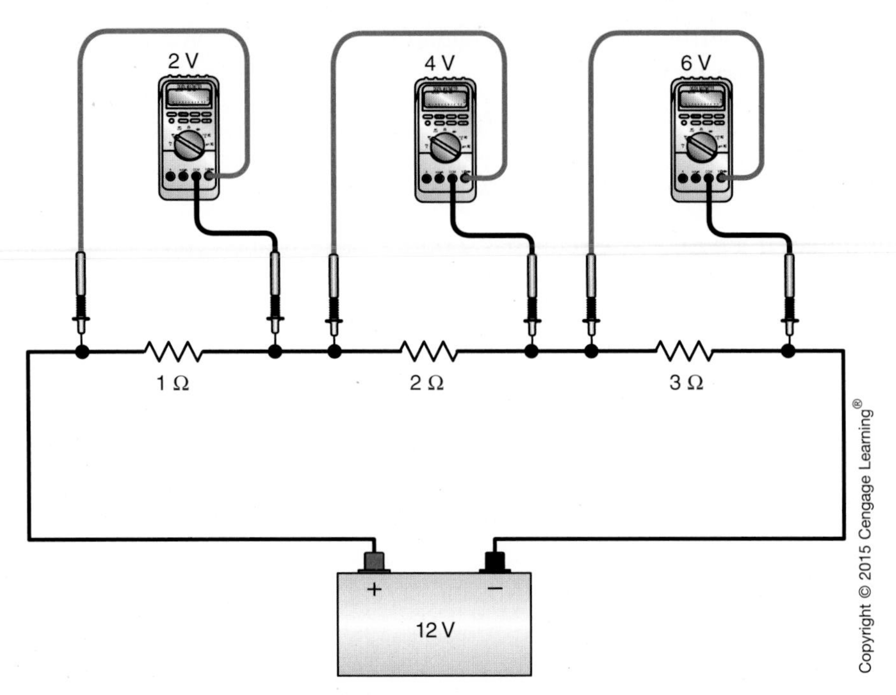

FIGURE 2-30 **The sum of the individual voltage drops equals the source voltage (2V+4V+6V=12V or source voltage).**

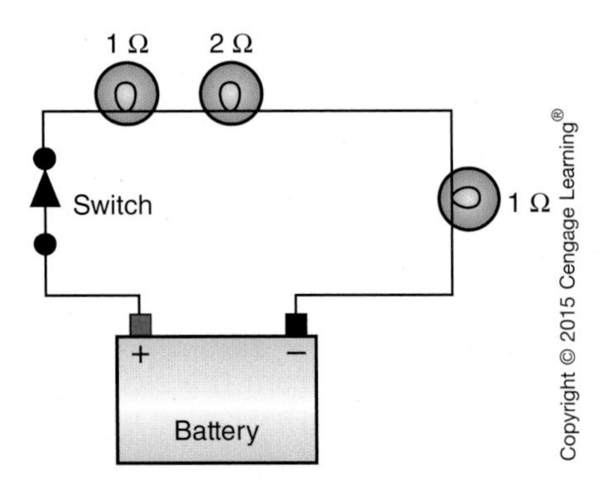

FIGURE 2-31 **Total circuit resistance is the sum of the individual circuit resistances, 1Ω + 2Ω + 1Ω = 4Ω of total circuit resistance.**

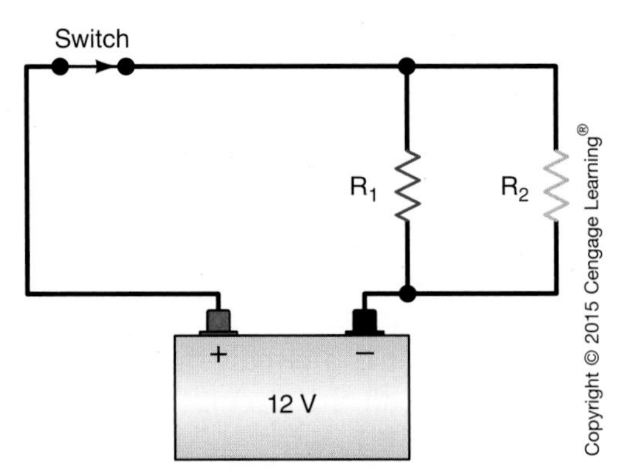

FIGURE 2-32 **There are multiple paths for current to flow in a parallel circuit. If R_1 were to fail R_2 would still work.**

Parallel circuit is a circuit that provides two or more paths for current flow through all the circuit's components and back to the source (Figure 2-32). In a parallel circuit each path has separate resistances that operate independently or in conjunction with one another depending on the design of the circuit. Current can flow through more than one branch at a time and voltage is the same across each branch of the circuit. In this type of circuit, the failure of a component in one branch does not affect the operation of the components in the other branches of the circuit.

Parallel Circuit Laws:

- Voltage is the same across each branch of the parallel circuit (Figure 2-33) and the voltage in each branch is used by the load(s) in that branch. The voltage dropped across each parallel branch will be the same; but, if the branch contains more than one resistor, the voltage drop across each of them will depend on the resistance of each resistor in the branch.

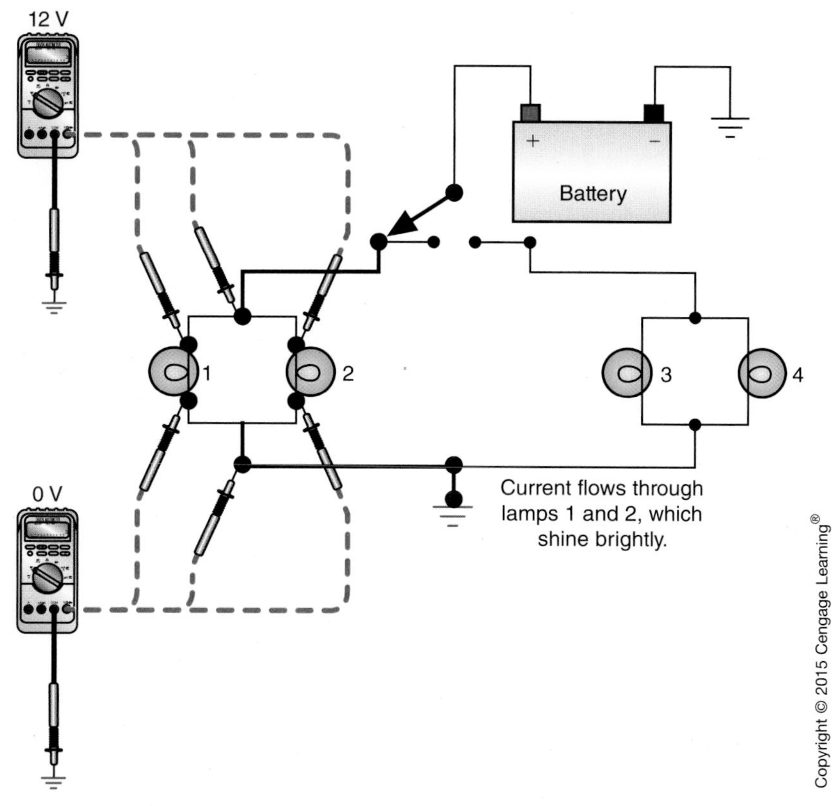

FIGURE 2-33 **Voltage is the same across each branch of the parallel circuit, and the voltage in each branch is used by the load(s) in that branch.**

- Total current in a parallel circuit is equal to the sum of the individual branch currents (Figure 2-34).
- Total resistance in a parallel circuit is always less than the smallest resistive branch (Figure 2-35).

$$R_T = \frac{1}{1/R_1 + 1/R_2 + 1/R_3 \ldots 1/R_{10}}$$

$$R_T = \frac{1}{1/3\Omega + 1/6\Omega + 1/1\Omega}$$

$$R_T = \frac{1}{0.333 + 0.166 + 1}$$

$$R_T = \frac{1}{1.5}$$

$$R_T = 0.667\Omega$$

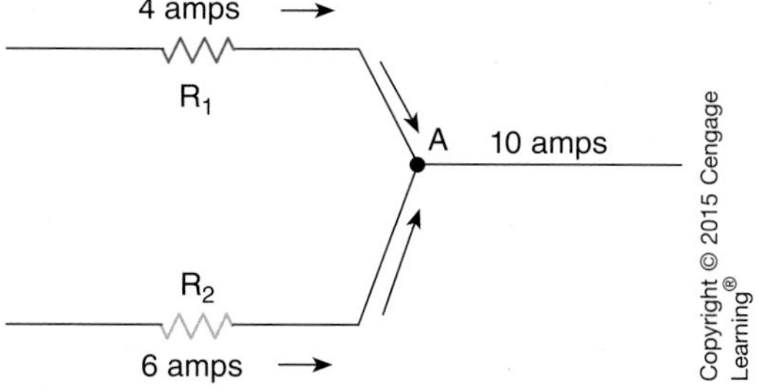

FIGURE 2-34 **Total current in a parallel circuit is equal to the sum of the individual branch currents.**

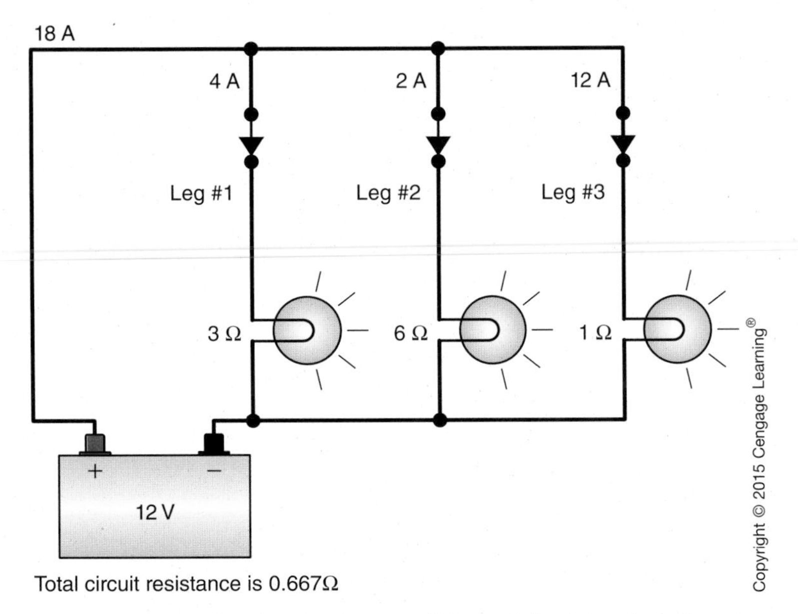

FIGURE 2-35 **Total resistance in a parallel circuit is always less than the smallest resistive branch.**

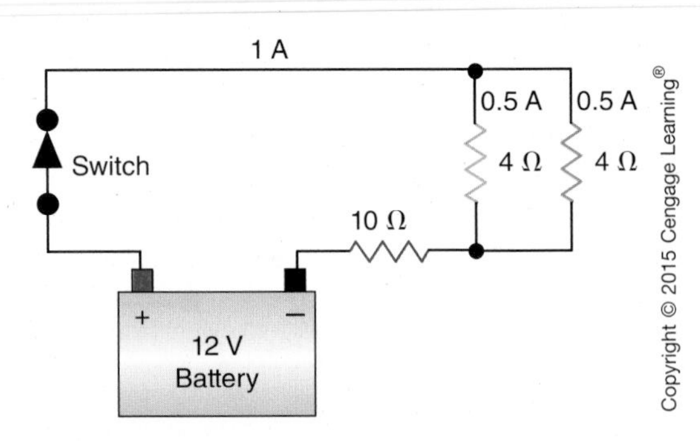

FIGURE 2-36 **The series-parallel circuit has some loads that are in series and some that are in parallel with each other.**

The series-parallel circuit has some loads that are in series and some that are in parallel with each other (Figure 2-36).

A voltmeter can be used to check for available voltage at the battery, terminals of any component or connectors. It can also be used to test voltage drops across electrical circuits, component loads, connectors, and switches. A voltmeter is connected in parallel with the circuit being tested (Figure 2-37). In Figure 2-38 voltage is being tested in a closed 12V series circuit with two loads. At test point A, the voltage should be the source voltage of 12V. At point B, the 1Ω resistor would have dropped half the voltage (there are two 1Ω loads in the circuit) and the meter should read 6V. At test point C, all the voltage should have been used up by the two loads in the circuit and the meter reading should be 0V. These readings would indicate normal circuit operation.

One of the most useful tests to perform is the voltage drop test. In a circuit all of the voltage provided by the source power is used (dropped) by the circuit with nothing left over. The voltage is used by resistance in wiring, connectors, switches, and loads. This loss or use of voltage is called a voltage drop and is the amount of electrical energy

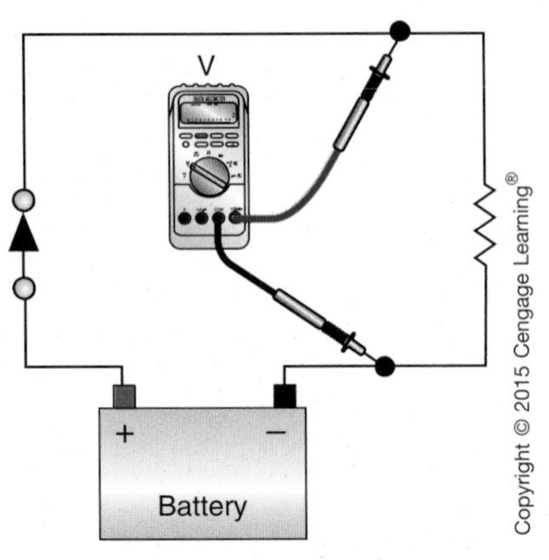

FIGURE 2-37 **A voltmeter is connected in parallel with the circuit being tested.**

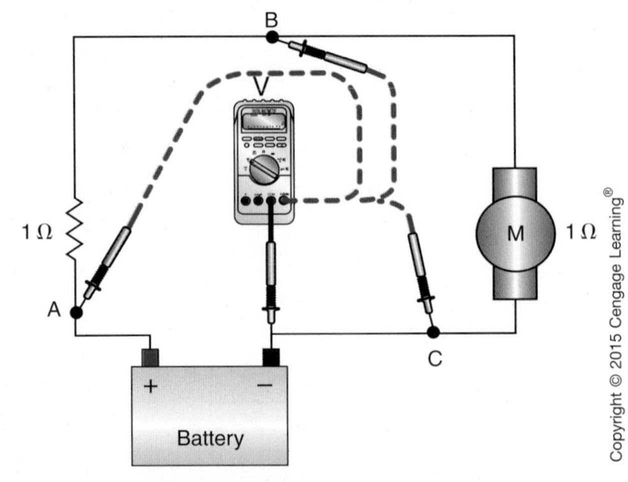

FIGURE 2-38 **A voltmeter testing for voltage at various points in the series circuit.**

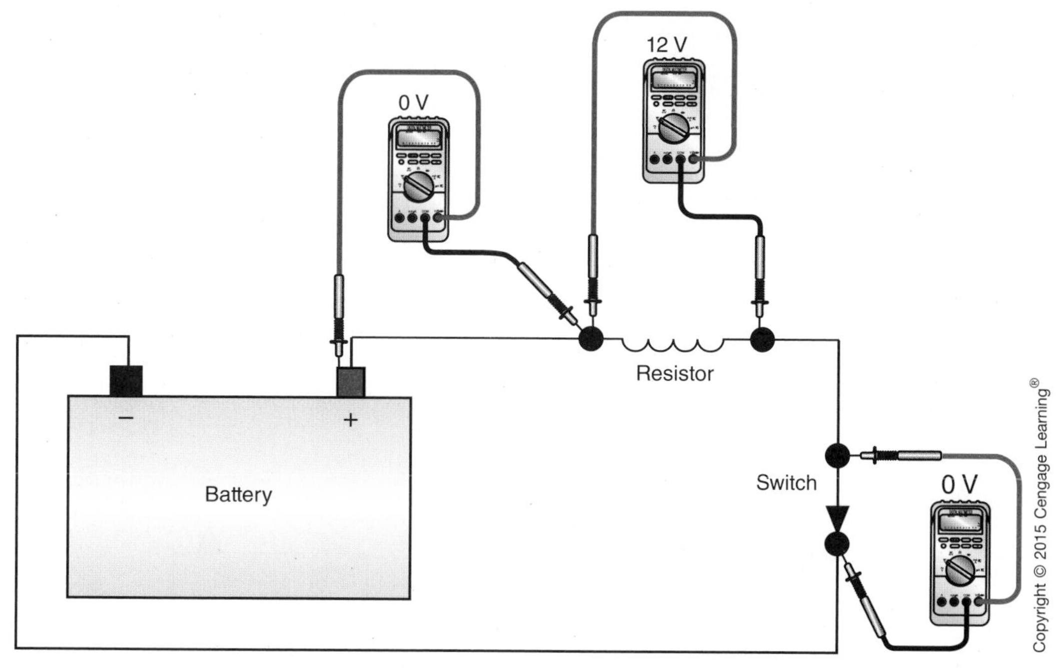

FIGURE 2-39 **Only the load will drop source voltage, the wiring and the switch should not drop significant voltage.**

converted into another form of energy. Voltage dropped in wiring, connectors and switches is converted into heat energy. When a circuit or branch of a circuit has only one load (resistance) source voltage is dropped across that load (Figure 2-39). If there is more than one load in a circuit each load will use a portion of the voltage. The total of all voltage drops in a circuit should equal source voltage. All of the voltage must be used by the circuit. You should verify that the circuit is turned on and that source voltage is available at the load and that the load drops source voltage. If the component (load) that is in the

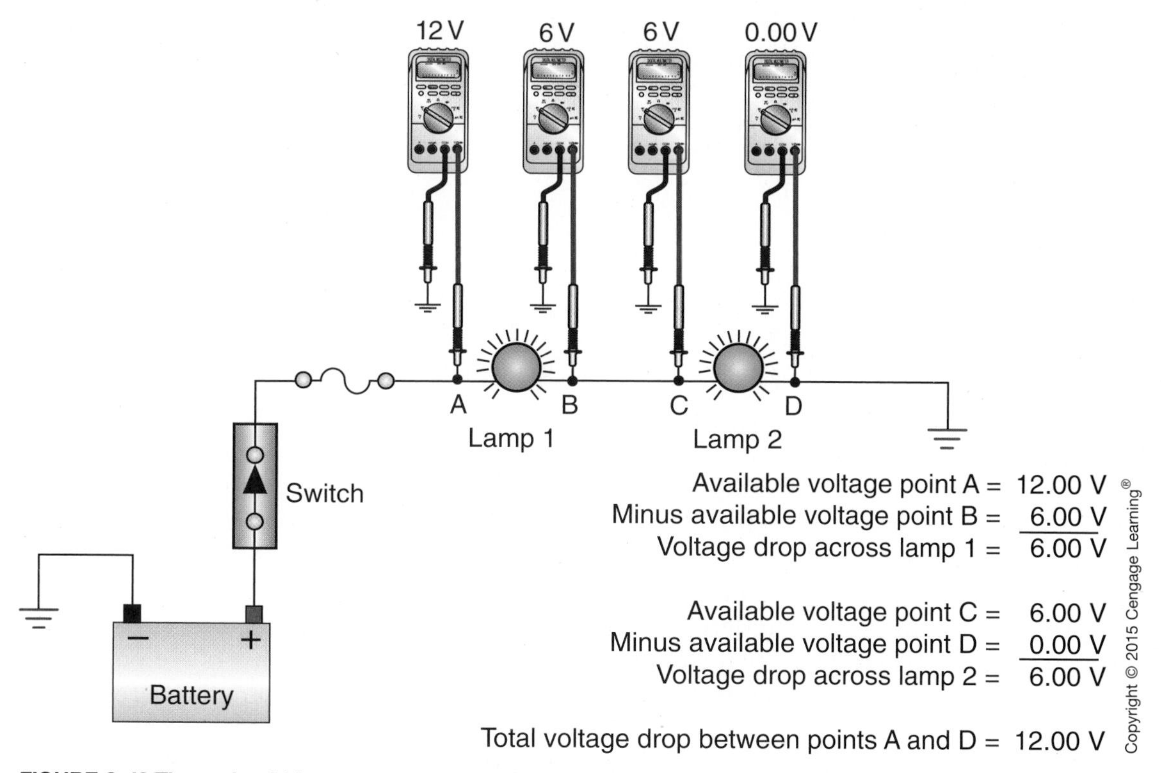

FIGURE 2-40 **There should be source voltage before the first load. If there are two loads in a series circuit each load will drop a portion of the source voltage. There should be no voltage after the last load.**

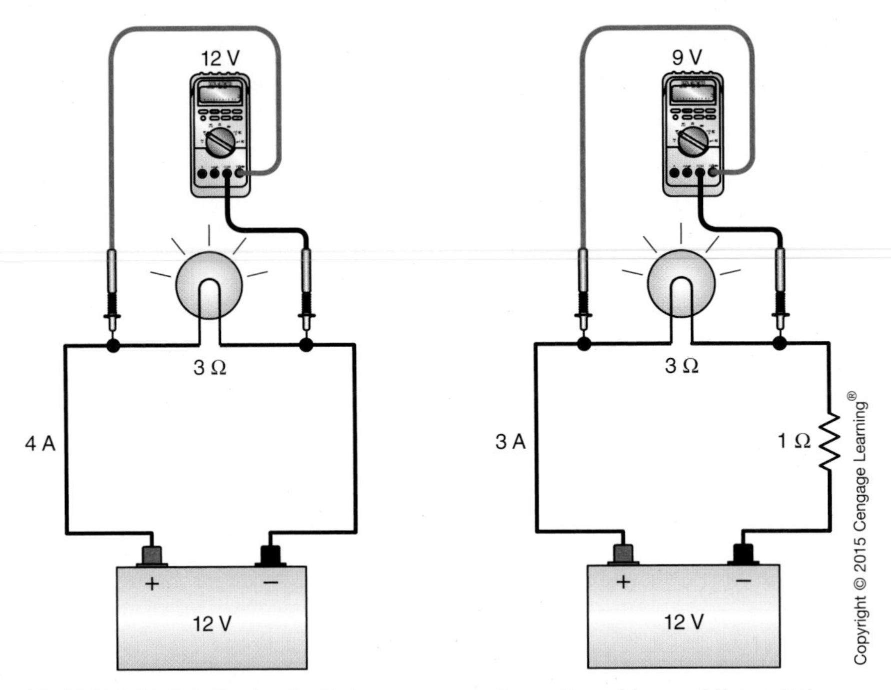

FIGURE 2-41 **Only the load will drop source voltage, the wiring and the switch should not drop significant voltage. But if there is unwanted resistance in a circuit, source voltage will not be available for the light bulb and the light bulb would be dimmer than normal or may not light at all.**

circuit drops source voltage then the component is faulty. In Figure 2-40, if both lamp 1 and lamp 2 are the same resistance value (size) they will both share the source voltage equally. Lamp 1 will use 6V and Lamp 2 will use 6V. If there is unwanted resistance in a circuit the load in the circuit will not drop all source voltage (Figure 2-41). There is some allowable voltage drop by circuit components other than the load, but it is generally limited to a maximum of:

0.2V (200 mV) for wires and cables
0.3V (300 mV) for a switch
0.1V (100 mV) for a ground
0V for a connection or connectors

Common circuit faults:

- Short circuits
- Short to ground
- Opens in a circuit
- High resistance, may be caused by corrosion or poor connections
- Low voltage

Table 2-1 shows common testing units and expected results when diagnosing the above common circuit faults. When using an ammeter to measure amperage, be sure to first turn power off in the circuit before connecting the multimeter. The circuit must be opened (disconnected) from the load being tested. The multimeter is placed in series in the circuit, recompleting the disconnected circuit (Figure 2-42). Always verify that the multimeter is capable of handling the highest expected amperage in the circuit being tested. What size fuse protects the circuit being tested? Many multimeters are only internally protected to 10 amperes by their internal fuse and many automotive circuits can provide 20 or more amps. In these cases, an inductive amp probe is the best choice of tools to use to avoid multimeter damage (Figure 2-43). The inductive probe eliminates the

TABLE 2-1 CIRCUIT TEST CHART

Type of Defect	Test Unit	Expected Results
Open	Ohmmeter	∞ infinite resistance between conductor ends
	Test light	No light after open
	Voltmeter	Ø volts at end of conductor after the open
Short to Ground	Ohmmeter	Ø resistance to ground
	Test light	Lights if connected across fuse
	Voltmeter	Generally not used to test for ground
Short	Ohmmeter	Lower than specified resistance through load component Ø resistance to adjacent conductor
	Test light	Light will illuminate on both conductors
	Voltmeter	A voltage will be read on both conductors
Excessive Resistance	Ohmmeter	Higher than specified resistance through circuit
	Test light	Light illuminates dimly
	Voltmeter	Voltage will be read when connected in parallel over resistance

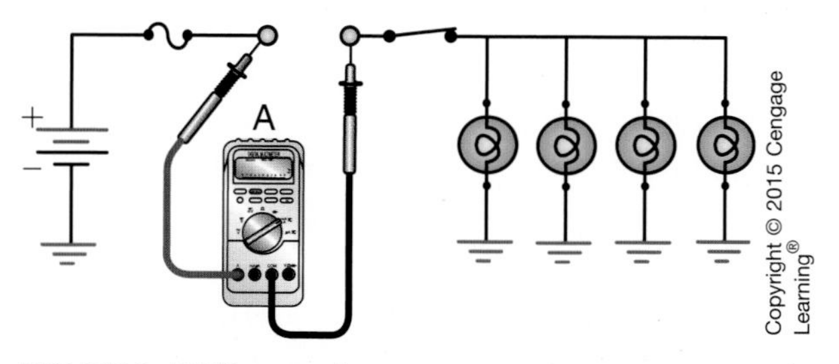

FIGURE 2-42 **When testing amperage draw the circuit must be opened (disconnected) from the load being tested the multimeter is placed in series in the circuit.**

FIGURE 2-43 **A multimeter with an inductive amp probe is a safe noninvasive method of measuring amperage in a circuit.**

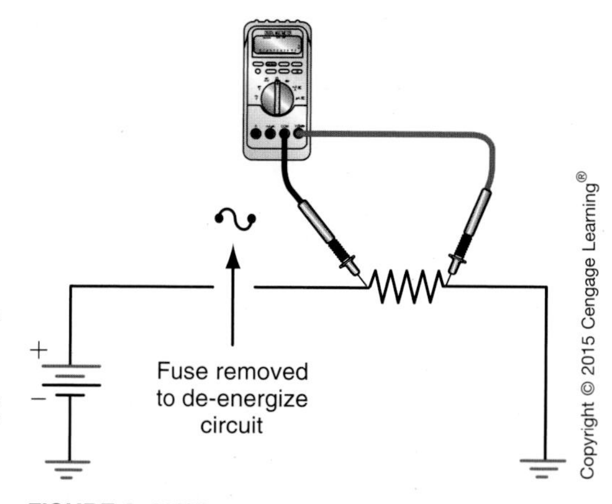

FIGURE 2-44 **When measuring resistance with an ohmmeter, the meter is connected in parallel with power removed from the circuit and circuit or component isolated.**

need to connect the multimeter in series and is a safe noninvasive method of measuring amperage in a circuit.

When using an ohmmeter to measure resistance, the power from the circuit must be removed and circuit or component should be isolated (Figure 2-44). Ohmmeter leads are placed across or in parallel with component or circuit being tested.

SUMMARY

TERMS TO KNOW

Ampere (A)
Atmospheric pressure
Compressible
Dynamic balance
Energy
Force
Friction
Horsepower
Inertia
Mass
Momentum
Ohms (Ω)
Ohm's law
Static balance
Torque
Vacuum
Voltage (V)
Volume
Weight
Work

- Work is the result of applying a force.
- Force is measured in pounds and distance.
- Energy is the ability to do work; there are six basic types of energy.
- Inertia is the tendency of an object at rest to remain at rest, or the tendency of an object in motion to remain in motion.
- An object gains momentum when force overcomes static energy and moves the object.
- Friction is the resistance to motion when one object is moved over another object.
- Mass is a measurement of an object's inertia.
- Weight is a measurement of the earth's gravitational pull on an object.
- Volume is the length, width, and height of a space occupied by an object.
- Power is a measurement for the speed at which work is done.
- A rolling, tilted wheel tends to move in the direction of the tilt.
- If the pivot point at the tire centerline is behind the centerline where the vehicle weight is projected on the road surface, the wheel tends to remain in the straight-ahead position.
- Weight must be distributed equally around the center of wheel rotation viewed from the side to obtain static balance.
- Weight must be distributed equally on both sides of the tire centerline viewed from the front of the tire to maintain proper dynamic balance.
- Torque is a twisting force that does work.
- Atmospheric pressure is the total weight of the earth's atmosphere.
- Vacuum is the absence of atmospheric pressure.
- Voltage (V) or electromotive force (EMF) is the electrical pressure and is measured in volts (V); it may be either direct current (DC) or alternating current (AC).
- Current is the flow or rate of flow of electrons under pressure in a conductor between two points having a difference in potential and is measured in ampere (A) or amperage.
- Resistance is defined as the opposition to current flow and is measured in ohms (Ω).
- Ohm's law defines the relationship between voltage, current, and resistance. It is the basic electrical law.

REVIEW QUESTIONS

Short Answer Essays

1. Describe Newton's first law of motion, and give an application of this law in automotive theory.
2. Explain Newton's second law of motion, and give an example of how this law is used in automotive theory.
3. Describe Newton's third law of motion, and give an example of how this law applies to an automotive suspension system.
4. Describe six different forms of energy.
5. Describe four different types of energy conversion.
6. Explain the difference between static and dynamic inertia.
7. Explain why a rotating, tilted wheel moves in the direction of the tilt.
8. Explain why the front wheel of a bicycle tends to remain in the straight-ahead position as the bicycle is driven.
9. Define static and dynamic balance.
10. Briefly define the term *current.*

Fill-in-the-Blanks

1. Improper static wheel balance causes wheel ________________ when driving the vehicle.
2. If the tire pivot point is behind the tire and wheel centerline where the vehicle weight is projected against the road surface, the wheel tends to remain in the ________________ position when driving the vehicle.
3. Work is calculated by multiplying ________________ × ________________.
4. ________________ is the flow or rate of flow of electrons under pressure in a conductor between two points having a difference in potential and is measured in ________________.
5. When one object is moved over another object, the resistance to motion is called ________________.
6. ________________ is the electrical pressure it may be either direct current (DC) or alternating current (AC).
7. Torque is a force that does work with a ________________ action.
8. Power is a measurement for the rate at which ________________ is done.
9. To obtain proper dynamic balance, the weight must be distributed equally on both sides of a wheel and tire centerline viewed from the ________________ of the tire.
10. Vacuum is defined as the absence of ________________.

Multiple Choice

1. When an engine is running, the alternator converts:
   - A. Thermal energy to mechanical energy.
   - B. Electrical energy to mechanical energy.
   - C. Chemical energy to electrical energy.
   - D. Mechanical energy to electrical energy.
2. Work is accomplished during all of these conditions EXCEPT:
   - A. When a mechanical force starts an object in motion.
   - B. When a mechanical force is applied to an object, but the object does not move.
   - C. When a mechanical force stops an object in motion.
   - D. When a mechanical force redirects an object in motion.
3. All these statements about Newton's Laws of Motion are true EXCEPT:
   - A. For every action there is an equal and opposite reaction.
   - B. A body in motion remains in motion unless an outside force acts on it.
   - C. A body's acceleration is directly proportional to the force applied to it.
   - D. A body moves in an arc away from the force acting upon the object.
4. Torque is calculated by:
   - A. Multiplying the force and the radius from the force to the object.
   - B. Dividing the radius by the force.
   - C. Adding the force and the distance of force movement.
   - D. Subtracting the radius from the force to the object and the weight of the object.
5. When working with gases and liquids:
   - A. Gases are not compressible.
   - B. Liquids are not compressible.
   - C. Temperature does not affect gas volume.
   - D. Pressure is applied unevenly to the inside of a container surface.
6. When applying the principles of work and force:
   - A. Work is accomplished when force is applied to an object that does not move.
   - B. In the metric system the measurement for work is cubic centimeters.
   - C. No work is accomplished when an object is stopped by mechanical force.
   - D. If a 50-pound object is moved 10 feet, 500 ft.-lb of work are produced.
7. All of the following are true of voltage drops EXCEPT:
   - A. Voltage drop can be measured with a voltmeter.
   - B. All of the source voltage in a circuit must be dropped.
   - C. Corrosion in a circuit does not cause a voltage drop.
   - D. Voltage drop is the conversion of electrical energy into another form of energy.

8. A lever is 10 ft. (3.0 m) long and the fulcrum is positioned 1 ft. (0.304 m) from the end of the lever. A 5 lb (2.26 kg) weight is placed on the end of the lever nearest the fulcrum. The weight required on the opposite end of the lever to lift the 5 lb (2.26 kg) weight is:
   A. 0.368 lb (0.166 kg).
   B. 0.555 lb (0.251 kg).
   C. 0.714 lb (0.323 kg).
   D. 0.748 lb (0.339 kg).

9. A tire and wheel assembly that does not have proper dynamic balance has:
   A. Weight that is not distributed equally on both sides of the tire centerline.
   B. Worn tread on the inside edge of the tire.
   C. A tire with improperly positioned steel belts.
   D. Weight that is not distributed equally around the center of the tire and wheel.

10. When a car is driven on the road above 50 mph (80 km/h), the left front tire has a tramping action. The MOST likely cause of this condition is:
    A. Improper left front dynamic wheel balance.
    B. A bent left front wheel rim.
    C. Improper left front static wheel balance.
    D. Improper rim offset.

Chapter 3

Tires and Wheels

Upon completion and review of this chapter, you should be able to understand and describe:

- General tire function.
- Typical tire construction, and identify the purpose of each component in a tire.
- Three types of tire ply and belt designs.
- Tire ratings, and explain the meaning of each designation in the rating.
- The purpose of the tire performance criteria (TPC) rating.
- The difference between all-season tires and conventional tires.
- Two different types of tire load ratings.
- The Uniform Tire Quality Grading (UTQG) designations.
- The precautions to be observed when selecting replacement tires.
- Tire contact area, free tire diameter, and rolling tire diameter.
- The tire motion forces while a tire and wheel assembly is rotating on a vehicle.
- The importance of tire design quality as it relates to the tire motion forces.
- Wheel offset.
- Plus tire sizing
- Wheel tramp, and explain how static unbalance causes wheel tramp.
- Wheel shimmy, and describe how dynamic unbalance results in wheel shimmy.
- Various types of tire pressure monitoring systems.
- The advantage of nitrogen tire inflation.
- Noise, vibration, and harshness analysis.

Introduction

Although tires are often taken for granted, they contribute greatly to the ride and steering quality of a vehicle. Tires also play a significant role in vehicle safety. Improper types of tires, incorrect inflation pressure, and worn-out tires create a safety hazard. When tires and wheels are out of balance, tire wear and driver fatigue are increased, which can create a driving hazard. Tires serve the following functions:

1. Tires cushion the vehicle ride to provide a comfortable ride for the occupant.
2. Tires must firmly support the vehicle weight.
3. Tires must develop traction to drive and steer the vehicle under a wide variety of road conditions. In other words they must transmit traction and braking forces to the road.
4. Tires contribute to directional stability of the vehicle and must absorb all the stresses of accelerating, braking, and centrifugal force in turns.

Tire Design

Tires equipped on today's passenger cars and light trucks are of the radial ply design. Tire construction varies depending on the manufacturer and the type of tire. A typical modern tire contains these components (Figure 3-1):

The **bead wire** is a group of wire strands that retain the bead on the wheel rim.

The tire **liner** is made from synthetic gum rubber and seals the inside of the tire.

1. **Bead wire**
2. Bead filler
3. **Liner**, which functions as the air seal membrane
4. Sidewall with hard side compound
5. Rayon carcass plies
6. Steel belts
7. Jointless belt cover
8. Hard undertread compound
9. Hard high-grip tread compound

The bead wire contains several turns of bronze-coated steel wire in a continuous loop. This bead wire is molded into the tire at the inner circumference and wrapped in the cord plies. The bead wire anchors the tire to the wheel. The **bead filler** above the bead reinforces the sidewall and acts as a rim extender.

Shop Manual
Chapter 3, page 96

Tire **sidewalls** are made from a blend of rubbers, which absorbs shocks and impacts from road irregularities, prevents damage to the plies, and also contains antioxidants and other chemicals that are gradually released to the surface of the sidewall during the life of the tire. These antioxidants help keep the sidewall from cracking and protect it from ultraviolet radiation and ozone attack. Since the sidewall must be flexible to provide ride quality, minimum thickness of this component is essential. Tire manufacturers have reduced sidewall thickness by 40 percent in recent years to reduce weight and heat buildup, improve ride quality, reduce rolling resistance, and improve fuel economy. Some off road and performance tires are equipped with raised rubber that extents above the rim area and is referred to as a rim guard to offer added protection. A lettering and numbering arrangement for tire identification is located on the outside of the sidewall.

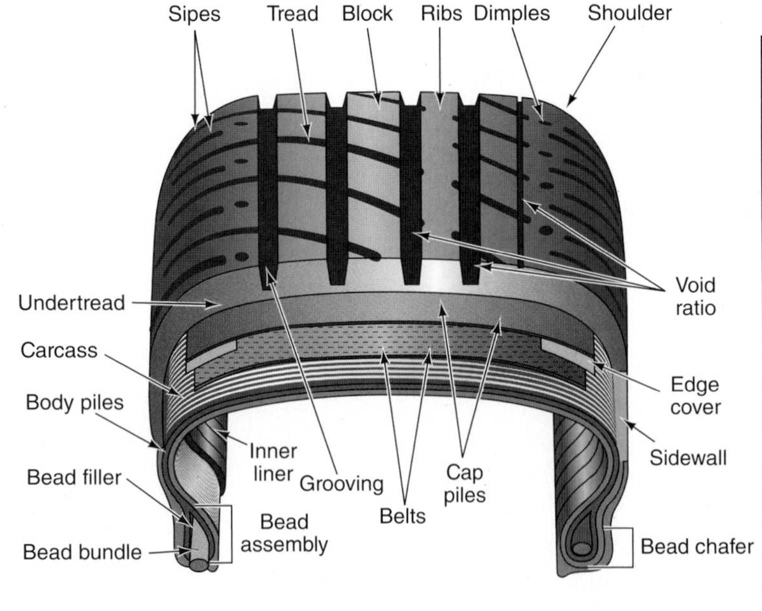

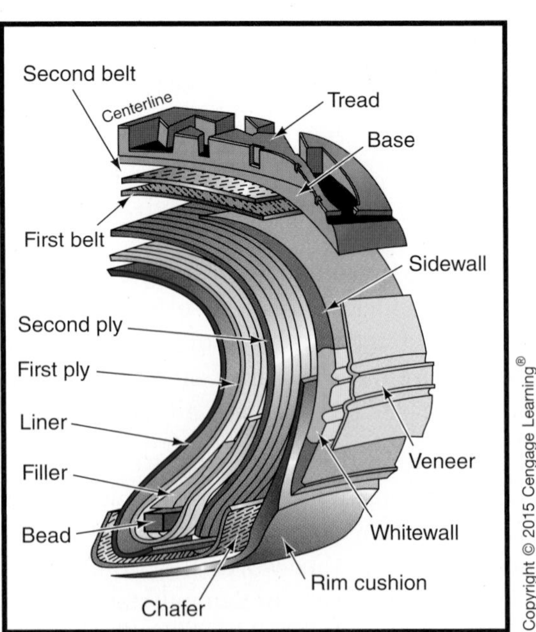

FIGURE 3-1 **The typical tire design and construction.**

The **cord plies** surround both beads and extend around the inner surface of the tire to enable the tire to carry its load. The plies are molded into the sidewalls. Each ply is a layer of rubber with parallel cords imbedded in its body. The load capacity of a tire may be increased by adding more cords in each ply or by installing additional plies. The most common materials in tire plies are polyester, rayon, and nylon. Passenger car tires usually have two-cord plies, whereas the tires of many trucks and recreation vehicles are rated to carry the heavier loads of these vehicles and are equipped with a higher ply-rated tire. In general, tires with more plies have stiffer sidewalls, which provide less cushioning and reduced ride quality.

Tire belts are positioned between the tread and the cord plies.

Tire treads are mounted between the sidewalls.

Steel is the most common material in **tire belts**, although other belt materials such as polyester have been used to some extent. Many tires contain two steel belts. The tire belts restrict ply movement and provide tread stability and resistance to deformation. This belt action provides longer tread wear and reduces heat buildup in the tire. Steel belts expand as wheel speed and tire temperature increase. Centrifugal force and belt expansion tend to tear the tire apart at high speeds and temperatures. Therefore, high-speed tires usually have a nylon jointless **belt cover**. This nylon belt cover contracts as it is heated and helps hold the tire together, providing longer tire life, improved stability, and better handling. Additionally a tire may have a layer of DuPont KEVLAR for added puncture proof protection under the tread layer.

Synthetic rubber compounds are developed in laboratories, whereas natural rubber is found in nature.

Belted bias-ply tires have steel or polyester belts between the cord plies and the tread.

Tire treads are made from a blend of rubber compounds that are very resistant to abrasion wear. Spaces between the tire treads allow tire distortion on the road without scrubbing, which accelerates wear. Modern automotive tires contain two layers of tread materials. The first tread layer is designed to provide cool operation, low rolling resistance, and durability. The outer layer, or tread, is designed for long life and maximum traction. Tread rubber is a blend of many different synthetic and natural rubbers. Tire manufacturers may use up to thirty different types of **synthetic rubber** and eight types of natural rubber in their tires. Manufacturers blend these synthetic and natural rubbers in both tread layers to provide the desired traction and durability. Tire treads must provide traction between the tire and the road surface when the vehicle is accelerating, braking, and cornering. This traction must be maintained as much as possible on a wide variety of road surfaces. For example, on wet pavement, tire treads must be designed to drain off water between the tire and the road surface. This draining action is extremely important to maintaining adequate acceleration, braking, and directional control. Lines cut across the tread provide a wiping action, which helps dry the tire-road contact area. Most tire manufacturers add a permablack compound to the rubber in their tires to maintain new-tire appearance throughout the life of the tire.

A BIT OF HISTORY

In 1834 Charles Goodyear was a hardware merchant from Philadelphia with a great interest in a new substance imported from Brazil called rubber. When rubber was first imported to the United States, many entrepreneurs were interested in manufacturing products from rubber. However, these entrepreneurs soon discovered that rubber became bone-hard in cold weather, and then turned to a glue-like substance in very hot weather, and these characteristics ended the rubber manufacturing business at that time.

The synthetic gum rubber liner is bonded to the inner surface of the tire to seal the tire. All modern passenger car and light truck tires are the tubeless type. In these tires, the tire bead must provide an airtight seal on the rim, and both the tire and the wheel rim must be completely sealed. The inner liner of the tire provides an air tight seal.

Some heavy-duty truck tires and off-road equipment have inner tubes mounted inside the tire. On tube-type tires, the air is sealed in the inner tube, and the sealing qualities of the tire and wheel rim are not important.

Designing tires is a very complex engineering operation. The average all-season tire contains the components (by weight) listed in Table 3-1. Tire design varies depending on the operating conditions and the load capacity of the tire. A tire designed for improved steering and handling characteristics has a nylon bead reinforcement and a hard bead filler with a slim tapered profile (Figure 3-2). This type of tire is suitable for sports car operation because the design stiffens the tire and reduces tire deflection during high-speed cornering. However, this type of tire may provide slightly firmer ride quality.

Tire Ply and Belt Design

The three basic tire construction designs that have been used are the bias, belted bias, and belted radial. In **bias-ply** or **belted bias-ply tires**, the cords crisscross each other. These

A BIT OF HISTORY
(continued)

Charles Goodyear persisted in experimenting with rubber! He experimented with mixing magnesia powder, nitric acid, and sulfur with rubber to reduce its stickiness. These substances improved the rubber quality, but it still was not perfected. Goodyear was often financially bankrupt and he had difficulty obtaining enough money to feed his children. He was even jailed on some occasions for failure to pay his bills, and ill health often plagued him.

By accident Goodyear discovered that extreme heat changed the rubber to a weatherproof, gum-like material. He found that pressurized steam applied to rubber for 4 to 6 hours at 270°F (132°C) provided the most uniform rubber. Goodyear was now able to obtain financial backing to produce rubber. Goodyear was plagued with patent right problems and those trying to steal his rubber patents. However, he did succeed in discovering modern rubber! Neither Charles Goodyear nor his family was ever connected with the company named in his honor, today's Goodyear Tire and Rubber Company.

Today there is a cultivated rubber tree for every two human beings on Earth, and three million people are employed collecting the rubber

TABLE 3-1 **TIRE COMPONENTS BY WEIGHT**

A typical P195/75-14 all-season tire contains:	
Synthetic rubber (30 types)	2.49 kg
Carbon Black (8 types)	2.27 kg
Natural Rubber (8 types)	2.04 kg
Chemicals, waxes, oils, pigments, etc. (40 types)	1.36 kg
Steel cord for belts	0.68 kg
Polyester and nylon	0.45 kg
Bead wire	0.23 kg
Total weight	9.42 kg

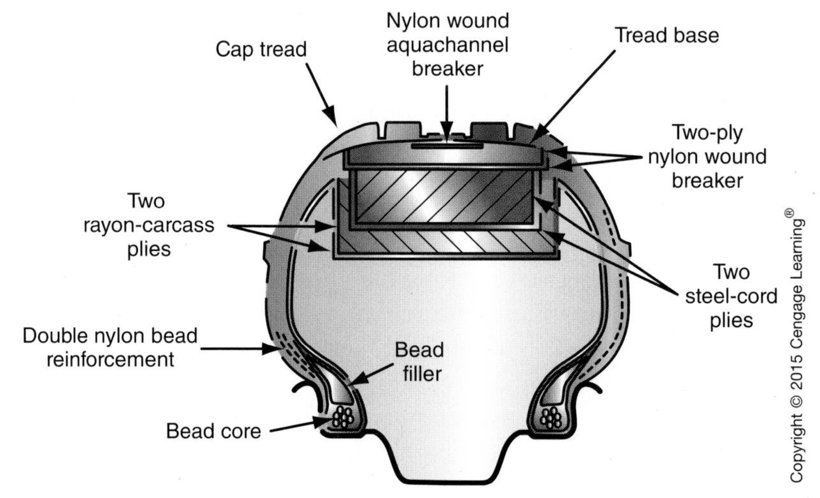

FIGURE 3-2 **Tires designed for improved steering and handling with a nylon bead reinforcement and a hard bead filler with a slim tapered profile.**

cords are usually at an angle of 25° to 45° to the tire centerline. The belt-ply cord angle is usually 5° less than the cord angle in the tire casing. Two plies and two belts are most commonly used, but four plies and four belts may be used in some tires. Compared to a bias-ply tire, a belted bias-ply tire has greater tread rigidity. The belts reduce tread motion during road contact. This action provides extended tread life compared to a bias-ply tire.

In **radial-ply tires**, the ply cords are arranged radially at a right angle to the tire centerline (Figure 3-3). Steel belts are most commonly used in radial tires, but other belt materials such as fiberglass, nylon, and rayon have also been used. The steel or fiberglass cords in the belts are crisscrossed at an angle of 10° to 30° in relation to the tire centerline. Many radial tires have two plies and two belts. Radial tires provide less rolling resistance, better steering characteristics, and longer tread life than bias-ply tires. As was stated earlier all modern vehicles are equipped with radial ply tires.

Regardless of the type of tire construction, the tire must be uniform in diameter and width. Radial runout refers to variations in tire diameter. A tire with excessive radial runout causes a tire thumping problem as the car is driven. When a tire has excessive variations in width, this condition is called lateral runout. A tire with excessive lateral runout causes the chassis to "waddle" when the car is driven.

The tire plies and belts must be level across the tread area. If the plies and/or belts are not level across the tread area, the tire is cone shaped. This condition is referred to as tire

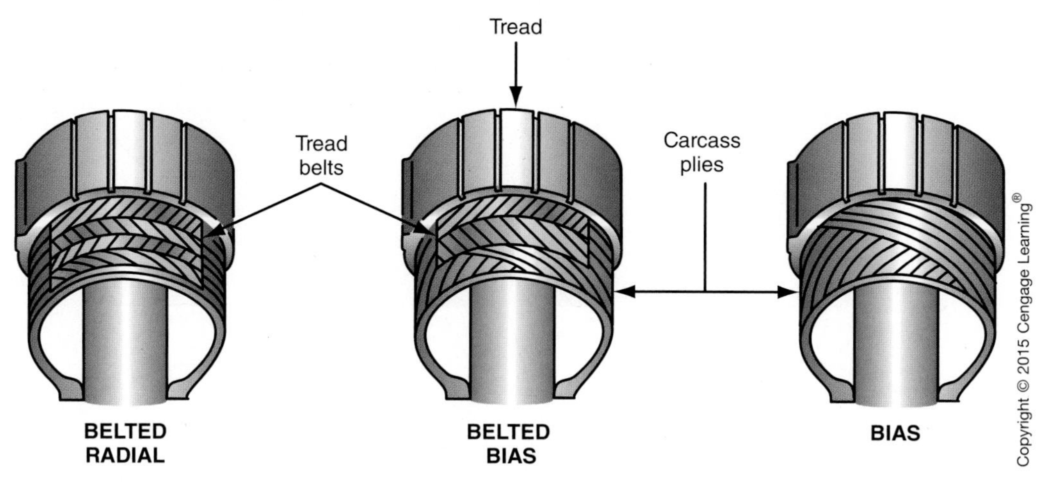

FIGURE 3-3 Three types of tire construction: bias-ply, belted bias-ply, and belted radial-ply.

A BIT OF HISTORY *(continued)*

from the trees in various countries. The United States imports approximately half of the rubber in the world and synthesizes an equal amount from petroleum. About 300,000 people in the United States are employed in the rubber industry, and they produce approximately $6 billion worth of products each year.

conicity. When a front tire has conicity, the steering may pull to one side as the car is driven straight ahead. This is referred to as a radial pull. A rear tire with conicity will not affect the steering as much as a front tire with **conicity**. If a tire pull is suspected the front tires can be cross rotated. If the pull shifts direction or goes away suspect a radial pull and further diagnosis will be required to determine which tire is affected.

Tire conicity refers to a condition where the plies and/or belts are not level across the tire tread. When a tire has conicity, the plies and/or belts are somewhat cone shaped and may cause the vehicle to pull in one direction when driven.

Tire Tread Design

Vehicle tires have many different tire treads designed to provide the tire performance desired by the tire and vehicle manufacturers. A typical modern tire tread has these features:

1. An interlocking tread pattern for improved tread gripping quality on icy or slick roads (Figure 3-4).
2. Deeply carved aquachutes to propel water off the tread and away from the tire to reduce the possibility of **hydroplaning** when driving on wet pavement (Figure 3-5).
3. Reinforced tread shoulders to improve tread gripping quality when turning corners on dry pavement (Figure 3-6).

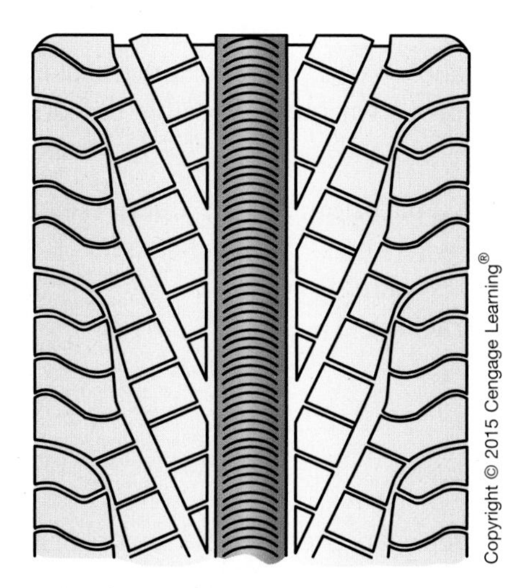

FIGURE 3-4 Interlocking tread pattern for improved traction on icy roads.

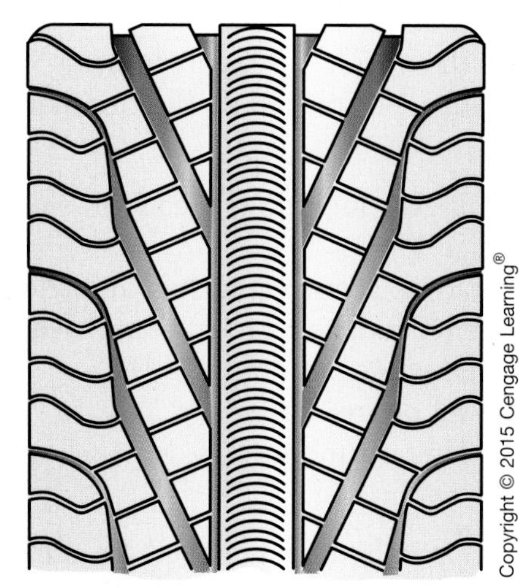

FIGURE 3-5 Aquachutes propel water off the tire tread.

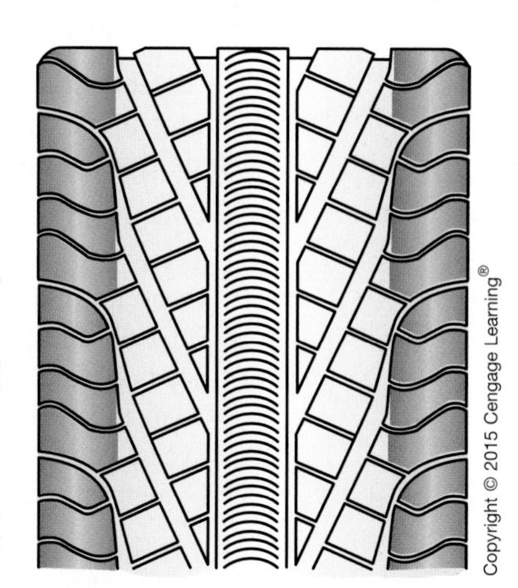

FIGURE 3-6 Reinforced tread shoulders improve tread gripping quality while cornering.

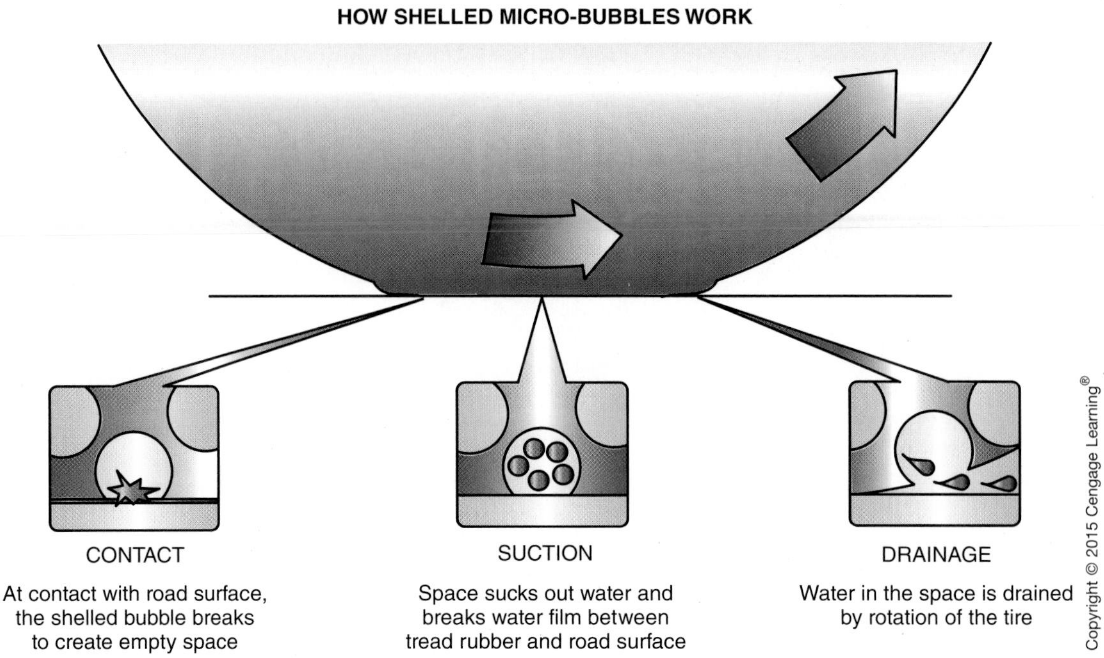

HOW SHELLED MICRO-BUBBLES WORK

CONTACT

At contact with road surface, the shelled bubble breaks to create empty space

SUCTION

Space sucks out water and breaks water film between tread rubber and road surface

DRAINAGE

Water in the space is drained by rotation of the tire

FIGURE 3-7 **Advantages of micro-bubbles in the tire tread.**

Hydroplaning occurs when water on the pavement is allowed to remain between the pavement and the tire tread contact area. This action reduces friction between the tire tread and the road surface, and can contribute to a loss of steering control.

One tire manufacturer installs 2 to 3 billion microscopic hollow shells or bubbles in the tread material on one brand of their tires. These hollow shells are installed to 60 percent of the tread depth, and add rigidity to the tread material. When these hollow shells contact the road surface they break open, and the shell edges provide a gripping effect to improve traction. When driving on wet pavement, each time a hollow shell contacts the road surface and breaks open, a small amount of water is pulled into the hollow shell (Figure 3-7). Because this action is occurring at the millions of hollow shells in contact with the road surface, water is removed from between the tread and the road surface. This action reduces the possibility of hydroplaning. As the tire continues to rotate, water is expelled from the hollow shells when they move out of contact with the road surface.

Tire Defects

Tire defects may be the result of manufacturing anomalies, road impact, or misuse. Manufacturing defects may be covered by the tire manufacturer's warranty. Possible defects include the following:

1. Separations and bulges, including tread or sidewall separation, or an open tread splice
2. Tread chunking, tearing, groove cracking, and shoulder cracking
3. Sidewall circumference fatigue, cracking, and weather cracking
4. Bead chafing, broken wires, or a pulled bead
5. Breaks in the sidewall or tread areas
6. Excessive radial or lateral runout
7. Conicity

If this inner liner is pinched or damaged air can leak between the tire plys forming a bubble on the side wall or under the tread of the tire. This is often the result of tire impact damage such as hitting a pothole or curb and is not a manufacturing defect. Inner liner damage can also result from an extreme underinflation condition that causes excessive heat buildup and flexing of the sidewall of the tire. It is critical that a tire suspected of being damaged or that has been driven under inflated be removed and inspected thoroughly before attempting to inflate or repair the assembly. A tire that is driven on while flat is almost always non-repairable and must be replaced.

Tire Ratings and Sidewall Information

A lot of essential information is molded into the sidewall of the average passenger car and light truck tire, including the tire rating. There are many dimensions related to tires and rims and not all this information is listed on the sidewall of the tire but is no less important (Figure 3-8). The tire rating is a group of letters and numbers that identify the tire type, section width, aspect ratio, construction type, rim diameter, load capacity, and speed symbol. The tire in Figure 3-9 has a P215/65R15 95H rating.

P indicates a passenger car tire, LT indicates a light truck tire, and T indicates a temporary tire most commonly a compact spare tire. The number 215 is the width of the tire in millimeters measured from sidewall to sidewall with the tire mounted on the recommended rim width.

The complete DOT code must be stamped on one side of the tire sidewall. This code is a combination of letters and numbers and indicates the manufacturer plant code where the tire was manufactured, the batch code, and the week and year the tire was manufactured (Figure 3-10). It is not a unique number to a specific tire but rather an identification for a batch of tires produced. Beginning in 2000 the week and year the tire was manufactured are indicated by the last 4 digits of the DOT number (Figure 3-11). The first 2 digits

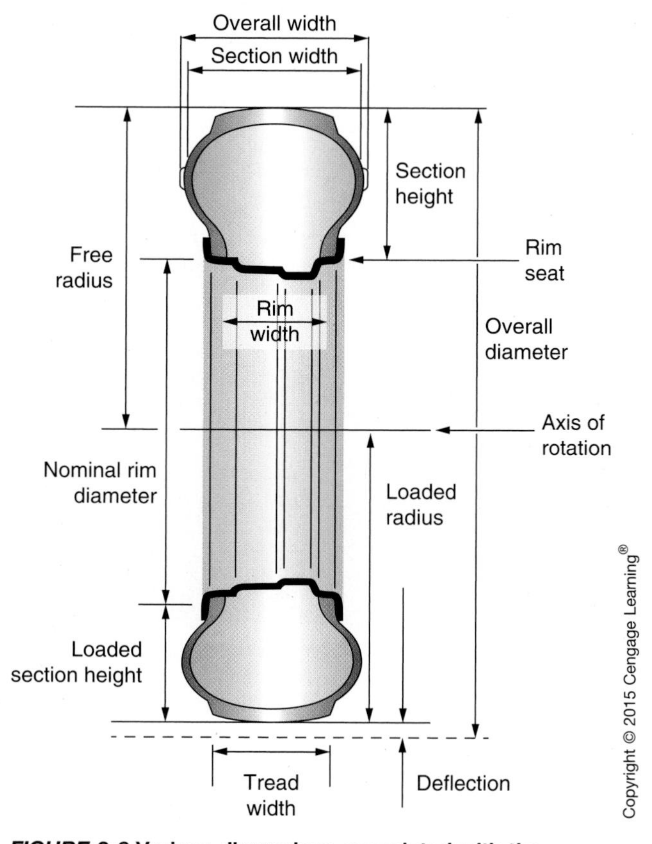

FIGURE 3-8 **Various dimensions associated with tire and rim.**

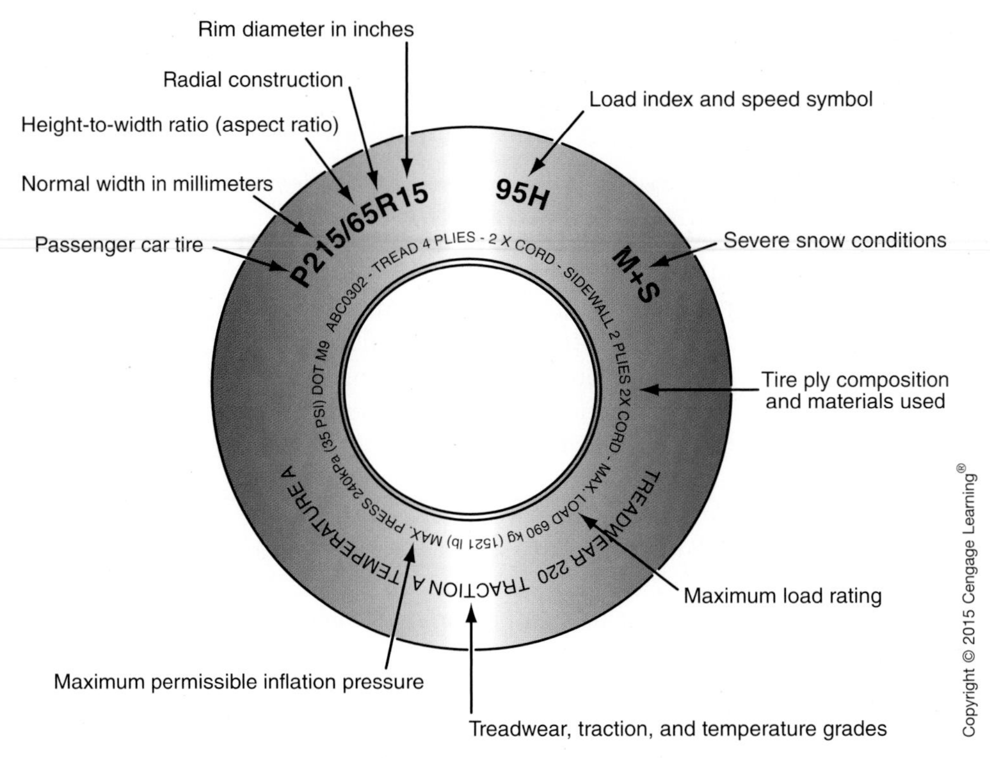

FIGURE 3-9 **Tire sidewall information.**

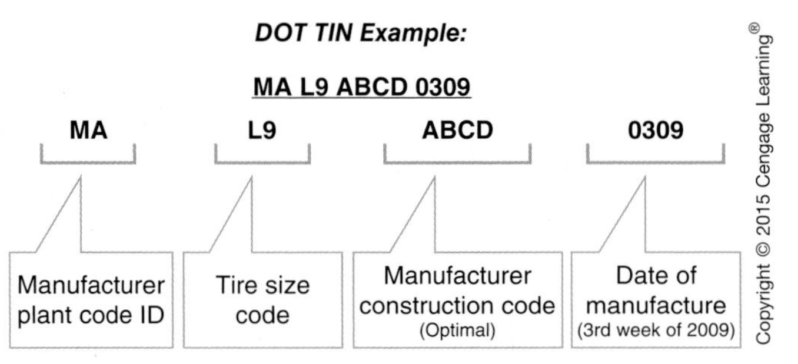

FIGURE 3-10 **The tire DOT code contains batch information about where and when a tire was manufactured.**

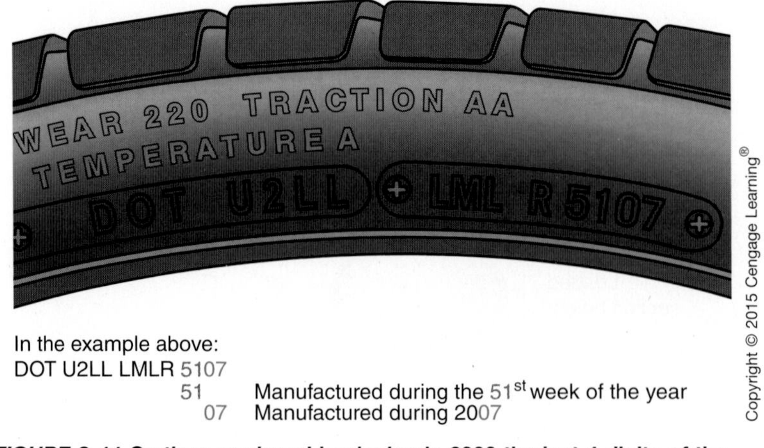

FIGURE 3-11 **On tires produced beginning in 2000 the last 4 digits of the DOT code indicates among other things the week and year the tire was produced.**

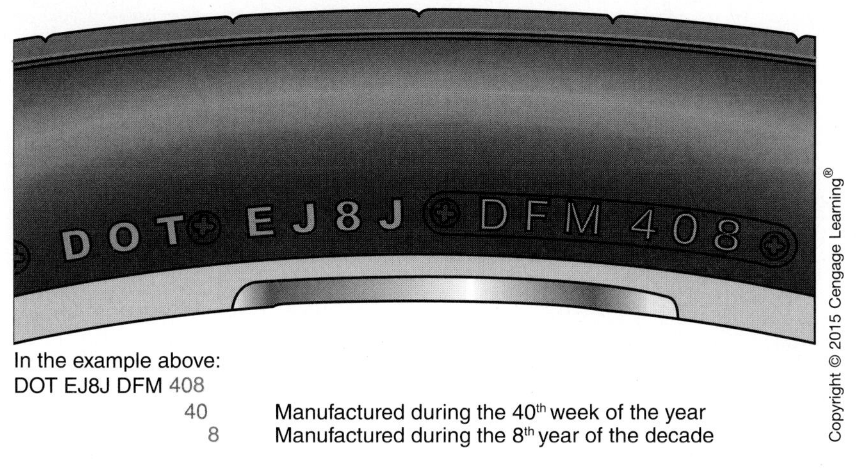

DOT EJ8J DFM 408
In the example above:
DOT EJ8J DFM 408
40 Manufactured during the 40th week of the year
8 Manufactured during the 8th year of the decade
Copyright © 2015 Cengage Learning®

FIGURE 3-12 **On tires produced before 2000 the last 3 digits of the DOT code indicates among other things the week and year the tire was produced.**

indicate the week while the last 2 digits indicate the year of production. As an example if the last 4 digits are 2514, then the tire was produced during the 25 week of 2014. On tires produced before 2000 the week and year the tire was manufactured are indicated by the last 3 digits of the DOT number (Figure 3-12). The first 2 digits indicate the week, while the last digit indicates the year of production. As an example if the last 3 digits are 254, then the tire was produced during the 25 week of 1994.

Aspect Ratio and Ply Design

The aspect ratio is a number that indicates the section height of the tire side wall. The number 65 indicates the aspect ratio, which is the ratio of the height to the width. With a 65 aspect ratio, the tire's height is 65 percent of its width (Figure 3-13). The letter R indicates radial-ply tire construction. Tires used on passenger cars and light trucks have been of radial ply construction since the late 1970s. If the tire construction was indicated by the letter B, the tire has a belted, bias-ply construction. The letter D represents diagonal bias-ply tire design. Bias belted tires can still be found on some utility and recreational trailers, but are not suitable for modern passenger vehicles.

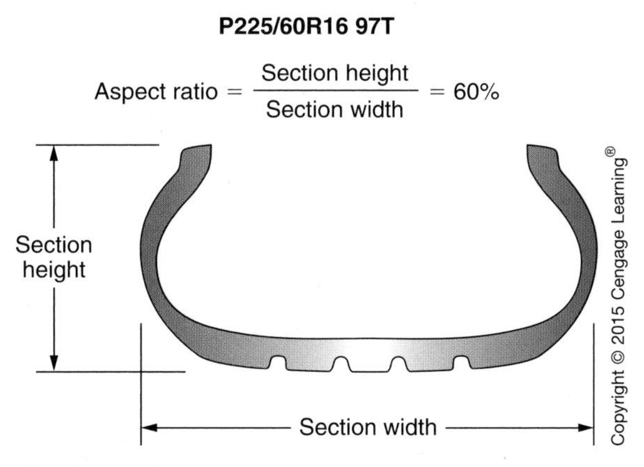

FIGURE 3-13 **The aspect ratio is determined by dividing section height by section width and is represented as a percentage.**

Load Index	Maximum Load (lbs.)	Load Index	Maximum Load (lbs.)
74	827	88	1235
75	853	89	1279
76	882	90	1323
77	908	91	1356
78	937	92	1389
79	963	93	1433
80	992	94	1477
81	1019	95	1521
82	1047	96	1565
83	1074	97	1609
84	1102	98	1653
85	1135	99	1709
86	1168	100	1764
87	1201		

FIGURE 3-14 **The load index number correlates to a number assigned to the maximum load the tire can carry.**

The **load rating** indicates the tire's load-carrying capability.

Aspect ratio is the percentage of a tire's height in relation to its width.

The aspect ratio may be referred to as the tire profile or series number.

Rim Diameter and Load Rating

The number 15 is the rim diameter in inches. The load index is represented by the number 95. The tire in Figure 3-9 has load rating of 1,521 pounds. The load index number correlates to the maximum rated load the tire can carry (Figure 3-14). A higher number represents the tire has a higher maximum load capacity. The maximum load is shown on the tire in pounds (lb) and kilograms (kg) together with the maximum inflation pressure in pounds per square inch (psi) and kilograms (kg). As an example MAX LOAD 730 kg (1609 lbs) and 240 kPa (35 psi) MAX PRESSURE COLD may be listed on the sidewall of the tire. This indicates the maximum load at a specified cold inflation pressure. Never exceed the maximum inflation pressure listed on the sidewall or tire failure could result. Some tire manufacturers have used the letters B, C, D, or E to indicate the **load rating** most commonly on LT tires. The letter B indicates the lowest load rating and the letter C represents a higher load rating. A tire with a D load rating is designed for light-duty trucks, and this tire will safely carry a load of 2,623 pounds when inflated to 65 psi. The ply rating is not the actual number of sidewall plies used but rather an equivalent strength compared to earlier bias ply tires. An E load range light truck tire may have the ply rating of a 10 ply tire but only have a 4 ply sidewall in actuality.

The **speed rating** indicates the tire's capability to withstand high speed.

Speed Rating

The letter H indicates the tire has a 130 miles per hour (mph) **speed rating**. Other speed ratings are:

Q – 99 mph
S – 112 mph
T – 118 mph
U – 124 mph
V – above 130 mph without service description
V – 149 mph with service description
Z – above 149 mph

W – 168 mph
Y – 186 mph

Tire speed ratings do not suggest that vehicles can always be driven safely up to the maximum designated speed rating, because many different road and weather conditions may be encountered. Tire manufacturers do not endorse the operation of a vehicle in an unlawful or unsafe manner. Speed ratings are based on laboratory tests, and these ratings are not valid if tires are worn out, damaged, altered, underinflated, or overloaded. The condition of the vehicle may also affect high-speed operation.

Department of Transportation (DOT) Tire Grading

The United States Department of Transportation (DOT) and the National Highway Traffic Safety Administration (NHTSA) developed the **Uniform Tire Quality Grading (UTQG)** system to provide customers with standardized information related to a tires tread wear, traction, and temperature capabilities to aid them with purchasing decisions. It is required that tire manufacturers grade most of their passenger car tires, except deep-treaded light truck tires, winter tires, temporary spare tires, and trailer tires. It should be noted that the DOT does not test the tires. It is left to the manufacturer to assign a grade based on their test results and those of independent testing companies they have hired.

Tread Wear Rating

The UTQG for **tread wear ratings** allows customers to compare tire tread life expectancies. The tread wear rating is based on the tire tread wear when tested under specific conditions on a regulated test track. Tire tread wear is monitored for a total of 7,200 miles and then the data is extrapolated out for total tread life. A baseline tire has a tread wear rating of 100. A tire with a 150 tread wear rating would last 1.5 times as many miles on the test track as the baseline tire with a 100 tread wear rating. Tread wear ratings are valid only for comparison within a manufacturer's product line and these ratings are not as valid for comparisons between tire manufacturers. This is one of the major flaws of the UTQG as it relates to tread wear. Though a tire with a tread wear number of 600 should be expected to last longer than one rated at 500.

Traction Rating

The UTQG for **traction ratings** indicates the tire's straight line wet coefficient of traction and braking on both wet asphalt and wet concrete surfaces. The test does not evaluate the dry braking performance, wet cornering, dry cornering, or high speed hydroplaning resistance. So while the information it provides is limited it does offer some ability to cross compare tires as long as the limits are understood.

To determine the traction rating a properly inflated tire is placed on the axle of a skid trailer that is electronically monitored. Test conditions are carefully controlled to maintain test uniformity. A trailer speed of 40 mph is maintained over both wet concrete and wet asphalt. The brakes are momentarily applied on both surfaces while sensors gather information related to the tire's coefficient of friction or braking g-force as the tire slides while a constant speed of 40 mph is maintained. By design this test places less emphasis on tread design and more emphasis on the tread compound.

The results of the skid tests are averaged, and the traction rating is designated as AA, A, B, or C (Figure 3-15). An AA traction rating indicates the best traction, while a C rating indicates acceptable traction.

Traction Grades	Asphalt g-Force	Concrete g-Force
AA	Above 0.54	0.41
A	Above 0.47	0.35
B	Above 0.38	0.26
C	Less Than 0.38	0.26

FIGURE 3-15 The UTQG Traction Grade is determined by the tire's average coefficient of brake traction on both wet asphalt and wet concrete at a constant 40 mph in a straight line.

Temperature Rating

The UTQG for **temperature rating** indicates the tire's ability to generate or dissipate heat during tire operation. A tire that is unable to dissipate heat efficiently or is unable to resist the destructive effects of internal heat buildup will have a reduced ability to run at higher speeds.

To obtain the temperature rating, tires are tested on a laboratory test wheel under a load. The temperature rating indicates the tires ability to operate at high speeds without failure. Extremely high temperatures may cause tire materials to degenerate and thus reduce tire life or cause tire failure. The temperature ratings indicate the tires' abilities to withstand heat at various speeds and are A for best, B for intermediate, and C for acceptable (Figure 3-16). Where a B-rated tire can withstand speeds between 100 to 115 mph. The test is similar to those used to confirm a tire's speed rating though it is not an exact test.

The UTQG is a good indicator of what can be expected from a particular tire. But it falls short of reflecting what actual real world performance will be. It should be viewed as a useful tool for comparison only.

DOT regulations also require tire manufacturers to place the following information on tire sidewalls:

1. Size
2. Load range
3. Maximum load
4. Maximum pressure
5. Number of plies under the tread and in the sidewalls
6. Manufacturer's name
7. Tubeless or tube-type construction
8. Type of carcass construction
9. DOT approval number, including the manufacturer's code number and date of manufacture

Some tires have a **tire performance criteria (TPC)** number molded on the sidewall (Figure 3-17). This number represents that the tire meets the car manufacturer's performance standards for traction, endurance, dimensions, noise, handling, and rolling

Temperature Grades	Speeds in mph
A	Over 115
B	Between 100 to 115
C	Between 85 to 100

FIGURE 3-16 The UTQG Temperature Grade indicates the tire's ability under a load to withstand heat build up at various speeds.

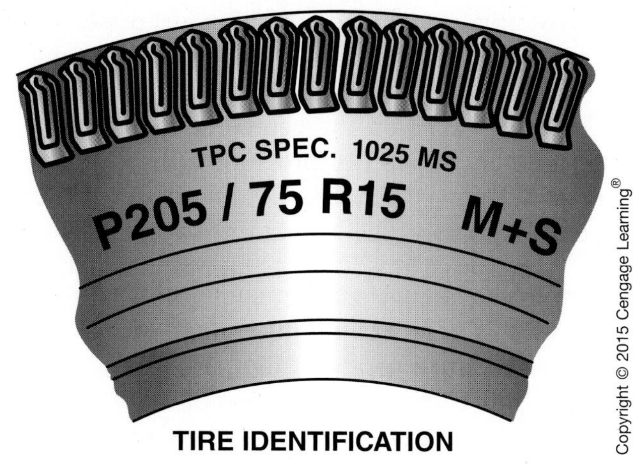

FIGURE 3-17 **Tire performance criteria (TPC) number.**

resistance. Most car manufacturers assign a different TPC number to each tire size. Replacement tires should have the same ratings and TPC number as the original tires.

Tire Contact Area

The **tire contact area** refers to the area of the tire that is in contact with the road surface when the tire is supporting the vehicle weight. The **tire free diameter** is the distance of a horizontal line through the center of the spindle and wheel to the outer edges of the tread. The **tire rolling diameter** is the distance of a vertical straight line through the center of the spindle to the outer edges of the tread when the tire is supporting the vehicle weight. The rolling diameter is always less than the free diameter. The difference between the free diameter and the rolling diameter is referred to as deflection. Tire tread grooves take up excess rubber and prevent scrubbing as the tire deflects in the contact area. The rolling diameter, free diameter, and contact area are shown in Figure 3-18.

Tire Placard and Inflation Pressure

The vehicle weight is supported by the correct air pressure exerted evenly against the entire interior tire surface, which produces tension in the tire carcass. Therefore,

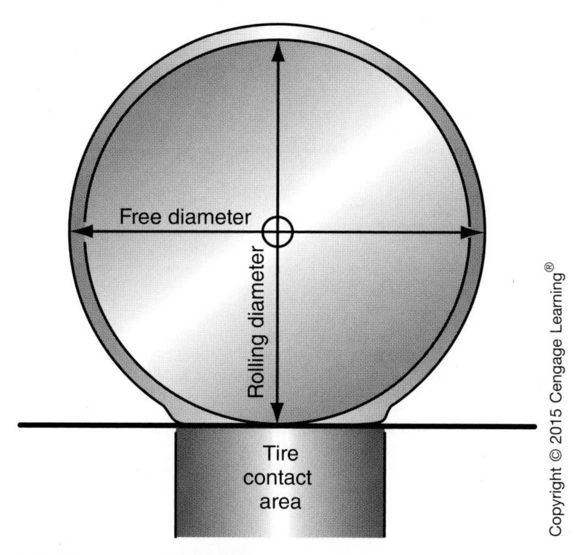

FIGURE 3-18 **Tire rolling diameter, free diameter, and contact area.**

Tire pressure is the amount of air pressure in the tire.

tire pressure is extremely important. Underinflation decreases the rolling diameter and increases the contact area, which results in excessive sidewall flexing and tread wear. Overinflation decreases the contact area, increases the rolling diameter, and stiffens the tire. This action results in excessive wear on the center of the tread. Tire pressure should be checked when the tires are cool. Since tire pressure normally increases at high tire temperatures, air pressure should not be released from hot tires. Excessive heat buildup in a tire may be caused by underinflation. This condition may lead to severe tire damage.

AUTHOR'S NOTE: There are many causes of excessive tire tread wear such as improper wheel balance or alignment. However, it has been my experience that one of the most common causes of excessive tire tread wear is improper inflation pressure. Many vehicle owners or drivers do not check tire pressures regularly.

On many vehicles, the **tire placard** is permanently attached to the rear face of the driver's door. This placard provides tire information such as maximum vehicle load; tire size, including spare; and cold inflation pressure, including spare (Figure 3-19).

Shop Manual
Chapter 3, page 108

Tire pressure is carefully calculated by the vehicle manufacturer to provide satisfactory tread life, handling, ride, and load-carrying capacity. Most vehicle manufacturers recommend that tire pressures be checked cold once a month or prior to any extended trip. The manufacturer considers the tires to be cold after the vehicle has sat for three hours, or when the vehicle has been driven less than one mile. The tires should be inflated to the pressure indicated on the tire placard. Tire pressures may be listed in metric or English system values. Conversion charts provide pressures in either of these systems (Table 3-2).

Nitrogen Tire Inflation

Some shops are presently equipped with **nitrogen tire inflation** equipment. When tires are inflated with compressed air, the air slowly passes through the tire walls and tread. Tires inflated with compressed air can lose as much as 12 psi in a 6-month period. When tire inflation pressure is lower than specified, tire tread wear is increased, while fuel economy and vehicle stability are decreased.

Nitrogen molecules are considerably larger than air molecules (Figure 3-20). Therefore, when a tire is inflated with nitrogen, the nitrogen passes through the tire walls and tread more slowly when compared with air. When tires are inflated with nitrogen, the tire inflation pressure remains more stable over a longer time period, which provides reduced tire

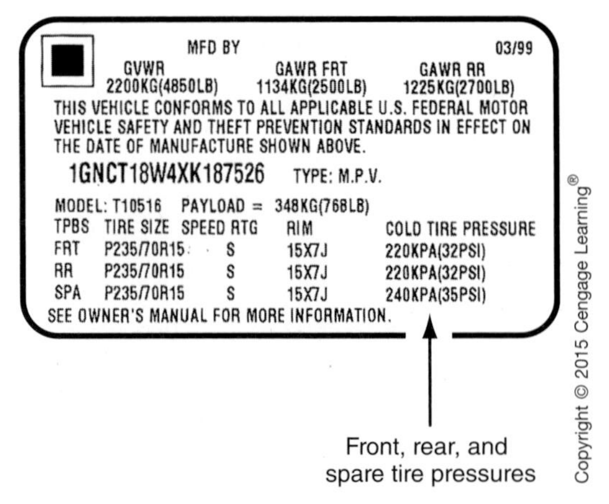

MFD BY
03/99
GVWR
2200KG(4850LB)
GAWR FRT
1134KG(2500LB)
GAWR RR
1225KG(2700LB)
THIS VEHICLE CONFORMS TO ALL APPLICABLE U.S. FEDERAL MOTOR VEHICLE SAFETY AND THEFT PREVENTION STANDARDS IN EFFECT ON THE DATE OF MANUFACTURE SHOWN ABOVE.
1GNCT18W4XK187526 TYPE: M.P.V.
MODEL: T10516 PAYLOAD = 348KG(768LB)
TPBS TIRE SIZE SPEED RTG RIM COLD TIRE PRESSURE
FRT P235/70R15 S 15X7J 220KPA(32PSI)
RR P235/70R15 S 15X7J 220KPA(32PSI)
SPA P235/70R15 S 15X7J 240KPA(35PSI)
SEE OWNER'S MANUAL FOR MORE INFORMATION.
Front, rear, and spare tire pressures

FIGURE 3-19 **Tire placard.**

TABLE 3-2 TIRE INFLATION PRESSURE CONVERSION CHART

Inflation Pressure Conversion Chart (Kilopascals to PSI)			
kPa	psi	kPa	psi
140	20	215	31
145	21	220	32
155	22	230	33
160	23	235	34
165	24	240	35
170	25	250	36
180	26	275	40
185	27	310	45
190	28	345	50
200	29	380	55
205	30	415	60

Conversion: 6.9 kPa = 1 psi

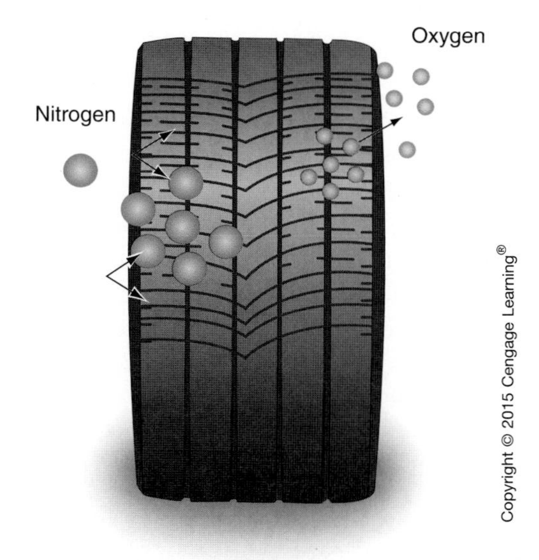

FIGURE 3-20 Oxygen and nitrogen leakage through tire walls and tread.

tread wear, improved fuel economy, and increased vehicle stability. Another benefit from using nitrogen to inflate tires is reduced aluminum rim corrosion because the nitrogen does not contain any oxygen or moisture (water molecules). When tires are filled with air, the oxygen and water vapor in the air tends to react with the aluminum in the rims to cause corrosion.

Specialty Tires

Sport Utility Vehicle (SUV) and 4 × 4 Tires

SUV tires may be classified for use on pavement or off-road. SUV tires have greater load-carrying capacity compared to passenger car tires, because of the extra loads that may be carried in an SUV. A typical pavement SUV tire has these features:

1. A silica tread compound that provides low noise levels and exceptional wet braking capability.

2. An enhanced casing system and a stable tire contact area that supplies even tread wear and responsive handling quality.
3. Larger cables in the steel belts to provide increased strength and durability.
4. Full-depth, interlocking sections in the tread that supply excellent wet and snow traction.

Compared to SUV tires for use on pavement, off-road SUV tires have stronger tread rubber to prevent cutting and chipping. Off-road SUV tires may also have thicker belt wire and more belt strands.

Tires designed for 4 × 4 vehicles may have some of the same features as SUV tires. A typical 4 × 4 tire has these additional features:

1. Two wide circumference grooves with a stepped profile to reduce hydroplaning.
2. Staggered shoulder blocks in the tread to improve lateral grip on slopes.
3. Gradual profile changes in the shoulder area of the tread to provide progressive break-away during hard cornering.
4. Additional rubber at the base of the tread to reduce the possibility of tire damage.

Puncture sealing tires are available as an option on certain car lines, and some rubber companies sell these tires in the replacement tire market. These tires contain a special rubber sealing compound applied under the tread area during the manufacturing process. When a nail or other object up to 3/16 inch (4.76 mm) in diameter punctures the tread area, it picks up a coating of sealant. If the object is removed, the sealant sticks to the object and is pulled into the puncture. This sealant completely fills the puncture and forms a permanent seal to maintain tire inflation pressure (Figure 3-21). Puncture sealing tires usually have a special warranty. These tires can be serviced with conventional tire changing and balancing equipment. When repairing tires, the maximum repairable puncture size is ¼ inch.

Mud and snow tires are available in various ply and belt designs. These tires provide increased traction in snow or mud compared to conventional tires. Mud and snow tires are identified with an MS suffix after the tire performance criteria (TPC) number on the tire sidewall. When snow tires are installed on a vehicle, these tires should be the same size and type as the other tires on the vehicle. In areas where snow is encountered, all-season tires have replaced snow tires to a large extent. Studded tires provide improved traction on ice, but these tires are prohibited by law in many states because their use resulted in road surface damage.

Many tires sold today are classified as **all-season tires**. These tires have a 37 percent higher average snow traction compared to non-all-season tires. All-season tires may have slightly improved performance in areas such as wet traction, rolling resistance, tread life, and air retention.

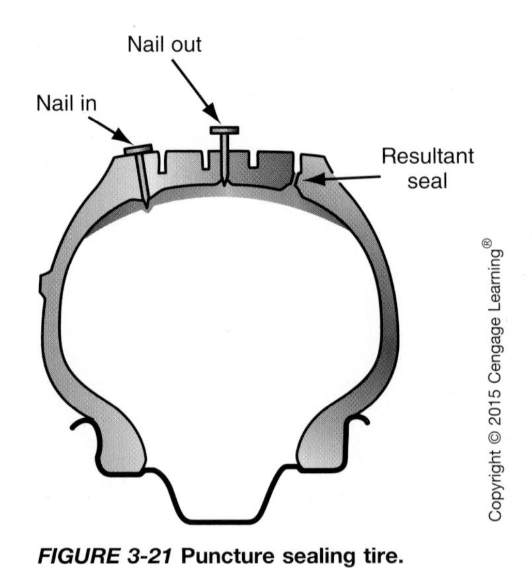

FIGURE 3-21 **Puncture sealing tire.**

Improvements in tread design and tread compounds provide the superior quality in all-season tires. All-season tires are usually identified by the letters MS (for mud and snow) on the sidewall.

Replacement Tires

Most tires have tread wear indicators built into the tread. When the tread wears down to 2/32″ (1.5 mm) the minimum allowable tread depth, the wear indicators appear as bands across the tread (Figure 3-22). Car manufacturers typically recommend tire replacement when the wear indicators appear in one or more tread grooves at three locations around the tire. It is generally advisable to recommend that tires be replaced before they reach the minimum of 2/32″ (1.5 mm) of tread depth.

If **replacement tires** have a different size or construction type than the original tires, vehicle handling, ride quality, speedometer/odometer calibration, and antilock brake system (ABS) operation may be seriously affected. When replacement tires are a different size than the original tires, the vehicle ground clearance and tire-to-body clearance may be altered. Steering and braking quality may be seriously affected if different sizes or types of tires are installed on a vehicle. This does not include the compact spare tire, which is intended for temporary use. The majority of vehicles manufactured are equipped with ABS. When different-sized tires are installed on these vehicles, the ABS operation is abnormal, which may result in serious braking defects. When selecting replacement tires, the following precautions must be observed to maintain vehicle safety:

1. Replacement tires must be installed in pairs on the same axle. Never mix tire sizes or designs on the same axle. If it is necessary to replace only one tire, it should be paired with the tire having tread depth within 4/32″ (3 mm) of the replacement tire.
2. The tire load rating must be adequate for the vehicle on which the tire is installed. Light-duty trucks, station wagons, and trailer-towing vehicles are examples of vehicles that require tires with higher load ratings than passenger car tires.
3. Snow tires should be the same type and size as the other tires on the vehicle.
4. A four-wheel-drive and all-wheel drive vehicle should have the same type and size of tires on all four wheels and tread depth between tires should be within manufacturer specifications.
5. Do not install tires with a load rating less than the car manufacturer's recommended rating.
6. Replacement tire ratings should be equivalent to the original tire ratings in all rating designations.
7. When combining different tires front to rear on a vehicle, consult the car manufacturer's or tire manufacturer's recommendations.

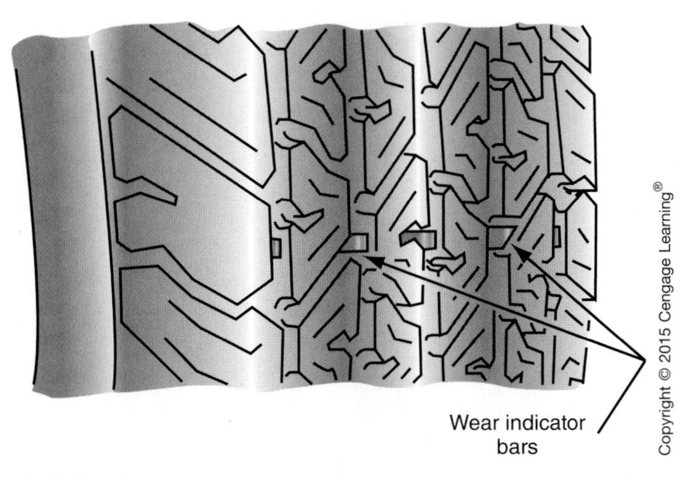

FIGURE 3-22 **Tread wear indicators.**

Vehicles that are 4-wheel drive or all-wheel drive, whether they are light trucks, SUVs, or passenger cars should have four matched tires. This includes tread depth. All tires should be within 2/32″ to 4/32″ (1.5–3 mm) of tread depth with one another depending on manufacturer recommendations. The reason for this is that four-wheel and all-wheel drive vehicles are equipped with viscous couplings or differentials that are designed for momentary differences in wheel speeds. If all drive tires are not exactly the same size the differentials and viscous couplings will be always forced to compensate for the different rolling diameters resulting in unwanted heat build-up and potential wear to expensive drivetrain components. If a tire requires replacement because it was damaged it may be necessary to replace all four tires due to too great a variation in tread depth created by installing one new tire. Refer to vehicle manufacturer service information for specific details and allowable tread depth differences.

Vehicles tires should be rotated every 6,000 miles as a general rule following manufacturer's recommendations for both frequency and rotation pattern (Figure 3-23). It is also

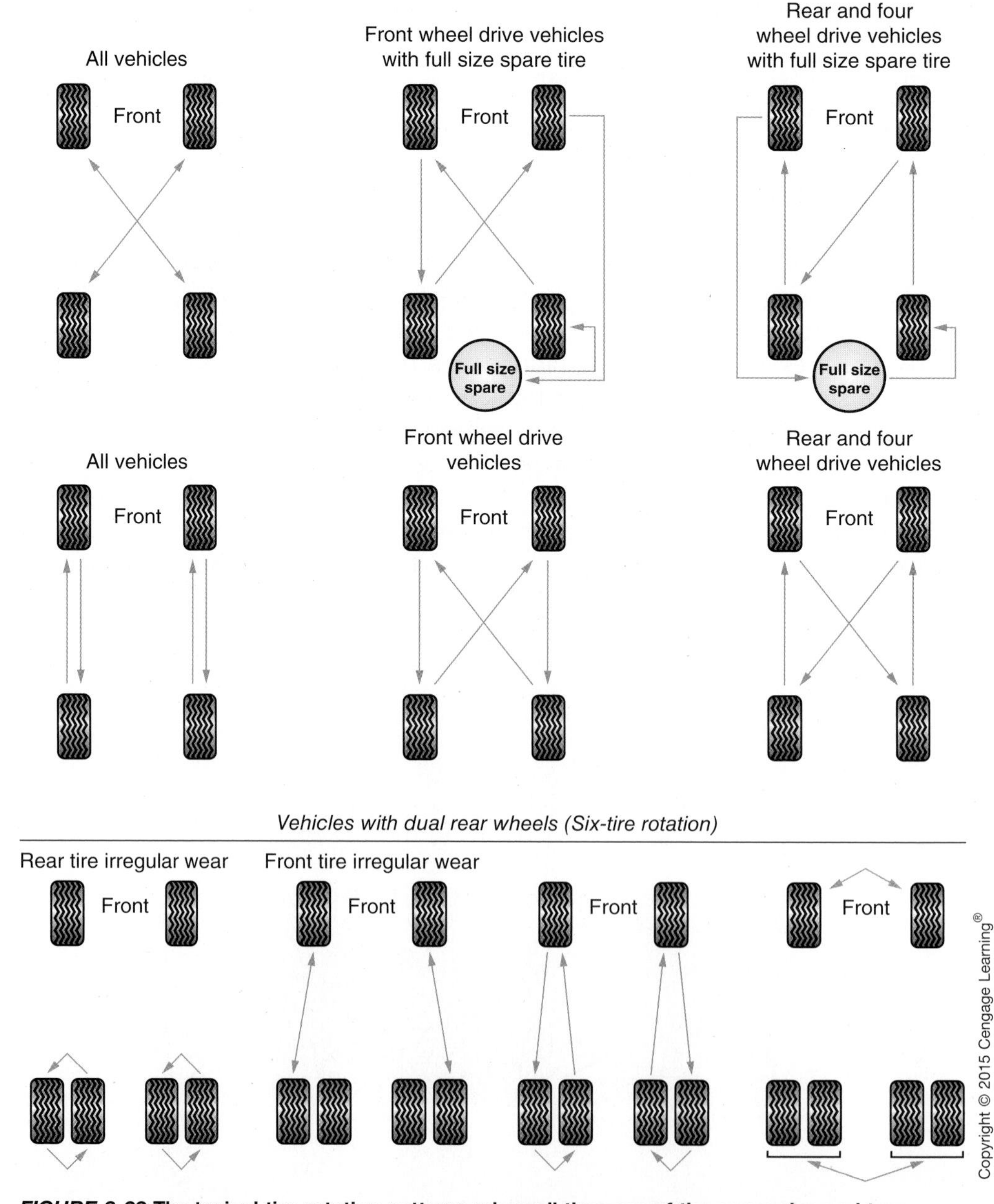

FIGURE 3-23 **The typical tire rotation patterns when all tires are of the same size and type.**

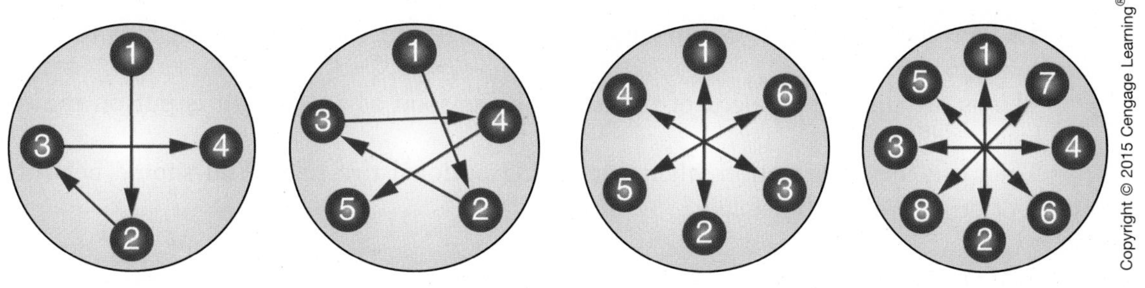

FIGURE 3-24 **The typical rim torqueing sequence.**

important to torque the wheel lug nuts to specification following the recommended torque sequence or star pattern (Figure 3-24).

Plus Sizing

Sometimes a customer does not like the appearance of their vehicle with the stock tire and wheel assembly and they would like to install a larger rim diameter to improve the looks and performance of the vehicle. This is referred to as plus sizing. It should be noted that a stock vehicle is designed to operate with a specific tire and wheel combination and deviating from allowed OEM combinations may adversely affect vehicle operation. Increasing a tire's contact patch may have a negative impact on wheel bearing loads, suspension component wear and cause harsher ride quality due to less road isolation caused by a narrower and stiffer tire side wall.

When the decision to plus size is made you must remember that the overall height of the tire and wheel assembly must be maintained (Figure 3-25). By using a larger diameter wheel with a lower profile, lower aspect number, it is possible to properly maintain the overall diameter of the tire and height of the assembly, keeping odometer and speedometer changes negligible. In addition, it is important to maintain the overall diameter of the tire to maintain the effectiveness of the antilock braking system, traction control, and vehicle stability system. It should be noted that when the decision is made to increase the rim diameter it may also be necessary to change rim width and wheel offset. Refer to the section on rims for more detail regarding rim dimensions.

Now that we know what the tire size numbers mean, we can calculate the overall height of a tire and wheel assembly. We multiply the tire width by the aspect ratio to get the height of the tire.

$$\begin{aligned}\text{Tire sidewall height} &= 235\text{ mm (width)} \times 0.75(\text{aspect ratio percentage})\\ &= 176.25\text{ mm}(64.94\text{ in})\end{aligned}$$

It will be necessary to convert rim diameter from inches to millimeters ($15\text{ in} \times 25.4 = 381$). Then we add twice the tire sidewall height to the rim diameter.

$$\text{Overall tire and wheel height} = 2 \times 176.25\text{ mm} + 381 = 733.5\text{ mm (28.9 in)}$$

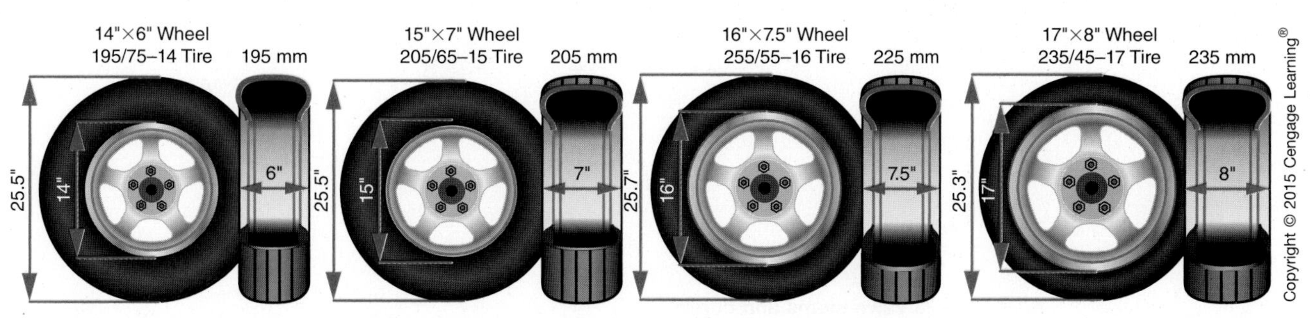

FIGURE 3-25 **Plus sizing must maintain the overall height of the tire and wheel assembly.**

To convert from inches to mm multiply by 25.4 and to convert from mm to inches divide by 25.4.

This is the unloaded height of the assembly; as soon as any weight is put on the tire, the height will decrease.

Then use this information when comparing optional tire sizes. You should be within 0.25″ of original assembly height.

Tire Valves

Shop Manual
Chapter 3, page 104

The **tire valve** allows air to flow into the tire, and it is also used to release air from the tire. The core in the center of the valve is spring loaded and allows air to flow inward while the tire is inflated (Figure 3-26). Once the tire is inflated, the valve core seats and prevents air flow out of the tire. The small pin on the outer end of the valve core may be pushed to unseat the valve core and release air from the tire. An airtight cap on the outer end of the valve keeps dirt out of the valve, and provides an extra seal against air leakage. A deep groove is cut around the inner end of the tire valve. When the valve assembly is pulled into the wheel opening, this groove seals the valve in the opening. Steel valve stems are installed in some wheels. The lower end of the valve stem is threaded, and a nut retains the valve stem in the wheel (refer to Figure 3-26). Steel washers and sealing washers are located on the valve stem on the inside and outside of the wheel. The sealing washers are positioned next to the wheel.

Wheel Rims

Wheel rims are circular devices on which the tires are mounted. Wheel rims are bolted to the wheel hubs or axles.

Wheel rims may be manufactured from stamped or pressed steel discs that are riveted or welded together to form the circular rim or may be cast or forged aluminum (Figure 3-27). Currently, less than 50 percent of the wheel rims installed by the original equipment manufacturers (OEMs) are stamped steel type.

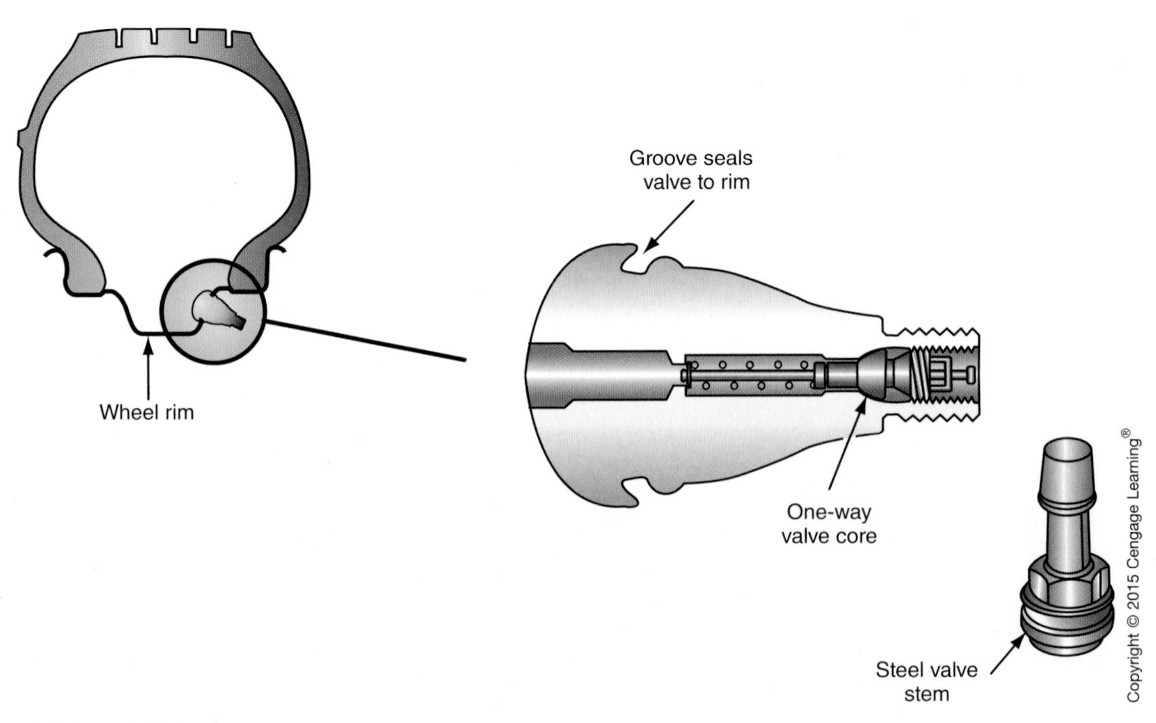

FIGURE 3-26 **Valve stems and cores.**

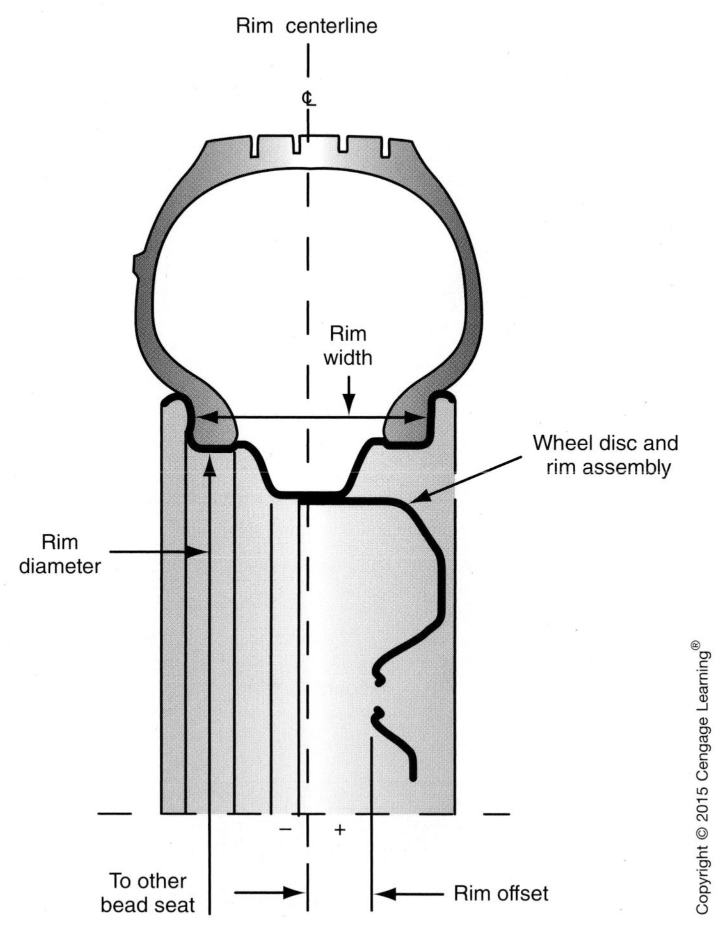

FIGURE 3-27 **Wheel rim design.**

A rim is equipped with a drop center, which allows tires to be more easily removed and installed (Figure 3-28). The closer the drop center is to the outside mounting flange the easier it is to remove and install the tire (Figure 3-29). Some performance rims and run flat tire rims are equipped with a bead lock ridge, which aids in keeping the tire bead locked or secured to the rim (Figure 3-30). This is important especially during corning maneuvers or in the case of a run flat tire when it has lost all air pressure. These bead lock rims also commonly have a drop center that is located in the center of the rim farther from the outer mounting flange. Rims of this design make tire service much more challenging (Figure 3-31).

There are many dimensions that are associated with the rim assembly (Figure 3-32) and all must be considered if a replacement rim of a different design is being considered. The rim offset is what locates the tire and wheel assembly in relationship to the suspension and the lines of force on the wheel bearing and spindle assembly (Figure 3-33). When the mounting face of the rim aligns directly with the rim's centerline then the rim has zero offset. If a rim is designed with positive offset, the rim centerline is inboard of the mounting face. Stated another way the mounting face is closer to the street side of the wheel. The rim with negative offset has the centerline outboard of the mounting face. Stated another way the mounting face is closer to the brake side of the wheel. Front-wheel drive vehicles tend to have positive offset rims to allow for proper clearance of the hub assembly within the vehicle wheel well. The rim offset affects front suspension loading and operation. As a general rule, if a replacement set of rims is being considered they should meet all the dimensional characteristics of the original equipment rim.

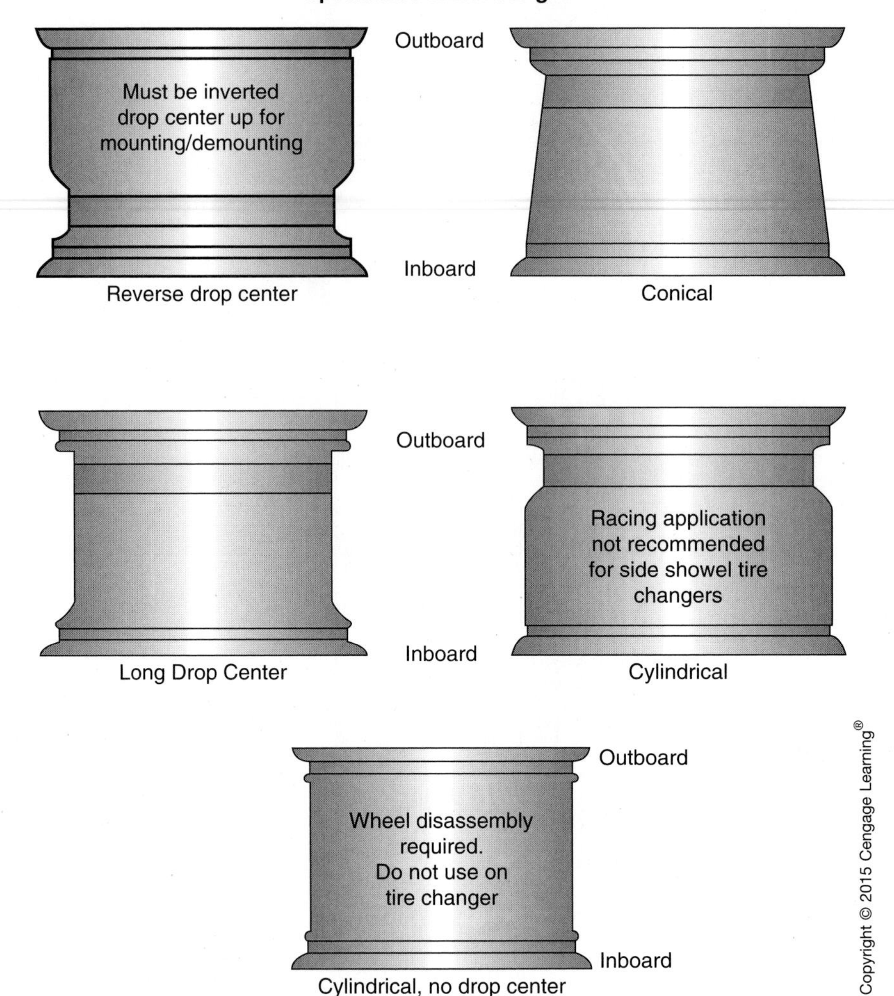

FIGURE 3-28 **Various rim drop center designs.**

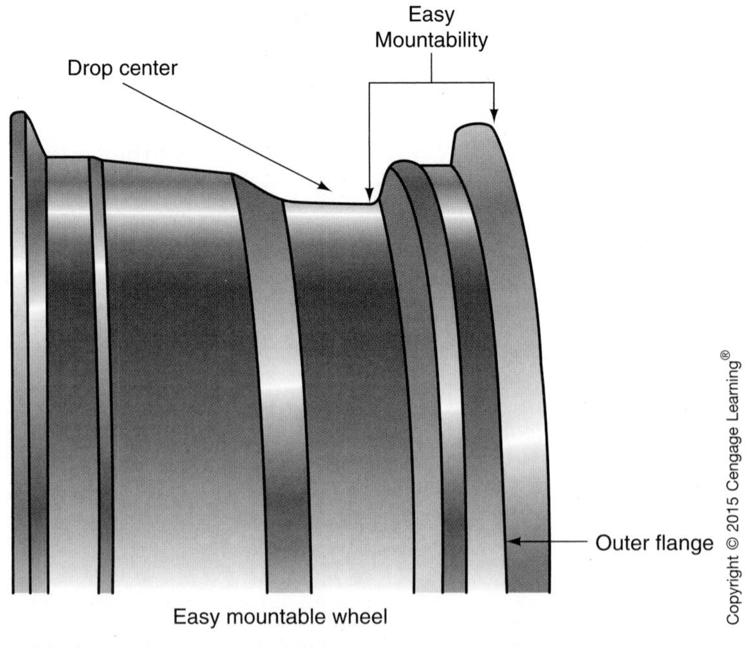

FIGURE 3-29 **The closer the drop center is to the outer mounting flange the easier it will be to dismount and mount tires on the rim assembly.**

FIGURE 3-30 **The bead lock on a rim assembly aids in keeping the tire bead secure to the rim.**

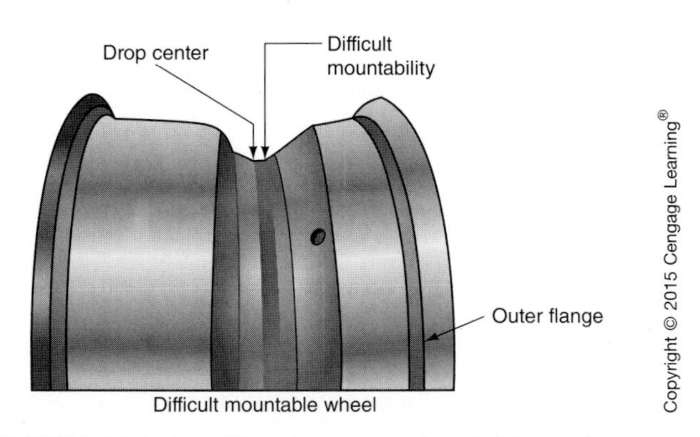

FIGURE 3-31 **A rim with a drop center located in or close to the center of the assembly can be very challenging to dismounting and mounting tires on the rim assembly.**

A large hole in the center of the rim fits over a flange on the mounting surface. The rim has a small hole for the valve stem. The wheel stud mounting holes in the rim are tapered to match the taper on the wheel nuts. Some late model Ford trucks have non-self-centering wheel nuts. On these nuts, a swivel washer is attached to the inner side of each nut (Figure 3-34). When the wheel and nuts are installed, the flat side of the swivel nut fits against the wheel rim. The wheels are hub-centered, and an O-ring mounted on the hub provides improved wheel centering (Figure 3-35). If this O-ring is missing or damaged, the wheel may not be properly centered, and this condition results in wheel vibrations. The O-ring also provides a seal to prevent corrosion.

Cars and light-duty trucks have four, five, six, or eight mounting stud openings and an equal number of matching studs in the hub or axle (Figure 3-36). The number of wheel studs depends on the vehicle weight and load. Heavier vehicles and/or increased vehicle load usually require more wheel mounting studs. These nuts usually have a taper on the inner side of the nuts, and this taper fits against a matching taper on the wheel stud openings (Figure 3-37). These tapers on the nuts and wheel openings center the wheel on the hub or axle; this is referred to as a stud piloted rim. Wheel stud openings are listed by the number of studs and the circle through the stud centers. For example, if wheel openings are listed as 5–4.5, there are five wheel studs and stud openings, and the centers of the wheel studs and wheel openings are on a 4.5 in. (11.4 cm) circle. If a wheel has an even

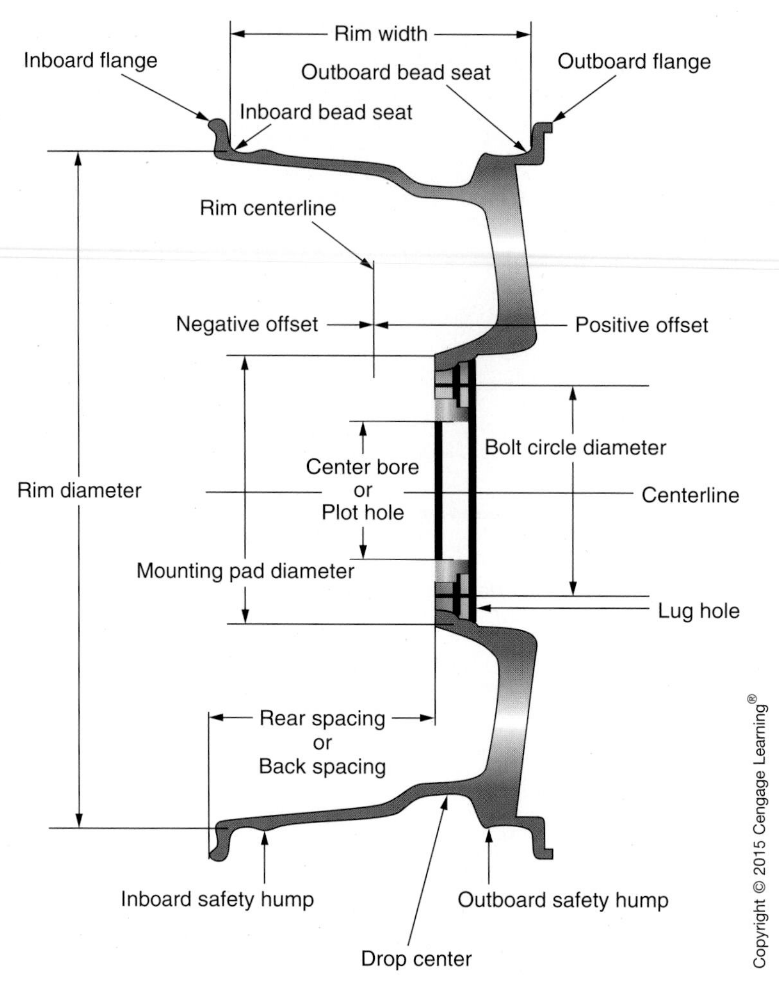

FIGURE 3-32 **Various dimensions associated with the wheel rim assembly.**

number of studs, measure the distance between opposite stud centers to check the stud position and wheel circle. On a five-bolt wheel, measure the distance between two adjacent stud centers. The specified distances between adjacent wheel stud centers and the corresponding wheel circles are the following:

4.5 in. wheel circle	2.64 in. between adjacent stud centers
4.75 in. wheel circle	2.79 in. between adjacent stud centers
5.0 in. wheel circle	2.93 in. between adjacent stud centers
5.5 in. wheel circle	3.23 in. between adjacent stud centers

Templates are available to measure the stud position on wheels that do not have an even number of studs.

The width of the wheel is measured between the vertical bead seats on the rim flanges. Rim diameter is the distance between the horizontal part of the bead seats measured through the rim center bottom. A drop center in the rim makes tire changing easier (Figure 3-38). Rims have a safety ridge or bead lock ridge behind the tire bead locations, which help prevent the beads from moving into the drop center area if the tire blows out. If the tire blows out and a bead enters the drop center area, the tire may come off the wheel.

Cast aluminum wheels are commonly used on many vehicles by OEMs. These wheels may be polished, chromed, or painted. A cast aluminum wheel costs about twice as much as a stamped steel wheel.

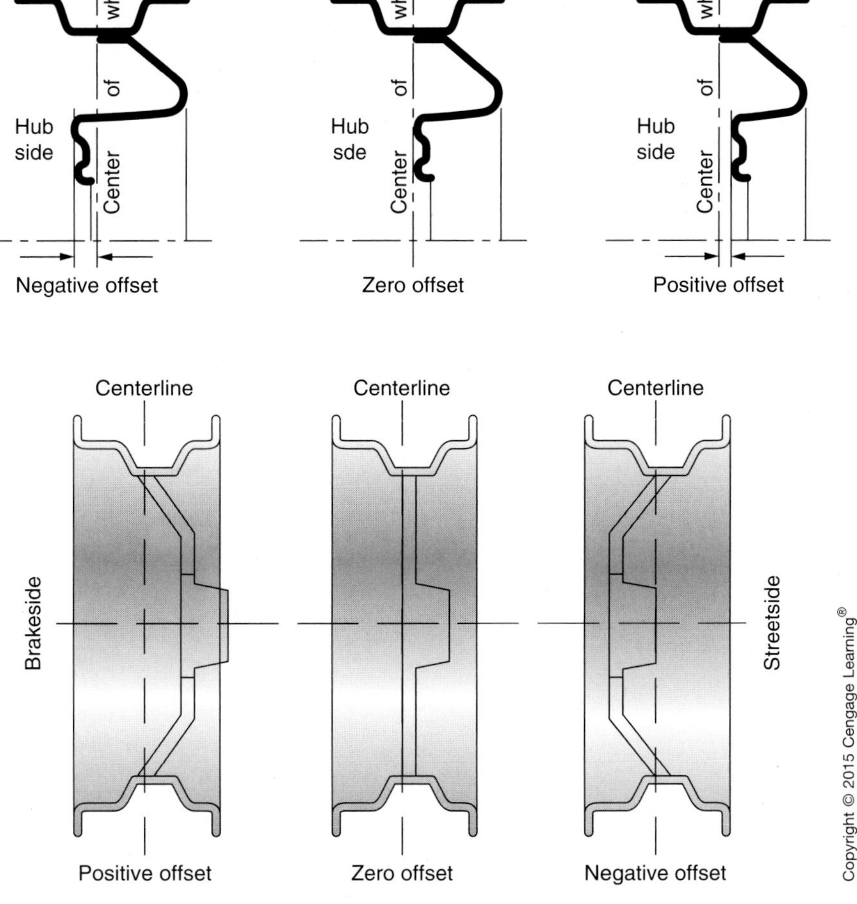

FIGURE 3-33 **If a rim is designed with positive offset, the rim centerline is inboard of the mounting face. The rim with negative offset has the centerline outboard of the mounting face.**

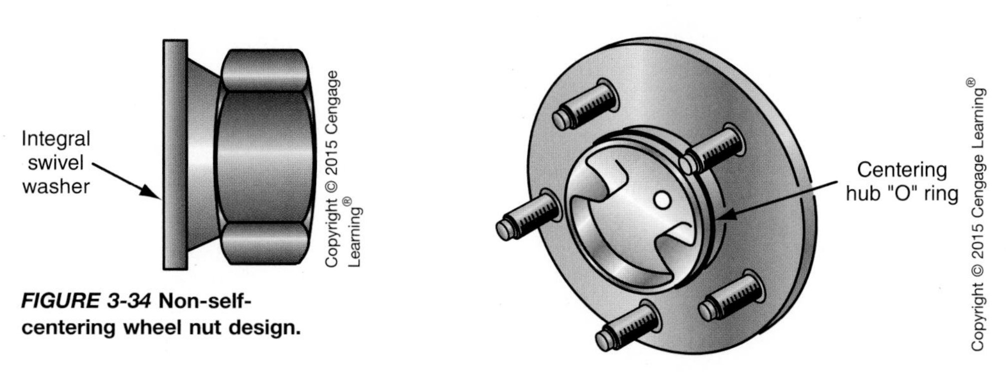

FIGURE 3-34 **Non-self-centering wheel nut design.**

FIGURE 3-35 **Hub O-ring that provides wheel centering.**

Forged aluminum wheels are installed on some vehicles. They are more durable than cast aluminum wheels and have a bright finish. Forged aluminum wheels are used on heavy-load applications such as trucks. Some expensive sports cars are also equipped with forged aluminum wheels. A forged aluminum wheel is lighter than a steel wheel, but the cost of a forged aluminum wheel is considerably higher than a stamped steel or cast aluminum wheel.

Some vehicles are equipped with pressure-cast aluminum wheels that are designed to be chrome plated. Pressure-cast wheels may also be called squeeze-cast wheels. During the manufacture of a pressure-cast wheel, the molten metal is squeezed under pressure

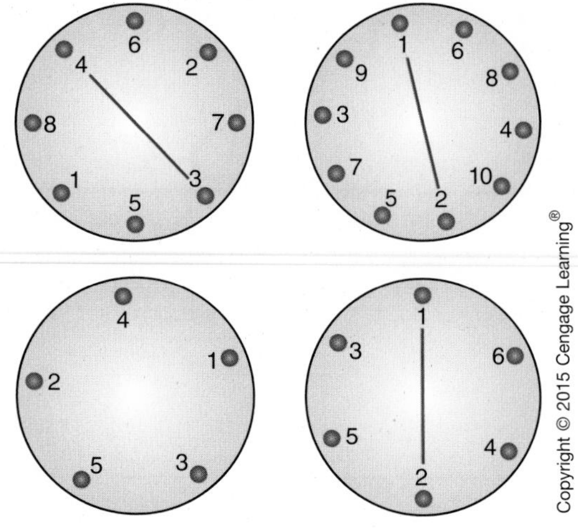

FIGURE 3-36 **Wheel stud configurations.**

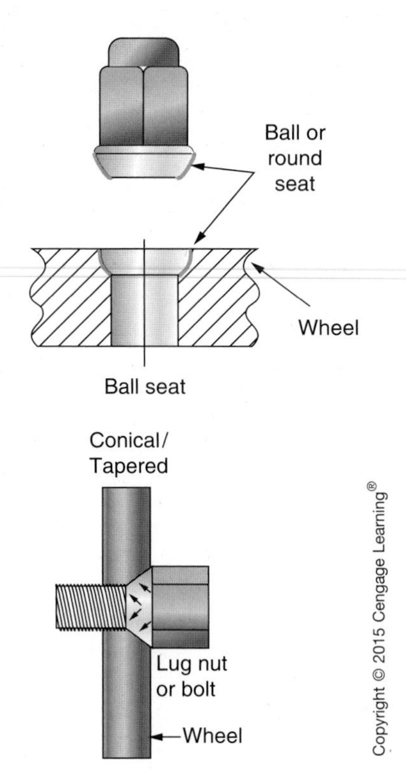

FIGURE 3-37 **Wheels lug nuts generally have a tapered or ball seat to locate and secure the rim to the hub assembly.**

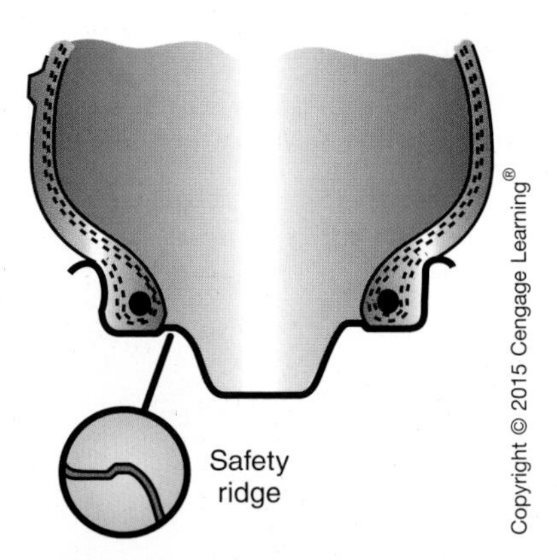

FIGURE 3-38 **Wheel rim with drop center and safety ridges.**

into the mold. This action squeezes all the air out of the metal and reduces the metal porosity, which provides an improved substrate for chrome plating. The chrome plating increases the cost of the pressure-cast aluminum wheel so it is considerably more expensive than other types of wheels.

Magnesium alloy wheels may be called "mag" wheels.

Race cars may be equipped with **magnesium alloy wheels**, which are lighter than either steel or aluminum wheels.

Replacement wheel rims must be the same as the original equipment wheels in load capacity, offset, width, diameter, and mounting configuration. An incorrect wheel can affect tire life, steering quality, wheel bearing life, vehicle ground clearance, tire clearance, and speedometer/odometer calibrations.

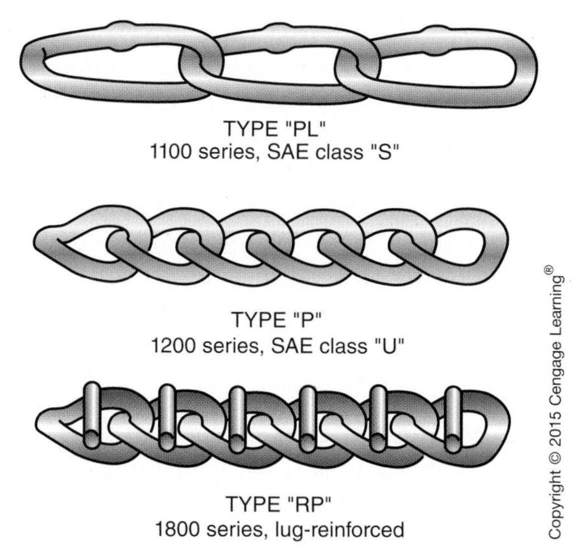

FIGURE 3-39 **Types of tire chains.**

Tire Chains

Tire chains may be used in emergency situations such as driving on snow-covered or ice-covered mountain roads. Most tire manufacturers do not recommend the use of tire chains. Use only SAE class "S" tire chains unless specified by the vehicle and tire manufacturer. Chain types include class S, type P, and type RP (Figure 3-39). These chains must be the proper size and must be installed tightly with the ends secured. Always follow the chain manufacturer's recommended installation procedure. While using chains, driving speed should be reduced. If the chains are heard striking the vehicle body, stop immediately and tighten the chains to prevent body damage.

Tire chains are made from steel and wrap around the tire to improve traction.

Compact Spare Tires

Because cars have been downsized in recent years, space and weight have become major concerns for vehicle manufacturers. For this reason, many car manufacturers have marketed cars with **compact spare tires** to save weight and space. The high-pressure mini-spare tire is the most common type of compact spare (Figure 3-40), and the first letter in the tire size will be a T for temporary. This compact spare rim is usually four inches wide, but is one inch larger in diameter than the other rims on the vehicle. The compact spare rim should not be used with standard tires, snow tires, wheel covers, or trim rings. Any of these uses may result in damage to these items or other parts of the vehicle. The compact spare should be used only on vehicles that offer it as original equipment. Inflation pressure in the compact spare should

Compact spare tires save onboard storage space, but are designed for very limited operation.

FIGURE 3-40 **Compact, high-pressure, minispare tire.**

be maintained at 60 psi (415 kPa). The compact spare tire is designed for very temporary use until the conventional tire can be repaired or replaced. Limit driving speed to 50 mph (80 kph) when the high-pressure minispare is installed on a vehicle.

The space-saver spare tire must be inflated with a special compressor. Battery voltage is supplied to the compressor from the cigarette lighter. This type of compact spare should be inflated to 35 psi (240 kPa). After the tire is inflated, be sure there are no folds in the sidewalls.

The lightweight-skin spare tire is a bias-ply tire with a reduced tread depth to provide an estimated 2,000 miles (3,200 km) of tread life. Always inflate the lightweight-skin spare tire to the pressure specified on the tire placard.

Run-Flat Tires

Run-flat tires are designed to operate at zero air pressure for limited mileage. Run-flat tires may be called self-supporting, extended mobility, or zero-pressure tires.

Some tire manufacturers utilize **run-flat tires** on some of their vehicle platforms. Run-flat tires may be called extended mobility tires (EMT). These tires are standard equipment on the 1995 and later model Corvette. Run-flat tires eliminate the need for a spare tire and a jack, saving both weight and space. Run-flat tires are designed with stiffer sidewalls that will partially support the vehicle weight without air pressure in the tire.

Run-flat tires must provide acceptable levels of inflated performance in the areas of comfort, ride, handling, and adequate deflated mobility. Run-flat tires share these same basic design objectives:

1. Minimize the difference between run-flat tires and conventional tires when the tires are inflated.
2. Enhance the handling and riding capabilities of run-flat tires when inflated.
3. Provide acceptable handling when run-flat tires have zero pressure on various vehicles.
4. Enhance low-pressure and zero-pressure bead retention on run-flat tires.
5. Provide sufficient zero-pressure durability so the vehicle can be driven a reasonable distance to a repair facility.

Shop Manual
Chapter 3, page 104

Run-Flat Tires with Sidewall Reinforcements

One of the most important requirements for run-flat tires is bead retention when running with low or zero pressure. Run-flat tires have improved beads to meet this requirement (Figure 3-41). Some run-flat tires have sidewall reinforcements that may be manufactured from flexible, low-hysteresis rubber, thermal resistive materials, or metallic and/or textile tissues. Run-flat tires with a high aspect ratio require increased sidewall stiffness compared to run-flat tires with a low aspect ratio. Because of these sidewall reinforcements, run-flat

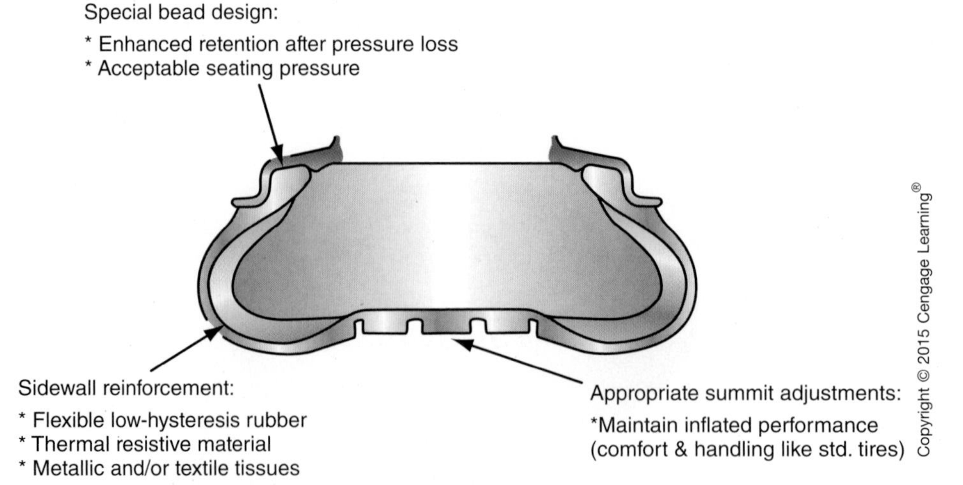

FIGURE 3-41 **Run-flat tire with sidewall reinforcement.**

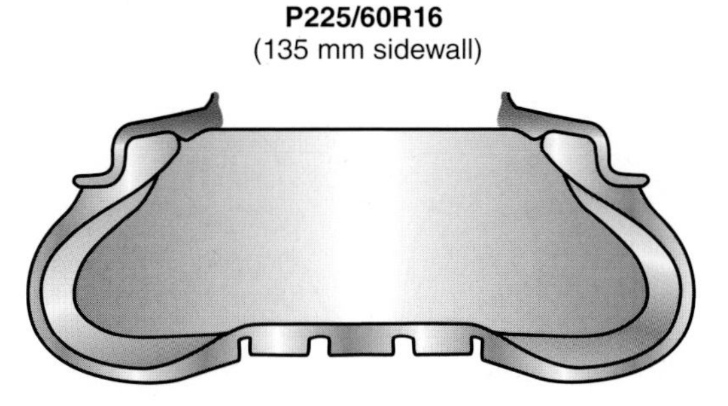

FIGURE 3-42 **Run-flat tire zero-pressure durability.**

tires are 20 to 40 percent stiffer compared to conventional tires. Therefore, run-flat tires may increase ride harshness and vertical firmness especially when the tire strikes a large road irregularity. This increase in ride harshness is not as noticeable when the tire contacts a smaller road irregularity. The increased stiffness in run-flat tires usually provides a small improvement in vehicle handling.

Run-flat tires with a low aspect ratio usually have the same or less rolling resistance compared to equivalent-size conventional tires. Because run-flat tires with a high aspect ratio have increased sidewall stiffness, these tires tend to have increased rolling resistance compared to conventional tires.

In moderate cornering and lane-change maneuvers, a run-flat tire with zero pressure provides a slight decrease in vehicle handling. The key word in the previous sentence is moderate. Run-flat tires are used with a tire pressure monitoring system, which has a warning light in the instrument panel to inform the driver when one tire has low pressure. Once the low tire pressure warning light is illuminated with the engine running, the driver should avoid high-speed driving and high-speed cornering, because run-flat tires provide reduced handling capabilities during high-speed cornering.

The zero-pressure durability of a run-flat tire varies depending on the vehicle weight, atmospheric temperature, and the tire aspect ratio. For example, a run-flat tire with a 40 aspect ratio supporting 880 lb (400 kg) provides 186 mi (300 km) of driving at 55 mph (89 km/h) (Figure 3-42). A run-flat tire with a 60 aspect ratio supporting 1,322 lb (600 kg) provides 50 mi (80 km) of driving at 55 mph (89 km/h).

CAUTION:
On a vehicle with PAX tires as original equipment, conventional tires and rims should not be substituted for the PAX tires and rims. This action will decrease ride quality and vehicle steering characteristics.

Run-Flat Tires with Support Ring

Michelin PAX run-flat tires were first introduced in 1998, but later discontinued in 2009. PAX run-flat tires were installed as original equipment on some Honda Odyssey, Toyota Sienna, Nissan Quest, and Acura RL models and Audi A8 and A4 models in Europe. The letters "PAX" translate to the values of peace of mind, safety, and the future. The technology never gained traction though on the world market and has now become part of automotive history.

The run-flat tires in the PAX system have a flexible support ring mounted on a special rim to support the tire if deflation occurs. PAX tires do not have a stiffer sidewall, and so they provide excellent ride quality. When PAX tires are installed on a vehicle as original equipment, all the suspension system components such as springs and shock absorbers are engineered to go with the PAX tires.

PAX run-flat tires have a vertical, mechanical locking bead system that "latches" the bead to the rim. The outer tire bead has a smaller diameter compared with the inside tire bead. Therefore, the outer edge of the rim has a smaller diameter compared with the inner edge (Figure 3-43). When mounted on the rim, the tire beads lock into vertical grooves in

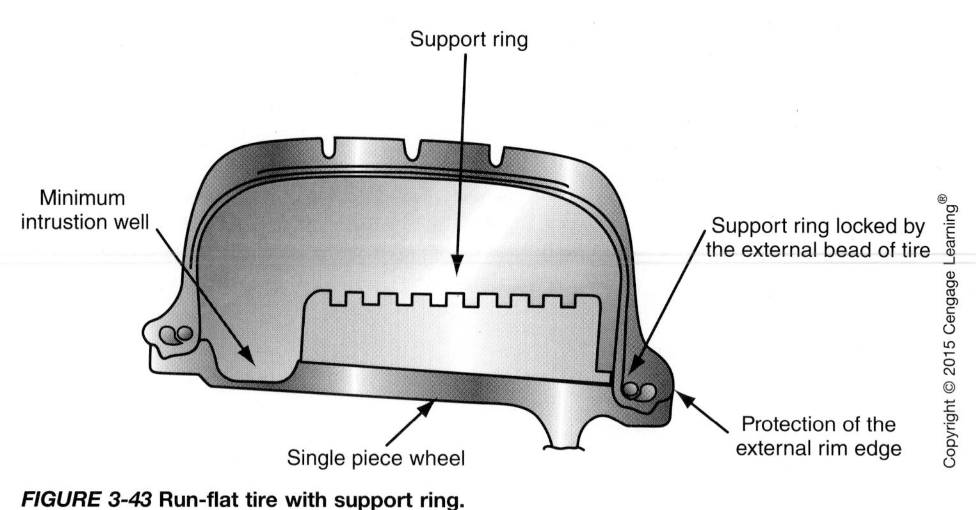

FIGURE 3-43 **Run-flat tire with support ring.**

CAUTION:
Using conventional tire changers that do not have PAX capabilities on PAX run-flat tires and rims may damage the tire and rim. Tire deflection is the difference between the free diameter and the rolling diameter of the tire.

the outer edges of the rim. The outer edges of the inner and outer tire beads extend slightly outside the wheel rim edges. This design helps protect the rim. PAX tires may be driven for 125 miles (201 km) at 55 mph (88 km/h) with zero air pressure. When mounting the tire on the rim, the tire is first positioned such that the tire beads are outside the rim. The outer (smaller diameter) bead is then mounted onto the rim, followed by the inner (larger diameter) bead. The disadvantages of this type of run-flat tire are the special rim that is required and the extra weight of the support ring. A tire pressure monitoring system is used with this type of run-flat tire.

Tire Pressure Monitoring Systems

The Transportation Recall Enhancement Accountability and Documentation Act passed by the U.S. government in October 2000 requires all new vehicles to have tire pressure monitoring systems no later than the 2008 model year. A tire pressure monitoring system illuminates a warning light in the instrument panel to inform the driver if one tire has low pressure; some systems also alert the driver if tire temperature is excessive.

Shop Manual
Chapter 3, page 108

Some tire pressure monitoring systems have a sensor strapped in the drop center on each rim (Figure 3-44). These sensors must be mounted on the rim directly opposite to the valve stem. Other tire pressure monitoring systems have a sensor mounted on top of each valve stem in place of the valve cap. Regardless of the sensor's mounting location, it transmits radio

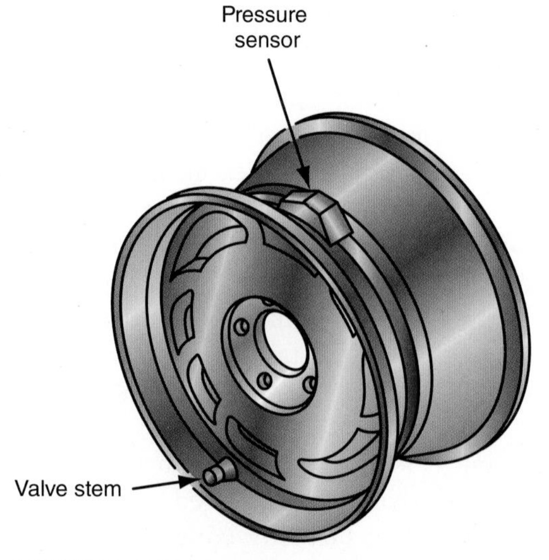

Location	Color code
Right front	Blue
Left front	Green
Right rear	Orange
Left rear	Yellow

WARNING: Pressure sensor inside tire. Avoid contacting sensor with tire changing equipment tools or tire bead.

Service note: Pressure sensor must be mounted directly across from valve stem.

FIGURE 3-44 **Tire pressure sensor.**

frequency (RF) signals. These RF signals will change if any tire is underinflated a specific amount. The RF signals are sent to a receiver that is usually mounted under the dash (Figure 3-45). On some systems the receiver illuminates the low tire pressure warning light in the instrument panel if the pressure in any tire drops below 25 psi (172 kPa). On other systems the tire pressure warning light is illuminated with a smaller reduction in tire pressure.

Some tire pressure monitoring systems provide continuous monitoring with the vehicle stopped and the ignition switch on. Other systems will not monitor tire pressure until the vehicle is moving above a specific speed such as 25 mph (40 km/h). Some tire pressure monitoring systems have two warning lights: the low/flat tire warning light informs the driver regarding low tire pressure, and the service low tire pressure warning system (LTPWS) warning light informs the driver if there is a defect in the system (Figure 3-46). On some vehicles the low tire pressure receiver is connected to the data link connector (DLC) under the dash. A scan tool may be connected to the DLC to diagnose the tire pressure monitoring system. If a defect occurs in the system, a diagnostic trouble code (DTC) is stored in the receiver memory. A scan tool may be connected to the DLC to obtain the DTCs from the receiver.

Other tire pressure monitoring systems use the wheel speed sensor signals on four-channel ABS to detect low tire pressure. A tire with low air pressure has a smaller diameter, and the wheel speed sensor generates a higher frequency signal.

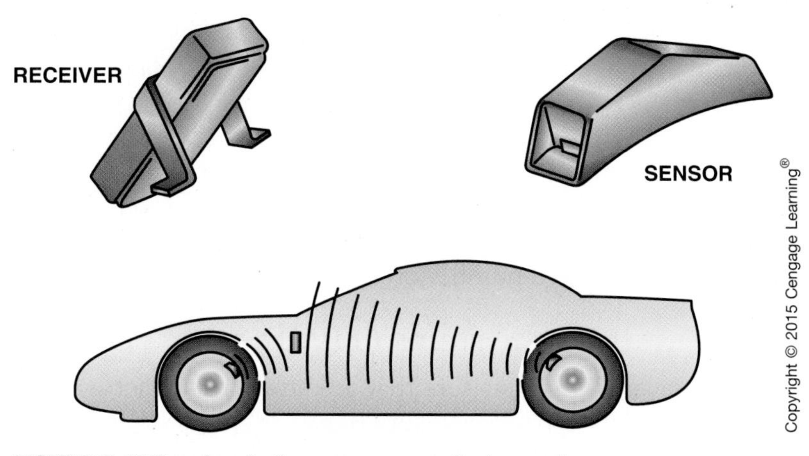

FIGURE 3-45 **Receiver in tire pressure monitoring system.**

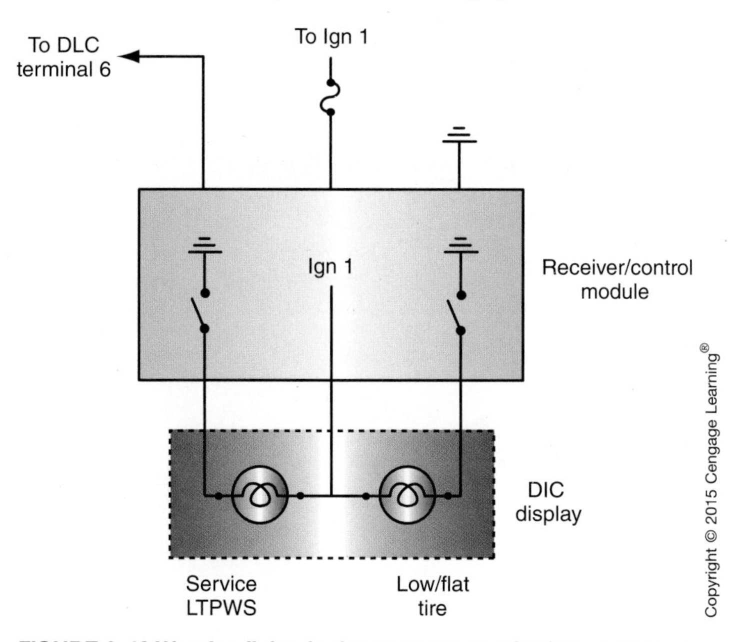

FIGURE 3-46 **Warning lights in tire pressure monitoring system.**

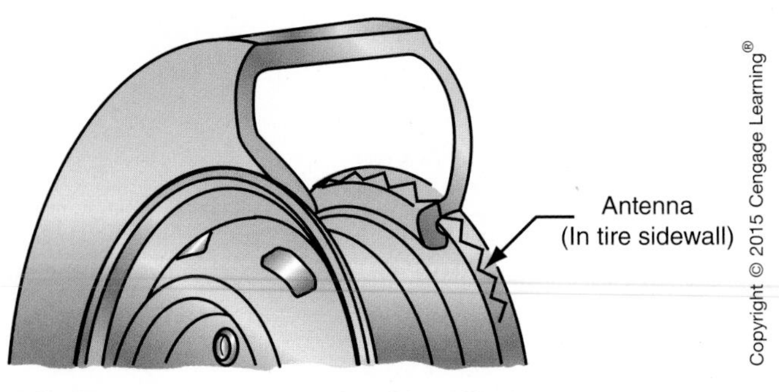

FIGURE 3-47 **Antenna in a tire sidewall.**

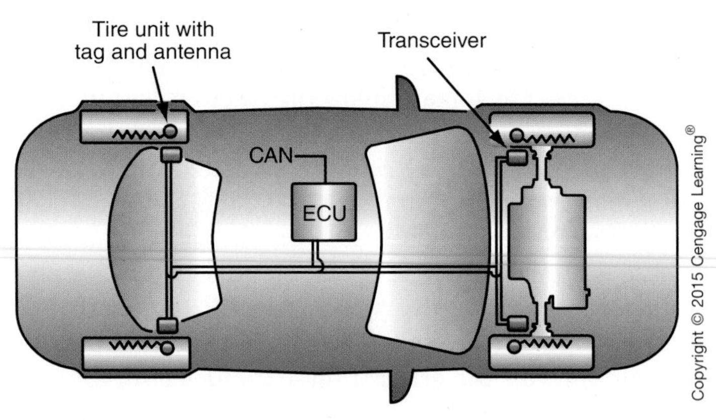

FIGURE 3-48 **Transceivers and ECU in a tire pressure monitoring system.**

Some tire pressure monitoring systems have a miniature pressure sensor and computer chip about the size of a watch battery. This sensor and chip assembly is imbedded in the tire. The pressure sensor senses the actual tire pressure and sends voltage signals in relation to the tire pressure to a transceiver via a 360° circumferential antenna mounted in the tire sidewall close to the rim (Figure 3-47). A transceiver is mounted in each wheel well. The pressure sensor and computer chip do not require a battery, because they receive energy from the transceiver-generated field. Each transceiver is connected by data links to a central electronic control unit (ECU) (Figure 3-48). The transceivers relay voltage signals from the pressure sensors to the ECU. The computer chip calculates the recommended tire pressure based on data such as air temperature, tire pressure, and vehicle speed. Vehicle speed information is transmitted via data links from the powertrain control module (PCM) to the ECU. If a low tire pressure signal is received by the ECU, the instrument panel displays a warning message together with a calculated number of miles of safe travel.

Tire Pressure Monitoring Systems (TPMS) with Valve Stem Sensors

Most current TPMS have pressure sensors that are attached to each valve stem inside the tire, and these sensors require a battery (Figure 3-49). Average battery life is approximately 10 years. A retention nut is threaded onto the valve stem on the outside of the wheel rim, and a sealing grommet is positioned between this nut and the rim (Figure 3-50). The internal threads on the retention nut are positioned near the bottom of the nut, and matching

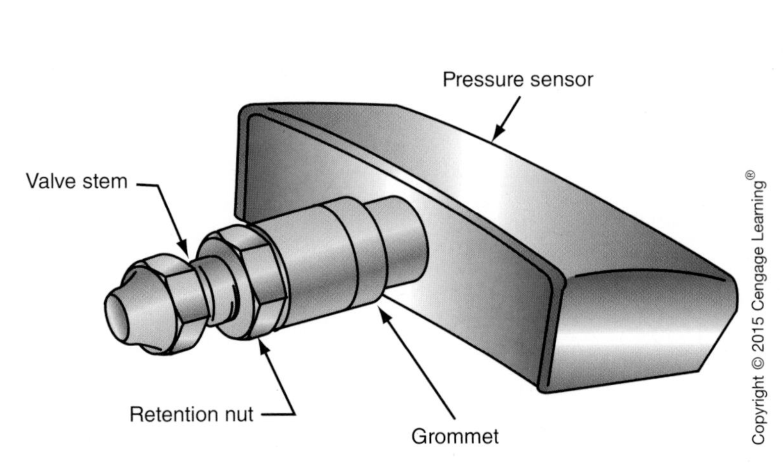

FIGURE 3-49 **Tire pressure sensor mounted on the valve stem.**

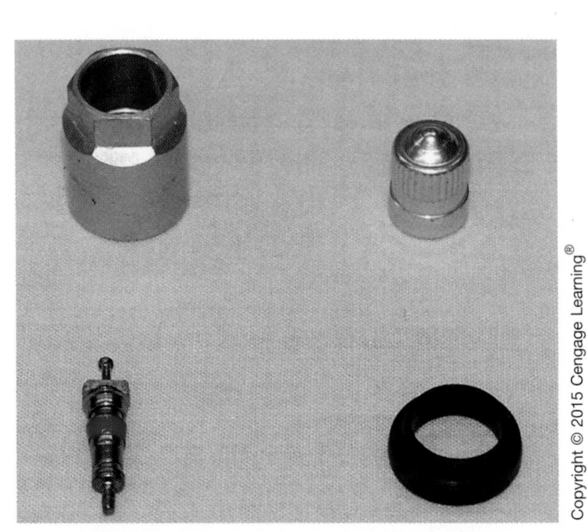

FIGURE 3-50 **Tire pressure sensor retention nut, valve cap, valve core, and grommet.**

threads are located near the bottom end of the valve stem. A nickel-plated valve core is threaded into the internal threads in the valve stem, and a special valve cap is threaded onto the external threads at the top of the valve stem.

Many vehicles have a TPMS sensor in the spare tire. The valve stem acts as a radio antenna, and the TPMS module is the receiver. Each sensor transmits radio signals to the TPMS module every 60 seconds. On some SUVs the TPMS module is located behind the right-hand C pillar. The TPMS module compares the data from each tire pressure sensor to the low tire pressure limits. When the TPMS module determines that any of the five tires has a pressure below the specified pressure, this module sends voltage signals through the data links to the vehicle message center. When the message center receives voltage signals indicating low pressure in one or more tires, a warning message is displayed in the message center, and the TPMS warning light is illuminated. The possible messages related to the TPMS are:

1. WARNING TIRE VERY LOW—On some systems this message is displayed if any tire has less than 25 psi pressure. An audible warning chime is heard when this message is displayed, and the TPMS warning light is illuminated.
2. CHECK TIRE PRESSURE—On some systems this message is displayed if any tire has less than 30 psi pressure. An audible warning chime is heard when this message is displayed, and the TPMS warning light is illuminated.
3. TIRE PRESSURE SENSOR FAULT—This message is displayed if one or more sensors are malfunctioning. If this message is displayed, the TPMS warning light flashes for 20 seconds.
4. TIRE PRESSURE MONITOR FAULT—This message is displayed if the TPMS module is defective or if all four sensors have failed. The TPMS warning light flashes for 20 seconds when this message is displayed.

Some TPMS have the capability to sense a loss or an increase in tire pressure, and these systems also allow the driver to display the individual tire pressures and their locations on the driver information center (DIC) while the vehicle is being driven. This type of TPMS has conventional radio frequency–transmitting pressure sensors in each of the four valve stems. These sensors transmit signals to the antenna module, dash integration module (DIM), instrument panel cluster (IPC), and the DIC via the serial data circuit. The sensor's pressure accuracy from 14°F to 158°F (10°C to 70°C) is plus or minus 1 psi (7 kPa).

When the vehicle speed is less than 20 mph (32 km/h), the system remains in the stationary mode. In this mode, the sensors transmit data every 60 minutes to provide longer sensor battery life. When the vehicle speed is 20 mph (32 km/h) or greater, centrifugal force closes a roll switch in each sensor, and this action causes the sensors to enter the drive mode in which the sensors transmit signals every 60 seconds. The antenna module receives and translates each sensor signal into sensor presence, sensor mode, and tire pressure. The temperature and speed ratings of the TPMS system vary depending on the vehicle manufacturer. Always consult the vehicle manufacturer's specifications. The antenna module then transmits this information from each sensor to the DIC via the serial data circuit. The DIC displays an overhead view of the vehicle in which tire locations and pressures are displayed. If the TPMS senses a specific pressure gain or loss in any tire, a CHECK TIRE PRESSURE warning message is displayed on the DIC, and the low tire pressure warning indicator is illuminated in the IPC. Any defect in the TPMS system causes a SERVICE TIRE MONITOR warning message to be displayed on the DIC.

TPMS with Electronic Vehicle Information Center (EVIC) Display

Some TPMS provide graphic tire pressure displays and warning messages on an EVIC display in the instrument panel. These systems may have 4-tire or 5-tire monitoring capabilities. A TPMS with 5-tire monitoring capabilities also monitors the spare tire. These systems have valve stem sensors that broadcast tire pressure once per minute when the vehicle is moving at 25 mph (40 km/h) or faster. If the vehicle has a 5-tire TPMS, the sensor in the spare tire transmits a signal every hour. Each valve stem sensor transmits a unique code so the EVIC module can determine the location of the sensor. If the wheels are rotated on the vehicle, or the spare tire and wheel are installed on the vehicle, the EVIC must be reprogrammed to recognize the new wheel and tire locations.

This type of TPMS provides a warning display in the EVIC if the tire pressure drops below a specific value or increases above a certain threshold. Typically, if the tire pressure in any tire decreases below 25 psi (172 kPa), the EVIC requests a chime warning, displays a LOW PRESSURE message, and indicates the location of the tire with low pressure. If the pressure in any tire exceeds 45 psi (310 kPa), the EVIC requests a chime warning and displays HIGH PRESSURE while indicating the location of the tire with high pressure. After a few seconds the EVIC reverts to a blinking display of the tire pressure. This blinking display continues for the entire ignition cycle or until an EVIC button is pressed. When an EVIC button is pressed, the blinking display returns after 60 seconds without the chime warning. If high or low pressure is detected in the spare tire on a 5-tire system, SPARE HIGH PRESSURE or SPARE LOW PRESSURE appears on the EVIC display for 60 seconds during each ignition cycle.

Shop Manual
Chapter 3, page 110

STATIC WHEEL BALANCE THEORY

Static balance refers to the balance of a wheel in the stationary position.

When a tire and wheel assembly has proper **static balance**, it has the weight equally distributed around its axis of rotation, and gravity will not force it to rotate from its rest position. If a vehicle is raised off the floor and a wheel is rotated in 120° intervals, a statically balanced wheel will remain stationary at each interval. When a wheel and tire are statically unbalanced, the tire has a heavy portion at one location. The force of gravity acting on this heavy portion will cause the wheel to rotate until the heavy portion is located near the bottom of the tire (Figure 3-51).

Results of Static Unbalance

Centrifugal force may be defined as the force that tends to move a rotating mass away from its axis of rotation. As we have explained previously, a tire and wheel are subjected to very strong acceleration and deceleration forces when a vehicle is in motion. The heavy portion of a statically unbalanced wheel is influenced by centrifugal force. This influence attempts to move the heavy spot on a tangent line away from the wheel axis. This action tends to lift the wheel assembly off the road surface (Figure 3-52).

Wheel tramp may be defined as rapid upward and downward wheel and tire oscillations.

The wheel-lifting action caused by static unbalance may be referred to as **wheel tramp** (Figure 3-53). Wheel tramp action allows the tire to slip momentarily when it is lifted vertically. When the wheel and tire move downward as the heavy spot decelerates, the tire strikes the road surface with a pounding action. This repeated slipping and pounding action causes severe tire scuffing and cupping (Figure 3-54).

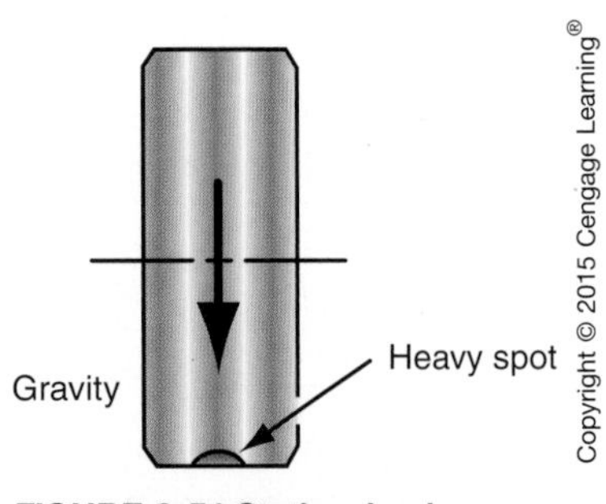

FIGURE 3-51 **Static wheel unbalance.**

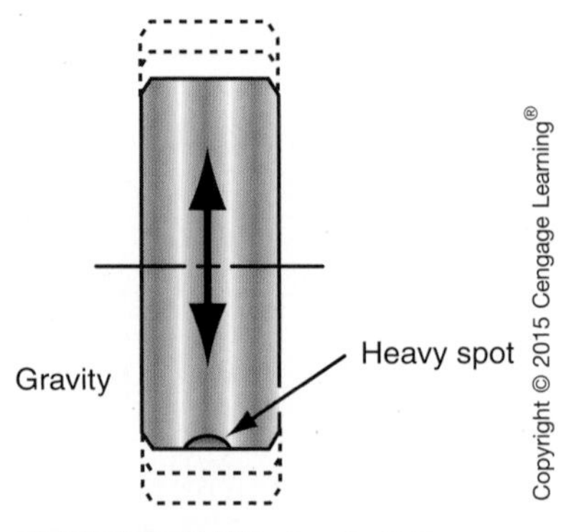

FIGURE 3-52 **Effects of static unbalance.**

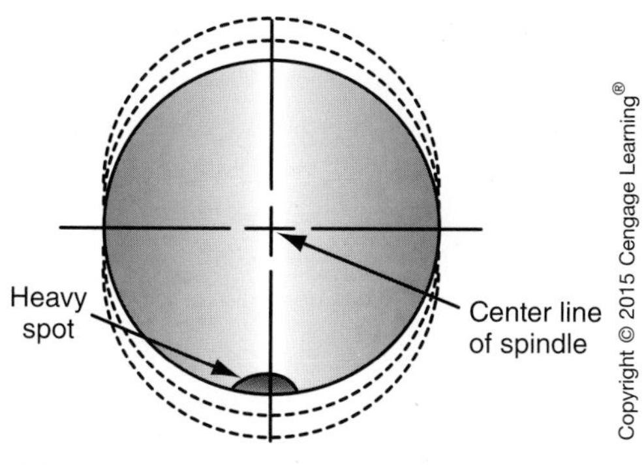

FIGURE 3-53 **Wheel tramp.**

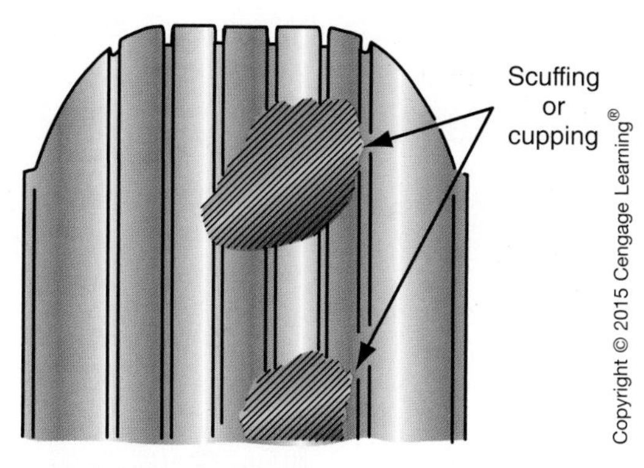

FIGURE 3-54 **Cupping tire wear caused by static unbalance.**

The vertical wheel motion from static unbalance is transferred to the suspension system and then absorbed by the chassis and body. This action causes rapid wear on suspension and steering components. The wheel tramp action resulting from static unbalance is also transmitted to the passenger compartment, which causes passenger discomfort and driver fatigue.

When a vehicle is traveling at normal highway cruising speed, the average wheel speed is 850 revolutions per minute (rpm). A statically unbalanced tire and wheel assembly is an uncontrolled mass of weight in motion. When a vehicle is traveling at 60 mph (97 km/h) and a tire has 2 ounces (oz), or 57 grams (g), of static unbalance, the resultant pounding force is approximately 15 pounds (lb), or 6.8 kilograms (kg), against the road surface.

Dynamic Wheel Balance Theory

Dynamic balance refers to the balance of a wheel in motion.

When a tire and wheel assembly has correct **dynamic balance**, the weight of the assembly is distributed equally on both sides of the wheel center viewed from the front. Dynamic wheel balance may be explained by dividing the tire into four sections (Figure 3-55). In Figure 3-55, if sections A and C have the same weight and sections B and D also have the same weight, the tire has proper dynamic balance. If a tire has dynamic unbalance, section D may have a heavy spot; thus, sections B and D have different weights (Figure 3-56).

From our discussion of dynamic balance, we can understand that a tire and wheel assembly may be in static balance, but have dynamic unbalance. Therefore, wheels must be in balance statically and dynamically. When a tire and wheel assembly is placed on a

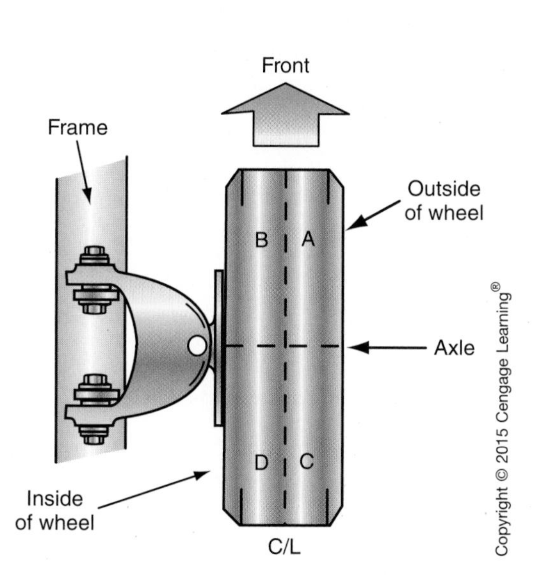

FIGURE 3-55 **Dynamic wheel balance theory.**

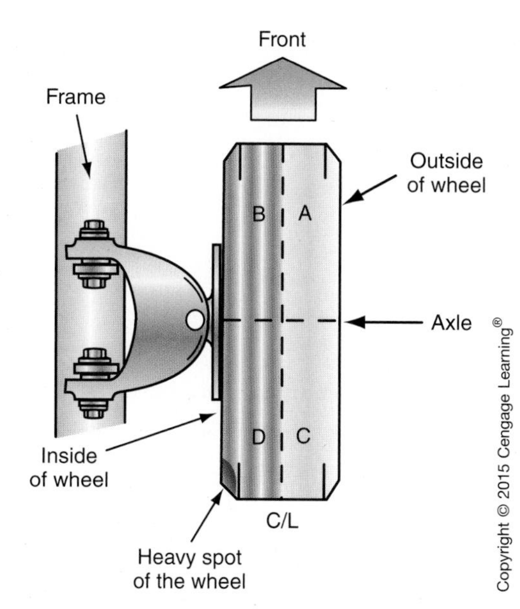

FIGURE 3-56 **Dynamic wheel unbalance.**

FIGURE 3-57 **Wheels weights come in various dimensions depending on wheel rim assembly design.**

computer wheel balancer to check and correct for an imbalance condition, it is generally necessary to attach a wheel weight to the rim assembly to offset this condition. Wheel weights are generally either clip on or stick on weights. Clip on wheel weights are attached to the outboard flange of the rim and come in an assortment of sizes and shapes depending on the shape of the rim flange (Figure 3-57).

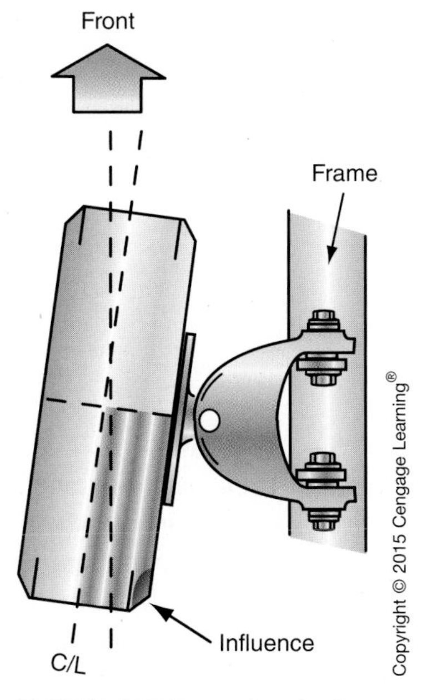

FIGURE 3-58 Dynamic wheel unbalance with heavy spot at the rear of the left front tire.

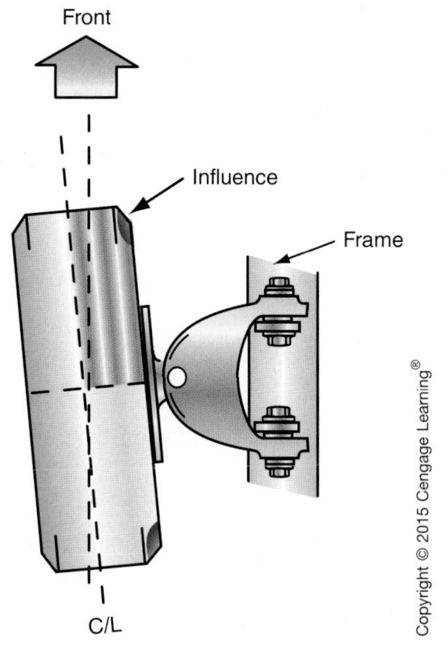

FIGURE 3-59 Dynamic wheel unbalance with heavy spot at the front of the left front tire.

Results of Dynamic Wheel Unbalance

When a dynamically unbalanced wheel is rotating, centrifugal force moves the heavy spot toward the tire centerline. The centerline of the heavy spot arc is at a 90° angle to the spindle. This action turns the true centerline of the left front wheel inward when the heavy spot is at the rear of the wheel (Figure 3-58). When the wheel rotates until the heavy spot is at the front of the wheel, the heavy spot movement turns the left front wheel outward (Figure 3-59).

From these explanations, we can understand that dynamic wheel unbalance causes lateral wheel shake, or shimmy (Figure 3-60). This action causes steering wheel oscillations at medium and high speeds with resultant driver fatigue and passenger discomfort. Wheel shimmy and steering wheel oscillations also cause unstable directional control of the vehicle.

Earlier in this chapter, when discussing wheel rotation, we mentioned that a tire stops momentarily where it contacts the road surface. A wheel with dynamic unbalance pivots on the contact area, which results in excessive tire scuffing and wear. Dynamic wheel unbalance causes premature wear on steering linkage and suspension components. Therefore, dynamic wheel balance is extremely important to provide normal tire life, reduce

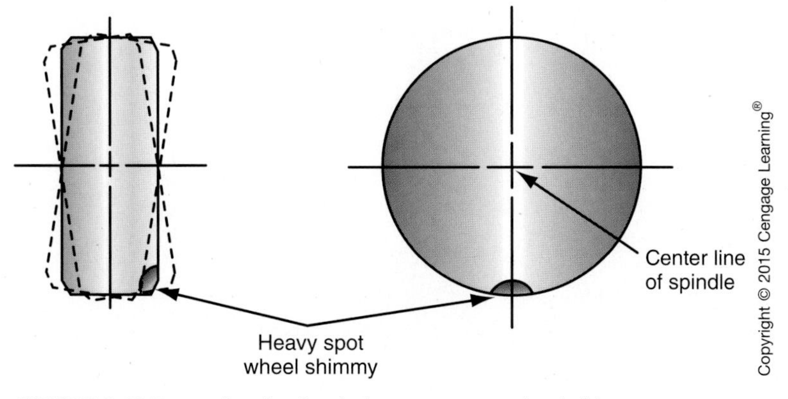

FIGURE 3-60 Dynamic wheel unbalance causes wheel shimmy.

steering and suspension component wear, increase directional control, and decrease driver fatigue. The main purposes of proper wheel balance are:

1. Maintains normal tire tread life.
2. Provides extended life of suspension and steering components.
3. Helps provide directional control of the vehicle.
4. Reduces driver fatigue.
5. Increases passenger comfort.
6. Helps maintain the life of body and chassis components.

Shop Manual
Chapter 3, page 117

Tire Motion Forces

When a vehicle is in motion, wheel rotation subjects the tires to centrifugal force. The tires are also subjected to accelerating and decelerating forces because of their path of travel. If a vehicle is traveling at 55 mph, or 88.5 km/h, the part of the tire exactly fore and aft of the spindle is also traveling at 55 mph (88.5 km/h). At the exact top of the tire, the tire speed is 110 mph (177 km/h). The tire speed actually drops to zero at the exact bottom of the tire where the arc of deceleration ends and the arc of acceleration begins. Since the tires are subjected to strong acceleration and deceleration forces, the tire construction must be uniform. A soft spot in a tire will deflect farther than the surrounding area, and this area will be subjected to rapid wear as it strikes the road surface (Figure 3-61).

Tires must have equal stiffness in the sidewalls. If a tire does not have equal stiffness in all areas around the tire sidewalls, it may cause a vibration when driving. This vibration is called **road force variation** and produces a vibration similar to the vibration caused by improper wheel and tire balance. Imagine the tire as a collection of springs around the circumference of the tire. If the sidewall stiffness is not uniform a varied force is exerted, which creates a vibration. Hunter road force wheel balancers have a roller that is forced against the tire tread during the balance procedure, and this type of balancer detects road force variation (Figure 3-62).

Road force variation is a condition that occurs when tire sidewalls do not have equal stiffness around their complete area.

The radial force variation can be determined by examining the radial run out. Loaded radial runout of one thousands of an inch is equivalent to approximately one pound of radial force variation. As an example at 50 mph a wheel imbalance of 1.5 ounces (42 grams) will result in the same amount of vibration as 0.30 in (0.76 mm) of loaded radial runout or 30 pounds of road force variation.

FIGURE 3-61 **Think of a tire as sets of springs around its diameter. The variation in sidewall stiffness is referred to as road force variation and if excessive can create harmonic vibration and tire wear.**

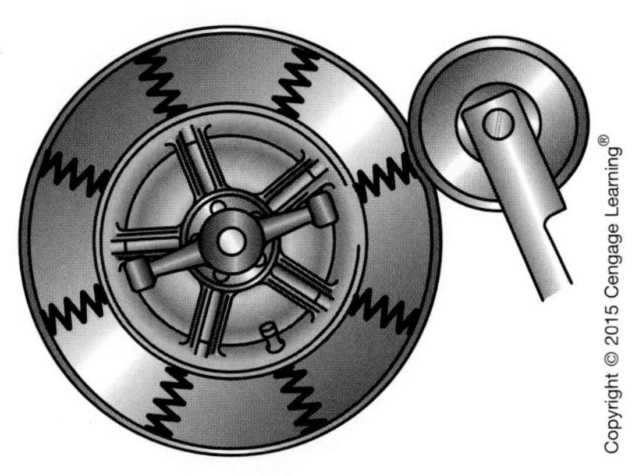

FIGURE 3-62 **The roller on some wheel balancers senses road force variation.**

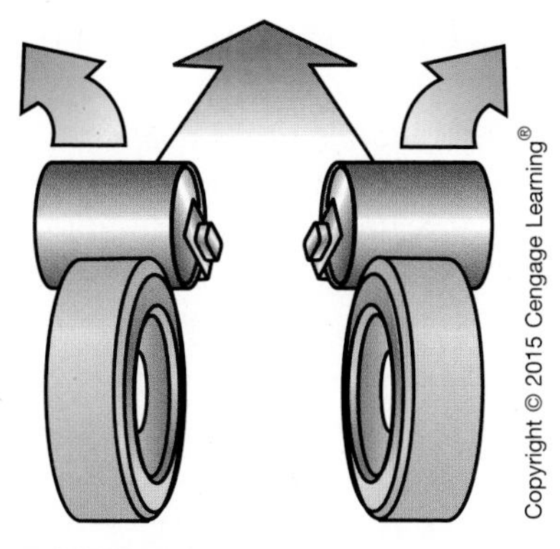

FIGURE 3-63 **The roller on some wheel balancers senses concentricity.**

If excessive road force variation is detected match mounting the rim and tire should be considered. In match mounting the high point on the tire is matched with the low point on the rim runout or vice versa to decrease the rolling vibration in the wheel assembly.

Tires must be manufactured so the cords are concentric with the center of the tire. Tire conicity is a term used to describe a tire in which the cords are off-center in relation to the tire center. If the tire cords are not concentric with the tire center, the tire may cause a steering pull condition when driving. Modern wheel balancers use a roller pressed against the tire tread to detect conicity during the balance procedure (Figure 3-63).

Noise, Vibration, and Harshness Analysis

OEMs have made significant improvements in reducing vehicle noise, vibration, and harshness (NVH) because many customers prefer a quieter-running vehicle with reduced vibration and ride harshness. The suspension system plays a significant role in reducing NVH. Many of the suspension features described in this chapter are designed to reduce NVH. These features include track bars and braces, proper car riding height, hydraulic suspension system mounts, independent rear suspension systems, and large rubber insulating bushings on suspension system mountings.

However, the technician must understand that reducing NVH involves many mechanical and body or chassis components on the vehicle. For example, some vehicles have an optimized body structure, which is the major reason for reduced NVH. In this body design, lateral tie bars that connect the front longitudinal rails provide a stiffer front end. At the rear of the car, one-piece side rings with integral quarters eliminate rear pillar seams and provide a more precise door fit. The instrument panel and steering column are integrated solidly into the body structure by a cast magnesium beam. The door hinges are through-bolted and thick spacer blocks are installed on these bolts to provide a very solid door attachment. Because the entire body is stronger and more rigid, the suspension can provide excellent ride quality and steering control without having to compensate for unwanted body flexing. In addition, appropriate body cavities like the dash panel are filled with expandable baffles to eliminate noise. The five-layer noise buffer in the dash panel contains these materials:

1. Fiberglass insulation mat
2. Viscoelastic energy-absorbing layer
3. Double steel panel
4. Single one-piece dash mat

A cast foam floor carpet system reduces noise transmitted through the floor pan and wheel wells. The door pillar and rocker panel cavities contain over 20 noise blockers.

Many engine refinements reduce vibration. For example, many engines now have a deep skirt block with the main bearings bolted through the sides of the block and vertically. Many engines now have cast oil pans and rocker covers rather than stamped steel components. The main bearing caps are contained in a one-piece casting on many engines to increase bottom end strength and reduce engine vibration. Some V-8 engines have a rubber intake manifold valley stuffer attached to the underside of the intake manifold to reduce vibration. Other vehicles have a slip yoke vibration damper mounted on the front of the drive shaft to reduce driveline vibration. Therefore, reducing NVH is a total vehicle concept.

Vibration Theory

Vibrations have these three elements:

1. Source—the cause of the vibration
2. Path—where the vibration travels through the vehicle
3. Responder—the component where the vibration is felt

For example, if the vehicle has an unbalanced tire, this is the vibration source. The vibration path is the steering and suspension system through which the vibration travels. The responder is the steering wheel because this component is where the customer feels the vibration (Figure 3-64). When diagnosing vibration problems, locate and correct the source of the vibration. In the previous example of the unbalanced tire, installing a rigid brace from the steering column to the instrument panel and chassis may reduce the vibration experienced by the customer, but this does not solve the problem. To eliminate the problem, also diagnose the unbalanced tire condition and then balance the tire and wheel assembly.

Vibration may also produce noise. If a vehicle has a broken or improperly positioned tailpipe hanger that allows the tailpipe to contact the chassis, the customer may complain about a vibrating noise. The floor panel acts as a large speaker and amplifies the vibrating noise. The vibration path is through the exhaust system and chassis to the floor panel. The responder is the chassis and floor panel. In this case, after diagnosing the broken or improperly positioned tailpipe hanger, replace or reposition it to eliminate the vibration transfer path.

A **cycle** begins and ends at the same point.

Clamp a yardstick to a table top with 18 in (45 cm) hanging over the edge of the table (Figure 3-65). If the outer end of the yardstick is pulled upward or downward and then released, the end of the yardstick repeatedly vibrates up and down. Each vibration **cycle** begins at the midpoint with the yardstick straight. From this point, the vibration cycle

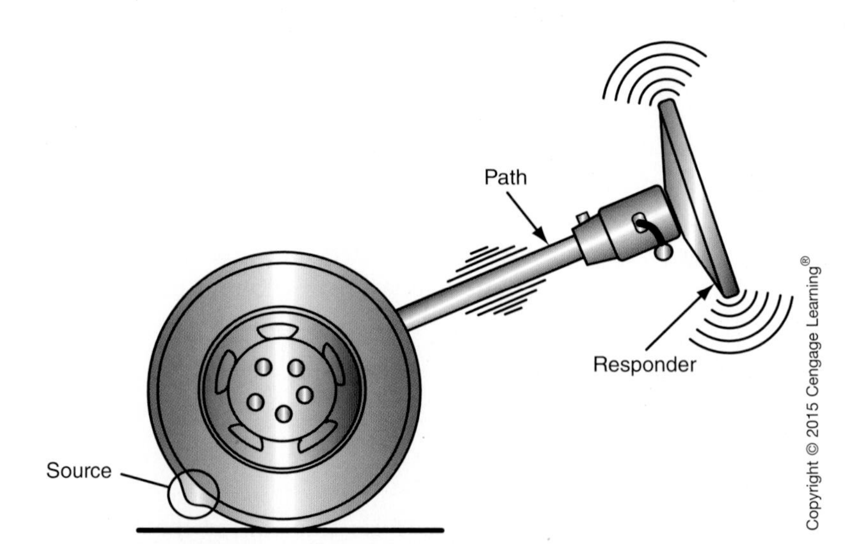

FIGURE 3-64 **Vibration source, path, and responder.**

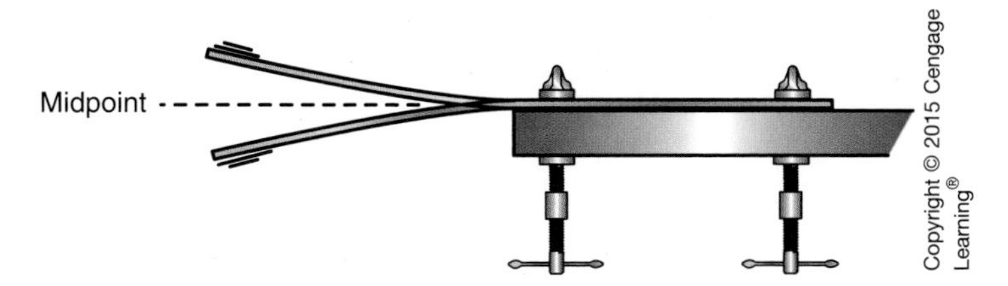

FIGURE 3-65 **Theory of vibration cycles.**

FIGURE 3-66 **Electronic vibration analyzer (EVA).**

continues to the lowest point of travel and then moves up through the midpoint to the highest point of travel. The vibration cycle then returns to the midpoint where it begins over again.

If the end of the vibrating yardstick completed 10 cycles per second, this is called the **frequency** of the vibration. To calculate the cycles per minute multiply the cycles per second by 60. In this example it is $10 \times 60 = 600$ cycles per minute.

The **frequency** of a vibration is the number of cycles per second or cycles per minute.

If the yardstick is clamped to the table top with an 8 in (20 cm) overhang, the end of the yardstick will vibrate much faster when the end of the yardstick is pulled upward or downward. Under this condition, the end of the yardstick may vibrate at a frequency of 30 cycles per second or 1,800 cycles per minute.

Many vehicle vibrations are caused by an out-of-balance rotating component or engine firing pulses improperly isolated from the passenger compartment. Customers usually complain about vibrations that are felt in the steering wheel, instrument panel, frame or chassis, or the front or rear seat.

Vehicle vibrations may be tested with an **electronic vibration analyzer (EVA)**. (Figure 3-66). The EVA has a vibration sensor mounted on the suspected vibration source. The EVA senses and records vibration cycles. A vibration cycle begins and ends at the same point and is continually repeated (Figure 3-67). Cycles per second are measured in **Hertz (Hz)**, and the Hz may be multiplied by 60 to obtain the cycles per minute. The **amplitude** of a vibration is the maximum value of the varying vibration. In Figure 3-68, 1 represents the maximum amplitude, 2 is the minimum amplitude, 3 is the zero-to-peak amplitude, and 4 indicates the peak-to-peak amplitude. The vibration amplitude may vary with the rotating speed of a component. For example, if a tire and wheel assembly is unbalanced, the amplitude of the resulting vibrations increases with wheel speed.

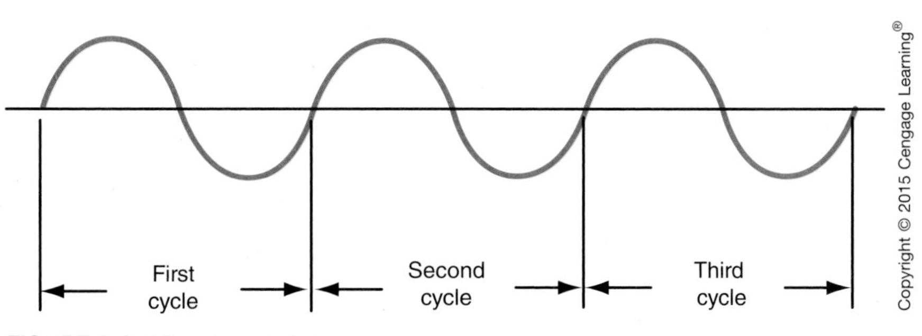

FIGURE 3-67 **Vibration cycles.**

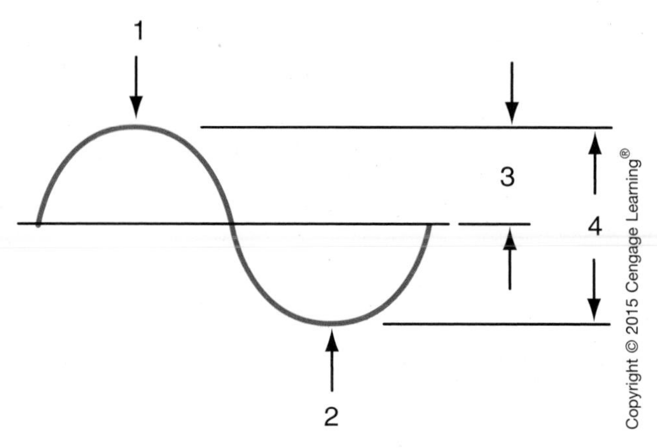

FIGURE 3-68 **Amplitude of vibration cycles.**

Natural vibration frequency is the frequency at which the object tends to vibrate.

Resonance is the point where the frequency of a vibration from a defective component intersects with the natural frequency of the component or system where the component is located.

All objects have **natural vibration frequencies**. The natural frequency of a front suspension system is 10 to 15 Hz. The suspension system design determines the natural frequency. The natural frequency of the suspension system is the same at all vehicle speeds.

The vibration frequency caused by an unbalanced tire and wheel assembly increases with wheel speed. When the vibration frequency caused by the wheel balance problem intersects with the natural frequency of the suspension system, the suspension system begins to vibrate. This intersection point is called the **resonance**.

Vibration Classifications and Terminology

Vibrations that can be felt are:

1. Shake
2. Roughness
3. Buzz
4. Tingling

Vibrations that result in noise may be classified as:

1. Boom
2. Moan or drone
3. Howl
4. Whine

Shake

A shake is a vibration with a low frequency of 5 to 20 Hz, and is sometimes experienced in the steering wheel, seat, or console. Customers may refer to shake as shimmy, wobble, shudder, waddle, or hop. Most cases of a shake vibration are caused by brake rotors and drums or unbalanced tire and wheel assemblies. These defects cause a shake-type vibration that is vehicle-speed sensitive. The engine, clutch, or transmission may cause a shake-type vibration that is engine-speed sensitive.

Roughness

Roughness is a vibration with a higher frequency of 20 to 50 Hz. Holding a jigsaw to cut a piece of wood produces a roughness-type vibration. A roughness vibration may be caused by a defective wheel bearing. Prior to causing a roughness vibration, the defective wheel bearing would cause a howl.

Buzz

A buzz has a faster frequency of 50 to 100 Hz. Holding a vibrator-type electric razor produces a feeling similar to a buzz vibration. This type of vibration is usually felt in the vehicle floor or the seat. A buzz-type vibration is usually caused by defects in the exhaust system hangers, A/C compressor, or the engine.

Shop Manual
Chapter 3, page 126

Tingling

A tingling vibration has a very high frequency much like a pins-and-needles sensation. Customers may complain that a tingling-type vibration puts their hands and feet to sleep. A tingling vibration may be caused by improper drive shaft balance in a rear-wheel-drive vehicle.

Boom

A boom noise has a low frequency of 20 to 60 Hz. A boom-type vibration produces a noise similar to a bowling ball rolling down a bowling alley. A customer may describe a boom-type vibration as droning, growling, humming, rumbling, roaring, or moaning. A boom-type noise and vibration may be caused by engine backfiring resulting from ignition defects.

Moan or Drone

A moan or drone is a tone with a higher frequency of 60 to 120 Hz. A moan or drone produces a noise similar to a bumblebee in flight. Moan or drone may be caused by defects in the exhaust system or defective engine or transmission mounts.

Howl

A howl is a noise with frequency of 120 to 300 Hz, and this sound is much like the wind howling. A howling noise may be caused by worn differential bearings.

Whine

A whine is a high-pitched sound with a frequency of 300 to 500 Hz, and this sound is similar to a vacuum cleaner. A whine problem may be caused by meshing gears in the transmission of differential. When diagnosing vibration problems, the technician should match the vibration frequency to the rotating speed of a component, to help locate the component responsible for the vibration.

SUMMARY

- Tires are extremely important because they provide ride quality, support the vehicle weight, provide traction for the drive wheels, and contribute to steering quality and directional stability.
- Tires have many design features and are carefully engineered to meet specific driving requirements.
- Tires may be bias-ply, belted bias-ply, or radial belted design.
- Tire ratings provide information regarding tire type, section width, aspect ratio, construction type, rim diameter, load capacity, and speed rating.
- Numeric tire ratings provide information about tire type, aspect ratio, construction type, and rim diameter.

TERMS TO KNOW

All-season tires
Amplitude
Bead wire
Bead filler
Belt cover
Belted bias-ply tires
Bias-ply tires
Compact spare tires
Conicity
Cord plies
Dynamic balance
Electronic vibration analyzer
Hertz
Hydroplaning
Liner
Load rating
Magnesium alloy wheels
Mud and snow tires
Natural vibration frequencies
Nitrogen tire inflation
Puncture sealing tires
Radial-ply tires
Replacement tires
Resonance
Road force variation
Run-flat tires
Sidewalls
Speed rating
Static balance
Synthetic rubber
Temperature rating
Tire belts
Tire chains
Tire contact area
Tire free diameter

TERMS TO KNOW (continued)

Tire performance criteria (TPC)
Tire placard
Tire pressure
Tire rolling diameter
Tire treads
Tire valve
Traction ratings
Tread wear ratings
Uniform Tire Quality Grading (UTQG) System
Wheel rims
Wheel tramp

- The tire performance criteria (TPC) number represents that the tire meets the car manufacturer's performance standards for traction, endurance, dimensions, noise, handling, and rolling resistance.
- The Uniform Tire Quality Grading (UTQG) designation includes tread wear, traction, and temperature ratings.
- Replacement tires must be the same type and size as the original tires to maintain vehicle safety.
- Replacement tires must have the same ratings as the original tires to maintain vehicle safety.
- The tire placard provides valuable information regarding the tires on the vehicle.
- Tires are subjected to severe acceleration and deceleration forces during normal operation.
- Replacement wheel rims must have the same width, diameter, offset, load capacity, and mounting configuration as the original rims to maintain vehicle safety.
- Static wheel unbalance causes wheel tramp and severe tire cupping.
- Dynamic wheel unbalance causes wheel shimmy, increased tire wear, unstable directional control, driver fatigue, and increased wear on suspension and steering components.

REVIEW QUESTIONS

Short Answer Essays

1. State general tire functions.
2. Interpret what each item of a tire size means, using P225/45R16 as your example.
3. What is the meaning of the service description 88H on the tire side wall following the tire size?
4. Describe the structural difference between a bias-ply and radial-ply tire.
5. List and define the three major uniform tire quality grading system ratings.
6. What is the tire DOT code and how do you know the date the tire was manufactured?
7. Explain the difference between an A and C tire temperature rating.
8. Define tire contact area and tire deflection.
9. Explain the purpose of the wheel rim drop center and safety ridges.
10. Describe the results of static unbalance.

Fill-in-the-Blanks

1. To calculate the tire aspect ratio, the tire section height is divided by the ________________.
2. Tires equipped on today's passenger cars and light trucks are of the ________________ ply design.
3. In ________________ ________________, the ply cords are arranged radially at a right angle to the tire centerline.
4. Replacement tires should be installed in pairs on the same ________________.
5. When a high-pressure minispare tire is installed on a vehicle, driving speed should not exceed ________________ mph.
6. Run-flat tires eliminate the need for a ________________ ________________ and a ________________.
7. Radial runout refers to variations in tire ________________.
8. Car manufacturers recommend that tire inflation pressures should be checked when the tires are ________________.
9. ________________ occurs when water on the pavement is allowed to remain between the pavement and the tire tread contact area.
10. ________________ ________________ vehicles tend to have positive offset rims to allow for proper clearance of the hub assembly within the vehicle wheel well.

Multiple Choice

1. All of the following statements about rim offset are correct EXCEPT:
   A. Zero offset, the rims centerline aligns with the mounting face of the rim.
   B. Positive offset, the rim centerline is inboard of the mounting face of the rim.
   C. Positive offset, the centerline is outboard of the mounting face of the rim.
   D. Negative offset, the centerline is outboard of the mounting face of the rim.

2. Tire pressure monitoring systems (TPMS) became required equipment on all vehicles sold in the United States beginning with what model year?
   A. 1996
   B. 2000
   C. 2006
   D. 2008

3. What is the maximum speed rating for a T rated tire?
   A. 112 mph
   B. 118 mph
   C. 124 mph
   D. 130 mph

4. Dynamic wheel unbalance can result in:
   A. Lateral wheel shimmy.
   B. Increased steering effort.
   C. Tire and wheel tramp.
   D. Normal tire tread life.

5. When a tire has a conicity problem:
   A. The tire has excessive lateral runout.
   B. The tire has too much radial runout.
   C. The tire is cone shaped and not level across the tread area.
   D. The tire has separation between the cord plies in the sidewall.

6. The Uniform Tire Quality Grading (UTQG) system provides consumers with useful information to help them purchase tires based on all of the following EXCEPT:
   A. Temperature Rating
   B. Traction Rating
   C. Speed Rating
   D. Tread wear rating

7. Nitrogen tire inflation provides:
   A. Increased tire rolling resistance.
   B. Improved puncture resistance.
   C. More stable tire pressure.
   D. Improved ride quality.

8. The tread wear indicator bar becomes visible when tread depth is less than?
   A. 4/32 inch
   B. 2/16 inch
   C. 2/32 inch
   D. 2/16 inch

9. Static wheel unbalance causes:
   A. Cupped tire tread wear.
   B. Even wear on one edge of the tire tread.
   C. Even wear on the center of the tire tread.
   D. Feathered tire tread wear.

10. When installing replacement tires:
    A. Different tire sizes can be installed on the same axle.
    B. Snow tires can be a different type or size than the original tires on the vehicle.
    C. If the tire size is different than the original tire size, antilock brake system operation can be adversely affected.
    D. A four-wheel-drive vehicle can have different size tires on the front and rear wheels.

Chapter 4

WHEEL BEARINGS

UPON COMPLETION AND REVIEW OF THIS CHAPTER, YOU SHOULD BE ABLE TO UNDERSTAND AND DESCRIBE:

- The purposes of a bearing.
- Three different types of bearing loads.
- The basic parts in ball bearings or roller bearings, and describe the purpose of each part.
- The action between the balls and the race when a ball bearing is rotating.
- The purposes of bearing snaprings, shields, and seals.
- The load-carrying capabilities of ball bearings, roller bearings, tapered roller bearings, and needle roller bearings.
- The advantage of tapered roller bearings compared to other types of bearings.
- Seal design and purpose.
- The purpose of the garter spring behind a lip seal.
- The purpose of flutes on seal lips.
- Two different types of rear axle bearings in rear-wheel-drive cars, and give the seal location for each bearing type.
- Grease classifications.

INTRODUCTION

Many different types of bearings are used in the automobile. A bearing may be defined as a component that supports and guides one of these parts:

1. Pivot
2. Wheel
3. Rotating shaft
4. Oscillating shaft
5. Sliding shaft

While a bearing is supporting and guiding one of these components, the bearing is designed to reduce friction and support the load applied by the component and related assemblies. Since the bearing reduces friction, it also decreases the power required to rotate or move the component. Bearings are precision-machined assemblies, which provide smooth operation and long life. When bearings are properly installed and maintained, bearing failure is rare.

> **AUTHOR'S NOTE: Only 21 percent of the power developed by a vehicle engine actually gets to the drive wheels. Much of the engine energy is lost in overcoming friction and wind resistance. The U.S. Department of Energy (DOE) is working with a major bearing manufacturer to explore the use and advantage of roller bearings to reduce friction in vehicle engines. The DOE estimates that if all vehicle engines used roller-bearing, low-friction technology, 100 million barrels of oil could be saved in a year.**

BEARING LOADS

Shop Manual
Chapter 4, page 148

When a bearing load is applied in a vertical direction on a horizontal shaft, it is called a **radial bearing load**. If the vehicle weight is applied straight downward on a bearing, this weight is a radial load on the bearing. A **thrust bearing load** is applied in a horizontal direction (Figure 4-1). For example, while a vehicle is turning a corner, horizontal force is applied to the front wheel bearings. When an **angular bearing load** is applied, the angle of the applied load is somewhere between the horizontal and vertical positions.

A **thrust bearing load** may be referred to as an axial load.

BALL BEARINGS

Front and rear wheel bearings may be **ball bearings** or roller bearings. Either type of bearing contains these basic parts:

1. Inner race, or cone
2. Rolling elements, balls, or rollers
3. Separator, also called a cage or retainer
4. Outer race, or cup

The **inner race** supports the inner side of the rolling elements in a bearing.

The **inner race** is an accurately machined component. The inner surface of the race is mounted on the shaft with a precision fit. The **rolling elements** are mounted on a very smoothly machined surface on the inner race. The surfaces of the rolling elements and the inner and outer races are case hardened to provide long bearing life. Positioned

Ball bearings have round steel balls between the inner and outer races.

The **rolling elements** are the precision-machined balls or rollers between the inner and outer bearing races.

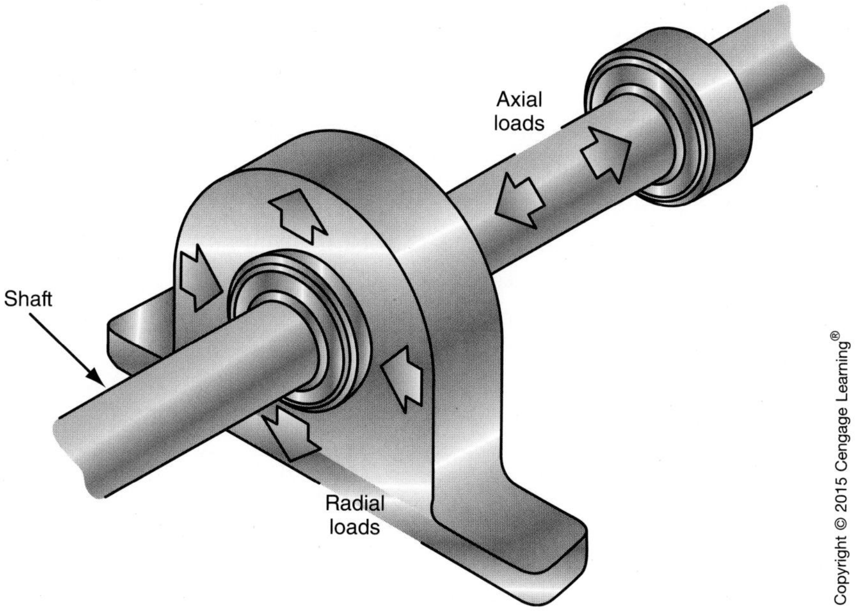

FIGURE 4-1 **Types of bearing loads.**

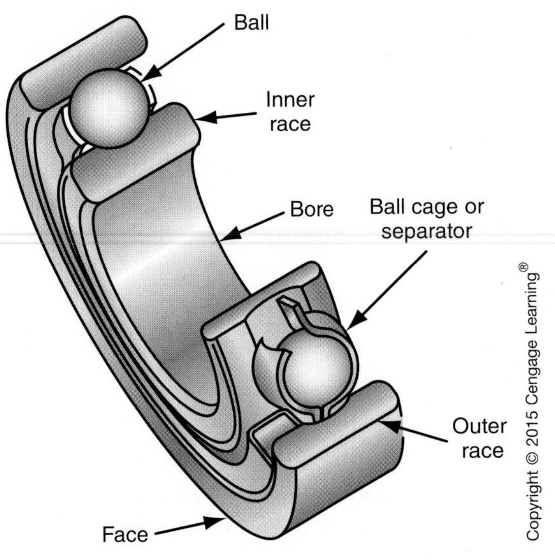

FIGURE 4-2 **Parts of a ball bearing.**

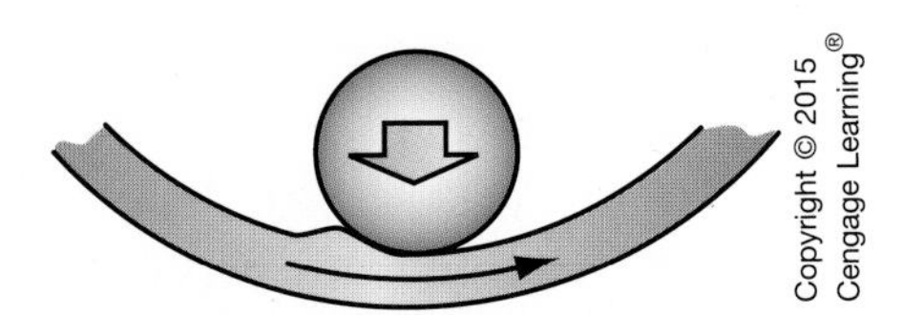

FIGURE 4-3 **When a load is applied to a ball bearing, the metal in the race bulges out in front of the ball and flattens out behind the ball.**

The **separator** keeps the rolling elements equally spaced in a bearing.

The **outer race** supports the outer side of the rolling elements and positions the bearing properly in the housing.

between the inner and outer races, the **separator** retains the rolling elements and keeps them evenly spaced. The rolling elements have precision-machined surfaces, and these elements are mounted between the inner and outer races. The **outer race** is the bearing's exterior ring. Both sides of this component have precision-machined surfaces. The outer surface of this race supports the bearing in the housing, and the inner surface is in contact with the rolling elements.

A **single-row ball bearing** has one row of balls between the inner and outer races.

A **single-row ball bearing** has a crescent-shaped machined surface in the inner and outer races in which the balls are mounted (Figure 4-2). When a ball bearing is at rest, the load is distributed equally through the balls and races in the contact area. When one of the races and the balls begin to rotate, the bearing load causes the metal in the race to bulge out in front of the ball and flatten out behind the ball (Figure 4-3). This action creates a certain amount of friction within the bearing, and the same action is repeated for each ball, while the bearing is rotating. If metal-to-metal contact is allowed between the balls and the races, these components will experience very fast wear. Therefore, bearing lubrication is extremely important to eliminate metal-to-metal contact in the bearing and reduce wear.

A ball bearing is designed primarily to handle radial loads. However, this type of bearing can also withstand a considerable amount of thrust load in either direction, even at high speeds. A maximum capacity ball bearing has extra balls for greater radial load-carrying capacity. Ball bearings are available in many different sizes for various applications (Figure 4-4).

Double-row ball bearings contain two rows of balls side by side. As in the single-row ball bearing, the balls in the double-row bearing are mounted in crescent-shaped grooves in the inner and outer races. The double-row ball bearing can support heavy radial loads and withstand thrust loads in either direction.

A **snapring** is made from spring steel and fits into a groove to retain a bearing on a shaft.

Bearing shields may be positioned between the inner and outer bearing races on the outside of the rolling elements.

Ball Bearing Seals, Shields, and Snaprings

For some applications, a ball bearing is held in place with a **snapring**. A groove is cut around the outside surface of the outer race, and the snapring is mounted in this groove. The snapring may fit against a machined housing surface, or the outer circumference of the snapring may be mounted in a groove in the housing. Ball bearings retained with a snapring are not used on wheels because they are not designed to withstand high thrust loads encountered by wheel bearings.

Bearing shields cover the space between the two bearing races on one, or both, sides of the bearing. These shields are usually attached to the outer race, but space is left

FIGURE 4-4 **A ball bearing.**

between the shield and the inner race. Bearing shields prevent dirt from entering the bearing, but excess lubrication can still flow through the bearing.

A **bearing seal** is a circular metal ring with a sealing lip on the inner edge. The seals are usually attached to the outer bearing race on each side of the bearing, and the lip surface contacts the inner race. The seal lip may have single, double, or triple lips made of synthetic or nonsynthetic rubber or elastomers. Lubricant is retained in the bearing by the seal, and the seal also keeps moisture, dirt, and contaminants out of the bearing. Some rear axle bearings on rear-wheel drive cars are sealed on both sides and retained on the axle with a retainer ring (Figure 4-5).

A **bearing seal** may be mounted on the outside of the rolling elements. The seal is attached to the outer race, and the seal lip contacts the inner race.

Shop Manual
Chapter 4, page 165

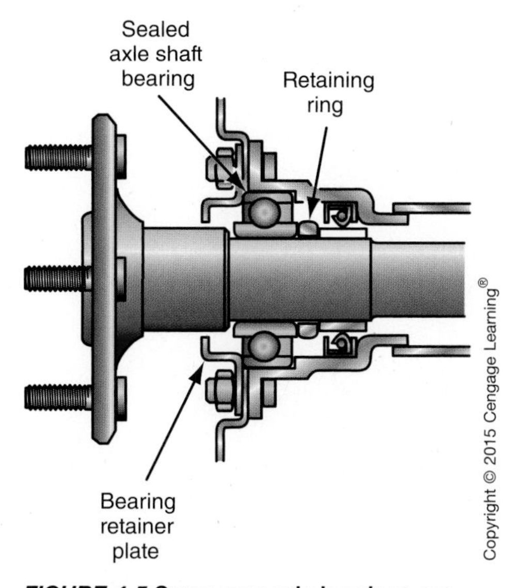

FIGURE 4-5 **Some rear axle bearings are sealed on both sides and retained on the axle with a retainer ring.**

A **roller bearing** is sometimes called a nontapered roller bearing.

Shop Manual
Chapter 4, page 148

A **tapered roller bearing** has tapered rollers between the inner and outer races. Tapered surfaces on the races match the tapered roller surfaces.

ROLLER BEARINGS

Roller Bearing

A **roller bearing** contains precision-machined rollers that have the same diameter at both ends. These rollers are mounted in square-cut grooves in the outer and inner races (Figure 4-6). In the roller bearing, the races and rollers run parallel to one another. Roller bearings are designed primarily to carry radial loads, but they can withstand some thrust load. Since ball bearings do not withstand high thrust loads, they are usually not used in wheel bearings.

Tapered Roller Bearing

In a **tapered roller bearing**, the inner and outer races are cone shaped. If imaginary lines extend through the inner and outer races, these lines taper and eventually meet at a point extended through the center of the bearing (Figure 4-7). The most important advantage of the tapered roller bearing compared to other bearings is an excellent capability to carry radial, thrust, and angular loads. In the tapered roller bearing, the rollers are mounted on cone-shaped precision surfaces in the outer and inner races. The bearing separator has an open space over each roller (Figure 4-8). Grooves cut in the side of the separator roller

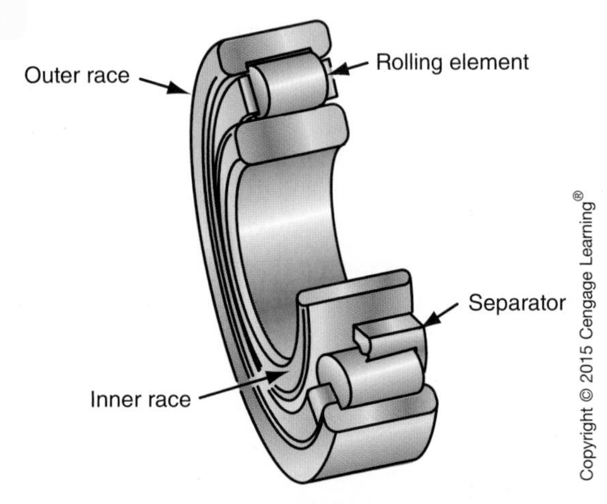

FIGURE 4-6 **Parts of a roller bearing.**

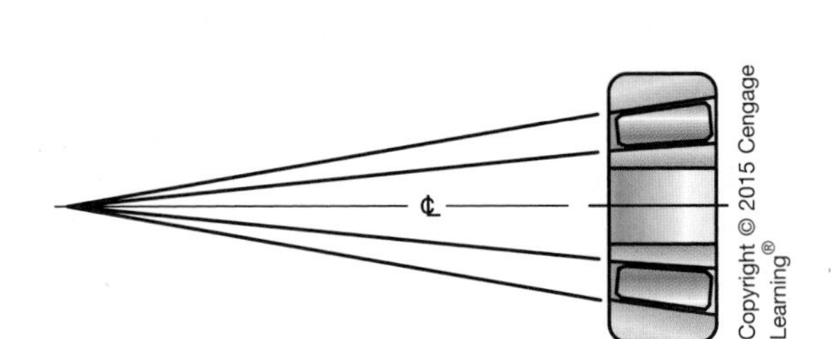

FIGURE 4-7 **Imaginary lines extending from tapered roller bearing races eventually meet at a point extending from the bearing center.**

FIGURE 4-8 **Tapered roller bearings.**

FIGURE 4-9 **Needle roller bearing.**

openings match the curvature of the roller. This design allows the rollers to rotate evenly without interference between the rollers and the separator. Lubrication and proper endplay adjustment are critical on tapered roller bearings. A tapered roller bearing may be called a cup and cone.

Needle Roller Bearings

A **needle roller bearing** contains many small-diameter steel rollers in a thin outer race. This type of bearing is very compact, and it is used in steering gears where mounting space is limited. Most needle roller bearings do not have a separator, but the steel rollers push against each other and maintain the roller position. Rather than having an inner race, a machined surface on the mounting shaft contacts the inner surface of the rollers (Figure 4-9). The needle roller bearing is designed to carry radial loads; it does not withstand thrust loads.

WARNING: Never strike a roller bearing with a steel hammer. This action may cause the bearing to shatter, resulting in severe personal injury.

WARNING: Spinning a roller bearing with compressed air may rotate the bearing at extremely high speed and cause the bearing to disintegrate, resulting in serious personal injury.

Seals

Seals are designed to keep lubricant in the bearing and prevent dirt particles and contaminants from entering the bearing. Static seals are used between two surfaces that do not move and dynamic seals are used between two surfaces where at least one surface moves. Wheel bearing seals are mounted in front and rear wheel hubs, and in rear axle housings on rear-wheel-drive cars. The metal seal case has a surface coating that resists corrosion and rust and acts as a bonding agent for the seal material. Seals have many different designs, including single lip, double lip, and fluted. A lip seal is usually encased in metal. The seal material is usually made of a synthetic rubber compound such as nitrile, silicon,

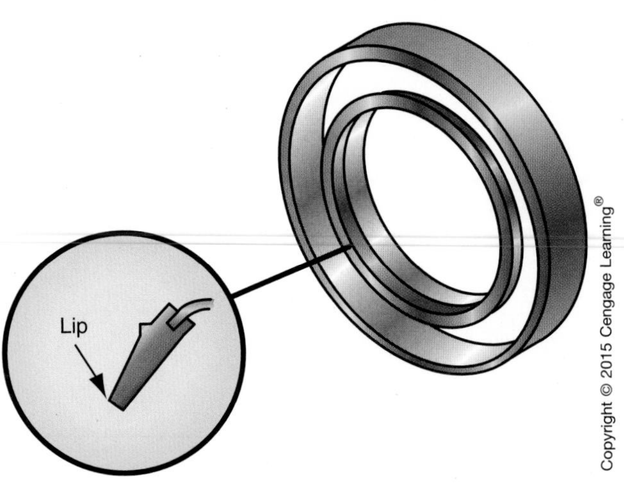

FIGURE 4-10 **Springless seal.**

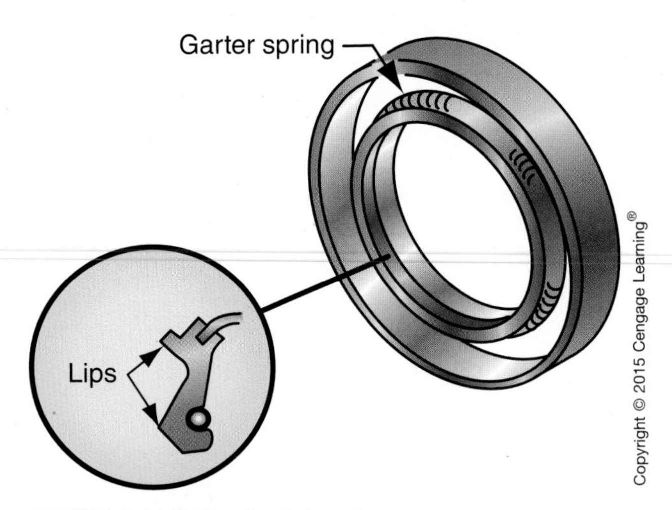

FIGURE 4-11 **Spring-loaded seal.**

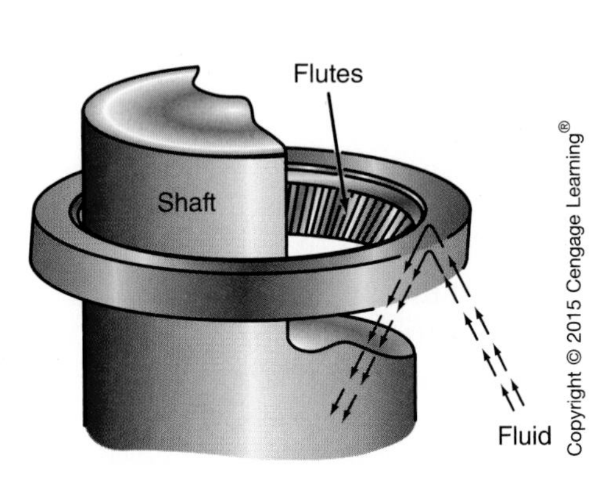

FIGURE 4-12 **Fluted lip seal redirects oil back into the housing.**

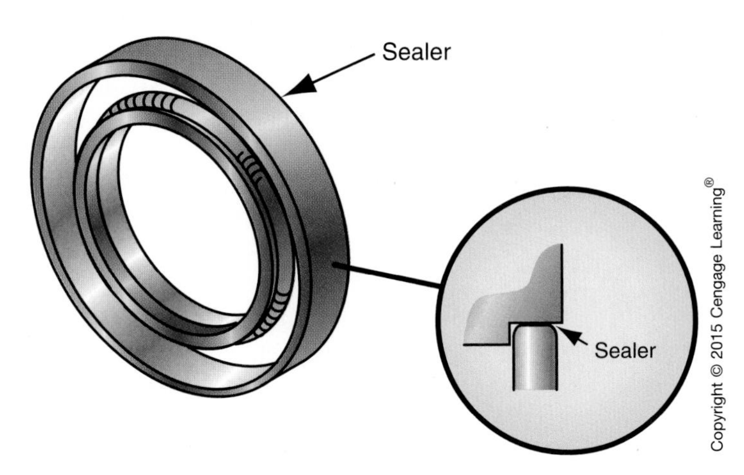

FIGURE 4-13 **Sealer painted on the seal case prevents leaks between the case and the housing.**

Springless seals do not have a garter spring behind the seal lip.

A **spring-loaded seal** has a garter spring behind the seal lip.

A **garter spring** is a circular coil spring mounted behind a seal lip.

A **fluted lip seal** has small ridges on the outer lip surface.

polyacrylate, or a fluoroelastomer such as Viton. The actual seal material depends on the lubricant and contaminants that the seal encounters. All seals may be divided into two groups, springless and spring loaded. **Springless seals** are used in some front or rear wheel hubs, where they seal a heavy lubricant into the hub (Figure 4-10).

In a **spring-loaded seal**, the **garter spring** behind the seal provides additional force on the seal lip to compensate for lip wear, shaft movement, and bore eccentricity (Figure 4-11). A **fluted lip seal** may be used to direct oil back into a housing. This seal design provides a pumping action to redirect the oil back into the housing (Figure 4-12).

Some seals have a sealer painted on the outside surface of the metal seal housing. When the seal is installed, this sealer prevents leaks between the seal case and the housing (Figure 4-13). Seals should be replaced whenever it is necessary to remove one during a service procedure.

Wheel Bearings

Some front-wheel-drive vehicles have front wheel bearing and hub assemblies that are bolted to the steering knuckles (Figure 4-14). The bearings are lubricated and sealed, and the complete bearing and hub assembly is replaced as a unit. The **bearing hub unit** is more compact than other types of wheel bearings mounted in the wheel hub. This type of bearing contains two rows of ball bearings with an angular contact angle of 32°

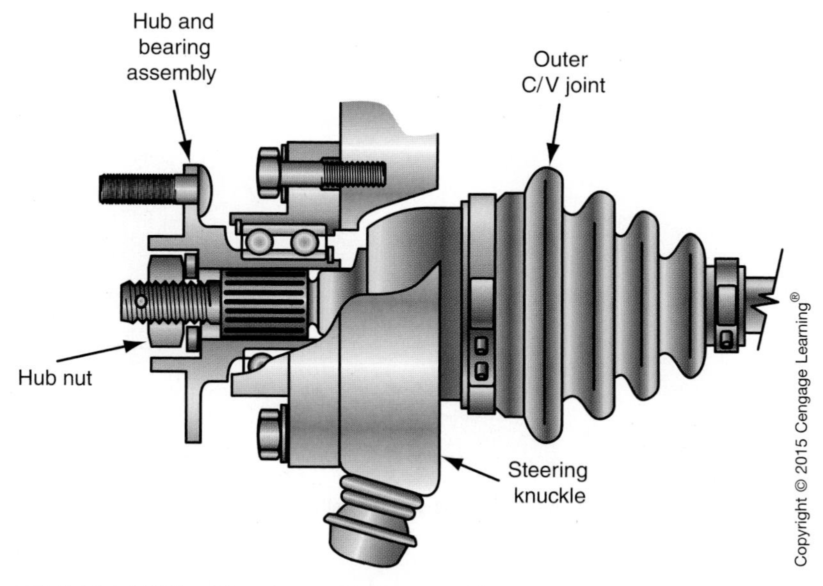

FIGURE 4-14 **Wheel bearing and hub assembly.**

(Figure 4-15). The inner bearing assembly bore is splined, and the inner ring extends to the outside to form a flange and spigot. The flange attached to the outer ring contains bolt holes, and bolts extend through these holes into the steering knuckle. This type of bearing attachment allows the bearing to become a structural member of the front suspension. Since the bearing outer ring is self-supporting, the main concern in knuckle design is fatigue strength rather than stiffness. The drive axle shaft transmits torque to the inner bearing race. This shaft is not designed to hold the bearing together. This type of wheel bearing is designed for mid-sized front-wheel-drive cars.

A **bearing hub unit** is a one-piece bearing and hub assembly that supports a front or rear wheel.

Each front drive axle has splines that fit into matching splines inside the bearing hubs (Figure 4-16). A hub nut secures the drive axle into the inner bearing race. Some wheel bearing hubs contain a wheel speed sensor for the antilock brake system (ABS).

Shop Manual
Chapter 4, page 161

FIGURE 4-15 **Double-row, sealed wheel bearing hub unit.**

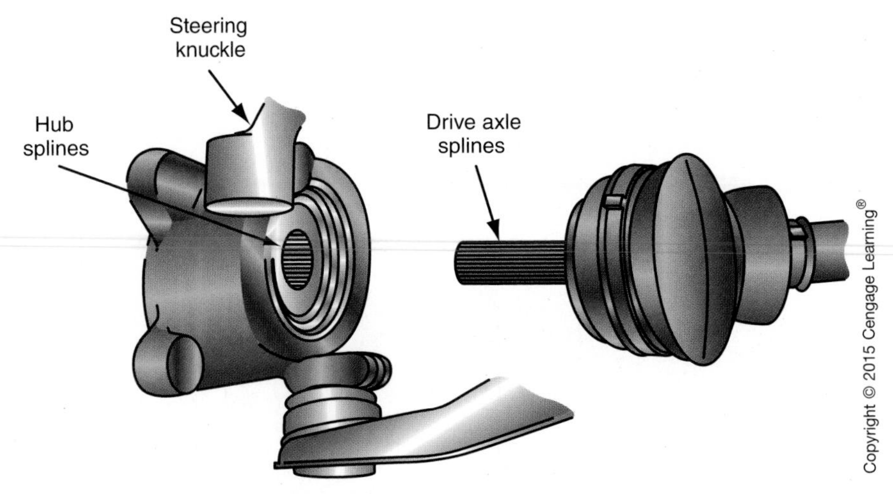

FIGURE 4-16 **Front drive shaft installed in wheel bearing hub.**

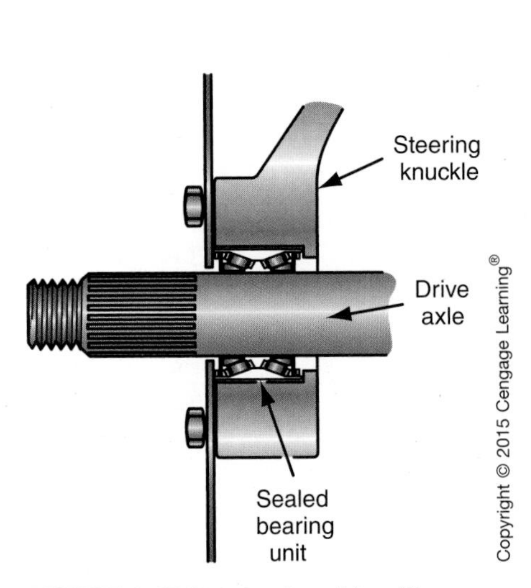

FIGURE 4-17 **Steering knuckle with pressed-in bearing.**

FIGURE 4-18 **Cutaway view of front wheel bearing that is pressed into the steering knuckle.**

A BIT OF HISTORY

Nearly all rear-wheel-drive cars have two serviceable tapered roller bearings or ball bearings in the front wheel hubs. Most front-wheel-drive cars have sealed bearing hubs on the front wheels, and some have these units on the rear wheels. These sealed bearing hub units have largely replaced the tapered roller bearings or ball bearings in the front wheel hubs.

Some front-wheel-drive vehicles have a sealed bearing unit that is pressed into the steering knuckle (Figure 4-17). The wheel hub is pressed into the inner bearing race, and the drive axle is splined into the hub. This type of bearing is designed for smaller front-wheel-drive cars. These bearings may contain two rows of ball bearings, or two tapered roller bearings and a split inner race (Figure 4-18). The bearing containing two tapered roller bearings has more radial load capacity than the double-row ball bearing. However, the tapered roller bearing is more sensitive to misalignment. Both sides of the bearing are sealed, and a seal is positioned behind the bearing in the steering knuckle to keep contaminants out of the bearing area.

Front Steering Knuckles with Two Separate Tapered Roller Bearings

Other front-wheel-drive vehicles have two separate tapered roller bearings mounted in the steering knuckles. The bearing races are pressed into the steering knuckle, and seals are located in the knuckle on the outboard side of each bearing (Figure 4-19). Correct bearing endplay adjustment is supplied by the hub nut torque. The wheel hub is pressed into the inner bearing races, and the drive axle splines are meshed with matching splines in the wheel hub.

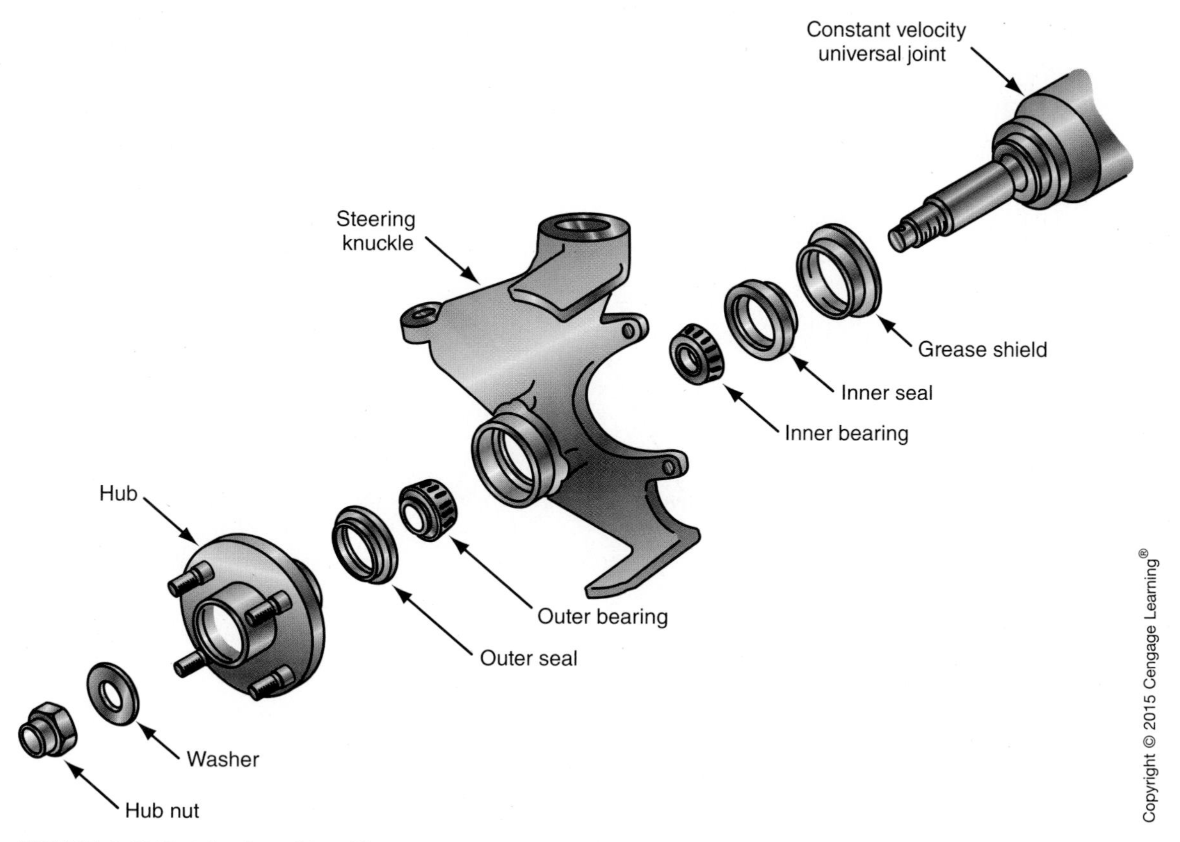

FIGURE 4-19 **Steering knuckle with two separate tapered roller bearings.**

Wheel Hubs with Two Separate Tapered Roller Bearings

Many rear-wheel-drive cars have two tapered roller bearings in the front hubs that support the hubs and wheels on the spindles. This type of front wheel bearing has the bearing races pressed into the hub. A grease seal is pressed into the inner end of the hub to prevent grease leaks and keep contaminants out of the bearings. The hub and bearing assemblies are retained on the spindle with a washer, adjusting nut, nut lock, and cotter pin. The adjusting nut must be adjusted properly to provide the correct bearing endplay. A grease cap is pressed into the outer end of the hub to prevent bearing contamination (Figure 4-20).

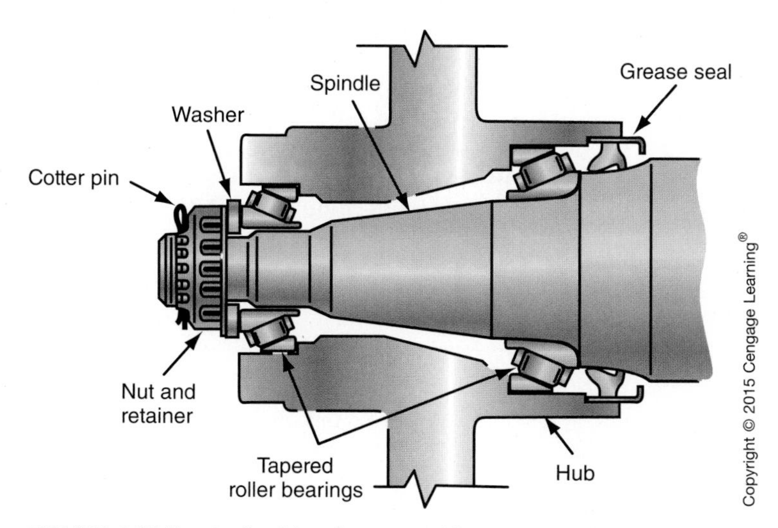

FIGURE 4-20 **Front wheel bearing assembly, rear-wheel-drive car.**

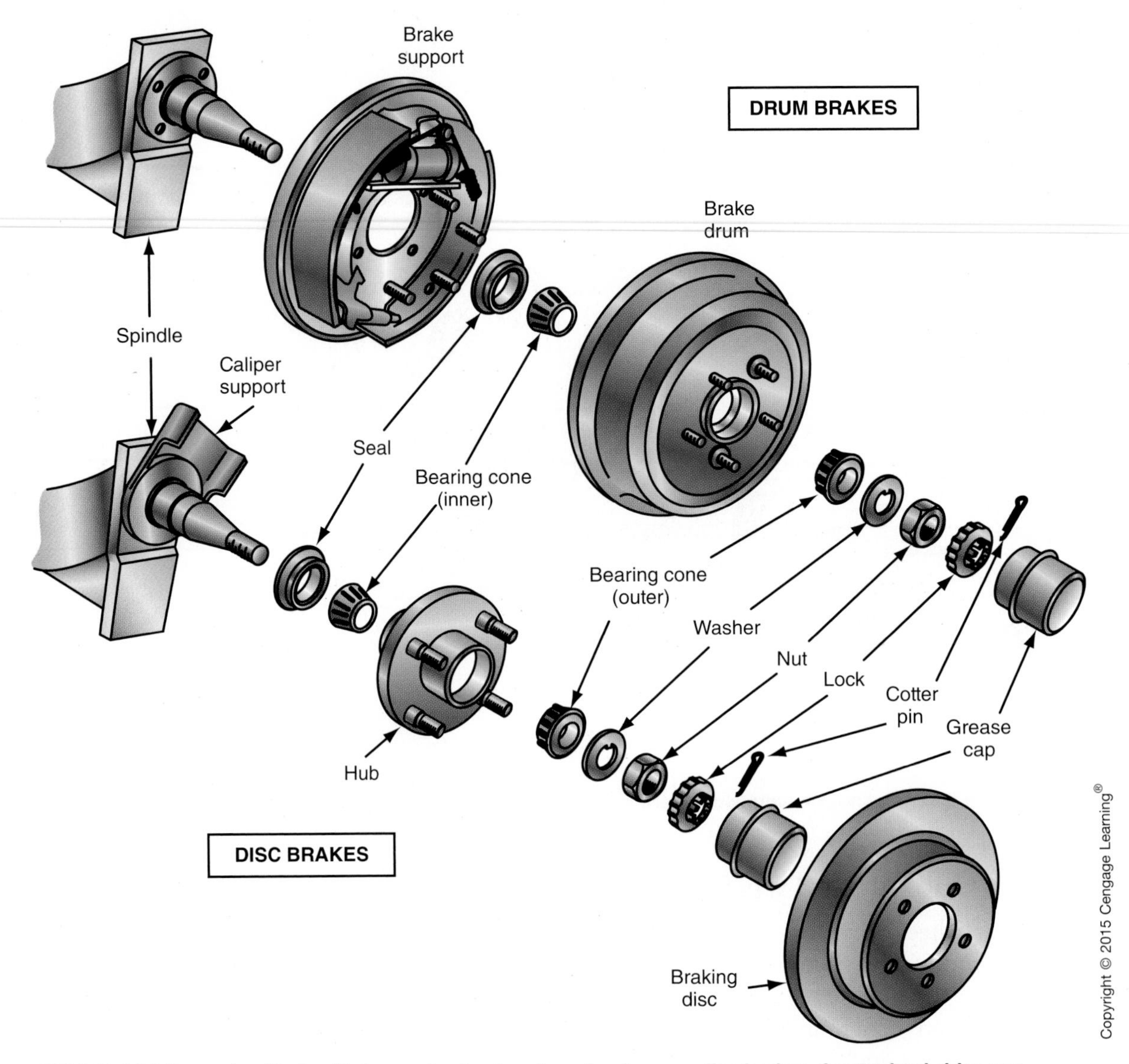

FIGURE 4-21 **Rear wheel hub with tapered roller bearings for drum or disc brakes, front-wheel-drive car.**

Some front-wheel-drive cars have two tapered roller bearings in the rear wheel hubs that are very similar to the front wheel bearings in Figure 4-20. The two tapered roller bearings in the rear wheel of a front-wheel-drive car are shown in Figure 4-21.

AUTHOR'S NOTE: An understanding of wheel bearings and related service procedures is critical to maintain vehicle safety! If the technician's knowledge of wheel bearings and appropriate service procedures is inadequate, bearing failure may occur. Wheel bearing failure may cause a wheel to fly off a vehicle with disastrous results!

Shop Manual
Chapter 4, page 163

REAR-AXLE BEARINGS

On many rear-wheel-drive cars, the rear axles are supported by roller bearings mounted near the outer ends of the axle housing. The outer bearing race is pressed into the housing,

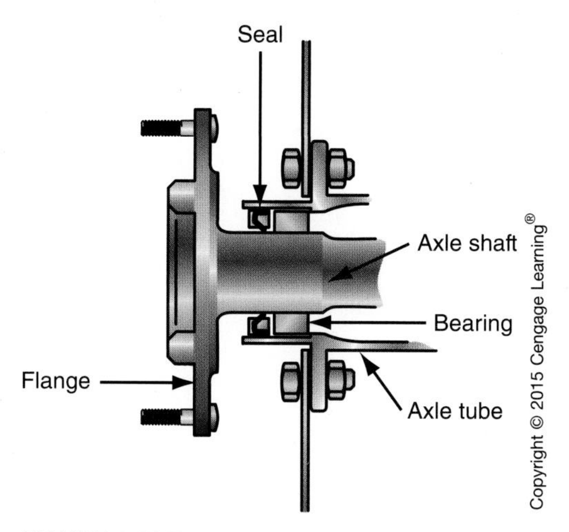

FIGURE 4-22 **Rear-axle bearing, rear-wheel-drive car.**

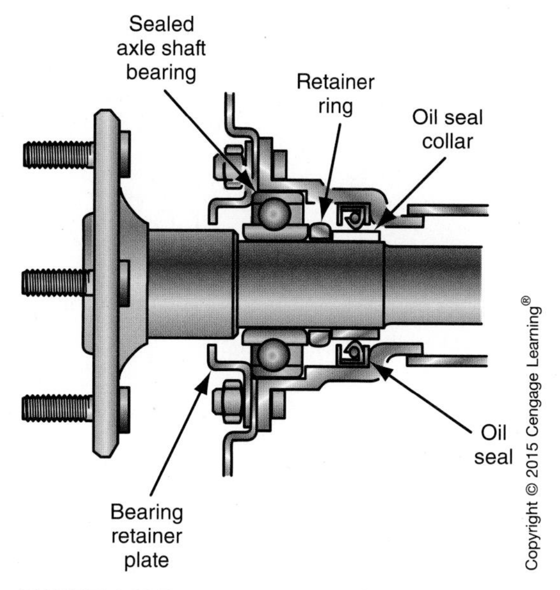

FIGURE 4-23 **Rear-axle bearing and retainer.**

and a machined surface on the axle contacts the inner roller surface. A seal is mounted in the axle housing on the outboard side of each bearing (Figure 4-22). This type of axle bearing is usually not sealed, and lubricant in the differential and rear axle housing provides axle bearing lubrication. The seals prevent lubricant leaks from the outer ends of the axle housing and keep dirt out of the bearings.

Other **rear-axle bearings** on rear-wheel-drive vehicles have sealed roller bearings pressed onto the rear axles. These axle bearings are sealed on both sides, and an adapter ring is pressed onto the axle on the inboard side of the bearing (Figure 4-23). The outer bearing race is mounted in the rear axle housing with a light press fit, and a seal is positioned in the housing on the inboard side of the bearing and adapter ring. A retainer plate is mounted between the bearing and the outer end of the axle. This retainer plate is bolted to the axle housing to retain the axle in the housing.

Rear-axle bearings support the rear axles in the rear axle housing on rear-wheel-drive vehicles.

Many wheel bearings require **lithium-based** or **sodium-based grease**.

Bearing Lubrication

Proper bearing lubrication is extremely important to maintain bearing life. Bearing lubricant reduces friction and wear, dissipates heat, and protects surfaces from dirt and corrosion. Sealed or shielded bearings are lubricated during the manufacturing process, *and no attempt should be made to wash these bearings or pack them with grease.*

Bearings that are not sealed or shielded require cleaning and repacking at intervals specified by the vehicle manufacturer. *Always use the bearing grease specified by the vehicle manufacturer.* Bearing lubricants may be classified as greases or oils. Grease is oil with a thickening agent added. Greases are named for the thickening agent:

- Aluminum
- Barium
- Calcium
- Lithium
- Sodium

A **lithium-based grease** has a specific amount of lithium mixed with the lubricant.

A **sodium-based grease** has a specific amount of sodium mixed with the lubricant.

ASTM International sets the standards for testing for grease. Classification ASTM D 4950 covers greases designed for the lubrication of wheel bearings and chassis components for passenger cars, trucks, and other vehicles. The National Lubricating Grease Institute

NLGI SYMBOLS

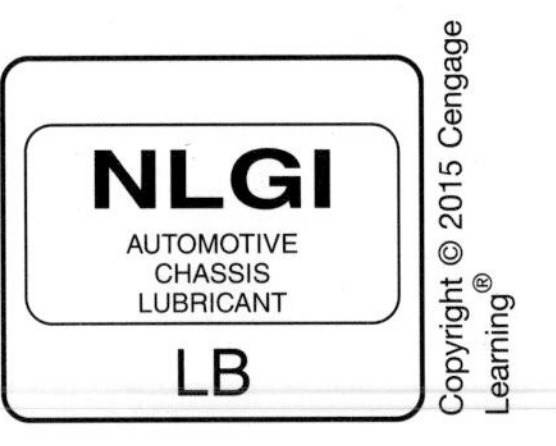

FIGURE 4-24 **The National Lubricating Grease Institute (NLGI) classifies automotive greases into two main groups, L and G.**

(NLGI) classifies automotive greases into two main groups. The prefix designation for wheel bearing grease is G, and the prefix designation for chassis lubrication grease is L (Figure 4-24).

Service category L chassis grease:

- LA- Mild duty, non-critical applications with frequent relubrication intervals of less than 2,000 miles (3,200 km). Performance offers oxidation resistance, shear stability, and corrosion and wear protection.
- LB- Mild to severe duty, high loads, and vibration with a usable temperature range of −40 to 248°F. Relubrication intervals of greater than 2,000 miles (3,200 km). Performance offers oxidation resistance, shear stability, and corrosion and wear protection even under heavy loads and in the presence of water contamination.

Service category G wheel bearing grease:

- GA- Mild duty
- GB- Mild to moderate duty, with a usable temperature range of −40 to 248°F and occasional spikes to 320°F. Performance offers oxidation and evaporation resistance, shear stability, and corrosion and wear protection.
- GC- Mild to severe duty, with a usable temperature range of −40 to 320°F and occasional spikes to 392°F. Performance offers oxidation and evaporation resistance, shear stability, and corrosion and wear protection.

The **Society of Automotive Engineers (SAE)** is responsible for the establishment of many automotive standards.

The **American Petroleum Institute (API)** is responsible for establishing standards related to oils and lubricants.

The most common grease designation used in automotive shops is GC and LB. A designation label GC/LB indicates that the grease is acceptable for bearings and chassis parts. Number 2 grease is the most common rating with the higher the number indicating the stiffer the consistency of the grease.

New bearings usually have a protective coating to prevent rust and corrosion. This coating should not be washed from the bearing. When rear-axle bearings are lubricated from the differential housing, the type and level of oil in the housing are important.

Vehicle manufacturers usually recommend a **Society of Automotive Engineers (SAE)** No. 90 or SAE No. 140 hypoid gear oil in the differential. In very cold climates, the manufacturer may recommend an SAE No. 80 differential gear oil. The **American Petroleum Institute (API)** classifies gear lubricants as GL-1, GL-2, GL-3, GL-4, and GL-5. The GL-4 lubricant is used for hypoid gears under normal conditions. The GL-5 lubricant is used in heavy-duty hypoid gears. Always use the vehicle manufacturer's specified differential gear oil.

CAUTION:
If a bearing is operated without proper lubrication, bearing life will be very short.

The differential should be filled until the lubricant is level with the bottom of the filler plug opening in the differential housing. If the differential is overfilled, the bearings and seals may have excessive lubricant. Under this condition, the lubricant may leak past the seal. When the lubricant level is low in the differential, the lubricant may not be available in the axle housings. When this condition exists, the bearings do not receive enough lubrication and bearing life is shortened.

SUMMARY

- A bearing reduces friction, carries a load, and guides certain components such as pivots, shafts, and wheels.
- Radial bearing loads are applied in a vertical direction.
- Thrust bearing loads are applied in a horizontal direction.
- Angular bearing loads are applied at an angle between the vertical and horizontal.
- The inner bearing race is positioned at the center of the bearing and supports the rolling elements.
- The rolling elements in a bearing are positioned between the inner and outer races.
- The bearing separator keeps the rolling elements evenly spaced.
- The outer bearing race forms the outer ring on a bearing.
- A cylindrical ball bearing is designed primarily to withstand radial loads, but these bearings can handle a considerable thrust load.
- A snapring can be mounted in a groove in the outer bearing race, and the snapring retains the bearing in the housing.
- A bearing shield prevents dirt from entering the bearing, but it is not designed to keep lubricant in the bearing.
- Bearing seals keep lubricant in the bearing and prevent dirt from entering the bearing.
- Roller bearings are designed primarily to carry radial loads, but they can handle some thrust loads.
- Tapered roller bearings have excellent radial, thrust, and angular load-carrying capabilities.
- Needle roller bearings are very compact and are designed to carry radial loads. They will not carry thrust loads.
- Springless seals are used for wheel bearing seals in some wheel hubs.
- The garter spring provides additional force on the seal lip to compensate for lip wear, shaft movement, and bore eccentricity.
- Flutes on seal lips provide a pumping action to direct oil back into a housing.
- Bearing hub units are compact compared to bearings that are mounted in the wheel hub. This compactness makes bearing hub units suitable for front-wheel-drive cars.
- Some bearing hub units are bolted to the steering knuckle; other bearing hub units are pressed into the steering knuckle.
- Some steering knuckles contain two separate tapered roller bearings.
- Rear-axle bearings are mounted between the drive axles and the housing on rear-wheel-drive cars.

TERMS TO KNOW

American Petroleum Institute (API)
Angular bearing load
Ball bearings
Bearing hub unit
Bearing seals
Bearing shields
Double-row ball bearings
Fluted lip seal
Inner race
Lithium-based grease
Needle roller bearing
Outer race
Radial bearing load
Rear-axle bearings
Roller bearing
Rolling elements
Separator
Single-row ball bearing
Snapring
Society of Automotive Engineers (SAE)
Sodium-based grease
Springless seals
Spring-loaded seal
Tapered roller bearing
Thrust bearing load

REVIEW QUESTIONS

Short Answer Essays

1. Define a radial bearing load.
2. Define a thrust bearing load, and give another term for this type of load.
3. Explain an angular bearing load.
4. Describe the main parts of a bearing, including the location and purpose of each part.
5. Describe the difference between a maximum-capacity ball bearing and an ordinary ball bearing.
6. Explain the design and purpose of bearing seals.

7. Explain the purpose of the sealer on the outside surface of a seal housing.
8. What does ASTM and NLGI stand for and what is their function?
9. Explain how the proper bearing endplay adjustment is obtained when two tapered roller bearings are mounted in the steering knuckle.
10. Describe two different types of rear-axle bearings on rear-wheel-drive cars.

Fill-in-the-Blanks

1. A bearing can be described as a component that supports a ________________, ________________, or ________________.
2. A bearing is designed to support a load and ________________.
3. An angular bearing load is applied at an angle between the ________________ and the ________________.
4. A ball bearing is designed primarily to withstand ________________ loads.
5. Greases are named for the ________________ agent used.
6. Lubrication and proper ________________ adjustment are important on tapered roller bearings.
7. A needle roller bearing is not designed to carry ________________ loads.
8. A springless seal may be used to seal a ________________ lubricant into a hub.
9. When a vehicle is turning a corner, the front wheel bearings must carry a ________________ load.
10. When a rear-wheel-drive vehicle has cylindrical roller bearings mounted in the rear axle housing, the inner surface on the rollers contacts a machined surface on the ________________.

Multiple Choice

1. A tapered roller bearing has:
   A. Rollers that have the same diameter at both ends.
   B. Excellent ability to carry radial, thrust, and angular loads.
   C. Horizontal inner surfaces of the inner and outer races.
   D. A separator that allows the rollers to lightly contact each other.
2. A spring-loaded seal compensates for all of these conditions EXCEPT:
   A. Seal lip wear.
   B. Shaft movement.
   C. Bore eccentricity.
   D. Wear on the seal bore.
3. When two separate tapered roller bearings are located in a front steering knuckle on a front-wheel-drive car, the wheel bearing endplay adjustment is provided by:
   A. The bearing race position.
   B. Wheel nut torque.
   C. Hub nut torque.
   D. Wheel hub position.
4. A rear-wheel hub contains two tapered roller bearings on a front-wheel-drive car. This type of wheel hub has:
   A. An adjusting nut to adjust the wheel bearing endplay.
   B. A lock washer to retain the adjusting nut.
   C. A staked-type nut lock.
   D. A seal in the outer end of the hub to prevent grease contamination.
5. All of the following statements are correct EXCEPT:
   A. Greases are named for their thickening agent.
   B. NLGI sets the standards for testing greases.
   C. The prefix GC is the highest designation for automotive wheel bearing greases.
   D. The prefix LB is the highest designation for automotive wheel chassis greases.
6. When inspecting and servicing bearings:
   A. Some rear-wheel-drive vehicles have rear-axle bearings that are sealed on both sides.
   B. In a sealed bearing, the seal is usually attached to the inner race.
   C. Endplay adjustment is not critical on tapered roller bearings.
   D. A roller bearing can be removed by striking it with a steel hammer.
7. All of these statements about bearings are true EXCEPT:
   A. Tapered roller bearings have excellent capability to carry radial, thrust, and angular loads.
   B. Ball bearings are used in most wheel bearings.
   C. A roller bearing has rollers with the same diameter at each end.
   D. In a tapered roller bearing, the openings in the separator grooves match the curvature of the rollers.

8. While inspecting and servicing wheel hub seals:
   A. Many wheel hub seal lips are made from a plastic compound.
   B. Springless seals may be used in applications where they must seal a light fluid.
   C. A spring-loaded seal helps compensate for wheel hub bore eccentricity.
   D. The sealer on the outer surface of a seal housing improves seal and hub bore alignment.

9. When inspecting and servicing front wheel bearings on front-wheel-drive cars:
   A. Some one-piece front wheel bearing hubs are bolted to the front strut.
   B. In a one-piece front wheel bearing hub, the outer end of the drive axle is splined to the outer bearing race.
   C. When the steering knuckle contains two tapered roller bearings, the torque on the drive axle nut does not affect bearing endplay.
   D. Some front wheel bearings mounted in the steering knuckle contain two tapered roller bearings with a split inner race.

10. When inspecting and servicing rear wheel bearings on rear-wheel-drive cars:
    A. On some rear wheel bearings, a threaded adapter ring retains the bearing on the axle shaft.
    B. Roller-type rear-axle bearings may be lubricated from the oil supply in the differential.
    C. A GL-2 lubricant is recommended in hypoid-type differentials.
    D. An SAE 50 gear oil is recommended in many differentials.

Chapter 5

Shock Absorbers and Struts

Upon completion and review of this chapter, you should be able to understand and describe:

- The three purposes of shock absorbers.
- How shock absorbers contribute to vehicle safety.
- Wheel jounce and rebound.
- Spring operation during wheel jounce and rebound.
- Shock absorber operation during wheel jounce.
- Shock absorber operation during wheel rebound.
- The advantages of nitrogen-gas-filled shock absorbers and struts.
- Shock absorber ratios.
- Travel-sensitive shock absorber operation.
- The operation of an adjustable shock absorber.
- The operation of load-leveling struts and shock absorbers.

Introduction

Shock absorbers are devices used to control, or dampen, spring oscillations.

Two front **shock absorbers** or **struts** are connected from the front suspension to the chassis, and two rear shock absorbers or struts are attached between the rear suspension and the chassis. The internal strut design is very similar to shock absorber design, and struts perform the same control functions as shock absorbers. Shock absorbers have three main purposes:

1. They control spring action and oscillations (bounce) to provide the desired ride quality and noise reduction.
2. They help prevent body sway (roll) and lean while cornering.
3. They reduce brake dive and acceleration squat, thereby reducing the tendency of a tire tread to lift off the road, which improves tire life, traction, and directional stability.

The kinetic energy of the suspension system spring action is converted to thermal energy as the shock absorbers piston travels though the hydraulic fluid. This heat is then dissipated into the air flowing surrounding the outer steel tube of the shock absorber. Because shock absorbers control spring action, spring oscillations, and chassis oscillations, they contribute to vehicle safety and passenger comfort. If the shock absorbers are worn out, excessive chassis oscillations may occur, particularly on rough road surfaces. These excessive chassis oscillations may result in loss of steering control. Worn-out shock absorbers also cause excessive body lean and sway while cornering, which may cause the driver to lose control of the vehicle and can contribute to increased stopping distance (Figure 5-1). Shock absorbers are extremely important to providing longer tire life and

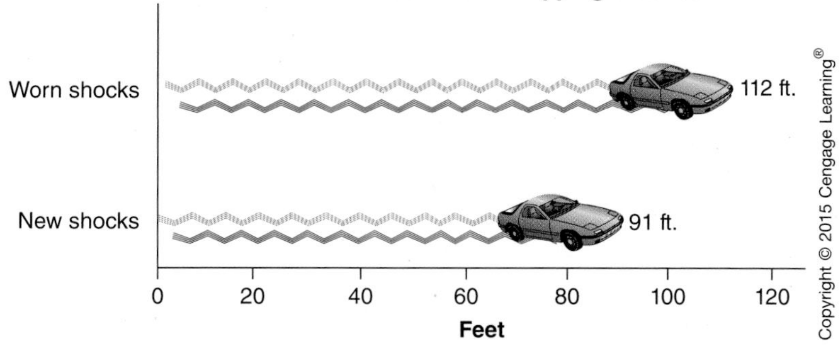

FIGURE 5-1 **Worn shock absorbers can increase stopping distance.**

improving vehicle handling, steering quality, and ride quality. A faulty shock absorber can cause tire cupping to occur. The two most common causes of tire cupping are a static imbalance condition of the wheel assembly and a weak hydraulic shock absorber.

Shock absorber design is matched to the deflection rate of the suspension spring to control the spring action. Several different types of shock absorbers are available for special service requirements.

Twin Tube Shock Absorber Design

Shop Manual
Chapter 5, page 180

The lower half of a shock absorber is a twin tube steel unit filled with hydraulic oil and nitrogen gas (Figure 5-2). In some shock absorbers, the nitrogen gas is omitted. The outer tube is the reservoir tube, which stores excess hydraulic fluid and the inner tube is the pressure or working tube. A relief valve is located in the bottom of the unit to control fluid movement during compression, and a circular lower mounting is attached to the lower tube. This mounting contains a rubber isolating bushing, or grommet, to reduce the transmitted road noise and suspension vibration. The rubber bushings are flexible to allow for minimal movement as road forces are transmitted through the assembly. The upper mount is connected to the vehicle frame and the lower mount is typically connected to the lower control arm or axle tube of the suspension system. A piston and rod assembly is connected to the upper half of the shock absorber. This upper portion of the shock absorber has a dust shield that surrounds the lower twin tube unit. The piston is precision fit in the inner cylinder of the lower unit. A piston rod guide and seal are located in the top of the lower unit. A circular upper mounting with a rubber bushing is attached to the top of the shock absorber. Shock absorber bore size refers to the diameter of the working piston. Generally the larger the piston diameter, the higher the potential energy and damping control level of the shock absorber because of larger piston displacement and pressure area. As a general rule the larger the piston area, the lower the internal operating temperature and pressure. To achieve optimal ride characteristics of stability and balance under a wide variety of driving conditions, ride engineers select piston size and valve rates for specific vehicle applications.

The advantages of the twin tube shock absorber are:

- Combines both comfort and control.
- Adjusts more rapidly than a single tube design shock absorber to changing weight and road conditions.

A disadvantage of the twin tube design is that it can only be mounted in one direction. The twin tube shock absorber is the most common original equipment manufacturer (OEM) design used on cars, SUVs, and light trucks.

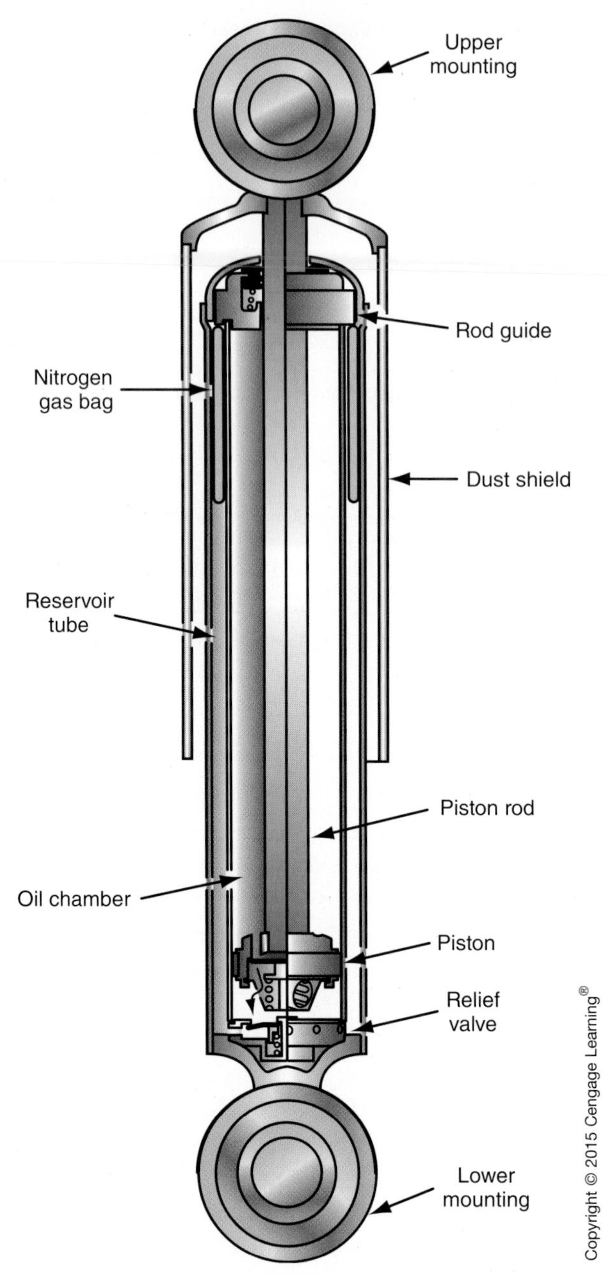

FIGURE 5-2 **Shock absorber filled with hydraulic oil and nitrogen gas.**

Shock Absorber Operation

Shock absorbers are usually mounted between the lower control arms and the chassis. When a vehicle wheel strikes a bump, the wheel and suspension move upward in relation to the chassis. Upward wheel movement is referred to as **jounce travel**. This jounce action causes the spring to deflect or compress. Under this condition, the spring stores energy and springs back downward with all the energy absorbed when it deflected upward. This downward spring and wheel action is called **rebound travel**. If this spring action were not controlled, the wheel would strike the road with a strong downward force, and the wheel jounce would occur again. Therefore, some device must be installed to control the spring action, or the wheel would bounce up and down many times after it hit a bump, causing passenger discomfort, directional instability, and suspension component wear along with severe tire cupping.

Shock absorbers are installed on suspension systems to control spring action. When a wheel strikes a bump and jounce travel occurs, the shock absorber lower tube unit is forced upward. This action forces the piston downward in the lower tube unit. Because oil cannot leak past the piston, the amount of resistance a shock absorber develops depends on the piston size and the number of holes passing through it. The oil in the lower unit is forced through the piston orifices or valves to the upper oil chamber. These valves provide precise oil flow control and control the upward action of the wheel and suspension, which is referred to as a shock absorber compression stroke (Figure 5-3). Modern shock absorbers are velocity-sensitive damping devices that increase resistance the faster the suspension moves in effect adjusting themselves to road conditions.

When the spring expands downward in rebound travel, the lower shock absorber unit is also forced downward. When this occurs, the piston moves upward in the lower tube unit, and hydraulic oil is forced through the piston orifices or valves from the upper oil chamber to the lower oil chamber. Because the valves restrict oil flow with precise control, the downward suspension and wheel movement is controlled (Figure 5-4).

When the shock absorber piston moves, oil is forced through the piston. Because the piston valves and orifices resist the flow of oil, friction and heat are created. The resistance of the oil moving through the piston must be calibrated as closely as possible to the spring's

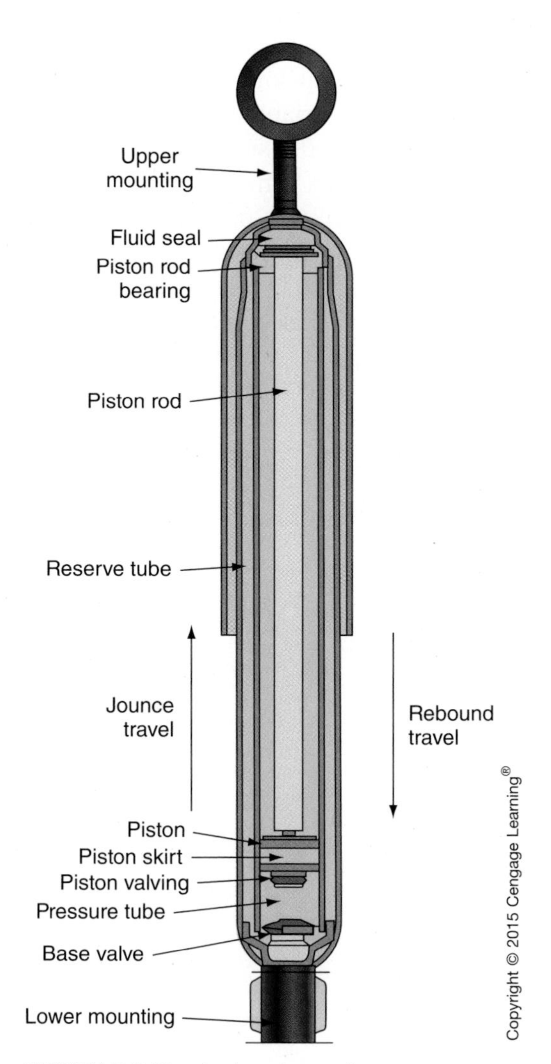

FIGURE 5-3 **Shock absorber action.**

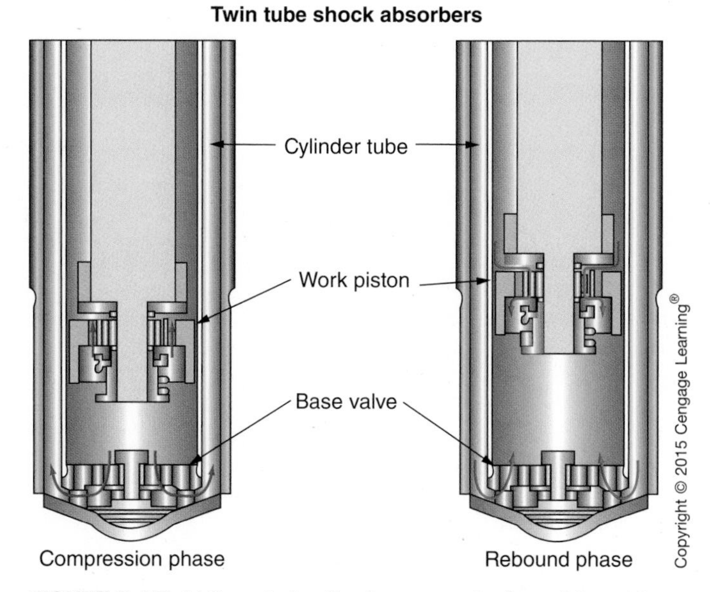

FIGURE 5-4 **Fluid flow during the jounce and rebound travel is restricted through valve passages in the working piston and base valve to achieve proper shock rate.**

deflection rate or strength. Wheels and suspension systems deflect at many different speeds, depending on the type and size of bump and the vehicle speed. The resistance of a shock absorber piston increases with the square of its speed. For example, if the wheel deflection speed increases four times, the piston resistance is sixteen times as great. Therefore, if a wheel strikes a large bump at high speed, the wheel deflection and rebound can be effectively locked by the shock absorber. Shock absorber engineers prevent this action by precisely designing shock absorber valves and orifices to provide enough friction to prevent the spring from overextending on the rebound stroke. These piston valves and orifices must not create excessive friction, which slows the wheel from returning to its original position.

Shock absorber pistons have many different types of valves and orifices. Valving is selected to achieve optimal ride characteristics of a specific vehicle platform. In some pistons, small orifices control the oil flow during slow wheel and suspension movements. Stacked steel valves control the oil flow during medium speed wheel and suspension movements. During maximum wheel and suspension movements, larger orifices between the piston valves provide oil flow control (Figure 5-5). On other shock absorber pistons, the stacked steel valves alone provide oil flow control (Figure 5-6). As a simple example, think of a shock absorber with a multistage damping stack of three steel valve plates or discs (Figure 5-7). The first stage valve is for light bumps experienced during normal flat road travel and is constantly opening and closing as the vehicle is driven. The second stage valve is for moderate suspension movement that may occur during corning maneuvers. The third stage valve is for extreme suspension movement such as large bumps, pot holes, speed bumps, and it does not open as frequently for each mile driven as the first and second stage valve. When enough force is built up inside the shock absorber each valve progressively opens to allow fluid to bypass the piston seal offering variable damping. A shock absorber piston assembly may make as many as 1,500 to 1,900 strokes per mile or 75 million strokes in 50,000 miles (80,467 km), which means that a first stage valve may open and close 37.5 million times in that same distance driven. In this design the constant flexing of the valve plate eventually causes metal fatigue of one or more of the valve plates effecting damping performance related to the pressure valve that has failed. Regardless of the piston orifice and valve design, the shock absorber must be precisely matched to absorb the spring's energy.

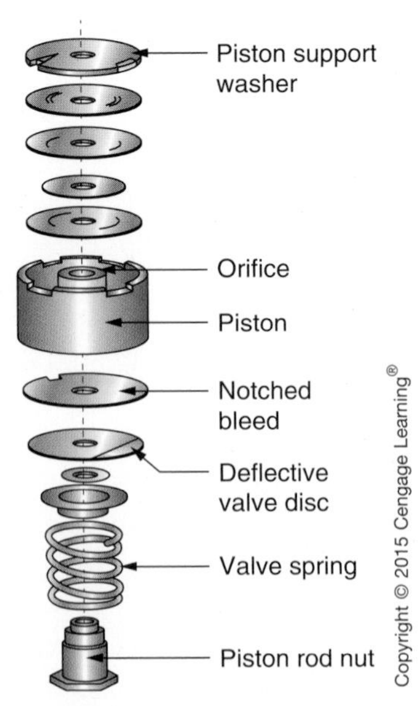

FIGURE 5-5 **A combination of valve discs and orifice passages are used on some shocks to achieve proper shock rate.**

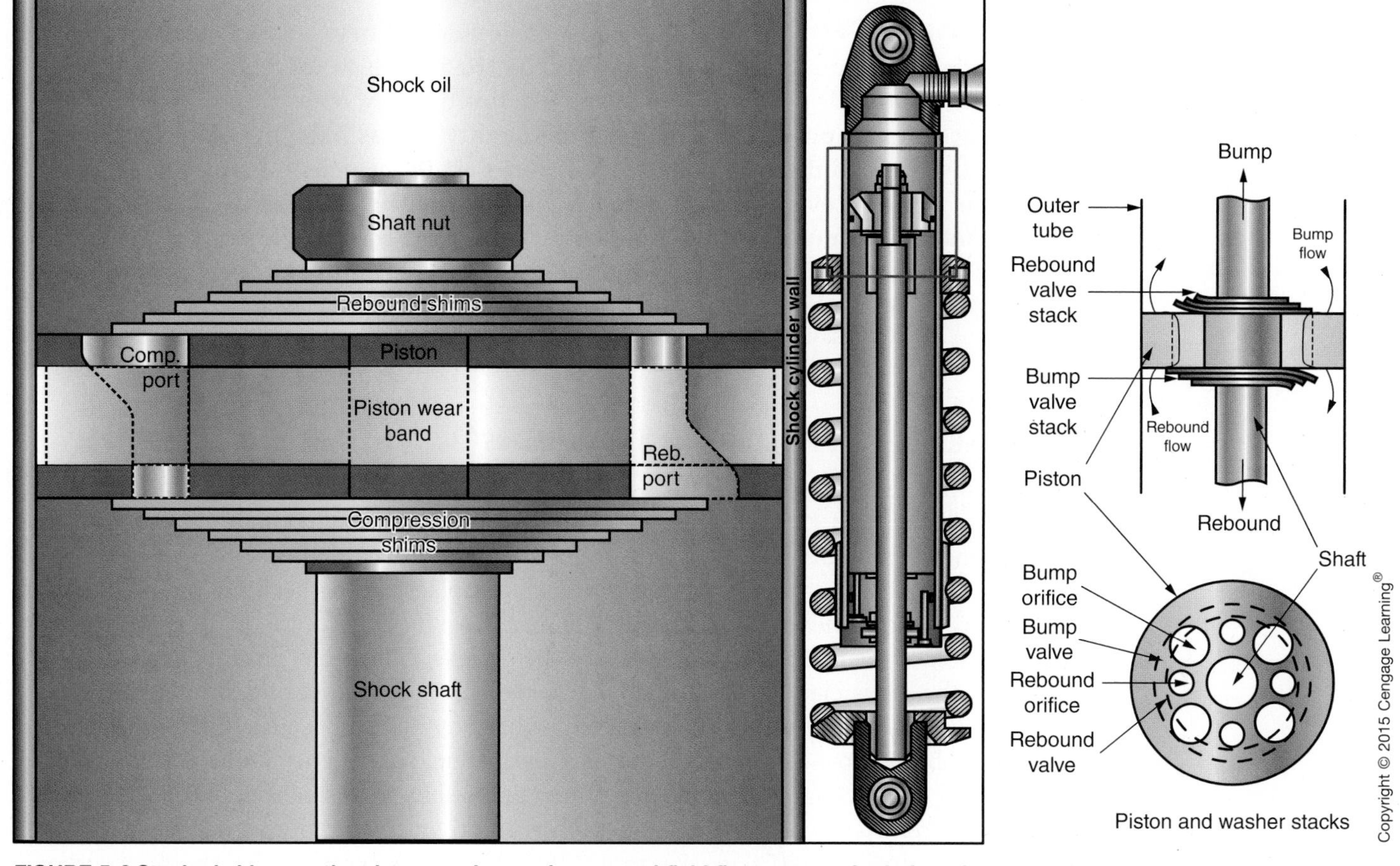

FIGURE 5-6 Stacked shims on the piston can be used to control fluid flow rates on both the rebound and compression cycles.

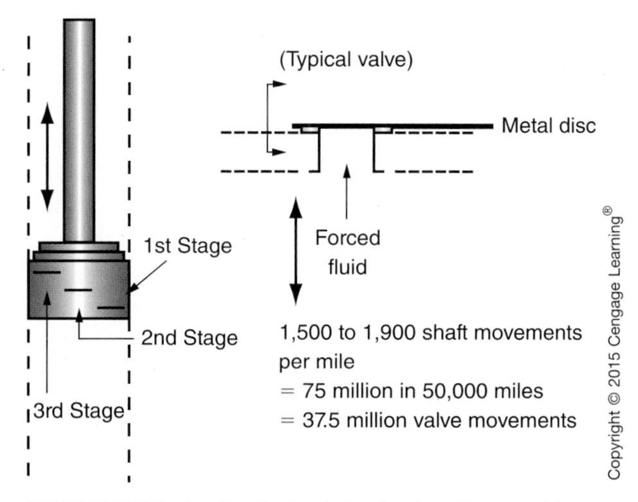

FIGURE 5-7 A simple stacked steel valve disc used to achieve proper shock damping rate.

During fast upward wheel movement on the compression stroke, excessive pressure in the lower oil chamber forces the base valve open and thus allows oil to flow through this valve to the reservoir. A rebound rubber is located on top of the piston. If the wheel drops downward into a hole, the shock absorber may become fully extended. Under this condition, the rebound rubber provides a cushioning action.

AUTHOR'S NOTE: Of all the suspension components, shock absorbers and struts contribute the most to ride quality. As the vehicle is driven over road irregularities, the shock absorbers or struts are continually operating to control the spring action and provide acceptable ride quality. It has been my experience that shock absorbers and struts usually wear out first in suspension systems, because they are working every time a wheel strikes a road irregularity. Therefore, you must understand not only shock absorber and strut operation but also how ride quality is affected if these components are not functioning properly. Thus, you must know shock absorber and strut diagnosis and service procedures.

GAS-FILLED SHOCK ABSORBERS AND STRUTS

Gas-filled units are identified with a warning label. Many shock absorbers and struts today contain a nitrogen gas charge. A gas charge shock absorber may contain either a low pressure in twin tube designs or a high pressure charge in a mono-tube design. The nitrogen gas provides a compensating space for the oil that is displaced into the reservoir on the compression stroke and when the oil is heated. The primary function of the gas charge is to minimize aeration of the hydraulic fluid. Because the gas exerts pressure on the oil, cavitation, or foaming of the oil, is eliminated, which in turn reduces fade or loss of damping capability. If foaming occurs, shock absorber performance would be reduced because of the compressibility of air bubbles suspended in the hydraulic fluid, reducing efficiency and transfer of energy. As you recall from Chapter 2, fluids are noncompressible. When oil bubbles are eliminated by the pressurized gas charge, the shock absorber provides continuous damping for wheel deflections as small as 0.078 in. (2.0 mm). By eliminating aeration the shock absorber is more responsive and predictable under a broader performance range. The gas charge also creates a mild boost in spring rate assisting in the reduction of body roll, acceleration squat, and brake dive. This mild boost in spring rate is achieved because of the increased surface area below the working piston than above (Figure 5-8). With an increased surface area below the piston as a result of the piston rod attachment to the upper piston surface more pressurized fluid contacts the bottom of the piston causing more force to be transferred. The force applied to the piston is equal to the pressure applied to the piston time's the surface area of the piston ($F = P \times A$). This is the reason that gas charged shocks will extend on their own. If a low pressure (100–150 psi) **gas-filled shock absorber** is removed and compressed to its shortest length, it should reextend when it is released. Failure to reextend indicates that shock absorber or strut replacement is necessary. A high pressure shock absorber should not be compressible by hand because it contains a 360 psi gas charge.

A **gas-filled shock absorber** contains a nitrogen gas charge to maintain pressure on the oil in the shock.

WARNING: New gas-filled shock absorbers are wired in the compressed position for shipping purposes. Exercise caution when cutting this wire strap because shock absorber extension may cause personal injury. After the upper shock absorber attaching bolt is installed, the wire strap can be cut to allow the unit to extend. Front gas-filled struts have an internal catch that holds them in the compressed position. This catch is released when the strut rod is held and the strut rotated 45 degrees counterclockwise.

WARNING: Do not throw gas-filled shock absorbers or struts in the fire or apply excessive heat or flame to these units. These procedures may cause the unit to explode, resulting in personal injury.

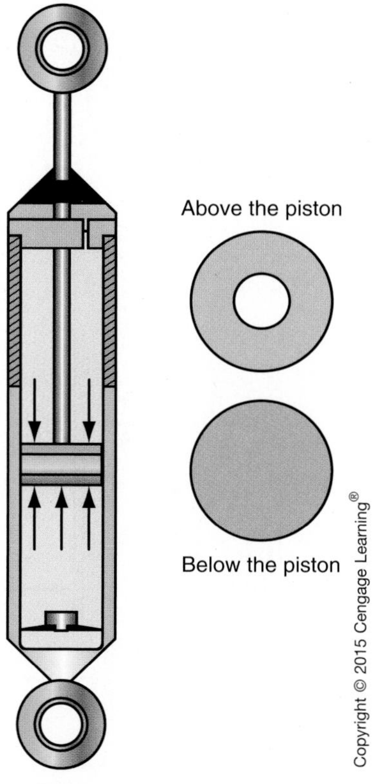

FIGURE 5-8 **The surface area of the shock absorber working piston is greater at the bottom than it is at the top because of the loss of area due to the attachment area of the piston rod. This causes applied force to be greater below the piston than above the piston. Though, this does not take valving and orifices into account.**

CAUTION:
When drilling worn-out shock absorbers or struts to relieve the gas pressure prior to disposal, drill the shock absorber only at the vehicle manufacturer's specified location.

WARNING: Never apply heat to a shock absorber or strut chamber with an acetylene torch. This action may cause a shock absorber or strut explosion resulting in personal injury.

Heavy-Duty Mono-Tube Shock Absorber Design

Some **heavy-duty mono-tube shock absorbers** have a dividing piston in the lower oil chamber. The area below this piston is pressurized with nitrogen gas to 360 pounds per square inch (psi), or 2,482 kilopascals (kPa). Hydraulic oil is contained in the oil chamber above the dividing piston. The other main features of the heavy-duty shock absorber are:

1. High-quality seal for longer life.
2. Single tube design to prevent excessive heat buildup. This design may be called a mono-tube shock absorber.
3. Rising rate valve to provide precise spring control under all conditions.

The operation of the heavy-duty shock absorber is similar to that of the conventional type (Figure 5-9). A dent in the oil chamber will affect shock absorber operation on this type of design.

Heavy-duty mono-tube shock absorbers have several design features that provide improved durability compared with conventional shock absorbers.

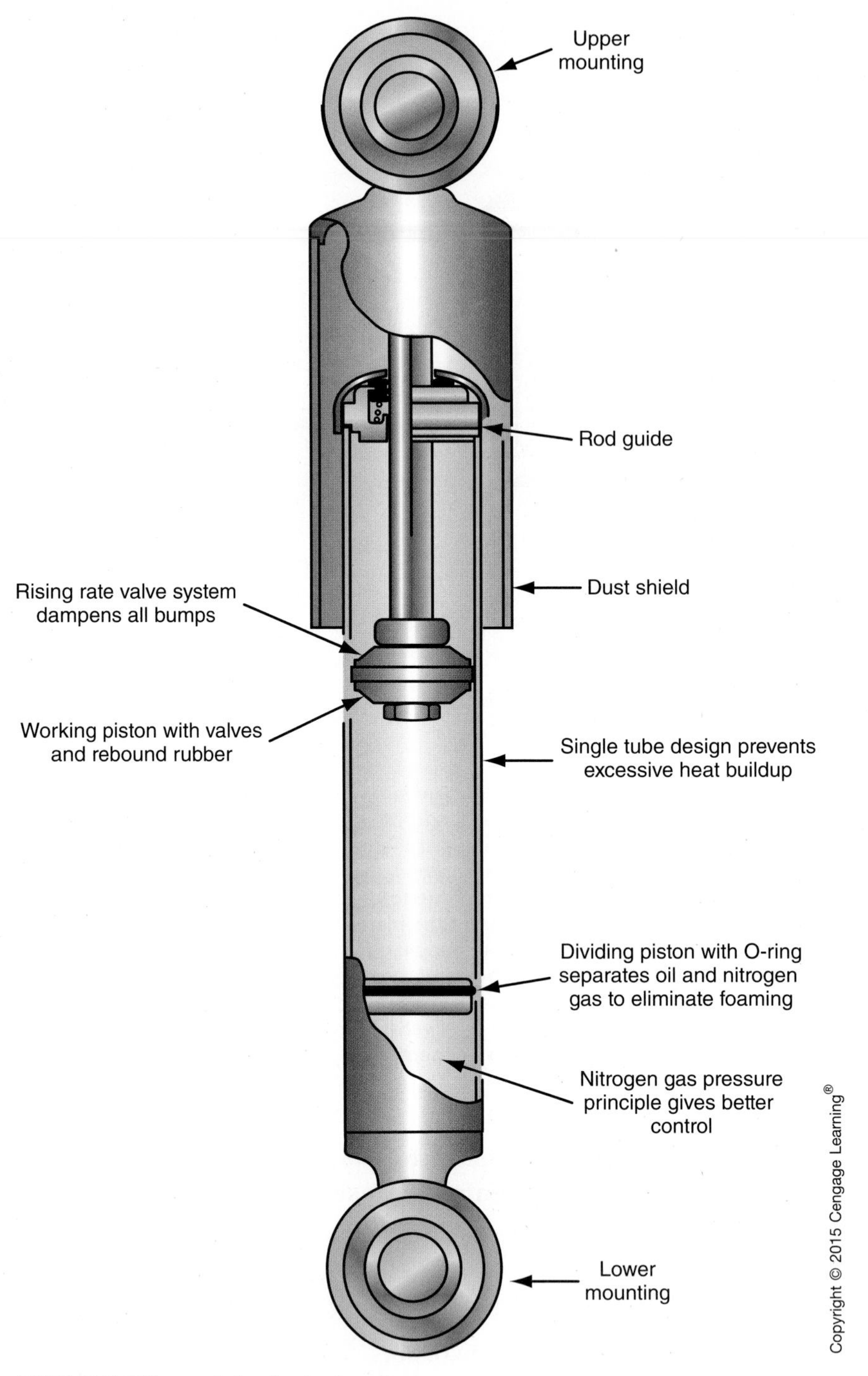

FIGURE 5-9 **Heavy-duty shock absorber.**

The advantages of the mono-tube shock absorber are:

- Heat may be dissipated more rapidly.
- The shock assembly may be mounted in any position including upside down, reducing unsprung weight.
- Larger piston diameter allows for lower working pressures.

The disadvantages of the mono-tube shock absorber are:

- Piston rod seal is subjected to internal damping pressure.
- Sufficient room must be provided by the suspension system layout to provide for its larger diameter.

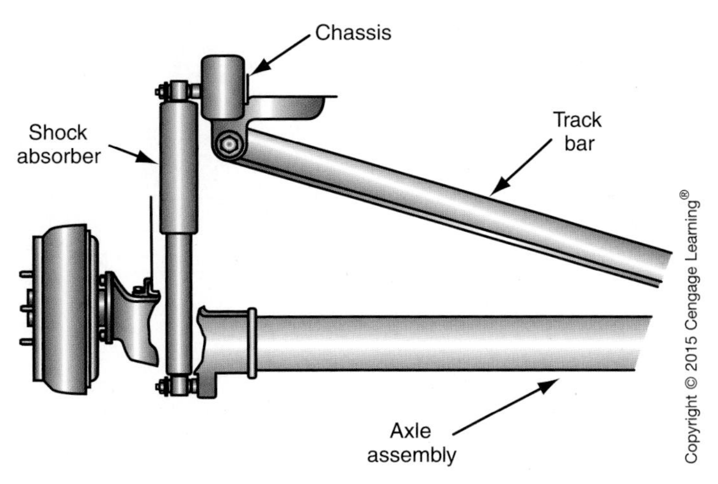

FIGURE 5-10 **Rear shock absorber mounting on a front-wheel-drive car.**

- The single wall working chamber is susceptible to damage from rocks and other road debris. A dent will destroy the unit.
- Larger than a twin tube design.

Shock Absorber Ratios

Most automotive shock absorbers are a double-acting-type that controls spring action during jounce and rebound wheel movements. The piston and valves in many shock absorbers are designed to provide more extension control than compression control. An average shock absorber may have 70 percent of the total control on the extension cycle, and thus 30 percent of the total control is on the compression cycle. Shock absorbers usually have this type of design because they must control the heavier sprung body weight on the extension cycle. The lighter unsprung axle, wheel, and tire weight are controlled by the shock absorber on the compression cycle. A shock absorber with this type of design is referred to as a 70/30 type. **Shock absorber ratios** vary from 50/50 to 80/20.

Shock absorber ratios indicate the amount of control on the extension and compression cycles.

A shock absorber is mounted between the rear axle and the chassis on a front-wheel-drive car (Figure 5-10). Mounting bolts extend through hangers on the rear axle and chassis. These bolts also pass through the isolating bushings on each of the shock absorbers. The isolating bushings are very important for preventing vibration and noise. Front shock absorbers may be mounted in a similar way between the lower control arms and the chassis.

Strut Design, Front Suspension

A strut-type front suspension is used on most front-wheel-drive cars and some rear-wheel-drive cars. Internal strut design is very similar to shock absorber design, and **struts** perform the same functions as shock absorbers. Some struts have a replaceable cartridge. In many strut-type suspension systems, the coil spring is mounted on the strut. The coil spring is largely responsible for proper curb riding height. A weak or broken coil spring reduces curb riding height and provides harsh riding. The lower end of the front suspension strut is bolted to the steering knuckle (Figure 5-11). An **upper strut mount** is attached to the strut, and this mount is bolted into the chassis strut tower. A lower spring seat is part of the strut assembly, and a lower insulator is positioned between the coil spring and the spring seat on the strut. Another **spring insulator** is located between the coil spring and the upper strut mount. The two insulators prevent metal-to-metal contact between the spring and the strut, or mount. These insulators reduce the transmission of

Struts are similar to shock absorbers, but struts are usually positioned between the knuckle and the chassis to provide knuckle support.

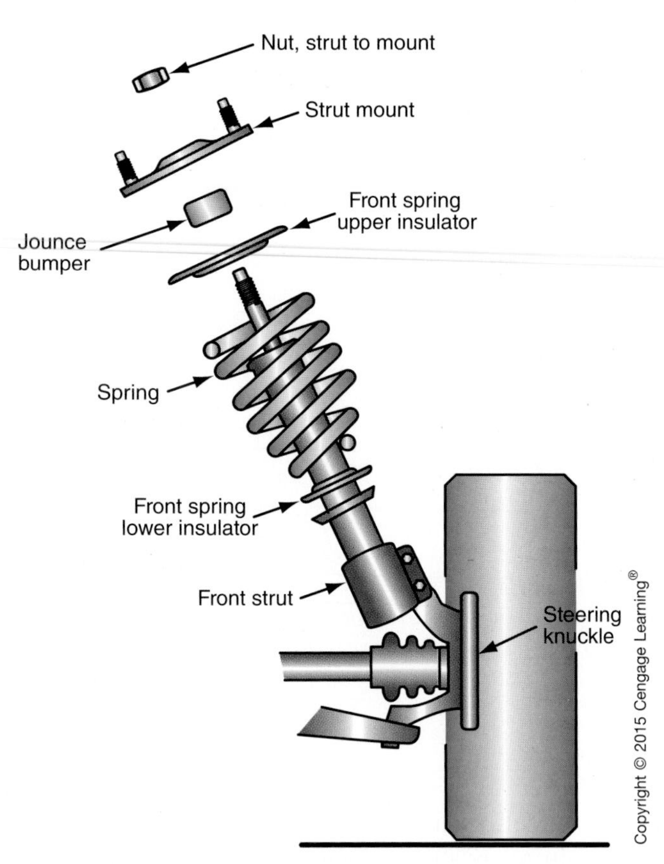

FIGURE 5-11 Front strut assembly.

FIGURE 5-12 Upper strut mount.

A BIT OF HISTORY

For many years, rear-wheel-drive cars were equipped with front and rear shock absorbers. Most front-wheel-drive cars are equipped with front struts, and some of these cars also have rear struts. Because massive numbers of front-wheel-drive cars have been introduced in the 1980s and 1990s, front and rear struts are now very common.

Shop Manual
Chapter 5, page 184

noise and harshness from the suspension to the chassis. A rubber spring bumper is positioned around the strut piston rod. When a front wheel strikes a large road irregularity and the strut is fully compressed, the jounce bumper provides a cushioning action between the top of the strut and the upper support. The jounce bumper stops the upward wheel and suspension movement before the spring is completely compressed. If the spring becomes completely compressed and the coils strike each other, ride quality is very harsh. Therefore, the jounce bumper in the strut improves ride quality. Most jounce bumpers are made from butyl rubber. Some late model vehicles have microcellular urethane (MCU) jounce bumpers, which are lighter than rubber and provide more progressive cushioning to improve ride quality. MCU jounce bumpers are also 20 to 40 percent lighter than rubber jounce bumpers, which reduces road noise transmission to the passenger compartment. In relation to temperature changes, MCU jounce bumpers remain more stable and provide improved ride quality regardless of the temperature. The upper strut mount contains a bearing, upper spring seat, and jounce bumper (Figure 5-12).

When the front wheels are turned, the front strut and coil spring rotate with the steering knuckle. The strut-and-spring assembly rotates on the upper strut mount bearing.

Some cars have a multilink front suspension with an upper link connected from the chassis to the steering knuckle. The strut is connected from the upper link to the strut tower (Figure 5-13). A bearing is mounted between the upper link and the steering knuckle, and the wheel and knuckle turn on this bearing and the lower ball joint. Therefore, the coil spring and strut do not turn when the front wheels are turned, and a bearing in the upper strut mount is not required (Figure 5-14).

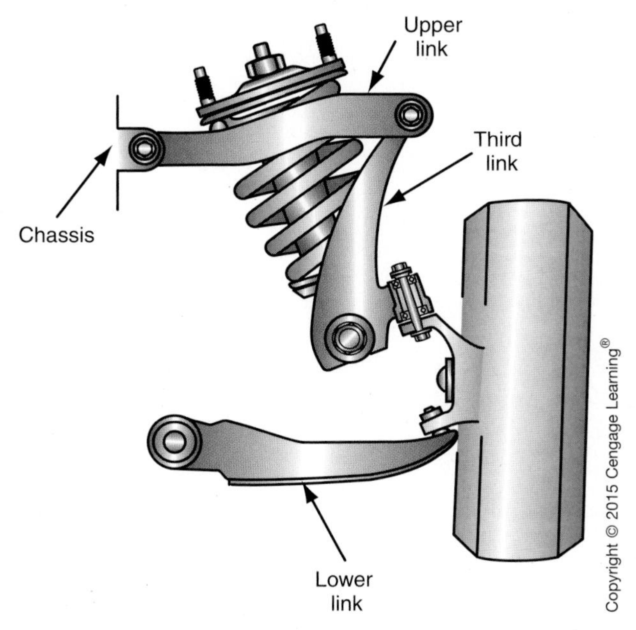

FIGURE 5-13 **Multilink front suspension with strut connected between the upper link and the strut tower.**

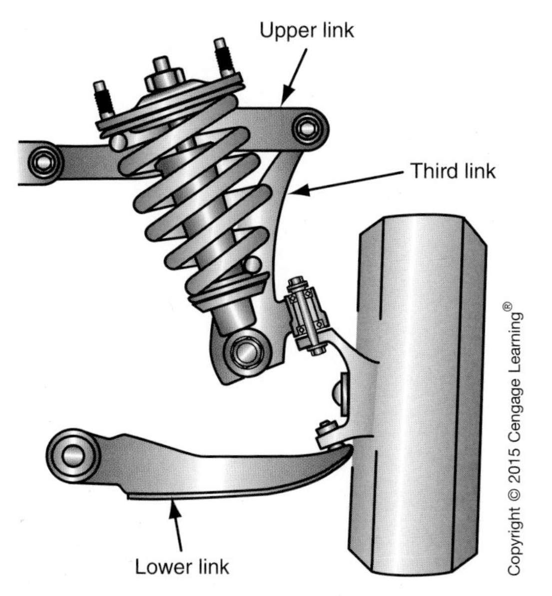

FIGURE 5-14 **Multilink front suspension with knuckle and wheel pivots at the upper link bearing and lower ball joint, and a nonrotating upper strut mount.**

Shock Absorber and Strut Design, Rear Suspension

In some rear suspension systems, the lower end of the strut is bolted to the spindle, and the top of the strut is connected through a strut mount to the chassis. The rear coil springs are mounted separately from the struts. These springs are mounted between the lower control arms and the chassis (Figure 5-15).

Shop Manual
Chapter 5, page 190

In other rear suspension systems, the coil springs are mounted on the rear struts (Figure 5-16). An upper insulator is positioned between the top of the spring and the upper spring support, and a lower insulator is located between the bottom of the spring and the spring mount on the strut. A rubber spring bumper is positioned on the strut piston rod. If a rear wheel strikes a severe road irregularity and the strut is fully compressed, the spring bumper provides a cushioning action between the top of the strut and the upper support.

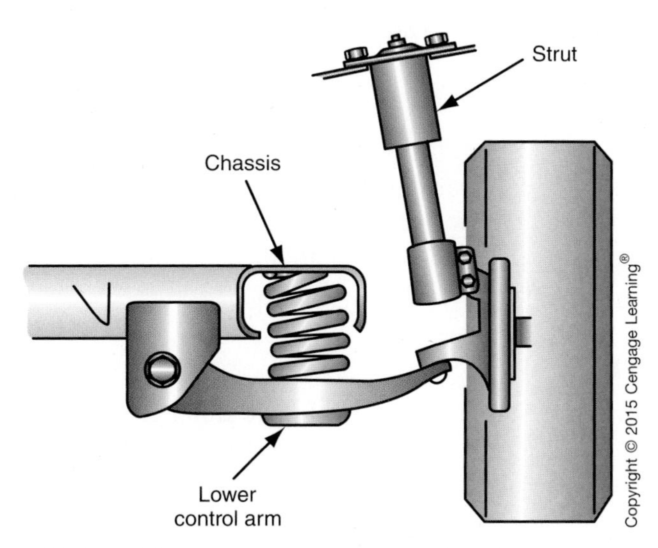

FIGURE 5-15 **Rear suspension system with coil springs mounted separately from the struts.**

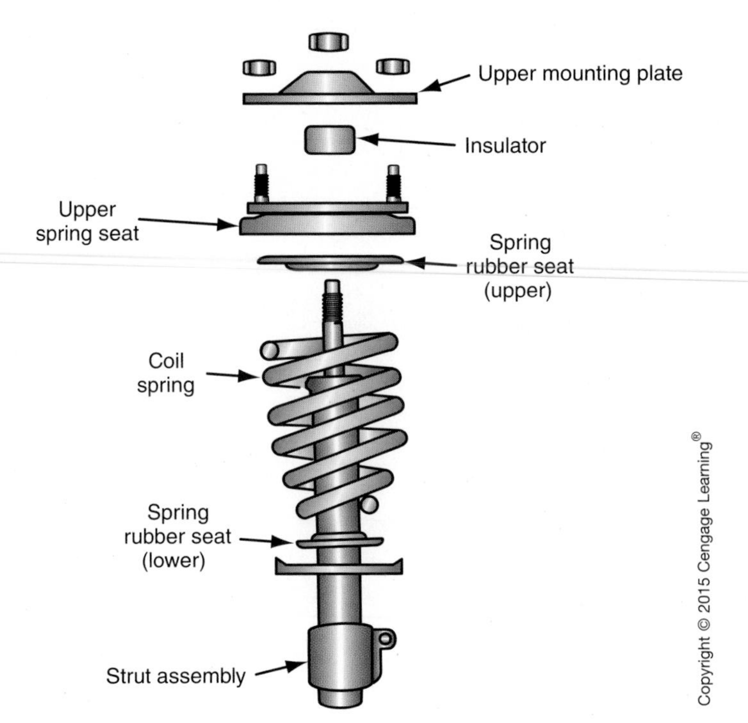

FIGURE 5-16 **Rear suspension system with coil springs mounted on the struts.**

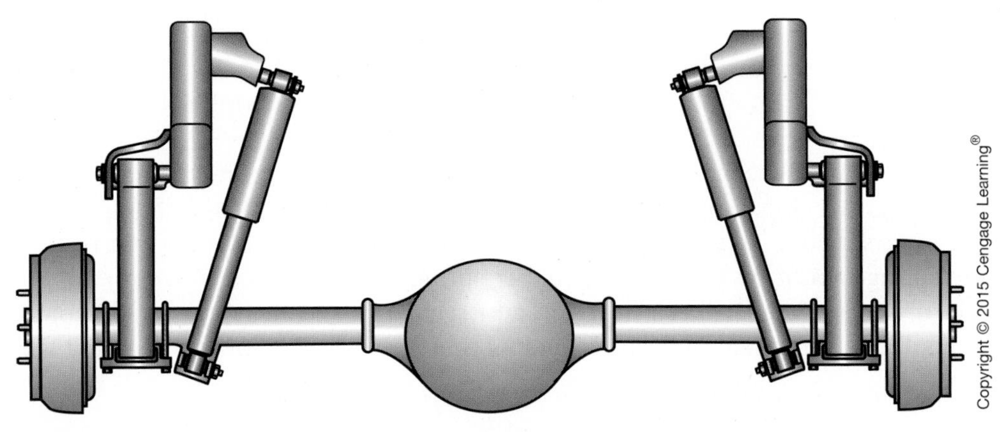

FIGURE 5-17 **Rear shock absorber mounting, sport utility vehicle (SUV).**

On some rear-wheel-drive light-duty trucks and sport utility vehicles (SUVs), the rear shock absorbers are slanted rearward and inward (Figure 5-17). On other light-duty trucks and SUVs, the rear shock absorbers are staggered on each side of the rear axle (Figure 5-18). Either of these shock absorber mountings improves ride quality and reduces noise, vibration, and harshness (NVH).

Travel-Sensitive Strut

Travel-sensitive struts vary the amount of strut control in relation to strut travel.

Some **travel-sensitive struts** contain narrow longitudinal grooves in the lower oil chamber (Figure 5-19). These grooves are parallel to the piston orifices, and some oil flows through the grooves as well as the orifices. Under normal driving and road conditions, the orifices and grooves are calibrated to provide normal spring damping and control. If the front wheel drops suddenly, such as when it strikes a large hole, the piston moves into the narrow portion of the oil chamber. Under this condition, all the oil must flow through the

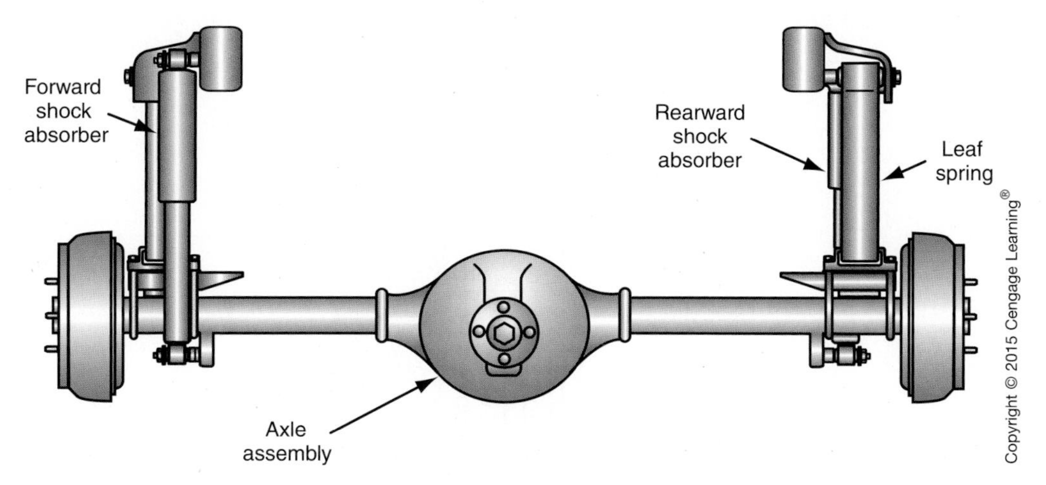

FIGURE 5-18 **Shock absorbers staggered on each side of the rear axle.**

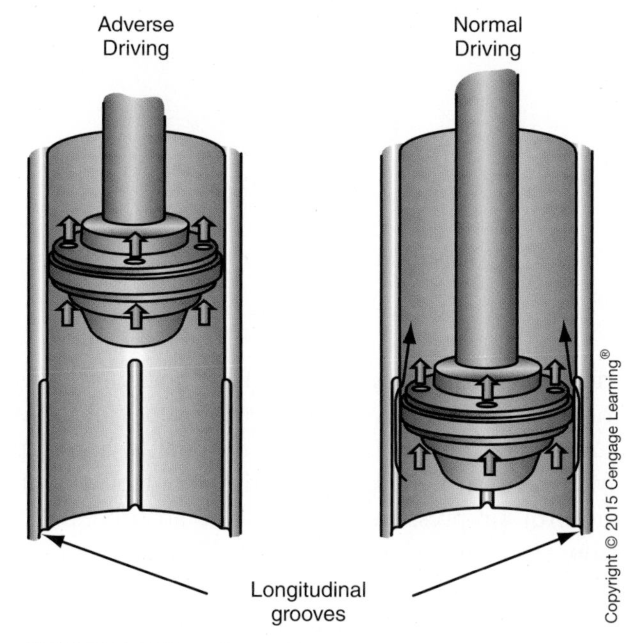

FIGURE 5-19 **Travel-sensitive strut.**

piston orifices, which greatly increases the strut's resistance to movement and the suspension damping action. This strut action prevents harsh impacts against the internal strut rebound rubber.

Adjustable Struts

Some adjustable struts have a manual adjustment that allows the vehicle owner or technician to adjust the struts to suit driving conditions (Figure 5-20). The strut adjusting knob varies the strut orifice opening. This knob has eight possible settings. The factory setting is No. 3, which provides average suspension control. The No. 1 setting provides reduced spring control and the softest ride, whereas a No. 8 adjustment gives increased spring control and the hardest ride. The adjustment knob is usually accessible without raising the vehicle.

Load-Leveling Shock Absorbers

Load-leveling rear shock absorbers or struts are used with an electronic height control system. An onboard air compressor pumps air into the rear shocks to raise the rear of the

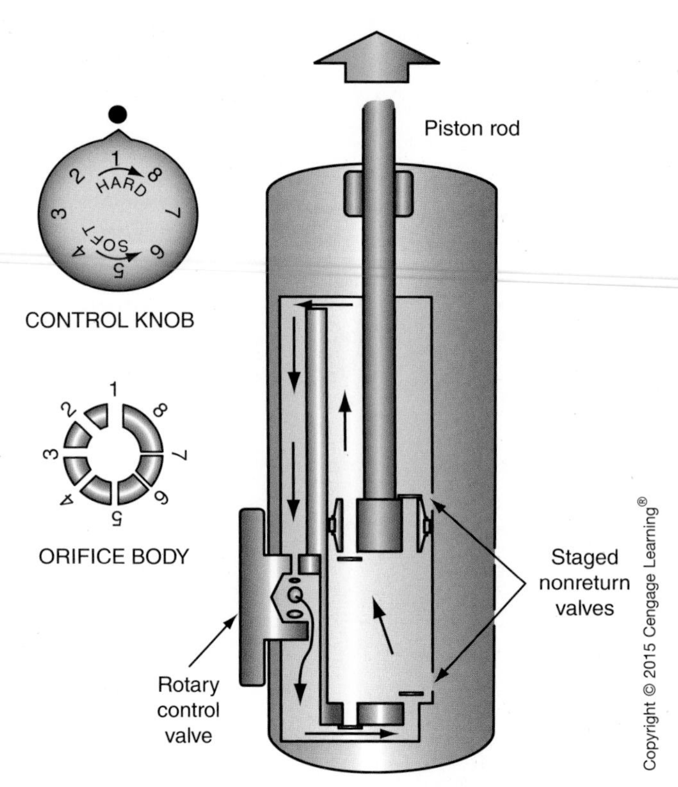

FIGURE 5-20 Manual adjustable strut.

vehicle, and an electric solenoid releases air from the shocks to lower the rear chassis. An electromagnetic height sensor may be contained in the shock absorber, or an external sensor may be used (Figure 5-21). This sensor sends a signal to an electronic control module in relation to the rear suspension height. The module controls the air compressor and the exhaust solenoid to control air pressure in the shock absorbers. This action maintains a specific rear suspension trim height regardless of the load on the rear suspension. If a heavy package is placed in the trunk, the vehicle chassis is forced downward. However, the **load-leveling shock absorbers** extend to restore the original rear suspension height.

Load-leveling shock absorbers use air pressure supplied to the shock absorbers to maintain rear suspension height. Load-leveling shock absorbers may be called air shocks.

Aftermarket air shock absorbers are available. These shock absorbers contain an air valve connection. A shop air hose may be used to supply the desired pressure in these air shock absorbers.

Aftermarket spring-assisted shock absorbers are also available. These are conventional shock absorbers with a small coil spring mounted over them. Upper and lower spring seats are attached near the top and bottom of these shock absorbers to support and retain the spring. The coil springs on the shock absorbers help the springs in the suspension system support the vehicle weight.

Shop Manual
Chapter 5, page 194

Electronically Controlled Shock Absorbers and Struts

Many cars are equipped with computer-controlled **active suspension systems**. In these systems, a computer-controlled actuator is positioned in the top of each shock absorber or strut (Figure 5-22). The shock absorber or strut actuators rotate a shaft inside the piston rod, and this shaft is connected to the shock valve. Many of these systems have two modes, soft and firm. In the soft mode, the actuators position the shock absorber valves so there is less restriction to the movement of oil. When the computer changes the actuators to the firm mode, the actuators position the shock valves so they provide more restriction to oil movement, which provides a firmer ride.

Computer-controlled suspension systems may be referred to as **active suspension systems**.

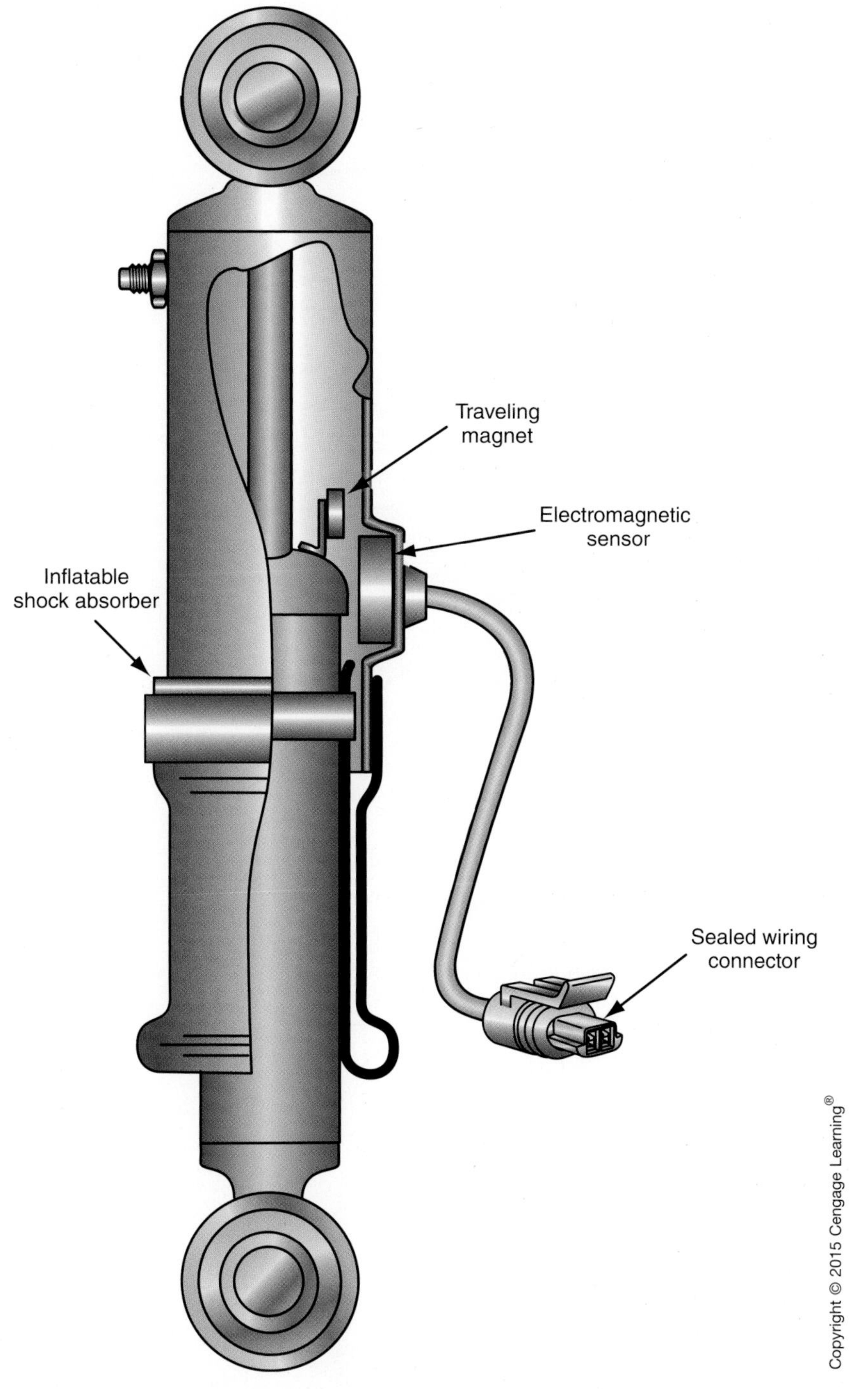

FIGURE 5-21 **Load-leveling strut.**

Some electronically controlled shock absorbers and struts contain a synthetic **magneto-rheological fluid** that contains numerous small, suspended metal particles. Delphi's trademarked name for the system is called MagneRide. There are no small moving parts or electromechanical valves to fail. The electro-rheological fluid viscosity changes when electrical current passes through the fluid itself, which contains magnetically soft particles of iron microspheres in a synthetic hydrocarbon based fluid. Each shock absorber contains a electromagnetic coil winding inside the piston itself that is energized by the suspension computer (Figure 5-23). Only the fluid in the oil passage of the piston is involved in the viscosity change. The wires to the coil are routed through the piston rod to a

Magneto-rheolgical fluid A fluid that when subjected to a magnetic field will increase viscosity as the magnetic field is increased, to the point of becoming an elastic solid.

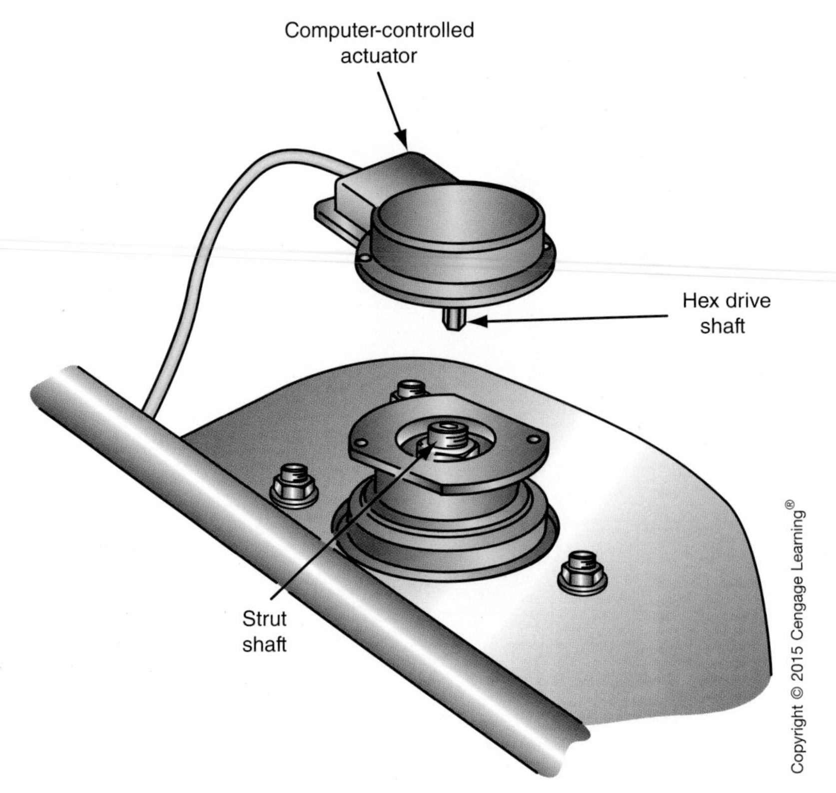

FIGURE 5-22 **Computer-controlled strut actuator.**

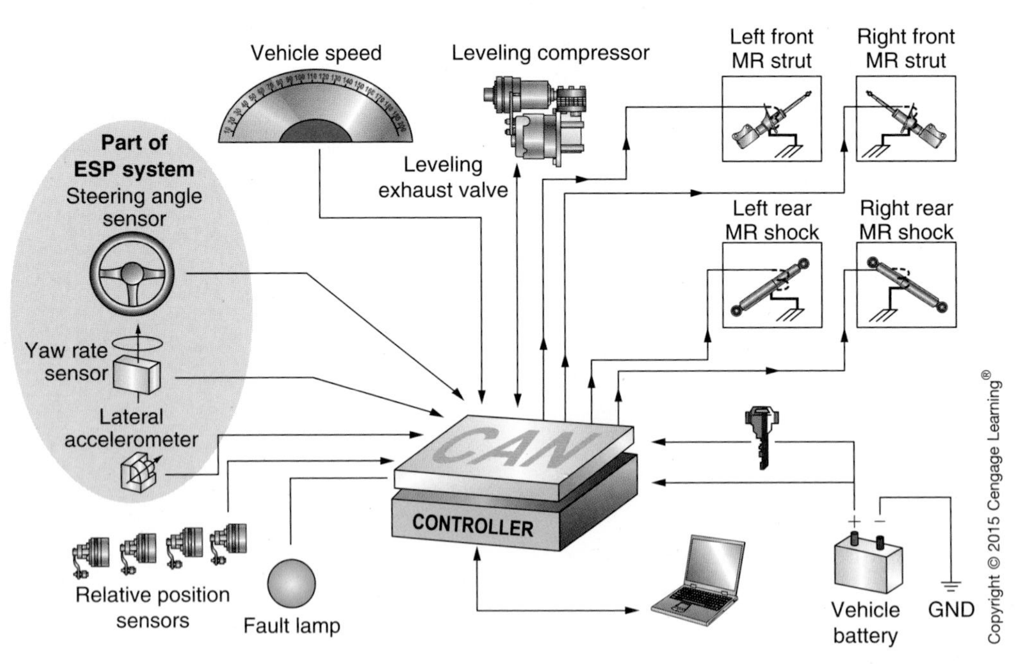

FIGURE 5-23 **Some of the major electrical components that are integrated into a magneto-rheological damping system.**

connector at the top of the shock housing. If the shock absorber winding is not energized, the metal particles in the fluid align randomly in the fluid. Under this condition, the fluid has a mineral oil-like consistency and the fluid moves easily through the shock absorber orifices. If the suspension computer energizes a shock absorber coil winding, the metal particles in the fluid are aligned into fibrous structures. When this occurs, the fluid has a

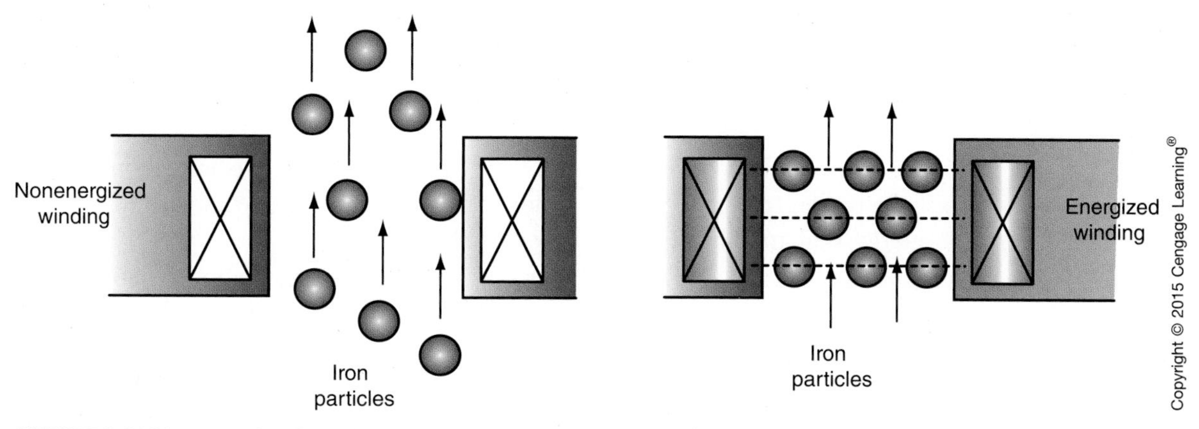

FIGURE 5-24 **Magneto-rheological fluid action in a strut or shock absorber.**

jelly-like consistency for a firm ride (Figure 5-24). The computer can change the shock absorber damping characteristics almost instantaneously, in 1 millisecond, and can also supply a varying amount of current through the shock absorber windings to provide a wide variety of shock absorber damping characteristics. Depending on the amount of current supplied, the fluid's viscosity or resistance to flow can be varied from thinner than water to a near plastic or solid state or any consistency in-between. This makes the fluid infinitely adjustable in viscosity and achieves continuously variable real-time damping to suit almost every driving condition instantly.

The advantages of magneto-rheological fluid-controlled shock absorber are:

- Ability to smooth out the action of each tire.
- Reduced noise, bounce, and vibration giving a flatter ride by controlling body motions.
- Exceptional roll control during evasive steering maneuvers.
- Excellent handling by controlling weight transfer during lateral and longitudinal maneuvers.
- Enhanced road isolation by reducing high-frequency road noise through the dampers.
- When integrated with ABS and traction control, ensures maximum stability and balance on slippery surfaces and gravel.

SUMMARY

- Shock absorbers or struts play a very important role in ride quality, steering control, and tire life.
- Tire and wheel jounce travel occurs when a tire strikes a bump in the road surface and the tire and wheel move upward.
- Rebound tire and wheel travel occurs when the tire and wheel move downward after jounce travel.
- When a spring is deflected upward during jounce travel, it stores energy. The spring then expands downward in rebound travel with all the energy it stored during the jounce travel. If the spring action is not controlled, the energy in the spring during rebound travel drives the tire against the road surface with excessive force. This action drives the tire and wheel back upward in jounce travel and the wheel continues oscillating up and down.
- The shock absorbers control spring action and prevent excessive tire and wheel oscillations.

TERMS TO KNOW

Active suspension systems
Adjustable struts
Gas-filled shock absorber
Heavy-duty shock absorbers
Jounce travel
Load-leveling shock absorbers
Magneto-rheological fluid

TERMS TO KNOW
(continued)

Rebound travel
Shock absorbers
Shock absorber ratios
Spring insulator
Struts
Travel-sensitive struts
Upper strut mount

- During jounce travel, the piston moves downward in the lower shock absorber chamber; during rebound travel, this piston moves upward. Because the lower shock absorber chamber is sealed and filled with oil, this oil must flow past the piston during any piston movement.
- Valves and openings in the shock absorber piston provide precision control of the oil flow past the piston to control spring action. Shock absorber valves are matched to the amount of energy that may be stored in the spring.
- A nitrogen gas charge is located in the oil reservoir of many shock absorbers and struts to prevent oil cavitation or foaming, which provides more positive shock absorber action.
- Shock absorber ratio refers to the difference between the shock absorber control on the compression and extension cycle. Many shock absorbers provide more control on the extension cycle.
- Internal design is similar in shock absorbers and struts, but struts also support the coil spring.
- Most front struts are connected between the steering knuckle and the upper strut mount.
- Many rear struts are connected between the spindle and the upper support.
- A travel-sensitive shock absorber provides increased resistance to piston movement as the shock absorber is extended.
- Adjustable shock absorbers and struts have a manual adjustment that allows the technician or owner to adjust the strut orifice opening.
- Load-leveling shock absorbers have air pumped into the shock absorbers from an onboard compressor to maintain a specific rear suspension height regardless of the rear suspension load.
- Most front struts rotate on the upper strut mount bearing as the front wheels are turned.

REVIEW QUESTIONS

Short Answer Essays

1. Describe spring action during jounce and rebound travel.
2. Describe uncontrolled spring action without a shock absorber.
3. Explain shock absorber operation.
4. Describe the vehicle safety hazards created by worn-out shock absorbers.
5. Explain shock absorber ratio.
6. Explain the purpose of the nitrogen gas charge in shock absorbers and struts.
7. Explain the differences between heavy-duty monotube and conventional twin tube shock absorbers.
8. Identify the purposes of an upper strut mount.
9. Explain the purpose of strut-and-spring insulators.
10. Describe the purpose of the rubber spring bumper on a strut piston rod.

Fill-in-the-Blanks

1. Shock absorbers control spring action and prevent excessive spring ________________.
2. Shock absorber design is matched to the ________________ rate of the spring.
3. Modern shock absorbers are ________________ ________________ damping devices that increase resistance the faster the suspension moves.
4. The nitrogen gas charge in a shock absorber prevents oil ________________ and ________________.
5. The single tube design in a heavy-duty shock absorber prevents excessive ________________ ________________.
6. A shock absorber with a 70/30 ratio provides more control on the ________________ cycle.
7. The lower end of many front struts is bolted to the steering ________________.

8. When a higher number is selected on the adjustment knob of an adjustable strut, the strut provides a ________________ ride.
9. Load-leveling shock absorbers may have an internal ________________ ________________.
10. Travel-sensitive shock absorbers provide increased spring control when the shock absorber is ________________.

Multiple Choice

1. The main purpose of shock absorbers and struts is to:
   - A. Control spring action.
   - B. Prevent fore-and-aft wheel movement.
   - C. Reduce lateral wheel movement.
   - D. Prevent wheel shimmy.
2. All of these statements about shock absorber design and operation are true EXCEPT:
   - A. The oil flow through the orifices and valves is matched to the spring's strength and deflection rate.
   - B. A typical shock absorber may have 70 percent of the total control on the extension cycle.
   - C. During jounce wheel travel the piston moves upward in the shock absorber lower tube unit.
   - D. A nitrogen gas charge in a shock absorber prevents oil foaming.
3. Compared to a conventional shock absorber, a heavy-duty shock absorber has:
   - A. Triple layers of steel in the lower tube unit.
   - B. Higher viscosity oil.
   - C. A larger diameter piston rod.
   - D. A high-quality seal for longer life.
4. All of these statements about gas-filled shock absorbers are true EXCEPT:
   - A. A hydrogen gas charge is used in most shock absorbers and struts.
   - B. The gas charge creates a mild boost in spring rate.
   - C. The primary function of the gas charge is to minimize aeration of the hydraulic fluid.
   - D. A gas-filled shock absorber provides continuous damping action with very little wheel movement.
5. The magneto-rheological fluid used in some shock absorbers and struts contains:
   - A. Transmission fluid.
   - B. Small, suspended metal particles.
   - C. Antifoaming agents.
   - D. A dye for lead detection purposes.
6. During normal strut operation:
   - A. Downward wheel movement is called wheel jounce.
   - B. The strut prevents excessive wheel oscillations.
   - C. When the wheel moves upward the strut piston also moves upward.
   - D. The strut prevents wheel shimmy.
7. Travel-sensitive struts:
   - A. Have an adjustment on the outside of the strut.
   - B. Provide more resistance to oil movement when compressed.
   - C. Provide more resistance to oil movement when driving on a smooth road surface.
   - D. Bypass some of the oil past the piston when compressed.
8. Adjustable struts:
   - A. May be adjusted with a control knob in the instrument panel.
   - B. Provide five different settings.
   - C. Provide softest ride with a No. 1 setting.
   - D. Rotate the upper part of the strut to provide adjustable ride quality.
9. Load-leveling shock absorber systems:
   - A. May be pressurized from an onboard air pressure tank.
   - B. May have a suspension height sensor.
   - C. Maintain lower rear suspension height when driving on smooth road surfaces.
   - D. Prevent suspension bottoming on rough road surfaces.
10. Electronically controlled shock absorbers and struts:
    - A. Contain an electronically controlled actuator in the top of each strut.
    - B. Usually have 12 different modes of strut operation from soft to very firm.
    - C. Have an electronically controlled actuator that rotates the strut piston rod.
    - D. Have an actuator that positions the strut valves to provide more oil restriction in the soft mode.

Chapter 6

Front Suspension Systems

Upon completion and review of this chapter, you should be able to understand and describe:

- The causes of coil spring failure.
- The design and spring rate of a linear-rate coil spring.
- The difference between a regular-duty coil spring and a heavy-duty coil spring.
- The functions of three different types of coils in a variable-rate coil spring.
- Basic torsion bar action as a front wheel strikes a road irregularity.
- How friction and noise problems are reduced in a multiple-leaf spring.
- The advantages of a front suspension system with ball joints compared to earlier I-beam front suspension systems.
- Two types of load-carrying ball joints, and explain the location of the control arm in each type.
- How you would recognize a worn ball joint from a visual inspection of the ball joint wear indicator.
- The mounting location and purpose of a stabilizer bar.
- The purpose of a strut rod.
- The two steering knuckle pivot points in a MacPherson strut front suspension system.
- The advantage of a short-and-long arm front suspension system compared with a suspension system with equal-length upper and lower control arms.
- Two methods of attaching the spindles to the I-beams in a twin I-beam front suspension.
- The effect of sagged front springs on caster angle and directional stability.
- The effect of sagged front springs on camber angle.
- The advantages of hydraulic control arm bushings.
- The advantages of aluminum control arms.

Introduction

The front and rear suspension systems must perform several extremely important functions to maintain vehicle safety and owner satisfaction. The suspension system must supply steering control for the driver under all road conditions. Vehicle owners expect the suspension system to provide a comfortable ride. The suspension, together with the frame, must maintain proper vehicle tracking and directional stability. Another important purpose of the suspension system is to provide proper wheel alignment and minimize tire wear.

The impact of the front tires striking road irregularities must be absorbed and dissipated by the front suspension system. These impacts are distributed throughout the suspension, and this action isolates the vehicle passengers from road shock. The vehicle's ride characteristics are determined by the amount of impact energy that the suspension can absorb and by the rate at which the suspension dissipates these tire impacts. Ride characteristics are designed into the suspension system, and these characteristics are not adjustable. For example, the front suspension may be designed to provide a very soft ride in which all the tire impacts are absorbed and dissipated quickly by the suspension system. This type of suspension system provides a very comfortable ride, but it also allows excessive body lean during cornering, which reduces high-speed cornering and handling capabilities. Such vehicles as sports cars and sport utility vehicles (SUVs) are usually designed with a suspension system that provides a firm ride and absorbs and dissipates tire impacts more slowly. Although high-speed cornering and handling capabilities are improved, this type of suspension may transfer some road shock to the passenger compartment.

When the vehicle is driven over road irregularities, the front suspension system must allow the front wheels to move vertically while maintaining the tire's proper horizontal position in relation to the road surface. To provide this wheel action, the steering knuckle must be mounted between the upper and lower control arms on short-and-long arm (SLA) suspension systems or between the lower control arm and the lower end of the strut on MacPherson strut suspension systems. The upper and lower control arms must be pivoted on the inner ends; the steering knuckle pivots on the ball joints in SLA suspensions or on the lower ball joint and upper strut mount on MacPherson strut suspensions.

Springs absorb much of the shock from tire impacts with the road surface. When a front wheel strikes a road irregularity and moves upward, the coil spring compresses and absorbs energy during this movement. The coil spring immediately dissipates this energy as the spring moves back to its original state. Shock absorbers are installed in the front suspension system to dampen the oscillations of the coil springs.

A stabilizer bar is mounted on rubber insulating bushings and bolted to the chassis. The outer ends of the stabilizer bar are attached through links to the lower control arms. The stabilizer bar controls the amount of independent lower control arm movement. When one front wheel strikes a road irregularity and moves upward, the stabilizer bar transfers part of the lower control arm movement to the opposite lower control arm, which reduces and stabilizes body roll. Therefore, the stabilizer bar helps to define the suspension characteristics related to body roll.

All suspension components, including the frame and chassis, are designed or "tuned" to provide the ride and handling qualities that the manufacturer believes the average driver will desire. For example, the late model Explorer Sport Trac frame has been stiffened 40 percent more than the Explorer frame. To achieve this increased stiffness, the frame side rails are thicker than the ones on the Explorer, and a new tubular crossmember has been added to the frame. **Gussets** have also been welded into the corners where the crossmembers meet the side rails. This frame design is matched to the torsion bar suspension to provide improved vehicle agility on and off the road.

Gussets are pieces of metal welded into the corner where two pieces of metal are attached together, which provide increased strength and stiffness.

Suspension System Components

Coil Springs

Shop Manual
Chapter 6, page 228

The coil spring is the most commonly used spring for front and rear suspension systems. Coil springs are actually a coiled-spring steel bar. When a vehicle wheel strikes a road irregularity, the coil spring compresses to absorb shock, and then recoils back to its original installed height. Many coil springs contain a steel alloy that contains different types of steel mixed with other elements such as silicon or chromium. Coil springs may be manufactured by a cold or hot coiling process. The hot coiling process includes procedures for tempering and hardening the steel alloy. Coil springs are designed to carry heavy loads, but they must be light in weight. Many coil springs have a vinyl coating, which increases corrosion resistance and reduces noise.

Coil spring failures may be caused by these conditions:

1. Constant overloading
2. Continual jounce and rebound action
3. Metal fatigue
4. A crack or nick on the surface layer or coating

Coil springs do not have much ability to resist lateral movement. However, when coil springs are used on the drive wheels, the suspension usually has special bars to prevent lateral movement.

Linear-Rate Coil Springs. Coil springs are classified into two general categories: linear rate and variable rate. **Linear-rate coil springs** have equal spacing between the coils and one basic shape with a consistent wire diameter. When the load is increased on a linear-rate spring, the spring compresses and the coils twist or deflect. As the load is removed from the spring, the coils unwind, or flex, back to their original position. The spring rate is the load required to deflect the spring 1 inch. Linear-rate coil springs have a constant spring rate, regardless of the load. For example, if 200 pounds deflect the spring 1 inch, 400 pounds deflect the spring 2 inches. The spring rate on linear springs is usually calculated between 20 and 60 percent of the total spring deflection.

Variable-Rate Coil Springs. Variable-rate coil springs have a variety of wire sizes and shapes. The most common variable-rate coil springs have a consistent wire diameter with a cylindrical shape and unequally spaced coils (Figure 6-1).

Inactive coils are positioned at each end of a coil spring.

Transitional coils are positioned between the inactive coils and the active coils in the center of the spring.

The active coils are located in the center of a coil spring.

The **inactive coils** at the end of the spring introduce force into the spring when the wheel strikes a road irregularity. When the **transitional coils** are compressed to their point of maximum load-carrying capacity, these coils become inactive. The **active coils** operate during the complete range of spring loading. When a stationary load is applied to a variable-rate coil spring, the inactive coils theoretically support the load. If the load is increased, the transitional coils support the load until they reach maximum load-carrying capacity, and the active coils carry the remaining overload. This spring action provides automatic load adjustment while maintaining vehicle height.

Some variable-rate coil springs have a tapered wire in which the active coils have the larger diameter, and the inactive coils have the smaller diameter. Other variable-rate spring designs include truncated cone, double cone, and barrel shape. A variable-rate spring does not have a standard spring rate. This type of spring has an average spring rate based on the load at a predetermined spring deflection. *It is impossible to compare variable spring rates and linear spring rates because of this difference in spring rates.* Variable-rate coil springs usually have more load-carrying capacity than linear-rate springs in the same application.

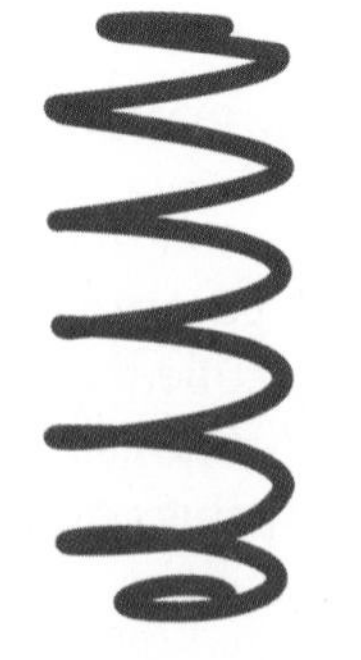

FIGURE 6-1 **Variable-rate coil springs have consistent wire diameter and unequally spaced coils.**

Light-Weight Coil Springs

A few sports cars are presently equipped with titanium coil springs. This type of coil spring reduces the weight on the front springs by 39 percent and 28 percent on the rear springs compared to steel coil springs. Decreasing the weight of the coil springs reduces the **unsprung weight** and improves ride control. The unsprung vehicle weight tends to force the wheel downward during rebound wheel travel. Higher unsprung weight drives the wheel downward with greater force and increases the impact force between the tire and the road surface, resulting in a harsh ride. To compensate for the high unsprung weight, the shock absorber damping rate must be increased. This suspension design reduces ride quality. When the unsprung weight is reduced, the wheel is forced downward with less force, reducing the shock absorber damping rate and improving ride quality.

Unsprung weight is the vehicle weight that is not supported by the coil springs, and sprung weight is the vehicle weight that is supported by the coil springs.

Heavy-Duty Coil Springs. Heavy-duty coil springs are designed to carry 3 to 5 percent greater loads than regular-duty coil springs. The wire diameter may be up to 0.100 inch greater in a heavy-duty spring than in a regular-duty coil spring. This larger-diameter wire increases the load-carrying capacity of the spring. The free height of a heavy-duty coil spring is shorter than a regular-duty coil spring for the same application (Figure 6-2).

Selecting Replacement Springs. When replacement coil springs are required, the technician must select the correct spring. The original part number is usually on a tag wrapped around one of the coils. However, this tag may have fallen off if the spring has been in service for very long. Some aftermarket suppliers stamp the part number on the end of the coil spring. If the original part number is available, the replacement springs may be ordered with the same part number. Most vehicle manufacturers recommend that both front or rear springs be replaced at the same time. The replacement springs must have the same type of ends as the springs in the vehicle. Coil spring ends may be square tapered, square untapered, or tangential (Figure 6-3).

Full-wire open-end springs have the ends cut straight off, and sometimes these ends are flattened, squared, or ground to a D-shape. **Taper-wire closed-end** springs are ground to a taper and wound to ensure squareness. **Pigtail spring ends** are wound to a smaller diameter. Springs are generally listed for front or rear suspensions.

Regular-duty coil springs are a close replacement for the original spring in the vehicle, and these springs may replace several different original equipment (OE) springs in the same vehicle. Linear-rate coil springs are usually found in regular-duty, heavy-duty, and sport suspension packages. Heavy-duty coil springs are required when the vehicle is carrying a continuous heavy load, such as trailer towing.

Variable-rate coil springs are generally used when automatic load leveling is required under increased loads. Variable-rate coil springs maintain the correct vehicle height under various loads and provide increased load-carrying capacity compared to heavy-duty coil springs. The technician must select the correct spring to meet the requirements of the vehicle and load conditions.

If a wheel moves upward in relation to the chassis, this action is referred to as jounce travel. Downward wheel movement is called rebound travel.

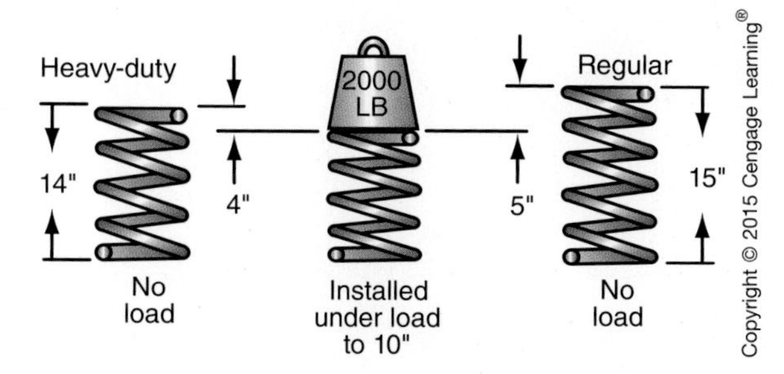

FIGURE 6-2 **Comparison of heavy-duty and regular-duty coil springs.**

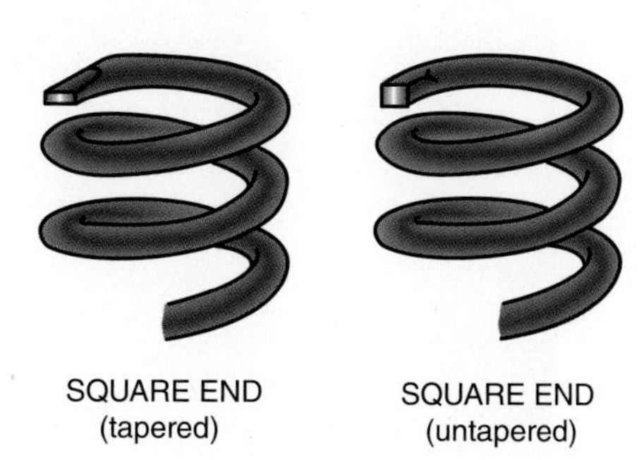
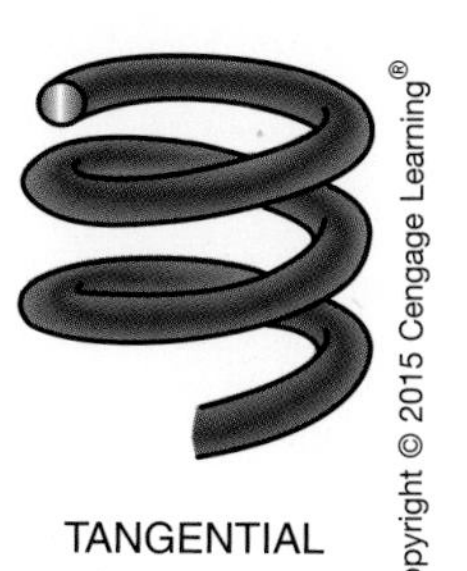

FIGURE 6-3 **Types of coil spring ends.**

WARNING: A very large amount of energy is stored in a compressed coil spring. Always follow the spring service procedures in the vehicle manufacturer's service manual to avoid personal injury.

WARNING: Never disconnect a suspension component that quickly releases the tension on a compressed coil spring. This action may cause personal injury.

Torsion Bars

In some front suspension systems, torsion bars replace the coil springs. During wheel jounce, the torsion bar twists. During wheel rebound, the torsion bar unwinds back to its original position. A torsion bar may be thought of as a straight, flattened coil spring. One end of the heat-treated alloy steel torsion bar is attached to the vehicle frame, and the opposite end is connected to the lower control arm. A few vehicles have the end of the torsion bars connected to the upper control arm. Some light-duty trucks and SUVs are presently equipped with longitudinal torsion bars in the front suspension (Figure 6-4). Transversely mounted torsion bars were used in some front suspensions on older cars.

Because the lower control arm moves up and down as the wheel strikes road irregularities, this control arm action twists the torsion bar. The bar's natural resistance to twisting causes it to return to its original position. During the manufacturing process, torsion bars are prestressed to provide fatigue strength. These bars are directional. Torsion bars are marked right or left, and they must be installed on the appropriate side of the vehicle. Left and right on a vehicle is always viewed from the driver's seat.

A torsion bar is capable of storing a higher maximum energy compared with a loaded coil or leaf spring. Shorter, thicker torsion bars have increased load-carrying capacity compared with longer, thinner bars. Since torsion bars require less space compared with coil or leaf springs, they are usually found on front suspensions. However, a few rear suspensions have torsion bars. Torsion bars have a riding height adjustment screw at the end where they are attached to the frame.

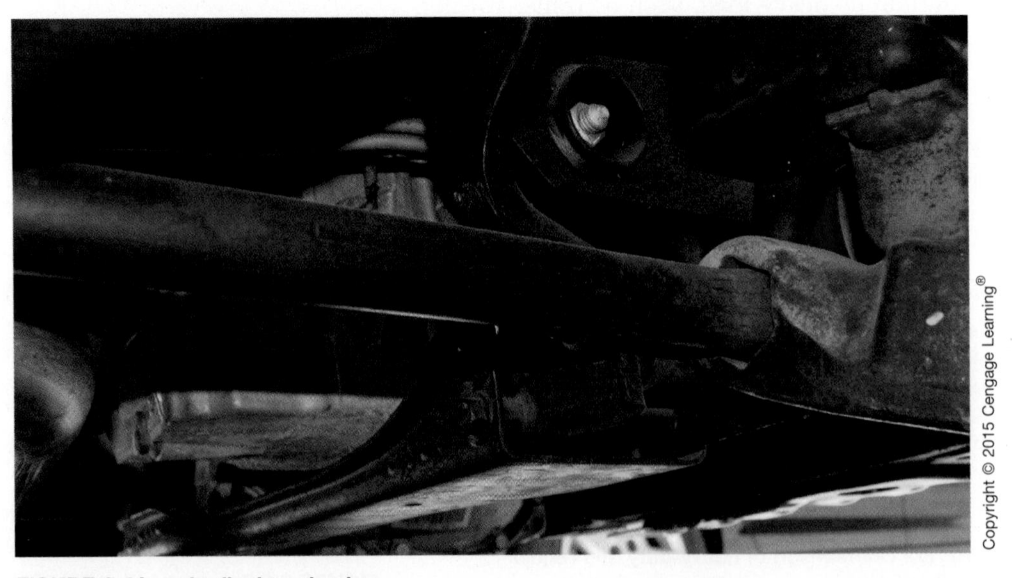

FIGURE 6-4 **Longitudinal torsion bar.**

Multiple-Leaf Springs

Leaf springs may be multiple-leaf or mono leaf. **Multiple-leaf springs** have a series of flat steel leaves of varying lengths that are clamped together. A center bolt extends through all the leaves to maintain the leaf position in the spring. The upper leaf is called the main leaf, and this leaf has an eye on each end. An insulating bushing is pressed into each main leaf eye. The front bushing is attached to the frame, and the rear bushing is connected through a shackle to the frame. The shackle provides fore-and-aft movement as the spring compresses (Figure 6-5).

The main leaf is the longest leaf in the spring, and the other leaves get progressively shorter. Each spring leaf is curved in the manufacturing process. If this curve were doubled, it would form an ellipse. Therefore, leaf springs are referred to as semielliptical or quarter elliptical. Most leaf springs are semielliptical. The ellipse designation refers to how much of the ellipse the spring actually describes.

As a leaf spring compresses, it becomes progressively stiffer. When a leaf spring is compressed, the length of the leaves changes, and the leaves slide on each other. This sliding action could be a source of noise and friction. These noise and friction problems are reduced by interleaves, or spacers, made from zinc and plastic placed between the steel leaves. The head on the spring center bolt fits into an opening in the axle to position the axle properly and provide proper vehicle tracking. If the center bolt is broken, the axle position may shift and alter vehicle tracking and alignment. Leaf springs are usually mounted at right angles to the axle. They provide excellent resistance to lateral movement.

Mono Leaf Springs

Some leaf springs contain a single steel leaf, and these springs may be referred to as **mono leaf springs**. The single leaf is thicker in the center and becomes gradually thinner toward the outer ends. This design provides a variable spring rate for a smooth ride and adequate load-carrying capacity. Mono leaf springs do not have a friction and noise problem as the spring compresses. Some cars, such as the Corvette and front-wheel-drive Oldsmobile Cutlass Supreme, use a fiberglass-reinforced plastic mono leaf spring in place of a steel

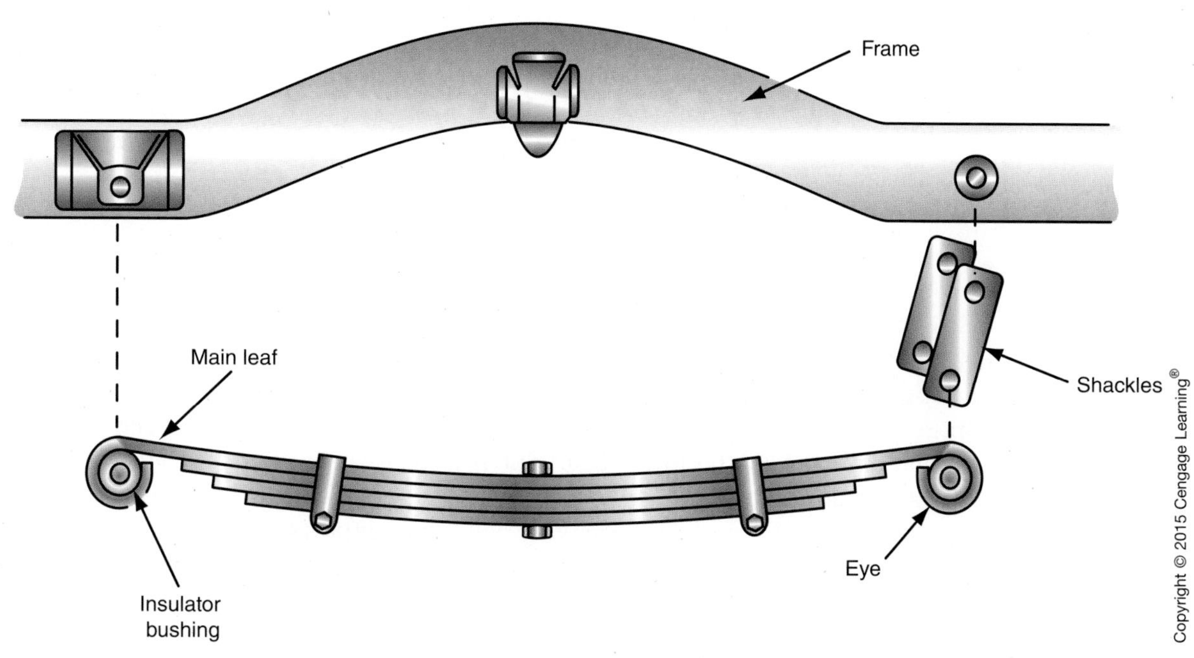

FIGURE 6-5 **Leaf spring design.**

A BIT OF HISTORY

The transverse leaf spring front and rear suspension is one of the oldest suspension systems in the automotive industry. It was used on Ford products from the early 1900s until 1948.

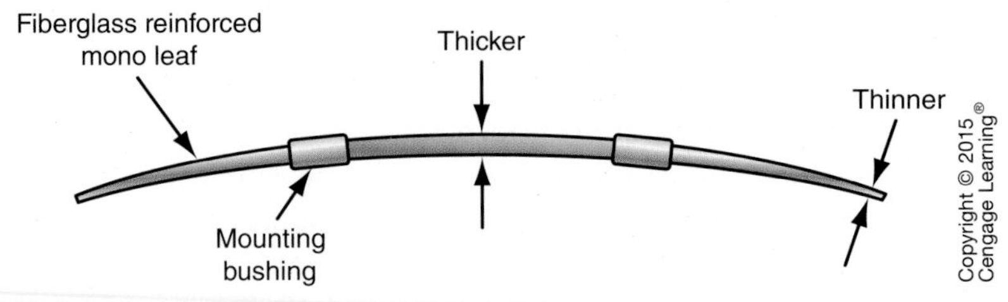

FIGURE 6-6 **Transversely mounted fiberglass mono leaf spring.**

spring to reduce weight (Figure 6-6). The mono leaf spring may be mounted longitudinally or transversely, and this type of spring may be used in front or rear suspensions. Older model Chevrolet Astro vans have longitudinally mounted fiberglass mono leaf rear springs.

Ball Joints

Shop Manual
Chapter 6, page 213

The ball joints act as pivot points that allow the front wheels and spindles, or knuckles, to turn between the upper and lower control arms. Compared with the earlier I-beam and kingpin-type front suspension systems, ball-joint suspension systems are much simpler. A front suspension with ball joints reduces the number of load-carrying bearing surfaces. Compared with an I-beam front suspension with kingpins, a front suspension with ball joints has these advantages:

1. Reduced space requirements
2. Reduced unsprung weight
3. Easier alignment
4. More dependable steering control
5. Improved safety
6. Improved tire life
7. Reduced steering effort
8. Improved ride quality
9. Simplified service—no kingpin reaming or honing
10. Simplified lubrication

WARNING: Worn ball joints may suddenly pull apart, resulting in loss of steering control and possibly a collision! Refer to the procedures in the Shop Manual Chapter 6 for measuring ball joint wear.

Load-Carrying Ball Joint

Ball joints may be grouped into two classifications, load carrying and non-load carrying. Ball joints may be manufactured with forged, stamped, cold-formed, or screw-machined housings. The coil spring is seated on the control arm to which the load-carrying ball joint is attached. For example, when the coil spring is mounted between the lower control arm and the chassis, the lower ball joint is a load-carrying joint (Figure 6-7). In a torsion bar suspension, the **load-carrying ball joint** is mounted on the control arm to which the torsion bar is attached. A load-carrying ball joint supports the vehicle weight.

In a load-carrying ball joint, the vehicle weight forces the ball stud into contact with the bearing surface in the joint. Load-carrying ball joints may be compression loaded or tension loaded. If the control arm is mounted above the lower end of the knuckle and rests on the knuckle, the ball joint is **compression loaded**. In this type of ball joint, the

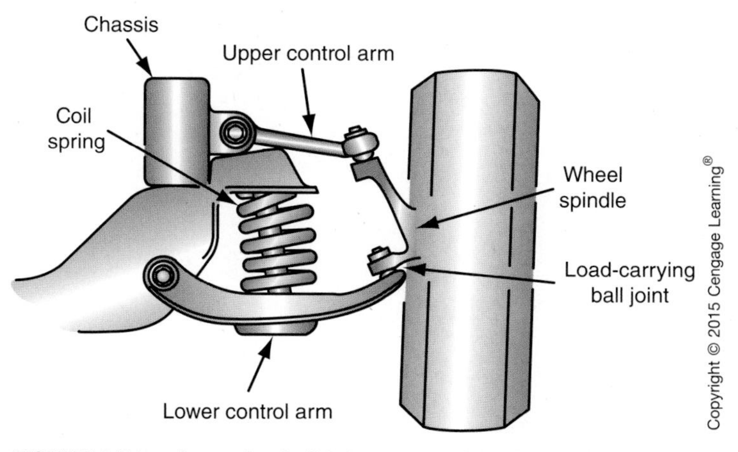

FIGURE 6-7 **Load-carrying ball joint mounted on the control arm on which the spring is seated.**

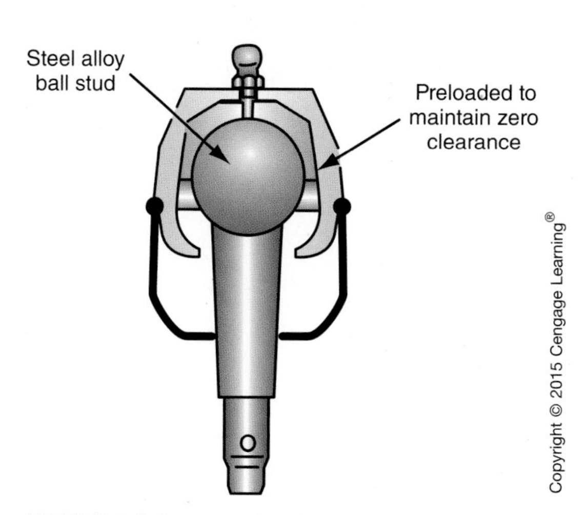

FIGURE 6-8 **Compression-loaded ball joint.**

vehicle weight is pushing downward on the control arm. This weight is supported on the tire and wheel, which are attached to the steering knuckle. Since the ball joint is mounted between the control arm and the steering knuckle, the vehicle weight squeezes the ball joint together (Figure 6-8). In this type of ball-joint mounting, the ball joint is mounted in the lower control arm and the ball joint stud faces downward (Figure 6-9).

When the lower control arm is positioned below the steering knuckle, the vehicle weight is pulling the ball joint away from the knuckle (Figure 6-10). This type of ball joint mounting is referred to as **tension loaded**. This type of ball joint is mounted in the lower control arm with the ball-joint stud facing upward into the knuckle (Figure 6-11).

Since the load-carrying ball joint supports the vehicle weight, this ball joint wears faster compared with a non-load-carrying ball joint. Many load-carrying ball joints have built-in **wear indicators**. These ball joints have an indicator on the grease nipple surface that recedes into the housing as the joint wears. If the ball joint is in good condition, the grease nipple shoulder extends a specified distance out of the housing. If the grease nipple shoulder is even with or inside the ball-joint housing, the ball joint is worn and replacement is necessary (Figure 6-12).

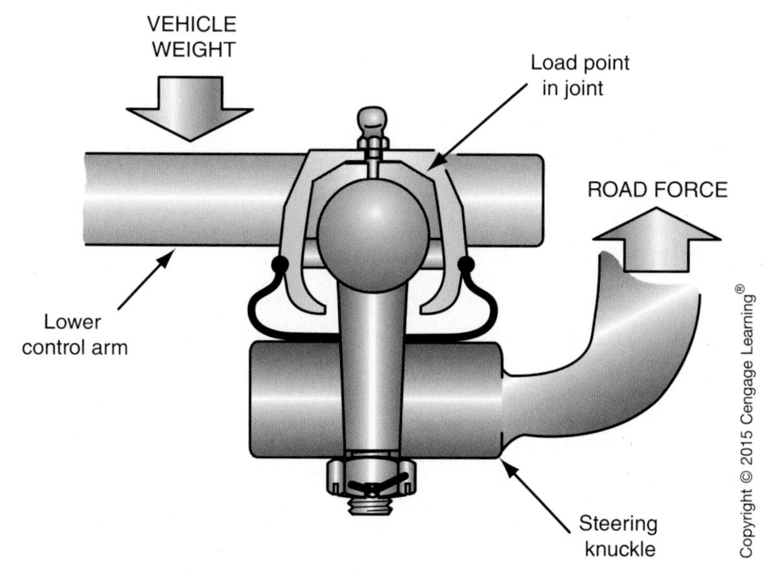

FIGURE 6-9 **Compression-loaded ball joint mounting.**

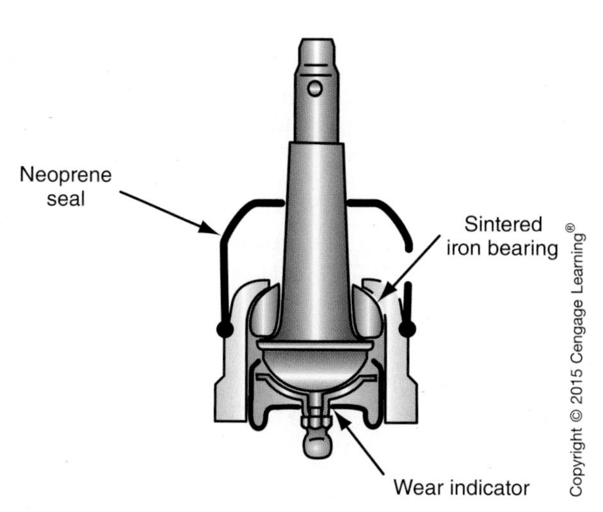

FIGURE 6-10 **Tension-loaded ball joint.**

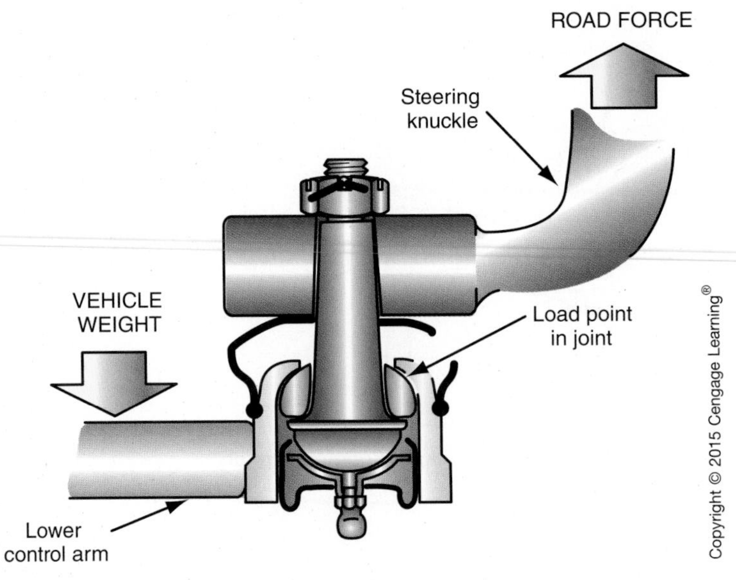

FIGURE 6-11 **Tension-loaded ball joint mounting.**

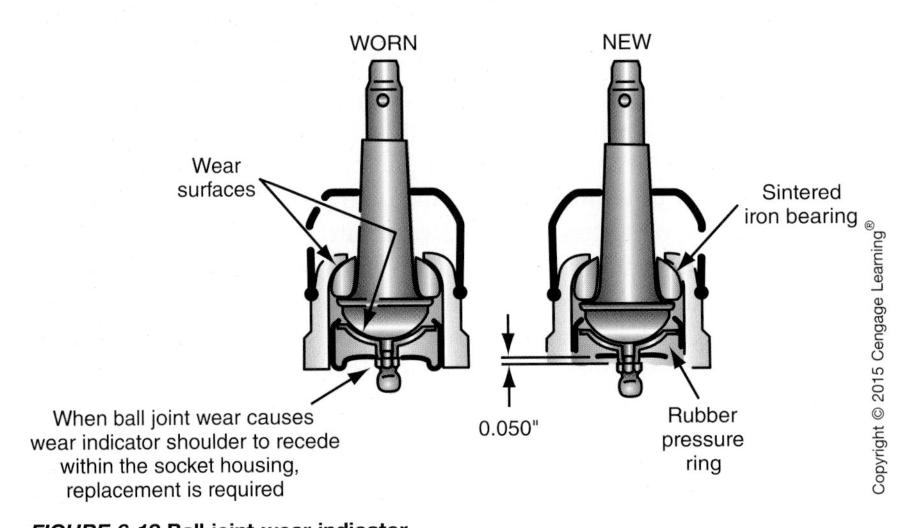

FIGURE 6-12 **Ball joint wear indicator.**

Most states have safety inspection procedures for testing ball joints and other safety-related components. These safety inspection procedures include specifications. Always follow the guidelines when performing safety inspections.

A **non-load-carrying ball joint** maintains knuckle position, but this ball joint does not support the vehicle weight.

Non-Load-Carrying Ball Joint

A **non-load-carrying ball joint** may be referred to as a stabilizing or follower ball joint. A non-load-carrying ball joint is designed with a preload, which provides damping action (Figure 6-13). This ball joint preload provides improved steering quality and vehicle stability.

Shop Manual
Chapter 6, page 214

Low-Friction Ball Joints

Low-friction ball joints are standard equipment on many vehicles. Low-friction ball joints provide precise low-friction movement of the ball socket in the ball joint. Compared to conventional ball joints, two-thirds of the internal friction is eliminated in a low-friction

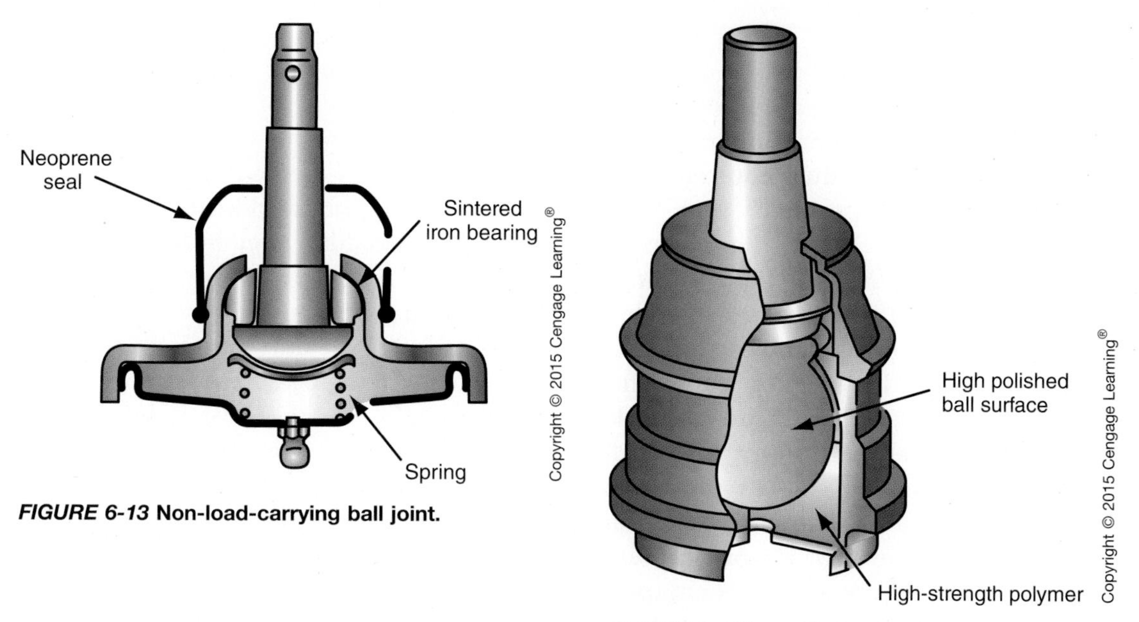

FIGURE 6-13 **Non-load-carrying ball joint.**

FIGURE 6-14 **Low-friction ball joint.**

ball joint. The smooth ball socket movement in a low-friction ball joint provides improved steering performance, better steering wheel return, and longer ball joint life. Low-friction ball joints have a highly polished ball socket surface surrounded by a high-strength polymer bearing (Figure 6-14).

Strut Rod

On some front suspension systems, a strut rod is connected from the lower control arm to the chassis. The strut rod is bolted to the control arm, and a large rubber bushing surrounds the strut rod in the chassis opening. The outer end of the strut rod is threaded and steel washers are positioned on each side of the strut rod bushing. Two nuts tighten the strut rod into the bushing (Figure 6-15). The strut rod prevents fore-and-aft movement of the lower control arm. In some suspension systems, the position of the strut rod nuts provides proper front wheel adjustment.

Shop Manual
Chapter 6, page 233

Some strut rods are presently manufactured from tubular steel to reduce strut rod and unsprung weight.

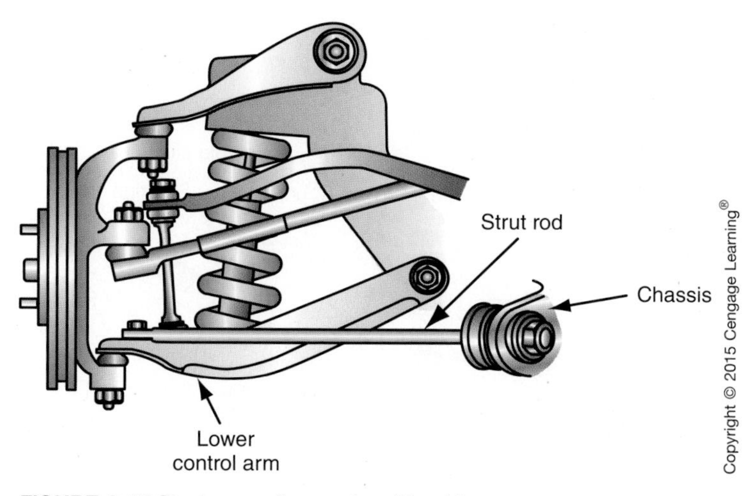

FIGURE 6-15 **Strut, or radius, rod and bushings.**

AUTHOR'S NOTE: **It has been my experience that the most important item when diagnosing automotive problems is your knowledge of automotive systems and components. You must understand the operation of automotive systems and the individual components within these systems. Second, you must be familiar with the problems that certain worn, defective, or misadjusted components can cause. When diagnosing an automotive problem, you must identify the exact problem, and then think of all the causes of the problem. When you are familiar with all the causes of the problem, you will know the system and component where diagnosis should begin. The cause of an automotive problem may not be in the system where the problem seems to appear. For example, a problem of steering pull during braking would usually be caused by a defect in the brake system. However, a worn front strut rod bushing may allow the front suspension on one side to move rearward during braking, which changes the front suspension alignment angles and causes steering pull when braking. Therefore, if you know that a worn radius arm bushing may cause steering pull when braking, you will inspect this component in the diagnosis as well.**

SHORT-AND-LONG ARM FRONT SUSPENSION SYSTEMS

Upper and Lower Control Arms

Many years ago, trucks and cars were equipped with I-beam front suspension systems designed with kingpins and longitudinally mounted leaf springs. A few trucks still used this type of suspension until the late 1980s. The kingpins were retained in a vertical opening in the ends of the axle with a lock bolt. Upper and lower bushings in the steering knuckle pivoted on the upper and lower ends of the kingpins (Figure 6-16). In this type of front suspension, if one front wheel moves upward or downward, some movement is transferred to the other end of the axle. This action tends to result in reduced ride quality, steering instability, and tire wear.

As automotive technology evolved, the control arm front suspension system replaced the I-beam front suspension system. This type of front suspension system has coil springs with upper and lower control arms. Since wheel jounce or rebound movement of one front wheel does not directly affect the opposite front wheel, the control arm suspension is an independent system. Many rear-wheel-drive cars have control arm front suspension systems. Early front suspension systems had equal-length upper and lower control arms. On

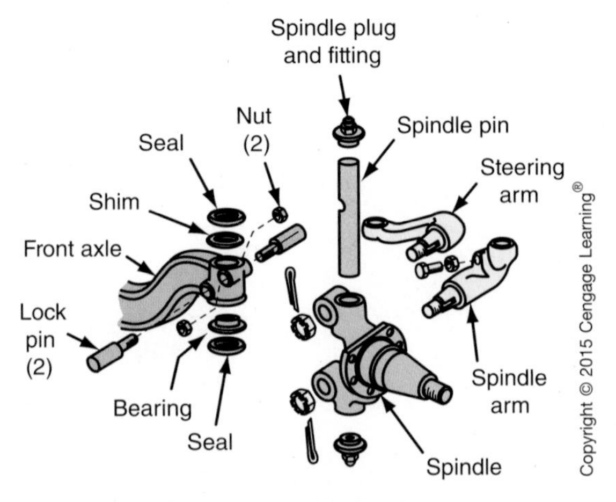

FIGURE 6-16 **I-beam front suspension with kingpins.**

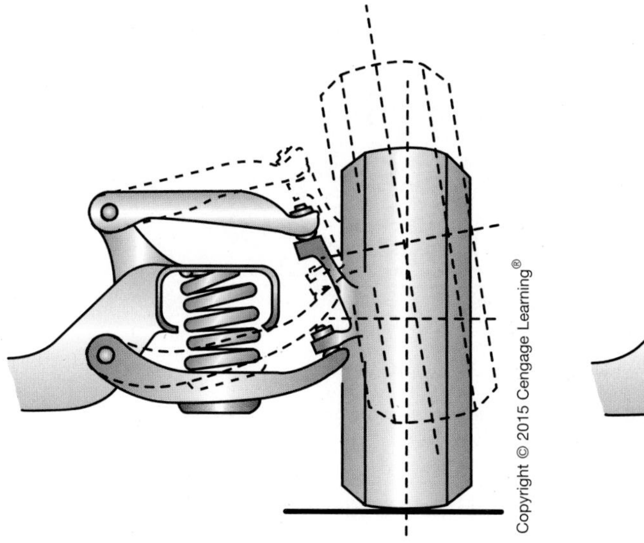

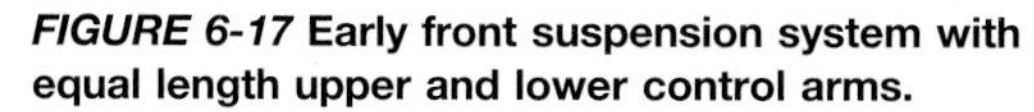

FIGURE 6-17 **Early front suspension system with equal length upper and lower control arms.**

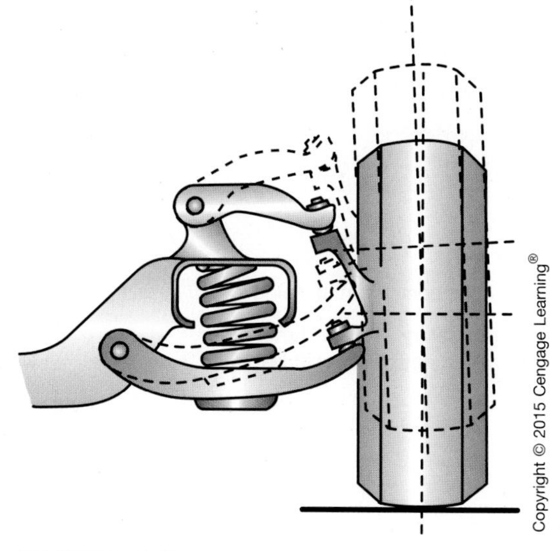

FIGURE 6-18 **Short-and-long arm front suspension system.**

these early suspension systems, the bottom of the tire moved in and out with wheel jounce and rebound travel. This action constantly changed the tire tread width and caused tire scuffing and wear problems (Figure 6-17).

An independent suspension system is one in which wheel jounce or rebound travel of one wheel does not directly affect the movement of the opposite wheel.

In later **short-and-long arm front suspension systems**, the upper control arm is shorter than the lower control arm. During wheel jounce and rebound travel in this suspension system, the upper control arm moves in a shorter arc than the lower control arm. This action moves the top of the tire in and out slightly, but the bottom of the tire remains in a more constant position (Figure 6-18). This short-and-long arm front suspension system provides reduced tire tread wear, improved ride quality, and better directional stability compared to I-beam suspension systems and suspension systems with equal-length upper and lower control arms.

The inner end of the control arm contains large rubber insulating bushings, and the ball joint is attached to the outer end of the control arm that attaches to the steering knuckle (Figure 6-19). A shaft or bolt passes through the eye of the bushing to connect it to the frame of the vehicle. The rubber bushings are used to separate and insulate parts while providing up-and-down pivot points for the suspension. The rubber cushions reduce

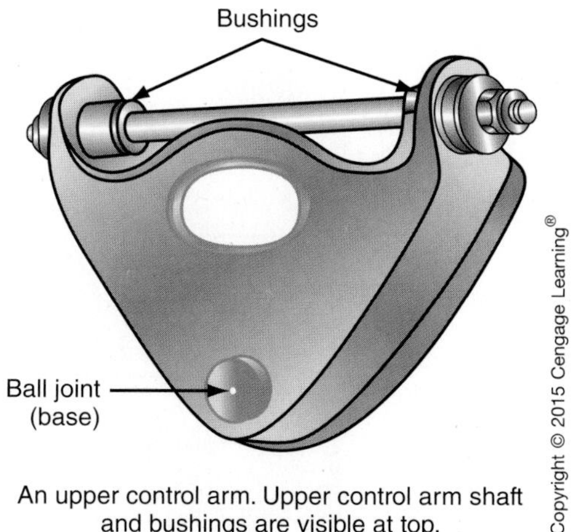

FIGURE 6-19 **The inner end of the control arm contains large rubber insulating bushings, and the ball joint is attached to the outer end of the control arm.**

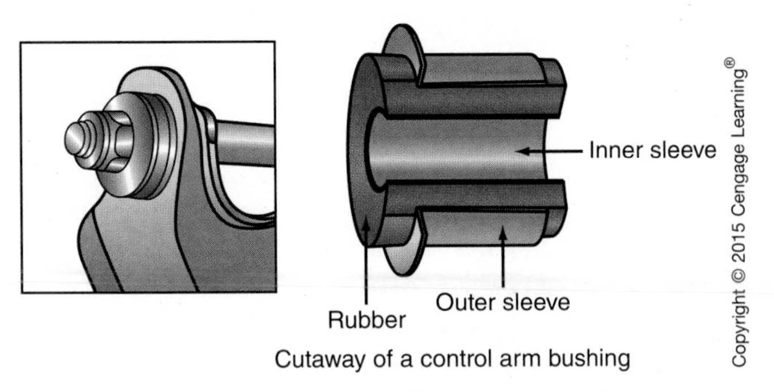

FIGURE 6-20 **The control arm bushings have a metal inner sleeve and outer shell with rubber sandwiched in between.**

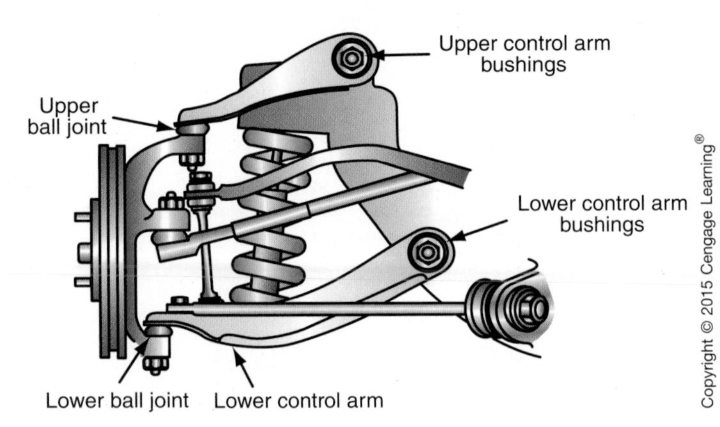

FIGURE 6-21 **Ball joints and complete short-and-long arm front suspension system.**

noise and vibration from being transmitted to the vehicle body. The control arm bushings have a metal inner sleeve and outer shell with rubber sandwiched in between (Figure 6-20). The bushings are usually pressed into the control arms. The inner sleeve is bolted to the frame and remains stationary, whereas the outer shell of the bushing moves with the control arm. Thus the control arm moves as the rubber is twisted and flexed allowing for up and down movement and some shock absorption. As was seen in other areas of the suspension system, bushings are found on shock absorbers, strut rods, leaf springs, and sway bars to name a few.

The lower control arm is bolted to the front crossmember, and the attaching bolts are positioned in the center of the lower control arm bushings (Figure 6-21). The ball joint may be riveted, bolted, pressed, or threaded into the control arm (Figure 6-22). A spring

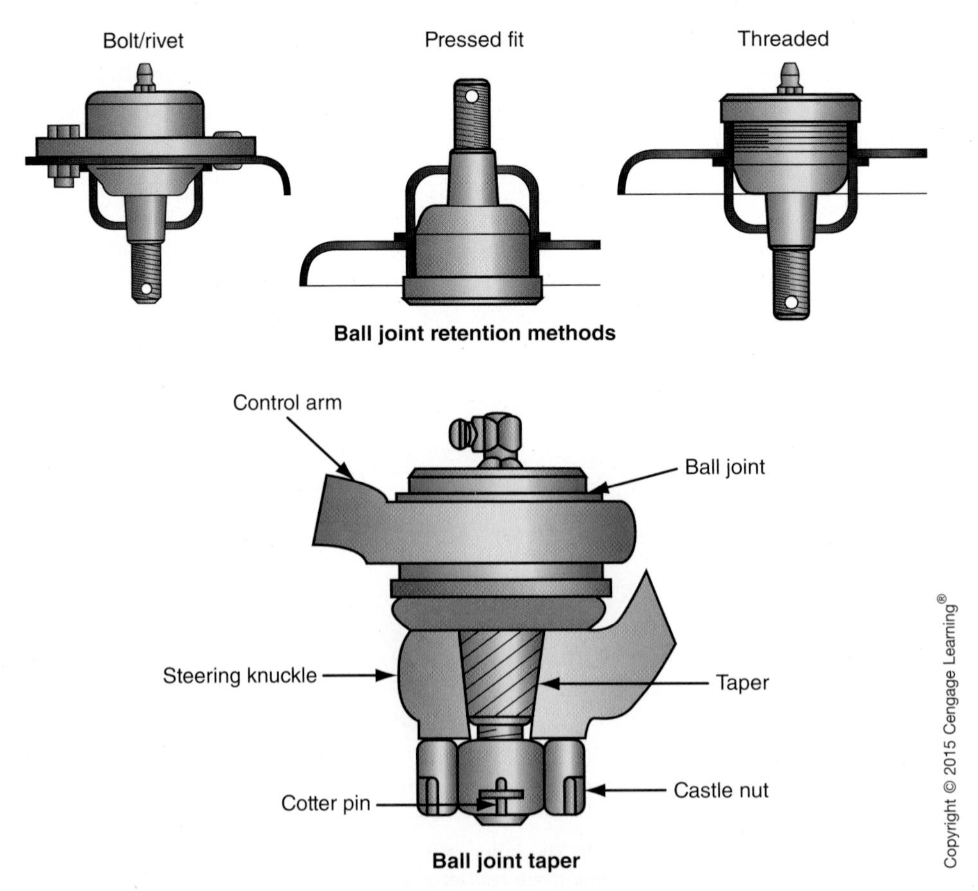

FIGURE 6-22 **The ball joint may be riveted, bolted, pressed, or threaded into the control arm.**

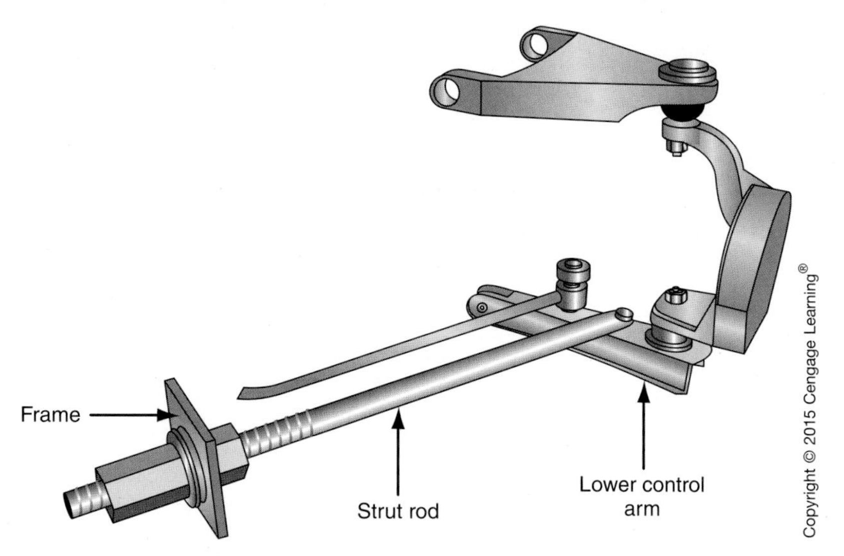

FIGURE 6-23 **A strut rod is used with a control arm that uses only one anchor point at the frame to provide the second anchor point.**

seat is located in the lower control arm. An upper control arm shaft is bolted to the frame, and rubber insulators are located between this shaft and the control arm.

On some suspension systems that utilize a lightweight lower control arm that has a single anchor point at the frame instead of two, a strut rod is used to provide the second anchor point (Figure 6-23).The strut rod provides stability to the control arm. This second anchor point is required to keep the control arm from moving backwards (aft) or forwards (fore) as a result of acceleration and braking forces. The strut rod is bolted to the lower control arm and the frame. At least one end of the strut rod is insulated with a bushing. But the other end may be rigidly attached to either the control arm or the frame or both ends may contain a bushing. On some platforms the strut rod provides adjustment for wheel alignments settings.

Some current vehicles have hydraulic control arm bushings on the front suspension. Hydraulic control arm bushings do a superior job of preventing road shocks and vibrations supplied to the control arms from reaching the body. This action improves ride quality.

Many current vehicles are equipped with aluminum upper and lower control arms and/or steering knuckles to reduce unsprung weight and improve ride quality. Any significant weight reduction also helps to improve fuel economy and vehicle performance and reduce CO_2 omissions. When servicing aluminum control arms, the use of the wrong tools may damage these components. Always use the tools specified by the vehicle manufacturer.

On some short-and-long arm front suspension systems, the coil spring is positioned between the upper control arm and the chassis (Figure 6-24). In these suspension systems, the upper ball joint is compression loaded.

Shop Manual
Chapter 6, page 229

Steering Knuckle

The upper and lower ball joint studs extend through openings in the steering knuckle. Nuts are threaded onto the ball joint studs to retain the ball joints in the knuckle, and the nuts are secured with cotter pins. The wheel hub and bearings are positioned on the steering knuckle extension, and the wheel assembly is bolted to the wheel hub. When the steering wheel is turned, the steering gear and linkage turn the steering knuckle. During this turning action, the steering knuckle pivots on the upper and lower ball joints. The upper and lower control arms must be positioned properly to provide correct tracking and wheelbase between the front and rear wheels. The control arm bushings must be in satisfactory condition to position the control arms properly.

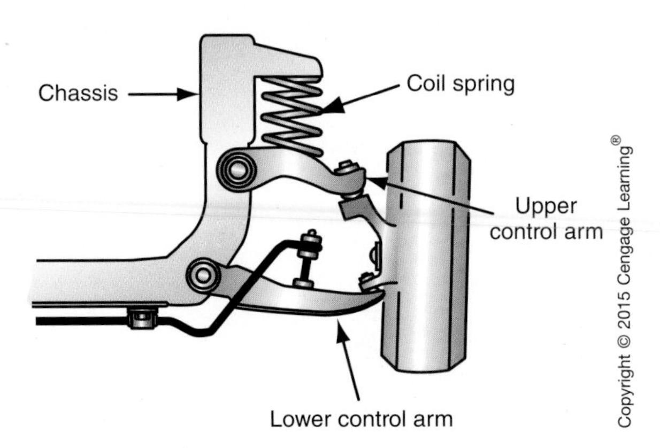

FIGURE 6-24 **Short-and-long arm front suspension with the coil spring between the upper control arm and the chassis.**

Coil Spring and Shock Absorber

The coil spring is positioned between the lower control arm and the spring seat in the frame. A spring seat is located in the lower control arm, and an insulator is positioned between the top of the coil spring and the spring seat in the frame. The shock absorber is mounted in the center of the coil spring, and the lower shock absorber bushing is bolted to the lower control arm. The top of the shock absorber extends through an opening in the frame above the upper spring seat. Washers, grommets, and a nut retain the top of the shock absorber to the frame. Side roll of the front suspension is controlled by a steel stabilizer bar, which is mounted to the lower control arms and the frame with rubber bushings.

On some later model short-and-long arm front suspension systems, the lower end of the coil spring is mounted on a seat attached to the shock absorber. A rubber bushing containing a metal bar is installed in the lower end of the shock absorber. This bushing is bolted to the lower control arm (Figure 6-25 and Figure 6-26). The upper end of the coil spring is seated on an upper shock absorber mount. The rod in the center of the shock

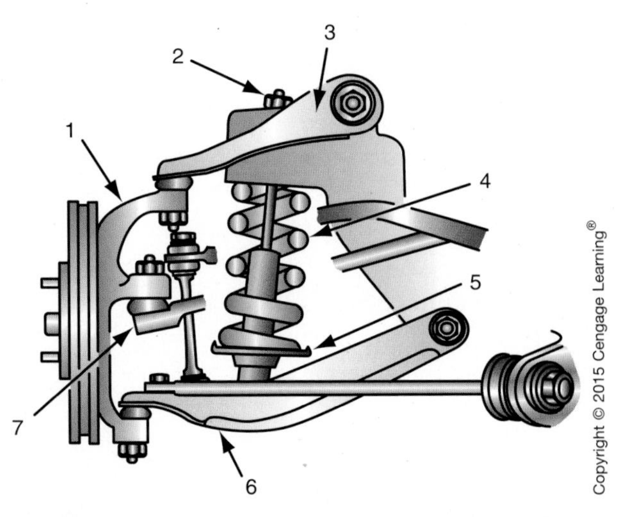

FIGURE 6-25 **Short-and-long arm front suspension system with coil spring mounted on the shock absorber.**

ITEM	DESCRIPTION
1	Steering knuckle
2	Nut, shock absorber upper mount
3	Upper control arm
4	Coil spring
5	Coil spring seat
6	Lower control arm
7	Tie rod end

FIGURE 6-26 **Component identification in short-and-long arm suspension with the coil spring mounted on the shock absorber.**

absorber extends through this mount, and a nut on the threaded end of the shock absorber rod retains the mount on the shock absorber. Bolts in the top of the upper shock absorber mount extend through the upper control arm.

The upper control arm is mounted high in the suspension system, and the upper end of the knuckle has a "goose neck" shape. The lower control arm is made from stamped steel to reduce weight. The rear lower control arm bushing is mounted vertically and carries only fore-and-aft loads. This mounting allows the use of a softer rear bushing in the lower control arm. The horizontal front lower control arm bushing and the lower shock absorber mounting are aligned with the wheel center. This provides a direct path for lateral cornering loads. This design allows the use of a hard front lower control arm bushing.

When servicing this suspension system, a special spring compressing tool must be used to compress all the spring tension before loosening the nut on top of the shock absorber rod (Figure 6-27 and Figure 6-28). After this nut is removed, the compressing tool is operated to gradually release the spring tension, and then the upper mount may be removed. (Refer to Chapter 5 in the Shop Manual for strut and coil spring service.)

In the short-and-long arm front suspension system on the Trailblazer SUV, the coil spring is mounted over the shock absorber; the coil spring tension is applied against the upper mount and the lower spring seat on the shock absorber (Figure 6-29). A special yoke attaches to the lower end of the shock absorber, and the lower end of this yoke pivots on a bolt in the lower control arm (Figure 6-30). This type of suspension system provides a more

CAUTION:
Never loosen the nut on top of the shock absorber rod until a coil spring compressing tool is used to compress all the spring tension. Failure to follow this procedure may cause the spring tension to suddenly release, resulting in personal injury.

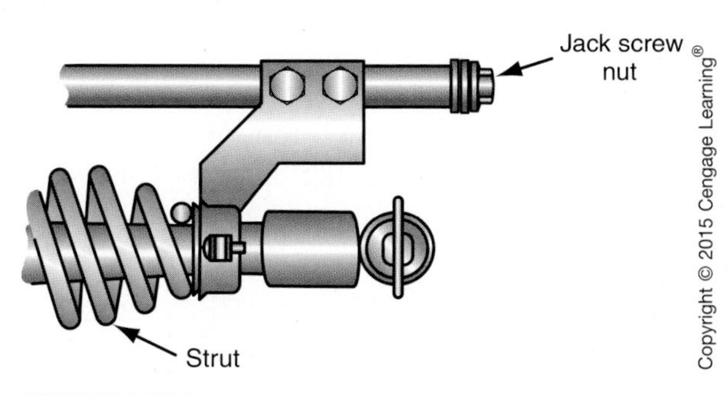

FIGURE 6-27 **Lower end of spring compressing tool.**

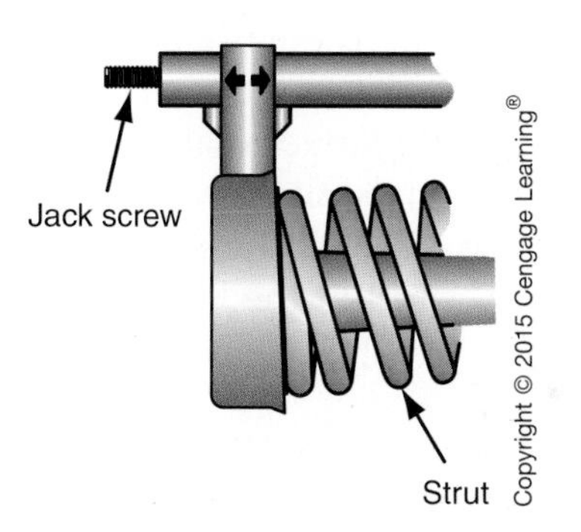

FIGURE 6-28 **Upper end of spring compressing tool.**

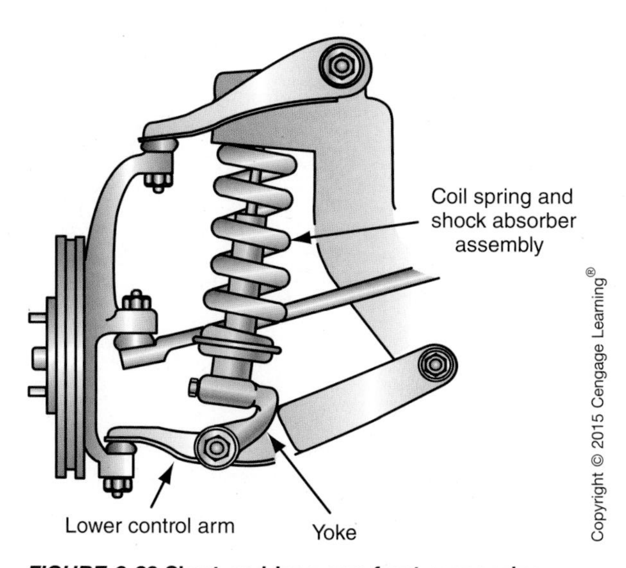

FIGURE 6-29 **Short-and-long arm front suspension with yoke-type shock absorber mounting.**

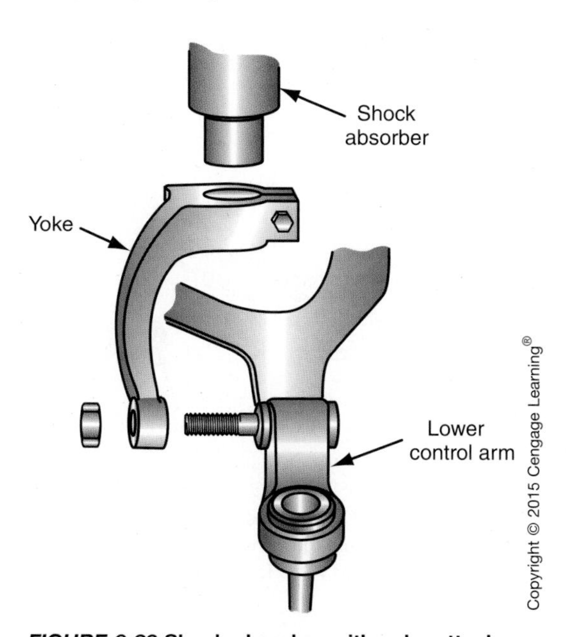

FIGURE 6-30 **Shock absorber with yoke attachment to lower control arm.**

positive shock absorber and spring mounting because the rubber bushing-type mounting between the shock absorber and the lower control arm is not required. The steering knuckle pivots on the upper and lower ball joints in the control arms. Therefore, the shock absorber and coil spring do not have to turn when the knuckle turns, which allow the use of a more rigid upper spring mount. This suspension design provides improved vehicle handling and ride characteristics.

Improved Designs in Short-and-Long Arm Front Suspension Systems

Some vehicles now have a reverse-L front suspension system. In these suspension systems, the rear attachment point on the lower control arm extends rearward and attaches to the engine cradle at a point farther toward the rear of the vehicle (Figure 6-31). The lower control arm has an L-shaped design. A firm bushing is installed at the location where the shorter forward leg of the lower control arm attaches to the cradle to control lateral control arm movement and quicken steering response. A more compliant bushing is installed between the longer rear leg and the rear attachment point to absorb the impact of longitudinal forces caused by road irregularities. This type of suspension separates the control of suspension loads into fore-and-aft control and side-to-side control to improve ride and steering quality.

Other modern short-and-long arm front suspension systems have high upper control arms that place the upper ball joints above the tires to provide suspension articulation that helps to keep the tire perpendicular to the road surface while cornering (Figure 6-32). This type of front suspension system has a lateral link and a tension strut to position the lower end of the knuckle and the lower ball joint rather than a lower control arm. The lateral link extends slightly forward from the lower ball joint to an attachment point on the chassis. Bushings are mounted in both ends of the lateral link. The tension strut extends rearward from the lower ball joint to the attachment location on the chassis. This type of

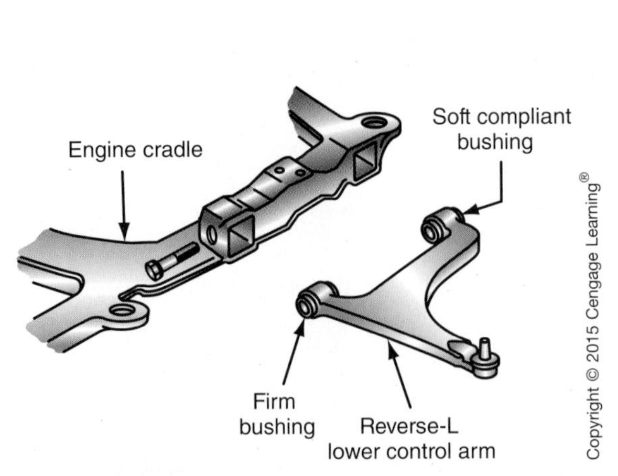

FIGURE 6-31 **Reverse-L front suspension system with firm, front lower control arm bushing and more compliant rear, lower control arm bushing.**

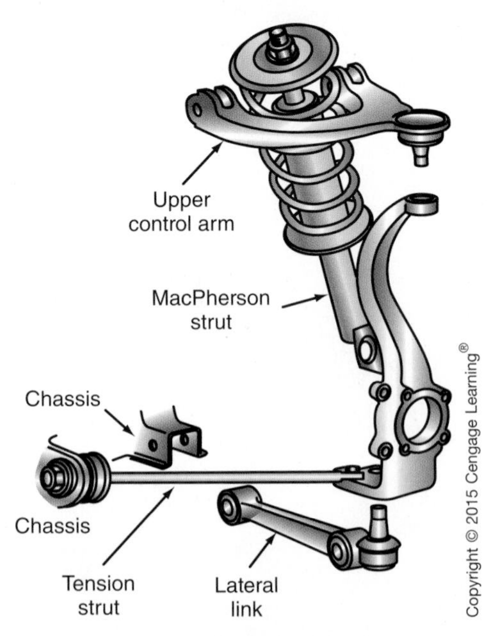

FIGURE 6-32 **Front suspension system with lateral link and tension strut in place of lower control arm.**

front suspension also separates the control of suspension loads to improve ride quality and steering control. This type of front suspension may be called a four-link suspension system.

Sway Bar

The sway bar prevents excessive body roll or lean by resisting centrifugal forces. During a turn the inside side of the suspension rises and the outside side of the suspension drops. The sway bar (stabilizer bar) is made of spring steel much like a torsion bar and is designed to absorb and transfer energy as it is twisted during suspension jounce and rebound. As the bar twists during a cornering maneuver the inside side of the suspension is forced down and the outside side of the suspension is forced up as energy is transmitted from one side to the other. In this manner, body roll is reduced and the vehicle maintains a flatter profile with improved vehicle control and maneuverability during turns. If a sway bar were to fail or if one was not present, the inside tire of the vehicle would have less weight on it during a turn as the weight was transferred to the outside of the vehicle. This would reduce the ability of the driver to control the vehicle during the turn, thereby reducing the stability of the vehicle. A sway bar may be found on both the front and rear suspension system of many vehicles. The front sway bar on a short arm long arm suspension system connects both lower control arms to one another by sway bar links (Figure 6-33). Generally, each sway bar link has four bushings and a link pin and is a frequent area of wear and failure especially on higher mileage vehicles. The sway bar is connected to the frame or unibody of the vehicle with one piece sway

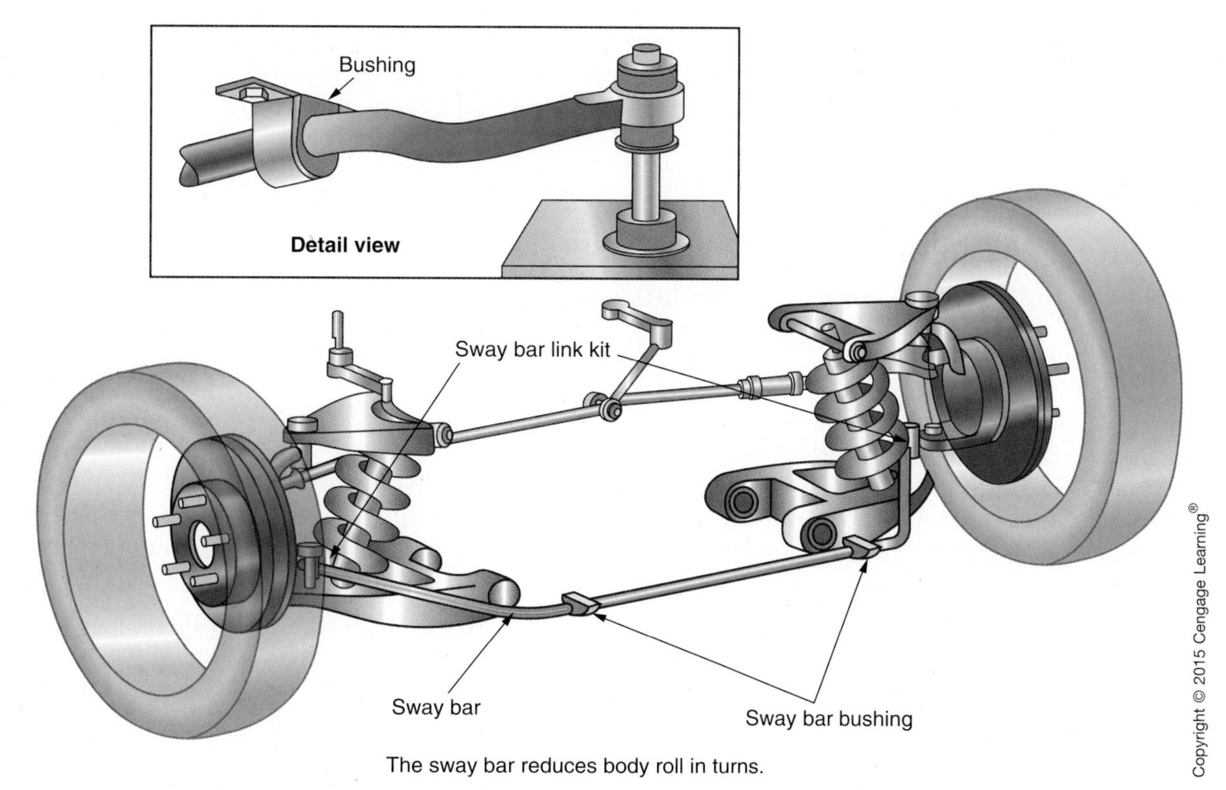

FIGURE 6-33 **The sway bar (stabilizer bar) reduces body roll and is attached to the vehicle with mounting bushings and link pin bushings.**

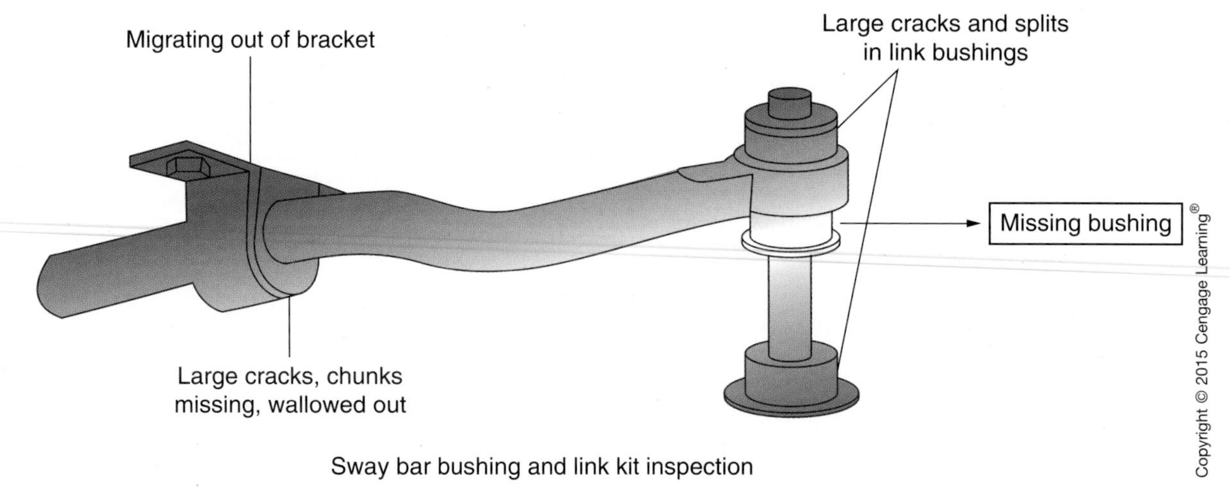

FIGURE 6-34 **The sway bar (stabilizer bar) mounting bushings and link pin bushings are common areas of wear and damage on higher mileage vehicles.**

bar mounting bushings and brackets. Some vehicles may exhibit a squeak or squawking noise over bumps especially on cold mornings as a result of faulty sway bar mounting bushings (Figure 6-34). Shock absorbers and struts also help in reducing body roll so it is important to inspect all suspension components when excessive body role is suspected.

AUTHOR'S NOTE: Driving over speed bumps is very useful to diagnose the presence of suspension noises. It is also helpful to drive the vehicle first thing in the morning when suspension components are cold and the suspension has settled from sitting for a long period especially for intermittent noises.

MACPHERSON STRUT FRONT SUSPENSION SYSTEM DESIGN

Lower Control Arms and Support

When smaller front-wheel-drive cars became popular, most of these cars had **MacPherson strut front suspension systems**. In these suspension systems, the lower end of the strut is bolted to the top of the steering knuckle, and the lower end of the knuckle is attached to the ball joint in the lower control arm (Figure 6-35). An upper strut mount connects the top of the strut to the chassis. An upper control arm is not required in this type of suspension system, because the strut supports the top of the steering knuckle. Since the upper control arm is not required in these suspension systems, they are more compact and therefore very suitable for smaller cars.

A BIT OF HISTORY

Rear-wheel-drive cars usually have short-and-long arm or torsion bar front suspension systems. Since space and weight are important factors on today's smaller, more efficient front-wheel-drive cars, the lighter, more compact MacPherson strut front suspension is used on most of these cars.

On some MacPherson strut front suspension systems, a steel support is positioned longitudinally on each side of the front suspension. These supports are bolted to the unitized body. The inner ends of the lower control arms contain large insulating bushings with a bolt opening in the bushing center. The control arm retaining bolts extend through the center of these bushings and openings in the support (Figure 6-36).

Road irregularities cause the tire and wheel to move up and down vertically, and the lower control arm bushings pivot on the mounting bolts during this movement. When the vehicle is driven over road irregularities, vibration and noise are applied to the tire and

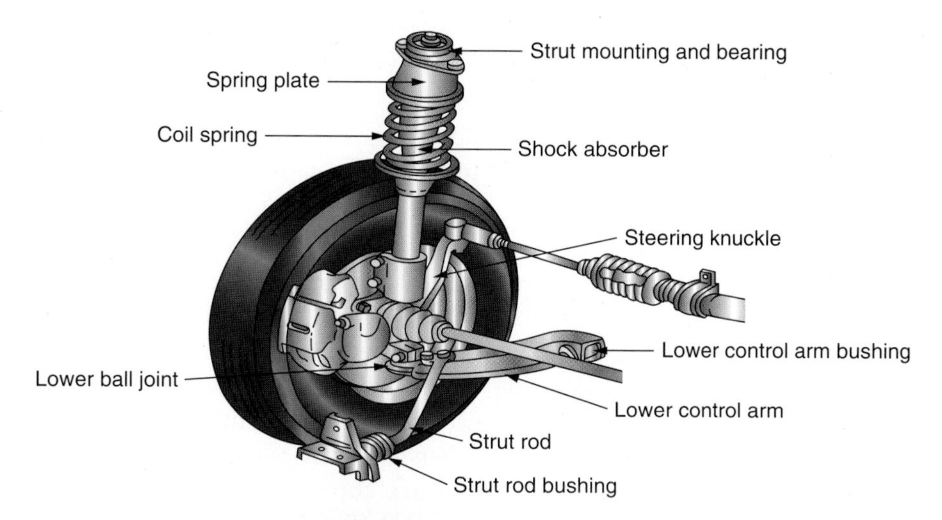

MacPherson strut suspension

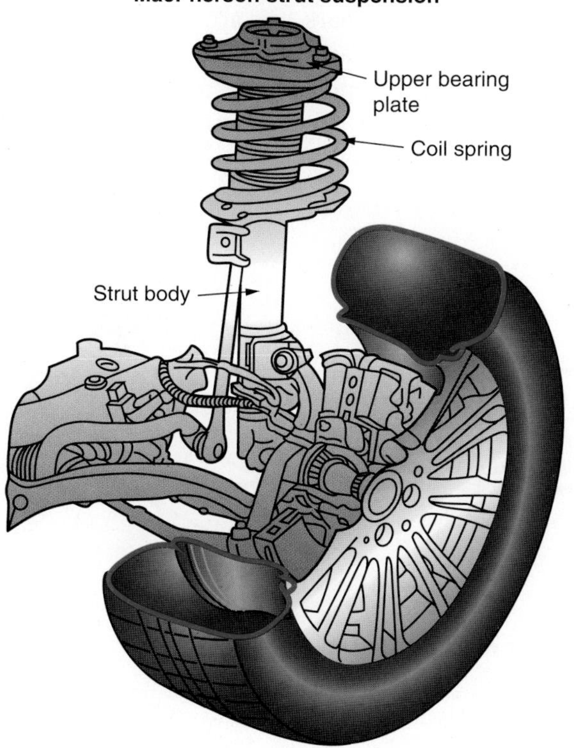

FIGURE 6-35 **The MacPherson strut suspension system replaces the upper control arm and combines the shock absorber, spring, and upper pivot point into one assembly.**

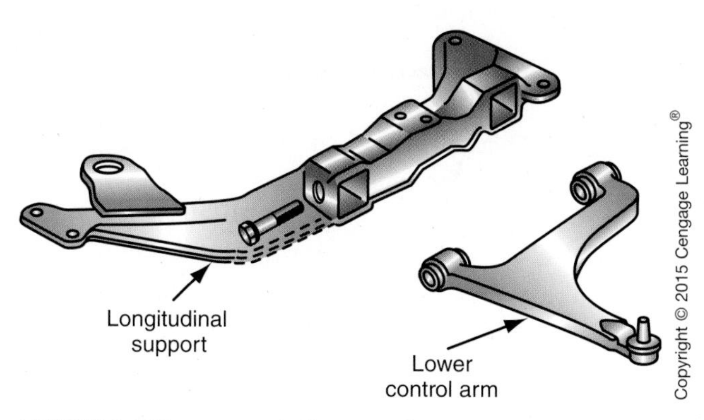

FIGURE 6-36 **Lower control arm and support.**

Shop Manual
Chapter 6, page 232

wheel. The control arm bushings help prevent the transfer of this noise and vibration to the support, the unitized body, and the passenger compartment. Proper location of the support and lower control arm is important to provide correct vehicle tracking. The supports also carry the engine and transaxle weight. Large rubber mounts are positioned between the supports and the engine and transaxle. These mounts absorb engine vibration.

Stabilizer Bar

A stabilizer bar may be called a sway bar.

The **stabilizer bar** is attached to the chassis and interconnects the lower control arms. Rubber insulating bushings are used at all stabilizer bar mounting positions. Some stabilizer bars are attached to the underside of the front crossmember, and the outer ends of the stabilizer bar are connected to the lower side of the front control arms (Figure 6-37). This type of torsion bar may be called "direct contact" because the outer ends of the bar are in direct contact with the lower control arms. Other stabilizer bars are attached to the upper side of the front longitudinal supports, and links are connected between the outer ends of the bar and the upper side of the front control arms (Figure 6-38). On some MacPherson strut front suspension systems, the stabilizer bars are attached to the upper side of the front subframe and the outer ends of the bar are linked to the front struts (Figure 6-39). Stabilizer bars with links between the outer ends of the bar and the suspension components may be called "indirect contact." Large rubber bushings with steel mounting caps

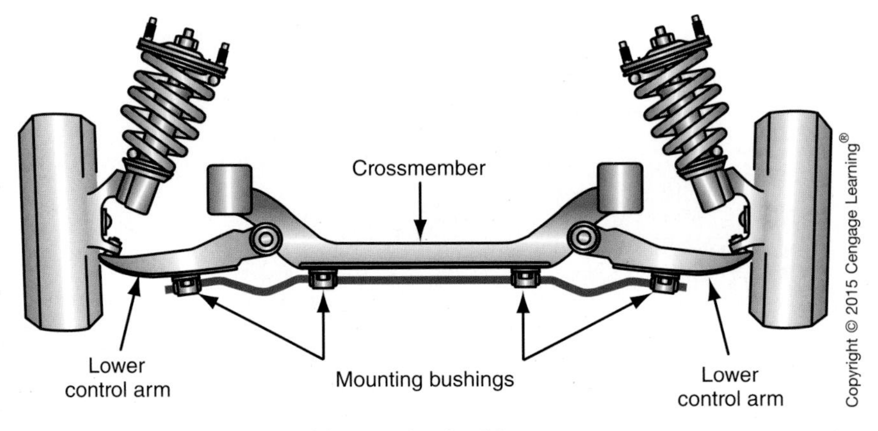

FIGURE 6-37 **Stabilizer bar with mounting bushings.**

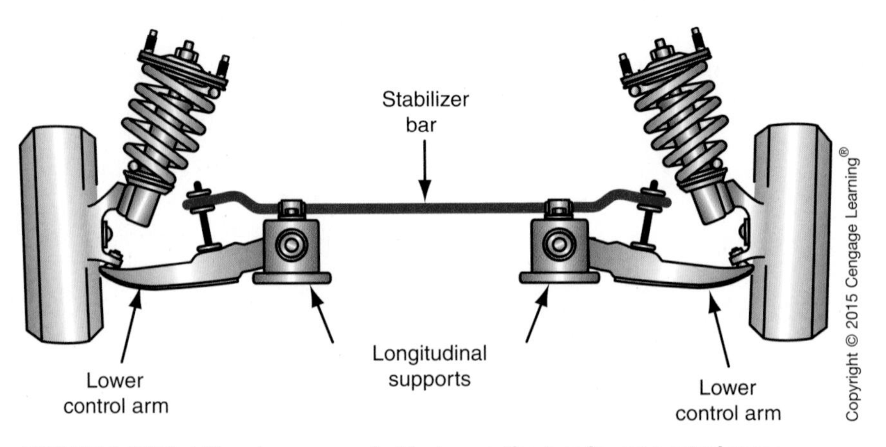

FIGURE 6-38 **Stabilizer bar connected between the two lower control arms.**

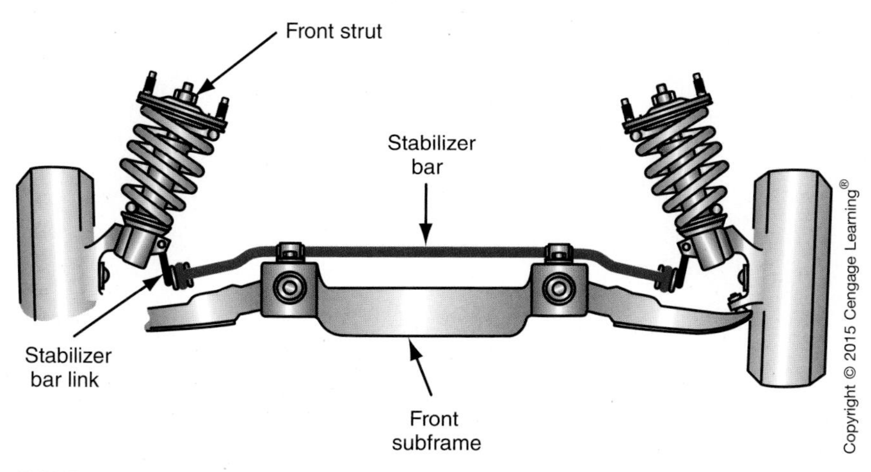

FIGURE 6-39 **Stabilizer bar connected to the front struts.**

attach the stabilizer bar to the chassis. The linkages at the outer ends of the stabilizer bar are connected to the control arms or struts with retaining bolts, small rubber bushings, steel washers, and sleeves.

The outer ends of the stabilizer bar move up and down with the control arm movement. When jounce or rebound wheel movement occurs, the stabilizer bar transmits part of this movement to the opposite front wheel to reduce and stabilize body roll. The rubber stabilizer bar mounting and linkage bushings prevent noise.

Some current vehicles have front and rear stabilizer bars with flat areas on the bars in the bushing contact areas. The bushings used with these bars have oval openings in the center of the bushing to match the flat areas on the bars. This stabilizer bar design reduces body roll while cornering or driving on irregular road surfaces, because more force is required to twist the stabilizer bar. Some vehicles are now equipped with aluminum stabilizer bars to reduce unsprung weight. Hollow stabilizer bars are used on some current vehicles for weight reduction.

An aftermarket electronically controlled stabilizer bar is available for off-road vehicle operation (Figure 6-40). In this type of stabilizer bar system the driver may press a switch in the instrument panel, and the electronic control disconnects one side of the stabilizer bar from the opposite side, thus making this component ineffective. When the stabilizer bar is disconnected electronically, full vertical wheel travel is allowed on rough terrain. The driver can press the control switch again to return the stabilizer bar to normal operation for stabilized control of body roll during on-road operation.

Lower Ball Joint

The lower ball joint is attached to the outer end of the lower control arm. Methods used to attach the ball joint to the control arm include bolting, riveting, pressing, and threading. A threaded stud extends from the top of the lower ball joint. This stud fits snugly into a hole in the bottom of the steering knuckle. When the ball joint stud is installed in the steering knuckle opening, a nut and cotter pin retain the ball joint (Figure 6-41).

Steering Knuckle and Bearing Assembly

The front wheel bearing assembly is bolted to the outer end of the steering knuckle, and the brake rotor and wheel rim are retained on the studs in the wheel bearing assembly.

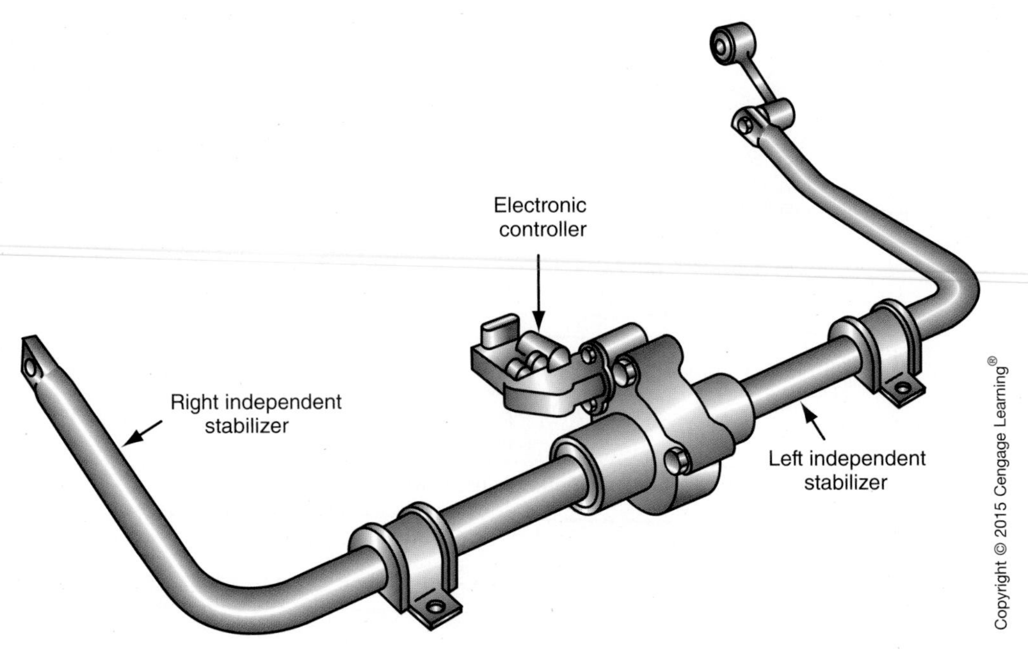

FIGURE 6-40 **Electronically controlled stabilizer bar.**

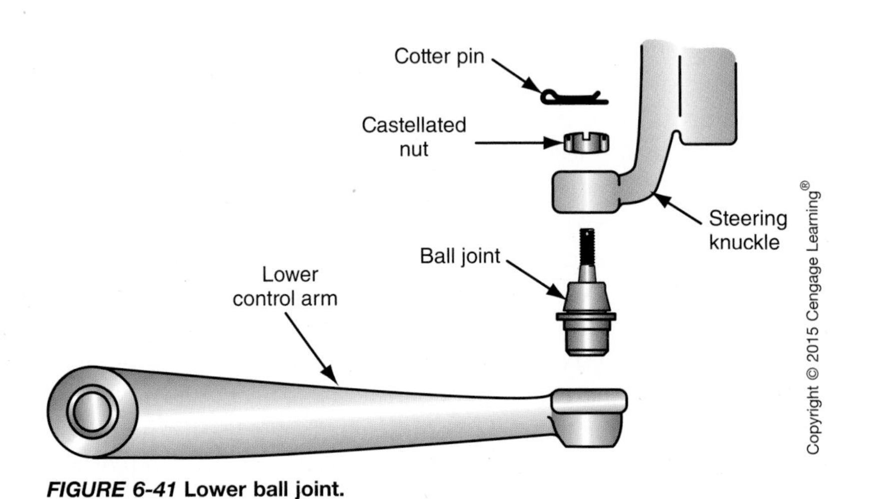

FIGURE 6-41 **Lower ball joint.**

This front wheel bearing assembly is a complete, non-serviceable, sealed unit. The front drive shaft is splined into the center of the wheel bearing hub. Thus, drive axle torque is applied to the front wheel. A tie rod end connects the steering linkage from the steering gear to the steering knuckle. The top end of the steering knuckle is bolted to the lower end of the strut (Figure 6-42). Many steering knuckles are manufactured from metal, but some current vehicles have aluminum steering knuckles to reduce vehicle weight and unsprung weight. Reducing unsprung weight improves ride quality and vehicle performance and reduces fuel consumption and CO_2 emissions.

Strut and Coil Spring Assembly

The strut is the shock absorber in the front suspension. The lower spring seat is attached near the center area of the strut. An insulator is located between the lower spring seat and the bottom of the coil spring. An upper strut mount is retained on top of the strut with a

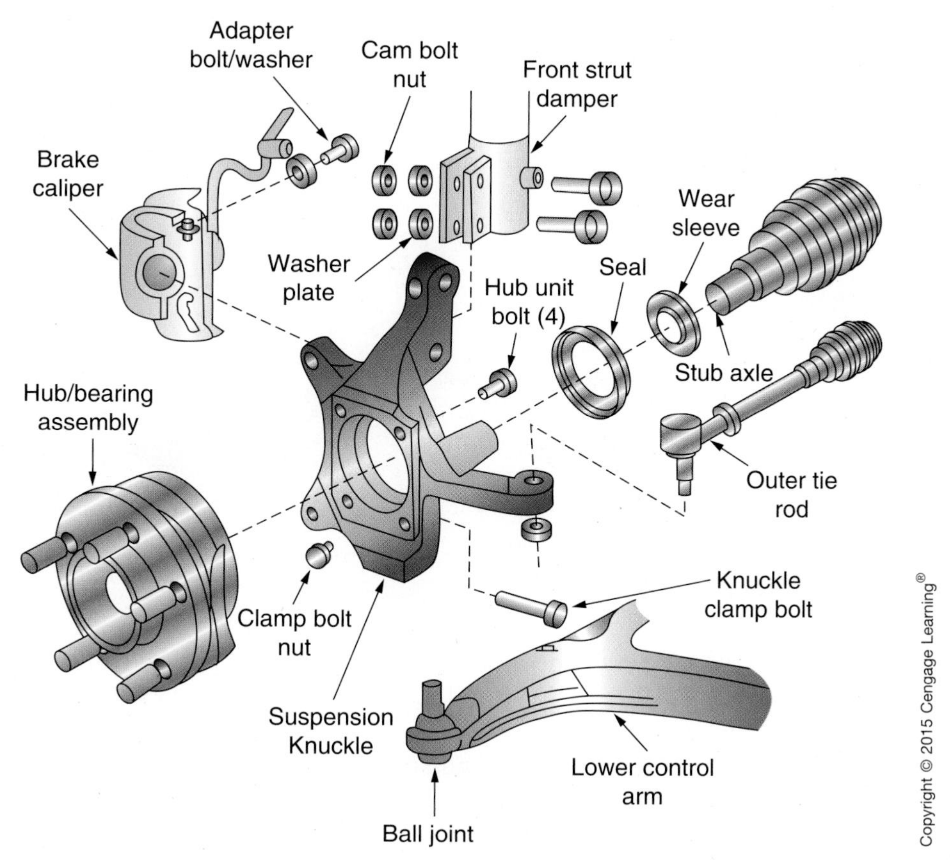

FIGURE 6-42 **The front wheel bearing assembly is bolted to the outer end of the steering knuckle and the strut assembly is attached to the upper portion of the steering knuckle where the upper ball joint would have been located on a short-arm-long-arm suspension system.**

nut threaded onto the upper end of the strut rod. The upper strut mount contains a bearing and upper spring seat, and an insulator is positioned between the top of the coil spring and the seat. The upper and lower insulators help prevent the transfer of noise and vibration from the spring to the strut and body.

A bumper is located on the upper end of the strut rod. This bumper reduces harshness while driving on severe road irregularities. During upward wheel movement, the bumper strikes the upper spring seat before the coils in the spring hit each other. Therefore, this bumper reduces harshness when the wheel and suspension move fully upward. The spring tension is applied against the upper and lower spring seats and insulators. However, the nut on top of the upper mount holds the spring in the compressed position between the upper and lower spring seats. When the steering wheel is turned, the steering linkage turns the steering knuckles to the right or left. During this front wheel turning action, the strut-and-spring assembly pivots on the lower ball joint and the upper strut mount bearing.

All the suspension-to-chassis mounting devices such as the lower control arm bushings and the upper strut mount must be positioned properly and be in satisfactory condition to provide correct vehicle tracking and the same wheelbase on both sides of the vehicle.

The purpose of the main components in a MacPherson strut front suspension system may be summarized as follows:

1. Lower control arm—controls lateral (side-to-side) movement of each front wheel.
2. Stabilizer bar—reduces body roll when a front wheel strikes a road irregularity.
3. Coil springs—allow proper setting of suspension ride heights and control suspension travel during driving maneuvers.

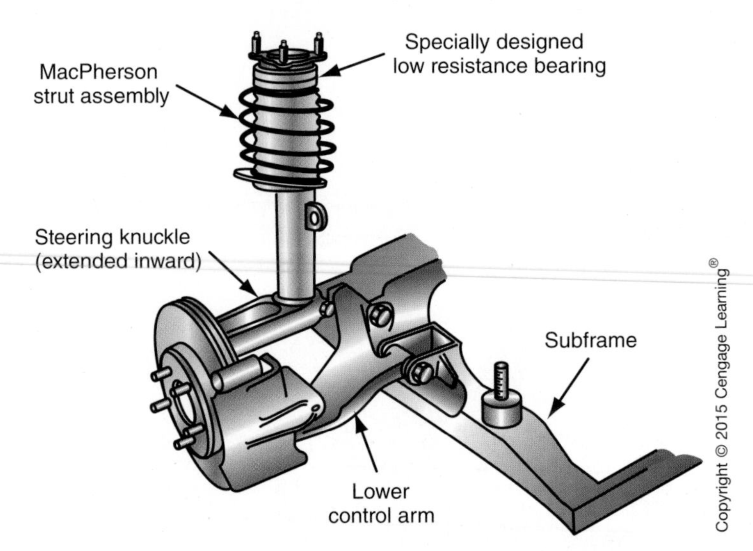

FIGURE 6-43 **MacPherson strut front suspension system on Jaguar X-type car.**

4. Shock absorber struts—provide necessary suspension damping and limit downward wheel movement with an internal rebound stop and upward wheel movement with an external jounce bouncer.
5. Strut upper mount—insulates the strut and spring from the body and provides a bearing pivot for the strut-and-spring assembly.
6. Ball joint—connects the outer end of the lower control arm to the steering knuckle and acts as a pivot for the strut, spring, and knuckle assembly.

MacPherson strut front suspension systems are all similar in design, but some vehicle manufacturers provide unique differences in their suspension systems. For example, the MacPherson strut front suspension system on the new Jaguar X-type car has two significant differences: an arm extends several inches inward from the top of the steering knuckle and the lower end of the strut is attached to the inner end of this arm (Figure 6-43). The upper strut mount contains a specially designed bearing that allows the strut-and-spring assembly to rotate freely regardless of the forces supplied to the front suspension. This upper strut mount design reduces friction within the struts and provides very smooth steering action.

Modified MacPherson Strut Suspension

A modified MacPherson strut front suspension is used on some older vehicles. This type of suspension has MacPherson struts with coil springs positioned between the lower control arms and the frame (Figure 6-44). The struts in these systems may be gas-filled or oil-filled.

The **multilink front suspension** may be referred to as a double wishbone suspension because of the link design. It has upper and lower links, and a third link connects the upper link to the top of the knuckle through a bearing.

High-Performance Front Suspension Systems

Multilink Front Suspension System

High-performance suspension systems are usually installed on sports cars. In these cars, the driver expects improved steering quality, especially when driving and cornering at higher speeds.

In a **multilink front suspension**, a short upper link is attached to the chassis with a bracket, and the outer end of this upper link is connected to a third link. Large rubber insulating bushings are mounted in each end of the upper link. The lower end of the third link is connected through a heavy pivot bearing to the steering knuckle (Figure 6-45). The lower link is similar to the conventional lower control arm. A rubber insulating bushing connects the inner end of the lower link to the front crossmember, and a ball joint is connected from

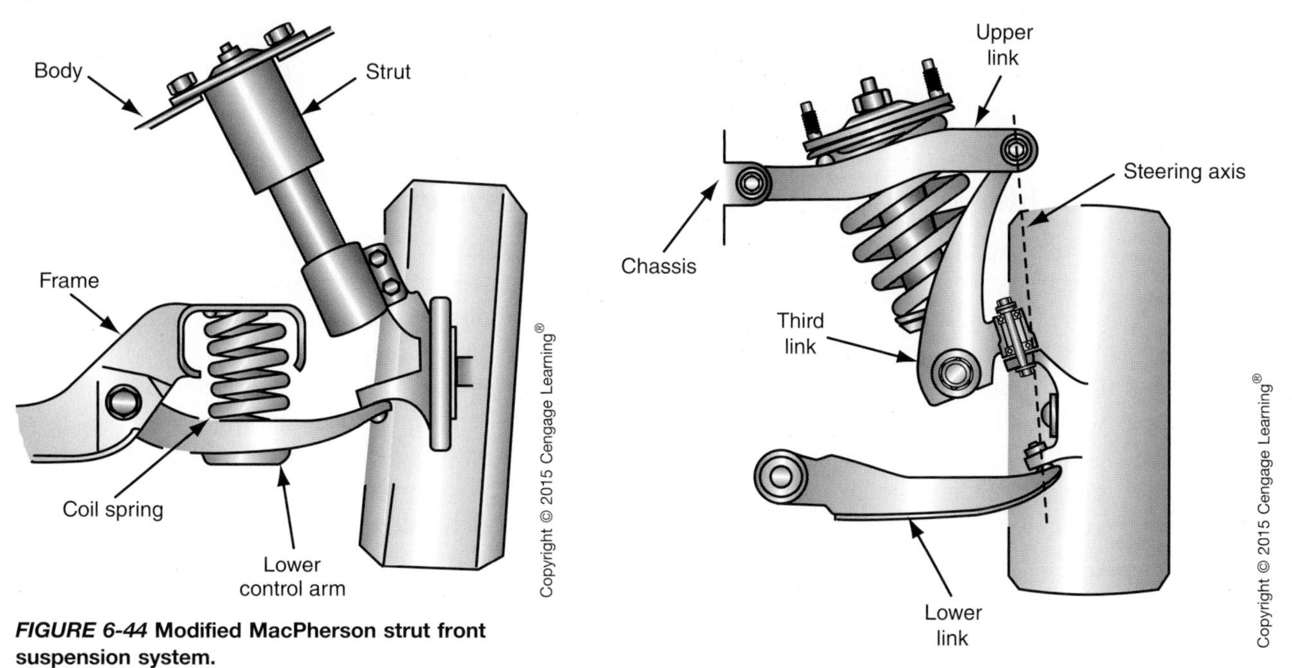

FIGURE 6-44 **Modified MacPherson strut front suspension system.**

FIGURE 6-45 **Multilink front suspension system.**

the outer end of the lower link to the steering knuckle. In the multilink suspension system, the ball joint axis extends through the lower ball joint and upper pivot bearing, but the ball joint axis is independent from the upper and third links. The extra links in a multilink suspension system maintain precise wheel position during cornering to provide excellent directional stability and steering control while minimizing tire wear.

The shock absorbers are connected from the lower end of the third link to the fender reinforcement. A coil spring seat is attached to the lower end of the shock absorber, and the upper spring seat is located on the upper shock absorber mounting insulator (Figure 6-46). Since the steering knuckle pivots on the lower ball joint and the upper pivot bearing, the coil spring and

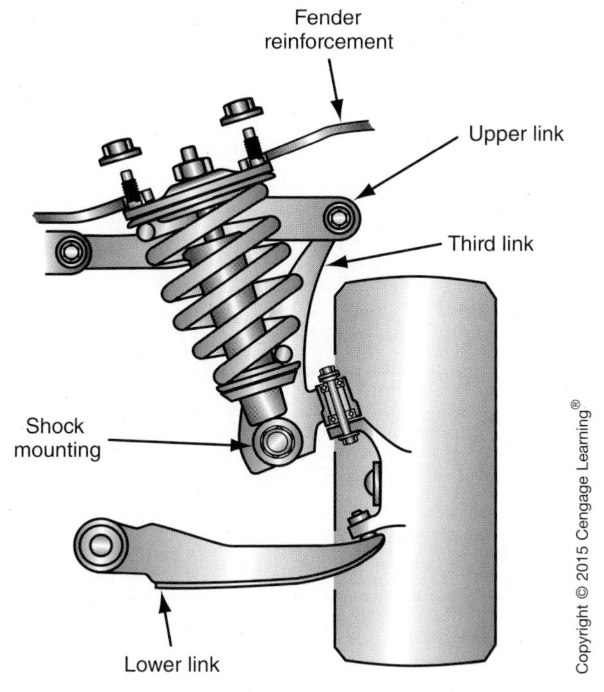

FIGURE 6-46 **Complete multilink suspension system.**

shock absorber do not rotate with the knuckle as they do in a MacPherson strut suspension. Tension or strut rods are connected from the lower links to tension rod brackets attached to the chassis. A stabilizer bar is mounted on rubber insulating bushings in the tension rod brackets, and the outer ends of this bar are attached to the third link.

Double Wishbone Front Suspension System

Double wishbone suspension systems provide increased suspension rigidity and maintain precise wheel position under all driving conditions to supply improved directional stability and steering control. In the double wishbone front suspension system, the upper and lower control arms are manufactured from lightweight, high-strength aluminum alloys designed for maximum strength and rigidity. These lighter control arms decrease the unsprung weight of the vehicle, which improves traction and ride quality. Since the upper and lower control arms have a wishbone shape, the term *double wishbone* is used for this type of suspension. Suspension rigidity is also increased by positioning the ball joints and steering knuckle inside the wheel profile (Figure 6-47). On each side of the car, the front suspension is attached to the chassis by a cast aluminum subframe. This design also reduces vehicle weight. The double wishbones are attached to the chassis at the most efficient locations to maintain precise wheel position and provide improved ride quality.

The front ends of the upper and lower control arms are attached to a compliance pivot assembly. Bushings are mounted in the front ends of the upper and lower control arms, and these bushings are bolted to the compliance pivot (Figure 6-48). When one of the front wheels is subjected to rearward force by hard braking or a road irregularity, the coil spring is compressed and the ride height is lowered. This rearward force on the front wheel twists the compliance pivot, allowing both control arms to pivot slightly. Under this condition, the upper and lower control arm movement allows the front wheel to move rearward a small amount, and this wheel movement absorbs energy to significantly improve ride quality (Figure 6-49). During this upper and lower control arm movement, track width

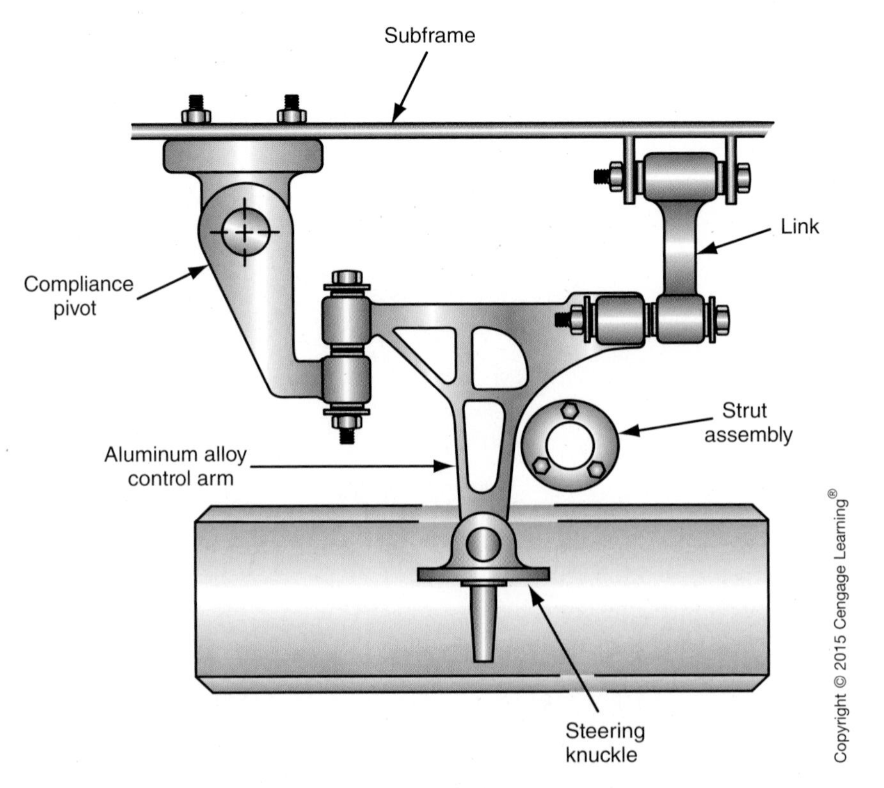

FIGURE 6-47 **Double wishbone front suspension.**

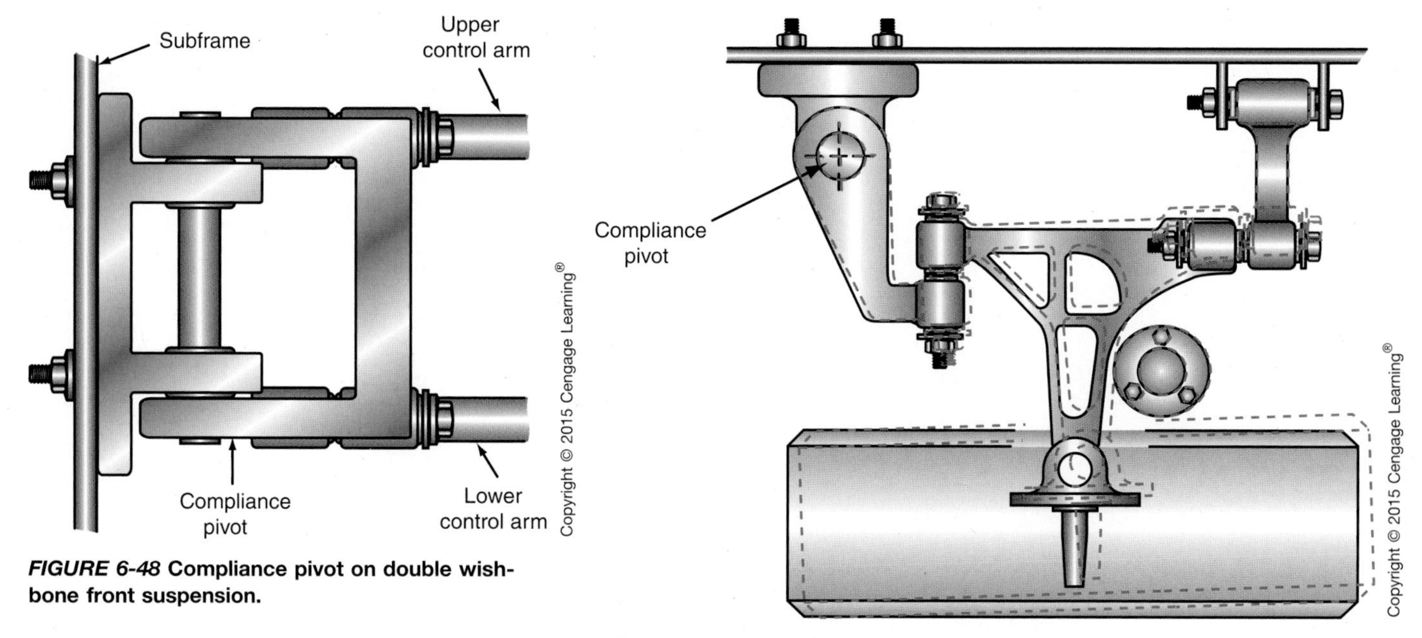

FIGURE 6-48 **Compliance pivot on double wishbone front suspension.**

FIGURE 6-49 **Compliance pivot action.**

and wheel geometry changes are minimal and do not affect steering control. While cornering, the compliance pivot does not move, and lateral suspension stiffness is maintained to supply excellent steering control.

Multilink Front Suspension with Compression and Lateral Lower Arms

Some lateral multilink front suspensions have compression and lateral lower arms. The lateral arm prevents front wheel movement, and the compression arm prevents fore-and-aft front wheel movement. A rubber insulating bushing in the inner end of the lateral arm is bolted to the chassis. A second rubber insulating bushing near the outer end of the lateral arm is bolted to the damper fork. The upper end of the damper fork is bolted to the front strut (Figure 6-50 and Figure 6-51). A ball joint in the outer end of the lateral arm is bolted

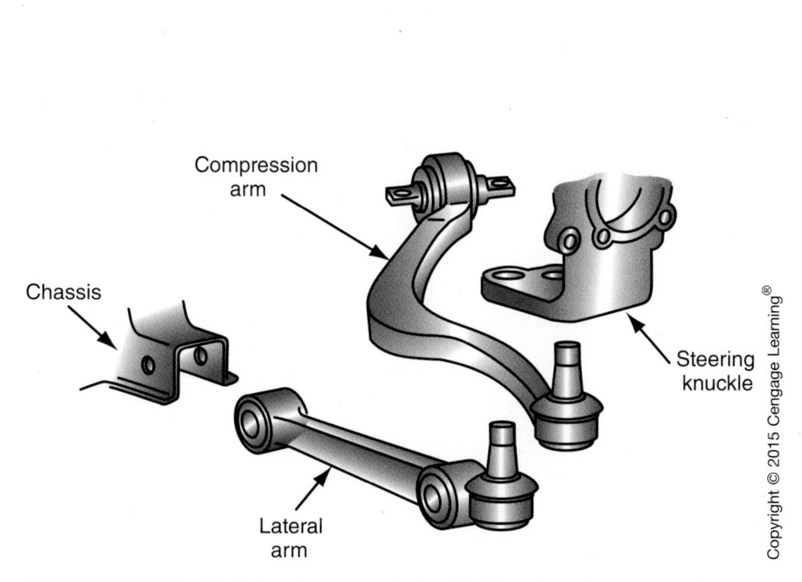

FIGURE 6-50 **Multilink front suspension with lateral and compression lower arms.**

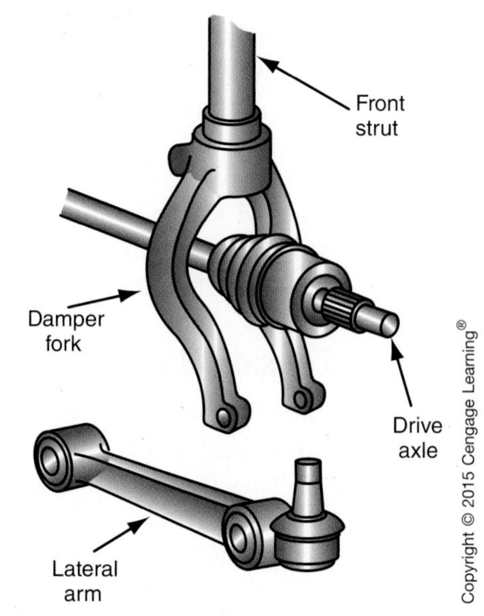

FIGURE 6-51 **Multilink front suspension damper fork.**

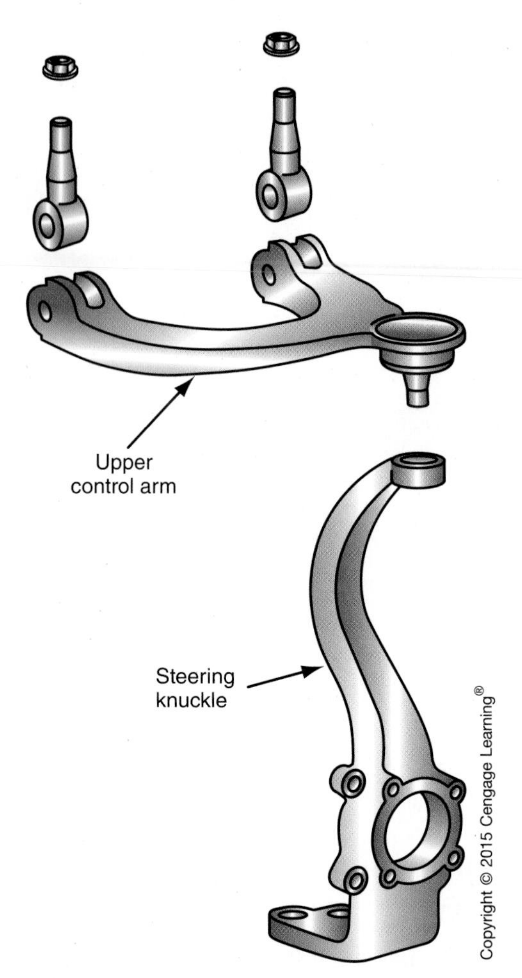

FIGURE 6-52 **Multilink front suspension upper control arm.**

into the steering knuckle. A rubber insulating bushing in the inner end of the compression arm is bolted to the chassis, and a ball joint in the outer end of this arm is bolted into the steering knuckle.

The upper control arm is mounted higher so it is above the front tire. The higher upper control arm and the lateral and compression lower arms provide excellent suspension stability and steering control, especially during high-speed cornering or when driving on irregular road surfaces. A ball joint in the outer end of the upper control arm is attached to the top of the knuckle, and two shafts in the inner end of this arm are bolted into the strut tower (Figure 6-52). There are no provisions for camber or caster adjustments on this multilink front suspension.

Torsion Bar Suspension

Shop Manual
Chapter 6, page 235

Some light-duty trucks and SUVs have torsion bar front suspension systems. These suspension systems have longitudinally mounted torsion bars. The front and rear ends of the torsion bars have a hex shape. The front end of each torsion bar is anchored in the upper or lower control arm and the rear end of the torsion bars are anchored to a chassis crossmember (Figure 6-53). Since the twisting of the torsion bars supports the vehicle weight, the torsion bars replace the coil springs.

Torsion bar front suspension systems are often used on four-wheel-drive trucks, because the absence of coil springs allows more space for the front drive axles. On some

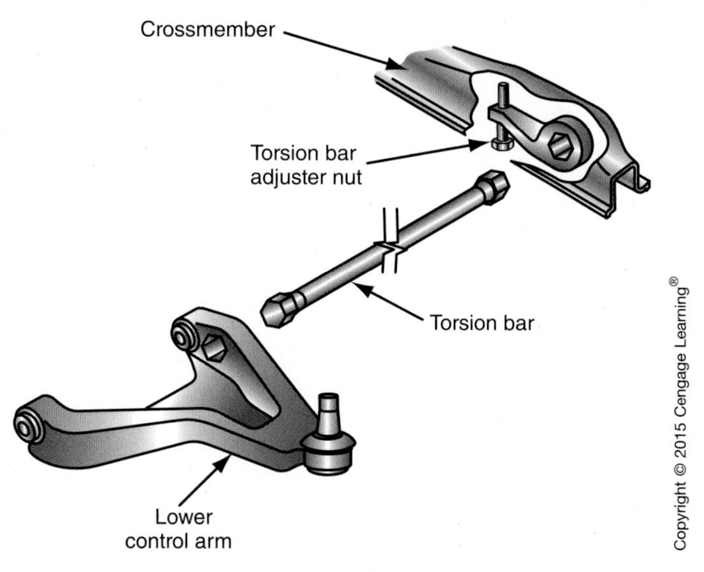

FIGURE 6-53 **Torsion bar mounting.**

torsion bar front suspension systems the torsion bars are anchored into the upper control arms rather than the lower control arms.

Vehicle ride height is controlled by the torsion bar anchor adjusting bolts in the crossmember. Front suspension heights must be within specifications for correct wheel alignment, tire wear, satisfactory ride, and accurate bumper heights. A conventional stabilizer bar is connected between the lower control arms and the crossmember. Ball joints are located in the upper and lower control arms, and both ball joints are bolted into the steering knuckle. The shock absorbers are connected between the lower control arms and the crossmember support, and the inner ends of the lower control arms are bolted to the crossmember through an insulating bushing (Figure 6-54).

Twin I-Beam Suspension Systems

Some Ford trucks are equipped with twin I-beam front suspension systems. In this type of suspension system, each front wheel is connected to a separate I-beam. The outer ends of the I-beam are connected to the spindles, and the inner ends of the beams are connected

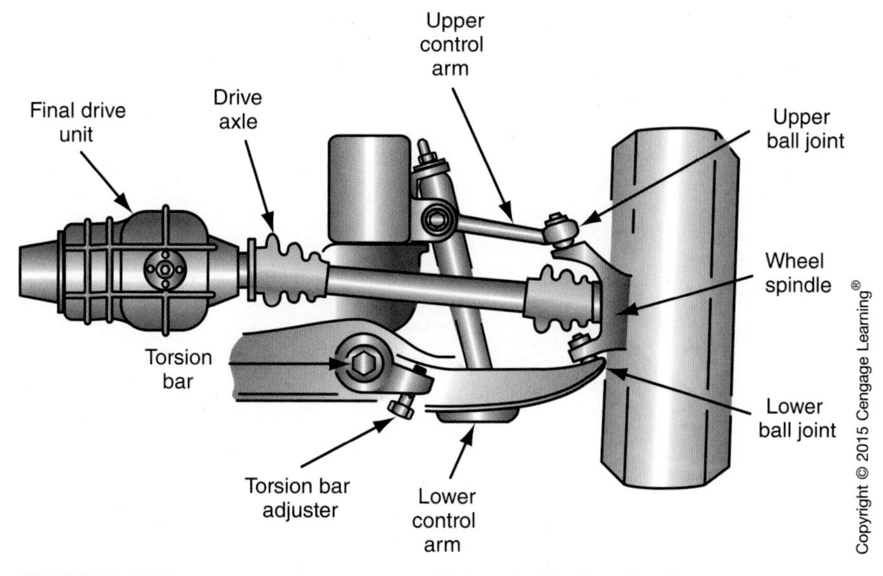

FIGURE 6-54 **Front suspension system with longitudinal torsion bars.**

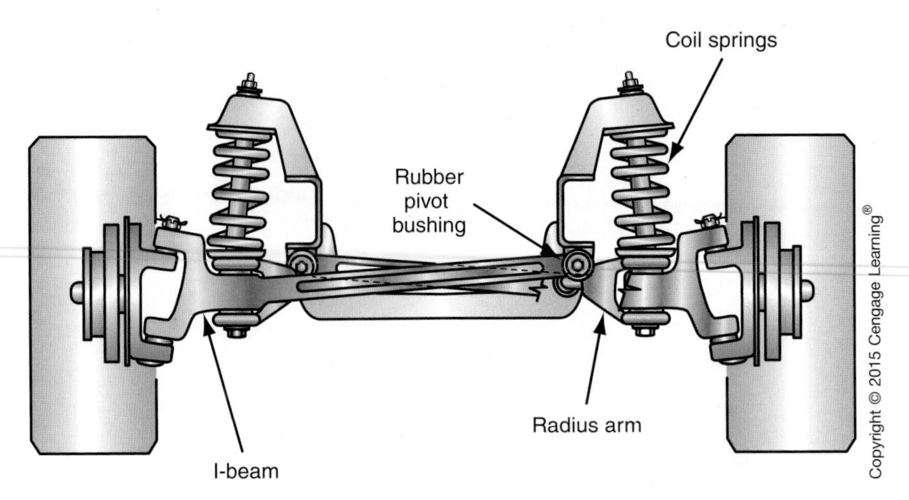

FIGURE 6-55 **Twin I-beam front suspension.**

through a rubber pivot bushing to the chassis (Figure 6-55). Coil springs are positioned between the I-beams and the chassis to support the vehicle weight. Radius arms are connected rearward from each I-beam to the chassis to prevent longitudinal wheel movement. Since each front wheel can move independently in a twin I-beam suspension system, the problems associated with straight I-beam systems are greatly reduced.

In some I-beam front suspension systems, kingpins are used to attach the I-beams to the spindles. In other I-beam suspension systems, ball joints connect the I-beams to the spindles.

Light-Duty Four-Wheel-Drive Truck Front Suspension Systems

Ford light-duty four-wheel-drive trucks have twin I-beam front suspension systems. F-150 four-wheel-drive models have a coil spring twin I-beam front suspension (Figure 6-56). The lower ends of the coil springs are seated on the twin I-beams, and the upper spring seat is positioned on the chassis. Heavy radius arms are bolted to the twin I-beams, and the other ends of the radius arms are mounted in a bushing and frame bracket (Figure 6-57). The radius arms prevent axle movement. The bushings in the inner ends of the twin I-beams are mounted on pivots attached to the front crossmember. Since each front wheel can move independently, twin I-beam suspensions are independent suspension systems. Upper and lower ball joints are pressed in the ends of the steering knuckle. The studs on these ball joints extend through openings in the outer ends of the twin I-beams. The ball joint studs are retained in the twin I-beams with nuts and cotter pins. Universal joints in the outer ends of

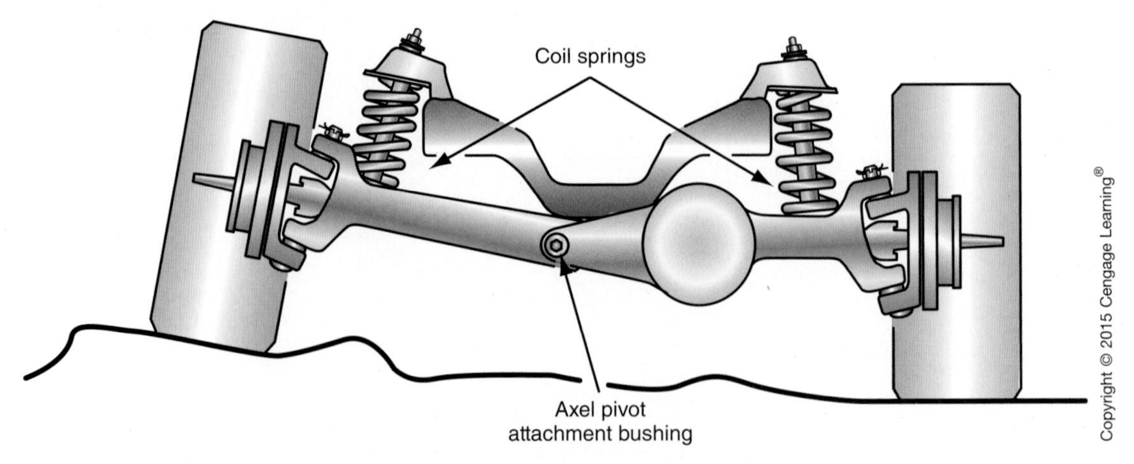

FIGURE 6-56 **Four-wheel-drive twin I-beam front suspension with coil springs.**

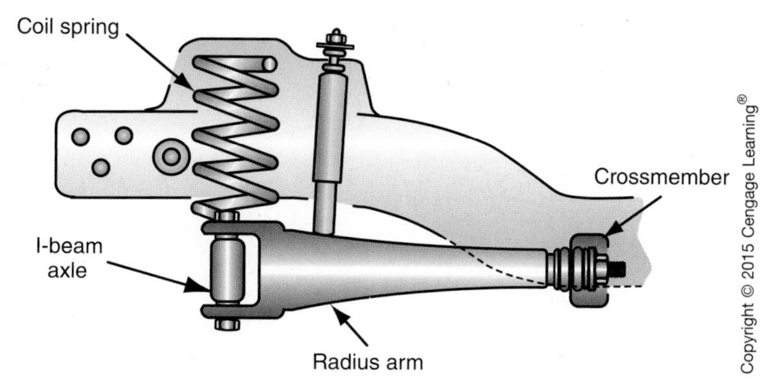

FIGURE 6-57 **Radius arms on a twin I-beam front suspension system with coil springs.**

the drive axles allow simultaneous wheel rotation and wheel turning to the right or left. The right-side drive axle also has an inner universal joint near the differential.

Ford F-250 and F-350 four-wheel-drive trucks have a leaf spring twin I-beam front suspension system (Figure 6-58). U-bolts retain the leaf springs to the twin I-beams. A bushing in the front of the leaf spring eye is bolted into a frame bracket. The bushing in the rear spring eye is connected to the frame through a conventional spring shackle (Figure 6-59). Since the leaf springs maintain the axle position, the radius arms are not required.

Some light-duty four-wheel-drive trucks have a straight front drive axle housing and a coil spring suspension system (Figure 6-60). Upper and lower radius rods control fore-

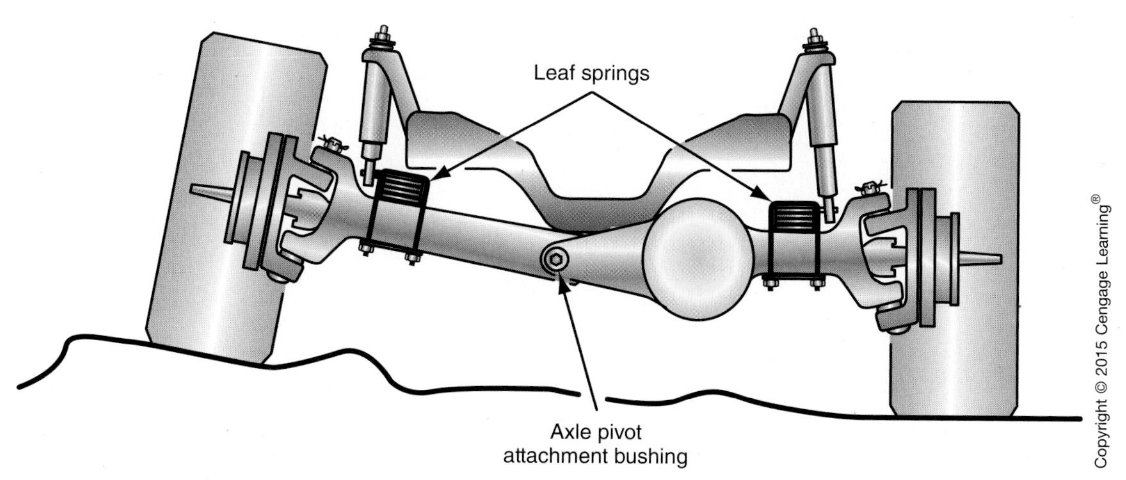

FIGURE 6-58 **Four-wheel-drive twin I-beam front suspension system with leaf springs.**

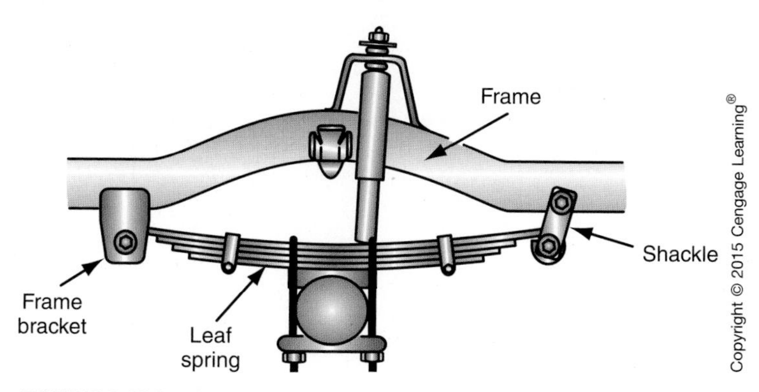

FIGURE 6-59 **Leaf spring mounting, four-wheel-drive twin I-beam front suspension.**

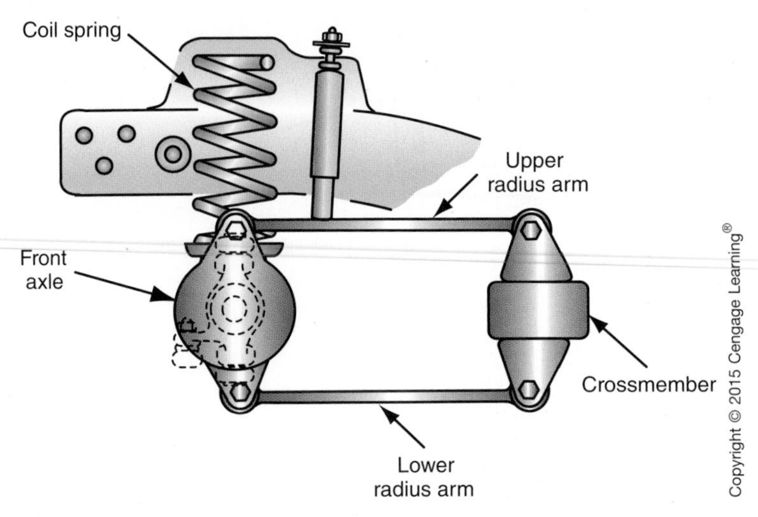

FIGURE 6-60 **Four-wheel-drive front suspension with straight drive axle and coil springs.**

and-aft drive axle movement, and a track bar controls lateral axle movement. Ball joints connect the steering knuckles to the outer ends of the drive axle housing.

Other light-duty four-wheel-drive trucks have a torsion bar front suspension system with upper and lower control arms (Figure 6-61 and Figure 6-62). The front of the torsion bars has hex-shaped ends that are mounted in a matching hex in the lower control arm. The hex-shaped rear ends of the torsion bars are mounted in adjusting arms that are retained in the torsion bar crossmember. An adjusting bolt in this crossmember contacts the outer end of each adjusting arm (Figure 6-63). Rotation of this adjusting bolt changes the tension on the torsion bar to adjust the curb riding height. On this type of torsion bar front suspension system, the front wheel bearing hubs are a prelubricated, sealed, non-serviceable assembly (Figure 6-64).

Twin I-beam front suspension systems have recently been replaced with short-and-long arm or torsion bar suspension systems, because these systems are lighter weight and provide improved ride quality.

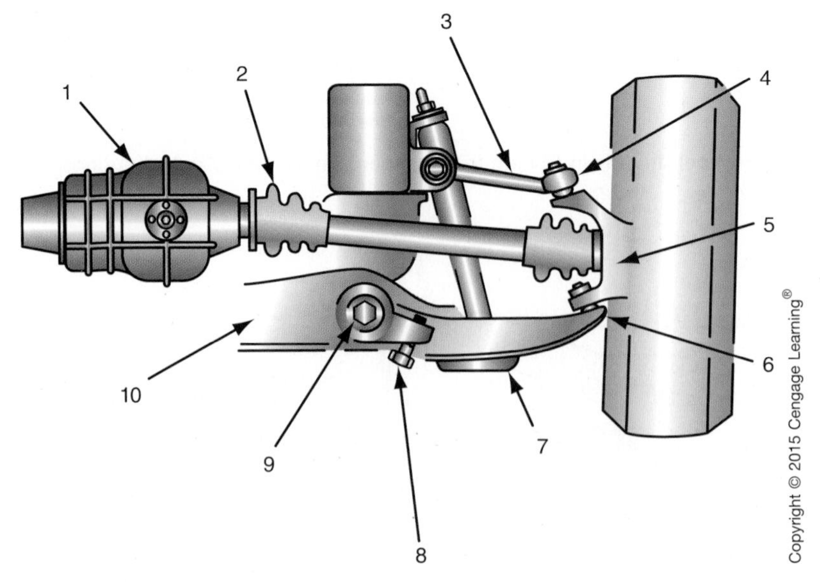

FIGURE 6-61 **Four-wheel-drive light-duty truck torsion bar front suspension.**

ITEM	DESCRIPTION
1	Final drive unit
2	Drive axle
3	Upper control arm
4	Upper ball joint
5	Steering knuckle
6	Lower ball joint
7	Lower control arm
8	Torsion bar adjuster
9	Torsion bar
10	Crossmember

FIGURE 6-62 **Component identification light-duty four-wheel-drive truck torsion bar front suspension.**

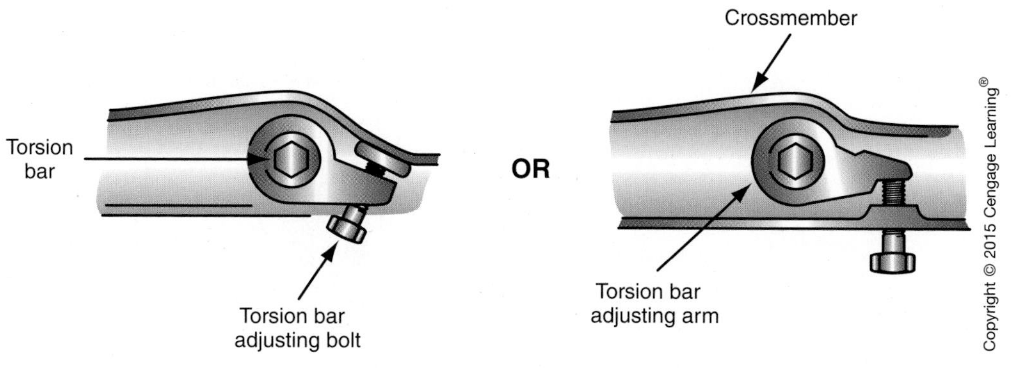

FIGURE 6-63 **Torsion bar adjusting arm and adjusting bolt.**

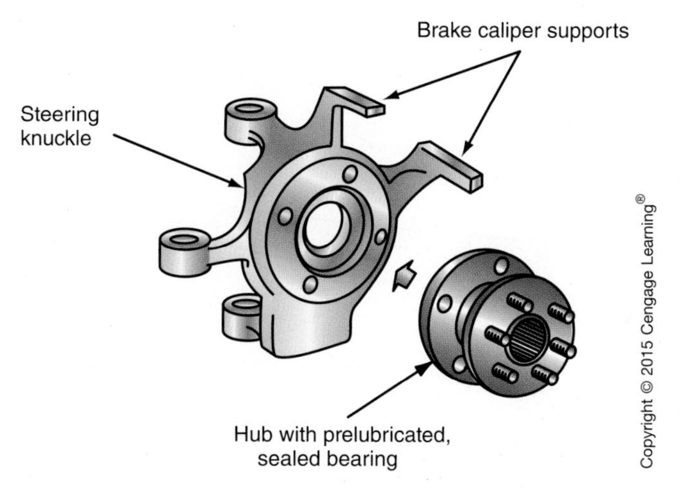

FIGURE 6-64 **Four-wheel-drive torsion bar front suspension with sealed non-serviceable front wheel bearing hubs.**

Caster angle is the tilt of an imaginary line through the center of the upper strut mount and ball joint in relation to the true vertical line through the center of the wheel and spindle as viewed from the side.

Directional stability refers to the tendency of the steering to remain in the straight-ahead position with very little correction by the driver when the vehicle is driven on a smooth, level road surface. Certain suspension defects reduce directional stability and cause the steering to pull to one side.

Curb Riding Height

Regular inspection and proper maintenance of suspension systems are extremely important to maintain vehicle safety. The curb riding height is determined mainly by spring condition. Other suspension components such as control arm bushings will affect curb riding height if they are worn. Since incorrect curb riding height affects most of the other suspension angles, this measurement is critical. Sagged springs change the normal operating arc of the lower ball joint. This action causes excessive lateral movement of the tire during wheel jounce and rebound with resulting tire wear (Figure 6-65).

The curb riding height must be measured at the vehicle manufacturer's specified location, which varies depending on the type of suspension system. When the vehicle is on a level floor or an alignment rack, measure the curb riding height from the floor to the manufacturer's specified location on the chassis.

Front Spring Sag, Curb Riding Height, and Caster Angle

Sagged springs cause insufficient curb riding height. Therefore, the distance is reduced between the rebound bumper and its stop. This distance reduction causes the bumper to hit the stop frequently with resulting harsh ride quality. The caster angle is the number of degrees between the true vertical centerline of the tire and wheel, and an imaginary line through the center of the upper strut mount and lower ball joint. Positive caster angle occurs when the caster line is tilted toward the rear of the vehicle. Negative caster angle is present when the caster line is tilted toward the front of the vehicle (Figure 6-66).

When both rear springs are sagged, the caster angle tilts excessively toward the rear of the vehicle (Figure 6-67). This caster angle results in increased steering effort and rapid steering wheel return after a turn.

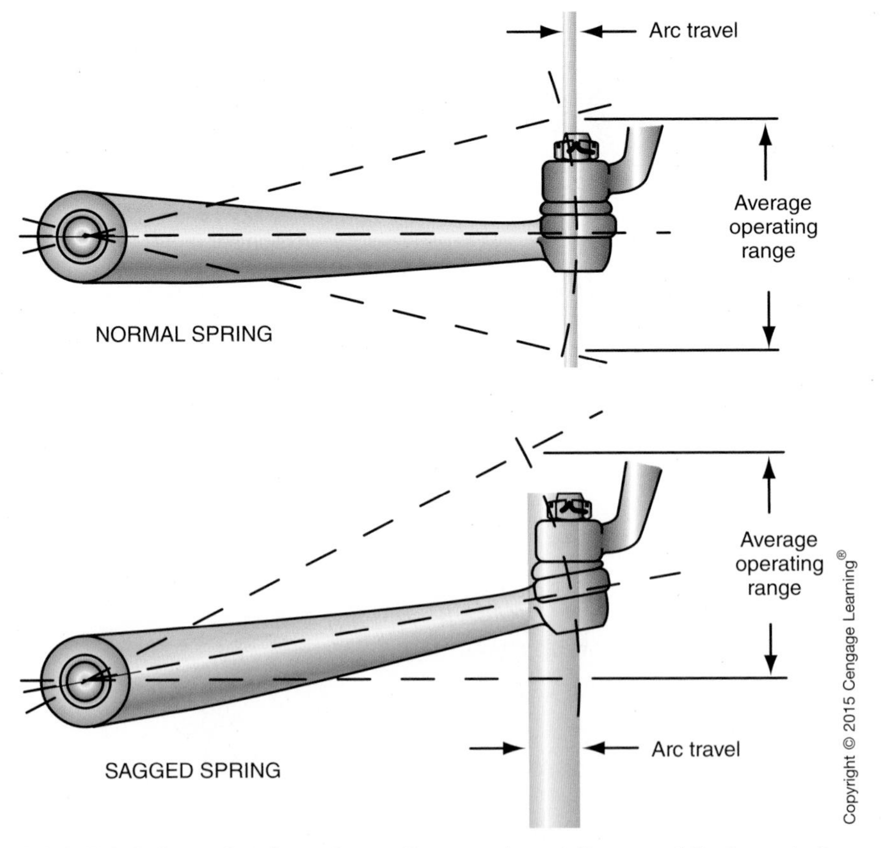

FIGURE 6-65 **Sagged springs change the normal operating arc of the lower ball joint and cause excessive lateral tire movement during wheel jounce and rebound with resulting tire wear.**

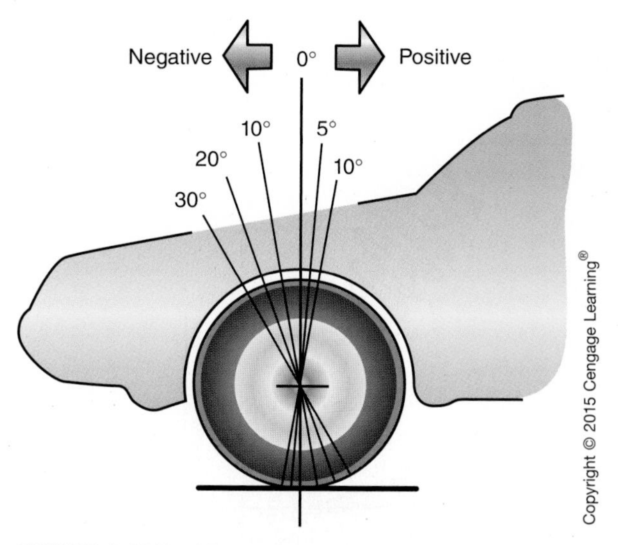

FIGURE 6-66 **Positive and negative caster.**

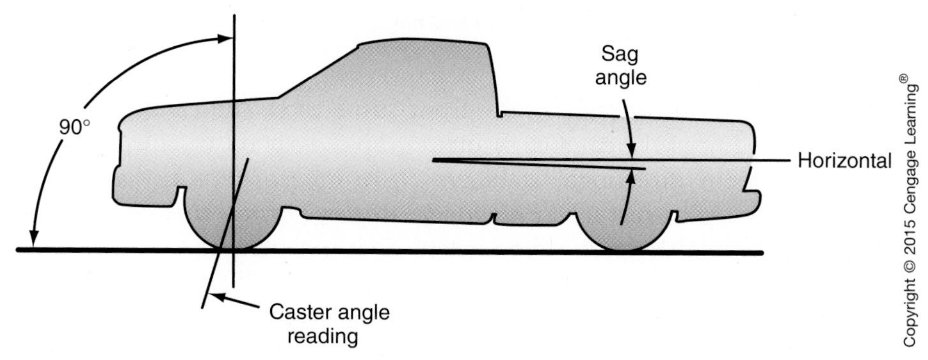

FIGURE 6-67 **Effects of rear spring sag and incorrect curb riding height on caster angle.**

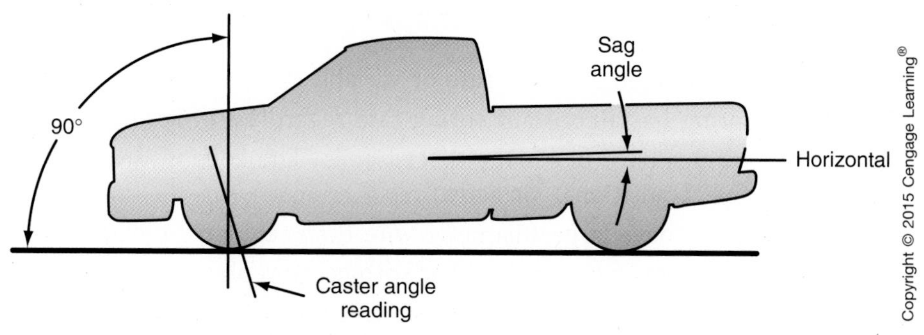

FIGURE 6-68 **Effects of front spring sag on caster angle.**

Shop Manual
Chapter 6, page 211

If both front springs are sagged, the caster angle tilts excessively toward the front of the vehicle (Figure 6-68). When the front springs are sagged and caster is negative, directional stability is decreased, and the front wheels tend to wander.

Spring Sag, Curb Riding Height, and Camber Angle

Camber angle is the tilt of a line through the center of the wheel in relation to the vertical centerline viewed from the front of the wheel.

Positive camber occurs when the camber line is tilted outward from the vertical centerline of the wheel. If the camber line is tilted inward from the vertical centerline of the wheel, the camber is said to be negative (Figure 6-69).

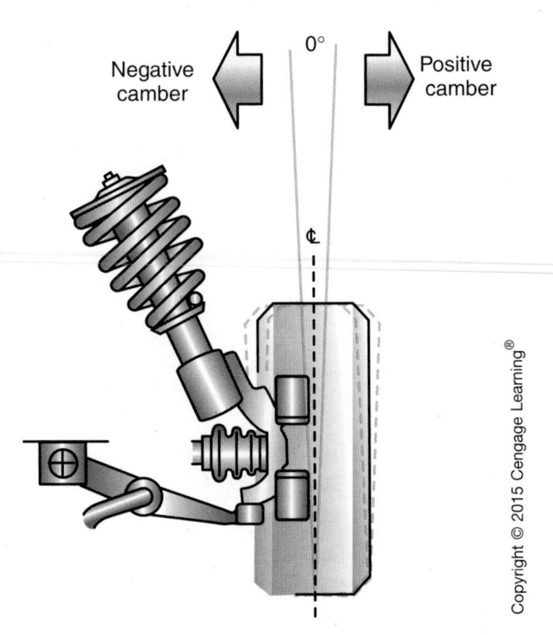

FIGURE 6-69 **Positive and negative camber.**

If a spring is sagged on one side of the front suspension, the camber angle is negative on that side. Excessive negative camber results in rapid wear on the inside edge of the tire tread and a decrease in directional stability. *Therefore, curb riding height is extremely important in maintaining correct wheel alignment angles, normal tire wear, and satisfactory directional stability.* When the curb riding height does not meet the vehicle manufacturer's specifications, check for bent control arms and worn control arm bushings or mounting bolts. If these components are satisfactory, spring replacement is required to correct the curb riding height.

SUMMARY

TERMS TO KNOW

Compression loaded
Full-wire open-end springs
Gussets
Heavy-duty coil springs
Inactive coils
Linear-rate coil springs
Load-carrying ball joint
MacPherson strut front suspension systems
Mono leaf springs
Multilink front suspension
Multiple-leaf springs
Non-load-carrying ball joint

- Coil springs may be classified as linear rate or variable rate.
- A linear-rate coil spring has a constant spring rate regardless of spring load.
- Variable-rate coil springs have unequally spaced coils. The coils in these springs may be referred to as inactive, transitional, or active.
- Heavy-duty coil springs have larger-diameter wire than regular-duty coil springs.
- Most vehicle manufacturers recommend that springs be replaced in pairs.
- In a torsion bar suspension system, the torsion bars replace the coil springs.
- Torsion bars are directional, and they are marked right or left.
- In a leaf spring, zinc and plastic spacers between the leaves reduce noise and friction problems.
- Leaf springs may be mono leaf or multiple-leaf design.
- Spring leaves may be manufactured from steel or fiberglass.
- Although many leaf springs are mounted longitudinally, some of these springs are mounted transversely.
- Ball joints may be classified as load-carrying or non-load-carrying.
- A load-carrying ball joint may be compression loaded or tension loaded.
- The load-carrying ball joint wears faster than the non-load-carrying ball joint.

- Many ball joints have wear indicators that provide visual ball joint wear inspection.
- A stabilizer bar reduces body roll or sway when one front wheel strikes a road irregularity.
- A strut rod prevents fore-and-aft lower control arm movement.
- In many MacPherson strut front suspension systems, the lower end of the strut is bolted to the steering knuckle, and the upper end of the strut is connected through an upper mount to the fender reinforcement.
- Torsion bars may be transversely or longitudinally mounted in a front suspension system.
- In a short-and-long arm suspension system, the coil springs may be mounted between the lower control arm and the frame or between the upper control arm and the chassis.
- In a twin I-beam suspension system, the spindles may be connected to the I-beams with kingpins or ball joints.
- Curb riding height is extremely important for maintaining normal tire wear and proper suspension alignment angles.

TERMS TO KNOW (continued)

Pigtail spring ends
Regular-duty coil springs
Short-and-long arm front suspension systems
Stabilizer bar
Taper-wire closed-end
Tension loaded
Transitional coils
Unsprung weight
Variable-rate coil springs
Wear indicators

REVIEW QUESTIONS

Short Answer Essays

1. Explain the meaning of constant spring rate.
2. Describe the main difference in the design of heavy-duty coil springs compared with regular-duty coil springs, and compare the free diameter of each type of spring.
3. Explain the type of load condition that requires variable-rate coil springs, and describe the load-carrying advantage of these springs.
4. Describe the design of taper-wire closed-end springs.
5. Explain torsion bar action during wheel jounce and rebound.
6. Describe the action of a leaf spring as it compresses.
7. Explain the position of the lower control arm when the lower ball joint is compression loaded.
8. Describe the position of the wear indicator when a ball joint is worn.
9. Explain the basic purpose of the lower control arms.
10. Describe the effect of sagged front springs on caster angle and directional stability.

Fill-in-the-Blanks

1. Heavy-duty coil springs are designed to carry ________________ to ________________ percent greater loads than regular-duty coil springs.
2. A variable-rate coil spring provides automatic ________________ adjustment.
3. In a variable-rate spring, the ________________ coils operate during the complete range of spring loading.
4. The load-carrying ball joint is attached to the control arm on which the ________________ is seated.
5. If the lower control arm is mounted above the steering knuckle and rests on the knuckle, the lower ball joint is ________________.
6. When the grease nipple shoulder is extended from the ball joint housing, the ball joint is ________________.
7. When one front wheel strikes a road irregularity, the stabilizer bar reduces ________________.
8. The strut rod prevents ________________ and ________________ lower control arm movement.
9. When the front springs are sagged, the positive caster is ________________.
10. If a front spring is sagged, the camber angle moves toward a ________________ position.

Multiple Choice

1. A 400-pound load deflects a linear-rate coil spring 1 inch. An 800-pound load will deflect this coil spring:
   A. 1.5 inches.
   B. 2 inches.
   C. 3 inches.
   D. 4 inches.

2. Heavy-duty coil springs have:
   A. The same wire diameter as a regular-duty coil spring.
   B. A shorter free height compared to a regular-duty coil spring.
   C. A 10 percent increase in load carrying capacity.
   D. A high aluminum content in the spring material.

3. A car's rear coil springs keep breaking. This problem is likely because of:
   A. Continual driving on rough road surfaces.
   B. Continual driving on curved roads.
   C. Excessive air pressure in the rear tires.
   D. Constant overloading of the rear suspension.

4. Aluminum control arms have all of these advantages EXCEPT:
   A. Reduce unsprung weight.
   B. Provide improved road feel.
   C. Provide improved ride quality.
   D. Contribute to improved fuel economy.

5. While discussing a torsion bar front suspension system:
   A. One end of the torsion bar is attached to the upper control arm.
   B. A suspension height adjustment is positioned on the end of the torsion bar connected to the control arm.
   C. Torsion bars may be mounted longitudinally or transversely in a front suspension system.
   D. Torsion bars eliminate the need for shock absorbers in the front suspension system.

6. While discussing leaf springs:
   A. A fiberglass mono leaf spring is heavier than a steel mono leaf spring.
   B. Some mono leaf springs are mounted transversely.
   C. In a multiple-leaf spring, the lower leaf is called the main leaf.
   D. A multiple-leaf spring has the same stiffness regardless of how much it is compressed.

7. All these statements about ball joints are true EXCEPT:
   A. A load-carrying ball joint wears faster than a non-load-carrying ball joint.
   B. A non-load-carrying ball joint is designed with a preload.
   C. Wear indicators may be positioned in the side of the ball-joint housing.
   D. Low-friction ball joints have a highly polished ball socket surrounded by a high-strength polymer bearing.

8. While discussing MacPherson strut suspension systems:
   A. During a turn, the strut, spring, and knuckle rotate on the upper strut mount.
   B. The lower end of the coil spring is seated on the lower control arm.
   C. The strut is welded into the knuckle and these components are replaced as an assembly.
   D. The inner ends of the lower control arms are bolted to the strut.

9. While discussing short-and-long arm front suspension systems:
   A. The coil springs are positioned between the upper and lower control arms.
   B. Compared with a suspension system with equal length control arms, the short-and-long design reduces track width change and tire wear.
   C. The outer ends of the stabilizer bar may be attached to the steering knuckles.
   D. The coil spring is retained in the lower control arm with a clamp and bolt.

10. While discussing sagged springs:
    A. Sagged front springs cause harsh riding.
    B. Sagged front springs cause excessive positive caster on the front wheels and increased directional stability.
    C. Sagged rear springs cause excessive negative caster on the front wheels.
    D. Sagged rear springs cause excessive positive camber on the front wheels.

Chapter 7

Rear Suspension Systems

UPON COMPLETION AND REVIEW OF THIS CHAPTER, YOU SHOULD BE ABLE TO UNDERSTAND AND DESCRIBE:

- A live axle rear suspension system.
- The advantages and disadvantages of a live axle leaf-spring rear suspension system.
- The movement of the rear axle housing during vehicle acceleration.
- How the differential torque is absorbed in a live axle coil spring rear suspension system.
- The purpose of a tracking bar in a live axle coil spring rear suspension system.
- The difference between a semi-independent and an independent rear suspension system.
- How individual rear wheel movement is provided in a semi-independent rear suspension system.
- The difference between a MacPherson strut and a modified MacPherson strut rear suspension.
- The advantage of attaching the differential housing to the chassis in an independent rear suspension system.
- How differential and suspension vibration, noise, and shock are insulated from the chassis in a multilink independent rear suspension system.
- How the top of the knuckle is supported in a multilink independent rear suspension system.
- The effect of sagged rear springs on caster angle and steering.

Introduction

The **rear suspension system** plays a very important part in ride quality and in the control of suspension and differential noise, vibration, and shock. Although the front wheels actually steer the vehicle, the rear suspension is also vital to steering control. The rear suspension must also provide adequate tire life and maintain tire traction on the road surface. Rear suspension systems described in this chapter include live axle, semi-independent, and independent. **Live axle rear suspension systems** are found on rear-wheel-drive (RWD) trucks and vans, a few RWD cars, and some four-wheel-drive (4WD) cars. Most front-wheel-drive (FWD) vehicles have semi-independent or independent rear suspensions. Independent rear suspensions are also found on RWD cars and 4WD cars.

A **rear suspension system** with two longitudinal leaf springs and a one-piece rear axle housing may be called a Hotchkiss drive.

A **live axle rear suspension system** may be defined as one in which the differential axle housing, wheel bearings, and brakes act as a unit.

Live Axle Rear Suspension Systems

Leaf-Spring Rear Suspension

A leaf spring is mounted longitudinally on each side of the rear suspension on some rear-wheel-drive cars and trucks (Figure 7-1). These relatively flat springs provide excellent lateral stability and reduce side sway, which contribute to a well-controlled ride with very

A BIT OF HISTORY

The leaf-spring rear suspension was one of the first widely used rear suspension systems. Because this type of rear suspension system has weight and ride quality disadvantages, it has been replaced on many vehicles with independent or semi-independent rear suspension systems.

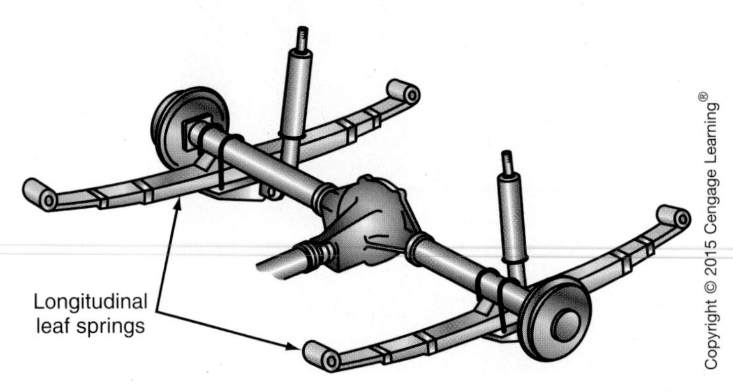

FIGURE 7-1 **Rear suspension system with long torque arm and track bar.**

good handling characteristics. However, leaf-spring rear suspension systems have a lot of **unsprung weight**, and leaf springs require a considerable amount of space.

Unsprung weight refers to the weight that is not supported by the springs, which includes the weight of the suspension system.

The **semielliptical springs** have steel leaves and zinc or plastic interleaves to reduce corrosion, friction, and noise. A large rubber bushing is installed in the front eye of the main spring leaf, and a bolt retains this bushing to the front spring hanger (Figure 7-2). The rear spring shackle is bolted to a rubber bushing in the rear main leaf eye, and the upper shackle bolt extends through a similar rubber bushing in the rear spring hanger (Figure 7-3). Some rear spring shackles contain threaded steel bushings or a slipper mount in which the end of the spring slides through the shackle. Shackle insulating bushings help prevent the transfer of noise and road shock from the suspension to the chassis and vehicle interior. When a rear wheel strikes a road irregularity, the spring is compressed and the spring length changes. The rear shackle provides fore-and-aft movement with variations in spring length.

Semielliptical springs have individual leaves stacked with the shortest leaf at the bottom and the longest leaf atthe top.

Because the differential axle housing is a one-piece unit, jounce and rebound travel of one rear wheel affects the position of the other rear wheel. This action increases tire wear and decreases ride quality and traction.

The differential axle housing is mounted above the springs, and a spring plate with an insulating clamp and U-bolts retains the springs to the rear axle housing (Figure 7-4). The shock absorbers are mounted between the spring plates and the frame.

Shop Manual
Chapter 7, page 261

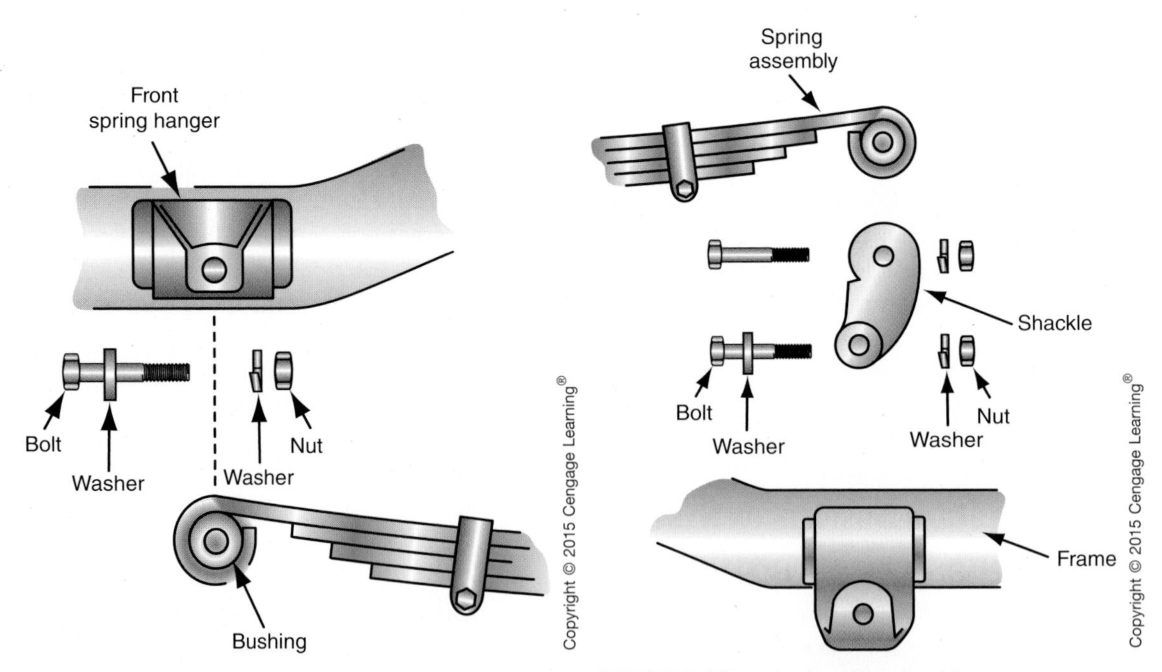

FIGURE 7-2 **Rear leaf-spring eye bushing.**

FIGURE 7-3 **Rear leaf-spring shackle.**

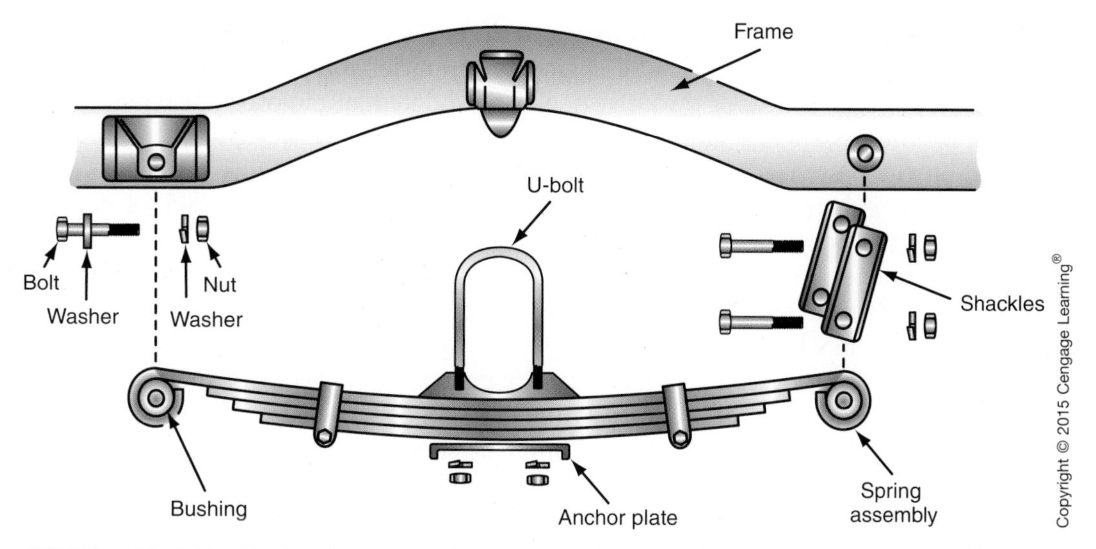

FIGURE 7-4 **Individual leaf-spring suspension components.**

The vehicle **sprung weight** is supported by the springs through the rear axle housing and wheels. When the vehicle accelerates, the rear wheels turn counterclockwise when viewed from the left vehicle side. One of Newton's laws of motion states that for every action there is an equal and opposite reaction. Therefore, when the wheels turn counterclockwise (when viewed from the left), the rear axle housing tries to rotate clockwise. This rear axle torque action is absorbed by the rear springs and the chassis moves downward (Figure 7-5). Engine torque supplied through the driveshaft to the differential tends to twist the differential housing and the springs. This twisting action may be referred to as **axle windup**. Many leaf springs have a shorter distance from the center bolt to the front of the spring compared to the distance from the center bolt to the rear of the spring. This type of leaf spring is referred to as an **asymmetrical leaf spring**, and the shorter distance from the center bolt to the front of the spring resists axle windup. A **symmetrical leaf spring** has the same distance from the center bolt to the front and rear of the spring.

Sprung weight refers to the weight carried by the springs, which includes the chassis and all components attached to the chassis.

When braking and decelerating, the rear axle housing tries to turn counterclockwise. This rear axle torque action applied to the springs lifts the chassis. This action may be called **braking and deceleration torque**.

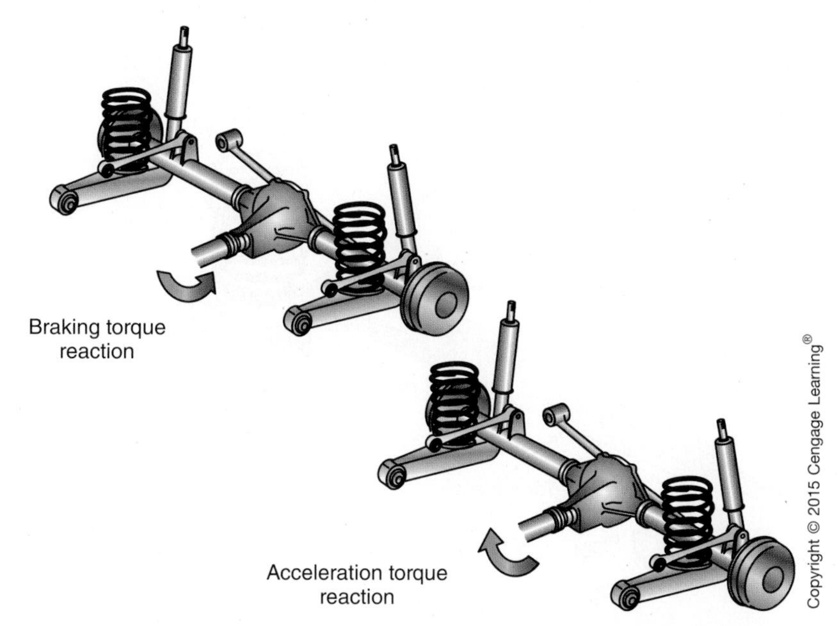

FIGURE 7-5 **Rear axle torque action during acceleration and deceleration.**

During hard acceleration, the entire power train twists in the opposite direction to engine crankshaft and drive shaft rotation. The engine and transmission mounts absorb this torque. However, the twisting action of the drive shaft and differential pinion shaft tends to lift the rear wheel on the passenger's side of the vehicle. Extremely hard acceleration may cause the rear wheel on the passenger's side to lift off the road surface. Once this rear wheel slips on the road surface, engine torque is reduced, and the leaf spring forces the wheel downward. When this rear tire contacts the road surface, engine torque increases and the cycle repeats. This repeated lifting of the differential housing is called **axle tramp**, and this action occurs on live axle rear suspension systems. Axle tramp is more noticeable on live axle leaf-spring rear suspension systems in which the springs have to absorb all the differential torque. For this reason, only engines with moderate horsepower were used with this type of rear suspension. Rear suspension and axle components such as spring mounts, shock absorbers, and wheel bearings may be damaged by axle tramp. Mounting one rear shock absorber in front of the rear axle and the other rear shock behind the rear axle helps reduce axle tramp.

AUTHOR'S NOTE: Leaf-spring rear suspension systems are still used on many light-duty trucks because of their load-carrying capability. However, today's design engineers have improved the ride quality of these suspension systems compared with past models. Ride quality in these leaf-spring suspension systems has been improved by installing longer leaf springs and using larger, improved rubber insulating bushings in the spring eye and shackle. Ride quality has also been improved by maximizing the shock absorber mounting location and matching the shock absorber design more closely to the leaf-spring jounce and rebound action. Optimizing the rear axle mounting position on the leaf springs also improves ride quality.

In some cars with higher torque engines, a long torque arm is bolted to the rear axle housing (Figure 7-6). This torque arm helps prevent differential rotation during hard acceleration

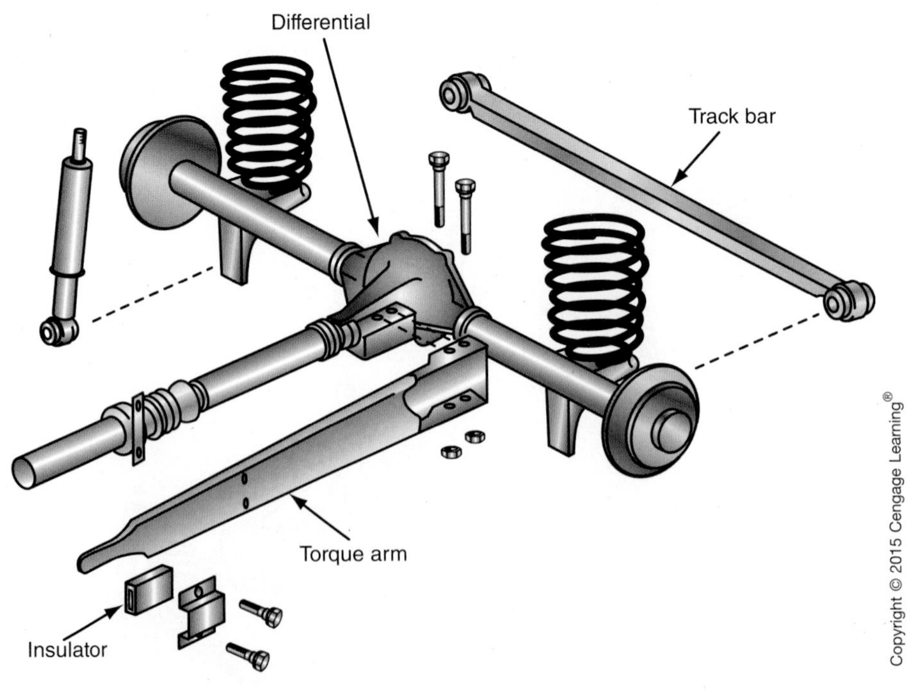

FIGURE 7-6 **Rear suspension system with long torque arm and track bar.**

and braking. The front of this torque arm is mounted in a rubber insulator and bracket that is bolted to the back of the transmission housing. This long torque arm helps prevent differential rotation when high torque is delivered from the engine to the differential. The center bearing assembly on the drive shaft is bolted to the long torque arm. This rear suspension system has a track bar (tie rod) connected from the left side of the rear axle housing to the chassis to help prevent lateral rear axle movement. A track bar brace is connected from the chassis end of the track bar to the other side of the chassis to provide extra rigidity.

Coil Spring Rear Suspension

WARNING: Compressed coil springs contain a large amount of energy. Never disconnect any suspension component that suddenly releases the coil spring tension. This action may result in personal injury and vehicle damage.

WARNING: Always follow the vehicle manufacturer's recommended rear suspension service procedures in the service manual to avoid personal injury.

Some rear-wheel-drive cars have a coil spring rear suspension. Upper and lower suspension arms with insulating bushings are connected between the differential housing and the frame (Figure 7-7). The upper arms control lateral movement, and the lower trailing control arms absorb differential torque. In some rear suspension systems, the upper arms are replaced with strut rods. The front of the upper and lower arms contains large rubber bushings. When strut rods are used in place of the upper arms, both ends of these rods contain large rubber bushings to prevent noise and vibration transfer from the suspension to the chassis. The coil springs are usually mounted between the lower suspension arms and the frame, whereas the shock absorbers are mounted between the back of the suspension arms and the frame.

Some rear suspension systems have a **track bar** connected from one side of the differential housing to the chassis to prevent lateral chassis movement. Large rubber insulating bushings are positioned in each end of the track bar.

A **track bar** may be referred to as a Panhard rod or Watts rod.

Some late model sport utility vehicles (SUVs) have a rear suspension system with coil springs and shock absorbers mounted separately from the springs (Figure 7-8). Notice that

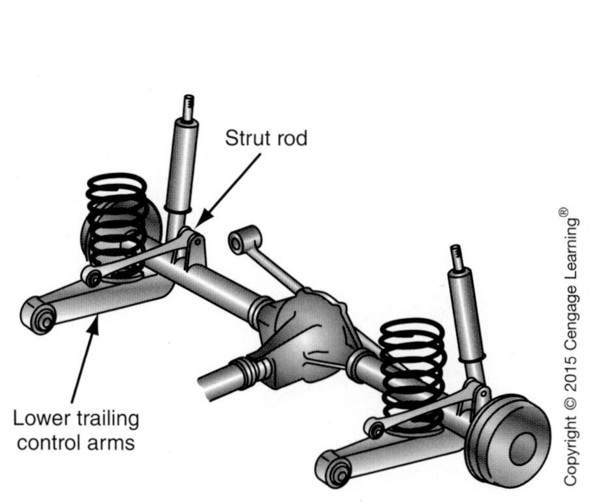

FIGURE 7-7 **Live axle coil spring rear suspension.**

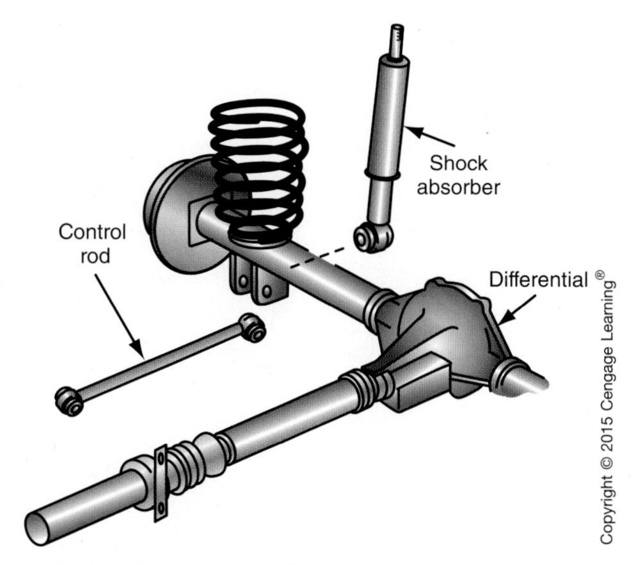

FIGURE 7-8 **Rear suspension system with coil springs and shock absorbers mounted separately.**

a control rod is mounted in rubber insulating bushings on the bottom of the rear axle housing, and the outer end of this bar is connected through links to the chassis. This rear suspension has an upper and lower control rod on each side of the suspension. Each lower control rod is connected from a bracket on the lower side of the axle housing to a frame bracket. Rubber insulating bushings are located in both ends of the lower control rods. The frame bracket is located ahead of the rear axle. The upper control rod is connected from a bracket on top of the axle housing to a frame bracket near the lower control rod frame bracket (Figure 7-9). Both ends of the upper control rod contain rubber insulating bushings. The upper and lower control rods prevent rear axle windup during hard acceleration.

This SUV rear suspension also has a track bar connected from a bracket on the rear of the axle housing to a frame bracket on the opposite side of the vehicle. The track bar prevents lateral rear axle movement. A rear axle brace is also connected from the track bar frame bracket to another frame bracket on the opposite side of the vehicle (Figure 7-10). The rear axle brace prevents any movement of the track bar frame bracket during

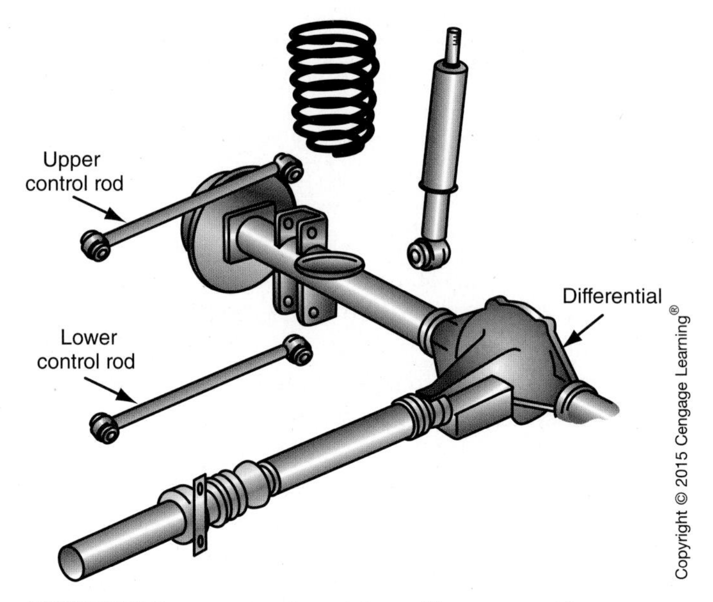

FIGURE 7-9 **Rear suspension system with upper and lower control rods.**

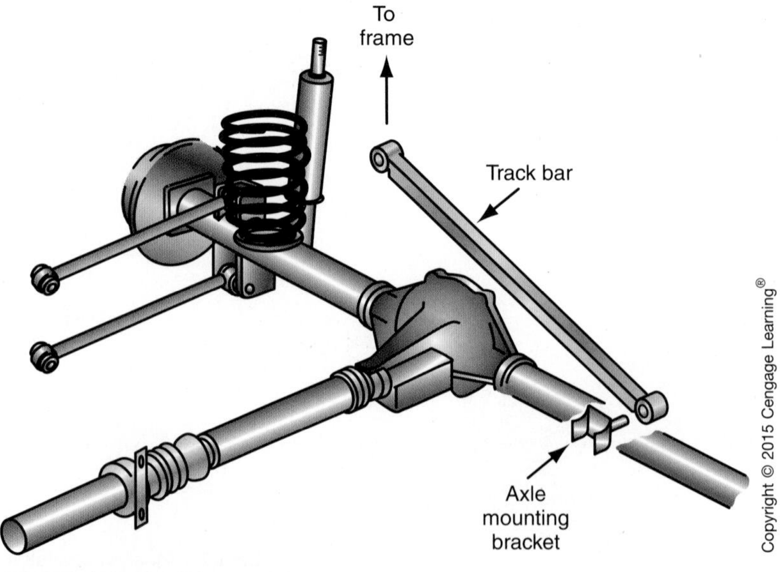

FIGURE 7-10 **Track bar and brace.**

off-road or severe driving conditions. The dual rear axle rods on each side of this suspension with the track bar and brace provide a very stable rear axle position during hard acceleration and severe driving conditions. This rear axle stability improves tire life, ride comfort, and steering control.

Semi-Independent Rear Suspension Systems

Shop Manual
Chapter 7, page 262

Many front-wheel-drive vehicles have a **semi-independent rear suspension** that has a **solid axle beam** connected between the rear trailing arms (Figure 7-11). A solid axle beam is usually a transverse inverted U-section channel connected between the rear wheels in a semi-independent rear suspension system. When one rear wheel strikes a bump, this beam twists to allow some independent wheel movement. Some of these rear axle beams are fabricated from a transverse inverted U-section channel.

A **semi-independent rear suspension** allows some individual rear wheel movement when one rear wheel strikes a bump.

In some rear suspension systems, the inverted U-section channel contains an integral tubular stabilizer bar. When one rear wheel strikes a road irregularity and the wheel moves upward, the inverted U-section channel twists, which allows some independent rear wheel movement. The trailing arms are connected to chassis brackets through rubber insulating bushings. In some semi-independent rear suspension systems, the coil springs are mounted on the rear struts, the lower spring seat is located on the strut, and the upper spring seat is positioned on the upper strut mount.

In other semi-independent rear suspension systems, the coil springs are mounted separately from the shock absorbers. Coil spring seats are located on the trailing arms, and the shock absorbers are connected from the trailing arms to the chassis. A crossmember connected between the trailing arms provides a twisting action and some independent rear wheel movement (Figure 7-12).

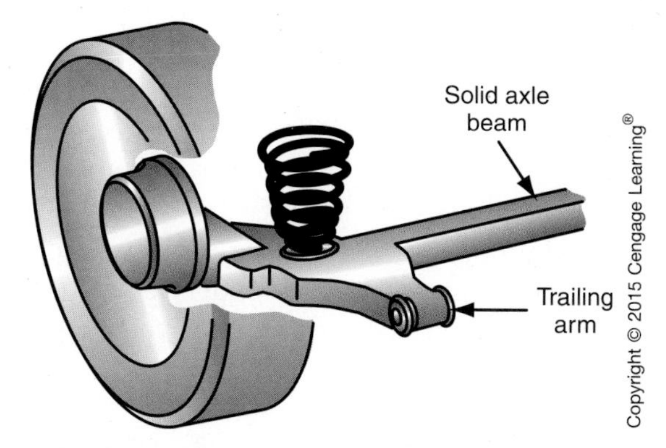

FIGURE 7-11 **Semi-independent rear suspension system.**

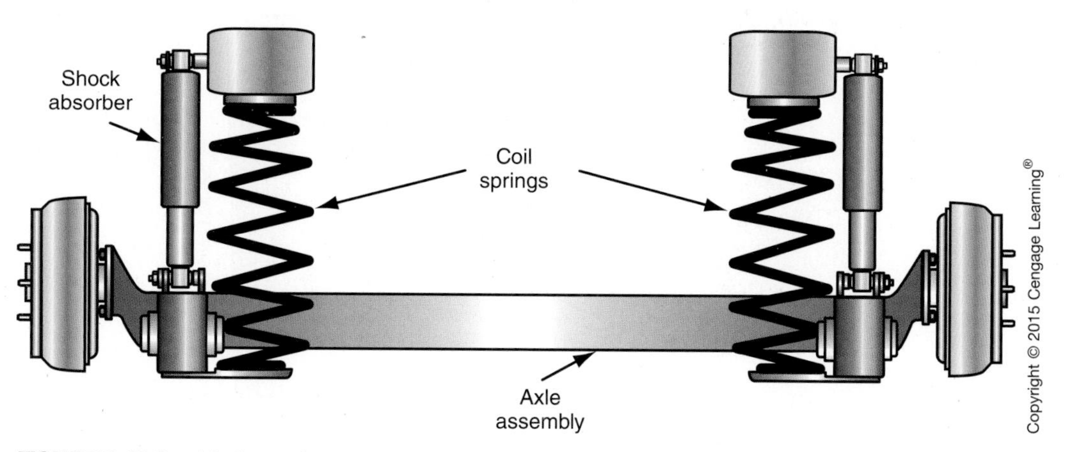

FIGURE 7-12 **Semi-independent rear suspension system with coil springs and shock absorbers mounted separately.**

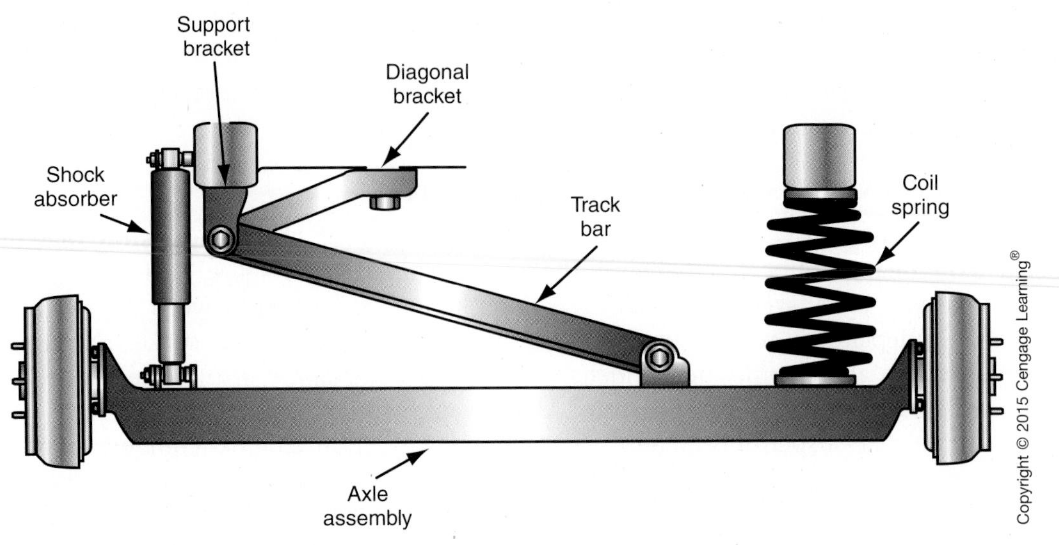

FIGURE 7-13 **Semi-independent rear suspension with track bar and brace.**

Some semi-independent rear suspension systems have a track bar connected from a rear axle bracket to a chassis bracket. In some applications, an extra brace is connected from this chassis bracket to the rear upper crossmember (Figure 7-13). The track bar and the brace prevent lateral rear axle movement.

Shop Manual
Chapter 7, page 258

Independent Rear Suspension Systems

MacPherson Strut Independent Rear Suspension System

In an **independent rear suspension** system, each rear wheel can move independently from the opposite rear wheel. Independent rear suspension systems may be found on front-wheel-drive and rear-wheel-drive vehicles. When rear wheel movement is independent, ride quality, tire life, steering control, and traction are improved. In a MacPherson strut rear suspension system, the coil springs are mounted on the rear struts. A lower spring seat is located on the strut, and the upper spring seat is positioned on the upper strut mount. This upper strut mount is bolted into the inner fender reinforcement. Dual lower control arms on each side of the suspension are connected from the chassis to the lower end of the spindle (Figure 7-14).

In an **independent rear suspension**, vertical movement of one rear wheel does not affect the opposite rear wheel.

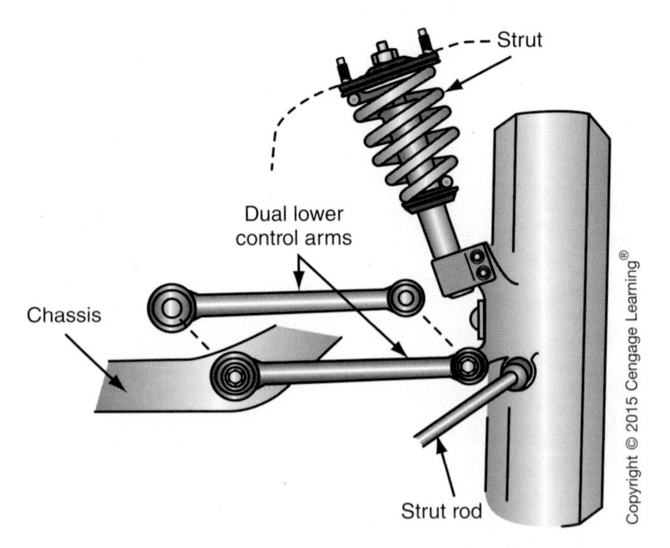

FIGURE 7-14 **MacPherson strut independent rear suspension system.**

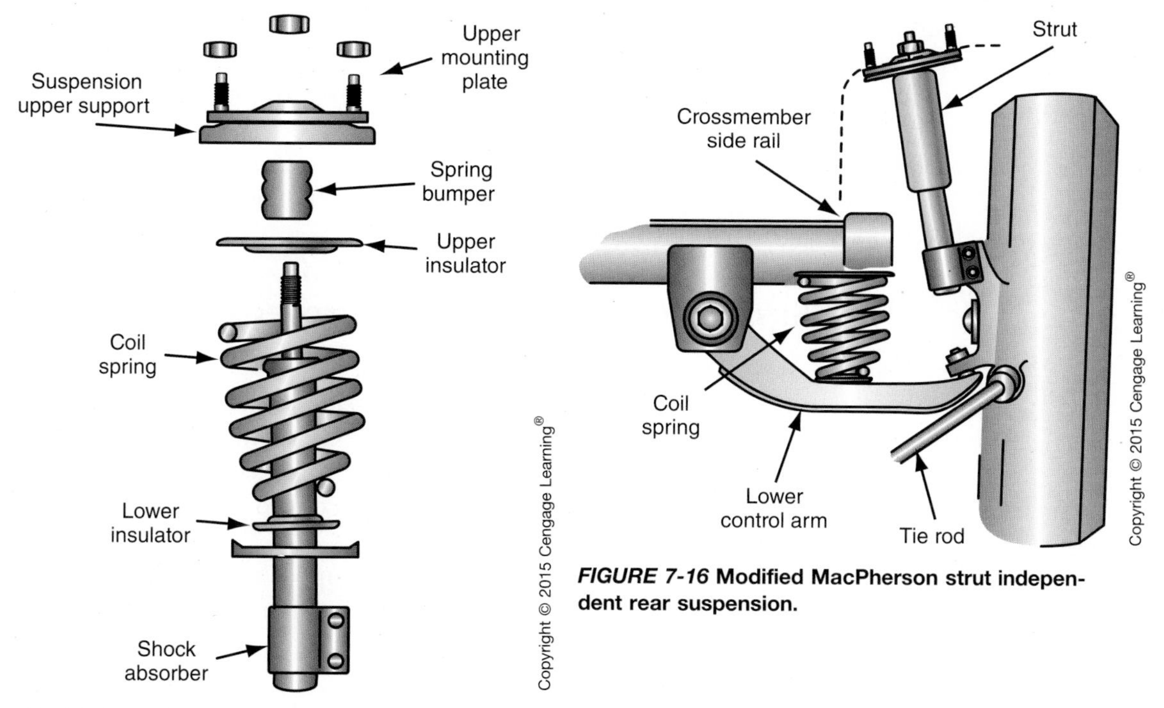

FIGURE 7-15 **Upper and lower spring insulators, MacPherson strut independent rear suspension.**

FIGURE 7-16 **Modified MacPherson strut independent rear suspension.**

The lower end of each strut is bolted to the spindle. Two strut rods are connected forward from the spindles to the chassis. Rubber insulating bushings are located in both ends of the strut rods. A stabilizer bar is mounted in rubber bushings connected to the chassis, and the ends of this bar are linked to the struts.

Insulators are mounted between the lower end of the coil spring and the lower spring seat, and the top of the coil spring and the upper spring support (Figure 7-15). These insulators help prevent the transfer of spring noise and vibration to the chassis and passenger compartment.

Modified MacPherson Strut Independent Rear Suspension System

Some front-wheel-drive vehicles have a modified MacPherson strut independent rear suspension. Each side of the rear suspension has a shock strut, lower control arm, tie rod, forged spindle, and coil spring mounted between the lower control arm and crossmember side rail (Figure 7-16).

The shock absorber strut has a rubber isolated top mount with a one-piece jounce bouncer dust shield. This top mount is attached to the body side panel and the lower end of the strut is bolted to the spindle. The stamped lower control arms are bolted to the crossmember and the spindle. A tie rod is connected from the spindle to the underbody. The purpose of each rear suspension component may be summarized as follows:

1. Stamped lower control arm—controls the lateral (side-to-side) wheel movement and contains the lower spring seat.
2. Tie rod—controls fore-and-aft wheel movement and positions the spindle properly.
3. Shock absorber strut—reacts to braking forces and provides suspension damping. A strut internal rebound stop provides rebound control, and an external jounce bumper supplies jounce control.
4. Coil spring—controls suspension travel, provides ride height control, and acts as a metal-to-metal jounce stop.

5. Forged spindle—supports the wheel bearings and attaches to the lower control arms, tie rod, brake assembly, and strut.
6. Suspension bushings—insulate the chassis and passenger compartment from road noise and vibration.
7. Suspension fasteners—connect components such as the spindle and strut. These fasteners must always be replaced with equivalent quality parts, and each fastener must be tightened to the specified torque.

Independent Rear Suspension with Lower Control Arm and Ball Joint

Some front-wheel-drive cars have an independent rear suspension system with a ball joint pressed into the outer end of the lower control arm. The ball joint contains a conventional wear indicator. The upper end of the ball joint stud is bolted into the knuckle (Figure 7-17). The inner end of the lower control arm is connected to the chassis through two rubber insulating bushings (Figure 7-18).

The lower end of the strut is bolted to the knuckle, and the upper strut mount is bolted to the inner fender reinforcement (Figure 7-19). A stabilizer bar is mounted in bushings

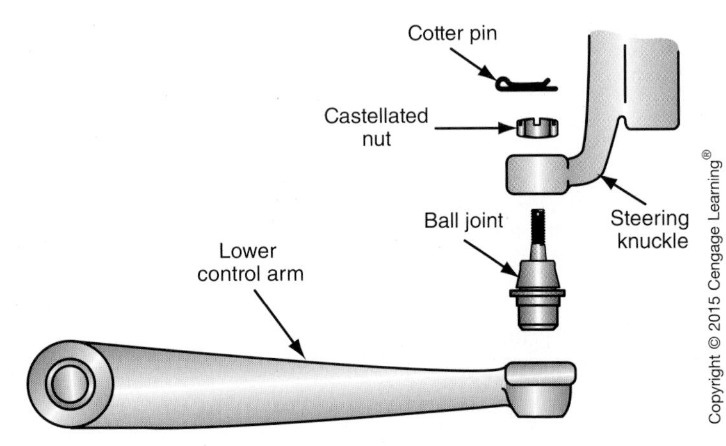

FIGURE 7-17 **Lower control arm and ball joint.**

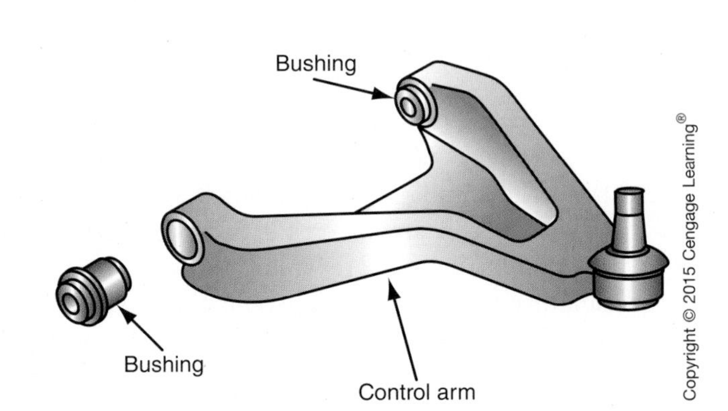

FIGURE 7-18 **Lower control arm bushings.**

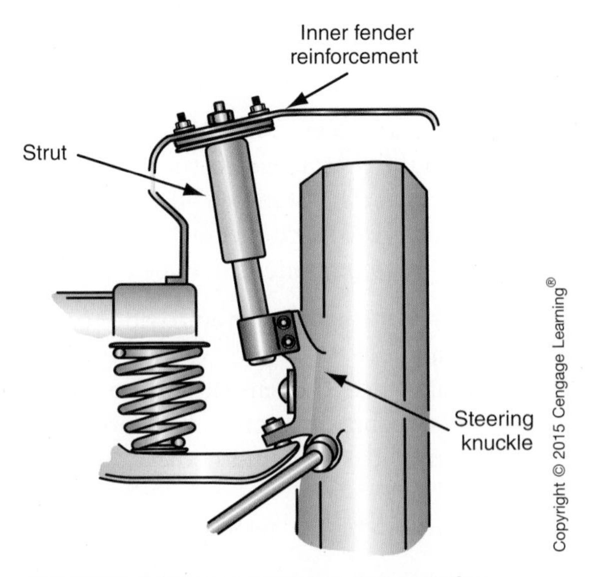

FIGURE 7-19 **Upper and lower strut mounting.**

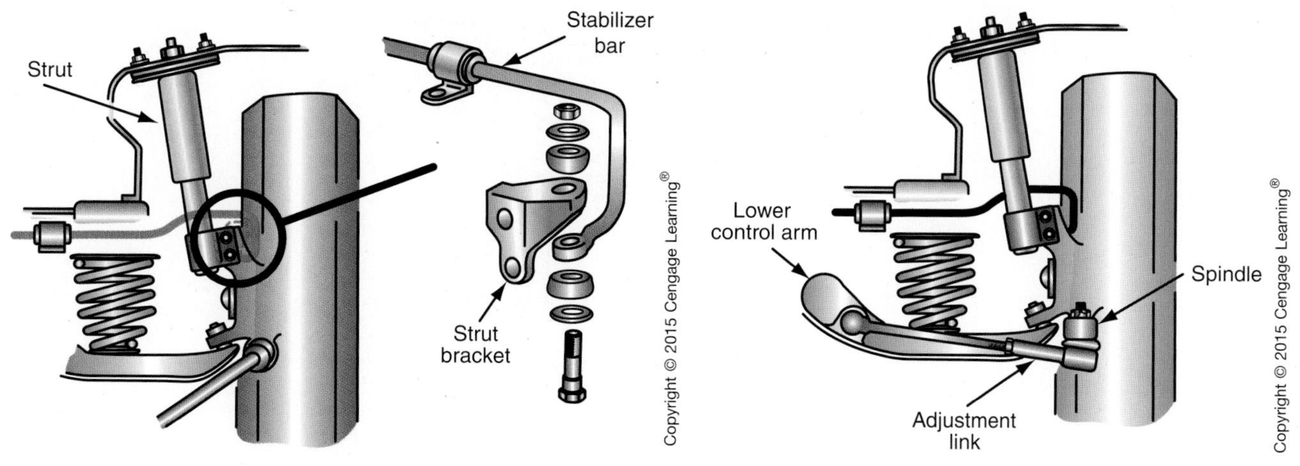

FIGURE 7-20 **Stabilizer bar mounting.**

FIGURE 7-21 **Suspension adjustment link.**

attached to the chassis, and the outer ends of this bar are linked to brackets connected to the strut (Figure 7-20).

A suspension adjustment link is connected from the lower control arm to the knuckle (Figure 7-21). This link provides a **rear wheel toe** adjustment. A coil spring seat is located in the lower control arm and the coil spring is mounted between this seat and an upper seat in the chassis (Figure 7-22). Upper and lower insulators are mounted between the coil spring and the seats.

Rear wheel toe is the variation between the distance measured between the tires at the front edge and the rear edge. If the distance between the front edge of the tires is less than the distance between the rear edge of the tires, the rear wheels have toe-in.

Independent Short-and-Long Arm Rear Suspension System

Some late model SUVs have a short-and-long arm (SLA) independent rear suspension system (Figure 7-23). This type of suspension system has upper and lower control arms, and the coil springs are mounted on the shock absorbers. The lower end of the shock absorber is mounted to the lower control arm with a rubber bushing. An upper rubber-insulated mount is positioned between the top of the shock absorber and the chassis. A toe link is connected from the spindle to the chassis, and the inner end of this link may be rotated to adjust the rear wheel toe. The inner end of the upper control arm is secured to the

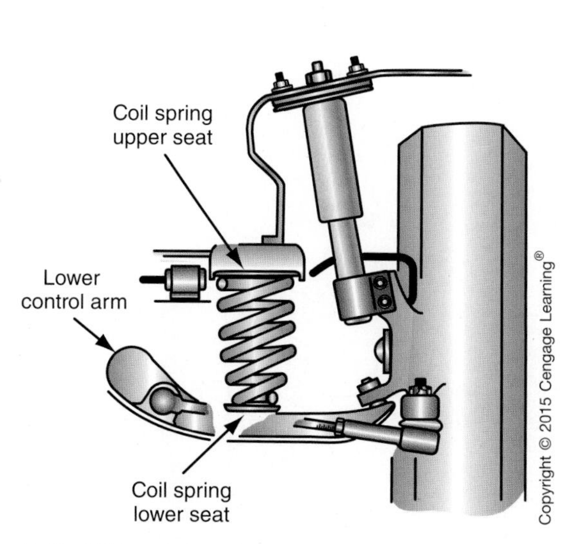

FIGURE 7-22 **Coil spring mounting.**

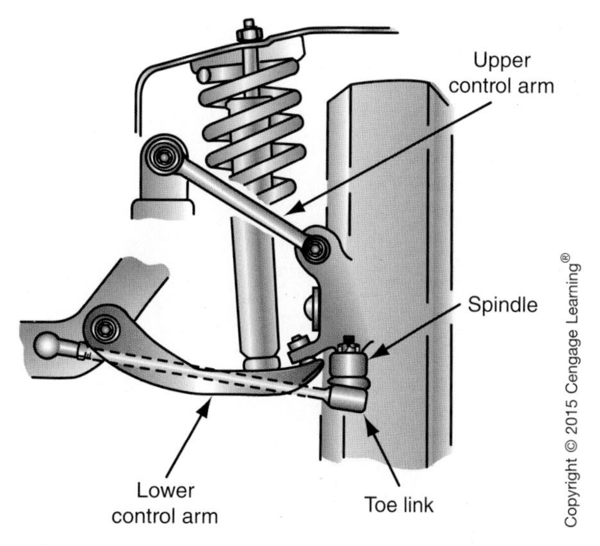

FIGURE 7-23 **Short-and-long arm independent rear suspension system.**

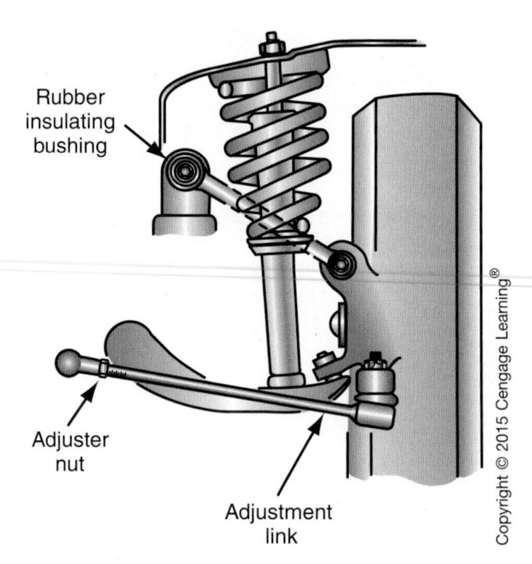

FIGURE 7-24 **Upper control arm, short-and-long arm independent rear suspension system.**

chassis with rubber insulating bushings (Figure 7-24). The SLA independent rear suspension system design allows the rear floor in the SUV to be lowered 5 inches. This body design allows extra cargo space and more headroom in the rear of the vehicle if the optional third seat is installed.

Independent Rear Suspension System with Rear Axle Carrier

Some import rear-wheel-drive cars have an independent rear suspension with a large rear axle carrier extending across the width of the chassis. A large extension on the rear axle carrier is bolted to the top of the differential, and the outer ends of this carrier are connected through heavy insulating bushings to chassis brackets (Figure 7-25).

A large final drive mount extends from the rear of the differential housing, and this mount is bolted to the chassis. When the differential is attached to the chassis, rather than being connected to the suspension, the unsprung weight is reduced. Two rubber insulating bushings connect the large trailing arms to the rear axle carrier. The rear axle carrier provides very stable differential and wheel position, which improves steering control, ride quality, and tire life. Lower coil spring seats are located on the trailing arms, and upper coil spring seats are positioned in the chassis.

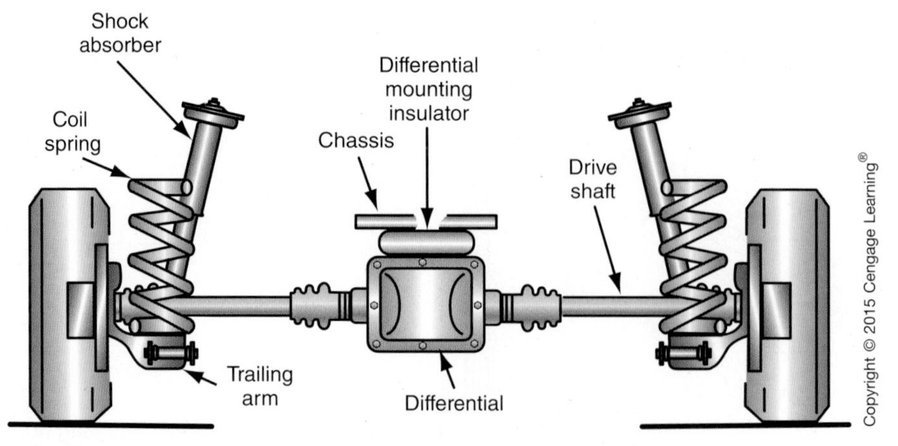

FIGURE 7-25 **Independent rear suspension with rear axle carrier.**

Shock absorbers are connected from the back of the trailing arms to the chassis. The rear wheel bearings are mounted in the outer ends of the trailing arms. Drive axles with inner and outer drive joints are connected from the differential to the rear wheels. When the drive axles have inner and outer joints, rear wheel camber change is minimized during wheel jounce and rebound. On some early independent rear suspension systems in rear-wheel-drive cars, the drive axles had only inner joints. With this type of rear suspension and drive axle, camber change was excessive during wheel jounce and rebound, and this action caused wear on the tire edges.

A stabilizer bar is bolted to the rear axle carrier through rubber insulating bushings. The outer ends of this bar are linked to the trailing arms.

Multilink Independent Rear Suspension System

Many vehicles have a **multilink independent rear suspension system**. A typical multilink rear suspension system has upper and lower control arms and additional links connected from the rear knuckles to the chassis to stabilize the rear wheel position. The suspension components are attached to the rear frame. The lower control arm is connected from the rear frame to the lower end of the knuckle. A large bushing is mounted in the inner end of the control arm at the frame attachment location (Figure 7-26). The coil springs are mounted between the lower control arms and the rear chassis, and insulators are positioned on the upper and lower ends of the springs. The upper control arms are connected from the rear frame to the top of the knuckle. The inner end of each upper control arm has two attachment points to frame, and large insulating bushings are installed at these

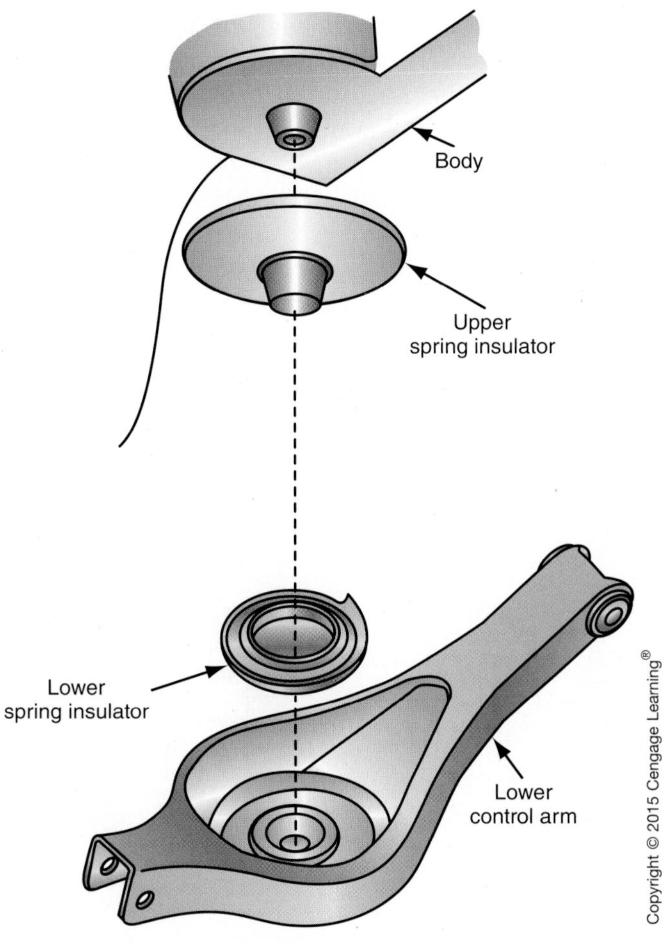

FIGURE 7-26 **Lower control arm with upper and lower spring insulators.**

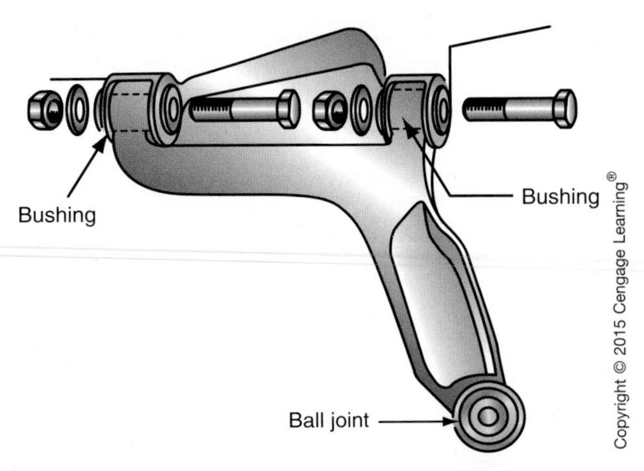

FIGURE 7-27 **The upper control arm is connected from the frame to the top of the knuckle.**

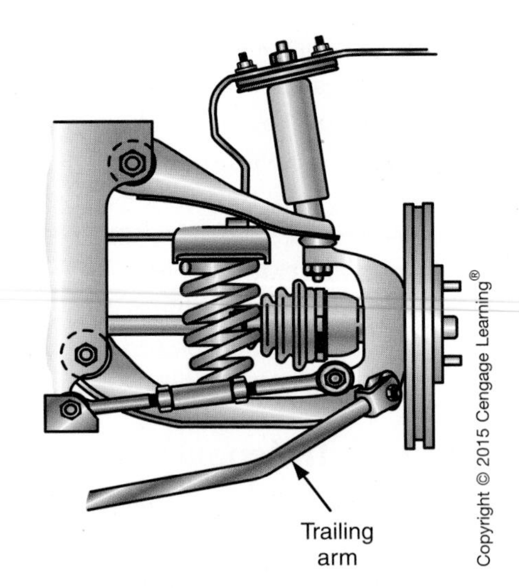

FIGURE 7-28 **The trailing arm is connected from the lower end of the knuckle to the frame.**

attachment locations (Figure 7-27). A ball joint in the outer end of each upper control arm connects the control arm to the top end of the knuckle. Many current rear suspension systems have aluminum components such as upper and lower control arms. Aluminum components reduce weight and improve fuel mileage, which in turn reduces emissions.

A trailing arm is connected from the lower end of the knuckle to the frame. Large insulating bushings are mounted at each trailing arm attachment location (Figure 7-28). The trailing arm prevents fore-and-aft wheel movement. An adjustment link is also connected from the lower end of the knuckle to the frame through insulating bushings (Figure 7-29). The adjustment link and lower control arm prevent lateral wheel movement, and this link also provides a method of rear wheel toe adjustment. This type of rear suspension may be called a **five-link suspension** because there are five attachments points between each knuckle and the frame. The lower control arm, trailing arm, and adjustment link have one attachment location, and the upper control arm has two attachment points. A five-link rear suspension provides a very stable rear wheel position, which improves steering control and rear tire tread life.

A **five-link suspension** has five attachment locations between each side of the suspension and the frame.

The shock absorbers are connected from the rear knuckles to the chassis (Figure 7-30). Two bolts retain the upper insulating shock absorber mount to the chassis. In this rear-wheel-drive car, the top of the differential is bolted to the upper part of the rear frame.

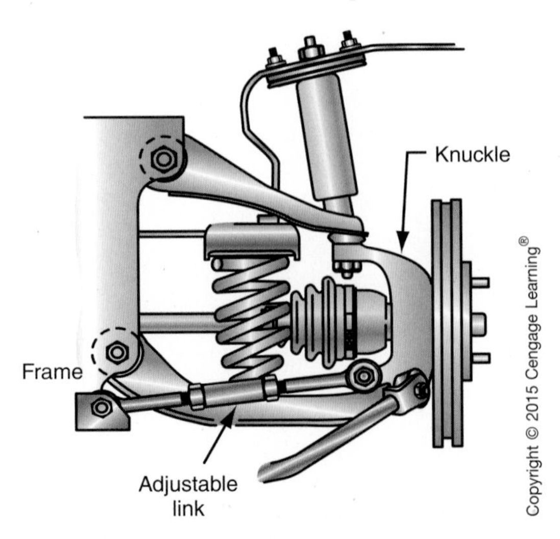

FIGURE 7-29 **Adjustment link.**

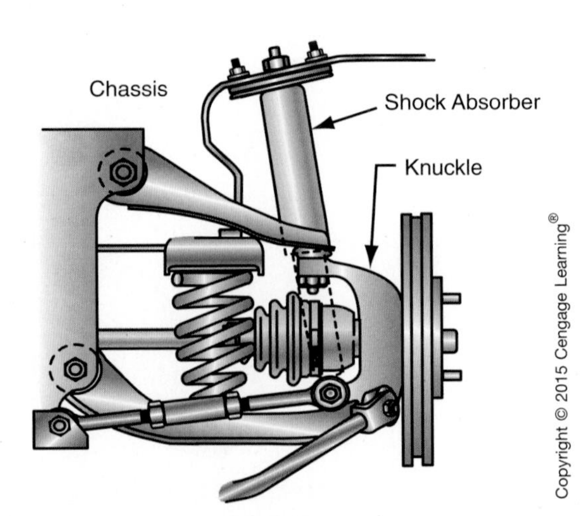

FIGURE 7-30 **Shock absorber mounting.**

Two large insulating bushings are mounted between the differential and the frame (Figure 7-31). Mounting the differential to the frame reduces the unsprung weight, which improves ride quality, because the unsprung weight forces the wheels downward during wheel rebound, which contributes to harsh ride quality.

In some current multilink rear suspension systems on rear wheel drive vehicles, four large rubber mounts are positioned between the rear suspension member or frame and the chassis. These rubber mounts are designed such that they are soft in fore and aft movement but stiff in lateral movement. These rubber mounts reduce the transfer of road vibrations and shocks from the rear suspension system to the body. The stiff lateral movement in these bushings maintains lateral rear wheel position to reduce rear tire wear and improve vehicle stability.

Some multilink rear suspension systems have four hydraulic mounts mounted between the rear suspension member or frame and the chassis. The differential is bolted to the rear frame. These hydraulic mounts contain silicone oil that helps prevent noise, vibration, and shock transfer from the differential and suspension to the chassis and vehicle interior (Figure 7-32).

The rear suspension member is connected to the outer shell of the hydraulic mounts, and the inner shell is attached to the chassis. Silicone oil fills the area between the inner

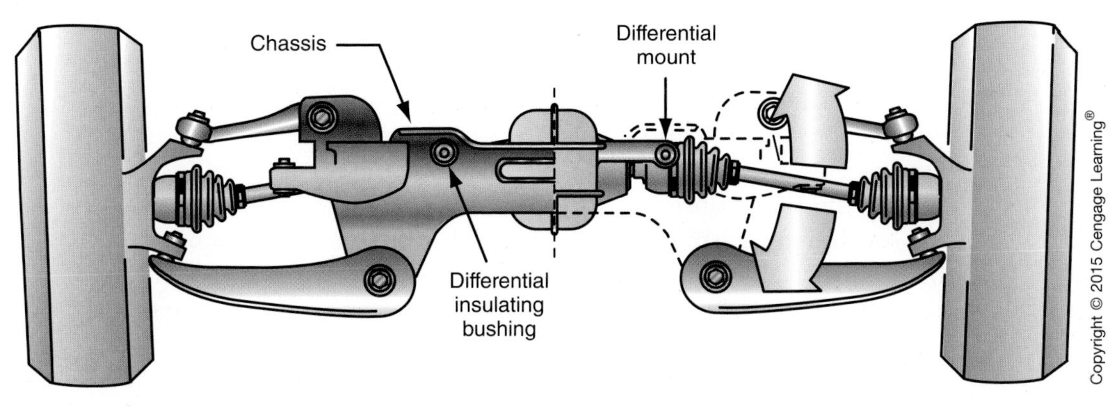

FIGURE 7-31 **Differential to frame mounting bushings.**

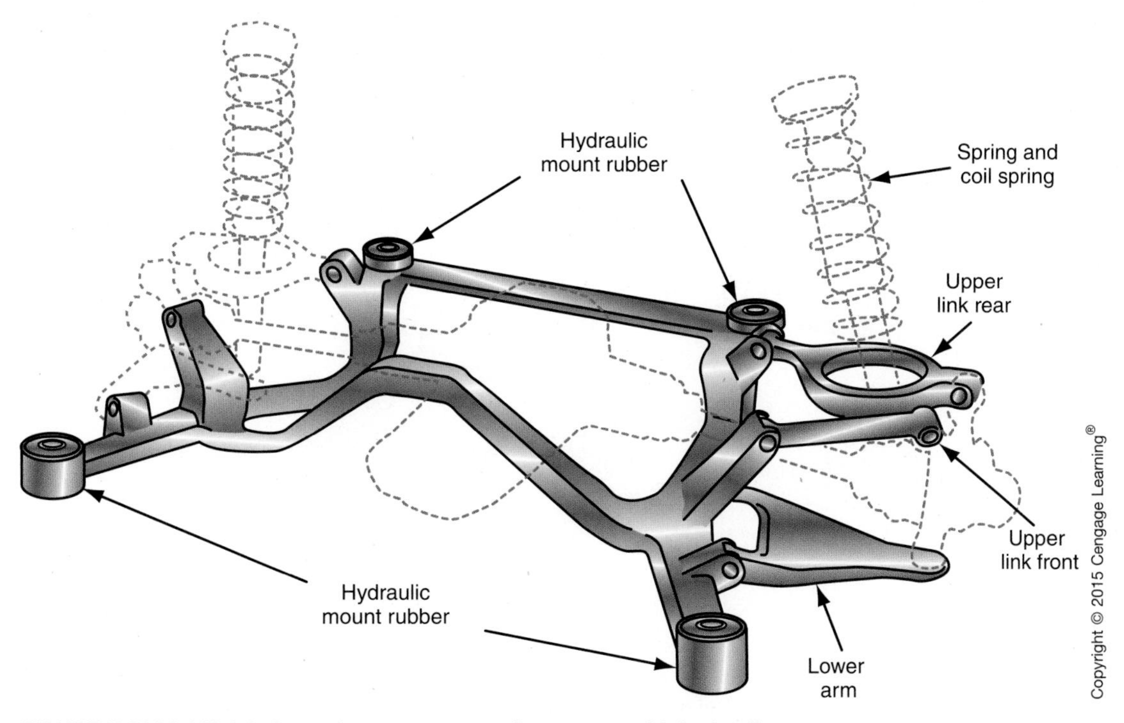

FIGURE 7-32 **Multilink independent rear suspension system with hydraulic mounts.**

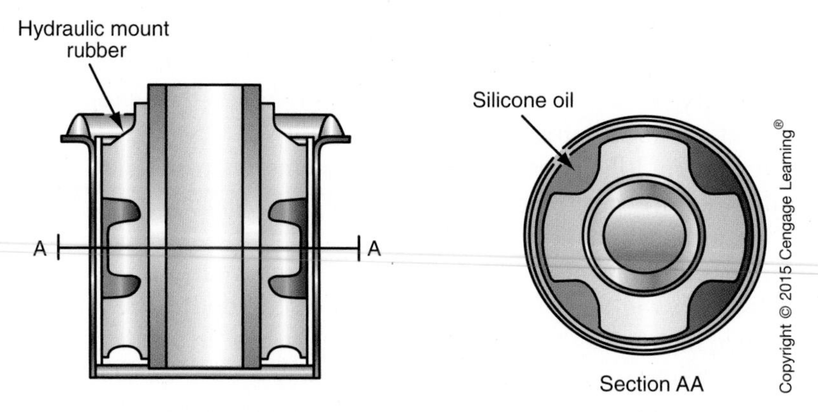

FIGURE 7-33 **Hydraulic mount design.**

and outer shells in each mount (Figure 7-33). Noise, vibration, and shock are transferred from the differential and suspension to the rear suspension member, but the silicone oil in the hydraulic mounts prevents the transfer of these undesirable forces to the chassis and vehicle interior. These hydraulic mounts have superior noise and vibration dampening characteristics compared to rubber bushings.

Shop Manual
Chapter 7, page 264

A rear upper link is connected from each side of the rear suspension member to the top of the knuckles. Both ends of this rear upper link contain rubber insulating bushings. The lower end of the shock absorber strut extends through a circular opening in the rear upper link. A front upper link is also connected from the rear suspension member to the knuckle. The top of the knuckle is supported by the front and rear upper links, rather than being supported by the shock absorber strut. The coil springs are mounted on the shock absorber struts, and the lower spring seat is attached to the strut. An upper spring seat is attached to the top of the shock absorber strut, and this upper seat is bolted into the inner fender reinforcement. The lower end of the shock absorber strut is connected to the back of the lower control arm. Lower control arms are connected from the rear suspension member to the lower end of the knuckles.

Some multilink rear suspension systems have a lower control arm and a toe control arm parallel to the lower control arm (Figure 7-34). A bolt in the outer end of the lower control arm extends through a bushing in the lower end of the steering knuckle. A ball

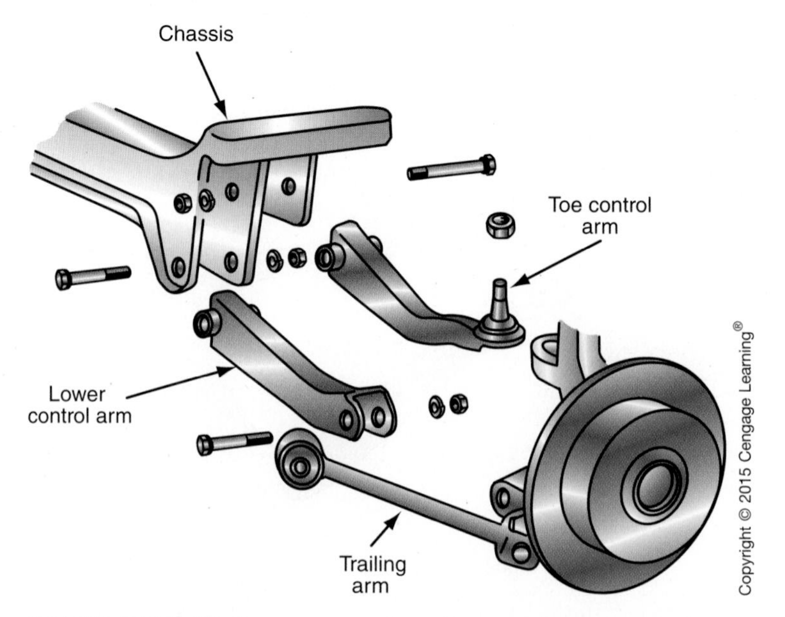

FIGURE 7-34 **Multilink rear suspension with parallel lower control arm and toe control arm.**

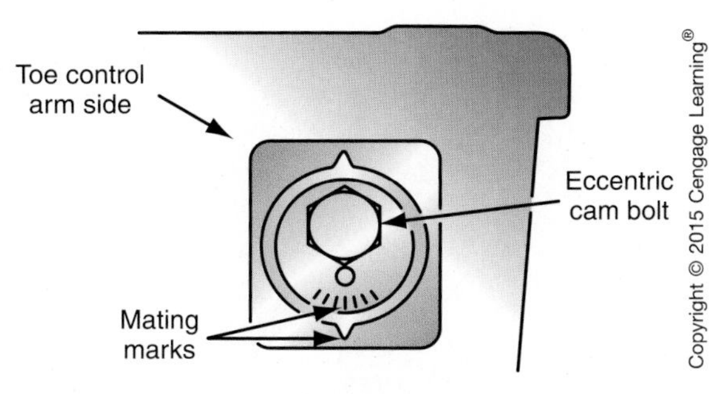

FIGURE 7-35 Eccentric cam bolt on inner end of toe control arm.

joint in the outer end of the toe control arm is bolted into the lower end of the knuckle. Rubber insulating bushings are pressed into the inner ends of the toe control arm and lower control arm. Bolts extend through these bushings and openings in the chassis. The rear end of the trailing arm is bolted to a bushing in the knuckle, and a bushing in the front end of this arm is bolted to the chassis. The trailing arm, lower control arm, and toe control arm prevent lateral and fore-and-aft wheel movement and provide excellent suspension rigidity when cornering or driving on irregular road surfaces. An eccentric cam bolt on the inner end of the toe control arm provides a rear wheel toe adjustment (Figure 7-35).

A bushing in the upper end of the knuckle is bolted to the outer end of the upper control arm (Figure 7-36). Front and rear bushings in the inner ends of the upper control arm are bolted to brackets that are in turn bolted to the chassis.

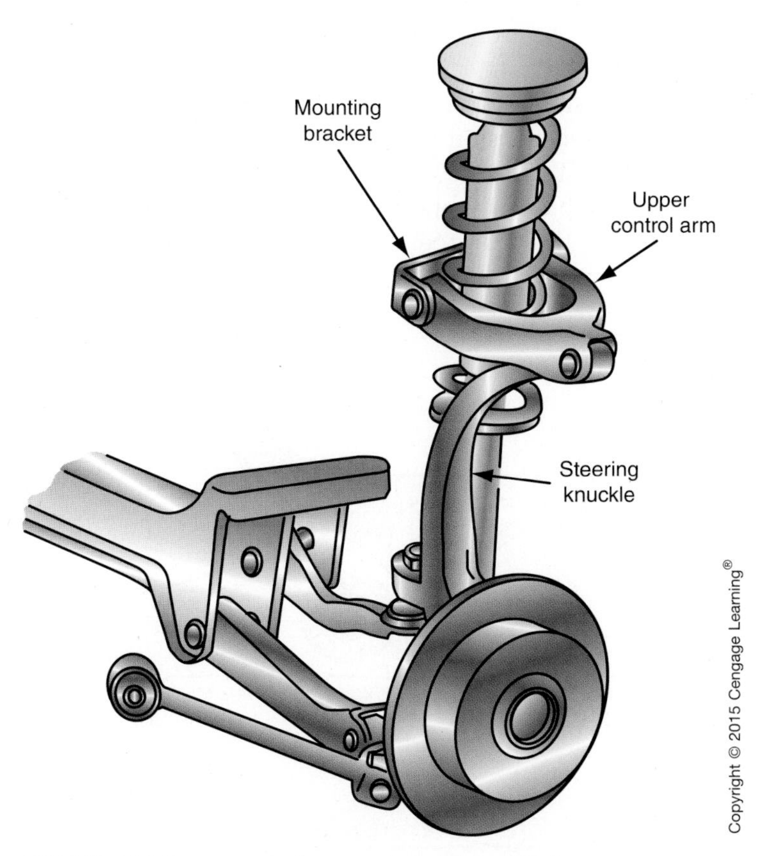

FIGURE 7-36 Upper control arm and strut in multilink rear suspension.

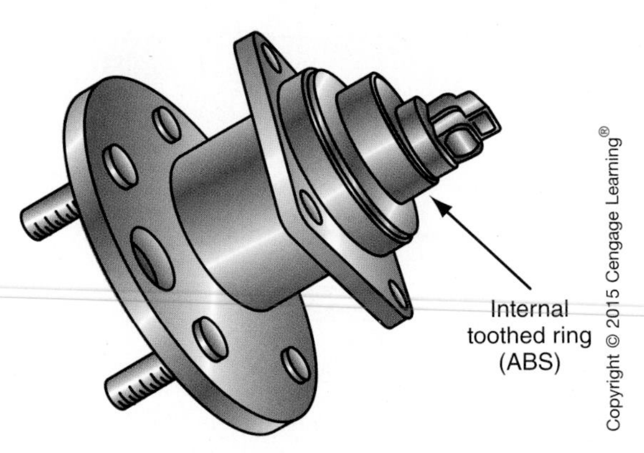

FIGURE 7-37 **One-piece rear wheel bearing assembly.**

The one-piece rear wheel bearing assemblies are bolted to the knuckles (Figure 7-37). If the vehicle has an antilock brake system, the toothed ring is attached to the inner end of each rear wheel bearing and stub axle shaft, and a wheel speed sensor is mounted in the knuckle. The rear wheel bearings assemblies are non-serviceable.

Double Wishbone Rear Suspension

In the **double wishbone rear suspension system**, the upper and lower control arms are manufactured from lightweight, high-strength aluminum alloys designed for maximum strength and rigidity. These lighter control arms decrease the unsprung weight, which helps improve traction and ride quality. Suspension rigidity is also increased by positioning the ball joints and steering knuckle inside the wheel profile (Figure 7-38). Since the upper and lower control arms have a wishbone shape, the term *double wishbone* suspension system is used for this type of rear suspension. On each side of the car, the rear suspension is attached to the chassis by a cast aluminum subframe (Figure 7-39). This design also helps reduce vehicle weight and transmits suspension loads to the chassis at the most efficient locations.

Some independent rear suspension systems experience undesirable toe changes during wheel jounce and rebound. These toe changes cause vehicle instability during cornering and acceleration. In the double wishbone rear suspension system, the control arm design and the pivot locations on the rear toe control arm provide minimal change in toe-in

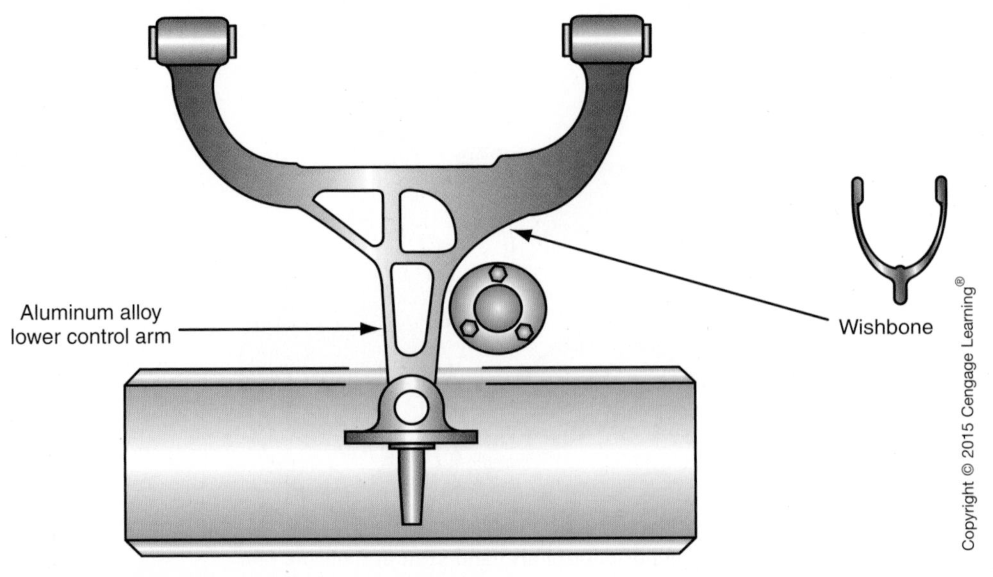

FIGURE 7-38 **Double wishbone rear suspension.**

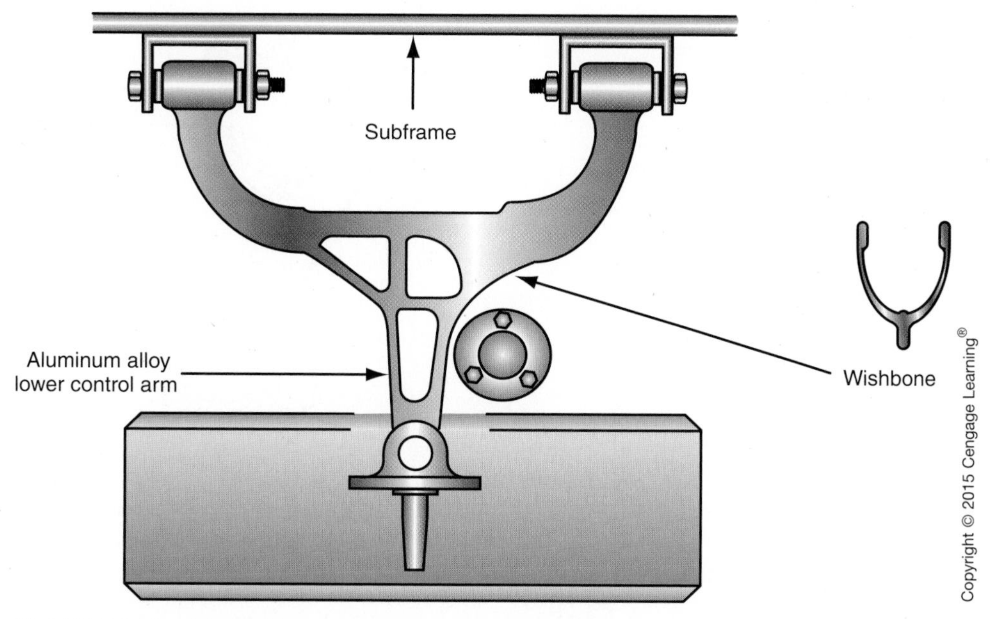

FIGURE 7-39 **Double wishbone rear suspension and subframe.**

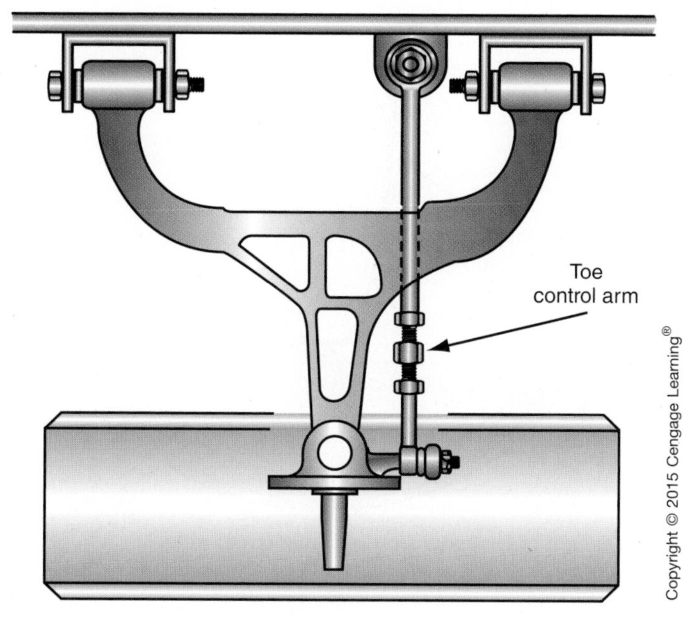

FIGURE 7-40 **Double wishbone rear suspension and toe control arm.**

during wheel jounce (Figure 7-40). This action results in extremely stable steering while cornering, accelerating, or driving on irregular road surfaces. The toe control arm may be lengthened or shortened to adjust rear wheel toe. An eccentric cam on the rear upper control arm bolt provides a rear wheel camber adjustment (Figure 7-41 and Figure 7-42).

Independent Rear Suspension with Transverse Leaf Spring

Some cars have an independent rear suspension system with a transverse mono leaf fiberglass spring. This type of spring is compact, lightweight, and corrosion free. Dual trailing arms are connected rearward from the chassis to the knuckle on each side of the suspension, and spindle support rods are attached from the center of the suspension to the bottom of the knuckle. Tie rods are connected from the rear of the knuckle to the center of the suspension (Figure 7-43).

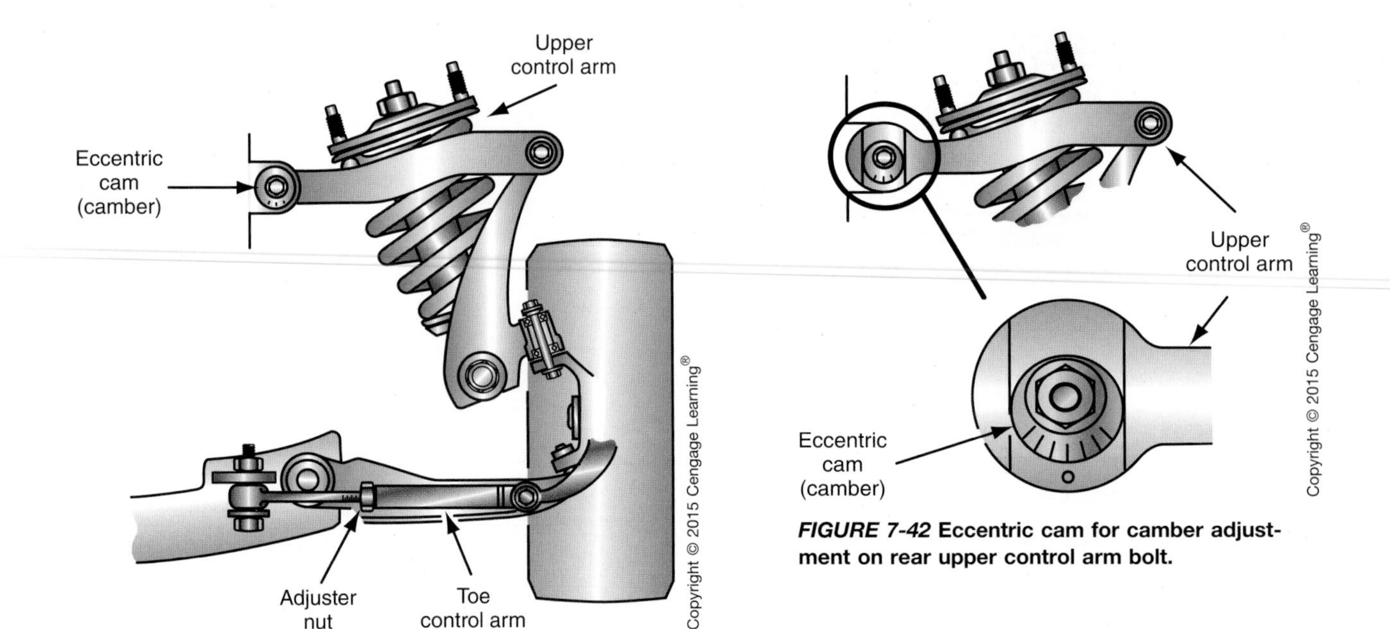

FIGURE 7-41 **Rear suspension camber and toe adjustments.**

FIGURE 7-42 **Eccentric cam for camber adjustment on rear upper control arm bolt.**

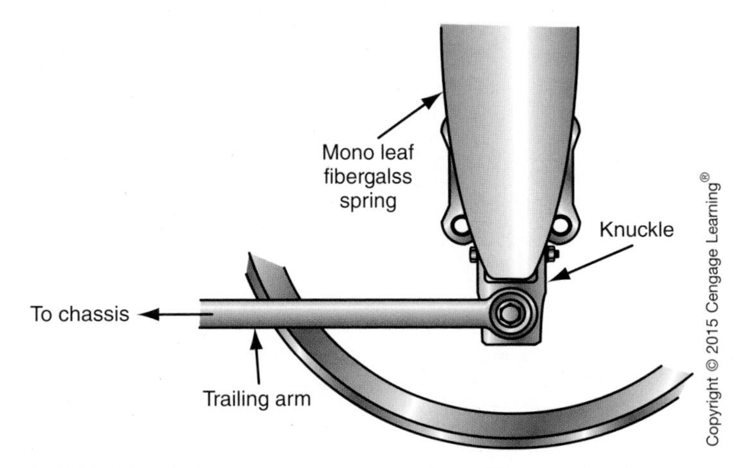

FIGURE 7-43 **Independent rear suspension with mono leaf transverse fiberglass spring.**

Shop Manual
Chapter 7, page 250

Other independent rear suspension systems have a multiple-leaf transversely mounted rear spring. In these suspension systems, heavy control arms extend rearward from the chassis to the knuckles, and strut rods are connected from the bottom of the knuckles to the center of the suspension. The shock absorbers are connected from the lower end of the knuckles to the chassis. A suspension member connects the differential housing to the chassis. All suspension component mounting locations are insulated with rubber bushings.

Curb Riding Height

Regular inspection and proper maintenance of suspension systems is extremely important to maintaining vehicle safety. *The curb riding height is determined mainly by spring condition.* Other suspension components such as control arm bushings will affect curb riding height if they are worn. Since incorrect curb riding height affects most of the other suspension angles, this measurement is critical. The curb riding height must be measured at the

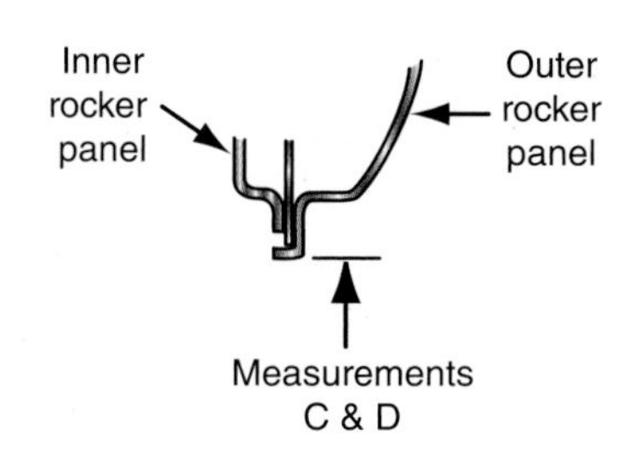

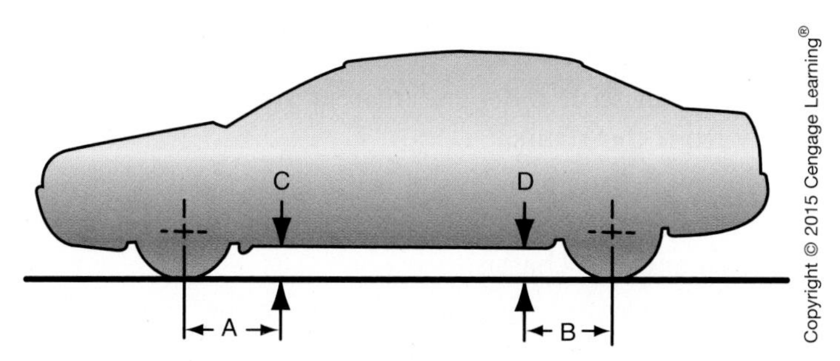

FIGURE 7-44 **Front and rear curb ride height measurement locations.**

vehicle manufacturer's specified location, which varies depending on the type of suspension system. When the vehicle is on a level floor or an alignment rack, measure the curb riding height from the floor to the manufacturer's specified location (Figure 7-44).

Caster angle is the tilt of an imaginary line through the center of the upper strut mount and ball joint in relation to the true vertical line through the center of the wheel and spindle when viewed from the side.

Spring Sag, Curb Riding Height, and Caster Angle

Sagged springs cause insufficient curb riding height. Therefore, the distance is reduced between the rebound bumper and its stop. This distance reduction causes the bumper to hit the stop frequently with resulting harsh ride quality.

When both rear springs are sagged, the **caster angle** tilts excessively toward the rear of the vehicle. This type of angle is called positive caster. Rear spring sag and excessive positive caster increase steering effort and cause rapid steering wheel return after a turn is completed.

SUMMARY

- A live axle rear suspension is one in which the differential housing, wheel bearings, and brakes act as a unit.
- A live axle rear suspension system provides good lateral stability, sway control, and steering characteristics, but this type of rear suspension causes increased tire wear with decreased ride quality and traction compared with other types of rear suspensions.
- Compared with other rear suspensions, the live axle leaf-spring suspension is more subject to axle tramp problems.
- In a live axle leaf-spring rear suspension system, the leaf springs absorb differential torque and provide lateral control.
- In a live axle coil spring rear suspension system, the lower control arms absorb differential torque and the upper arms control lateral movement.

TERMS TO KNOW

Asymmetrical leaf spring
Axle tramp
Axle windup
Braking and deceleration torque
Caster angle
Double wishbone rear suspension system
Five-link suspension

TERMS TO KNOW
(continued)

Independent rear suspension
Live axle rear suspension systems
Multilink independent rear suspension system
Rear suspension system
Rear wheel toe
Semielliptical springs
Semi-independent rear suspension
Solid axle beam
Sprung weight
Symmetrical leaf spring
Track bar
Unsprung weight

- Live axle coil spring rear suspension systems may have a tracking bar to control lateral movement.
- A semi-independent rear suspension has a limited amount of individual rear wheel movement provided by a steel U-section channel or crossmember.
- In an independent rear suspension, each rear wheel can move individually without affecting the opposite rear wheel.
- Compared with a live axle rear suspension system, an independent rear suspension provides improved ride quality, steering control, tire life, and traction.
- In a MacPherson strut independent rear suspension, the coil springs are mounted on the struts.
- In a modified MacPherson strut independent rear suspension, the coil springs are mounted separately from the struts.
- In some independent rear suspension systems, the knuckle is positioned by a ball joint on the lower end and the strut on the upper end, and the coil spring is positioned between the lower control arm and the chassis.
- In a multilink independent rear suspension, the top of the knuckle is positioned by a rear upper link and a front upper link, and the lower end of the knuckle is positioned by a lower arm.
- Some RWD vehicles have an independent rear suspension with the differential connected through insulating mounts to the chassis.
- Some FWD and RWD vehicles have an independent rear suspension with a mono leaf or multiple-leaf transversely mounted leaf spring.

REVIEW QUESTIONS

Short Answer Essays

1. Describe a live axle rear suspension system.
2. Explain the disadvantages of a live axle leaf-spring rear suspension.
3. Describe the purpose of a leaf-spring shackle.
4. Explain the differential torque action during acceleration, and describe how this torque is absorbed in a live axle coil spring rear suspension.
5. Define axle tramp.
6. Describe the purpose of a tracking bar.
7. Explain the difference between a semi-independent and an independent rear suspension system.
8. Explain the advantages of an independent rear suspension compared with a live axle rear suspension.
9. Describe the advantage of mounting the differential to the chassis in a RWD car with an independent rear suspension.
10. Describe the components that position the upper end of the knuckle in a multilink rear suspension.

Fill-in-the-Blanks

1. A live axle leaf-spring rear suspension provides excellent ________________ stability.
2. During braking and deceleration, the front of the differential is twisted ________________.
3. Axle tramp occurs during hard ________________.
4. In many live axle coil spring rear suspensions, the differential torque is absorbed by the trailing lower ________________ ________________.
5. In a modified MacPherson strut independent rear suspension, the coil spring seat is located on the ________________ ________________.
6. In a multilink independent rear suspension, the rear suspension member supports the suspension and the ________________.
7. In some multilink rear suspension systems, the rubber mounting bushings between the suspension member and the chassis are ________________ in the fore and aft direction and ________________ in the lateral direction.
8. A fiberglass mono leaf rear spring is compact, lightweight, and ________________ free.

9. When the rear springs are sagged, the caster angle on the front suspension becomes more ______________.

10. If the rear springs are sagged, the ______________ is increased.

Multiple Choice

1. Rear axle tramp is caused by:
   A. Irregular road surfaces.
   B. A bent rear control arm.
   C. Engine torque transmitted through the drive shaft.
   D. Improper rear wheel alignment.

2. All of these statements about a live axle leaf-spring rear suspension are true EXCEPT:
   A. During acceleration the front of the differential twists upward.
   B. The differential torque in a live axle rear suspension is applied to the springs.
   C. While decelerating and braking, the front of the differential twists downward.
   D. This type of rear suspension has a small amount of unsprung weight.

3. Rear axle tramp occurs during:
   A. Hard acceleration.
   B. Deceleration.
   C. High speed driving.
   D. Cornering at high speed.

4. In a semi-independent rear suspension system, some individual wheel movement is provided by:
   A. The U-section channel and integral stabilizer bar.
   B. The struts.
   C. The trailing arms.
   D. The track bar and brace.

5. In a semi-independent rear suspension system:
   A. The track bar and brace absorb differential torque.
   B. The trailing arms prevent lateral wheel movement.
   C. The lower coil spring seats may be on the trailing arms.
   D. The lower end of the shock absorbers may be attached to the track bar.

6. In a MacPherson strut independent rear suspension system:
   A. The spring seats are positioned on the struts.
   B. The upper spring seat is positioned in a chassis support.
   C. The lower end of each strut is attached to the strut rod.
   D. The ends of the coil spring are in direct contact with the spring seats.

7. All of these statements about modified MacPherson strut suspension systems are true EXCEPT:
   A. The lower control arms prevent lateral wheel movement.
   B. The tie rods control fore-and-aft wheel movement.
   C. The strut is bolted to the lower end of the spindle.
   D. The upper strut mount is retained on top of the strut.

8. In an independent rear suspension with a lower control arm and ball joint:
   A. The ball joint is pressed into the tie rod.
   B. The strut is bolted to the top of the knuckle.
   C. The lower coil spring seat is mounted on the strut.
   D. The suspension adjustment link is connected from the lower control arm to the chassis.

9. Sagged rear springs may cause:
   A. Slow steering wheel return after turning a corner.
   B. Decreased steering effort.
   C. Decreased positive caster on the front wheels.
   D. Harsh ride quality.

10. The adjustment link in a multilink rear suspension:
    A. Provides a rear wheel caster adjustment.
    B. Provides a rear wheel toe adjustment.
    C. Reduces fore-and-aft rear wheel movement.
    D. Absorbs engine torque transmitted through the drive shaft.

Chapter 8

STEERING COLUMNS AND STEERING LINKAGE MECHANISMS

UPON COMPLETION AND REVIEW OF THIS CHAPTER, YOU SHOULD BE ABLE TO UNDERSTAND AND DESCRIBE:

- How the steering column provides driver safety during a frontal collision.
- Two methods of steering column movement to protect the driver in a frontal collision.
- The purpose of a clock spring electrical connector.
- The mechanism that locks the steering wheel and gear shift when the ignition switch is in the lock position.
- A parallelogram steering linkage and explain the advantage of this type of linkage.
- The components in a parallelogram steering linkage.
- The purposes of the pitman arm and the idler arm.
- Two possible rack and pinion steering gear mountings.
- The rack and pinion steering linkage, and explain the advantages of this type of linkage.
- The design and operation of an active steering column.
- The design and operation of a driver protection module.

INTRODUCTION

Steering columns play a significant part in steering control, safety, and driver convenience. The steering column connects the steering wheel to the steering gear. The column components must be in satisfactory condition to minimize free play and to provide adequate steering control. If steering column components such as universal joints are worn, column free play is excessive and steering control is reduced.

Most steering columns provide some method of column collapse during a collision. Some vehicles have plastic pins that shear off in the column jacket, gearshift tube, and steering shaft if the driver hits the steering wheel in a frontal collision. This shearing action of the plastic pins allows the column to collapse away from the driver. In other vehicles, the column-to-instrument-panel mounting is designed to allow column movement if the driver hits the steering wheel in a collision. This action helps prevent driver injury.

All air-bag-equipped vehicles have a driver's side air bag located in the upper side of the steering wheel, and most vehicles also have a passenger's side air bag mounted in the instrument panel. Some vehicles now have seat belt pretensioners that tighten the seat belts and hold the driver or passengers back against the seat if the vehicle is involved in a collision. Many vehicles now have side air bags mounted in the outer edge of the seat back, or an air bag curtain mounted just above the front and rear door openings. The side air bags have a separate module and sensors, and deploy only when the vehicle is involved in a side collision.

A recent development on some vehicles is the installation of **pre-safe systems** that react during the few milliseconds before a collision occurs to increase driver and passenger safety. The input signals to the pre-safe system include vehicle speed, braking torque, brake pedal application speed, wheel slip, vehicle acceleration around the vertical axis, spring compression and rebound travel, steering wheel rotational speed, and tire pressure. The pre-safe system recognizes and acts in three crucial situations:

1. Sideways skidding
2. Avoidance maneuvers
3. Emergency braking beyond ABS operation

The pre-safe system can differentiate between a drama and a crisis. For example, if the vehicle is driven over a patch of ice and the brakes are applied lightly without any vehicle skidding, the ABS system can operate to prevent wheel lockup, but the pre-safe system recognizes this condition as a drama and remains inoperative. The pre-safe system also remains inoperative during mild avoidance maneuvers.

When the inputs indicate an impending crisis driving situation, the pre-safe system performs these functions:

1. Pretensions the driver and front passenger seat belts so the driver and front passenger are optimally restrained in their seats during emergency braking or a skid prior to a collision. When a collision occurs, pyrotechnic seat belt tensioners provide increased seat belt tension. If a collision does not occur, the seat belts return to normal tension.
2. If the front passenger seat is moved too far forward, the pre-safe system moves the seat rearward before a collision occurs. If appropriate, the front passenger seat cushion and backrest are moved to a position that provides increased passenger protection if a collision occurs. For example, a steeply inclined seat backrest or a flat seat cushion may impair passenger restraint during a collision.
3. The electronically adjustable, individual rear seat cushions are moved to the best possible angle to improve rear passenger restraint.
4. The sliding sunroof is closed to reduce the risk of occupant injury if a rollover occurs.

The brake assist system (BAS) also plays a crucial role in the pre-safe system operation. The BAS system operates beyond the ABS and electronic stability program (ESP) parameters. If the BAS detects an unstable driving condition such as severe vehicle skidding, it uses brake intervention and engine power reduction to correct the situation.

When the vehicle is involved in a frontal collision above a specific speed, the air bag sensor signals deploy the air bag or bags in a few milliseconds. As the driver is thrust forward during the frontal collision, the inflated air bag prevents the driver from striking the steering wheel, windshield, or instrument panel. The air bag deflates very quickly so it does not block the driver's vision in case the driver is still attempting to steer the vehicle. Since the driver's side air bag and some connecting components are mounted in the steering column, this column is a very important part of the vehicle safety equipment.

Shop Manual
Chapter 8, page 274

Steering columns may be classified as nontilting, tilting, and tilting or telescoping. Tilt steering columns facilitate driver entry to and exit from the front seat. These columns also allow the driver to position the steering wheel to suit individual comfort requirements (Figure 8-1). A tilting or telescoping steering column allows the driver to tilt and extend or retract the steering wheel. In this type of steering column, the driver has more steering wheel position choices.

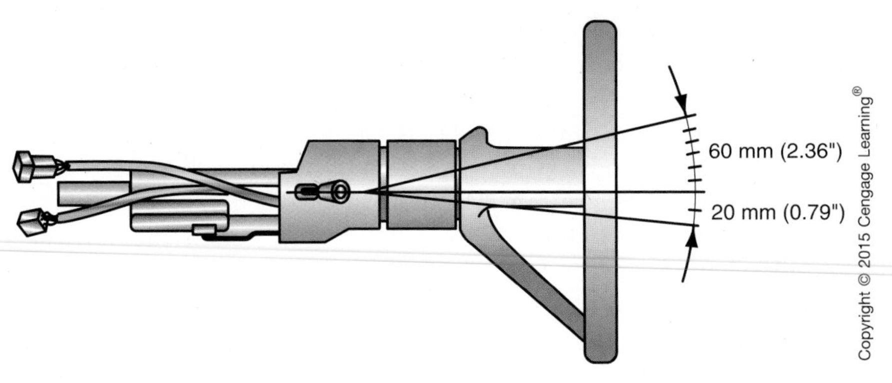

FIGURE 8-1 **Tilt steering column.**

The **toe plate** surrounds the steering column and covers the opening where the column extends through the vehicle floor.

The **silencer** is mounted with the toe plate and helps to reduce the transmission of engine noise to the passenger compartment.

Shop Manual
Chapter 8, page 279

Conventional Nontilt Steering Column

Design

Many steering columns contain a two-piece steering shaft connected by two universal joints. A jacket and shroud surround the steering shaft. The upper shaft is supported by two bearings in the jacket. A **toe plate**, seal, and **silencer** surround the lower steering shaft and cover the opening where the shaft extends through the floor (Figure 8-2). The lower steering shaft is surrounded by a shield underneath the toe plate.

The lower universal joint couples the lower shaft to the stub shaft in the steering gear. In some steering columns, a flexible coupling is used in place of the lower universal joint (Figure 8-3).

Studs and nuts retain the steering column bracket to the instrument panel support bracket. The steering column is designed to protect the driver if the vehicle is involved in a frontal collision. An **energy-absorbing lower bracket** and lower plastic adapter are used to connect the steering column to the instrument panel mounting bracket. This bracket allows the column to slide down if the driver is thrown forward into the wheel in a frontal collision. The mounting bracket is also designed to prevent rearward movement toward the driver in a collision.

In some steering columns, the outer column jacket is a two-piece unit retained with plastic pins (Figure 8-4). In this type of column, the lower steering shaft is a two-piece sliding unit retained with plastic pins (Figure 8-5). When the driver is thrown against the steering wheel in a frontal collision, the plastic pins shear off in the lower steering shaft and outer column

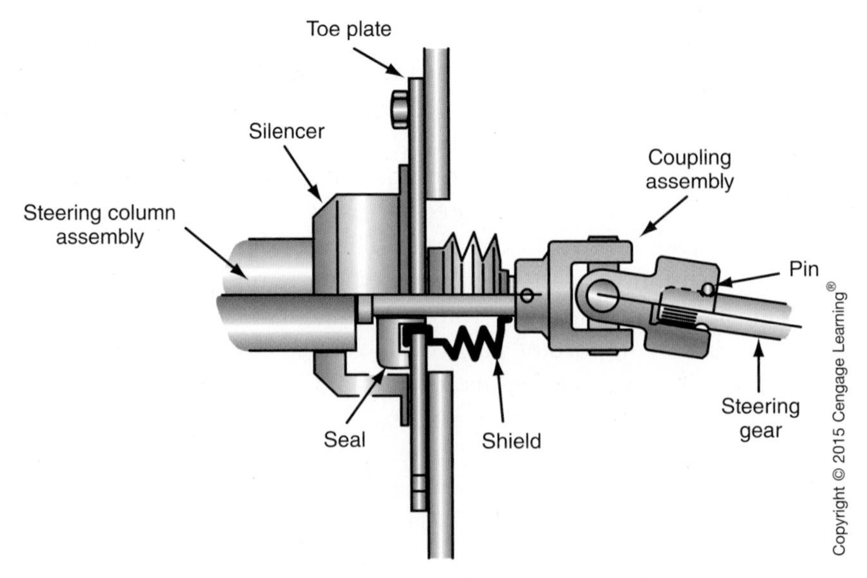

FIGURE 8-2 **Toe plate and silencer.**

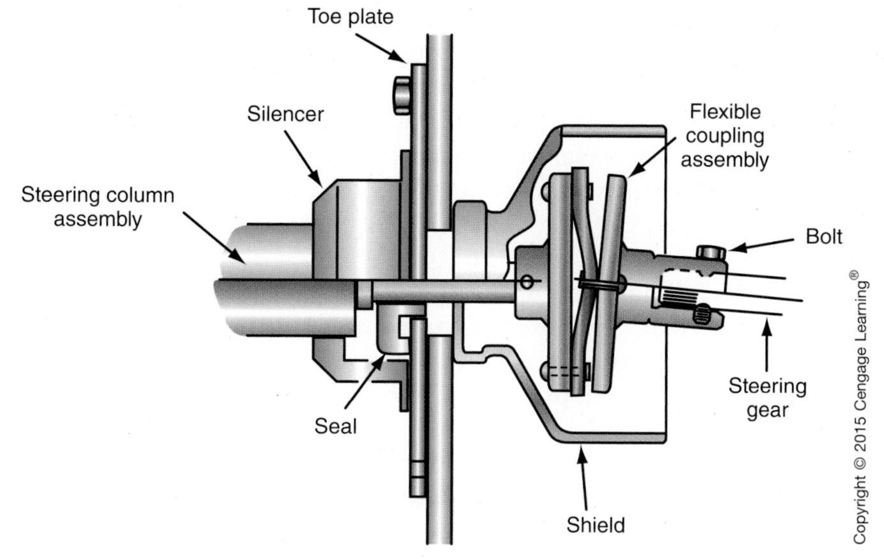

FIGURE 8-3 **Flexible coupling.**

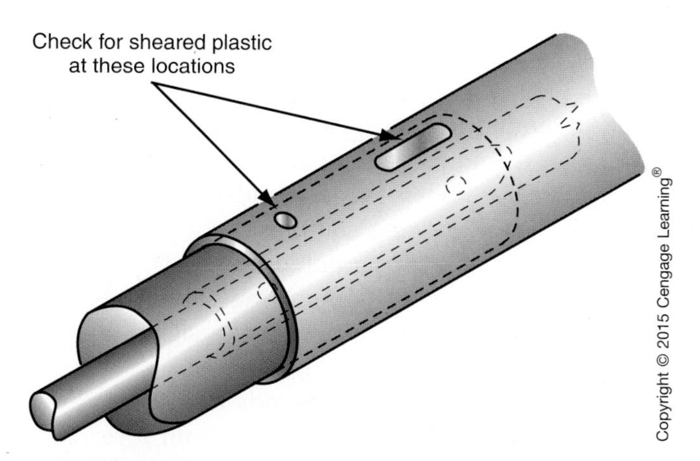

FIGURE 8-4 **Injection plastic in collapsible outer steering column jacket.**

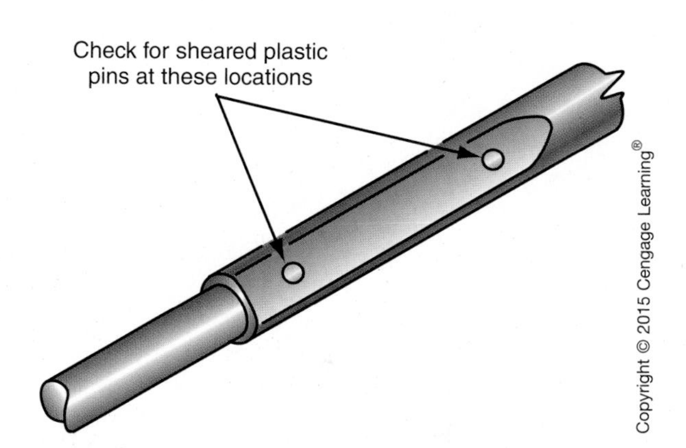

FIGURE 8-5 **Injection plastic in collapsible lower steering shaft.**

jacket. The shearing action of the plastic pins allows the steering column to collapse away from the driver, which reduces the impact as the driver hits the steering wheel.

A few current vehicles have steering columns with magnesium jackets or housings to reduce vehicle weight. Vehicle weight reduction improves fuel economy and reduces emissions.

An adaptive steering column has recently been introduced to the automotive market on some new models. The adaptive steering column collapses at two different speeds based on information received about the driver. Many vehicles have sensors in the lower part of a front seat to determine the weight of the driver and the weight and the presence of a passenger. The driver's weight signal, and other input signals are used by the restraints control module (RCM) to vary the air bag deployment force. The driver's weight input signal also determines how quickly the steering column should collapse. The adaptive steering column tailors the steering column collapse load to the driver's mass, safety belt use, and seat track position.

An adaptive steering column contains an energy-absorbing steel that buckles between the upper and lower portions of the steering column. The RCM receives input signals from various crash sensors indicating the severity of the crash, and this module determines the speed of steering column collapse. If input signals indicate a faster steering column

A **pyrotechnic** device contains an explosive and an ignition source.

An **air bag deployment module** contains the air bag and the inflator device.

The clock spring maintains positive electrical contact between the air bag module and the air bag electrical system regardless of steering wheel position. A clock spring electrical connector may be called a coil, **spiral cable**, or cable reel, depending on the manufacturer.

The combination turn signal, hazard warning, wipe or wash, and dimmer switch are sometimes called a smart switch.

CAUTION:
When servicing a vehicle equipped with an air bag or bags, follow all service precautions in the vehicle manufacturer's service manual. Failure to follow these precautions may result in an expensive, accidental air bag deployment.

collapse is required, the RCM fires a **pyrotechnic** device in the steering column that pulls a pin in the column and allows the energy-absorbing steel to buckle and provide faster column collapse. The result is a softer impact between the driver and the steering wheel. The adaptive steering column is designed to operate with the driver's air bag.

When the driver is thrown against the steering wheel during a collision, many steering wheels are designed to deform away from the driver to reduce the force on the driver's body.

WARNING: Small amounts of sodium hydroxide are a by-product of an air bag deployment. Sodium hydroxide is a caustic chemical that causes skin irritation and eye damage. Always wear eye protection and gloves when servicing and handling a deployed air bag.

On many cars, the **air bag deployment module** is mounted in the top of the steering wheel (Figure 8-6). A **clock spring electrical connector**, or **spiral cable**, is mounted under the steering wheel. This component contains a ribbon-type conductor that maintains constant electrical contact between the air bag module and the air bag electrical system during steering wheel rotation.

The steering wheel splines fit on matching splines on the top of the upper steering shaft, and a nut retains the wheel on the shaft. Most steering wheels and shafts have matching alignment marks that must be aligned when the steering wheel is installed.

An ignition switch cylinder is usually mounted in the upper right side of the column housing, and the ignition switch is bolted on the lower side of the housing (Figure 8-7). An operating rod connects the ignition switch cylinder to the ignition switch. Ignition switches are integral with the lock cylinder in some steering columns.

The turn signal switch and hazard warning switch are mounted on top of the steering column under the steering wheel. Lugs on the bottom of the steering wheel are used to cancel the signal lights after a turn is completed. On many vehicles, the signal light lever also operates the wipe or wash switch and the dimmer switch (Figure 8-8).

If the gear shift is mounted in the steering column, a tube extends from the gear shift housing to the shift lever at the lower end of the steering column. This shift lever is connected through a linkage to the transaxle or transmission shift lever. A lock plate is attached to the upper steering shaft, and a lever engages the slots in this plate to lock the steering wheel and gear shift when the gear shift is in Park and the ignition switch is in the Lock position (Figure 8-9).

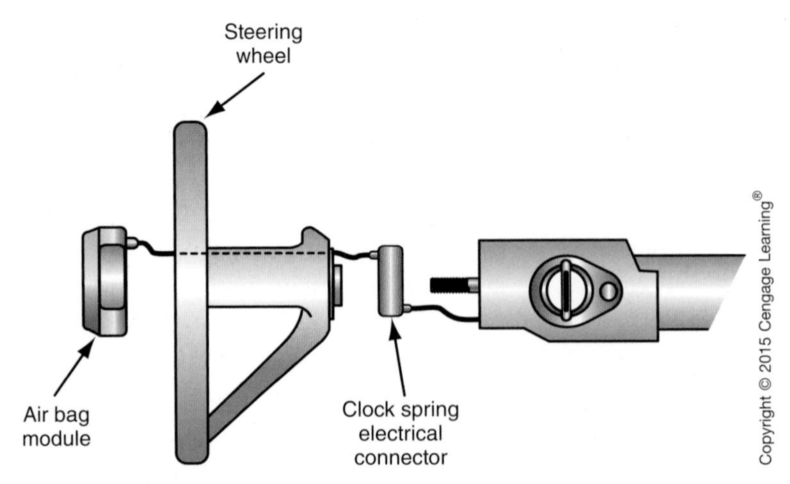

FIGURE 8-6 **Air bag inflator module mounted in the steering wheel and clock spring electrical connector located under the steering wheel.**

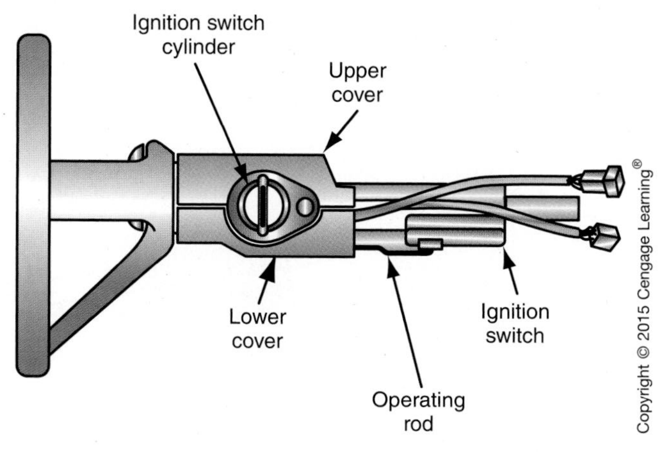

FIGURE 8-7 **Ignition switch and ignition switch cylinder mounted in steering column.**

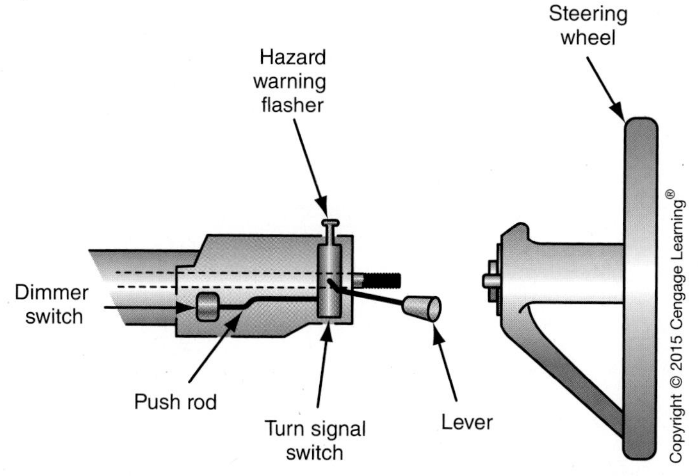

FIGURE 8-8 **Turn signal switch, hazard warning switch, dimmer switch, and wipe or wash switch mounted in steering column.**

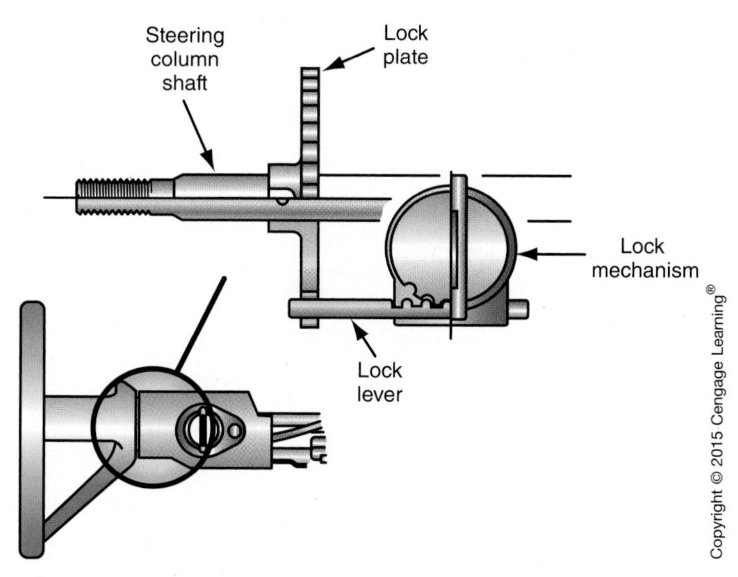

FIGURE 8-9 **Upper steering column with locking plate and lever.**

Tilt Steering Column

Design

Shop Manual
Chapter 8, page 280

Tilt steering columns have a short upper steering shaft connected to the steering wheel in the usual manner. This upper steering shaft is connected through a universal joint to the lower steering main shaft. An upper column tube surrounds the upper steering shaft, and the lower steering main shaft is supported on bushings in the lower column tube (Figure 8-10).

When the steering wheel is tilted, the wheel and upper steering column tube pivot on two bolts connected between the upper and lower column tubes. During the steering wheel tilting motion, the upper steering shaft pivots on the universal joint connected between the upper steering shaft and the lower steering main shaft. A release lever on the side of the steering column must be activated to allow the wheel and upper column to tilt. This steering wheel action allows the driver to position the steering wheel for greater comfort and easier movement in and out of the driver's seat.

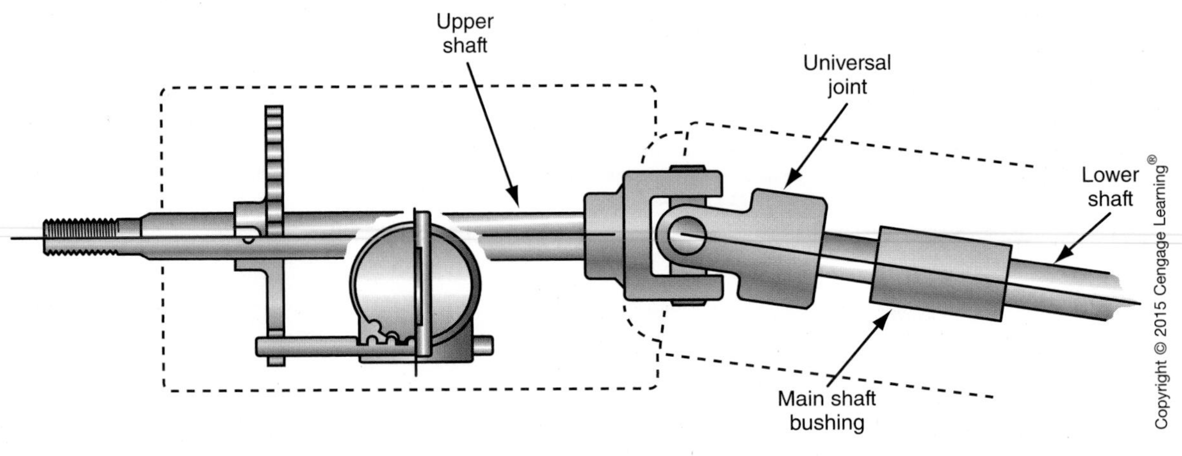

FIGURE 8-10 **Tilt steering column components.**

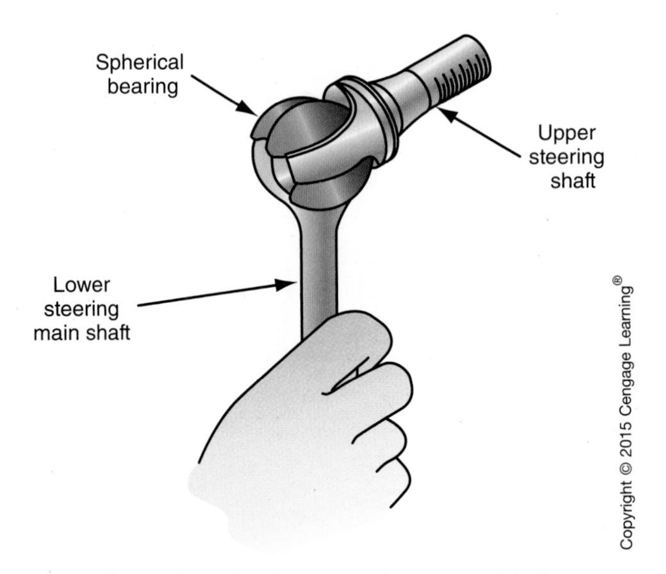

FIGURE 8-11 **Spherical bearing acts as a pivot between the upper steering shaft and the lower steering main shaft in some tilt steering columns.**

In some steering columns, a **spherical bearing** acts as a pivot between the upper steering shaft and the lower steering main shaft (Figure 8-11).

Four bolts attach the lower column support tube to the instrument panel. The ignition lock cylinder and switch are mounted in an upper bracket that is clamped to the lower column tube. A universal joint is connected between the lower steering mainshaft and the intermediate shaft assembly, which is attached to the stub shaft on the steering gear (Figure 8-12).

A steering wheel pad containing the air bag deployment module is mounted in the top of the steering wheel. Electrical contact between the air bag deployment module and the air bag electrical system is maintained by a clock spring electrical connector, or spiral cable, mounted directly under the steering wheel (Figure 8-13). A combination switch is mounted directly under the clock spring electrical connector. This switch contains the signal light switch, hazard light switch, dimmer switch, and wipe or wash switch. The upper side of the steering wheel also contains the horn switch contacts.

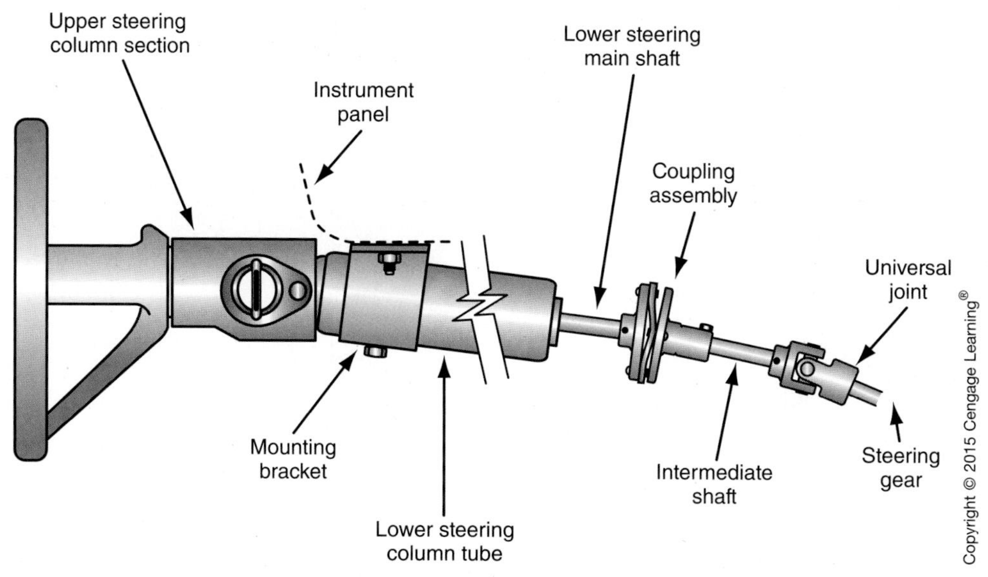

FIGURE 8-12 **Tilt steering column with universal joint, intermediate shaft, combination switch, and steering wheel.**

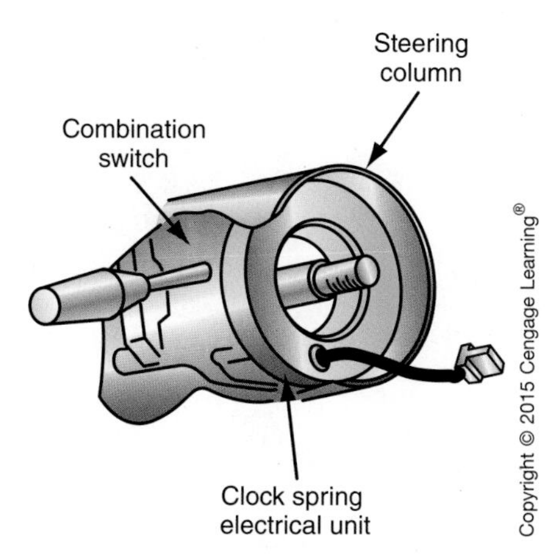

FIGURE 8-13 **Clock spring electrical connector, or spiral cable, maintains electrical contact between the air bag inflator module and the air bag electrical system.**

Electronic Tilt and Telescoping Steering Column

Some vehicles are equipped with an electronically controlled tilt and telescoping steering column. A driver-operated switch mounted in the steering column below the signal light lever controls the tilt and telescoping functions (Figure 8-14). A tilt and telescoping motor is mounted in the steering column power assembly. A potentiometer is mounted on the side of the tilt and telescoping motor, and this potentiometer is a steering column position sensor (Figure 8-15). This potentiometer sends a voltage signal to the driver position module (DPM) in relation to the motor and steering wheel position. The steering column position sensor signal is sent to the DPM, and the DPM uses this signal when storing and recalling steering column memory settings. A short tilt cable assembly is connected between the tilt and telescoping motor and the tilt mechanism in the steering column (Figure 8-16).

When the tilt and telescoping switch is operated by the driver, it sends voltage input signals to the DPM. If the driver operates the tilt and telescoping switch to request the

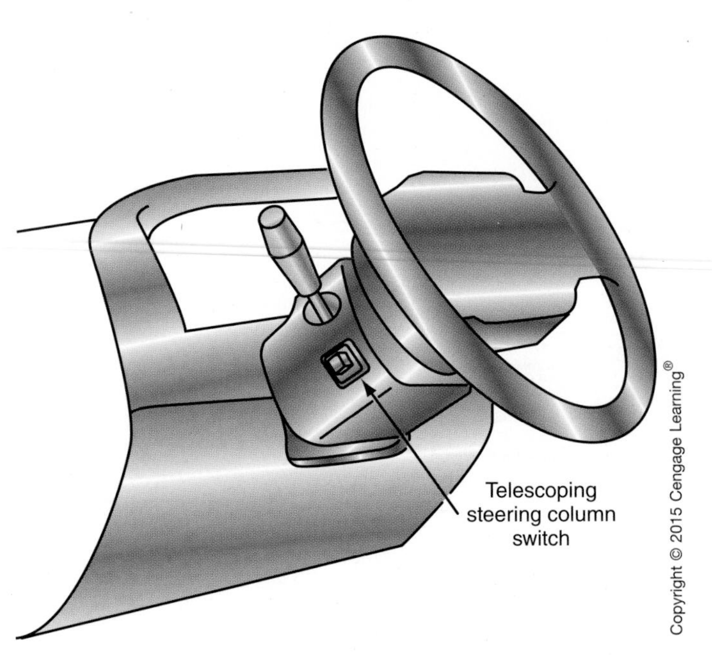

FIGURE 8-14 **Tilt and telescoping steering column switch.**

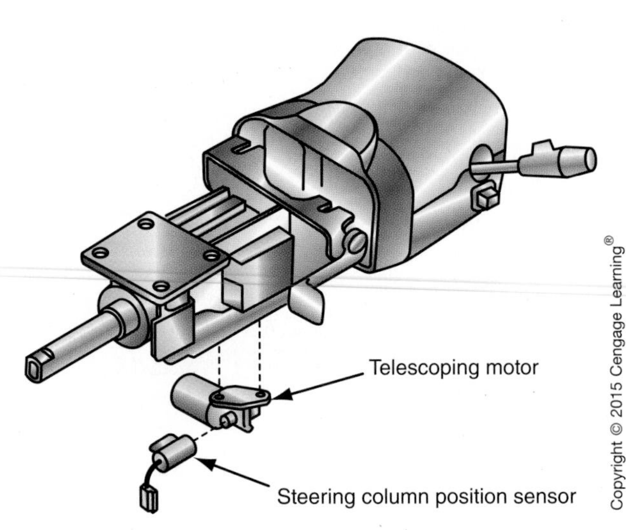

FIGURE 8-15 **Tilt and telescoping motor and steering column position sensor.**

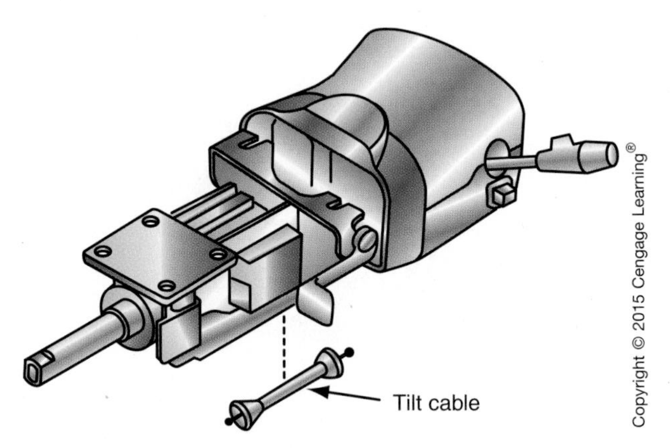

FIGURE 8-16 **Tilt cable for tilt and telescoping steering column.**

telescoping function, the DPM operates the tilt and telescoping motor to move the steering wheel toward or away from the driver depending on the switch request. When the driver operates the tilt and telescoping switch to request the tilt function, the DPM operates the tilt and telescoping motor to tilt the wheel upward or downward depending on the switch request. The DPM is interconnected via data links to some of the other on-board computers. Therefore, the DPM can receive input signals from other computers such as the driver door switch (DDS) module via the data links. When the ignition switch is turned off and the driver pushes the unlock button on the driver's door, the DPM tilts the steering column to a position that allows easier exit from the driver's seat. When the driver enters the vehicle and turns on the ignition switch, the DPM moves the steering column back to the previous setting.

Active Steering Column

The active steering column in some vehicles actively adjusts the energy-absorbing capability of the steering column using a pyrotechnic actuator. This type of steering column has a section of energy-absorbing steel that holds the upper and lower halves of the steering column together (Figure 8-17). The energy-absorbing steel section is designed to control the collapse of the steering column during impacts.

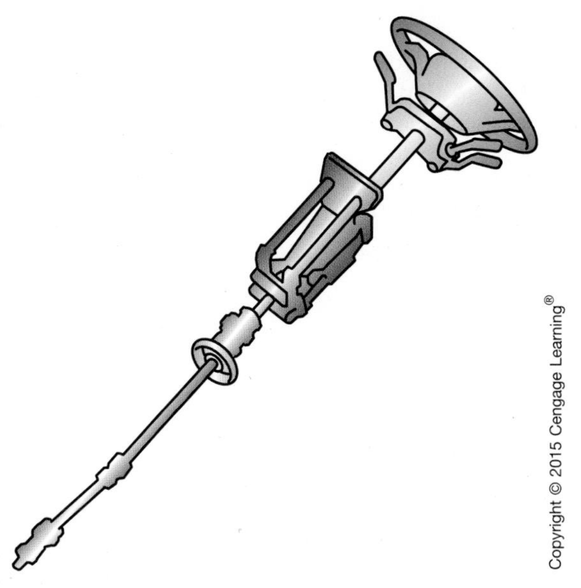

FIGURE 8-17 **Active steering column.**

A restraints control module (RCM) operates a pyrotechnic actuator in the steering column. Input sensors inform the RCM regarding driver weight, front seat position, driver seat belt usage, and crash severity. When the RCM receives input signals indicating softer steering column collapse is necessary to protect the driver, the RCM fires the pyrotechnic device in the steering column. This action pulls a pin out of the column, and reduces the columns' resistance to collapse. The steering column collapse is designed to operate with the air bag level that is being activated. The ease of steering column collapse is different for a belted or unbelted driver. The active steering column helps the vehicle manufacturers to meet enhanced federal safety regulations that now require crash protection for 5th percentile, smaller female drivers, and 50th percentile, large male drivers with and without seat belts.

Driver Protection Module

In the driver protection module, the steering column, knee bolster, and pedals are mounted in a module that allows these components to move away from the driver in a controlled manner during a vehicle crash (Figure 8-18). The driver protection module contains steel tubes supported by aluminum extrusions. This design allows the steering column, knee bolster, and pedals to move along the trajectory of the driver during a severe vehicle crash. This action helps to maintain air bag position. During a vehicle crash, the driver protection module

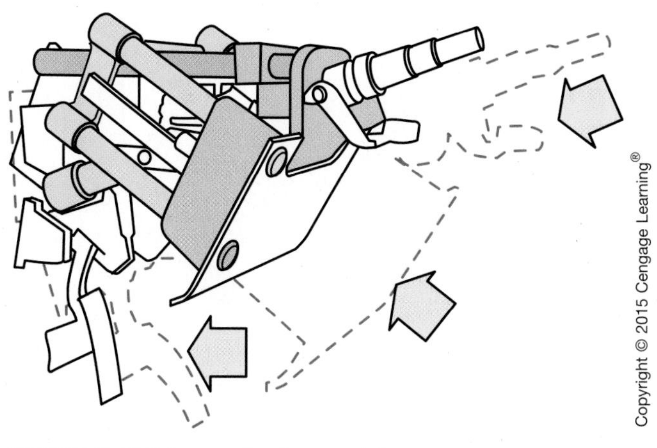

FIGURE 8-18 **Driver protection module.**

movement may be controlled actively by a pyrotechnic device operated by an electronic module. The driver protection module will provide adequate crash protection for drivers from the 5th percentile to the 95th percentile. It is expected that this level of protection will be required by federal legislation in the future.

Steering Linkage Mechanisms

Parallelogram Steering Linkage

Shop Manual
Chapter 8, page 291

Steering linkage mechanisms are used to connect the steering gear to the front wheels. A **parallelogram steering linkage** may be mounted behind the front suspension (Figure 8-19) or in front of the front suspension (Figure 8-20). The parallelogram steering linkage must not interfere with the engine oil pan or chassis components.

A **parallelogram steering linkage** may be defined as one in which the tie rods are mounted parallel to the lower control arms.

WARNING: Always remember that customers' lives may depend on the condition of the steering linkages on their vehicles. State safety inspections play a very important role in maintaining suspension, steering, and other vehicle systems in safe driving condition and saving lives. During undercar service, always make a quick check of the steering linkage condition.

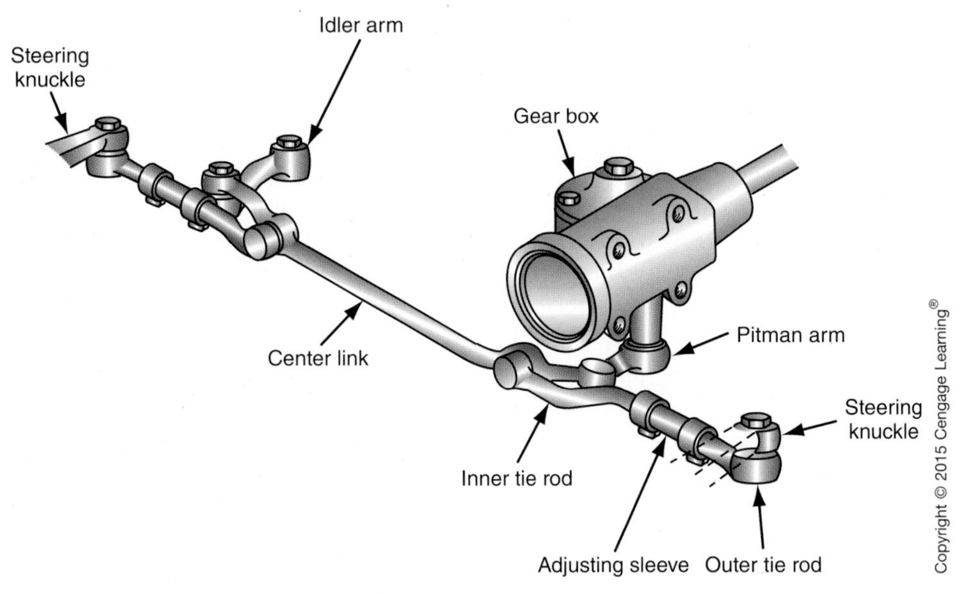

FIGURE 8-19 **Parallelogram steering linkage behind the front suspension.**

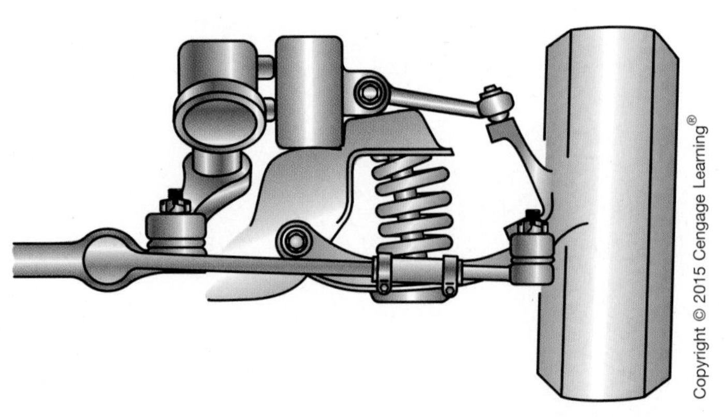

FIGURE 8-20 **Parallelogram steering linkage in front of the front suspension.**

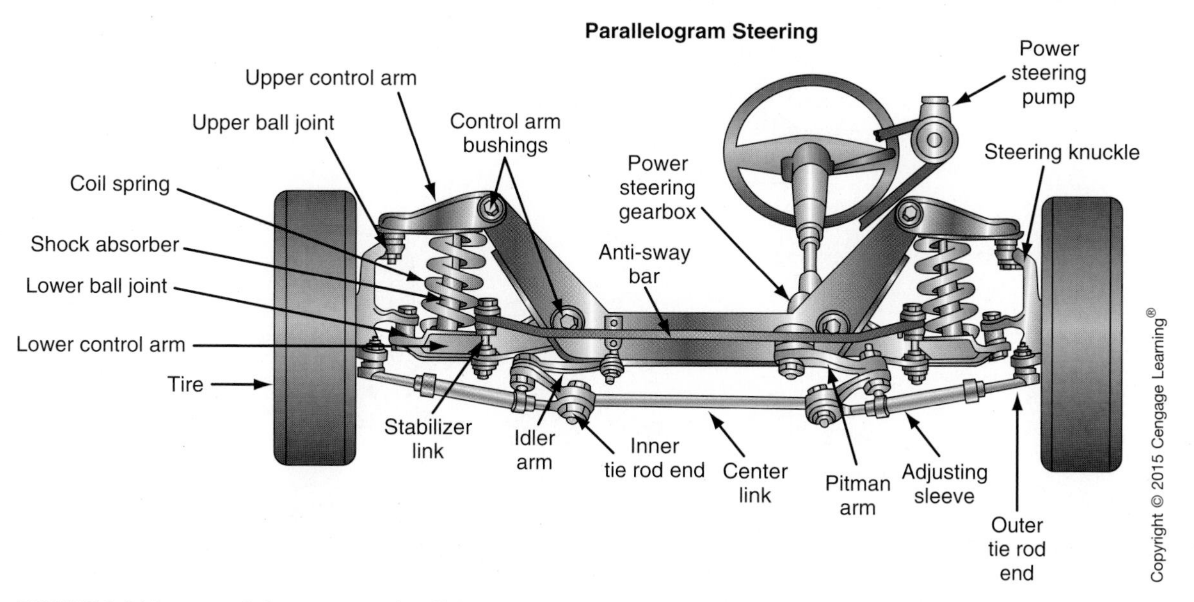

FIGURE 8-21 **In a parallelogram steering linkage, the tie rods are connected parallel to the lower control arms.**

Regardless of the parallelogram steering linkage mounting position, this type of steering linkage contains the same components. The main components in this steering linkage mechanism are:

1. Pitman arm
2. Center link
3. Idler arm assembly
4. Tie rods with sockets
5. Tie rod ends

Parallelogram steering linkages are found on independent front suspension systems (Figure 8-21). In a parallelogram steering linkage, the tie rods are connected parallel to the lower control arms. Road vibration and shock are transmitted from the tires and wheels to the steering linkage, and these forces tend to wear the linkages and cause steering looseness. If the steering linkage components are worn, steering control is reduced. Since loose steering linkage components cause intermittent toe changes, this problem increases tire wear. The wear points in a parallelogram steering linkage are the tie rod sockets and ends, idler arm, and center link end.

Tie Rods

The **tie rod** assemblies connect the center link to the steering arms, which are bolted to the front steering knuckles. In some front suspensions, the steering arms are part of the steering knuckle; in other front suspension systems, the steering arms are bolted to the knuckle. A ball socket is mounted on the inner end to each tie rod, and a tapered stud on this socket is mounted in a center link opening. A castellated nut and cotter pin retain the tie rods to the center link. A threaded sleeve is mounted on the outer end of each tie rod, and a tie rod end is threaded into the outer end of this sleeve (Figure 8-22).

Some outer tie rod ends have a ball stud that is surrounded by an upper hardened steel bearing and a high-strength polymer lower bearing seat (Figure 8-23). The hardened steel upper bearing provides strength and durability, and the polymer lower bearing seat provides smooth rotation of the ball stud in the tie rod end. An internal spring between the polymer lower bearing seat supplies self-adjusting action and constant tension on this seat. A seal in the upper part of the ball joint housing seals the ball stud to prevent contaminants from entering the tie rod end. These tie rod ends are installed on some original

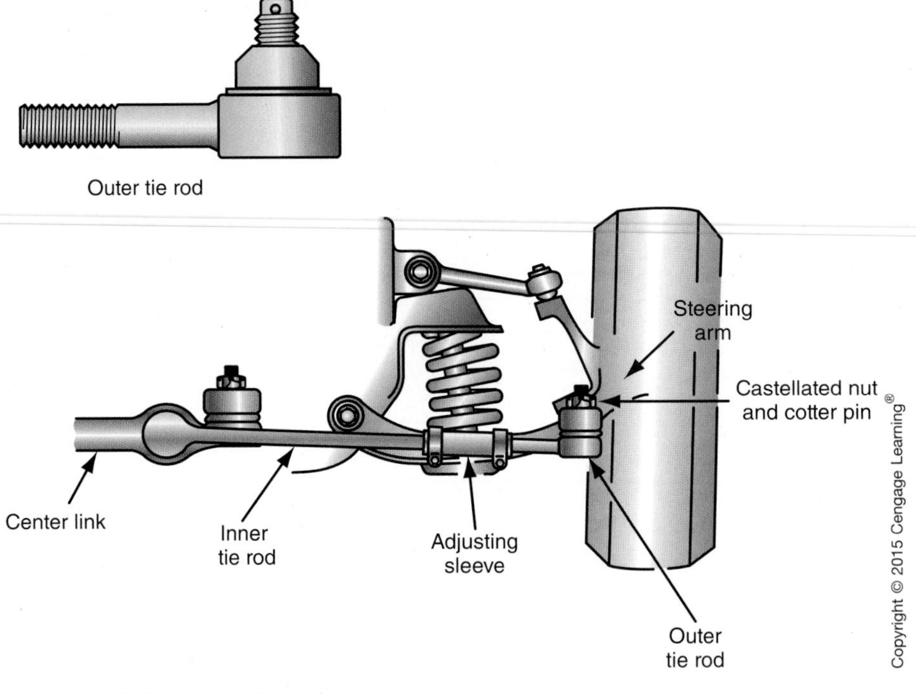

FIGURE 8-22 **Tie rod design.**

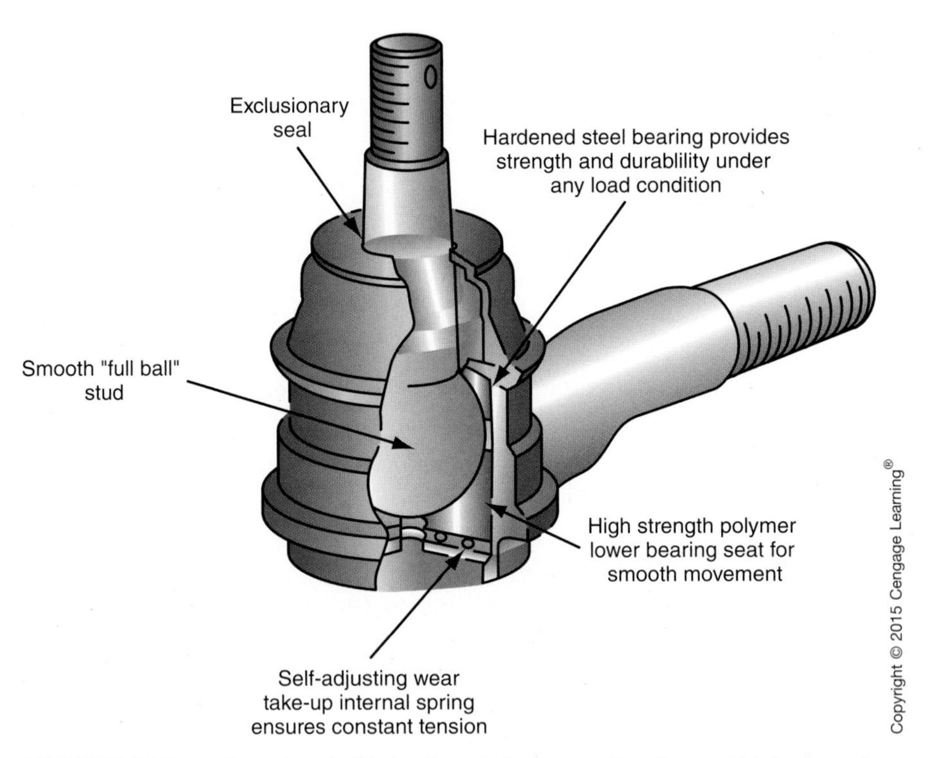

FIGURE 8-23 **Outer tie rod end with hardened steel upper bearing and high-strength polymer lower bearing.**

equipment manufacturers' vehicles, and they are available as replacement tie rod ends on most vehicles.

Some Ford cars have rubber-encapsulated outer tie rod ends in which a rubber bushing surrounds the lower end of the ball stud (Figure 8-24). Special service procedures required on these tie rod ends are explained in the Shop Manual. Similar outer tie rod ends are used on parallelogram steering linkages and rack and pinion steering linkages.

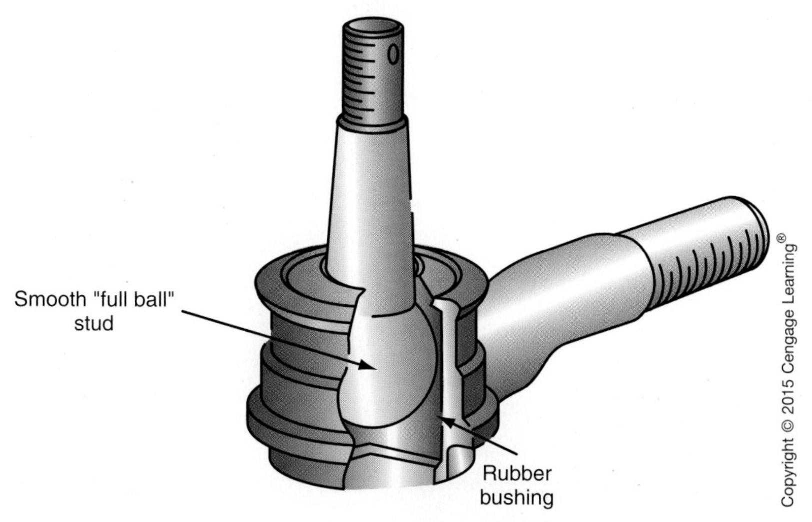

FIGURE 8-24 **Outer tie rod end with rubber-encapsulated ball stud.**

Each tie rod sleeve contains a left-hand and a right-hand thread where it is threaded onto the tie rod end and the tie rod. Therefore, sleeve rotation changes the tie rod length and provides a toe adjustment. Clamps are used to tighten the tie rod sleeves. The clamp opening must be positioned away from the slot in the tie rod sleeve. The design of the steering linkage mechanism allows multiaxial movement, since the front suspension moves vertically and horizontally. Ball-and-socket-type pivots are used on the tie rod assemblies and center link.

Multiaxial movement refers to movement in any direction around a pivot point.

If the front wheels hit a bump, the wheels move up and down and the control arms move through their respective arcs. Since the tie rods are connected to the steering arms, these rods must move upward with the wheel. Under this condition, the inner end of the tie rod acts as a pivot, and the tie rod also moves through an arc. This arc is almost the same as the lower control arm arc because the tie rod is parallel to the lower control arm. Maintaining the same arc between the lower control arm and the tie rod minimizes toe change on the front wheels during upward and downward wheel movement. This action improves the directional stability of the vehicle and reduces tread wear on the front tires.

AUTHOR'S NOTE: It has been my experience that the most common causes of premature wear on outer tie rod ends and other pivot points in a steering linkage are lack of lubrication or contamination that enters through broken seals. If the tie rod ends and other pivot points have grease fittings, lubrication at the manufacturer's recommended interval is important to maintain component life. However, you should not over lubricate the steering linkage pivot points with a high-pressure grease gun, because this may rupture the seals. The steering linkage components are exposed to a large amount of water and dirt contamination. If the seals are leaking on any of the linkage pivot points, the pivots will soon be contaminated with moisture and dirt that acts as an abrasive to shorten component life.

Pitman Arm

The **pitman arm** connects the steering gear to the center link. This arm also supports the left side of the center link. Motion from the steering wheel and steering gear is transmitted to the pitman arm, and this arm transfers the movement to the steering linkage. This pitman arm movement forces the steering linkage to move to the right or left, and the linkage

FIGURE 8-25 **Pitman arm design.**

moves the front wheels in the desired direction. The pitman arm also positions the center link at the proper height to maintain the parallel relationship between the tie rods and the lower control arms.

Wear-type pitman arms have ball sockets and studs at the outer end, and this stud fits into the center link opening (Figure 8-25). The ball stud and socket are subject to wear, and pitman arm replacement is necessary if the ball stud is loose in the pitman arm. A nonwear pitman arm has a tapered opening in the outer end. A ball stud in the center link fits into this opening. The nonwear pitman arm only needs replacing if the arm is damaged or bent in a collision. The opening in the inner end of both types of pitman arms has serrations that fit over matching serrations on the steering gear shaft. A nut and lock washer retain the pitman arm to the steering gear shaft.

Idler Arm

Steering wheel free play is the amount of steering wheel movement before the front wheels start to move to the right or left.

An idler arm support is bolted to the frame or chassis on the opposite end of the center link from the pitman arm. The **idler arm** is connected from the support bracket to the center link. Two bolts retain the idler arm bracket to the frame or chassis. In some idler arms, a ball stud on the outer end of the arm fits into a tapered opening in the center link (Figure 8-26), whereas in other idler arms, a ball stud in the center link fits into a tapered opening in the idler arm.

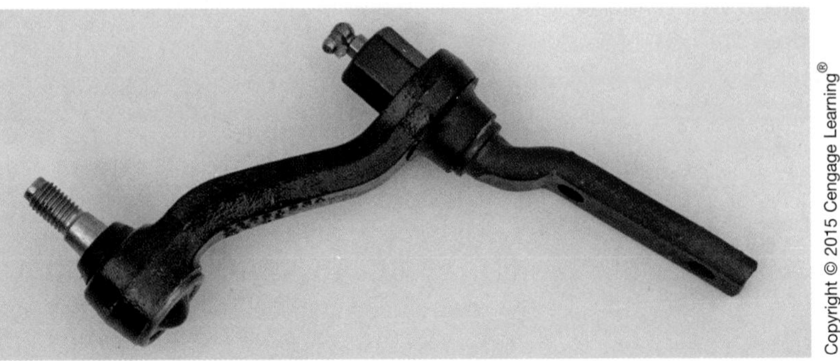

FIGURE 8-26 **Idler arm design.**

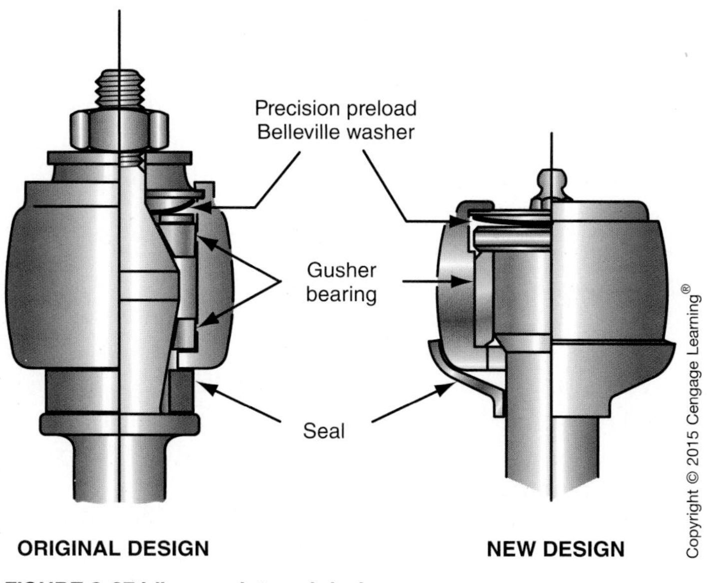

FIGURE 8-27 **Idler arm internal design.**

Bracket
Seal
Bearing
Conical socket design has 25% more bearing surface area.
Steel coil spring

FIGURE 8-28 **Idler arm with conical bearing.**

The idler arm supports the right side of the center link and helps maintain the parallel relationship between the tie rods and the lower control arms. The outer end of the idler arm is designed to swivel on the idler arm bracket, and this swivel is subject to wear. A worn idler arm swivel causes excessive vertical steering linkage movement and erratic toe. This action results in excessive steering wheel free play with reduced steering control and front tire wear.

A **center link** may be referred to as a drag link, steering link, or intermediate link.

Some idler arms contain an upper and lower gusher bearing that surrounds the bearing surface on the lower end of the bracket. A newly designed idler arm has a one-piece powdered metal gusher bearing that extends the full length of the friction surface on the lower end of the bracket (Figure 8-27). Compared with previous designs, this new design has 50 percent more bearing surface, which provides extended service life. A Belleville washer installed below the gusher bearing maintains a precision preload on the thrust washer, bracket, and gusher bearing.

Other idler arms have a conical machined surface on the lower end of the bracket. This conical surface is seated on a matching conical surface on the bearing (Figure 8-28). A coil spring between the lower end of the bracket and the housing maintains constant upward pressure on the bracket to maintain minimal endplay and constant turning torque as the bearing wears. Since this type of idler arm maintains minimal endplay, more precise front suspension alignments are possible.

A BIT OF HISTORY

For many years, rear-wheel-drive vehicles were equipped with parallelogram steering linkages. The smaller, lighter front-wheel-drive cars introduced in large numbers in the late 1970s and 1980s required a more compact, lighter steering gear. The rack and pinion steering gear is used almost exclusively on front-wheel-drive cars because of its reduced weight and space requirements.

Center Links

The **center link** controls the sideways steering linkage and wheel movement. The center link together with the pitman arm and idler arm provides the proper height for the tie rods, which is very important to minimizing toe change on road irregularities. Some center links have tapered openings in each end, and the studs on the pitman arm and idler arm fit into these openings. This type of center link may be called a taper end, or nonwear, link. Other wear-type center links have ball sockets in each end with tapered studs extending from the sockets (Figure 8-29). These tapered studs fit into openings in the pitman arm and idler arm, and they are retained with castellated nuts and cotter pins.

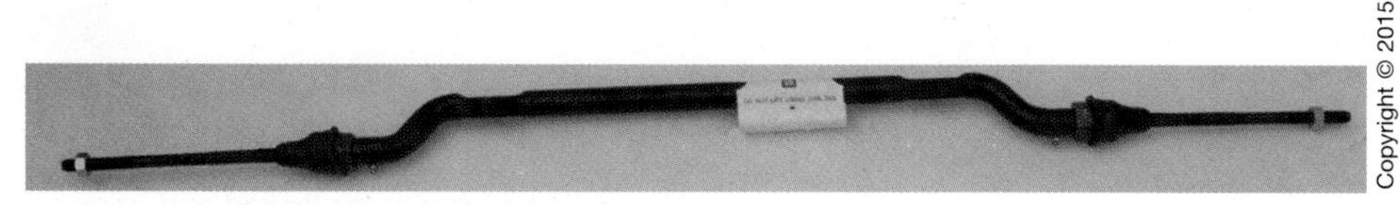

FIGURE 8-29 **Center link design.**

Rack and Pinion Steering Linkage

The **rack and pinion steering linkage** is used with rack and pinion steering gears (Figure 8-30). In this type of steering gear, the rack is a rod with teeth on one side. This rack slides horizontally on bushings inside the gear housing. The rack teeth are meshed with teeth on a pinion gear, and this pinion gear is connected to the steering column. When the steering wheel is turned, the pinion rotation moves the rack sideways. Tie rods are connected directly from the ends of the rack to the steering arms. The tie rods are similar to those found on parallelogram steering systems. An inner tie rod end connects each tie rod to the rack, and bellows boots are clamped to the gear housing. The bellows boots keep dirt out of these joints (Figure 8-31). The inner tie rod end contains a spring-loaded ball socket. The outer tie rod ends connected to the steering arms are basically the same as those in parallelogram steering linkages (Figure 8-32). Some inner tie rod ends contain a bolt and bushing. These tie rod ends are threaded onto the rack (Figure 8-33). Since the rack is connected directly to the tie rods, the rack replaces the center link in a parallelogram steering linkage.

Some inner tie rod ends have a mirror-finished ball and a high-strength polymer bearing to ensure low torque, minimal friction, and extended life (Figure 8-34). A hardened

Rack and Pinion Steering

Upper mounting plate and bearing
Coil spring
Macpherson strut
Steering knuckle
Anti-sway bar
Bellows
Rack and pinion unit
Tire
Outer tie rod end
Ball joint
Inner socket assembly (inside bellows)
Control arm
Control arm bushing
Rack and pinion bushings

FIGURE 8-30 **The rack and pinion steering linkage is used with rack and pinion steering gears.**

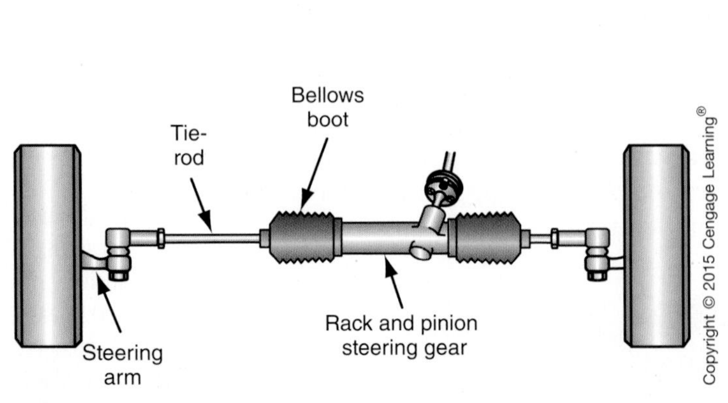

FIGURE 8-31 **Rack and pinion steering gear.**

FIGURE 8-32 **Inner tie rod and outer tie rod end, rack and pinion steering.**

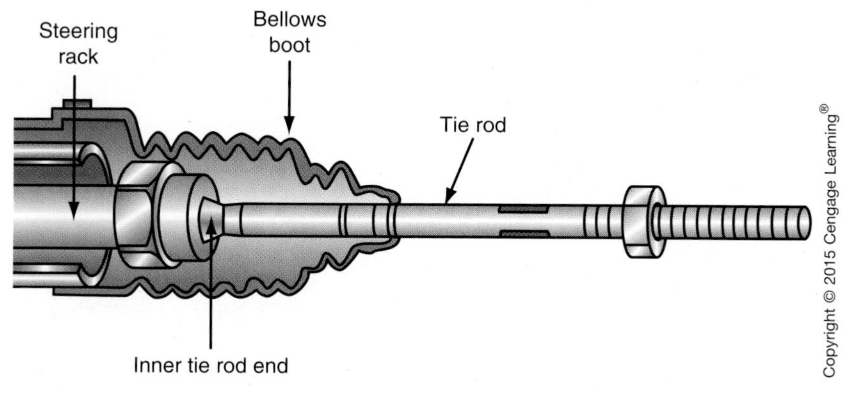

FIGURE 8-33 **The inner tie rod connects directly to the rack and pinion steering gear and is protected by a bellows boot.**

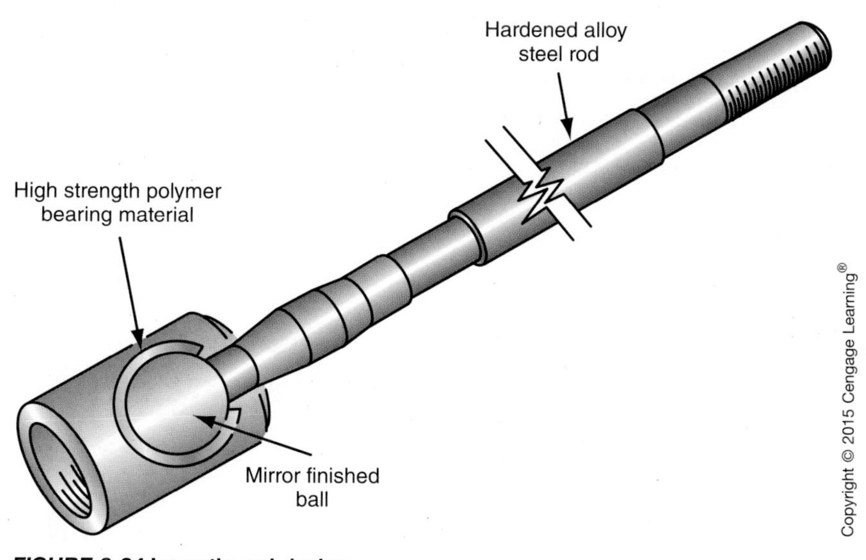

FIGURE 8-34 **Inner tie rod design.**

alloy steel rod extends from the ball to the outer tie rod end and provides maximum strength and durability.

The rack and pinion steering gear may be mounted on the front crossmember (Figure 8-35) or attached to the cowl behind the engine (Figure 8-36). Rubber insulating bushings surround the steering gear, and these bushings are clamped to the crossmember or cowl. The rack and pinion steering gear is mounted at the proper height to position the tie rods and lower control arms parallel to each other. The number of friction points is reduced in a rack and pinion steering system, and this system is light and compact. Most front-wheel-drive unibody vehicles have rack and pinion steering. Since the rack is linked directly to the steering arms, this type of steering linkage provides good road feel.

Steering Damper

A steering damper, or stabilizer, may be found on some parallelogram steering linkages and a few rack and pinion steering systems. The steering damper is similar to a shock absorber. This component is connected from one of the steering links to the chassis or frame (Figure 8-37). When a front wheel strikes a road irregularity, a shock is transferred from the front wheel to the steering linkage, steering gear, and steering wheel. The steering damper helps absorb this road shock and prevent it from reaching the steering wheel. Heavy-duty steering dampers are available for severe road conditions such as those sometimes encountered by four-wheel-drive vehicles.

Shop Manual
Chapter 8, page 298

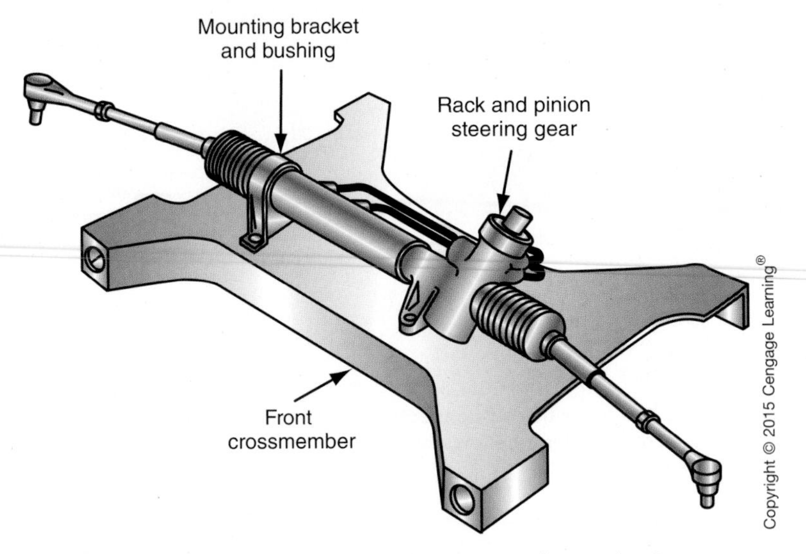

FIGURE 8-35 **Rack and pinion steering gear mounting on front crossmember.**

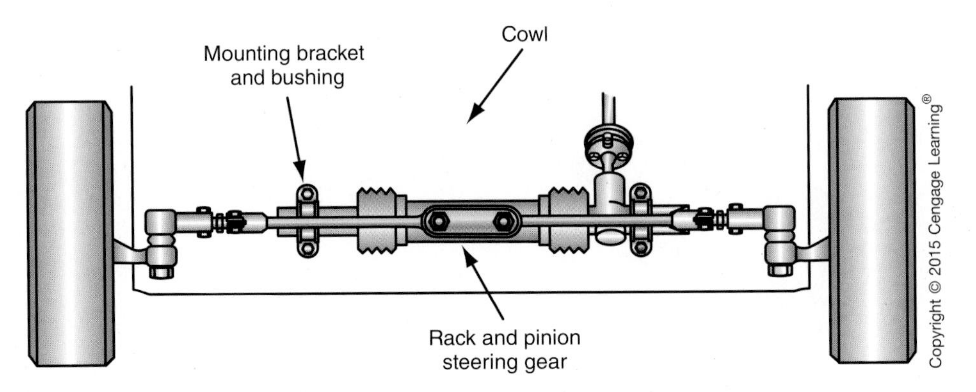

FIGURE 8-36 **Rack and pinion steering gear mounted on cowl.**

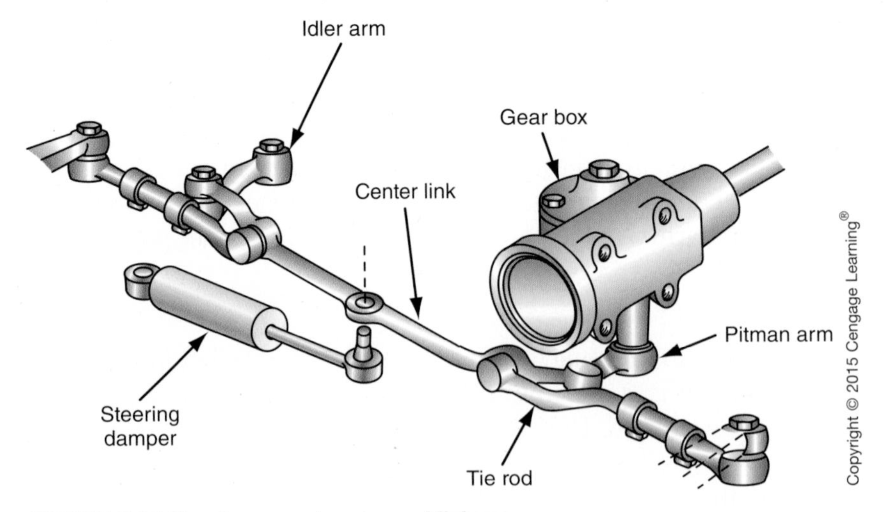

FIGURE 8-37 **Steering gear damper and linkage.**

SUMMARY

- Steering columns help provide steering control, driver convenience, and driver safety.
- Many steering columns provide some method of energy absorption to protect the driver during a frontal collision.
- Steering wheels and columns now contain an air bag deployment module to protect the driver in a frontal collision.
- Tilt steering columns increase driver comfort and ease while driving or getting in or out of the driver's seat.
- A clock spring electrical connector supplies positive electrical contact between the air bag module in the steering wheel and the air bag electrical system.
- The ignition switch, dimmer switch, signal light switch, hazard switch, and wipe or wash switch may be mounted in the steering column.
- When the ignition switch is in the Lock position, a locking plate and lever in the upper steering column locks the steering wheel and the gear shift.
- In some tilt steering columns, the upper column housing pivots on two bolts, and the upper steering shaft pivots on a universal joint.
- In a parallelogram steering linkage, the tie rods are parallel to the lower control arms.
- The parallelogram steering linkage minimizes toe change as the control arms move up and down on road irregularities.
- A rack and pinion steering linkage has reduced friction points; it is lightweight and compact compared with a parallelogram steering linkage.

TERMS TO KNOW

Air bag deployment module
Center link
Clock spring electrical connector
Energy-absorbing lower bracket
Idler arm
Parallelogram steering linkage
Pitman arm
Pre-safe systems
Pyrotechnic
Rack and pinion steering linkage
Silencer
Spherical bearing
Spiral cable
Tie rod
Toe plate

REVIEW QUESTIONS

Short Answer Essays

1. Explain how a collapsible steering column protects the driver in a frontal collision.
2. Explain how the driver's side air bag protects the driver in a frontal collision.
3. Describe the purpose of a clock spring.
4. List the switches commonly found in a steering column.
5. Describe the type of mechanism used to lock the steering wheel and gear shift when the ignition is in the Lock position.
6. Describe the pivot points in the upper shaft and upper column tube in a tilt steering wheel.
7. List the wear points in a parallelogram steering linkage.
8. List the five main components in a parallelogram steering linkage, and explain the purpose of each component.
9. Describe the basic design of a rack and pinion steering linkage.
10. Explain the advantages of a rack and pinion steering linkage compared with a parallelogram steering linkage.

Fill-in-the-Blanks

1. In some collapsible steering columns, _______________ in the outer column jacket and steering shaft shear off if the driver is thrown against the steering wheel in a frontal collision.
2. The driver's side air bag protects the driver if the vehicle is involved in a _______________ collision above a specific speed.
3. In an air bag system, a _______________ in the steering column maintains positive electrical contact between the air bag module and the air bag electrical system as the steering wheel is rotated.
4. In a parallelogram steering linkage, the center link connects the pitman arm to the _______________.
5. In a parallelogram steering linkage, the pitman arm and the _______________ position the center link and tie rods at the correct height.

6. In a wear-type pitman arm, a ________________ is positioned in the outer end of the pitman arm.
7. In a parallelogram steering linkage, the tie rods are parallel to the ________________.
8. The clamp opening in a tie rod sleeve must be positioned away from the ________________ in the tie rod sleeve.
9. In a rack and pinion steering system, the rack is connected directly to the ________________.
10. Compared with a parallelogram steering linkage, a rack and pinion steering linkage has a reduced number of ________________ points.

Multiple Choice

1. A typical air bag deployment time is:
   A. 1.5 minutes.
   B. 1 minute.
   C. 30 seconds.
   D. 40 milliseconds.
2. Many collapsible steering columns have:
   A. Plastic pins in the two-piece outer jacket.
   B. A collapsible bellows in the two-piece outer jacket.
   C. Steel pins in the two-piece lower steering shaft.
   D. A rubber spacer in the two-piece lower steering shaft.
3. The clock spring electrical connector:
   A. Maintains electrical contact between the air bag inflator module and the air bag electrical system.
   B. Is mounted above the steering wheel.
   C. Contains three spring-loaded copper contacts.
   D. Provides electrical contact between the signal light switch and the signal lights.
4. An active steering column has all of these components EXCEPT:
   A. A pyrotechnic actuator.
   B. A section of energy-absorbing steel.
   C. A telescoping cylinder.
   D. A pull-out pin that allows easier column collapse.
5. All of these statements about rack and pinion steering linkages are true EXCEPT:
   A. The tie rods are parallel to the lower control arms.
   B. The tie rod position depends on the steering gear mounting.
   C. The tie rods are connected to the pinion in the steering gear.
   D. The outer tie rod ends connect the tie rods to the steering arms.
6. All these statements about parallelogram steering linkages are true EXCEPT:
   A. Tie rod sleeves have the same type of thread in both ends of the sleeve.
   B. Loose steering linkages may cause excessive tire tread wear.
   C. Loose steering linkage causes excessive steering wheel free play.
   D. The pitman arm helps maintain the proper center link and tie rod height.
7. While discussing idler arms:
   A. Some idler arms contain a tapered roller bearing.
   B. The idler arm bracket is bolted to the upper control arm.
   C. A worn idler arm has no effect on front wheel toe.
   D. A partially seized idler arm bearing increases steering effort.
8. In a parallelogram steering linkage, the tie rods are parallel to the lower control arms to:
   A. Improve ride quality.
   B. Provide longer steering linkage life.
   C. Extend shock absorber and spring life.
   D. Reduce toe change during upward and downward front wheel movement.
9. A rack and pinion steering gear:
   A. Has tie rods that connect the rack directly to the steering arms.
   B. May be bolted to the vehicle frame.
   C. Has inner tie rod ends that are pressed onto the rack.
   D. Has more friction points compared with a parallelogram steering linkage.
10. While discussing steering linkages and dampers:
    A. In a rack and pinion steering gear, the rack positions the tie rods parallel to the lower control arms.
    B. A rack and pinion steering gear is used on most rear-wheel-drive cars.
    C. A steering damper is used on many front-wheel-drive vehicles.
    D. A defective steering damper may cause excessive steering effort.

Chapter 9

Four-Wheel Alignment, Part 1 Primary Angles

Upon completion and review of this chapter, you should be able to understand and describe:

- The effect of excessive positive camber on front suspension systems.
- Camber changes during front wheel jounce and rebound travel.
- The relationship between camber and vehicle directional stability.
- How front suspension camber and caster may be adjusted to compensate for road crown.
- The effects of positive and negative caster on directional control and steering effort.
- Positive and negative caster as they relate to ride quality.
- How higher or lower than specified front or rear suspension height affects front suspension caster.
- Toe-in and toe-out on front suspension systems.
- The toe-in settings required on front-wheel-drive and rear-wheel-drive vehicles.
- Tire tread wear caused by excessive toe-in.
- The customer complaints that may arise from incorrect rear wheel toe and thrust line adjustments.
- Suspension and chassis defects that may cause improper rear wheel toe, thrust line, or camber.
- The variables that affect wheel alignment.
- Wheel alignment, and explain five reasons for performing a wheel alignment.
- Why four-wheel alignment is essential on front-wheel-drive unitized body cars.
- Thrust line and geometric centerline and describe the effect of an improper thrust line on vehicle steering.
- The result of improper rear wheel toe and its effect on the thrust line.
- The causes of improper rear wheel toe.
- The causes of rear axle offset.
- Rear axle sideset, setback, and dog tracking.
- Front and rear tire tread wear caused by inaccurate rear wheel toe and thrust line settings.
- Geometric centerline alignment and explain the problem with this type of alignment.
- Thrust line alignment and describe the shortcoming of this type of alignment.
- Total four-wheel alignment and explain the advantage of this type of alignment.
- The safety hazards created by incorrect wheel alignment or worn suspension and steering components.

Introduction

Automotive engineers design suspension and steering systems that provide satisfactory vehicle control with acceptable driver effort and road feel. The vehicle should have a tendency to go straight ahead without being steered. This tendency is called **directional stability**. A vehicle must have predictable directional control, which means the steering must provide a feeling that the vehicle will turn in the direction steered. The wheels must be reasonably easy to turn and tire wear should be minimized. These steering qualities and tire conditions are achieved when front and rear **wheel alignment** angles are within the vehicle manufacturer's specifications.

Wheel alignment may be defined as an adjustment and refitting of suspension parts to original specifications that ensure design performance.

The condition of suspension system components and wheel alignment is extremely important to maintaining driving safety and normal tire wear. Worn suspension components such as tie rod ends, ball joints, and control arms can suddenly fall apart and cause complete loss of steering. This disastrous event may result in not only some very expensive property damage but also the loss of human life. When alignment angles are incorrect, an uncontrolled vehicle swerve or skid may occur during hard braking, resulting in a serious accident. Severe misalignment may reduce tire life to one-third of the normal expected tire life with correct alignment. After suspension components such as ball joints or control arms are replaced, wheel alignment is essential.

Technicians must be familiar with the symptoms that indicate incorrect rear wheel alignment. The rear wheel alignment procedures required to correct improper rear wheel alignment must be understood.

Camber Fundamentals

Shop Manual
Chapter 9, page 338

Camber

Camber is defined as the inward or outward tilt of the tire and wheel assembly as viewed from the front of the vehicle (Figure 9-1). A reference line begins at the base (ground) center of the tire and wheel and extends upward at a 90 degree angle to the road. Put another

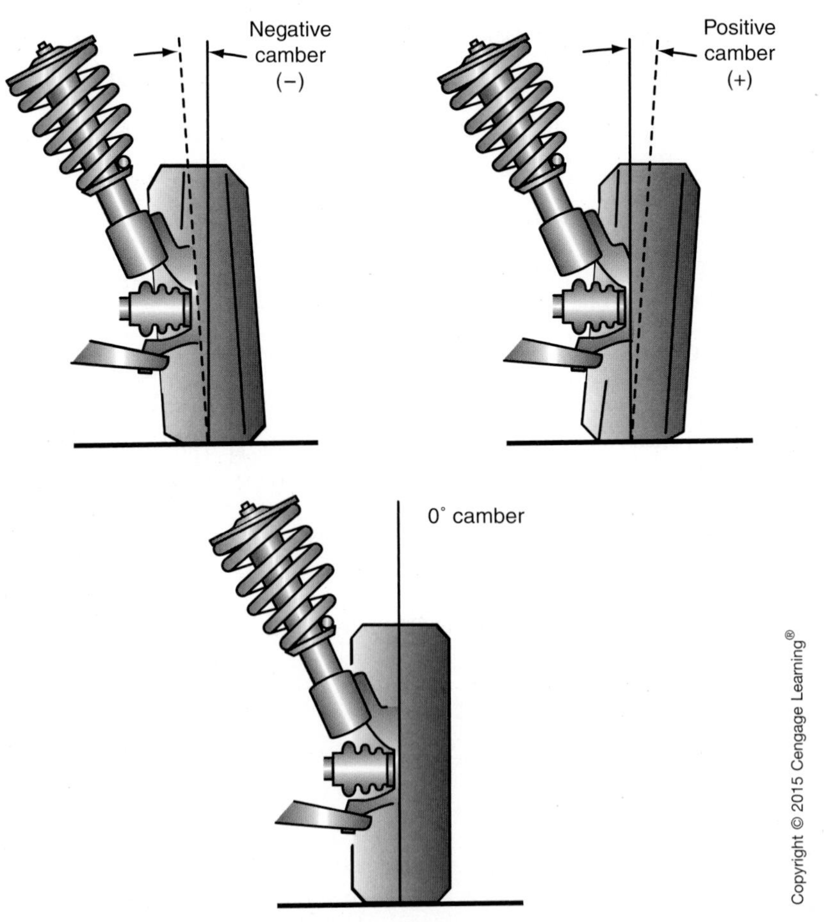

FIGURE 9-1 **Negative and positive camber angles on a MacPherson strut front suspension system.**

way, camber refers to the tilt of a line through the tire and wheel centerline in relation to the true vertical centerline of the tire and wheel. **Positive camber** is obtained when the top of the tire and wheel is tilted outward, away from the true vertical line of the wheel assembly. **Negative camber** occurs when the tire and wheel centerline tilts inward in relation to the wheel assembly true vertical centerline. The camber angle is measured and referred to the same for both the front and rear tires, meaning negative camber is the inward tilt of the tire at the top whether it is in the front or the rear of the vehicle.

Camber angle is displayed in degrees on an alignment computer. Camber is an angle that can affect both tire wear and directional stability, such as a pull or drift. If the side-to-side difference in front camber is greater than 0.5 to 0.75 degree, the vehicle may exhibit a pull or drift to the most positive side (Figure 9-2) as a general rule. This left side to right side difference is referred to as **cross camber** and is shown on the alignment computer's readings screen (Figure 9-3). Some manufactures specify a maximum allowable amount of cross camber. Other manufactures do not specify an acceptable amount and the general assumption as stated above is that a vehicle may exhibit a drift if the cross camber difference is between 0.5 and 0.75 degrees and a pull if the cross camber difference is greater than 0.75 degree toward the side with the higher camber reading. Rear camber differences do not cause vehicles to pull or drift but do cause tire wear just as errors in the front.

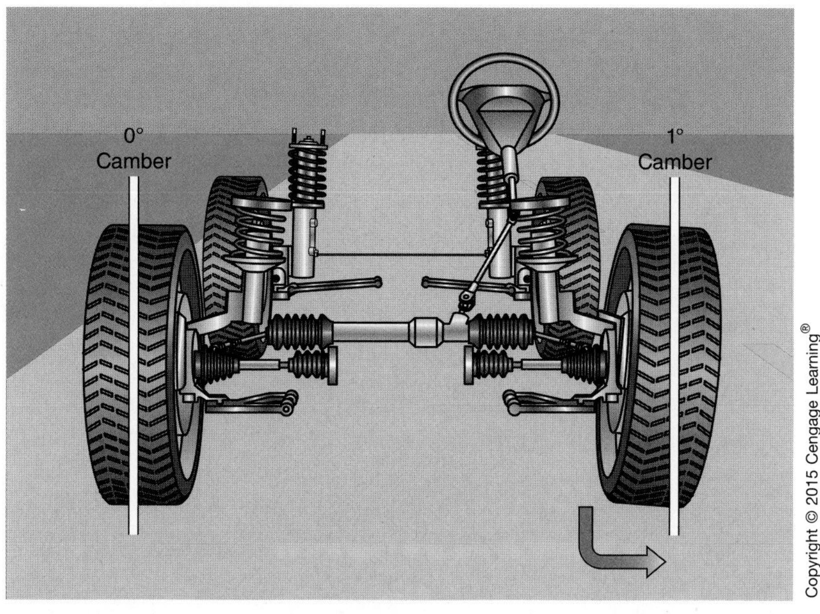

FIGURE 9-2 **Cross camber is the side-to-side difference in camber measurements. This vehicle will pull right toward the most positive side.**

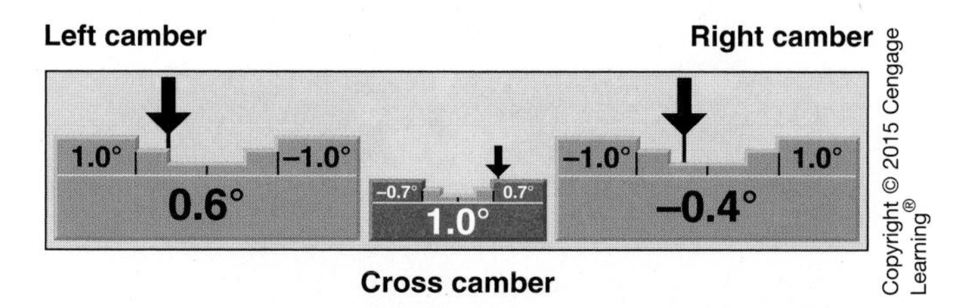

FIGURE 9-3 **The left and right camber preferred specification is 0 degree and the tolerance is + or –1.0 degree in this example. The actual left camber reading is positive 0.6 degree and the actual right camber reading is negative 0.4 degree. The maximum allowable cross camber is 0.7 degree with an actual cross camber of 1.0 degree.**

FIGURE 9-4 **Tire shoulder wear because of excessive camber.**

FIGURE 9-5 **Tire shoulder wear because of excessive positive camber.**

Camber can also cause smooth, tapered tire wear to either the outer or inner edge of the tire tread around the entire tire's circumference and is typically isolated to the tread shoulder area of the tire (Figure 9-4). Too much positive camber will cause outside edge wear to the tire shoulder (Figure 9-5). Too much negative camber will cause inside edge wear to the tire shoulder.

A camber specification and tolerance is specified by the vehicle manufacture for both the front and rear (Figure 9-6). The specification is the preferred angle and the tolerance is the permitted amount that can be increased (+) or decreased (–) from the preferred angle. The tolerance is a means provided by the manufacturer to fine tune the suspension system to specific driving habits. It also allows for some change and wear that may naturally occur between alignments without a noticeable impact on directional stability or excessive tire wear.

During an alignment, camber is considered a live measurement because the computer is continually updating the sensor information on the computer measurement screen. Some alignment machines will allow you to adjust camber with the weight on or off the

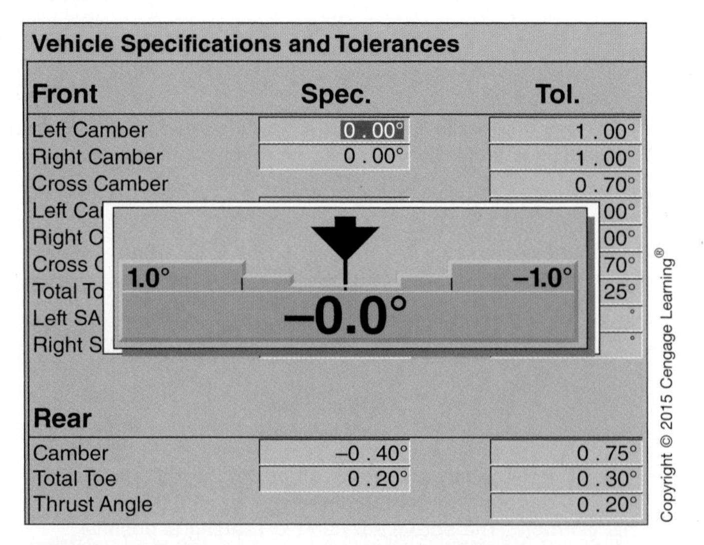

FIGURE 9-6 **Typical vehicle specification and tolerance data sheet with an alignment screen bar graph superimposed for camber.**

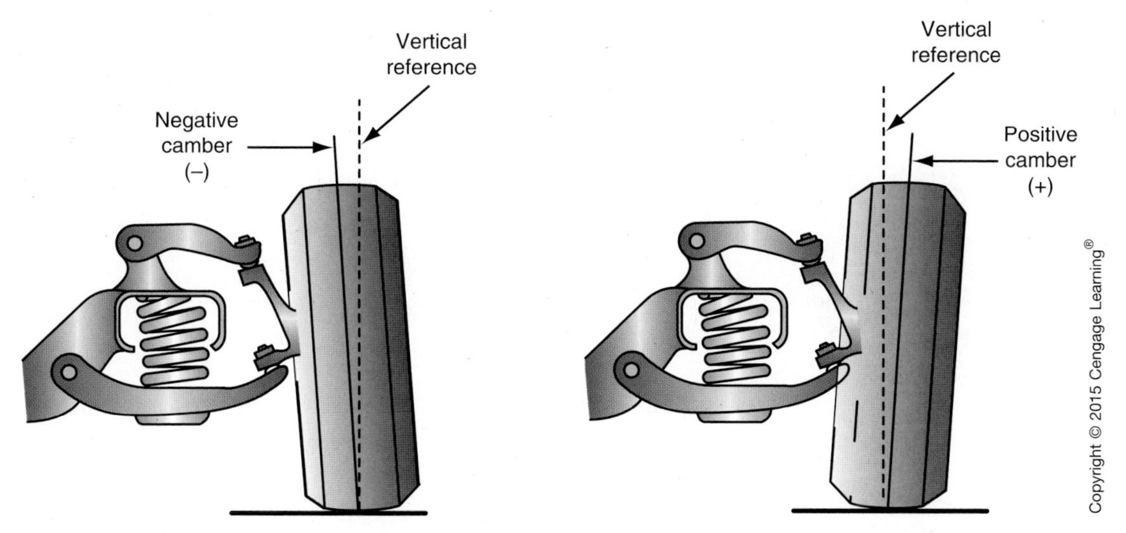

FIGURE 9-7 **Negative and positive camber angles on a short-and-long arm front suspension system.**

wheels using a program designed to allow the vehicle to be lifted. Hunter Engineering calls this program "Jack-Up Selected Axle."

The camber may be adjusted with the cam on the upper strut-to-steering knuckle bolt on some MacPherson strut front suspension systems. Negative and positive camber angles are the same on a short-and-long arm front suspension system as they are on a MacPherson strut suspension system (Figure 9-7).

Front cradle position is extremely important to provide proper front wheel camber on a front-wheel-drive vehicle. For example, if the vehicle is involved in a side collision and most of the collision force is supplied to the front wheel and chassis, the cradle may be pushed sideways. On the side of the vehicle impacted during the collision, the front suspension and cradle may be pushed inward, and the other side of the cradle may be forced outward. When the front suspension and cradle are moved inward, the camber on that side of the vehicle becomes more positive. When the cradle, suspension, and bottom of the wheel on the opposite side are moved outward, the camber is moved toward a negative position. Often this is noted as an equal and opposite difference in side-to-side camber.

Driving Conditions Affecting Camber

Jounce and Rebound

Upward wheel movements are referred to as **jounce**, whereas downward wheel movements are termed **rebound**. On most modern suspension systems during wheel jounce, the top of the wheel moves outward and creates a more positive camber angle (Figure 9-8). During wheel rebound, it is desirable to have very little camber change to minimize tire tread wear. A short-and-long arm front suspension system will have a slightly negative camber during wheel rebound travel (Figure 9-9).

Cornering Forces

During hard cornering at high speeds, centrifugal force attempts to move the vehicle sideways to the outside of the turn. Under this condition, more vehicle weight is transferred to the side of the vehicle on the outside of the curve. Therefore, the front suspension on the outside of the curve is forced downward, whereas the front suspension on the inside of the curve is lifted upward. When this action occurs, the camber on the inside wheel becomes less positive and the inside edge of the tire grips the road surface to help prevent sideways skidding (Figure 9-10).

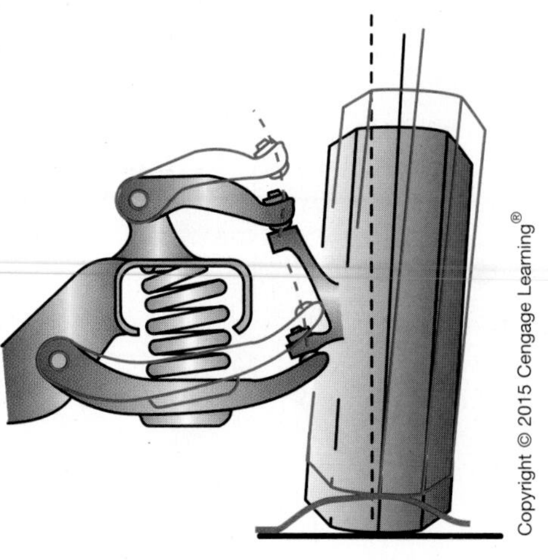

FIGURE 9-8 **Camber change during wheel jounce.**

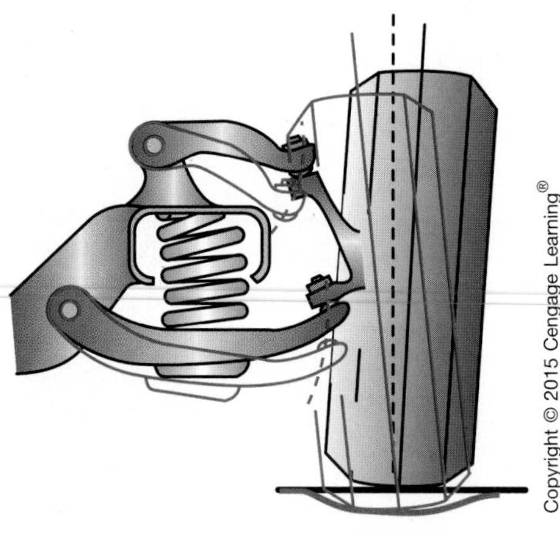

FIGURE 9-9 **Camber change during wheel rebound.**

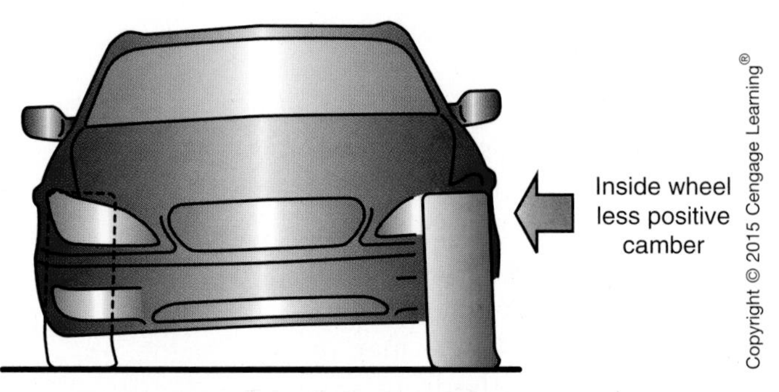

FIGURE 9-10 **While cornering at high speed, the camber on the inside front wheel becomes less positive and the inside edge of the tire grips the road surface to resist lateral skidding.**

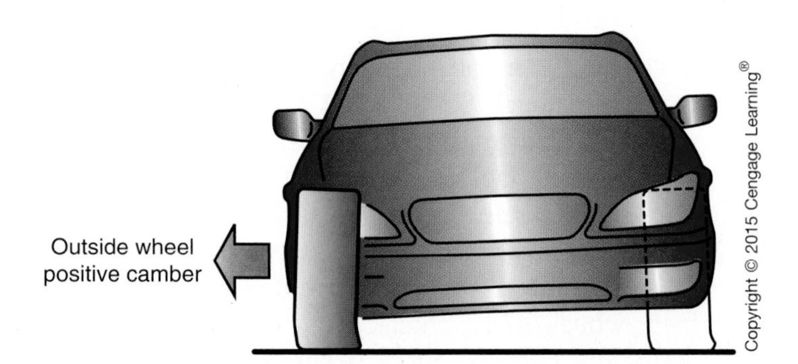

FIGURE 9-11 **While cornering at high speed, the camber on the outside front wheel moves to a slightly more positive position and the outside edge of the tire grips the road surface to resist lateral kidding.**

Simultaneously, the camber on the outside wheel moves to a slightly more positive position. This action on the outside wheel causes the outside edge of the tire to grip the road surface, which helps prevent sideways skidding (Figure 9-11). Frequent, hard, high-speed cornering causes wear on the edges of the tire treads.

Tire Tread Wear

The camber angle may be referred to as one of the tire wear alignment angles. When the front wheels are adjusted to the manufacturer's specified camber setting, the front wheels will remain at, or very close to, the 0 degree camber position during average driving conditions. Therefore, maximum tire tread life and directional stability are maintained.

If a front wheel has excessive positive camber, the wheel is tilted outward and the vehicle weight is concentrated on the outside edge of the tire. Under this condition, the outside edge of the tire has a smaller diameter than the inside tire edge. Therefore, the outside tire edge has to complete more revolutions to travel the same distance as the inside tire edge. Because both edges are on the same tire, the outside edge must slip and scuff on the road surface as the tire and wheel revolve (Figure 9-12).

Shop Manual
Chapter 9, page 339

Excessive negative camber tilts the wheel inward and concentrates the vehicle weight on the inside edge of the tire. This condition causes wear and scuffing on the inside edge of the tire tread (Figure 9-13). Therefore, correct camber adjustment is extremely important to providing normal tire tread life.

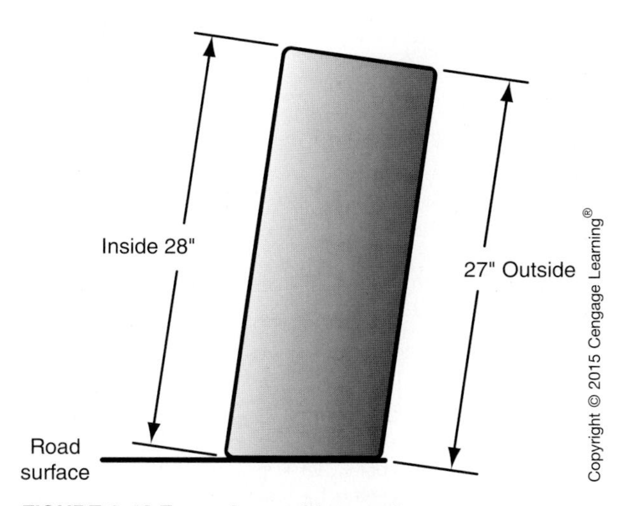

FIGURE 9-12 **Excessive positive camber causes a smaller diameter on the outside edge of the tire compared to the inside edge of the tire.**

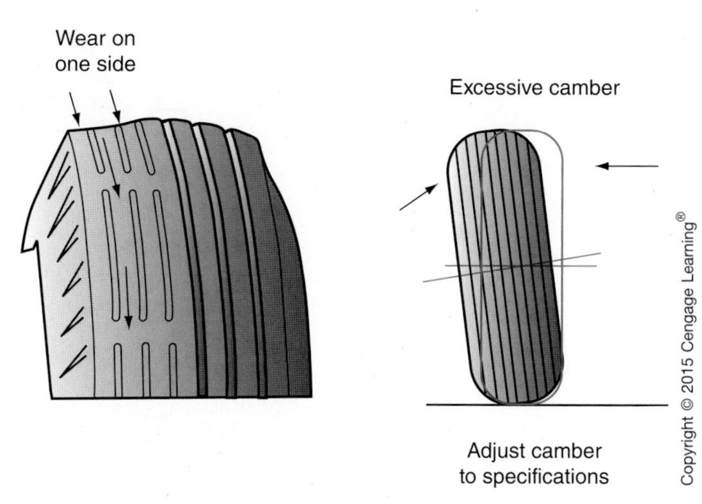

FIGURE 9-13 **Tire tread wear caused by incorrect camber adjustment.**

WARNING: Improper camber adjustment may result in rapid tire tread wear and steering pull, which is a vehicle safety hazard.

Road Crown

A wheel that is tilted tends to steer in the direction it is tilted (Figure 9-14). For example, a bicycle rider tilts the bicycle in the direction he or she wishes to turn, making the turning process easier.

When camber angles are equal on both front wheels, the camber steering forces are equal, and the vehicle tends to maintain a straight-line position. If the camber on the front wheels is significantly unequal, the vehicle will drift to the side with the greatest degree of positive camber.

Crowned highway design prevents water buildup on the driving surface. When a vehicle is driven on a crowned road, it is actually driven on a slight slope, which causes the vehicle steering to pull toward the right (Figure 9-15). Some car manufacturers use the pulling effect of camber to offset this pull to the right caused by road crown. In these vehicles, the left front wheel may have 1/4 to 1/2 degree more positive camber than the right front wheel.

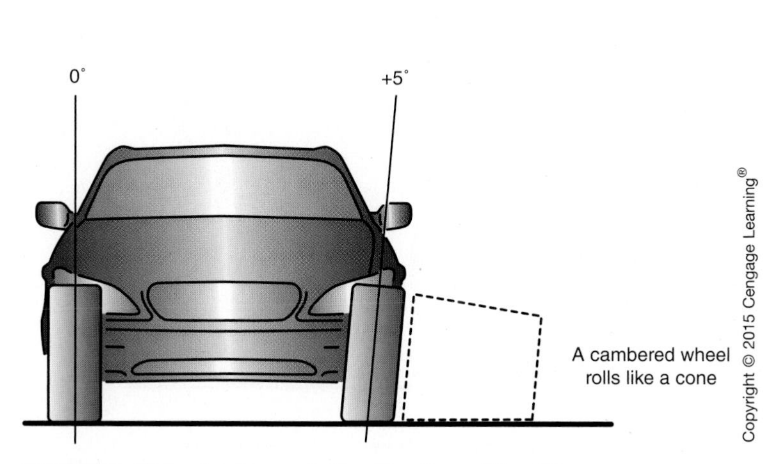

FIGURE 9-14 **A wheel turns in the direction of the tilt.**

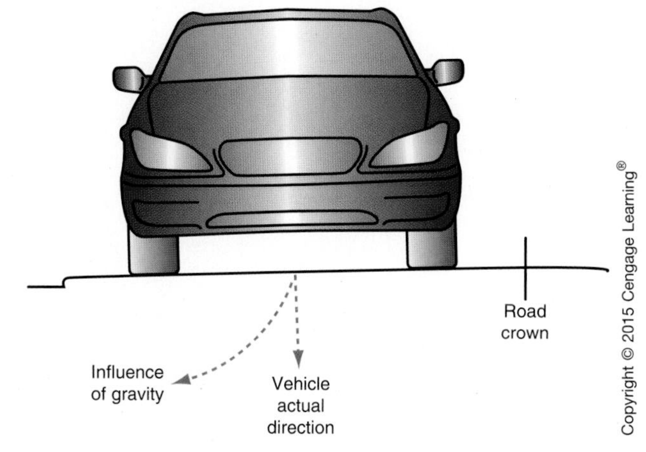

FIGURE 9-15 **A crowned road surface causes the vehicle steering to pull to the right.**

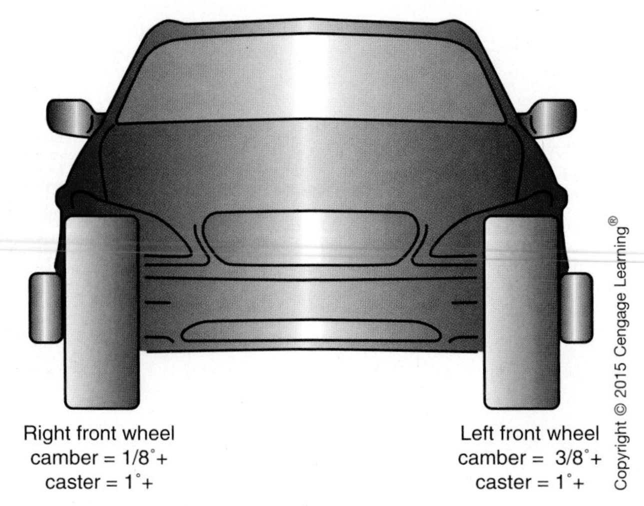

FIGURE 9-16 **Camber setting to offset road crown.**

When the camber on the front wheels is adjusted to the manufacturer's specifications, vehicle directional stability is maintained. If the camber adjustment on the left front wheel is used to offset road crown, the caster on both front wheels must be the same (Figure 9-16).

Caster Fundamentals

Caster Definition

When viewed from the side, caster is the angle formed when a line is drawn through the upper and lower pivot points in reference to true vertical. On a MacPherson strut suspension system, the caster line intersects the center of the upper strut mount and lower ball joint (Figure 9-17). While on a short-and-long arm suspension system, the caster line intersects the center of the upper and lower ball joints (Figure 9-18). **Positive caster** occurs when the caster line is tilted backward toward the rear of the vehicle in relation to the vertical centerline of the spindle and wheel viewed from the side (Figure 9-19). **Negative caster** occurs when the caster centerline is tilted toward the front of the vehicle in relation to the spindle and wheel vertical centerline.

Caster is not considered a direct tire wear angle but is a directional stability angle and is used to improve cornering and steering wheel returnability. Caster is measured in degrees and positive caster is specified on consumer production vehicles. Cross caster is the difference in caster between the right and left side of the vehicle (Figure 9-20). If one front wheel has more positive caster than the other front wheel, the steering pulls toward

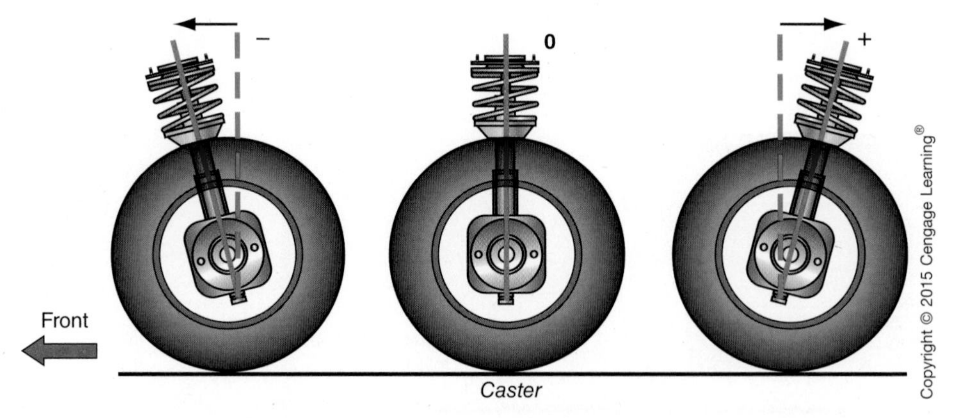

FIGURE 9-17 **On a vehicle with MacPherson strut front suspension, the caster line intersects the upper strut mount and lower ball joint.**

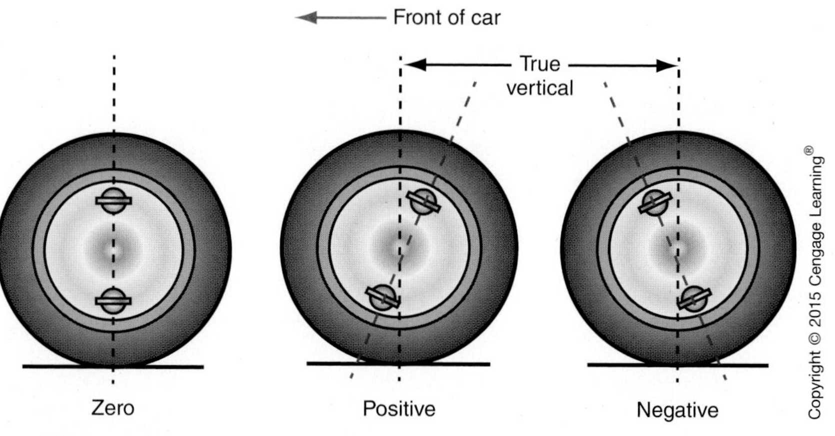

FIGURE 9-18 **On a vehicle with upper and lower ball joints, the caster line intersects the upper and lower ball joint.**

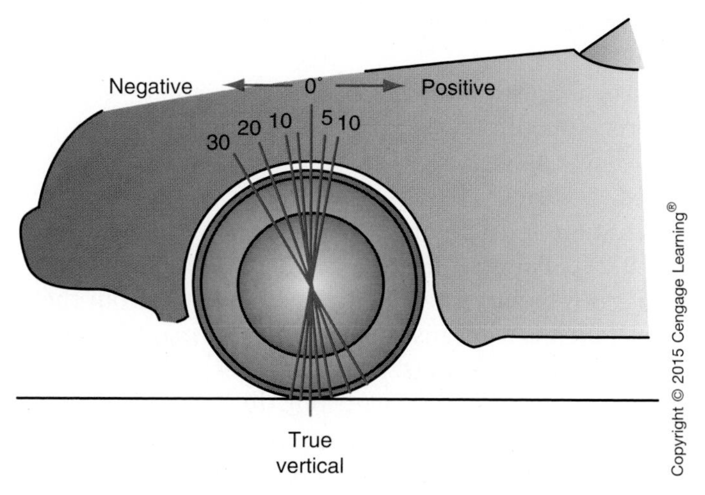

FIGURE 9-19 **Positive and negative caster.**

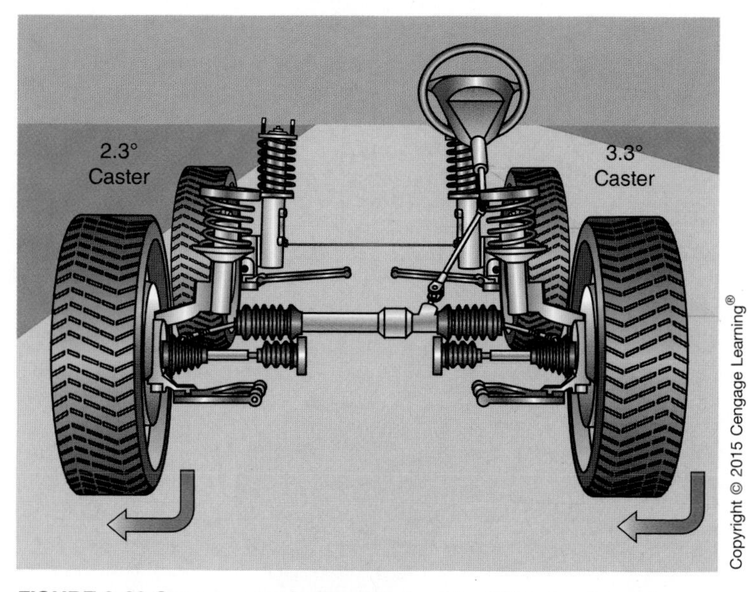

FIGURE 9-20 **Cross caster is the side-to-side difference in caster and may cause a drift or pull to the side with the least positive caster.**

Front	Spec.	Tol.
Left Camber	−0 . 50°	0 . 75°
Right Camber	−0 . 50°	0 . 75°
Cross Camber		0 . 75°
Left Caster	5 . 50°	0 . 75°
Right Caster	5 . 50°	0 . 75°
Cross Caster		0 . 75°
Total Toe	0 . 15°	0 . 25°
Left SAI	°	°
Right SAI	°	°

Rear	Spec.	Tol.
Camber	−0 . 50°	0 . 50°
Total Toe	0 . 12°	0 . 25°
Thrust Angle		°

FIGURE 9-21 **Typical alignment specification sheet for a vehicle. Actual specifications vary between year make and model of vehicle.**

the side with the least amount of positive caster. In general, a vehicle may exhibit a drift if the cross caster difference is of between 0.5 and 1.0 degrees and a pull if the cross caster difference is greater than 1.0 degree toward the side with the least positive caster. Rear caster is not measured.

The caster specification and tolerance is specified by the vehicle manufacturer for the front wheels (Figure 9-21). The specification is the preferred angle and the tolerance is the permitted amount that can be increased (+) or decreased (−) from the preferred angle. Like camber the caster tolerance is a means provided by the manufacture to fine tune the suspension system to specific driving habits. Many vehicles today do not allow for a change in the caster angle as part of a wheel alignment. But this angle should always be measured as part of a wheel alignment and if out of tolerance, the root cause should be determined and appropriate repairs should be recommended to the customer.

Effects of Positive Caster

If a piece of furniture mounted on casters is pushed, the casters turn on their pivots to bring the wheels into line with the pushing force applied to the furniture. Therefore, the furniture moves easily in a straight line (Figure 9-22).

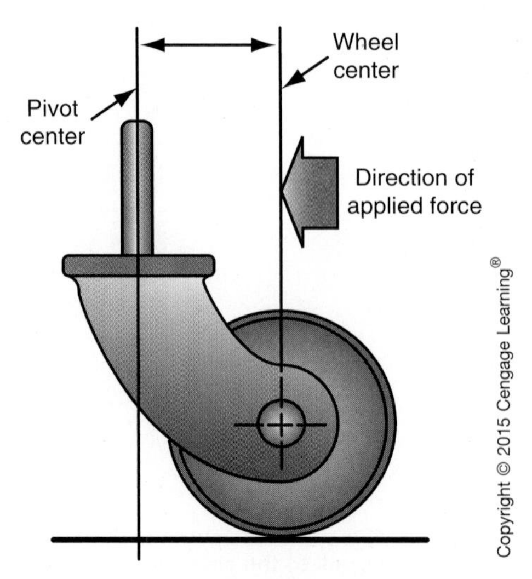

FIGURE 9-22 **Caster action on furniture.**

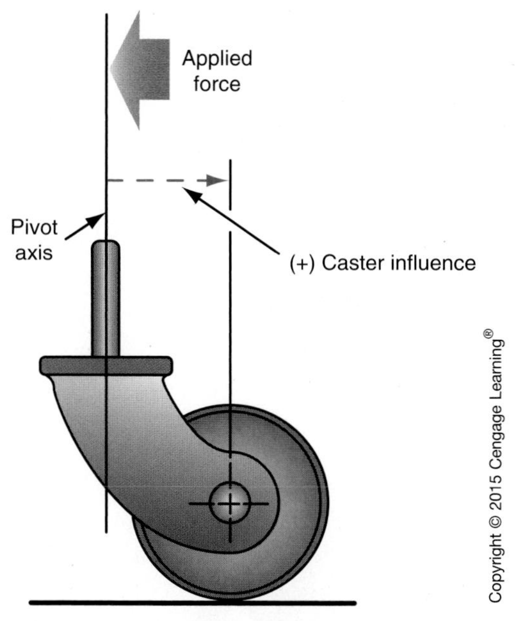

FIGURE 9-23 **Furniture caster wheel aligned with the force on the pivot.**

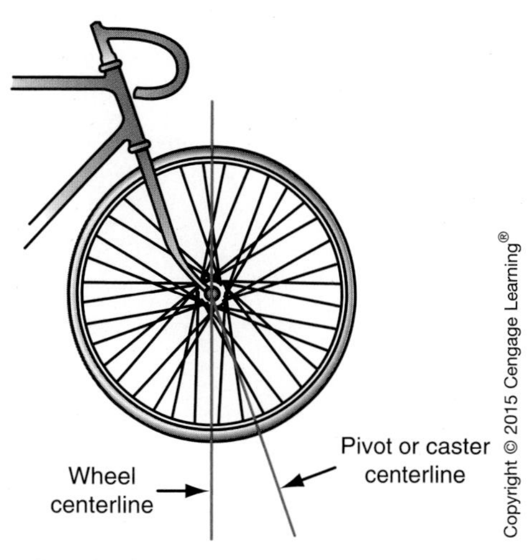

FIGURE 9-24 **Caster line on a bicycle.**

Any force exerted on the pivot causes the wheel to turn until it is lined up with the force on the pivot, because the weight on the wheel results in resistance to wheel movement (Figure 9-23).

Most bicycles are designed with positive caster. The weight of the bicycle and rider is projected through the bicycle front forks to the road surface. The tire pivots on the vertical centerline of the spindle and wheel when the handle bars and front wheel are turned. Notice that the caster line through the center of the front forks is tilted rearward in relation to the vertical centerline of the spindle and wheel as viewed from the side (Figure 9-24).

Because the pivot point is behind the caster line where the bicycle weight is projected against the road surface, the front wheel tends to return to the straight-ahead position after a turn. The wheel also tends to remain in the straight-ahead position as the bicycle is driven. Therefore, the caster angle on a bicycle front wheel provides the same action as the caster angle on a piece of furniture.

Positive caster projects the vehicle weight ahead of the wheel centerline, whereas negative caster projects the vehicle weight behind the wheel centerline. Because positive caster causes a larger tire contact area behind the caster pivot point, this large contact area tends to follow the pivot point. This action tends to return the wheels to a straight-ahead position after a turn. It also helps maintain the straight-ahead position. Positive caster increases steering effort because the tendency of the tires to remain in the straight-ahead position must be overcome during a turn. The returning force to the straight-ahead position is proportional to the amount of positive caster. Positive caster helps maintain vehicle directional stability. Excessive positive caster is undesirable because it increases steering effort and creates a very rapid steering wheel return.

Caster causes camber to change as the wheels are turned left or right. This effect is referred to as camber roll and is the effect that positive caster has on camber as the wheels are turned (Figure 9-25). If the caster angle were 0 degree, the front spindles would rotate horizontally in relation to the road surface. However, a positive caster angle causes the left front spindle to tilt toward the road surface during a left turn (Figure 9-26). Caster is the angle causing camber to increase or decrease when the wheels are turned but camber is the angle causing the tire wear. This effect means that caster can be an indirect tire wear angle if the caster angle is excessively out of specifications. If caster is severely out of

Copyright © 2015 Cengage Learning®

FIGURE 9-25 **Camber roll is the effect that positive caster has on camber during a turn.**

During a
left turn
the spindle
moves
downward
Copyright © 2015 Cengage Learning®

FIGURE 9-26 **Left front spindle movement during a left turn.**

specifications the tires may develop wear on both the inner and outer edges. Because the front wheels are the drive wheels on a front-wheel-drive vehicle, camber roll wear does occur on the front tires as part of the normal tire wear pattern on the front tires but the wear is minimal. Inner and outer edge wear becomes very noticeable on the front tires of a front-wheel-drive car if they are not rotated regularly. The severity of front tire wear because of camber roll is directly proportional to the amount of positive caster present.

This downward spindle movement tends to drive the tire into the road surface. Because this action cannot take place, the left side of the suspension and chassis is lifted. When the driver begins to return the wheel to the centered position, gravity forces the vehicle weight to its lowest position, which helps return the steering wheel to the straight-ahead position. Excessive positive caster increases the left front spindle downward tilt during a left turn, which increases the suspension and chassis lift. Therefore, excessive positive caster increases steering effort. The same action occurs at the right front spindle, but this spindle tilts downward during a right turn.

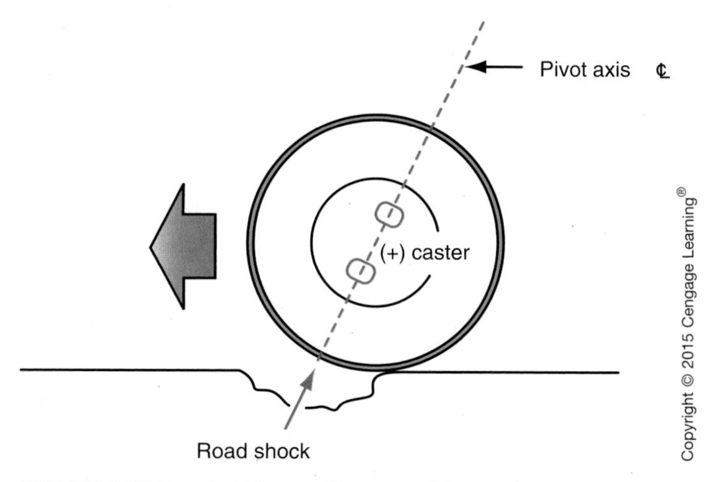

FIGURE 9-27 **Harsh riding quality caused by excessive positive caster.**

Harsh riding may be caused by excessive positive caster because the caster line is actually aimed at some road irregularities (Figure 9-27). Excessive positive caster may cause the front wheels to shimmy from side to side at low speeds.

When the caster line is aimed directly at the road irregularity, road shock is transmitted through the ball joint and upper strut mount to the suspension and chassis. A small degree of positive caster or a negative caster setting allows the front wheel to roll into a road depression without the caster line being aimed at the hole in the road. Therefore, this type of caster line improves ride quality.

Shop Manual
Chapter 9, page 353

Improper front wheel caster may be caused by an improperly positioned front cradle on a front-wheel-drive car. For example, if one or both sides of the cradle are driven rearward in a front-end collision, the front wheel caster is moved from a positive position toward a negative position. This condition may cause steering wander, reduced directional control, and slower steering wheel return.

Some car manufacturers recommend less positive caster on the left front wheel than on the right front wheel to offset the steering pull to the right caused by the road crown. If the caster adjustment is used to compensate for road crown, the camber on both front wheels should be the same. The most important facts about positive caster are the following:

1. Positive caster helps the front wheels return to the straight-ahead position after a turn.
2. Correct positive caster provides improved directional stability of a vehicle.
3. Excessive positive caster produces harsh riding quality.
4. Excessive positive caster promotes sideways front wheel shimmy.
5. The left front wheel may be adjusted with less positive caster than the right front wheel to compensate for road crown.

Rapid side-to-side wheel movement may be called **wheel shimmy**.

Effects of Negative Caster

Negative caster moves the centerline of the upper strut mount and lower ball joint behind the vertical centerline of the spindle and wheel at the road surface. If this condition is present, the friction of the tire causes the tire to pivot around the point where the centerline of the upper strut mount and ball joint meets the road surface. When this occurs, the wheel is pulled away from the straight-ahead position, which decreases directional stability.

Negative caster reduces steering effort. Because excessive positive caster increases road shock transmitted to the suspension and chassis, negative caster reduces this shock and improves ride quality. This improvement occurs because the front wheel rolls into a road depression without the caster line being aimed at the hole in the road. But, if caster is below manufacturers' specifications, poor steering wheel return may be experienced as well as a wondering feeling when driving straight ahead because of reduced directional stability that may result from a low or negative caster setting. Caster is never specified as a negative number. This is why we say that caster may pull to the side with the "least" positive caster because it is always specified to be a positive setting.

Effects of Suspension Height on Caster

When the rear springs become sagged or overloaded, the caster on the front wheels becomes more positive. This action explains why front wheel shimmy may occur when a trunk is severely overloaded.

If the rear suspension height is above the vehicle manufacturer's specification, the caster on the front wheels becomes less positive. Under this condition, the front wheel caster may change from positive to negative. This explains why a vehicle may have reduced directional stability and control when the rear suspension height is raised.

On many short-and-long arm front suspension systems, the inner front end of the upper control arm is higher than the inner rear side of this arm where it is attached to the frame. This design causes the front wheel caster to become more positive if the front suspension height is lowered. When the front suspension height is above the vehicle manufacturer's specification, the front wheel caster becomes less positive.

Most front-wheel-drive cars have a MacPherson strut front suspension system with a slightly positive caster setting. Therefore, the top strut mount is tilted rearward. If the front suspension height is lowered on this type of front suspension system, the front wheel caster becomes less positive. Conversely, if the suspension height is raised, positive caster increases. The most important facts about negative caster include the following:

1. Negative caster does not help return the front wheels to the straight-ahead position after a turn.
2. Negative caster contributes to directional instability and reduced directional control.
3. Negative caster does not contribute to front wheel shimmy.
4. Negative caster reduces road shock transmitted to the suspension and chassis.

Safety Factors and Caster

Directional Control

As explained earlier in this chapter, positive caster provides increased directional stability and control, whereas negative caster reduces directional stability. Therefore, front wheel caster must be adjusted to the manufacturer's specifications to maintain vehicle directional control and safe handling characteristics.

Suspension Height

We have already explained how incorrect front or rear suspension height results in changes in front wheel caster. Therefore, abnormal suspension heights may contribute to reduced directional control and unsafe steering characteristics.

WARNING: Improper caster adjustment may result in decreased directional stability of the vehicle and reduced driving safety.

Toe Definition

Total toe is the measurement difference between the distances between the fronts of the tires compared to the distance between the rears of the same tires (Figure 9-28). When the distance between the rear inside tire edges is greater than the distance between the front inside tire edges (Figure 9-29), the front wheels have a **toe-in** setting and is more commonly referred to as positive toe and is displayed on an alignment computer in degrees (Figure 9-30). **Toe-out** occurs when the distance between the inside front tire edges exceeds the distance between the inside rear tire edges and is more commonly referred to as negative toe and is displayed on an alignment computer as a negative number in degrees (Figure 9-31).

A total toe specification and tolerance is specified by the vehicle manufacture for both the front and rear (Figure 9-32). The specification is the preferred angle and the tolerance is the permitted amount that can be increased (+) or decreased (–) from the preferred angle. Manufacturers generally publish the toe specifications as total toe. On most vehicles individual right and left toe is adjustable independently by changing the length of right and left tie rod assemblies. The individual toe specification is half (1/2) the total toe specification and may be measured in either degrees or inches. Individual toe is the difference between the front of one tire and a reference line and the rear of the tire and the same reference line. The reference line is either the center line or the thrust line of the vehicle,

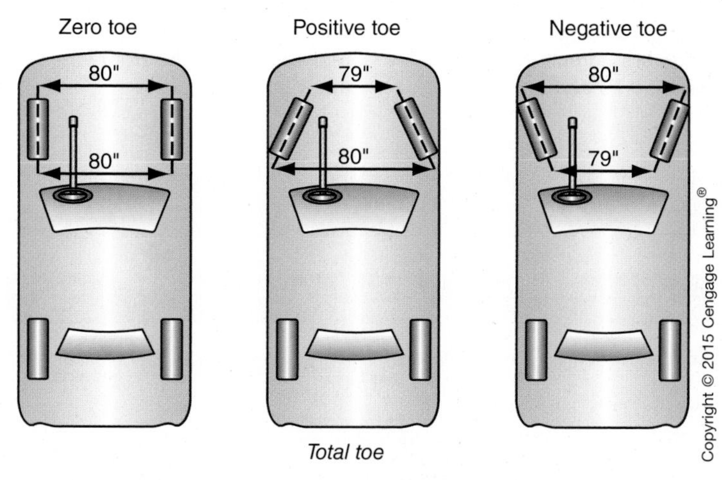

FIGURE 9-28 **Total toe is the difference in distance measurements between the front measurement compared to the rear measurement.**

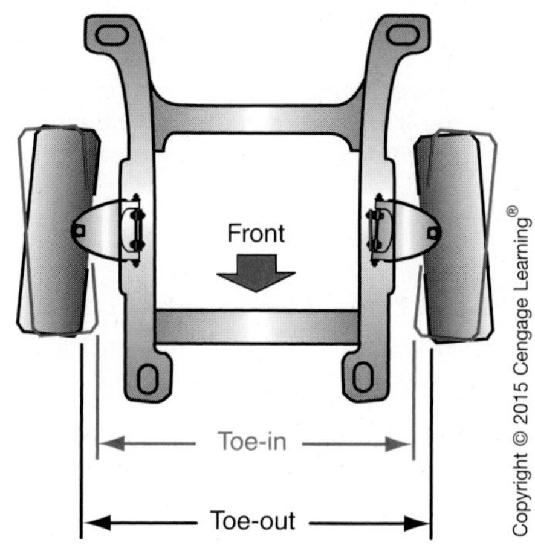

FIGURE 9-29 **Toe-in and toe-out.**

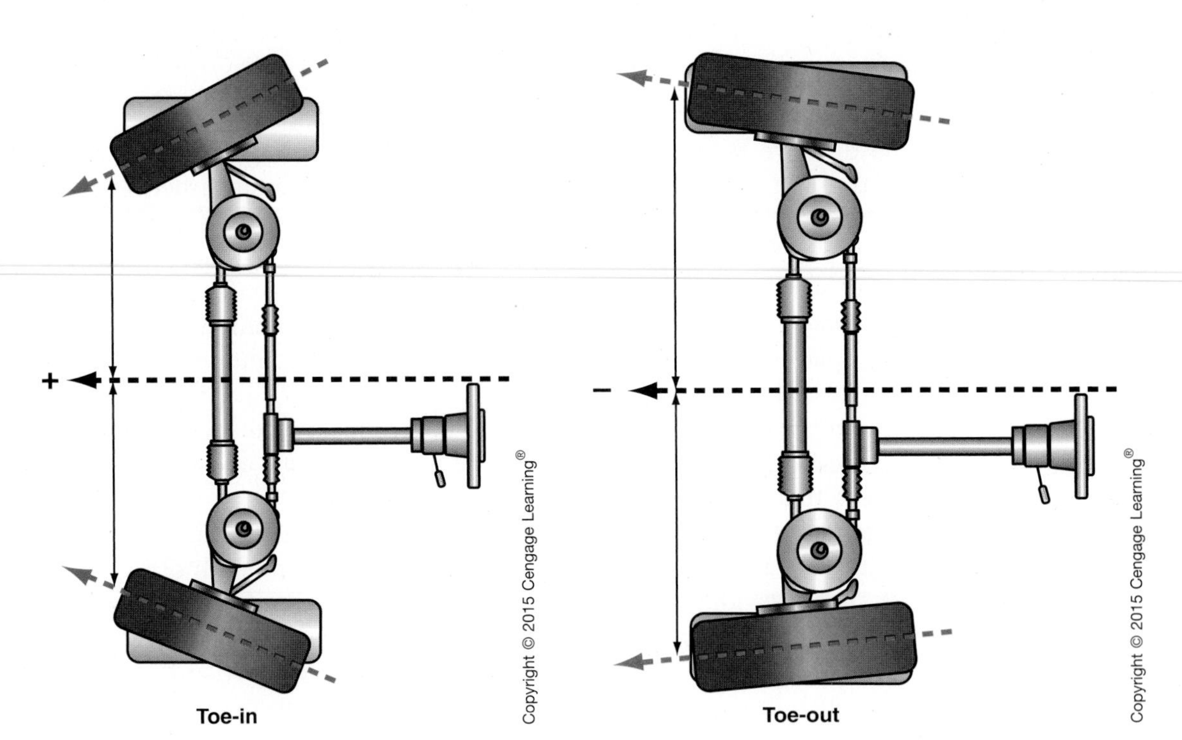

FIGURE 9-30 **If distance between the rear inside tire edges is greater than the distance between the front inside tire edges, the front wheels have a positive (toe-in) setting.**

FIGURE 9-31 **If distance between the inside front tire edges exceeds the distance between the inside rear tire edges, the front wheels have a negative (toe-out) setting.**

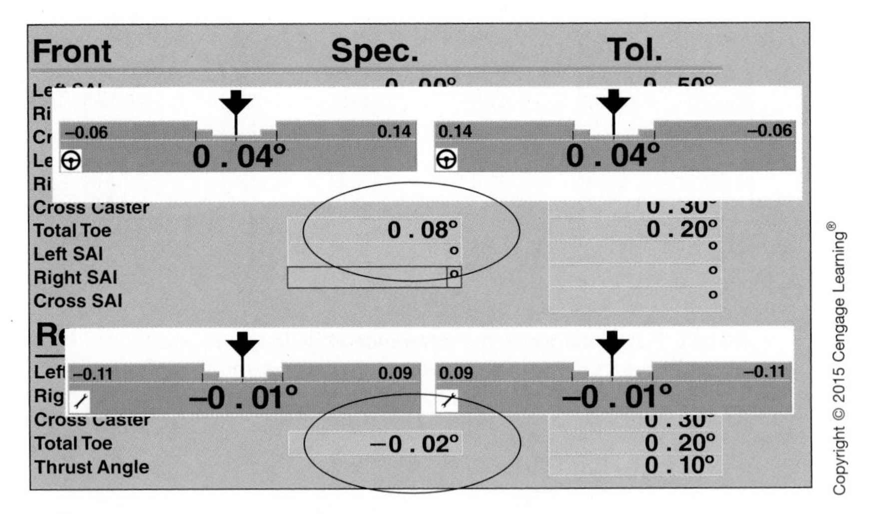

FIGURE 9-32 **A total toe specification and tolerance is specified by the vehicle manufacturer for both the front and rear. The individual toe specification is half (1/2) the total toe specification and may be measured in either degrees or inches.**

depending on the type of alignment being performed (Figure 9-33). Because the front tires are the steering wheels, front individual toe will attempt to equalize while driving. If one front tire's toe specification is out of adjustment, tire wear will occur on both front tires and the steering wheel will not be centered when driving straight ahead. Changes in front individual toe can occur as a result of road impacts such as curbs, pot holes, and severally rough roads or from loose or damaged components. Rear toe, unlike front toe, cannot equalize between the two rear tires because they are not steering wheels and the linkage

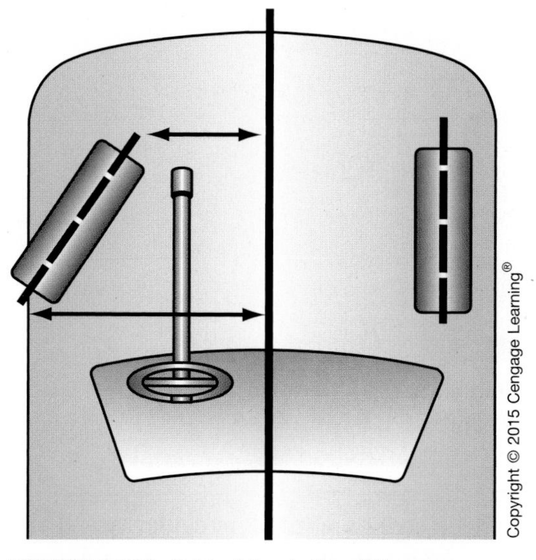

FIGURE 9-33 **Individual toe is the difference between the front of one tire and a reference line and the rear of the tire and the same reference line.**

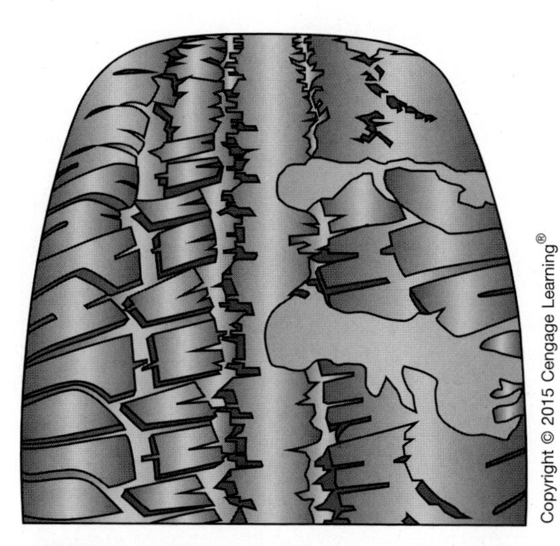

FIGURE 9-34 **When rear tire's toe angle error is extreme, it can cause a diagonal cupping often referred to as diagonal wipe on the tread surface.**

is independent of one another. A rear toe error will cause a dog track condition, which is discussed in further detail in the thrust angle alignment section, and tire wear to the effected wheel. On the rear tires, a toe angle error will cause the effected tire to squirm as it rolls down the road and when extreme can cause a diagonal cupping often referred to as diagonal wipe on the tread surface (Figure 9-34).

Toe is considered a major tire wear angle and incorrect total toe will wear tires extremely quickly. Toe angles that are outside the allowed tolerance will cause tire wear. Positive toe will wear the outside edge of the tire and negative toe will wear the inside edge of the tire. The wear will begin on the shoulder of the tire and taper toward the center of the tire.

Toe Setting for Front-Wheel-Drive and Rear-Wheel-Drive Vehicles

On a front-wheel-drive vehicle, the front drive axle torque forces the front wheels toward an increased positive (toe-in) position. Therefore, car manufacturers usually specify a slight toe-out position for the front wheels on these vehicles.

On rear-wheel-drive vehicles, front tire friction on the road surface moves the wheels toward the toe-out position when the car is driven. On this type of vehicle, manufacturers usually specify a slight toe-in on the front wheels. The front wheels are adjusted to a slight toe-in or toe-out with the vehicle at rest so the wheels will be parallel to each other when the vehicle is driven on the road. A slight amount of lateral movement always exists in steering linkages. Forces acting on the front wheels try to compress or stretch the steering linkages when the vehicle is driven. Whether a compressing or stretching action occurs on the steering linkage depends on whether the steering linkages are located at the rear edge or front edge of the front wheels.

Toe Adjustment and Tire Wear

When the front wheel toe is not adjusted to the manufacturer's specifications, front tire tread wear is excessive. If one of the tie rods is not parallel to the lower control arms, toe change on the front wheels is not equal during front wheel jounce and rebound. This

Shop Manual
Chapter 9, page 340

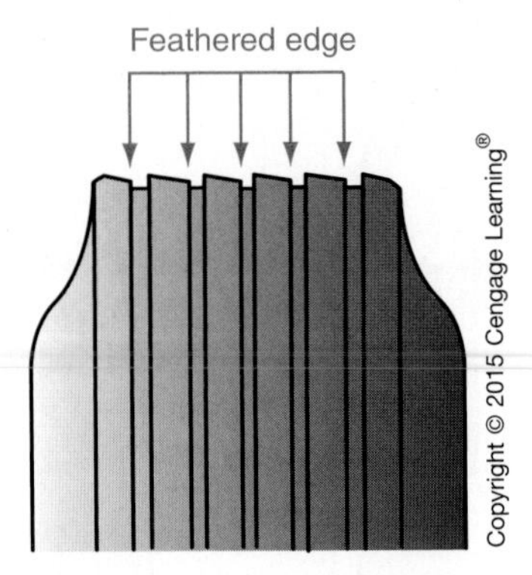

FIGURE 9-35 **Feathered tire tread wear caused by incorrect toe adjustment.**

action may result in bump steer when a front wheel strikes a road irregularity. While checking front wheel toe, improper toe change during suspension jounce and rebound should be checked to be sure the tie rods are parallel to the lower control arms. Technicians must understand front wheel toe and improper toe changes.

Ford Motor Company has calculated that a toe-in error of 1/8 in. (3.17 mm) is equivalent to dragging the tires crosswise for 11 ft. (3.3 m) for each mile the vehicle is driven. This crosswise movement causes severe feathered tire tread wear (Figure 9-35). Improper toe adjustment is the most common cause of rapid tire tread wear.

CAUTION:
Improper toe adjustment causes rapid tire tread wear, which may result in tire failure.

Excessive toe-out causes wear on the inside of the tire tread ribs and a sharp feathered edge on the outside of the tread ribs. If excessive toe-in is present, the tire tread wear is reversed.

Worn steering linkage components such as tie rod ends cause incorrect and erratic toe-in settings. If the front springs become weak, the front suspension height is lowered. When this occurs, the pitman arm and the idler arm move downward with the chassis. This action moves the tie rods to a more horizontal position that tends to push outward on the steering arms and increases front wheel toe-in. The toe-in change just described occurs when the steering linkage is located at the rear of the front wheels.

WARNING: Worn steering components may suddenly become disconnected, causing complete loss of steering control, collision damage, and personal injury.

Steering Terminology

Torque steer may be defined as the tendency of the steering to pull to one side during hard acceleration.

Bump steer is the tendency of the steering to veer suddenly in one direction when one or both of the front wheels strikes a bump.

In the automotive service industry, certain terms are used for specific steering problems. Some of these problems are related to camber and caster, others are caused by various suspension or steering defects. Technicians must be familiar with both the steering terminology and the causes of the problems.

A front-wheel-drive vehicle with unequal-length drive axles produces some **torque steer** on hard acceleration. Some front-wheel-drive vehicles, especially those with higher horsepower, have equal-length front drive axles to reduce torque steer. On a front-wheel-drive vehicle, torque steer is aggravated by different tire tread designs on the front tires or uneven wear on the front tires.

Bump steer occurs if the tie rods are not the same height, meaning one of the tie rods is not parallel to the lower control arm. This condition may be caused by a bent or worn idler arm or pitman arm on a parallelogram steering linkage. On a rack and pinion steering system, worn steering gear mounting bushings may cause this unparallel condition between one of the tie rods and the lower control arm.

Memory steer may be caused by a binding condition in the steering column or in the steering shaft universal joints. A binding upper strut mount may result in memory steer. Negative caster or reduced positive caster also causes memory steer.

Steering pull is the tendency of the steering to gradually pull to the right or left when the vehicle is driven straight ahead on a level road. Steering pull or drift may be caused by improper caster or camber adjustments. **Steering drift** is the tendency of the steering to slowly drift to the right or left when driving straight ahead on a level road.

When **steering wander** occurs, the vehicle tends to steer in either direction rather than straight ahead on a level road surface. Steering wander may be caused by improper caster adjustment.

When **memory steer** occurs, the vehicle doesn't want to steer straight ahead after a turn because the steering does not return to the straight-ahead position.

WHEEL ALIGNMENT THEORY

Road Variables

Shop Manual
Chapter 9, page 318

Vehicles are subjected to many **road variables** that affect wheel alignment. These variables must be counteracted by the suspension design and alignment, or steering would be very difficult. Some of the variables that affect wheel alignment and suspension design follow:

1. **Road crown** (the curvature of the road surface)
2. Bumps and holes
3. Natural crosswinds or crosswinds created by other vehicles
4. Heavy loads or unequal weight distribution
5. Road surface friction and conditions such as ice, snow, and water
6. Tire traction and pressure
7. Side forces while cornering
8. Drive axle forces in front-wheel-drive vehicles
9. Relationship between suspension parts as the front wheels turn and move vertically when road bumps and holes are encountered

Road variables are differences in road surface conditions, vehicle loads, and weather conditions.

Road crown refers to the high portion in the center of the road with a gradual slope to each side.

A desirable plan to reduce tire wear would be to place the front wheels and tires so they are perfectly vertical. The tires would then be flat on the road. However, if the wheels and tires are perfectly vertical, such variables as the driver entering the car, turning a corner, or adding weight to the luggage compartment would move the tire from its true vertical position. Therefore, tire wear and steering operation would be adversely affected.

Rather than allowing the variables to adversely affect tire wear and steering operation, the suspension and steering are designed with characteristics to provide directional stability, predictable directional control, and minimum tire wear. Wheel alignment angles are designed to provide these desired requirements despite road variables. Wheel alignment angles also control the **tracking** of the rear wheels in relation to the front wheels.

Vehicle **tracking** is the straightness of the rear wheels in relation to the front wheels.

A customer may enter your shop and request a wheel alignment for many reasons. The three most common reasons for customers requesting a wheel alignment are to straighten a steering wheel that is not centered (crooked), to prevent tire wear, or to address a vehicle handling issue such as a pull while driving straight ahead. In addition, a customer may request a wheel alignment for a steering wheel that is shaking. A shaking of the vehicle is not generally caused by a wheel alignment angle error. A tire or wheel balance or run out condition as well as bearing and axle faults are more likely causes of the steering wheel shake condition. Tire rotation and balancing are additional services that should be considered when a wheel alignment is performed. Because customers are often expecting a smooth vibration free ride after an alignment is performed.

After an alignment is performed, the benefits your customer can expect are:

- The steering wheel position is level and pointing straight down the road while driving the vehicle straight ahead.

- The steering wheel automatically returns to the straight-ahead position after making a right or left turning maneuver.
- The vehicle exhibits directional stability without pulling or wandering while driving straight ahead reducing driver fatigue.
- Improved tire tread life and fuel economy.
- Proper vehicle tracking whereby the front and rear wheels follow each other.
- Improved cornering performance, directional control, and feedback while maintaining optimum road isolation.

Types of Wheel Alignment

Geometric Centerline Alignment

In a **geometric centerline alignment**, the front wheel toe is adjusted using the geometric centerline as a reference.

Toe refers to the angle between the plane of a front or rear wheel and a reference line. Either the geometric centerline or the thrust line may be used as a toe reference line. When **geometric centerline alignment** is used as a reference for front wheel toe, the toe on each front wheel is adjusted to specifications using the geometric centerline as a reference (Figure 9-36). This type of wheel alignment has been used for many years. It may provide a satisfactory wheel alignment if the rear wheels are properly positioned and the thrust line is at the vehicle centerline. However, if the rear wheels are not positioned properly, the thrust line is not at the geometric centerline and steering problems will occur. Therefore, geometric centerline front wheel reference ignores rear wheel misalignment.

Thrust Line Alignment

If **thrust line alignment** is used, the thrust line created by the rear wheels is used as a reference for front wheel toe adjustment (Figure 9-37). When the front wheel toe is

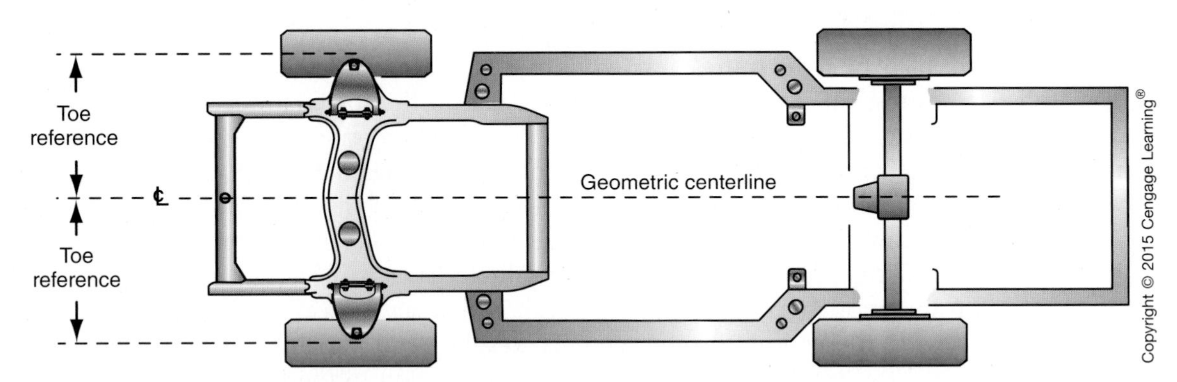

FIGURE 9-36 **Front wheel alignment with the front wheel toe referenced to the geometric centerline.**

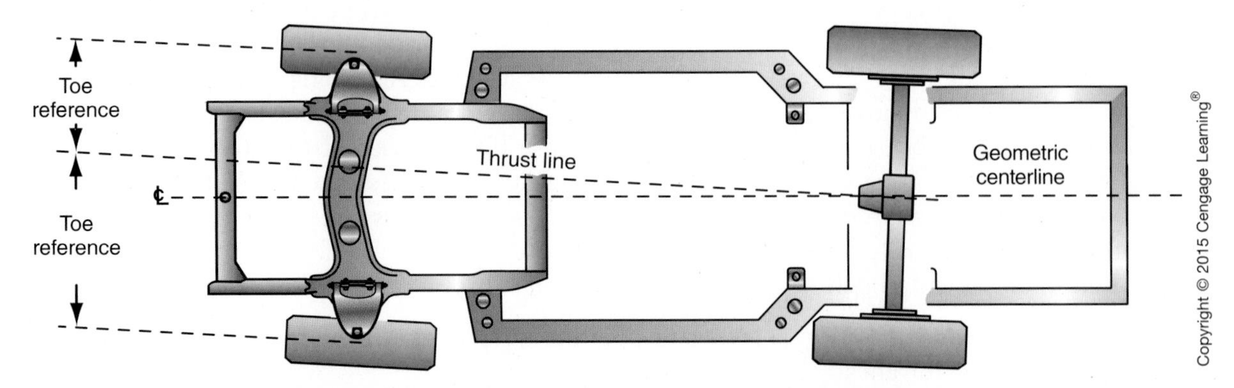

FIGURE 9-37 **Thrust line alignment with the front wheel toe referenced to the thrust line.**

adjusted with a thrust line reference, and this line is not at the geometric centerline, neither the front nor rear wheels are parallel to the geometric centerline. Under this condition, none of the four wheels is facing straight ahead when the vehicle is driven straight ahead. This action results in excessive wear on all four tire treads. This type of alignment ensures a centered steering wheel when the vehicle is driven straight ahead.

A positive thrust line (angle) is to the right of the vehicle center line when viewed from above and a negative thrust lie (angle) is to the left of the vehicle center line when viewed from above (Figure 9-38). If thrust angle error is excessive a visible dog track condition will be present (Figure 9-39). Dog tracking occurs when the front wheels follow the direction that the rear wheels are pointing. If excessive the vehicle will appear to be going down the road sideways. The next time you are driving on the highway in a straight-ahead path pay attention to the vehicles in front of you and look at the rear and front tires on the same side of the vehicle. Look for a dog track condition; if you can see both the rear and front tire when viewed from behind the vehicle is dog tracking. When viewing the actual alignment readings for a vehicle on a computer based alignment information screen pay

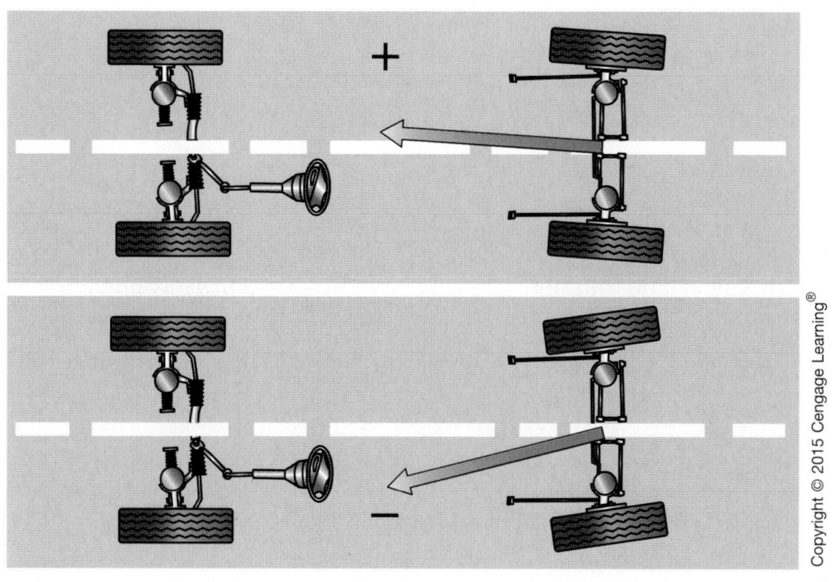

FIGURE 9-38 **Positive and negative thrust line referenced to the vehicle center line as viewed from above. A positive thrust line is to the right of the center line and a negative thrust line is to the left of the centerline.**

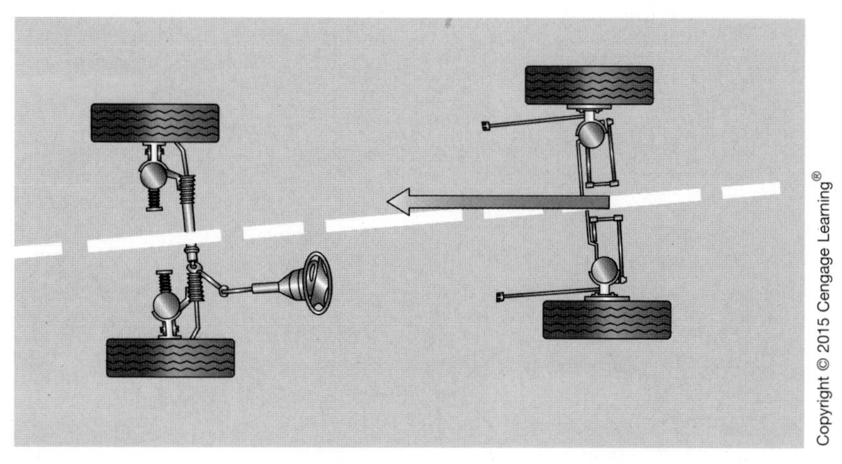

FIGURE 9-39 **If a severe thrust angle condition exists a vehicle will dog track.**

attention to whether the thrust angle is a positive or negative number. This information can aid your diagnosis when dealing with frame and structural damage. The maximum allowable thrust angle can vary between manufacturers but is generally not more than 0.125 degree.

Four-Wheel Alignment

When a **four-wheel alignment** is completed, the thrust line position is measured in relation to the geometric centerline. Individual rear wheel toe-in is measured and adjusted to manufacturer's specifications. This adjustment moves the thrust line to the geometric centerline. Thrust angle is measured to be sure the thrust line is positioned at the geometric centerline. Front wheel toe is measured using the common geometric centerline and thrust line as a reference (Figure 9-40). When a four-wheel alignment is completed, all four wheels are parallel to the geometric centerline and the steering wheel is centered as the vehicle is driven straight ahead. Wheel aligners have different alignment capabilities. Modern computer wheel alignment systems have four-wheel alignment capabilities, and this type of wheel alignment must be performed on today's vehicles to ensure proper steering control and vehicle safety.

When performing a wheel alignment the sequence that the angles are adjusted is critical, as a change in one angle may affect another angle. For example, a change in rear toe will change the position of the vehicles thrust angle, which is the reference line for front toe. So, a change in rear toe will also change the reading for front toe.

The sequence the wheel alignment angles must be adjusted in (if an angle adjustment is necessary) is:

1. Adjust rear camber (if a means of adjustment is available).
2. Adjust rear individual toe (if a means of adjustment is available).
3. Adjust front caster and camber (if a means of adjustment is available).
4. Adjust front individual toe.

The advantages of four-wheel alignment are the following:

1. *Improved fuel mileage.* After a four-wheel alignment, all four wheels are parallel, and this condition combined with proper tire inflation decreases rolling resistance, which improves fuel mileage.
2. *Longer tire life.* When all four wheels are aligned properly, tire tread wear is minimized.
3. *Improved vehicle handling.* When all four wheels are properly aligned and all steering and suspension components are in satisfactory condition, steering pulls, vibrations, and abnormal steering conditions are eliminated to ensure improved vehicle handling.
4. *Safer driving.* Proper alignment of all four wheels plus inspection and replacement of all worn or defective steering and suspension components improves vehicle handling, and this reduces the possibility of a collision and provides safer driving.

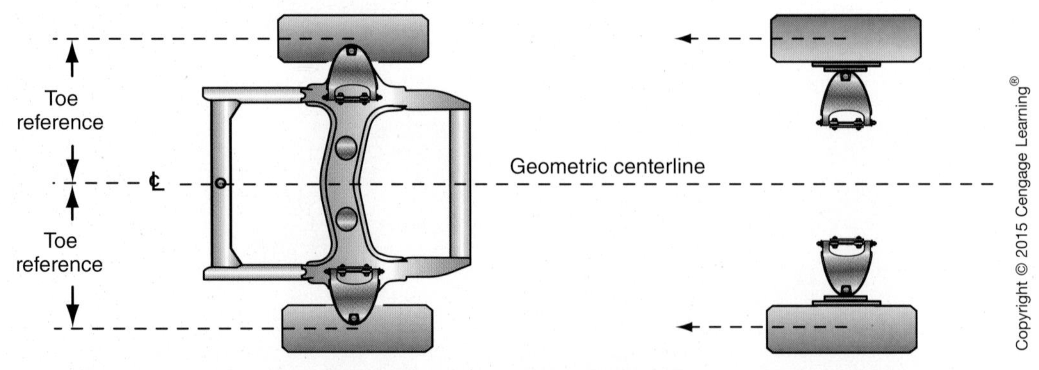

FIGURE 9-40 Four-wheel alignment with the thrust line adjusted so it is at the geometric centerline and the front wheel toe is referenced to the geometric centerline.

Importance of Four-Wheel Alignment

Until the late 1970s, most vehicles in the United States were rear-wheel drive. The majority of these vehicles had one-piece rear axle housings and frames. This type of vehicle design did not experience many rear wheel alignment problems and rear wheel alignment was usually not a concern in these years.

Shop Manual
Chapter 9, page 323

Beginning in the late 1970s, many domestic car makers began manufacturing front-wheel-drive cars with unitized bodies. The gasoline shortages in the 1970s and the introduction of federal corporate average fuel economy (CAFE) laws brought about a massive change to lighter weight, more fuel efficient front-wheel-drive cars. A significant number of these cars had four-wheel independent suspension systems.

A unitized body may be called a unibody.

Cars with unitized bodies and independent or semi-independent rear suspension systems are more likely to experience rear wheel alignment problems compared with rear-wheel-drive vehicles with frames and one-piece rear axle housings. This is especially true after collision damage. Therefore, with the introduction of unitized bodies in massive numbers, four-wheel alignment became a necessity.

The **thrust line** is an imaginary line at a 90 degree angle to the centerline of the rear wheels and projected forward.

Rear Wheel Alignment and Vehicle Tracking Problems

Result of Proper Rear Wheel Alignment

The driver uses the steering wheel to turn the front wheels and steer the vehicle in the desired direction. However, the rear wheels determine the direction of the vehicle to a large extent. When the **thrust line** is positioned at the **geometric centerline** and the front and rear wheels are parallel to the vehicle geometric centerline, the vehicle moves straight ahead with minimum guidance from the steering wheel (Figure 9-41).

The vehicle **geometric centerline** is an imaginary line through the exact center of the front and rear wheels.

Result of Improper Rear Wheel Alignment

Rear Axle Offset. A **rear axle offset** may occur on a rear-wheel-drive vehicle with a one-piece rear axle housing or on a front-wheel-drive car with a trailing arm rear suspension. If the rear axle is offset, the thrust line is no longer at the vehicle centerline (Figure 9-42). The **thrust angle** is the angle between the geometric centerline and the thrust line.

A **rear axle offset** refers to a condition where the complete rear axle housing has rotated slightly, moving one rear wheel forward and the opposite rear wheel backward. Under this condition, the rear wheels are no longer parallel to the geometric centerline of the vehicle.

With the rear axle offset problem shown in Figure 9-42, the thrust line is positioned to the left of the geometric centerline. Under this condition, the rear wheels steer the vehicle in a large clockwise circle. If the driver's hands are removed from the steering wheel, the steering pulls to the right. This steering action is similar to the rear wheels on a forklift. The rear wheels are used to steer the forklift because of the heavy load on the front wheels. If the rear wheels are pointed toward the left, the forklift turns to the right.

On a vehicle with this problem, the left front wheel toes out and the right front wheel toes in when the vehicle is driven straight ahead. This front wheel situation occurs because the front wheels try to compensate for the rear suspension defect. The front wheels are

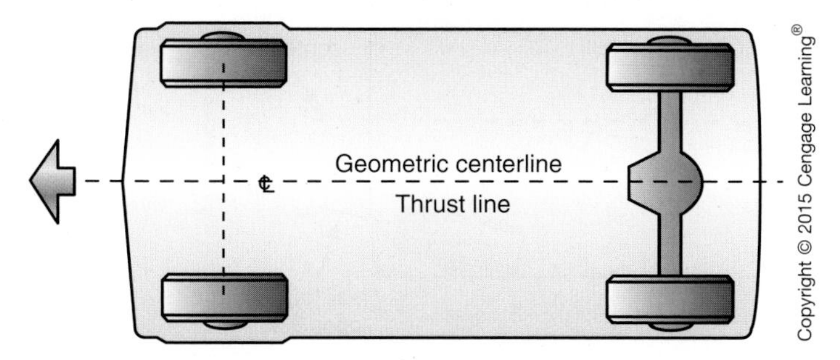

FIGURE 9-41 **Front and rear wheels parallel to the vehicle centerline.**

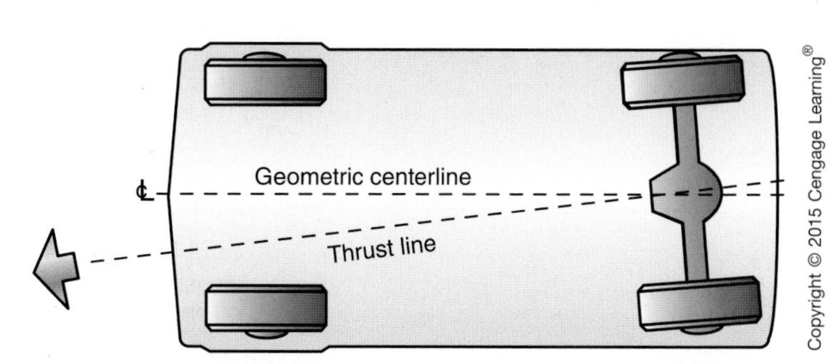

FIGURE 9-42 **Rear axle offset and improperly positioned thrust line.**

turned slightly to the left to compensate for the drift to the right caused by the rear suspension problem. Under this condition, both the front and rear tires may have feathered tire wear.

The ideal correction for this rear suspension problem is to reposition the rear suspension so the thrust line is at the vehicle centerline, then set the front wheel toe-in to the thrust line. The rear suspension problem that we have described also causes the steering wheel to be off-center when the vehicle is driven straight ahead.

Some of the causes of rear axle offset are:

1. A broken center bolt in rear leaf spring.
2. Worn shackles in rear leaf springs.
3. A bent frame.
4. A bent subframe or floor section, unitized body.
5. Worn trailing arm bushings.
6. Bent trailing arms.

Improper Rear Wheel Toe. If the left rear wheel has excessive toe-out, the thrust line is moved to the left of the geometric center line (Figure 9-43). This defect has basically the same effect on steering as rear axle offset. Improper toe on one rear wheel is most often encountered on vehicles with independent rear suspension.

Some of the causes of improper toe on one rear wheel are:

1. A bent one-piece rear axle.
2. A bent U section in trailing arm suspension.
3. A bent rear lower control arm.
4. A worn rear lower control arm bushing.
5. A bent rear spindle.
6. An improper rear toe adjustment.

AUTHOR'S NOTE: It has been my experience that the effect of rear wheel alignment on steering pull is sometimes ignored when diagnosing a customer complaint of steering pull. If the rear axle is offset or the rear wheel toe is incorrect, the rear wheel position actually pushes the front wheels away from the straight-ahead position and causes the steering to pull to one side. This problem is more likely to occur on front-wheel-drive cars, especially those with independent rear suspension and a unitized body design. Therefore, it is very important that you perform a four-wheel alignment and be sure the thrust angle is within specifications before aligning the front suspension.

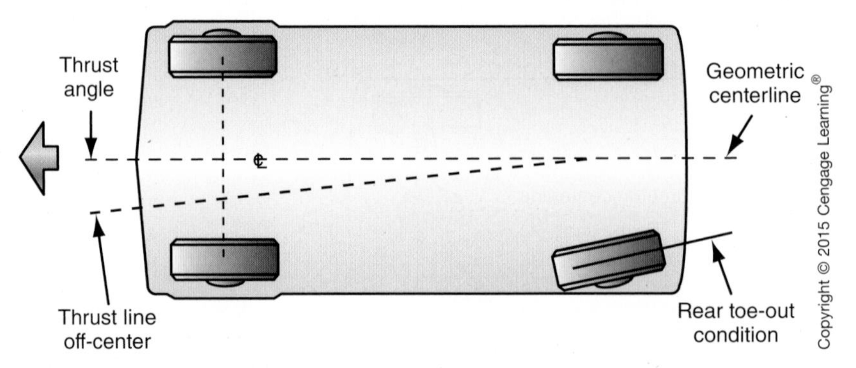

FIGURE 9-43 **Excessive toe-out on the left rear wheel moves the thrust line to the left of the geometric centerline.**

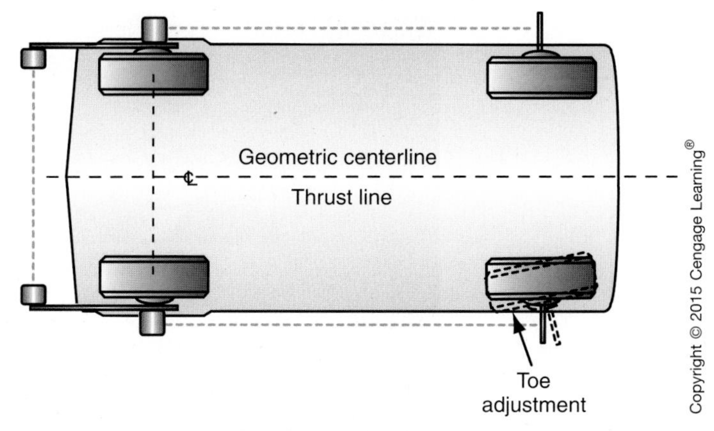

FIGURE 9-44 **Adjusting toe to specifications on the left rear wheel moves the thrust line to the geometric centerline.**

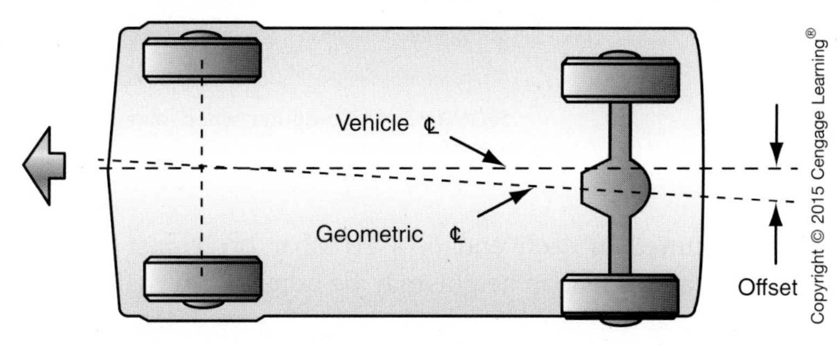

FIGURE 9-45 **Rear axle offset occurs when the rear axle assembly has moved straight sideways and the geometric centerline is shifted away from the proper vehicle centerline position.**

Shop Manual
Chapter 9, page 340

A BIT OF HISTORY

Early attempts at rear wheel alignment were slow and lacked precision. These attempts at rear wheel alignment included the use of a track bar and even backing the rear wheels of a car onto a front wheel aligner to align the rear wheels. To meet the need for fast, accurate front and rear wheel alignment, wheel alignment manufacturers designed computer wheel aligners. The technology in this equipment has greatly improved since the first models were introduced.

The necessary correction for this defect is to adjust the left rear wheel toe to the manufacturer's specifications. This adjustment moves the thrust line to the geometric centerline of the vehicle (Figure 9-44).

Rear Axle Sideset. When **rear axle sideset** occurs, the rear wheels are parallel to each other but the rear axle assembly has moved straight sideways so the geometric centerline is no longer at the proper vehicle centerline position (Figure 9-45). Because the rear wheels do not follow directly behind the front wheels when the vehicle is driven straight ahead, the vehicle dog tracks. Rear axle sideset does not affect steering pull as much as rear axle offset, but sideset will cause steering pull if it is severe.

Wheel Setback. Wheel setback is a condition where one front or rear wheel is moved rearward in relation to the opposite wheel (Figure 9-46). This condition is usually caused by front-end collision damage, and thus it is most likely to be a front suspension problem. However, setback may occur on independent rear suspension systems. A severe setback condition causes steering pull.

Computer Alignment Systems

Computer Wheel Aligner Features

Some computer wheel aligners have four high-resolution digital cameras that measure wheel target position and orientation. The front and rear wheel alignment angles are sensed by the digital cameras and wheel targets and then displayed on the wheel alignment monitor. The vehicle is raised to a comfortable working height on the aligner lift, and two

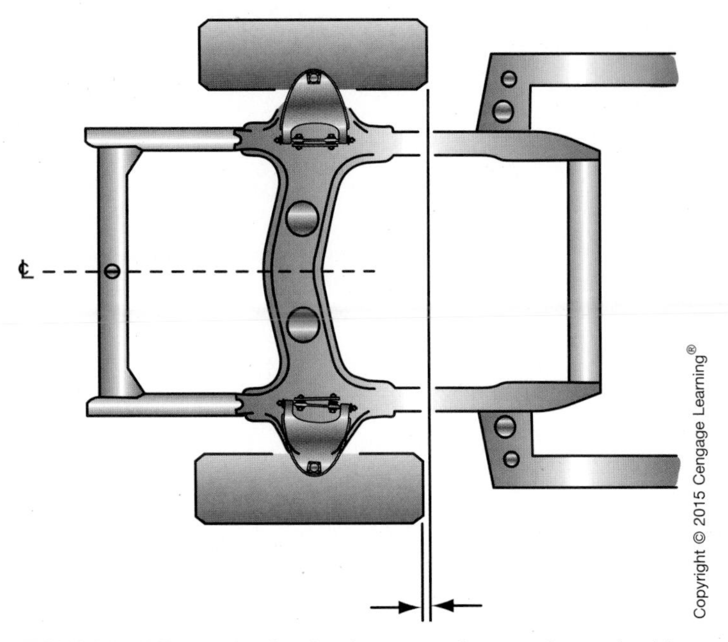

FIGURE 9-46 **Front wheel setback occurs when one front wheel is moved rearward in relation to the opposite front wheel.**

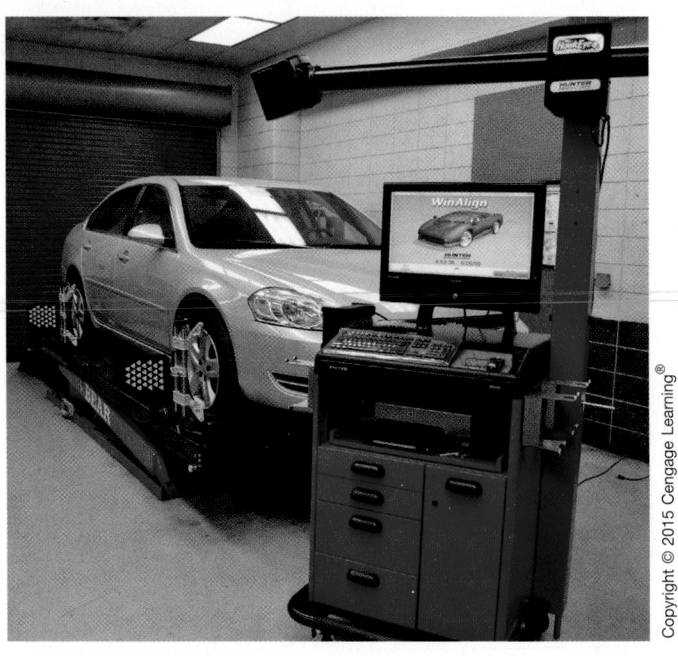

FIGURE 9-47 **Computer wheel aligner with digital cameras and wheel targets.**

digital cameras are mounted in each end of a crossbar on a post in front of the vehicle (Figure 9-47). The post and crossbar height may be adjusted to match the vehicle height.

One digital camera is aimed at each wheel target (Figure 9-48). A target is mounted on each wheel with a self-centering adapter. The adapters are adjustable to fit rims up to 24 in. (60 cm) in diameter. The targets contain a polished aluminum faceplate that acts as a mirror (Figure 9-49). The targets do not contain any electronics, glass, or cables, and they do not require calibration. The targets are also lightweight and do not require any maintenance.

FIGURE 9-48 **One digital camera is aimed at each wheel target.**

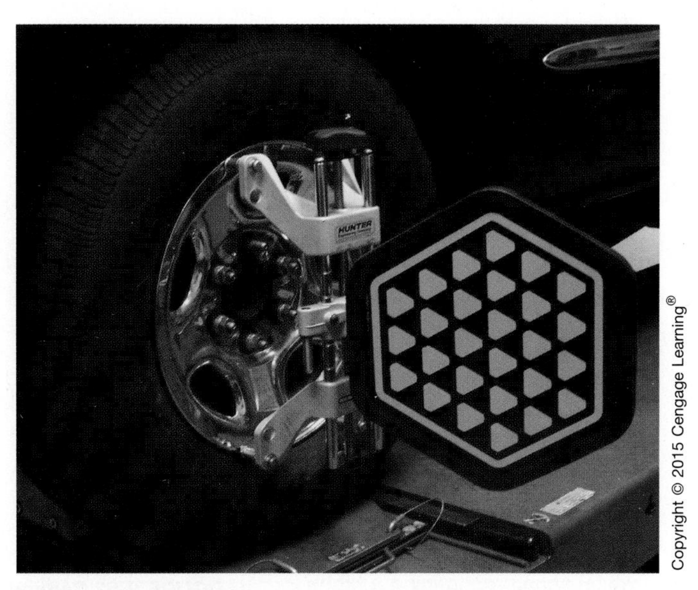

FIGURE 9-49 **Each target contains a polished aluminum face plate with no electronics, glass, or cables.**

Wheel sensors are mounted on each wheel, and these sensors transmit wheel alignment data to the computer wheel aligner.

A **high-frequency transmitter** is contained in many wheel sensors. This transmitter sends data to a receiver on top of the monitor.

This type of computer aligner is very simple to use. After the prealignment procedures and checks have been performed on the vehicle, and the vehicle is positioned at the desired height and position on the lift and turntables, the first step is to mount the wheel targets on the wheels. The second step is to roll the vehicle back a short distance until the on-screen indicators on the monitor turn green. The third step is to roll the vehicle forward and stop the vehicle so the front wheels are centered on the turntables. The fourth step is to observe all the front and rear camber and toe angles displayed on the monitor.

A remote control is wired into the monitor so the technician can perform the same functions when the monitor is not in the technician's view. A handheld remote control is used to measure the vehicle ride height (Figure 9-50). This remote is held against the lower edge of the fender well to measure the ride height, and the remote transmits the ride height measurement electronically to the console and monitor.

On some computer wheel aligners the **wheel sensors** contain a microprocessor and a **high-frequency transmitter** that acquire measurements and process data and then send

FIGURE 9-50 **A handheld remote control transmits ride height measurements electronically to the console and monitor.**

FIGURE 9-51 **Cordless wheel sensor containing a high-frequency transmitter.**

FIGURE 9-52 **Computer wheel aligner with data receiver on top of the monitor.**

A **receiver** on the computer wheel aligner receives signals from the wheel sensors.

these data to a **receiver** mounted on top of the wheel alignment monitor (Figure 9-51 and Figure 9-52). This type of wheel sensor does not require any cables connected between the sensors and the computer wheel aligner. The data from this type of wheel sensor are virtually uninterruptible, even by solid objects. When these wheel sensors are stored on the computer wheel aligner, a "docking station" feature charges the batteries in the wheel sensors. The front wheel sensors have optical arms that project ahead of the tires to provide front toe readings. Some rear wheel sensors also have optical arms that project behind the rear tires (Figure 9-53). This type of rear wheel sensor measures rear wheel setback.

On some computer alignment systems, the reference signals between the wheel sensors are provided by light-emitting diodes (LEDs) or electronic signals. Cables are connected from the wheel units to the computer aligner. Reference signals between the wheel units are provided by strings on some older model computer alignment systems (Figure 9-54). Most wheel sensors provide wheel runout compensation. Because the wheel unit is clamped to the rim, a bent rim will affect the alignment readings. On older computer wheel aligners, the technician had to perform a wheel runout check at each wheel. Newer wheel units use audible and visual prompts on the monitor screen to inform the technician if the wheel sensors require leveling or calibration. Wheel runout compensation may be

FIGURE 9-53 **Front and rear wheel sensors with optical arms.**

FIGURE 9-54 **Front wheel sensor with optical arm and string connected to the rear wheel sensor.**

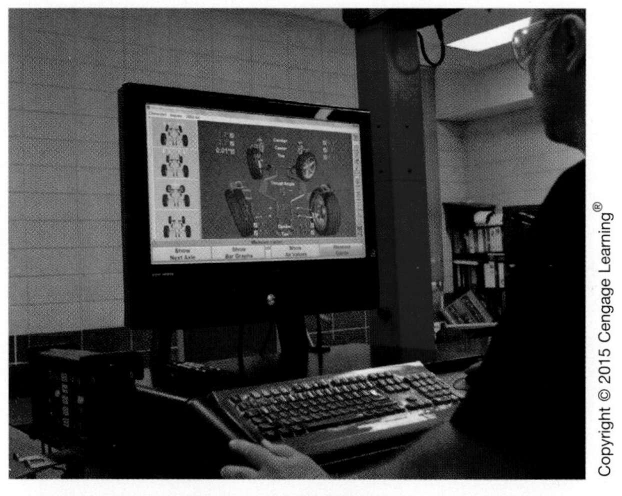

FIGURE 9-55 Computer wheel aligner.

Wheel Alignment Program

Wheel Alignment Specifications

FIGURE 9-56 CDs for wheel alignment program and specifications.

completed by pressing a button on the wheel sensor. Wheel sensor leveling is done by adjusting the sensor level control.

Computer alignment systems have the capability to measure thrust angle and setback as well as other front and rear alignment angles. A typical computer wheel aligner has the following features:

Shop Manual
Chapter 9, page 340

1. Color monitor and computer (Figure 9-55).
2. Computer software program stored on a hard disc drive. This program contains vehicle four-wheel alignment procedures and diagnostic drawings.
3. Compact disc (CD) or digital video disk (DVD) drive. The vehicle specifications and alignment program may be contained on CDs or DVDs, and updates are possible by replacing the CDs or DVDs (Figure 9-56). Wheel alignment specifications include domestic and imported vehicles for the current year and a minimum of 10 previous years.
4. The software in some computer wheel aligners includes a CD or DVD image database containing 2,100 digital photos of the adjustment and inspection points. In some computer aligners, a list of inspection points appears on the left of the screen. When the operator clicks on a listed inspection point, the matching photo appears, illustrating the selected inspection point (Figure 9-57). Special digital photos are available for inspection of the cradle and cradle-to-body alignment.

FIGURE 9-57 Point and click suspension inspection screen.

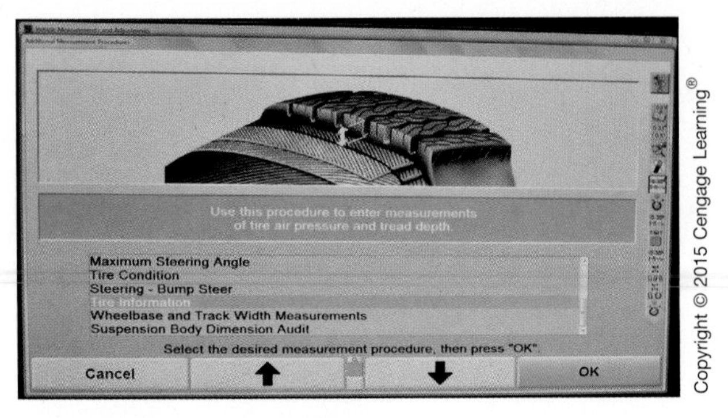

FIGURE 9-58 Tire inspection screen.

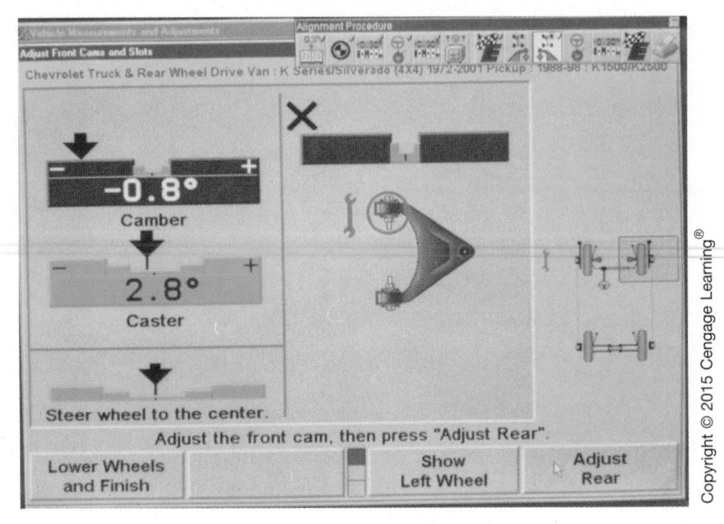

FIGURE 9-59 Wheel alignment adjustment screen.

5. Live action inspection videos may be accessed instantly from the inspection screen. These inspection videos are part of the 142-video training library.
6. A tire inspection screen illustrates various types of tire tread wear and the related causes (Figure 9-58). This screen may be printed out and shown to the customer.
7. Measurement and adjustment screens illustrate suspension measurements. If a measurement is within specifications, the measurement is illustrated with a green bar and an arrow indicates the measurement. A red measurement bar indicates the adjustment is not within specifications (Figure 9-59). In the adjustment mode, the measurement remains on the screen, and as the necessary suspension adjustment is performed, the adjustment bar turns green when the arrow moves within specifications.
8. An alignment procedure bar is available in some computer aligners. This vertical bar contains an icon for each step in an alignment procedure (Figure 9-60). These steps are arranged in the proper sequence. A check mark appears when a step has been completed, and the technician may select and click any step to move to that adjustment.
9. Correction kit videos may be accessed to illustrate the installation of aftermarket wheel alignment correction kits. Special tools videos are available in the computer wheel aligner software. These videos indicate the special wheel alignment adjustment tools

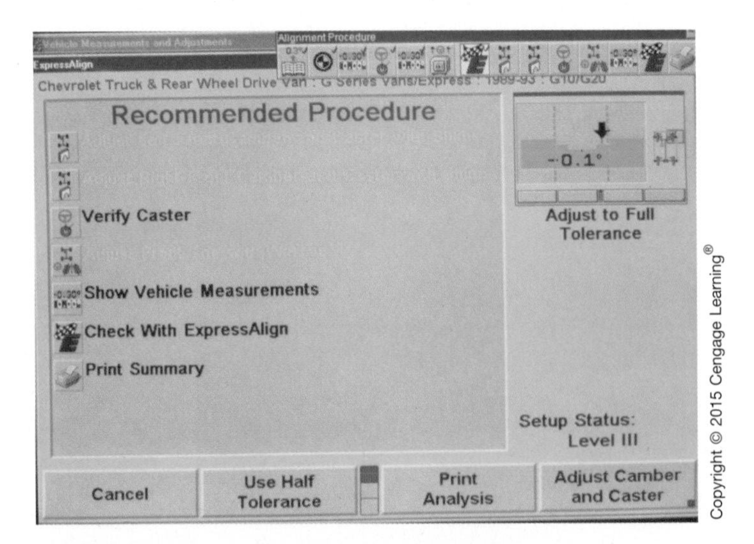

FIGURE 9-60 Wheel alignment procedure screen.

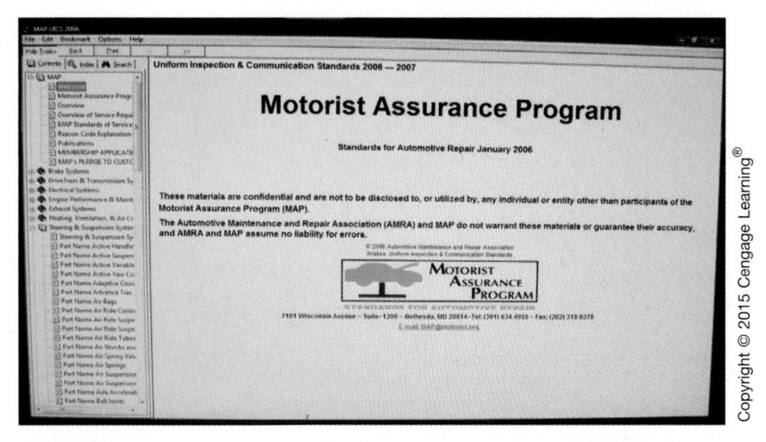

FIGURE 9-61 The Motorist Assurance Program (MAP) parts inspection guidelines may be accessed in computer wheel aligners.

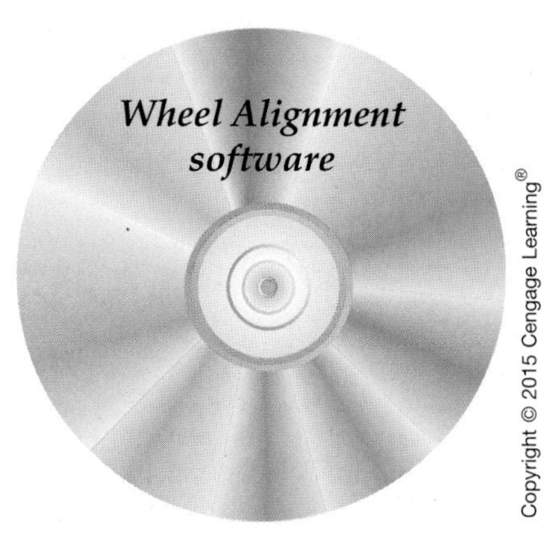

FIGURE 9-62 Wheel Alignment on CD or DVD for computer wheel aligners.

required for the vehicle being aligned. Operation videos may be accessed to show the technician how to operate the computer wheel aligner.

10. Some computer wheel aligners have multilingual capabilities. These capabilities usually include three languages; however, up to 20 languages may be available from the equipment manufacturer.
11. Full-function keyboard allows the technician to enter such information as the customer's name and address and the make and year of the vehicle.
12. Set of four self-centering adjustable rim clamps and four-wheel sensors.
13. Full-function remote control that allows the technician to control the alignment program and see the results of suspension adjustments while working under the vehicle.
14. Color or laser printer provides detailed wheel alignment reports for the customer and specific alignment messages and diagrams for the technician.
15. In some computer wheel aligners, the **Motorist Assurance Program (MAP)** uniform inspection guidelines may be accessed for many steering, suspension, and brake components (Figure 9-61). The MAP program establishes uniform parts inspection guidelines to improve customer satisfaction with the automotive industry.
16. A variety of software programs are available with some computer wheel aligners. Much of the software may be extra-cost options. This software may include:
    - Wheel alignment software on CD or DVD: This software provides the technician with a very extensive vehicle information database as well as many patented adjustment and productivity features (Figure 9-62).
    - Specific computer software provides service bulletins and other information on domestic and import vehicles (Figure 9-63 and Figure 9-64). Some computer software also provides labor estimates for automotive repairs.

CAUTION:
When using a computer wheel aligner, the equipment manufacturer's recommended alignment procedures must be followed to provide accurate readings and adjustments.

CAUTION:
Because wheel units contain sensitive electronic circuits, excessive shock such as dropping them on the floor can damage them.

Shop Manual
Chapter 9, page 323

Rear Wheel Alignment

Rear Wheel Alignment Diagnosis

Customer complaints that indicate rear wheel alignment problems are the following:

1. The front wheel toe is set correctly and the steering wheel is centered on the alignment rack, but the steering wheel is not centered when the vehicle is driven straight ahead.
2. The steering pulls to one side, but there are no worn suspension parts, defective tires, or improper front suspension alignment angles.
3. The vehicle is not overloaded and there are no worn suspension parts or improper front suspension alignment angles, but tire wear occurs.

The **Motorist Assurance Program (MAP)** improves automotive customer satisfaction by establishing uniform inspection guidelines.

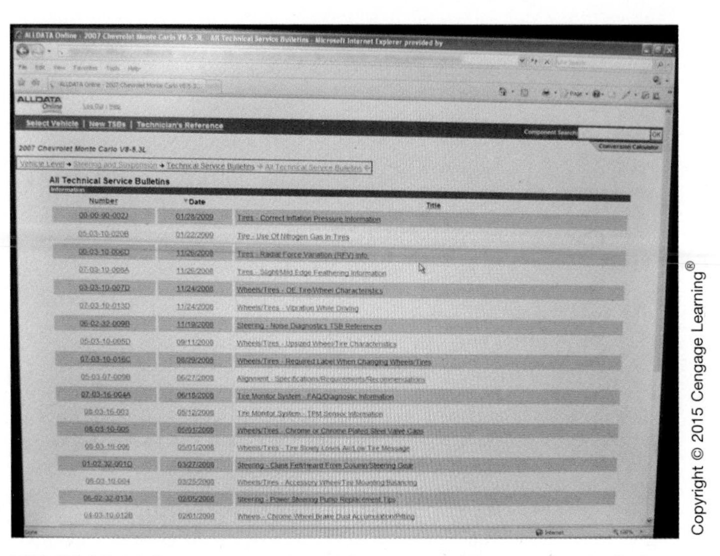

FIGURE 9-63 **Some computer software provides service bulletin information for suspension and steering problems.**

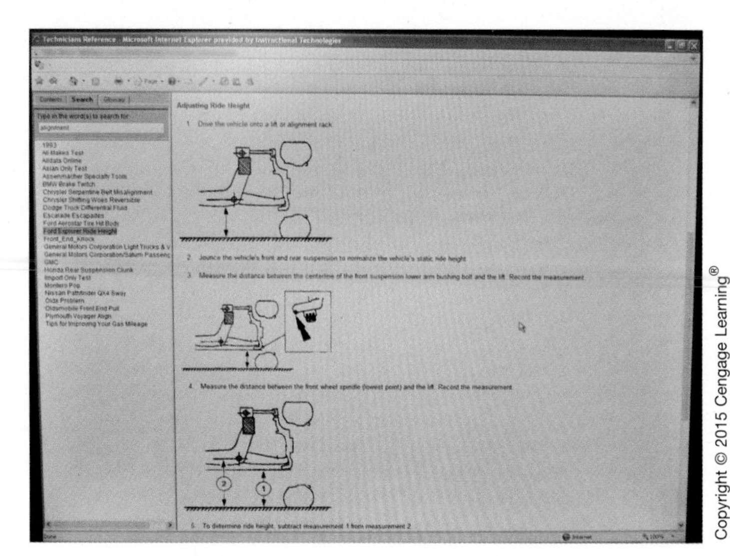

FIGURE 9-64 **Specific computer software provides diagnostic and service information in some computer wheel aligners.**

Defects That Cause Incorrect Rear Wheel Alignment

These suspension or chassis defects cause incorrect rear wheel alignment:

1. Collision damage that results in a bent frame or distorted unitized body
2. A leaf-spring eye that is unwrapped or spread open
3. Leaf-spring shackles that are broken, bent, or worn
4. Broken leaf springs or leaf-spring center bolts
5. Worn rear upper or lower control arm bushings
6. Worn trailing arm bushings or dislocated trailing arm brackets
7. Bent components such as radius rods, control arms, struts, and rear axles

Rear wheel camber is the tilt of a line through the center of the rear tire and wheel in relation to the true vertical centerline of the tire and wheel.

Rear Wheel Camber

Many front-wheel-drive vehicles have a slightly negative rear wheel camber that provides improved cornering stability. **Rear wheel camber** is basically the same as front wheel camber (Figure 9-65).

During rear wheel jounce on a multi-link rear suspension system, the camber becomes more positive, and the camber moves toward a negative position during wheel rebound. Because camber changes during wheel jounce, and rebound causes tire tread wear, the rear suspension is designed to minimize camber change during wheel jounce and rebound.

Rear wheel toe is the distance between the front edges of the rear tires in relation to the distance between the rear edges of the rear tires.

Rear Wheel Toe

On a front-wheel-drive vehicle, driving forces tend to push back the rear wheel spindles. Therefore these rear wheels are designed with zero toe-in or a slight toe-in depending on the vehicle (Figure 9-66). Correct **rear wheel toe** is important to obtain normal tire life.

Shop Manual
Chapter 9, page 340

On a multi-link rear suspension system with the toe adjustment link mounted in a level position, the toe moves toward a toe-in position during wheel jounce or rebound. These toe changes cause rear tire tread wear.

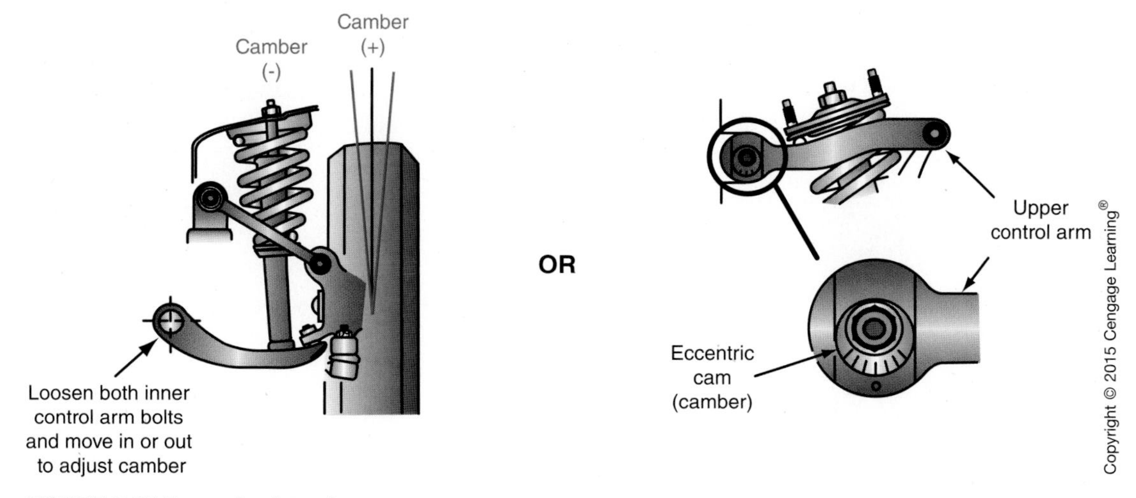

FIGURE 9-65 Rear wheel camber.

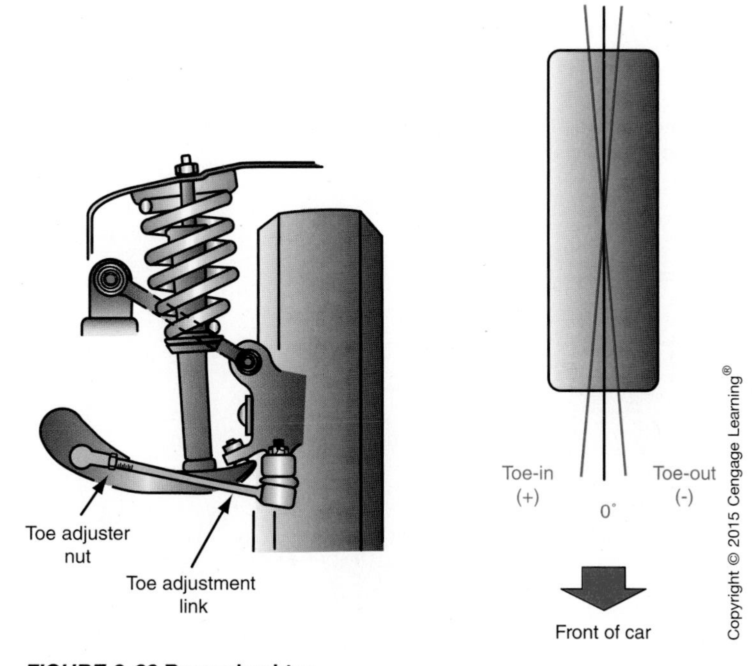

FIGURE 9-66 Rear wheel toe.

Steering Angle Sensor Reset

The last step of an alignment procedure that must be performed on some vehicle platforms is a reset procedure for the steering angle sensor (SAS) and other related sensors to update the vehicle geometry information stored in the vehicle's software system after an alignment (Figure 9-67). This procedure can be performed with an enhanced computer scan tool, depending on model, or your alignment equipment may incorporate this feature. Hunter Engineering has an interface called "CodeLink" that works with the alignment program to relearn the SAS and other related sensor information at the end of the alignment procedure (Figure 9-68). Reset requirements vary between vehicle manufacturer year, make,

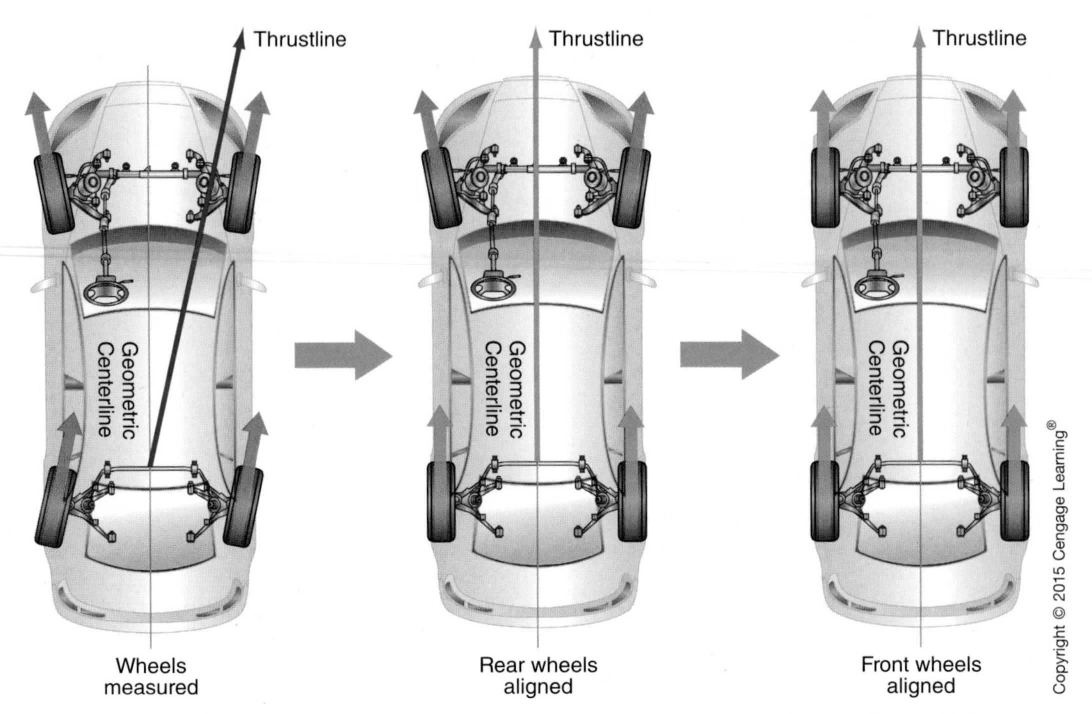

FIGURE 9-67 **On some vehicles the last step after an alignment procedure is to update vehicle geometry in the vehicle's software system.**

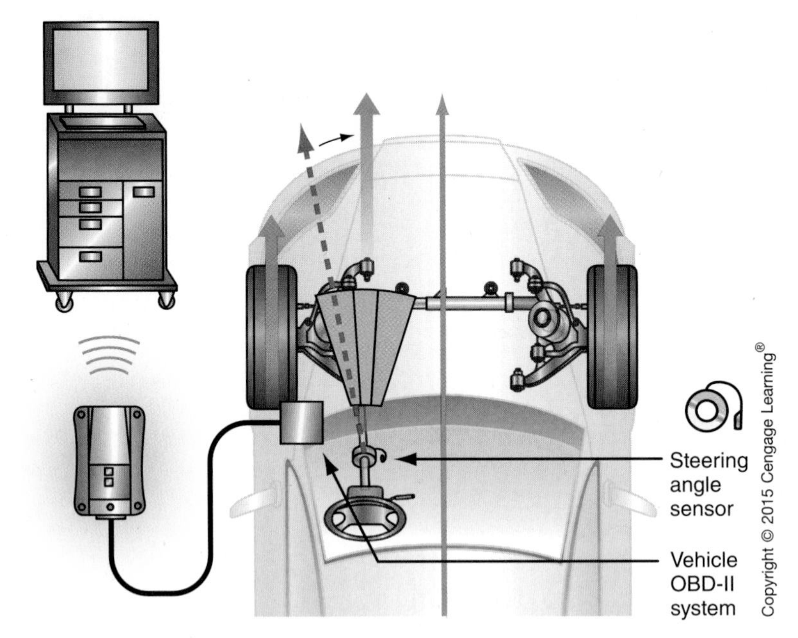

FIGURE 9-68 **To update vehicle geometry in the vehicles software system a scan tool or interface module is connected to the vehicle's data link connector.**

and model of the platform. Other sensors that may be part of this reset procedure are the yaw rate sensor, torque angle sensor, and deceleration sensor among others depending on vehicle manufacturer model. Refer to specific vehicle information for specific steps and requirements.

SUMMARY

- Front-wheel-drive cars with unitized bodies are more subject to rear wheel alignment problems compared with rear-wheel-drive cars with frames and one-piece rear axle housings.
- The vehicle geometric centerline is an imaginary line through the exact center of the front and rear wheels.
- The thrust line is an imaginary line at a 90 degree angle to the rear wheel centerline and projected forward.
- A rear axle offset occurs when the complete rear axle is rotated slightly, moving one rear wheel backward and the opposite rear wheel forward.
- Dog tracking is a term applied to a condition where the rear wheels are not directly following the front wheels.
- Rear axle offset or improper toe on one rear wheel causes steering pull to one side and tire tread wear.
- Rear axle sideset is a condition where the rear axle assembly has moved straight sideways and the geometric centerline is not positioned at the true vehicle centerline.
- Setback occurs when one front or rear wheel is moved backward in relation to the opposite wheel.
- In a geometric centerline front wheel alignment, the geometric centerline is used for a reference for front wheel toe. This type of alignment ignores thrust line position.
- In a thrust line front wheel alignment, the front wheel toe is checked using the thrust line as a reference, but the thrust line position may not be at the geometric centerline.
- In a total four-wheel alignment, the thrust line position is measured and the rear wheel toe is measured and adjusted as necessary so the thrust line is at the geometric centerline. The front wheel toe is measured using the common thrust line and geometric centerline as a reference.
- Computer alignment systems have a wheel unit mounted on a wheel clamp attached to each rim.
- Signals from the wheel sensors to the computer wheel aligner may be transmitted by a high-frequency transmitter in the wheel sensor.
- Computer alignment systems provide vehicle specifications plus diagrams of adjustment and inspection points.
- Directional stability is the tendency of a vehicle to travel straight ahead without being steered.
- Suspension and steering systems are designed to provide satisfactory vehicle control with acceptable driver effort and road feel and minimal tire tread wear.
- Proper wheel alignment and suspension component condition are extremely important to maintain vehicle driving safety.
- Many road variables such as bumps and holes, road crown, and heavy vehicle loads affect wheel alignment.
- Wheel alignment angles are designed to compensate for road variables.
- Camber is the tilt of a line through the center of the tire and wheel in relation to the vertical centerline of the tire and wheel.
- Positive camber is obtained when the centerline of the tire and wheel is tilted outward in relation to the vertical centerline of the tire and wheel.

TERMS TO KNOW

Bump steer
Directional stability
Four-wheel alignment
Geometric centerline
Geometric centerline alignment
High-frequency transmitter
Jounce
Memory steer
Motorist Assurance Program (MAP)
Negative camber
Negative caster
Positive camber
Positive caster
Rear axle offset
Rear axle sideset
Rebound
Receiver
Road crown
Road variables
Steering drift
Steering pull
Steering wander
Thrust angle
Thrust line
Thrust line alignment
Toe-in
Toe-out
Torque steer
Tracking
Wheel alignment
Wheel sensors
Wheel setback
Wheel shimmy

- Negative camber is present when the centerline of the tire and wheel is tilted inward in relation to the vertical centerline of the tire and wheel.
- During wheel jounce, the top of the wheel moves outward and the camber becomes more positive.
- During wheel rebound, the top of the wheel moves inward and the camber becomes less positive or moves to a slightly negative position.
- While cornering at high speeds, centrifugal force attempts to move the vehicle to the outside of the turn. This force raises the front suspension on the inside of the turn while lowering the front suspension on the outside of the turn.
- While cornering at high speeds, the front wheel on the inside of the turn moves to a less positive camber angle, and the camber angle becomes more positive on the outside front wheel.
- Excessive positive or negative camber concentrates the vehicle weight on one side of the front tire. The tire edge on which the weight is concentrated has a smaller diameter compared with the other side of the tire. Because the side of the tire with the smaller diameter makes more revolutions to go the same distance, this side of the tire becomes worn and scuffed.
- A wheel turns in the direction it is tilted.
- Road crown causes the vehicle steering to drift to the right.
- The camber on the left front wheel may be adjusted so it is slightly more positive than the right front wheel camber to compensate for steering pull to the right caused by road crown.
- Caster is the tilt of a line that intersects the center of the lower ball joint and the center of the upper strut mount in relation to a vertical line through the center of the spindle and wheel as viewed from the side.
- Positive caster occurs when the centerline of the lower ball joint and upper strut mount is tilted rearward in relation to the centerline of the spindle and wheel, viewed from the side.
- Negative caster is obtained when the centerline of the lower ball joint and upper strut mount is tilted forward in relation to the centerline of the spindle and wheel.
- Positive caster increases directional stability, steering effort, and steering wheel returning force.
- Excessive positive caster results in harsh ride quality.
- Excessive positive caster may cause front wheel shimmy.
- Negative caster reduces directional stability and steering effort while improving ride quality.
- If the caster is different on the two front wheels, the steering pulls toward the side with the least positive caster.
- The caster adjustment on the left front wheel may be adjusted so it is less positive than the caster on the right front wheel to compensate for road crown.
- If the rear suspension height is lowered, the front wheel caster becomes more positive.
- When the front suspension height is lowered, the front wheel caster becomes more positive.
- Proper caster adjustment is very important to maintain vehicle directional control and safety.
- Front wheel toe is the distance between the front edges of the tires compared with the distance between the rear edges of the tires.
- Toe-in is present when the distance between the front edges of the tires is less than the distance between the rear edges of the tires.
- Toe-out occurs when the distance between the front edges of the tires is greater than the distance between the rear edges of the tires.

- Most front-wheel-drive cars have a slight toe-out setting on the front wheels because drive axle forces tend to move the front wheels to a toe-in position.
- Most rear-wheel-drive cars have a slight toe-in setting because driving forces tend to move the front wheels to a toe-out position.
- The front wheel toe is adjusted with the vehicle at rest so the front wheels are straight ahead when the vehicle is driven.
- Improper toe adjustment results in feathered tire wear.

REVIEW QUESTIONS

Short Answer Essays

1. Explain why four-wheel alignment is essential on cars with semi-independent or independent rear suspension.
2. Explain why the rear wheel toe angle, thrust angle, and camber should be correct before adjusting the front suspension angles.
3. Describe the term total toe and what the difference is between a positive toe angle and a negative toe angle.
4. Explain what will happen if toe angles are outside the allowed tolerance.
5. Explain why excessive positive camber wears the outside edge of the tire tread.
6. Describe the type of tire tread wear caused by an improper toe setting.
7. Explain why positive caster provides increased directional stability.
8. Describe how positive caster provides increased steering wheel returning force.
9. Explain why positive caster causes harsh riding quality.
10. Define toe-in and toe-out.

Fill-in-the-Blanks

1. Excessive toe-out on the right rear wheel moves the thrust line to the ________________ of the geometric centerline.
2. If the thrust line is positioned to the left of the geometric centerline so the thrust angle is more than specified, the steering pulls to the ________________.
3. Directional stability refers to the tendency of a vehicle to travel straight ahead without being ________________.
4. When the front wheel camber is negative, the centerline of the tire and wheel is tilted ________________ in relation to the true vertical centerline of the tire and wheel.
5. Excessively ________________ camber on a front wheel assembly will cause the tire to wear on the inside edge.
6. Negative front wheel caster decreases ________________ ________________ and steering wheel ________________ ________________.
7. Raising the rear suspension height above the manufacturer's specification may change the front wheel caster from ________________ to ________________.
8. Front wheel shimmy may be caused by excessive ________________ caster.
9. Front wheel caster becomes more positive if the rear suspension height is ________________.
10. If the front suspension height is lowered on a MacPherson strut suspension, the front wheel caster becomes ________________ ________________.

Multiple Choice

1. While diagnosing front wheel toe problems:
   A. The front wheels are set to a straight-ahead position on most front-wheel-drive cars.
   B. Driving forces tend to move the front wheels toward a toe-in position on a rear-wheel-drive car.
   C. Improper toe adjustment may cause feathered wear on the front tire treads.
   D. Sagged front springs may increase the front wheel toe-out on a short-and-long arm suspension system.
2. "Feathering" type wear of a rear tire is likely caused by:
   A. Improper rear wheel camber alignment.
   B. Improper rear tire inflation.
   C. Improper rear wheel balance.
   D. Improper rear wheel toe alignment.

3. A front-wheel-drive vehicle with an independent rear suspension pulls to the right when driving straight ahead. All the front suspension alignment angles are within specifications. The most likely cause of this problem is:
   A. Excessive toe-out on the right rear wheel.
   B. Excessive negative camber on the right rear wheel.
   C. Excessive toe-out on the left rear wheel.
   D. Excessive positive camber on the left rear wheel.
4. All of these statements about the vehicle thrust line and steering pull are true EXCEPT:
   A. Excessive toe-out on the left rear wheel moves the thrust line to the left of the vehicle centerline.
   B. Excessive toe-in on the right rear wheel moves the thrust line to the right of the vehicle centerline.
   C. If the thrust line is moved to the left of the vehicle centerline, the steering tends to pull to the right.
   D. Excessive toe-out on the right rear wheel causes the steering to pull to the left.
5. While diagnosing suspension and wheel alignment problems:
   A. Road crown has no effect on vehicle steering or wheel alignment.
   B. Vehicle loads have no effect on wheel alignment.
   C. Steering angles are designed to reduce tire wear and provide directional control.
   D. Wheel alignment angles do not affect riding quality.
6. While diagnosing the vehicle geometric centerline and thrust line:
   A. The front and rear wheels should be parallel to the geometric centerline.
   B. The thrust angle is the difference between the front wheel camber and steering axis inclination (SAI) angles.
   C. If the thrust angle is more than specified, front wheel shimmy may occur.
   D. A bent front cradle may cause the thrust angle to be more than specified.
7. While driving straight ahead, a front-wheel-drive car pulls to the right. The most likely cause of this problem is:
   A. More positive camber on the left front wheel compared to the right front wheel.
   B. Sagged front springs and improper front wheel toe setting.
   C. Less positive caster on the right front wheel compared to the left front wheel.
   D. The SAI on the right front wheel is 1 1/2 degree more than the SAI on the left front wheel.
8. While adjusting front wheel camber and diagnosing camber-related problems:
   A. During front wheel jounce travel, the positive camber increases.
   B. Excessive positive camber on a front wheel causes premature wear on the inside edge of the tire tread.
   C. If the right front wheel camber is +1 1/2 degree and the left front camber is +1/2 degree, the steering pulls to the left.
   D. Excessive positive camber on both front wheels may cause front wheel shimmy.
9. A driver complains of harsh riding and the suspension height is normal. The most likely cause of this problem is:
   A. Excessive negative camber on both front wheels.
   B. Excessive positive caster on both front wheels.
   C. The left front wheel has more positive caster than the right front wheel.
   D. Both front wheels have negative camber and negative caster.
10. Excessive positive caster may cause all of these problems EXCEPT:
    A. Front wheel shimmy.
    B. Harsh ride quality.
    C. Excessive steering effort.
    D. Slow steering wheel return.

Chapter 10

Four-Wheel Alignment, Part 2 Diagnosic Angles and Frame Damage

Upon completion and review of this chapter, you should be able to understand and describe:

- The steering axis inclination (SAI) angle and the included angle.
- How SAI helps return the front wheels to the straight-ahead position.
- How SAI eliminates the need for excessive positive camber and caster.
- The effect that front suspension defects such as dislocated upper strut towers and bent struts or spindles have on SAI.
- Negative and positive scrub radius and the effect of each on steering quality.
- Setback on front suspension systems.
- How the front suspension system is designed to provide toe-out on turns.
- The purposes of a vehicle frame.
- Different types of frame construction.
- Different types of frame designs.
- Unitized body construction, and explain how this body design obtains its strength.
- Directional stability.
- Vehicle tracking, and explain how the four wheels on a vehicle must be positioned to obtain proper tracking.
- Wheelbase, and explain how the wheels on a vehicle must be positioned to provide correct wheelbase.
- Setback on a front wheel, and explain the effect that setback has on vehicle steering.
- Rear axle offset, and describe the effect of this problem on steering control.
- Rear axle sideset, and explain the effect of rear axle sideset on steering control.
- Side sway frame damage.
- Frame sag.
- Frame buckle.
- A diamond-frame condition.
- Frame twist.
- Aluminum space frame construction.

Introduction

The frame and unitized body play an important role in vehicle geometry and alignment angles. It is important to understand the vehicles structure elements and to be aware of how damage to the vehicles structure elements can affect alignment angle. Many of the angles discussed in this chapter are not adjustable on their own but are used as diagnostic angles. These angles are often used in combination with one another as an aid in determining the root cause for the vehicle geometry being out of specification.

Shop Manual
Chapter 10,
page 389

As an example, improper steering axis inclination (SAI) angles on either side of the front suspension may cause hazardous steering conditions while braking or accelerating. Therefore, technicians must be familiar with SAI and other related steering geometry if they are to diagnose no routine drivability and handling issues.

Steering Axis Inclination Definition

On front-wheel-drive vehicles with MacPherson strut front suspension systems, **steering axis inclination (SAI)** refers to the inward tilt of a line through the center of the top strut mount and the center of the lower ball joint in relation to the true vertical line through the center of the tire. These two lines are viewed from the front of the vehicle, and the SAI line always tilts inward in relation to the true vertical line (Figure 10-1).

Many rear-wheel-drive cars have a short-and-long arm front suspension system with unequal-length upper and lower control arms and a ball joint mounted in each control arm. The steering knuckle and spindle pivot on the ball joints as the wheels are turned. On this type of front suspension, the SAI line runs through the upper and lower ball joint centers. The **included angle** is the sum of the SAI angle and the camber angle (Figure 10-2).

If the camber angle is positive, this angle is added to the SAI angle to obtain the included angle. A negative camber angle must be subtracted from the SAI angle to calculate the included angle. On some truck front suspension systems, the steering knuckle pivots on a king pin that is mounted in an I-beam-type front axle. The SAI line on this type of suspension is referred to as a **king pin inclination (KPI)** line. This invisible line runs through the center of the king pin.

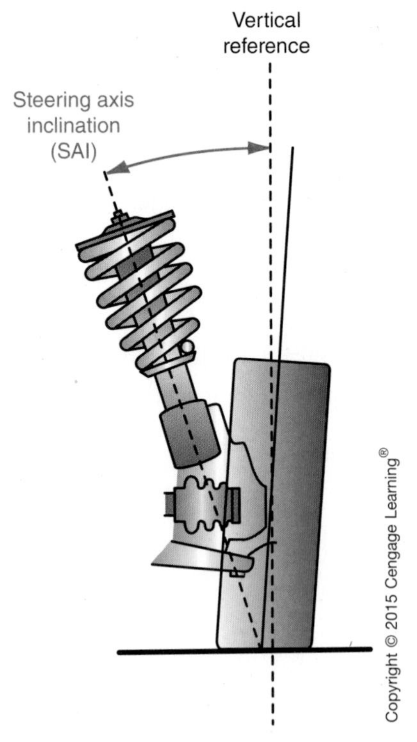

FIGURE 10-1 **Steering axis inclination angle.**

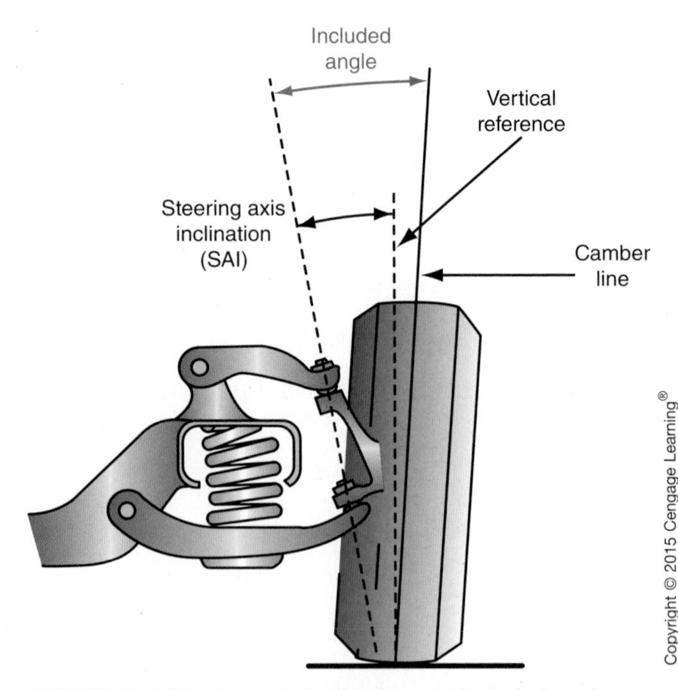

FIGURE 10-2 **Steering axis inclination and included angle.**

AUTHOR'S NOTE: It has been my experience that after front-end collision damage, a common problem for cars with a unitized body is improperly positioned strut towers. Even though the cosmetic appearance of the car body remains good, the strut towers may be out of position, resulting in unequal steering axis inclination (SAI) between the right and left front wheels. During severe braking, an unequal SAI causes steering pull and possible loss of steering control. Therefore, after a frontal collision on a unitized-body car, it is very important that you measure the strut tower location and be sure this measurement is within specifications. It is also critical that you check the front and rear end wheel alignment and adjust the alignment angles as necessary to bring them within specifications.

SAI Purpose

When the SAI angle is tilted toward the center of the vehicle and the wheels are straight ahead, the height of the spindle is raised closer to the chassis. This action lowers the height of the vehicle because of gravity. When the front wheels are turned, each spindle moves through an arc that tries to force the tire into the ground. Because this reaction cannot take place, the chassis lifts when the wheels are turned. When the steering wheel is released after a turn, the vehicle weight has a tendency to settle to its lowest point. Therefore, SAI helps return the wheels to the straight-ahead position after a turn, and it also tends to maintain the wheels in the straight-ahead position. However, SAI does increase steering effort because the chassis has to lift slightly on turns.

The vehicle weight is projected through the SAI line to the road surface. Let us assume that a front suspension is designed with a vertical 0 degree SAI line. Under this condition, the vehicle weight is projected through the SAI line a considerable distance inside the true vertical tire centerline. With this type of front suspension, severe tire scuffing would occur because the tire and wheel pivot around the SAI line during a turn (Figure 10-3).

With a 0 degree SAI line, greater steering effort is required during a turn and stress on the steering mechanism increases. This type of front suspension design causes excessive road shock and kick-back on the steering wheel during a turn, because the distance

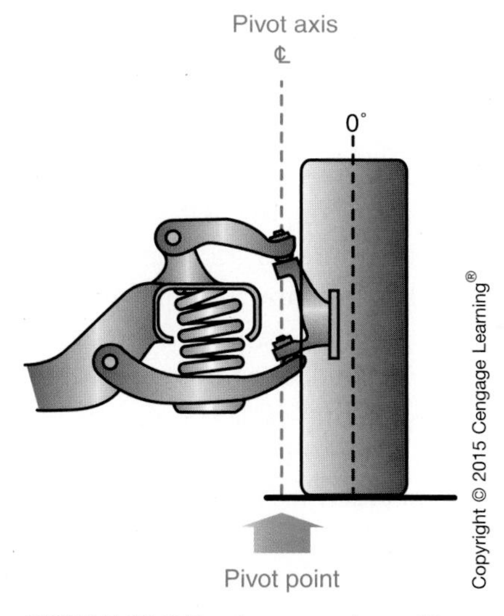

FIGURE 10-3 **Front suspension with a 0 degree SAI line.**

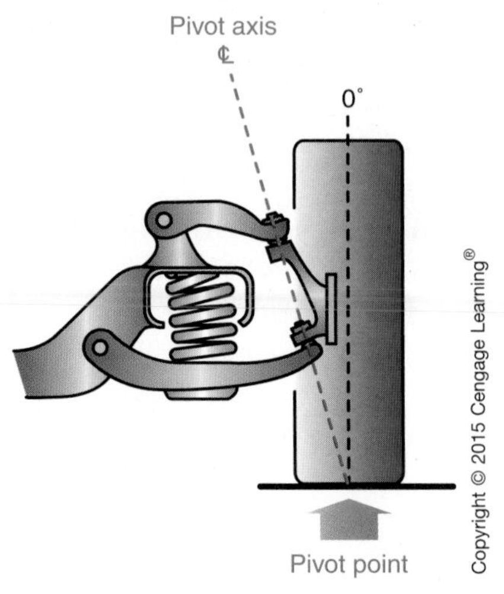

FIGURE 10-4 **If the SAI line meets the tire vertical centerline at the road surface, tire wear is reduced, steering effort is decreased, and stress on the steering components is diminished.**

between the SAI line and the tire vertical line returns the wheels to the straight-ahead position.

When the SAI line meets the tire vertical centerline at the road surface, tire wear is reduced, steering effort is decreased, and stress on steering components is diminished (Figure 10-4).

The point at which the SAI line and the tire vertical centerline intersect could be positioned closer together by designing the front suspension with excessive positive camber. However, this design would result in rapid wear on the outside edge of the tire tread. A correct SAI line reduces the need for excessive positive camber.

MacPherson strut front suspension systems have a much greater SAI angle (12 to 18 degrees) compared with short-and-long arm front suspensions in rear-wheel-drive cars (6 to 8 degrees). Front-wheel-drive cars require the higher SAI angle for directional stability. Positive caster also provides directional stability.

SAI and Safety Factors

On MacPherson strut suspension systems, an incorrect SAI angle may indicate that the upper strut mount is out of position, the lower control arm is bent, or the center cross member is shifted. Any of these defects may be caused by collision damage.

When the SAI angles are unequal on the left and right front suspension, serious handling problems may occur. These problems include torque steering during hard acceleration, steering pull during sudden stops, and bump steer. Torque steer is the tendency to pull to one side on hard acceleration because of unequal-length drive axles on a front-wheel-drive vehicle. Unequal SAI angles aggravate torque steer. Bump steer refers to unequal toe and/or camber changes that jerk the car to one side during front suspension jounce and rebound. Therefore, incorrect SAI angles may cause hazardous driving situations and contribute to serious accidents. Technicians must inspect the SAI angle during a wheel alignment.

If the SAI angle is correct and the included angle and camber are less than specified, bent components such as the strut or spindle are indicated. On the type of suspension that requires inward or outward upper strut movement to adjust camber, the SAI angle changes with this movement. When an eccentric bolt between the strut and steering knuckle is used for camber adjustment, the SAI angle will not change if the camber is adjusted.

It is very important to remember that a camber adjustment on many front suspension systems also changes the SAI angle. For example, if the upper control arm on a front suspension system with upper and lower ball joints is shimmed outward to increase positive camber, the SAI angle will change with the camber angle. Therefore, the included angle remains the same. The camber may be adjusted to specification, but the SAI angle and included angle could be out of specification. This is especially true on MacPherson strut suspension systems where collision damage may bend front struts or shift the upper strut towers. A service technician who inspects and adjusts the camber angle while ignoring the SAI and included angles may be overlooking serious and dangerous front suspension defects.

Scrub Radius

Scrub radius affects steering quality related to stability and returnability. However, scrub radius is not an alignment angle and it cannot be measured on conventional alignment equipment. Scrub radius is the distance from the point where the tire vertical line contacts the road to the location where the line through the upper strut center and the ball joint center meets the road surface. **Positive scrub radius** occurs when the line through the strut and ball joint meets the road surface inside the tire vertical centerline. A **negative scrub radius** is when the line through the strut and ball joint centers meets the road surface outside the tire and hub centerline (Figure 10-5).

Scrub radius is the distance between the SAI line and the true vertical centerline of the tire at the road surface.

Conventional short-and-long arm front suspension systems usually have positive scrub radius. Many front-wheel-drive vehicles have negative scrub radius.

When negative scrub radius is used in front-wheel-drive vehicles, straight-line braking is ensured and directional stability is maintained. As the vehicle moves forward, negative scrub radius tends to turn the front wheels inward. This action causes unequal forces applied to the steering to act inboard of the steering axis and pull the vehicle from any induced swerve. In a front-wheel-drive vehicle, a swerve to one side may be caused by one front wheel being on an ice patch while the other front wheel is on dry pavement. A failure of one-half the diagonal brake system, a sudden blowout of one front tire, or a grabbing brake on one front wheel will also cause a vehicle to swerve.

If a vehicle has positive scrub radius and the right front brake grabs, both the positive scrub radius and the grabbing brake tend to turn the right front wheel outward, and the vehicle pivots around the right front wheel. This action induces a swerve to the right.

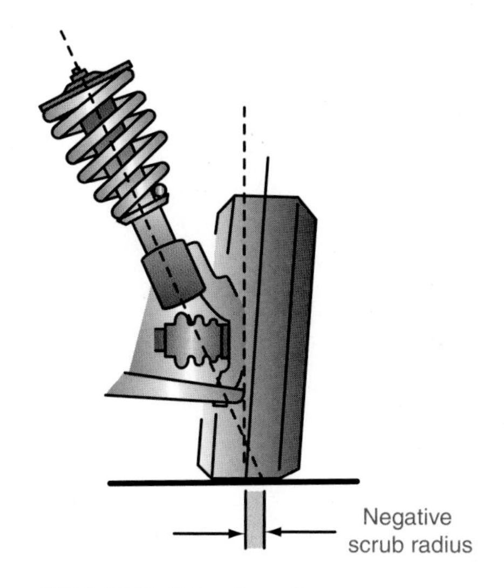

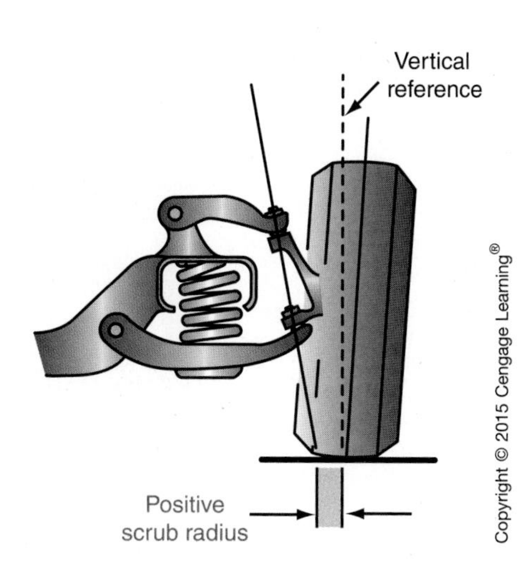

FIGURE 10-5 **Scrub radius.**

When a right front brake grab occurs with negative scrub radius, the brake grab causes the right front wheel to turn outward, and the vehicle tends to pivot around the right front wheel. However, the negative scrub radius tends to turn the right front wheel inward. The two forces on the right front wheel cancel each other to maintain directional stability.

Scrub Radius and Safety

If front tires that are larger in diameter than specified by the car manufacturer are installed, directional control may be affected. Large tires raise the chassis farther from the road surface, which changes the scrub radius. The installation of larger front tires could change a positive scrub radius to a negative scrub radius.

Reversing the front rims so they are inside out creates a significant scrub radius change and adversely affects directional control. This practice is not recommended by car manufacturers.

WARNING: Installing larger tires, or different rims than the ones specified by the vehicle manufacturer, changes the scrub radius, which may result in reduced directional control, collision damage, and personal injury.

WHEEL SETBACK

Wheel setback is a condition in which one wheel is moved rearward in relation to the other wheel (Figure 10-6). Setback will not affect handling unless it is extreme. Collision damage may drive one front strut rearward and cause extreme setback. Setback can also occur on rear wheels, but it is more likely to occur on front wheels because of collision damage. Some computer four-wheel aligners have setback measuring capabilities.

TURNING RADIUS

Front and Rear Wheel Turning Action

Turning radius may be referred to as cornering angle or toe-out on turns.

When a vehicle turns a corner, the front and rear wheels must turn around a common center with respect to the **turning radius** (Figure 10-7). On most front suspension systems, the front wheels pivot independently at different distances from the center of the turn, and therefore the front wheels must turn at different angles. The inside front wheel must turn at a sharper angle than the outside wheel. This is because the inside wheel is

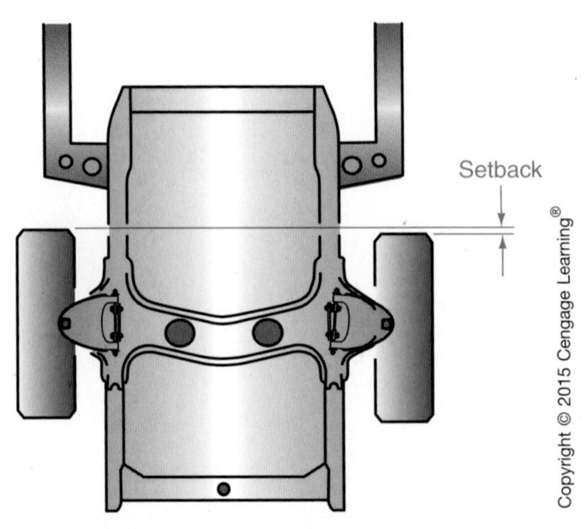

FIGURE 10-6 **Wheel bearing and hub assembly.**

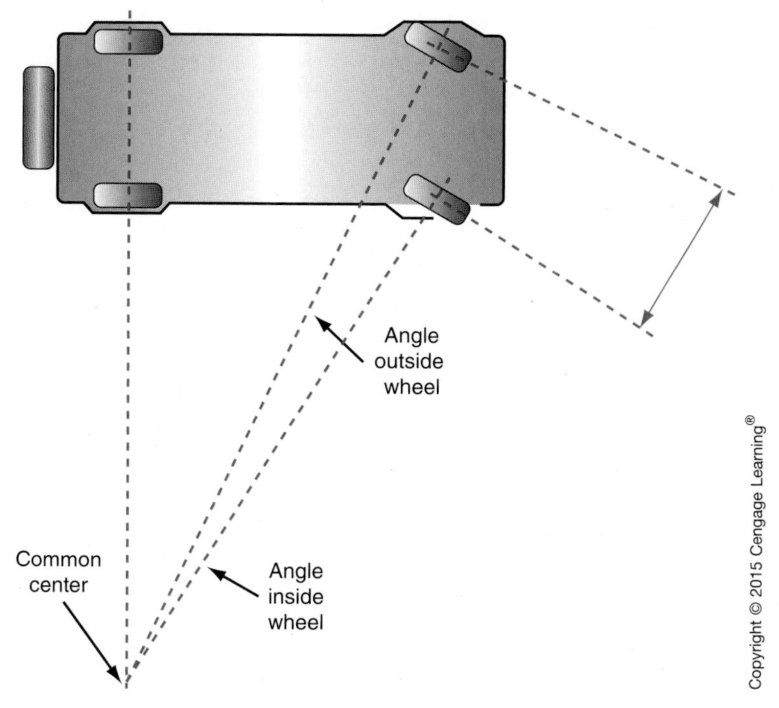

FIGURE 10-7 **Front and rear wheels turning around a common center.**

actually ahead of the outside wheel. When this turning action occurs, both front wheels remain perpendicular to their turning radius, which prevents tire scuffing.

The turning angles of the front wheels are determined by steering arm design, and these angles are not adjustable. If the steering angles are not correct, the steering arms may be bent.

Steering Arm Design

An understanding of a lever moving in a circle is necessary before an explanation of steering arm design and operation. If a lever moves from point A to B, it pivots around point O and moves through a horizontal distance A to B (Figure 10-8).

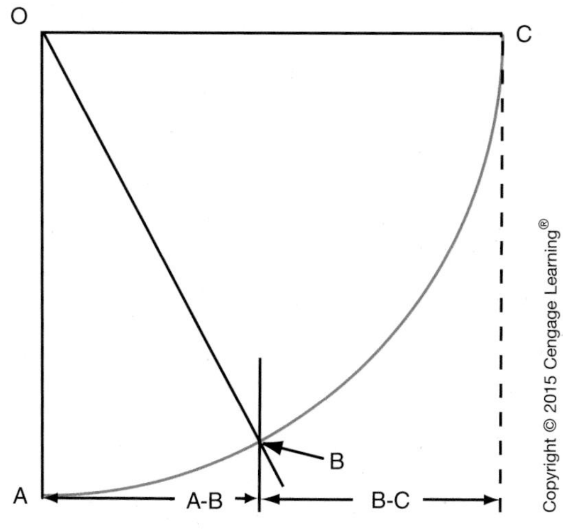

FIGURE 10-8 **Lever movement in a circle.**

Lever movement through arc B to C is much greater than the movement through arc A to B. However, during arc B to C, the lever moves through horizontal distance B to C and this distance is the same as horizontal distance A to B.

The steering arms are connected from the tie rod ends to the steering knuckles. Steering arms and linkages maintain the front wheels parallel to each other when the vehicle is driven straight ahead. However, the steering arms are not parallel to each other. If the steering linkage is at the rear edge of the front wheels, the steering arms are closer together at the point where the tie rods connect than at their spindle pivot point (Figure 10-9). When the steering linkage is positioned at the front edge of the front wheels, the steering arms are closer together at their spindle pivot point than at the tie rod connecting point.

Toe-out on turns is the angle of the inside wheel in relation to the angle of the outside wheel during a turn.

When the front wheels are turned on a vehicle with the steering linkage at the rear edge of the front wheels, the angle formed by the inside steering arm and linkage increases, whereas the angle of the outside steering arm and linkage decreases. The inside steering arm moves through the longer arc X, and the outside steering arm moves through shorter arc Z (Figure 10-10).

Therefore, the inside wheel turns at a sharper angle than the outside wheel. Because both steering arms are designed to have the same angle in the straight-ahead position, the inside front wheel always has a sharper angle regardless of the turning direction. The sharper inside wheel angle during a turn causes the inside wheel to toe out. Therefore, the term **toe-out on turns** is used for this steering action. If the front wheel turning angle is increased during a turn, the amount of toe-out on the inside wheel increases proportionally.

A BIT OF HISTORY

During a recent October Car Care Month, statistics were compiled from 65 inspection lanes in the United States. During these inspections, 29 percent of the cars failed the tire test. These figures indicate that almost one-third of the vehicles on the road have tire problems, and in many cases these problems create a safety hazard. These figures also tell us there is a tremendous business potential in tire sales, wheel balancing, and wheel alignment. Each time we perform even minor vehicle service the tires should be inspected. This inspection may not only help us increase tire sales and the volume of wheel balancing and alignment service work, but in some cases you may also save the customer's life!

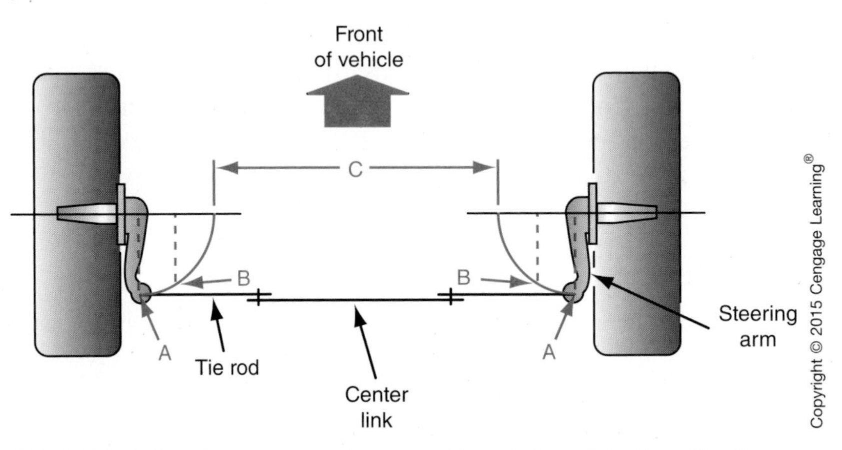

FIGURE 10-9 **Steering arms are closer together at the point where the tie rods connect than at their spindle pivot points.**

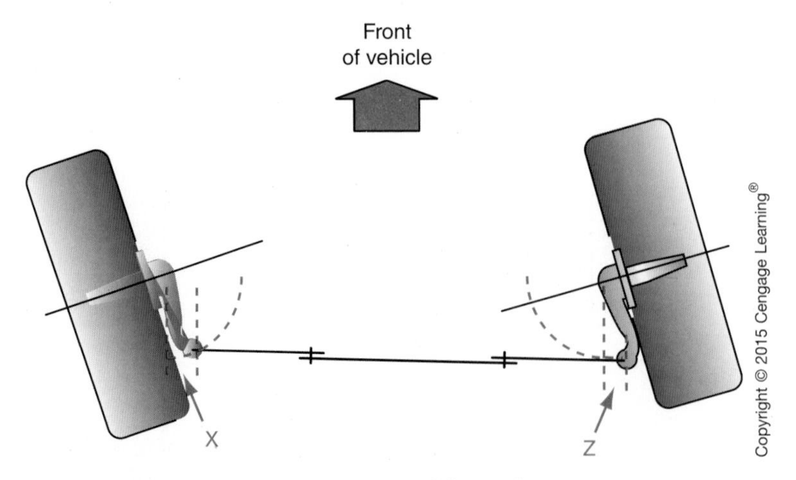

FIGURE 10-10 **Steering arm operation while turning.**

Slip Angle

During a turn, centrifugal force causes all the tires to slip a certain amount. The amount of tire slip increases with speed and the sharpness of the turn. This tire slip action causes the actual center of the turn to be considerably ahead of the theoretical turn center (Figure 10-11).

The **slip angle** on different vehicles varies depending on such factors as vehicle weight and type of suspension.

Slip angle is the actual angle of the front wheels during a turn compared to the turning angle of the front wheels with the vehicle at rest.

Frames and Frame Damage

The frame in a vehicle may be compared to the skeleton in the human body. Without the skeleton, a human body would not be able to stand erect. Likewise, if a vehicle did not have a frame, it could not support its own weight or the weight of its passenger or cargo load. The vehicle frame:

1. Enables the vehicle to support its total weight.
2. Enables the vehicle to absorb stress from road irregularities.
3. Enables the vehicle to absorb torque from the engine and drive wheels.
4. Provides a main member for attachment of body and other components.

The frame, together with the front and rear suspension systems, must position the wheels properly to minimize tire tread wear and provide accurate steering control.

In a **channel frame**, each side of the frame is made from a U-shaped steel channel.

In a **complete box frame**, each side of the frame forms a metal box.

A **tubular frame** member is formed in an oval shape or circle.

Types of Frames and Frame Construction

Frame Construction

Vehicle frame construction may contain three types of steel members: **channel** (partial box) **frame**, **complete box frame**, or **tubular frame** (Figure 10-12). On modern vehicles, most frames include all three types.

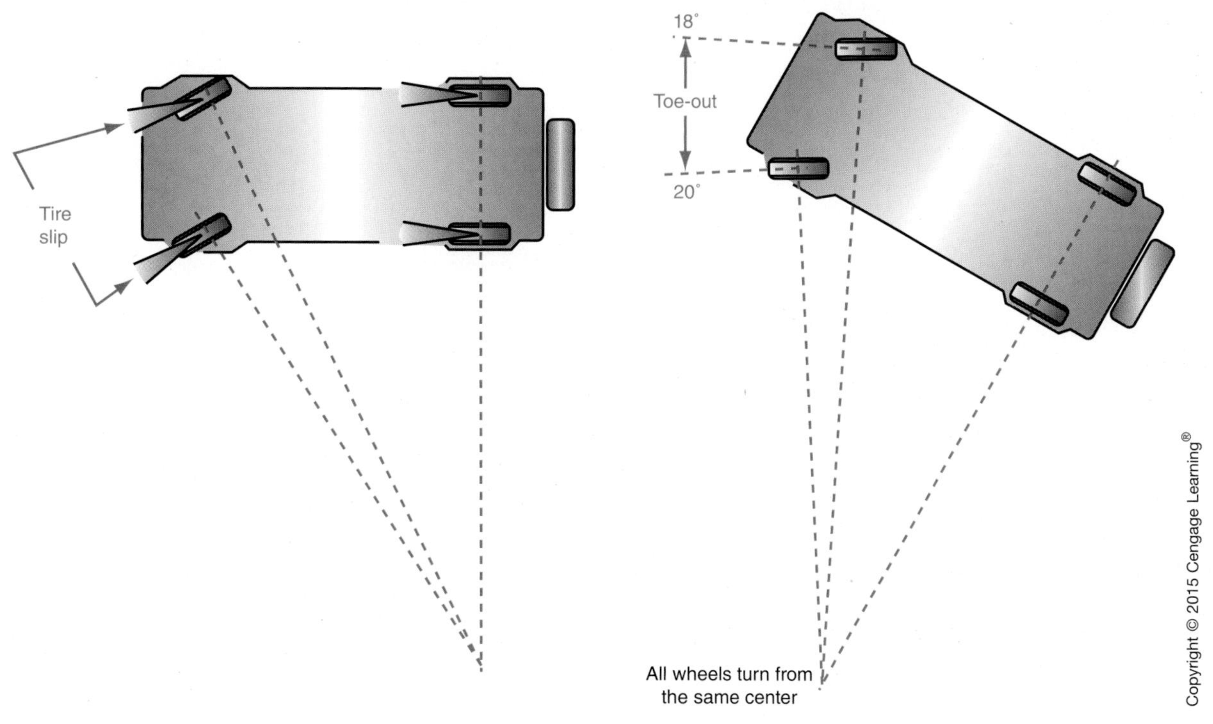

FIGURE 10-11 **Slip angle during a turn.**

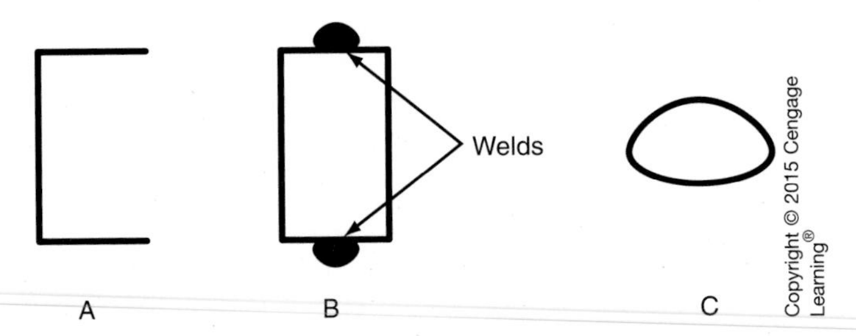

FIGURE 10-12 **Types of frame construction: (A) channel or partial box, (B) complete box, (C) tubular cross member.**

Compared to a channel frame or tubular frame, the complete box frame has increased torsional rigidity and improved crash performance. Therefore, in many frames the front section is a complete box design and the other part of the frame is a channel design. Some frames have X-shaped bracing at the rear of the frame for increased strength. Many vehicles now have frames or partial frames that are manufactured by a **hydro-forming** process. Hydro-formed frames are considerably more rigid than frames manufactured by heat-treating. Therefore, hydro-formed frames reduce the flexing of the frame and body components, which reduces squeaks and rattles to decrease noise, vibration, and harshness (NVH). Hydro-formed frames also provide more stable wheel position, which improves steering quality and reduces tire tread wear.

Hydro-forming is the process of using extreme fluid pressure to shape metal.

Ladder-Type Frames

The **ladder frame design** is used on trucks and full-size vans. In this type of frame, the side rails have very little offset and are in a straight line between the front and rear wheels. The ladder-type frame has more cross members for increased load-carrying capacity and rigidity (Figure 10-13).

In a **ladder frame design**, cross members are mounted between the sides of the frame, and the cross members are similar to rungs on a ladder.

Perimeter-Type Frames

The **perimeter-type frame** is used in some rear-wheel-drive cars. This type of frame forms a border around the passenger compartment. The frame rails are stepped inward at the

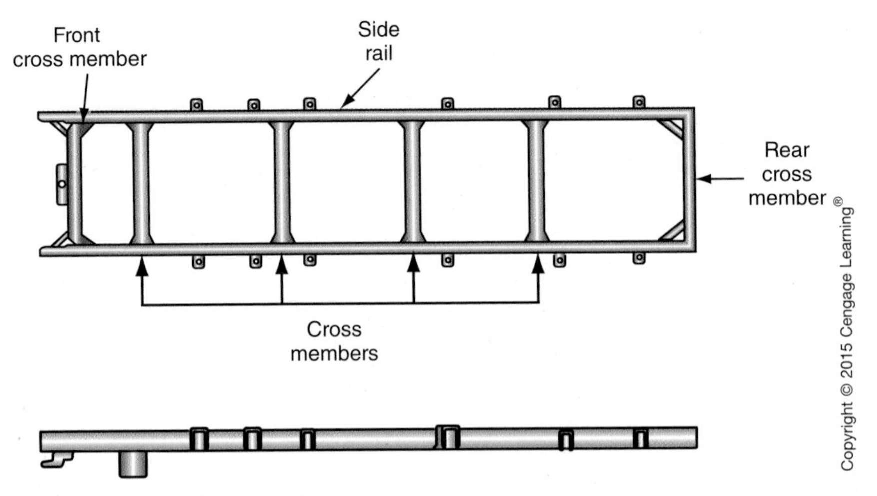

FIGURE 10-13 **Ladder-type frame.**

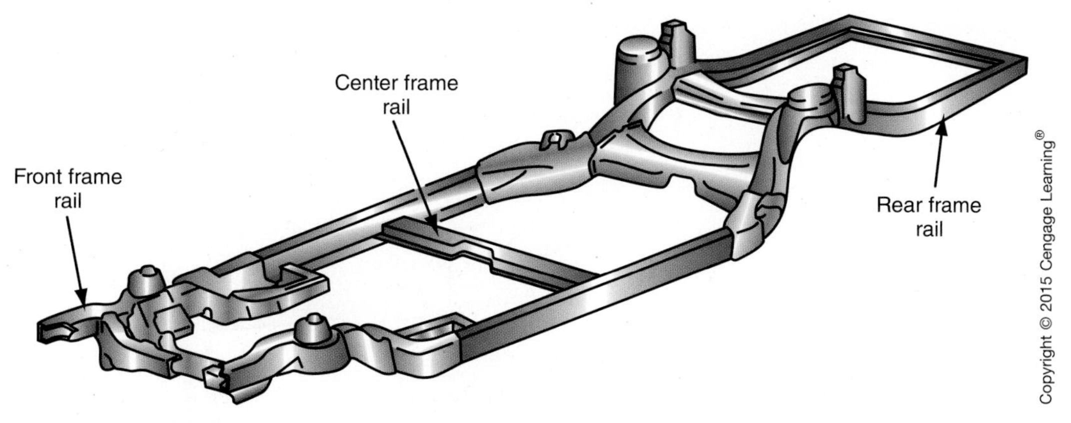

FIGURE 10-14 **Perimeter-type frame components.**

cowl area to provide increased strength, which supports the engine mounts and front suspension. This inward step at the cowl area also provides room for movement of the front wheels. Lateral support is provided by cross members welded between the frame rails near the front and rear of the frame. A transmission support member is welded or bolted between the frame rails at the back of the transmission. The rear frame kickup side rails support the rear suspension and the rear portion of the body weight. A front torque box is positioned just ahead of the transmission support member, and a rear torque box is located in front of the rear frame kickup side rails (Figure 10-14). These torque boxes are designed to absorb most of the impact during a side collision and thus reduce damage to other body components. The torque boxes also provide some protection for the vehicle occupants during a side collision.

Shop Manual
Chapter 10,
page 400

The body components are bolted to the frame, but rubber insulating bushings are positioned between the body and frame mounting locations. These bushings help prevent the transfer of road noise and vibration from the suspension and frame to the body and vehicle interior. Rubber insulating mounts are positioned between the engine and transmission and the frame mounting positions. These engine and transmission mounts reduce the transfer of engine vibration to the frame, body, and vehicle interior. Many suspension components are also connected to the frame through rubber insulating bushings to reduce the transfer of harshness and vibration from the suspension to the frame while driving over road irregularities.

During the manufacturing process, most manufacturers apply a special coating to the frame to help prevent rust and corrosion. For example, many light-duty truck frames are coated with epoxy or wax. An epoxy-coated frame is black and very smooth, whereas a wax-coated frame is gray and sticky. If these coatings are scratched or damaged, frame rusting and corrosion may occur.

Aluminum Spaceframes

Some sports cars are equipped with aluminum spaceframes (Figure 10-15). The aluminum spaceframe is manufactured using a hydroforming process. In theory, an aluminum frame must have three times the thickness of a steel frame to have the same strength as a steel frame. During the aluminum spaceframe–manufacturing process, a manufacturer adds thickness only where needed for adequate strength. As a result, the aluminum spaceframe has an average thickness of 1.9 times that of a comparable steel frame. This aluminum

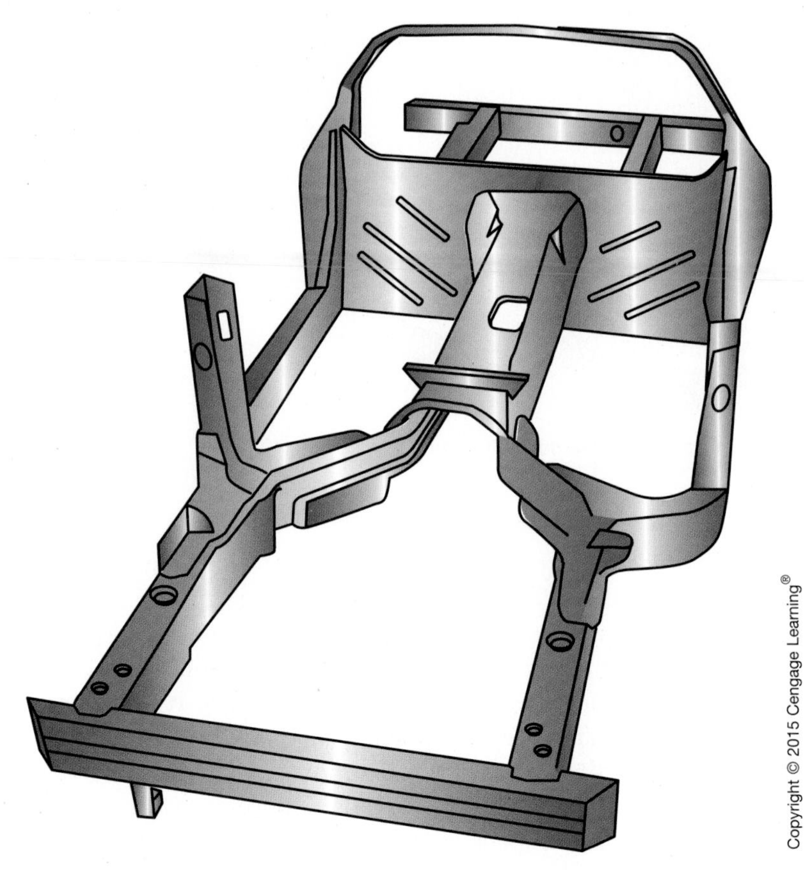

FIGURE 10-15 **Aluminum spaceframe.**

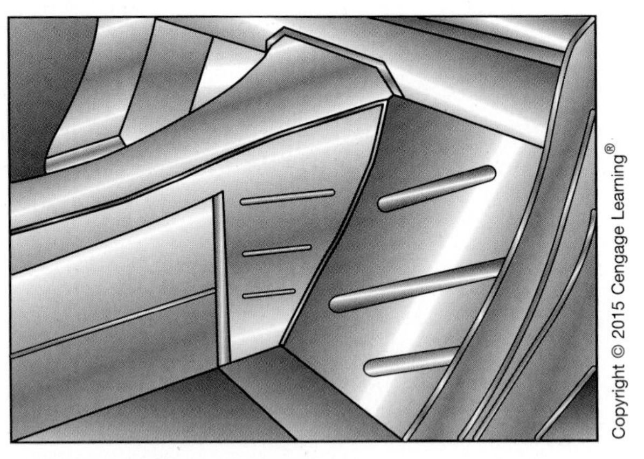

FIGURE 10-16 **Laser welding on aluminum spaceframe tunnel.**

Shop Manual
Chapter 10,
page 398

frame weighs 285 lb (129 kg), which is 136 lb (62 kg) lighter than a steel frame. Reducing vehicle weight improves fuel economy and vehicle performance.

Laser welding is used extensively on the tunnel in the aluminum spaceframe rather than spot welding (Figure 10-16). Laser welding adds stiffness and provides an improved seal compared with that provided by spot welding. Laser welding uses less heat per length of weld, which results in reduced aluminum distortion.

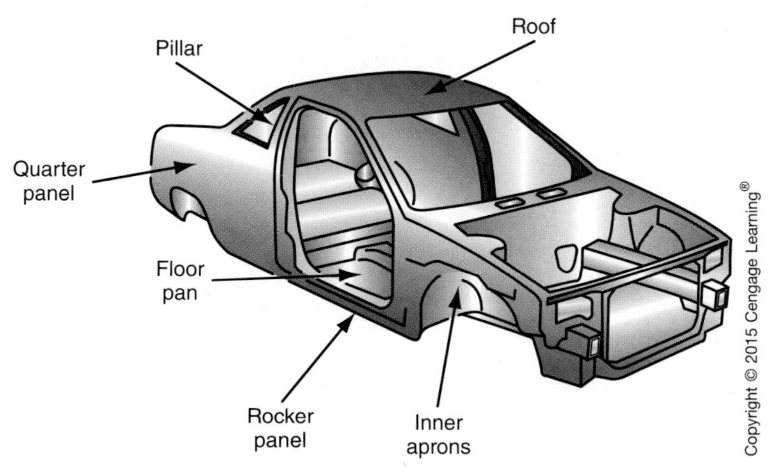

FIGURE 10-17 **Unitized body design.**

A BIT OF HISTORY

During the late 1940s, American Motors Corporation introduced the unitized body design in the United States. Some of these first-generation unitized bodies did not have partial frames. However, unitized bodies did not become popular until 1980 and 1981 with the introduction of General Motors X cars, Chrysler K cars, and other front-wheel-drive cars. These front-wheel-drive cars have partial frames and may be referred to as second-generation unitized bodies.

Unitized Body Design

Most front-wheel-drive cars have a **unitized body**. In this body design, the frame and body are combined as one unit, and the external frame assembly is eliminated. The strength and rigidity of the body is achieved by body design rather than by having a heavy steel frame to support the body. In the unitized body design, body sheet metal is fabricated into a box design. Body strength is obtained by shape and design in place of mass and weight of metal in vehicles with a separate frame.

All the members of a unitized body are load-carrying components. The floor pan, roof, inner aprons, quarter panels, pillars, and rocker panels are integrally joined to form a unitized body (Figure 10-17).

Most unitized bodies have bolt-on partial frames at the front and rear of the vehicle. Some of the components in a unitized body are made from **high-strength steels (HSS)** to provide additional protection in a collision. Ultra-HSS may be used in such unitized body components as door beams and bumper reinforcements. The unitized body has a complex design that spreads collision forces throughout the body to help protect the vehicle occupants.

Vehicle Directional Stability

Shop Manual
Chapter 10, page 392

The front and rear suspension systems are attached to the frame, partial frame, or unitized body. Therefore, the frame or unitized body must support the suspension systems properly to provide directional stability. Vehicle **tracking** is the parallel relationship between the front and rear wheels during forward vehicle motion. To provide proper tracking, each front wheel must be at the same distance from the vehicle centerline, and each rear wheel must be at an equal distance from the same centerline. The distance between the front wheels and the distance between the rear wheels does not necessarily have to be the same, but all four wheels must have a parallel relationship to provide proper tracking (Figure 10-18).

The vehicle **wheelbase** is the distance between the centers of the front and rear wheels. To provide an accurate wheelbase measurement, the centers of the front spindles and the centers of the rear axles must be square with the centerline of the vehicle (Figure 10-19). For this condition to exist, the front and rear axle centers must be at a 90 degree angle in relation to the vehicle centerline, and the wheelbase measurements must be equal on each side of the vehicle. Equal wheelbase measurements on each side of the vehicle and proper tracking are absolutely essential to providing **directional stability**.

Directional stability refers to the tendency of a vehicle to remain in the straight-ahead position when driven straight ahead on a reasonably smooth, straight road.

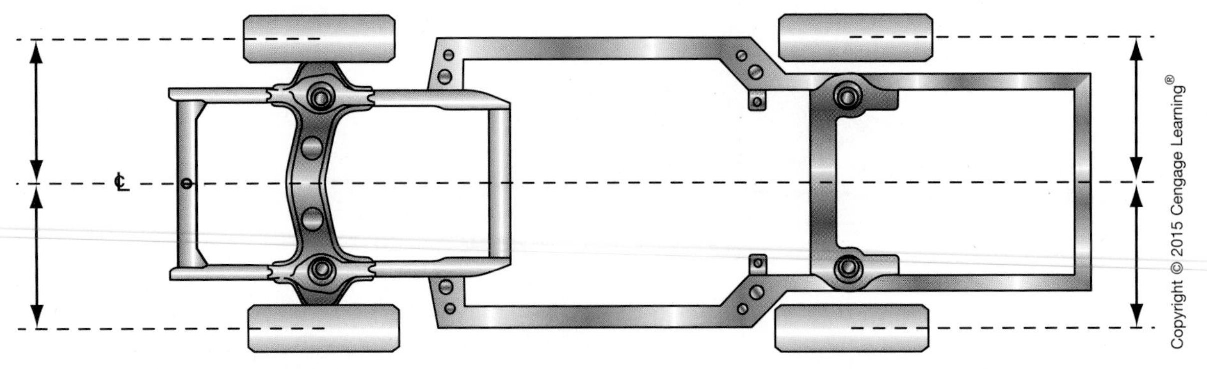

FIGURE 10-18 **Front and rear wheels must be at the same distance from the vehicle centerline to provide proper tracking.**

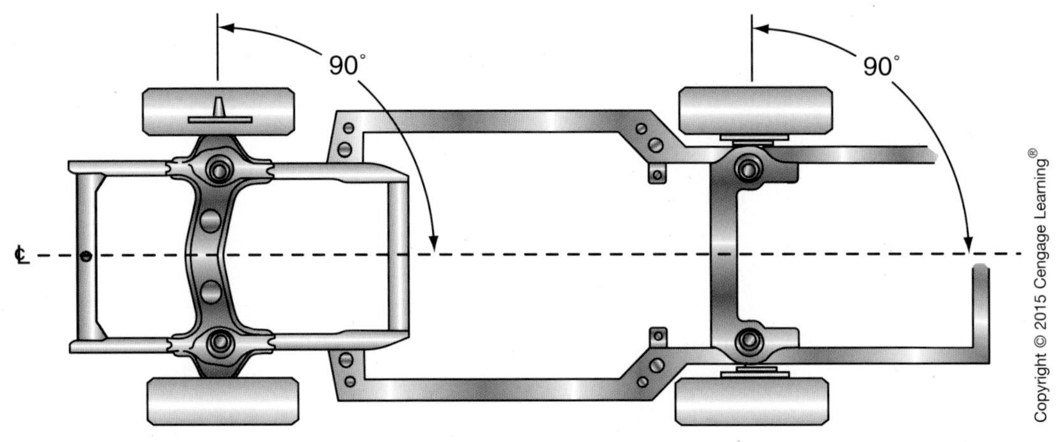

FIGURE 10-19 **To provide accurate wheelbase measurement, the front spindle centerlines and the rear axle centers must be at a 90 degree angle to the vehicle centerline.**

The term dog tracking may be applied to axle offset or sideset, because in either of these conditions the rear wheels are not parallel to the front wheels. Similarly, many dogs run down the street with their rear ends out of line with their front ends.

The **axle thrustline** is a line extending forward from the center of the rear axle at a 90 degree angle.

The **geometric vehicle centerline** refers to the front-to-rear centerline of the vehicle body. The centerlines of the front and rear axles should be positioned on this geometric centerline.

Vehicle Tracking

Wheel Setback

Front wheel setback occurs when the spindle on one front wheel is rearward in relation to the other front wheel. The centerline of each front wheel is still at a 90 degree angle to the vehicle centerline, but the front wheels no longer share the same centerline (Figure 10-20).

When left front wheel setback occurs, the vehicle has a tendency to steer to the left as it is driven straight ahead. Under this condition, the driver has to continually turn the steering wheel to the right to keep the vehicle moving straight ahead.

Axle Offset

If the rear axle is rotated, the **axle thrustline** is no longer at a 90 degree angle to the **geometric vehicle centerline** (Figure 10-21). This condition is referred to as **axle offset**. When the left side of the rear axle is rotated rearward, the steering pulls continually to the right. Under this condition, the driver has to turn the steering wheel to the left to keep the vehicle moving straight ahead.

Axle Sideset

When **axle sideset** occurs, the rear axle moves inward or outward, but the axle and vehicle centerlines remain at a 90 degree angle in relation to each other (Figure 10-22). Under this

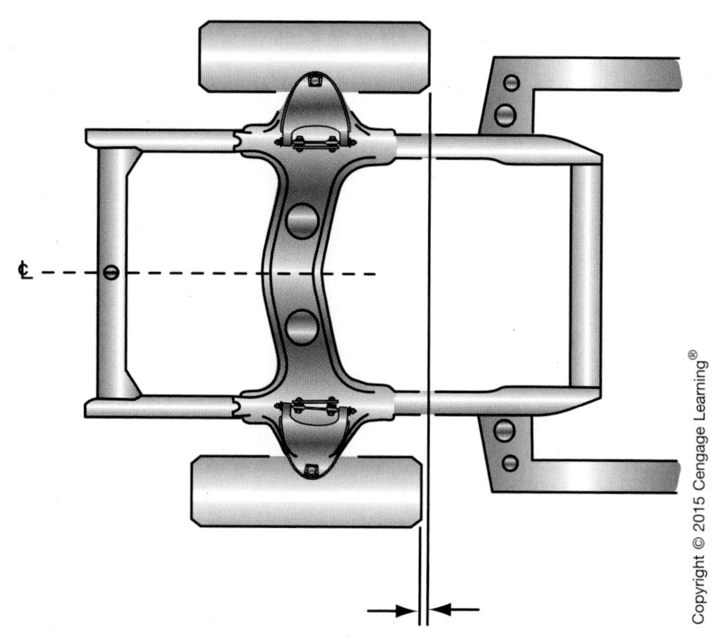

FIGURE 10-20 **Setback occurs when one front wheel is positioned rearward in relation to the other front wheel.**

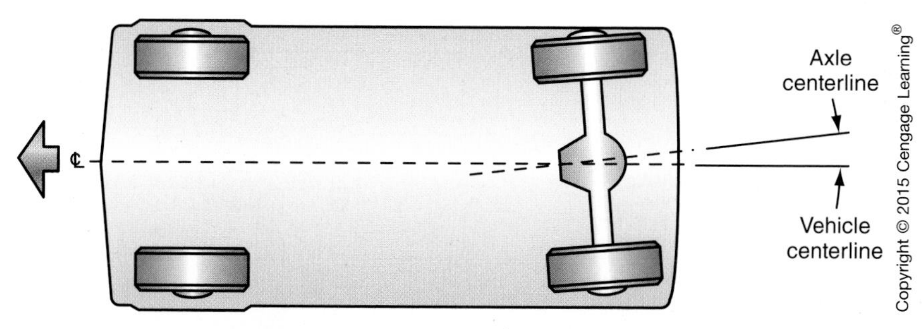

FIGURE 10-21 **Rear axle offset occurs when the rear axle is rotated so the axle centerline and the vehicle centerline are no longer at a 90 degree angle.**

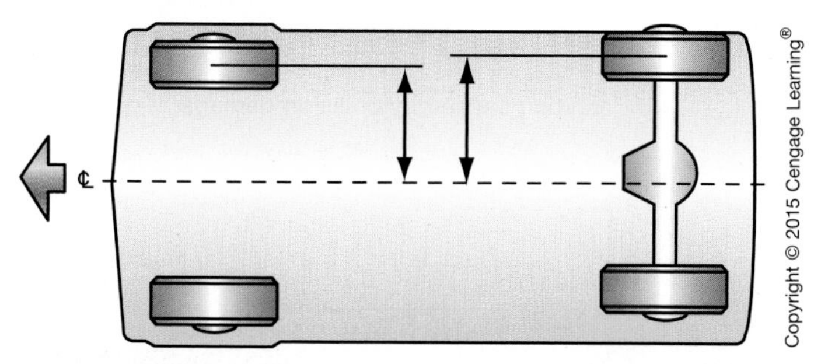

FIGURE 10-22 **Axle sideset occurs when the rear axle moves inward or outward and the axle centerline remains at a 90 degree angle in relation to the vehicle centerline.**

condition, the front-to-rear axle thrustline is no longer at the geometric vehicle centerline. This condition also causes steering pull. The vehicle frame or unitized body and the front and rear suspension systems must have proper tracking and equal wheelbase measurements on each side of the vehicle to provide directional stability and steering control.

Side sway is a lateral misalignment that affects the frame or body centerline.

CAUTION:
When attaching any type of additional equipment to a vehicle frame, always try to use the holes that already exist in the frame. Do not drill holes in a vehicle frame except at locations specified by the vehicle manufacturer, and do not attach additional equipment to a vehicle frame unless this procedure is approved by the vehicle manufacturer.

CAUTION:
Do not apply heat to a vehicle frame with an acetylene torch. When the metal is heated to a cherry red, the frame may be weakened.

Types of Frame Damage

Side Sway

AUTHOR'S NOTE: It has been my experience that frame damage is most commonly caused by abuse, and this problem is usually encountered on light-duty trucks or sport utility vehicles (SUVs). The frame damage may occur when the vehicle is overloaded and/or driven abusively on extremely rough terrain. Another common cause of frame damage is from a vehicle collision. In this case, the frame damage was likely ignored or overlooked during the body repairs. Regardless of the cause, frame damage usually results in excessive tire tread wear and steering complaints.

Side sway on the front or rear of a vehicle is usually the result of the vehicle being involved in a collision that pushes the front or rear frame sideways (Figure 10-23). Under this condition, the wheelbase on one side of the vehicle is longer than the opposite side. This side sway condition causes the steering to pull to the side with the shorter wheelbase.

Overall side sway occurs when the vehicle is hit directly on the side near the center in a collision. A vehicle frame is slightly V-shaped when it has overall side sway damage (Figure 10-24).

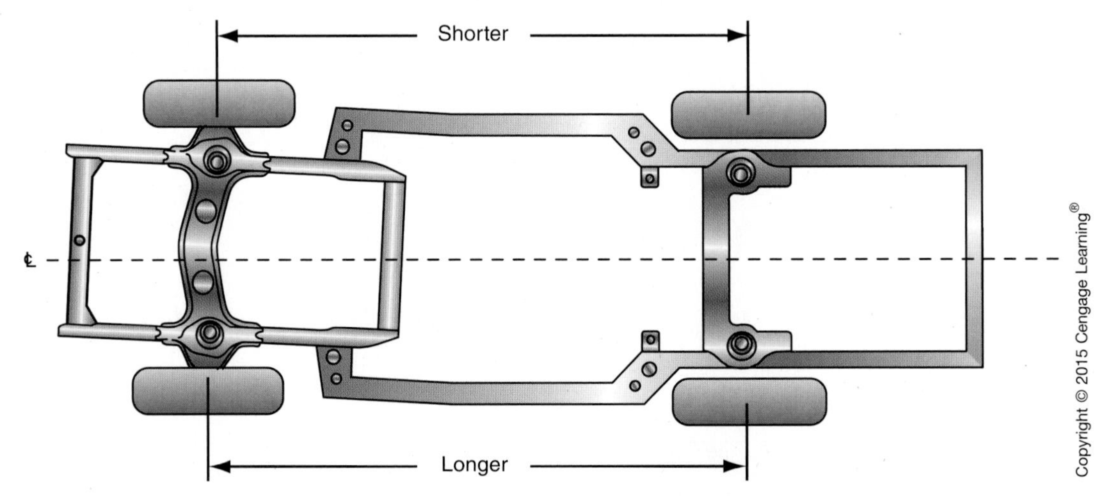

FIGURE 10-23 **Side sway frame condition caused by collision damage.**

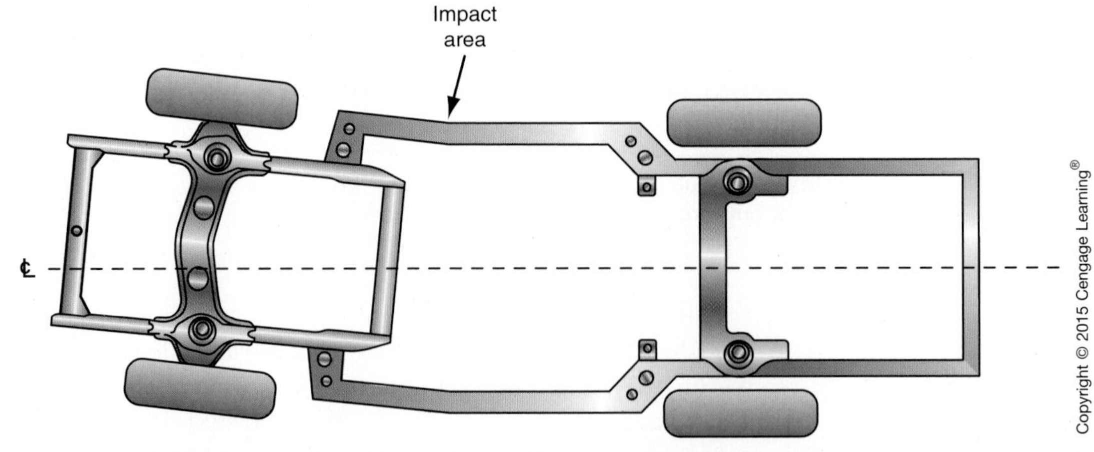

FIGURE 10-24 **When a frame has overall side sway damage, it is slightly V-shaped.**

WARNING: Some types of frame damage cause steering pull when driving straight ahead, and this steering pull is increased during hard braking. Therefore, frame damage can create a safety hazard that leads to a collision involving personal injury and vehicle damage.

Sag

Frame sag usually occurs when the vehicle is involved in a direct front or rear collision. When this condition is present, the front and/or rear frame rails are moved upward in relation to the center of the frame (Figure 10-25). If one side of the vehicle sustained more collision force than the opposite side, the left and right wheel base measurements will also likely be different.

The front cross member may receive sag damage in a collision. When this member is sagged, the upper control arms move closer together on a short-and-long arm suspension. If a MacPherson strut front suspension is sagged, the strut towers are moved closer together. In either type of front suspension, a sagged condition moves the top of the wheels inward to a **negative camber** position.

Negative camber occurs when the camber line through the center of the tire and wheel is tilted inward compared with the true vertical centerline of the tire and wheel.

Frame Buckle

A buckle condition exists when the distance from the cowl to the front bumper is less than specified, or the measurement from the rear wheels to the rear bumper is less than specified (Figure 10-26). Frame buckle is caused by a direct front or rear collision. In many cases of **frame buckle**, the wheelbase is reduced on one or both sides of the vehicle. This type of collision damage may cause the sides of the vehicle to bulge outward, especially on unibody cars. Under this condition, the side rails and door openings are distorted.

Frame buckle is accordion-shaped damage on the front or rear of the frame, which causes the distance to be reduced between the cowl and front bumper or between the rear wheels and rear bumper. Frame buckle may be called frame crush or mash.

Diamond-Frame Condition

A **diamond-frame condition** is present when collision damage causes a frame to be out of square. Under this condition, the frame is shaped like a parallelogram (Figure 10-27). If the

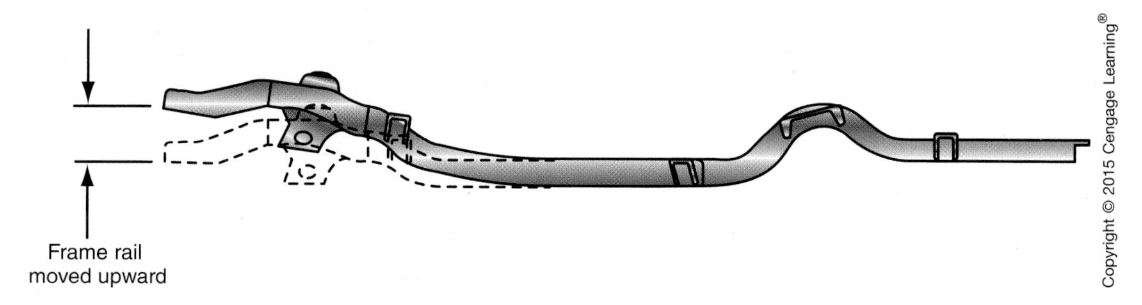

FIGURE 10-25 **Frame sag caused by direct front or rear collision damage.**

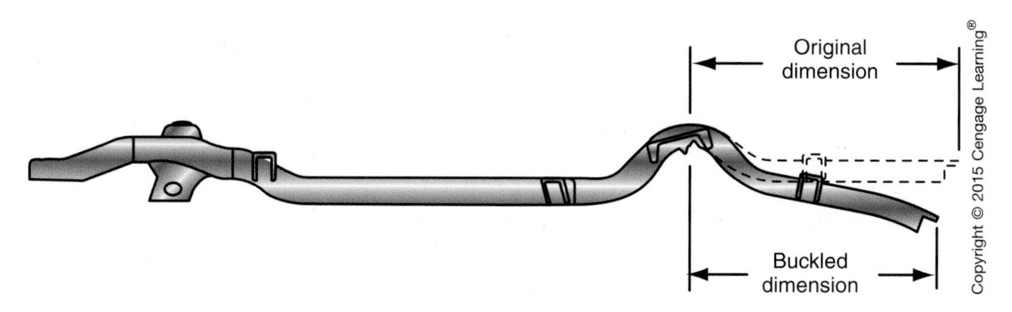

FIGURE 10-26 **Rear frame buckle.**

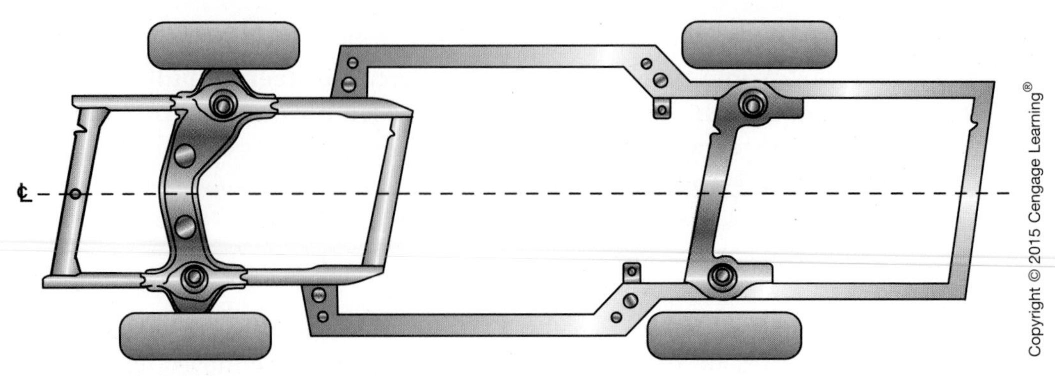

FIGURE 10-27 **Diamond-frame condition.**

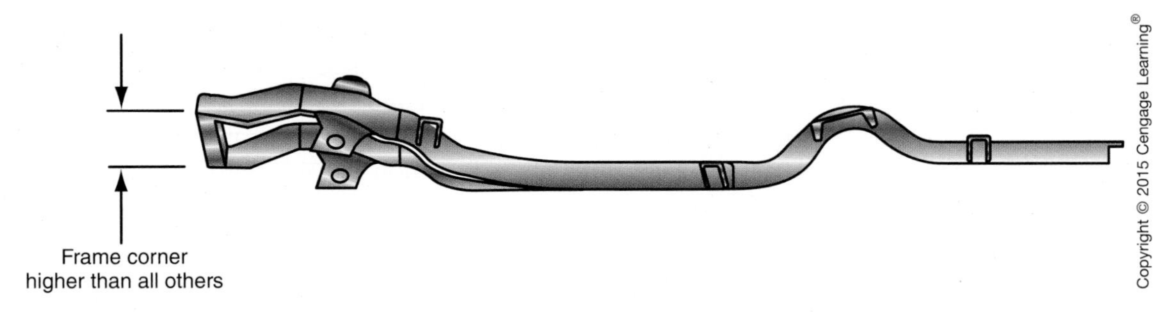

FIGURE 10-28 **Frame twist.**

right rear wheel is driven rearward in relation to the left rear wheel, the rear suspension steers the vehicle to the right, and this forces the front end of the vehicle to the left. Under this condition, the steering wheel must be held continually to the right to overcome the steering pull to the left. A diamond condition usually occurs on vehicles with frames. Vehicles with unitized bodies seldom have this type of condition.

Frame Twist

A **frame twist** condition exists when one corner of the frame is higher than the other corners. When frame twist is present, the front or rear chassis does not sit level in relation to the road surface (Figure 10-28). Frame twist is usually caused by vehicle rollover.

SUMMARY

TERMS TO KNOW

Axle offset
Axle sideset
Axle thrustline
Channel frame
Complete box frame
Diamond-frame condition
Directional stability
Frame buckle

- Steering axis inclination (SAI) is the angle of a line through the center of the upper strut mount and lower ball joint in relation to the true vertical centerline of the tire viewed from the front of the vehicle.
- On short-and-long arm front suspension systems, the SAI line runs through the center of the upper and lower ball joints.
- The SAI line is always tilted toward the center of the vehicle.
- The included angle is the sum of the SAI and positive camber angle.
- If the camber is negative, the camber must be subtracted from the SAI angle to obtain the included angle.
- SAI causes the front spindles to move through an arc when the front wheels are steered to the right or left.

- Because the front spindles move through an arc, the chassis lifts as the front wheels are turned. This lifting action helps return the front wheels to the straight-ahead position after a turn.
- SAI also helps maintain the front wheels in the straight-ahead position.
- SAI reduces the need for excessive positive caster.
- SAI is not adjustable. If the SAI does not equal the manufacturer's specifications and the other steering angles are correct, some suspension component, such as a strut tower, is out of place.
- A suspension system with a 0 degree SAI line would have increased tire wear, greater steering effort, increased stress on suspension and steering components, and excessive road shock and kickback on the steering wheel.
- When the SAI line intersects the true vertical tire centerline at or near the road surface, tire life is improved, stress on steering and suspension components is reduced, steering effort is decreased, and road shock and kickback on the steering wheel are minimized.
- Because excessive positive camber could be used to bring the SAI lines and tire vertical centerlines closer together, SAI reduces the need for excessive positive camber.
- If SAI angles on both sides of the front suspension are unequal, excessive torque steer may occur on hard acceleration, and severe steering pull may be present during hard braking.
- Adjusting camber on some front suspensions creates a corresponding change in SAI angle, whereas on other front suspension systems adjusting the camber does not change the SAI angle.
- Scrub radius is the distance between the SAI line and the true vertical tire centerline at the road surface.
- A front suspension has positive scrub radius when the SAI line contacts the road surface inside the true vertical tire centerline.
- A front suspension has negative scrub radius when the SAI line contacts the road surface outside the true vertical tire centerline.
- If the front tires are larger than the original tires specified by the vehicle manufacturer, a change occurs in scrub radius that may affect steering control.
- The installation of larger-than-specified front tires may change positive scrub radius to a negative scrub radius.
- Reversing the front rims so they are inside out results in a significant scrub radius change that adversely affects directional control.
- Wheel setback is a condition where one wheel is moved rearward in relation to the opposite wheel.
- Toe-out on turns is the turning angle of the wheel on the inside of the turn compared with the turning angle of the wheel on the outside of the turn.
- When the front wheel on the inside of a turn has turned 20 degrees outward, the front wheel on the outside of the turn may have turned 18 degrees.
- Turning radius is the amount of toe-out on turns.
- Toe-out on turns prevents tire scuffing. This angle is determined by the steering arm design.
- During a turn, centrifugal force causes all the tires to slip a certain amount depending on vehicle speed and the sharpness of the turn.
- Because the tires slip during a turn, the actual vehicle turning center is moved ahead of the theoretical turning center.

TERMS TO KNOW
(continued)

Frame sag
Frame twist
Geometric vehicle centerline
High-strength steels (HSS)
Hydro-forming
Included angle
King pin inclination (KPI)
Ladder frame design
Negative camber
Negative scrub radius
Perimeter-type frame
Positive scrub radius
Scrub radius
Side sway
Slip angle
Steering axis inclination (SAI)
Tracking
Tubular frame
Turning radius
Unitized body
Wheel setback
Wheelbase

- Slip angle is the actual angle of the front wheels during a turn compared with the turning angle of the front wheels with the vehicle at rest.
- The two rear wheel alignment angles are camber and toe.
- The frame enables a vehicle to support its weight and absorb stress and torque. It also provides a main member for attachment of other components.
- In a unitized body design, all body members are load-carrying components that are welded together.
- Proper tracking and wheelbase are essential to providing directional stability.
- Each front wheel and each rear wheel must be at an equal distance from the vehicle centerline to provide correct tracking.
- To provide proper wheelbase, the centers of the front and rear suspensions must be at a 90 degree angle to the vehicle centerline.
- Front wheel setback, rear axle offset, and rear axle sideset cause the steering to pull to one side.
- Regardless of the type of front or rear suspension system, the suspension system and the frame must position the wheels properly to provide correct tracking and wheelbase.
- Frame side sway occurs when the front suspension is forced sideways in a collision, and one front wheel is forced rearward in relation to the opposite front wheel. Side sway may also occur on the rear suspension.
- Frame sag occurs when the front or rear frame rails are bent upward in relation to the center of the frame.
- Frame buckle is accordion-shaped damage on the front or rear of the frame that causes the distance to be reduced between the cowl and front bumper or between the rear wheels and rear bumper.
- A diamond-frame condition is present when one side of the frame is driven rearward in relation to the opposite side of the frame, and the front and rear wheels on one side of vehicle are rearward in relation to the wheels on the other side.
- Frame twist occurs when one corner of the frame is bent up higher than the other frame corners.

REVIEW QUESTIONS

Short Answer Essays

1. Define steering axis inclination (SAI).
2. Explain the included angle.
3. Explain how SAI helps to return the steering wheel to the center position after a turn.
4. Define negative scrub radius and positive scrub radius, including the type of suspension system on which each condition is used.
5. Describe front wheel setback.
6. Explain the necessary wheel position to provide proper tracking.
7. Describe the necessary axle position to provide correct wheelbase.
8. Explain the turning angle of each front wheel during a turn.
9. Explain the effects of front wheel setback.
10. Describe the effects of rear axle offset.

Fill-in-the-Blanks

1. A 3 degree difference in the SAI angle on each side of the front suspension may cause ________________ during hard braking.
2. A 3 degree difference in the SAI angle on each side of the front suspension may cause increased ________________ during hard acceleration.
3. When a front suspension has a positive scrub radius, the SAI line meets the road surface ________________ the true vertical centerline of the tire.

4. A negative scrub radius tends to turn the front wheels ________________ when the car is driven.

5. Most rear-wheel-drive cars have a ________________ scrub radius.

6. Wheelbase is the distance between the front and rear wheel ________________.

7. Directional stability is the tendency of a vehicle to remain in the ________________ position when driven straight ahead on a level road.

8. During a turn, the ________________ front wheel turns at a sharper angle.

9. Tracking refers to the ________________ relationship between the front and rear wheels.

10. During a turn, centrifugal force causes all the tires to slip a certain amount, and the actual center of the turn is shifted ________________ of the theoretical center of the turn.

Multiple Choice

1. When measuring front wheel alignment angles, to calculate the included angle on the left front wheel when the camber on this wheel is positive:
   A. Add the camber to the toe setting.
   B. Add the camber to the SAI angle.
   C. Add the SAI to the caster angle.
   D. Subtract the SAI from the toe setting.

2. While diagnosing problems related to scrub radius:
   A. If the SAI line contacts the road surface inside the vertical tire and wheel centerline, the scrub radius is negative.
   B. Front-wheel-drive cars usually have a negative scrub radius and driving forces move the front wheels outward.
   C. If the SAI line contacts the road surface outside the tire and wheel vertical centerline, driving forces turn the front wheels outward.
   D. Larger than specified front tires may change the scrub radius from positive to negative.

3. While diagnosing a diamond-frame condition and frame twist:
   A. A diamond-frame condition causes the wheelbase to be unequal on the two sides of the vehicle.
   B. A diamond-frame condition does not affect steering pull and directional stability.
   C. Frame twist is usually caused when a vehicle is involved in a side collision.
   D. When frame twist occurs, the front or rear chassis does not sit level in relation to the road surface.

4. While diagnosing and adjusting turning radius:
   A. When a vehicle is making a left turn, the left front wheel turns at a sharper angle than the right front wheel.
   B. Improper turning radius may be caused by a bent tie rod on a short-and-long arm suspension system.
   C. Improper turning radius may be caused by an improperly positioned rack-and-pinion steering gear on a front-wheel-drive car.
   D. Improper turning radius is adjusted by turning an eccentric strut-to-steering knuckle bolt.

5. All of these statements about unitized body design are true EXCEPT:
   A. The frame and body are combined as one unit.
   B. The external frame assembly is eliminated.
   C. Some body members such as quarter panels do not contribute to body strength and rigidity.
   D. Body strength is obtained by body shape and design.

6. A front-wheel-drive car has an improper toe-out on turns setting, and a visual check indicates all the steering linkage and suspension components are satisfactory. The most likely cause of this problem is:
   A. A bent lower control arm.
   B. A bent front strut.
   C. A front strut tower that is out of position.
   D. A bent steering arm.

7. All of these statements about SAI and front spindle movement are true EXCEPT:
   A. When the steering wheel is turned, the front spindle movement is parallel to the road surface.
   B. When the SAI angle is increased, the steering wheel returning force is increased.
   C. The SAI angle tends to maintain the wheels in a straight-ahead position.
   D. Greater SAI angle is necessary on front-wheel-drive cars to provide directional stability.

8. Front wheel setback occurs when one front wheel is:
   A. Tilted inward from the true vertical position.
   B. Moved rearward in relation to the opposite front wheel.
   C. Tilted rearward from the true vertical position.
   D. Inward from its original position.

9. A light-duty truck with a one-piece rear axle housing and a leaf-spring rear suspension has excessive toe-out on the left rear wheel and too much toe-in on the right rear wheel. The most likely cause of this problem is:
   A. A broken center bolt in the left rear spring.
   B. Both rear springs are sagged.
   C. A bent rear axle housing.
   D. Worn-out rubber bushings in the shock absorbers.

10. While discussing scrub radius:
   A. Most front-wheel-drive cars have a positive scrub radius.
   B. If the SAI line meets the road surface outside the tire vertical centerline, the scrub radius is positive.
   C. Scrub radius is adjusted by shifting the upper strut tower on a MacPherson strut front suspension.
   D. Incorrect scrub radius may be caused by larger than specified front tires.

Chapter 11

COMPUTER-CONTROLLED SUSPENSION SYSTEMS

UPON COMPLETION AND REVIEW OF THIS CHAPTER, YOU SHOULD BE ABLE TO UNDERSTAND AND DESCRIBE:

- The types of integrated computer networks used on vehicles.
- The operation of controller area network (CAN) system.
- The conditions that cause a programmed ride control (PRC) system to switch from the normal to the firm mode.
- How the firm ride condition is obtained in PRC struts.
- The major components in an electronic air suspension system.
- How air is forced into and exhausted from the air springs in an air suspension system.
- How an electronic air suspension system corrects low suspension trim height.
- The operation of an electronic air suspension system while driving the car with the doors closed and the brake pedal applied.
- The normal operation of the warning lamp in an electronic air suspension system.
- The three modes in the air suspension system on some modern four-wheel-drive sport utility vehicles (SUVs).
- How unnecessary rear suspension height corrections are prevented on irregular road surfaces with an air suspension system.
- The design of the struts and air springs in an automatic air suspension system.
- The design of an electronic rotary height sensor.
- Speed-leveling capabilities and the advantage of this function in a suspension control module.
- The operation of an automatic ride control (ARC) system in relation to transfer case modes.
- The inputs in an electronic suspension control (ESC) system.
- The advantages of an ESC system with magneto-rheological fluid in the shock absorbers compared to other computer-controlled suspension systems.
- The operation of the rear electronic level control system that is combined with the road sensing suspension system.
- The operation of the speed sensitive steering system that is combined with the road sensing suspension system.
- The operation of a stability control system.
- The advantages of a traction control system.
- Various vehicle network systems.
- Active cruise control, lane departure warning, and collision-mitigation systems.

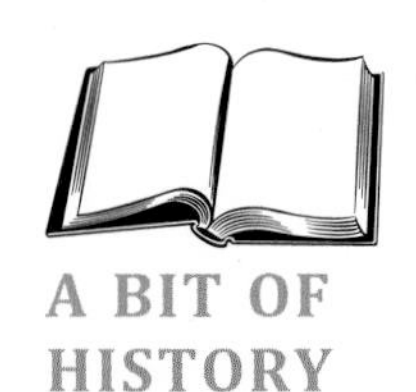

A BIT OF HISTORY

Suspension systems evolved slowly for many years. In the 1940s, many front suspensions were changed from I-beam to short-and-long arm. MacPherson strut suspensions replaced the short-and-long arm suspension systems on front-wheel-drive cars introduced in the late 1970s and 1980s. In the 2000s, we are experiencing an ever expanding use of electronics technology in automotive suspension systems. Some computer-controlled suspension systems now have the capability to react to road or driving conditions in one millisecond to improve ride and handling quality. The use of electronics technology will continue to improve suspension systems so they provide better handling characteristics and safety with longer tire life.

INTRODUCTION

We are all aware of the ever accelerating electronics revolution in the 1990s and early 2000s. Most industries have felt the impact of this revolution, and the automotive industry is no exception. Computers have greatly influenced the way vehicles are designed and built. Most systems on the automobile, including the suspension system, have been impacted by the computer. Many drivers like a soft, comfortable ride while driving normally on the highway. However, many of these same drivers prefer a firm ride during hard cornering, severe braking, or fast acceleration. A firm ride under these driving conditions reduces body sway and front end dive or lift. Prior to the age of electronics, cars were designed to provide either a soft, comfortable ride or a firm ride. Drivers who wanted a firm ride selected a sports car with a suspension designed to supply the type of ride and handling characteristics they desired. Car buyers who wanted a softer ride purchased a family sedan with a suspension designed to provide a softer, more comfortable ride.

Thanks to computer control, suspension system manufacturers can now provide a soft ride during normal highway driving, and then almost instantly switch to a firm ride during hard cornering, braking, fast acceleration, and high-speed driving. The computer-controlled suspension system allows the same car to meet the demands of both the driver who desires a soft ride, and the driver who wants a firm ride. Because computer-controlled suspension systems reduce body sway during hard cornering, these systems provide improved steering control.

Some computer-controlled suspension systems also supply a constant vehicle riding height regardless of the vehicle passenger or cargo load. This action maintains the vehicle's cosmetic appearance as the passenger and/or cargo load is changed. Maintaining a constant riding height also supplies more constant suspension alignment angles, which may provide improved steering control.

INTEGRATED ELECTRONIC SYSTEMS AND NETWORKS

Advantages of Integrated Electronic Systems and Networks

With the rapid advances in electronic technology, computer-controlled automotive systems have become integrated. Rather than having a separate computer for each electronic system, several of these systems may be controlled by one computer. Vehicles without any integrated electronic systems may have many individual modules and computers. Because computers must have some protection from excessive temperature changes, extreme vibration, magnetic fields, voltage spikes, and oil contamination, it becomes difficult for engineers to find a suitable mounting place for this large number of computers. Integration of several electronic systems into one computer solves some of these computer mounting problems and reduces the length of wiring harness. The ESC system explained in this chapter is an example of an integrated electronic system with suspension ride control, suspension level control, and speed sensitive steering controlled by one computer.

In a **network** several computers are interconnected by data links.

Another method of reducing the number of wires on a vehicle is to interconnect many of the on-board computers with data links. A data link system may be referred to as a **network**. Some input sensor signals may be required by several computers. For example, on some vehicles the VSS signal is required by the powertrain control module (PCM), suspension computer, transmission computer, cruise control module, and throttle control module. On many vehicles the VSS is hardwired to the PCM, and then the PCM relays the VSS signal to the other computers via the network.

Some vehicles now have a front control module mounted near the front of the vehicle and a rear control module mounted near the rear of the vehicle. These vehicles also have a body computer module (BCM). The BCM, PCM, front and rear control modules, and other modules are interconnected by a network. The headlight switch may be hard wired to the BCM. When the headlight switch is turned on, a voltage signal is sent to the BCM, and the BCM relays the appropriate LIGHTS ON message through the network to the front and rear control modules. These modules are hard wired to the exterior lights. When the front and rear control modules receive a specific LIGHTS ON message, these modules turn on the appropriate front and rear lights. Connecting the BCM and the front and rear control modules via a network reduces the number of wires between the light switch and the front and rear lights. The headlight switch in no longer just a switch but rather it is an input sensor to the BCM.

Some vehicles equipped with power windows and power door locks have a module in each door. These modules are hard wired to the window motor and door lock controls in each door. These door modules are connected by a network to the BCM that is usually mounted under the dash. The window and door lock control switches are hard wired to the BCM. When a WINDOW DOWN signal is sent from a window switch to the BCM, the BCM relays this signal through the network to the proper door module. When the door module receives the WINDOW DOWN signal, it supplies voltage to the window motor in the proper direction to roll the window down. Connecting the door modules to the BCM by a network greatly reduces the number of wires connected from the door switches into each door, and this design reduces wiring harness size and weight. Networks also reduce some of the problems associated with wiring harnesses. The previous examples are the type of networking that we see in a Controller Area Protocol (CAN) network described in more detail later in the chapter.

A BIT OF HISTORY

In 1996, the typical vehicle had six electronic control units (ECUs). In 2012, some luxury vehicles had up to 120 modules, 5 main networks, and 12 to 14 subnetworks depending on the on-board electronic equipment.

Types of Networks

One type of network system introduced in the early 1980s is the Chrysler Collision Detection (CCD) network. The CCD system has a twisted pair of wires connected between the PCM, BCM, transmission control module (TCM), air bag control module (ACM), electromechanical instrument cluster (MIC), and vehicle theft security system (VTSS) module. On some models the CCD system is also connected to the data link connector (DLC), allowing the computers in the system to communicate with a scan tool connected to the DLC. The network system may be called a data bus. The CCD system operates at 2.5V.

Another type of network system introduced in the 1980s is the universal asynchronous receive and transmit (UART) system. The UART data links are connected between various on-board computers and the DLC. The UART system operates at 5V and transmits data at 8.2 kilobits per second. When sending data, the UART system toggles the voltage from 5V to ground at a fixed bit pulse width. At rest the UART network system has 5V.

With the implementation of on-board diagnostic II (OBDII) systems in 1996, improved communication was required between the PCM, other computers, and the scan tool. Class 2 networks were installed to meet this demand. Class 2 networks transmit data at 10.4 kilobits per second, and transmit data by toggling the voltage from 0 to 7V. At rest this network system has 0V. The programmable communication interface (PCI) networks were also introduced on some vehicles to increase the data communication requirements. The PCI network system is a single wire system. The PCI system operates between 0V and 6 to 8V. Communication on a network system is accomplished by sending a group of 0 and 1 signals (Figure 11-1). A long voltage pulse at a high voltage and a short pulse at a low voltage represent a 0 signal. Conversely, a short pulse at a high voltage and a long pulse at a low voltage indicate a 1 signal (Figure 11-2).

Today's vehicles are now equipped with controller area network (CAN) systems. A local area network (LAN) system is similar to the CAN system, and the LAN system is used on a significant number of vehicles. Some vehicles have a low-speed LAN system

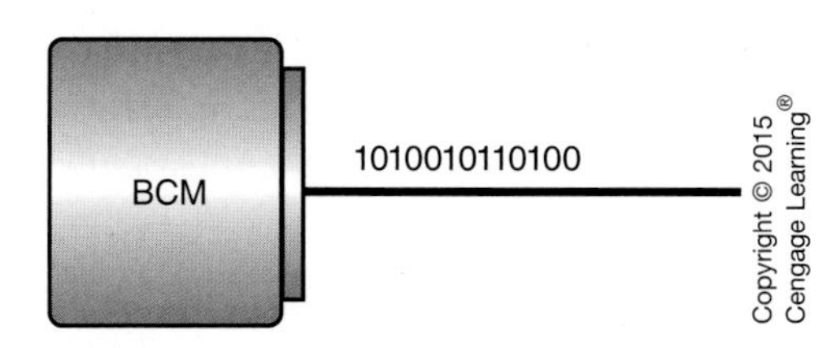

FIGURE 11-1 **A data link system transmits data by using a group of 0 and 1 signals.**

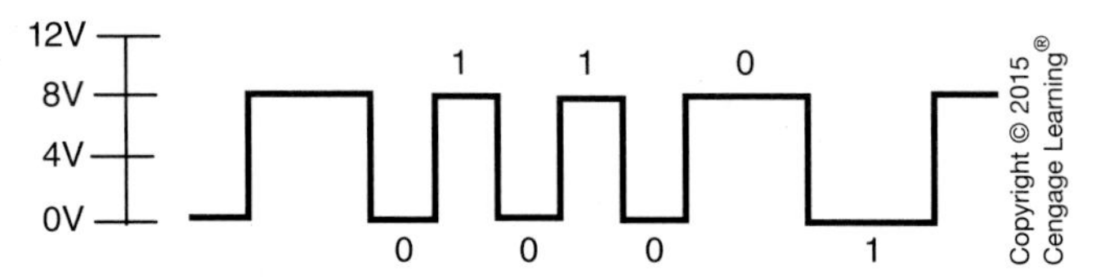

FIGURE 11-2 **Voltage level and duration required for 0 and 1 signals.**

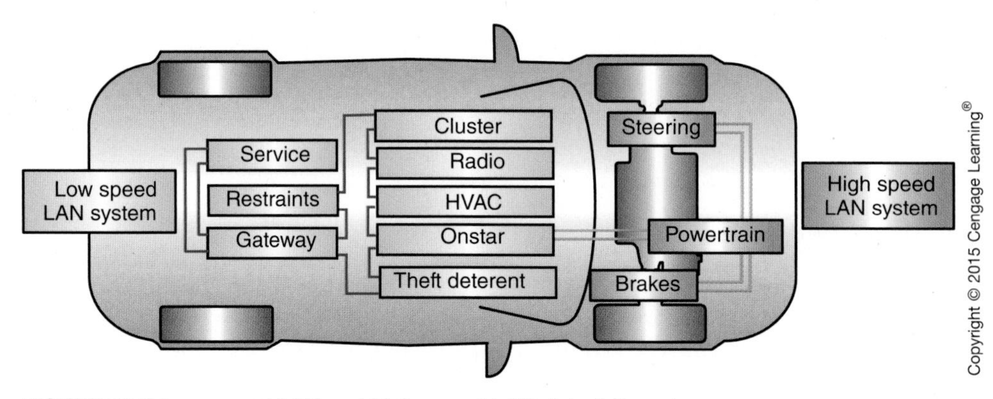

FIGURE 11-3 **Low-speed LAN and high-speed LAN data link systems.**

and a separate high-speed LAN system (Figure 11-3). The low-speed LAN system interconnects modules for applications such as door locks, window motors, HVAC, and radio. The low-speed LAN system is a single-wire system. The high-speed LAN system interconnects modules such as the PCM, transmission, antilock brake modules, and suspension modules. The high-speed LAN system is a two-wire system that operates at 500 kilobits per second. The greatly increased data transmission speed capabilities of the LAN system enhance the communication between various computers in the system and the scan tool. Other high-speed data link systems on modern vehicles include FlexRay and local interconnect network (LIN). The FlexRay data link system transmits data at 10 megabits per second (Mbps). The Byteflight network used on some luxury cars has much in common with the FlexRay network. The network systems on a current sport utility vehicle (SUV) are shown in Figure 11-4. Some networks, such as the one illustrated in Figure 11-4, have a gateway module. The gateway module changes and directs the signals to go to the appropriate network within the complete network system. The gateway module is often combined within one of the other network computers.

Some networks, such as CAN, contain terminators located at both ends of the network. The terminators are usually positioned inside some of the network computers. Terminators provide electrical resistance to absorb data and prevent this data from being transmitted back into the network.

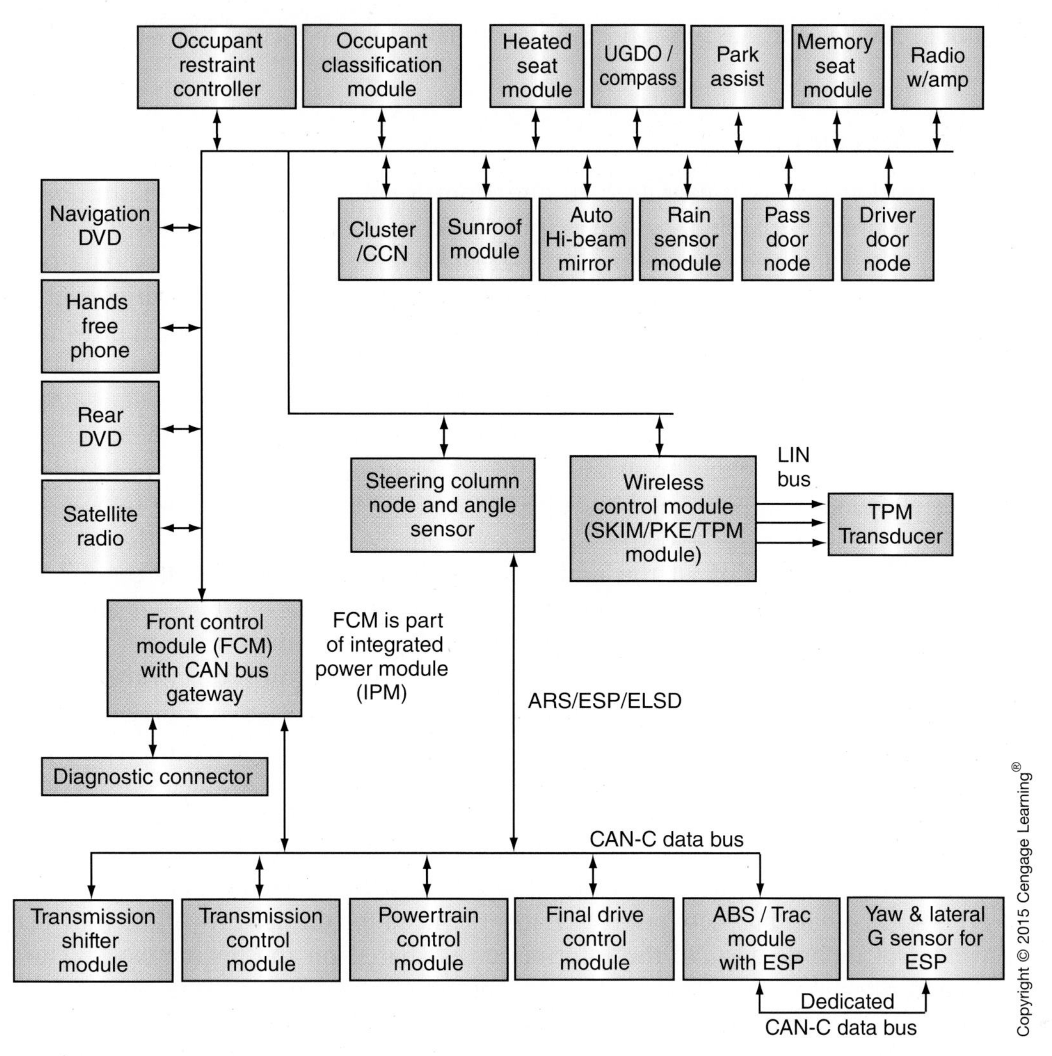

FIGURE 11-4 **Data link systems on a current SUV.**

Most luxury vehicles presently have a media oriented systems transport (MOST) network system in which the computers are interconnected by fiber-optic data links. The MOST system transmits data at 150 megabits per second. The data transmission rate depends on the model year of the vehicle and data link system. For example, early model MOST systems transmitted data at 22.6 Mbps and current MOST systems transmit data at 150 Mbps. This high data transmission speed is required on vehicles with navigation systems, CD changers, video systems, and satellite radios. The MOST system greatly reduces the number of wires in the wiring harnesses, but is more expensive compared to a network interconnected by wires.

Networks have collision resolution (CR) capabilities to prevent data collisions. The CR system varies depending on the network. In a single-wire CAN system, the system voltage is high when not transmitting data. When any computer wants to transmit data, it initiates a low voltage condition to begin transmission. As explained previously in this chapter, the data is a series of low and high voltages. The low voltage signal is the dominant voltage bit and the high voltage signal is the recessive voltage bit. The CR system uses the dominant and recessive voltage bits to determine transmission priority. If two computers transmit data at the same time, the CR system will recognize the computer with the most dominant bits and give priority to that computer. The low priority computer stops its communication, and the high priority computer continues data transmission. In a typical network, air bag, antilock brake, and suspension computers have high priority compared with audio computers.

Data Bus Network

Often information that is transmitted to and from a control module is in the form of a serial data stream. A serial data stream is a digitized code of ones and zeros that is known as a binary code. Each one or zero in a data stream is referred to as a binary digit or *bit* of information. A group of bits form a binary term or *byte*. The wire or wires over which serial data is transmitted is referred to as a **data bus network**. Often the data bus network is a twisted pair of wire (Figure 11-5). The twisted pair helps to eliminate the induction of electro-magnetic-interference (EMF), which could disrupt or cloud the data signal. Any microprocessor that communicates on the data bus network is referred to as a node. A node may only be capable of transmitting (send) information or may have bi-directional capabilities allowing it to both send and receive data on the network. The two wires on a CAN bus are the Can High and the CAN Low. On the CAN network, the entire CAN High and CAN Low pins of all the nodes are connected together. Both the CAN High and CAN Low wires transmit the same data across the network but in what amounts to a mirror image of the information on the two lines. In other words, they are a check and balance for one another. In the event of a network or connection problem the system will see a difference in the data packet over the two lines and set a Network diagnostic trouble code known as a U-code.

Control modules (nodes) that may be multiplexed on a data bus network allowing them to share information and sensor data between one another. Examples of control modules that may share data on a data bus network include the PCM, BCM, TCM, instrument panel cluster (IPC), and electronic brake control module (EBCM) to name a few (Figure 11-6). The data bus network eliminates the need to run hard wire from each sensor to each control module, instead the information is shared on the bus network. The DLC allows the connection of a diagnostic scan tool, which becomes a node on the network with bi-directional data communication (Figure 11-7).

In general, a twisted pair data bus sends data in a fixed pulse width data stream. Each of the data bits is the same length. Data bits that are strung together (0011011) in this way are referred to as a pulse width modulated (PWM) serial data stream (Figure 11-8).

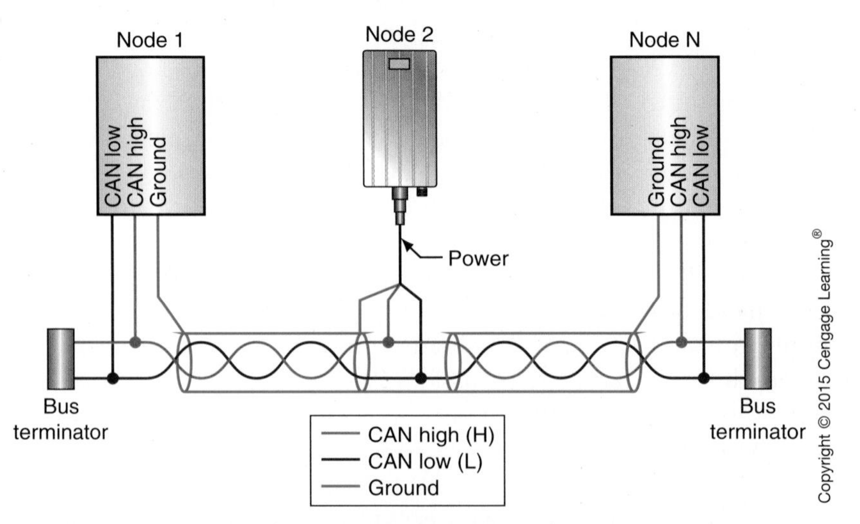

FIGURE 11-5 **Twisted wire pair used to share information between the controllers such as Node 2 the Powertrain Control Module (PCM), Node 1 the Body Control Module (BCM) and Node N the Traction Control Module.**

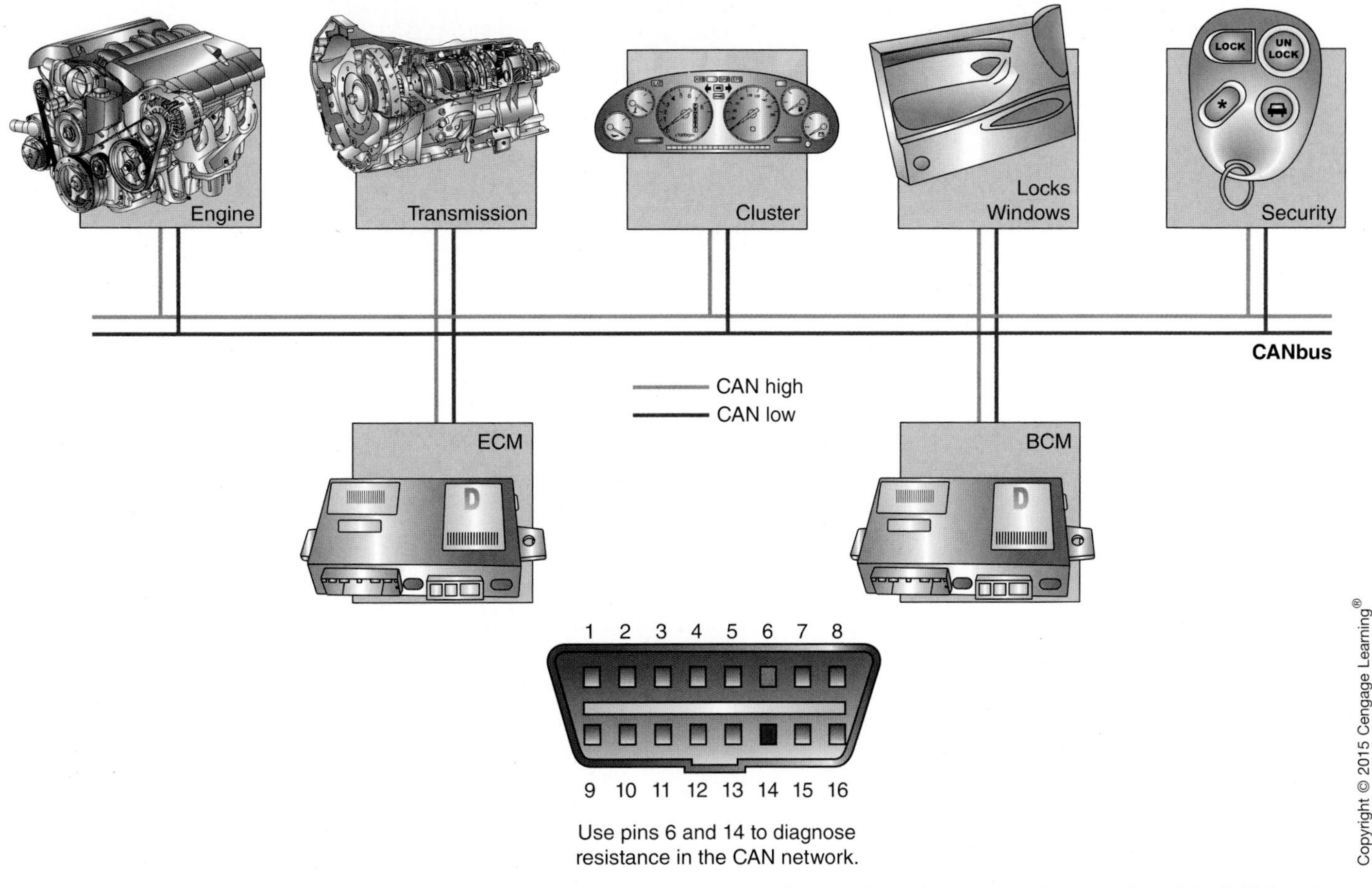

FIGURE 11-6 **Data bus network used to share information between the controllers such as the powertrain control module (PCM) and the transmission control module (TCM) as well as the DLC.**

Controller Area Network

The **Controller Area Network (CAN)** protocol is the latest serial bus communication network used on OBD II systems and offers real time control and is the predominate protocol in use. Mercedes-Benz first integrated CAN into their engine and transmission control units in 1992, with many other manufacturers integrating CAN into some of their new vehicle platforms in the early 2000s. It was mandated that by 2008 all DLC communicated on the CAN network, making CAN the standard protocol. The CAN protocol has been standardized by the International Standards Organization (ISO) as ISO 11898 standard for high speed and ISO 11519 for low speed data transfer. The speed of data transmission is expressed in bits per second (bps). The high speed version can operate at 1 Mega bit per second (Mbps) and is used for powertrain management systems, and operates in virtually real time data rate transfer speeds. The low speed version can operate at 125 kilobits per second (Kbps) and is used for body control modules and passenger comfort features. While the prefix kilo usually indicates a multiplier value of 1,000, in a serial data stream a kilobyte has a value of 1,024 bytes of data. This is the mathematical result of a base two numbering system (ones and zeros) carried to the tenth place. Additionally a megabyte has a value of 1,048,576 bytes of data, which is one kilobyte (1,024) squared. The CAN system has allowed for improved communication with on-board vehicle systems and is a true multiplexed network.

Terminal Assignment and Function	
PIN	**SAE/ISO**
1	Manufacturer discretionary
2	SAE J1850 (+)
3	Manufacturer discretionary
4	Chassis ground
5	Signal ground
6	ISO 15765-4 CAN-C (+)
7	ISO 9141-2 K-line ISO 1423-4 K-line
8	Manufacturer discretionary
9	Manufacturer discretionary
10	SAE J1850 (−)
11	Manufacturer discretionary
12	Manufacturer discretionary
13	Manufacturer discretionary
14	ISO 1565-4 CAN-C (−)
15	ISO 9141-2 L-line/ ISO 14230-4 L-line
16	Unswitched battery voltage

FIGURE 11-7 **DLC for SAE standard J1962 with CAN protocol pin assignments.**

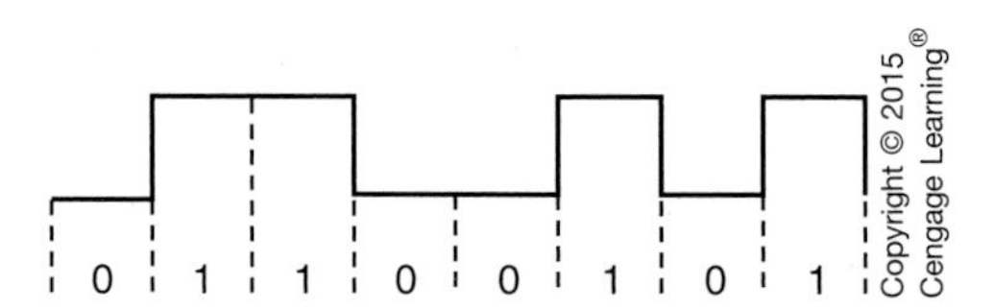

FIGURE 11-8 **Equal length bits of data strung together form a pulse width modulated serial data stream.**

The Society of Automotive Engineers (SAE) has divided the speed of serial data transfer for automotive applications into three classes. Class A is the slowest transmission rate with speeds less than 10 kbps. Class A networks are used for low priority data transmission; generally related to noncritical body control module functions such as memory seats. Class B networks are mid speed range networks with data transmission speeds between 10 kbps and 125 kbps; generally related to less critical devices such as HVAC, advanced lighting systems, and dash clusters. Class C networks have the fastest data transmission rate with speeds up to 1 bps. Class C networks are the most expensive to produce, and are used for "mission critical" data transmission that flow at "real-time" speeds. Examples of class C data include fuel control and ABS activation activity. The DLC is also connected to the class C network for improved on-board diagnostics.

CAN enables the use of enhanced diagnostics and more detailed DTCs. With CAN a scan tool is capable of communicating directly with sensors, independently of the PCM. The CAN protocol uses smart sensors. Each component contains its own control unit

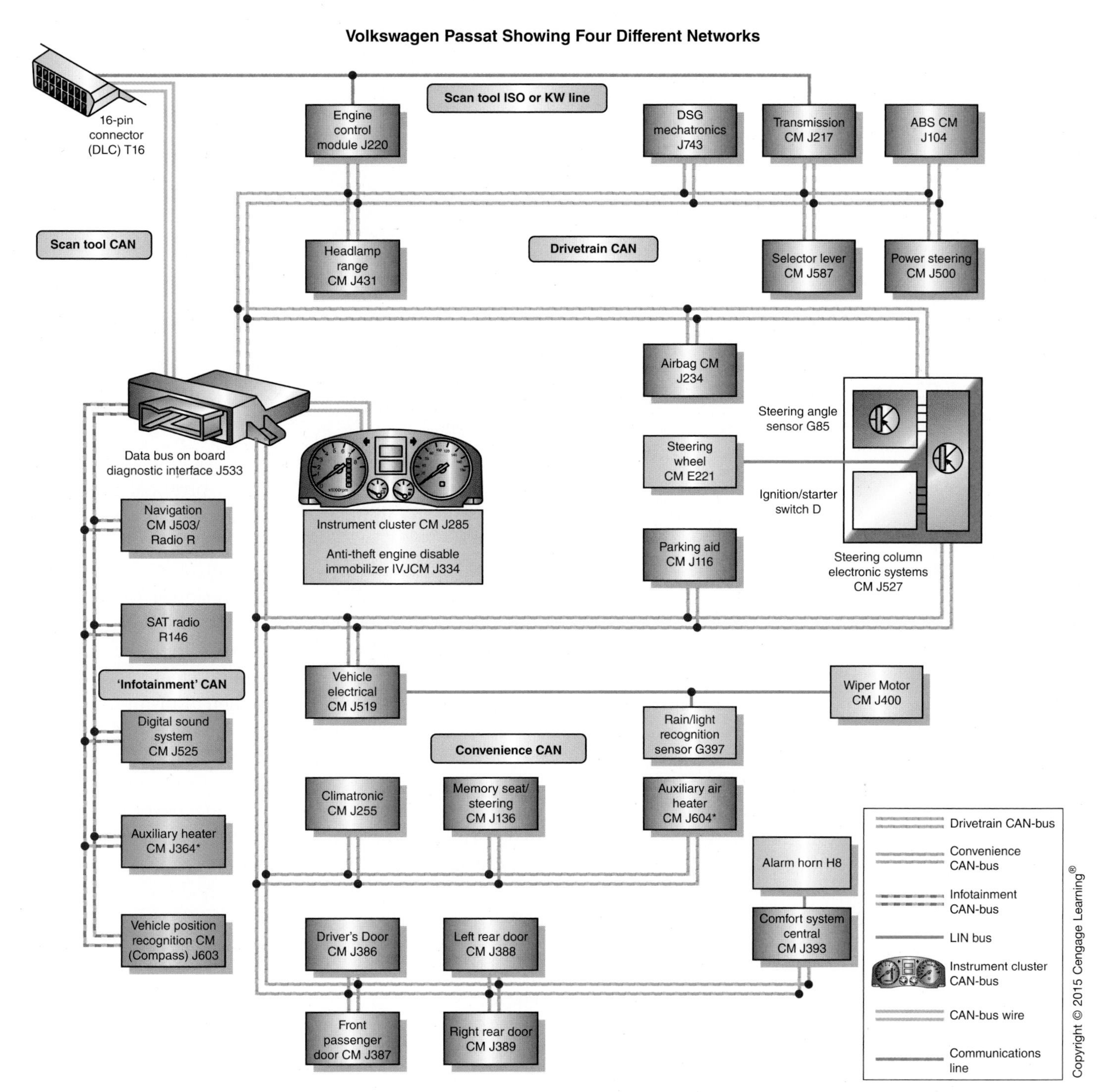

FIGURE 11-9 **Four different communication networks on one vehicle.**

(microprocessor) called a "node". Each node on the network has the ability to communicate over a twisted pair of wires or a single wire called a data bus with all the other nodes on the network (bi-directional communication) without having to go through a central processing unit (Figure 11-9) unlike other multiplexed systems used in the past for data sharing. Every component on the network is independently capable of processing and communicating data over a common transmission line. Nodes transmit information (messages) with an identifier that prioritizes the message. The messages transmitted from a node are a package of data bits, which include a beginning of message signal, component identifier, message (sensor output signal), and an end of message signal. Since this is a bi-directional communication network the control module receiving the data will send a signal back that

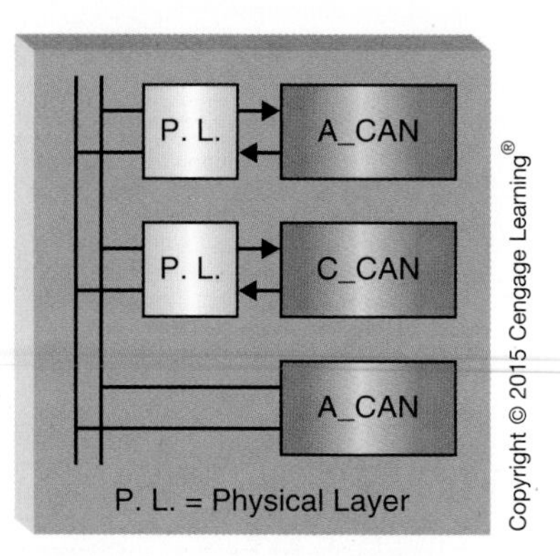

FIGURE 11-10 **When multiple nodes are sending data simultaneously to the control module, the CAN system segments transmissions based on the priority identifier.**

the information was received. For this sophisticated communication protocol to function, the data transmission package must be a set size (number of bits), format, and the information order must be consistent for all devices.

When multiple nodes need to send data simultaneously to the control module, the node will first see if the data bus is busy. The system uses collision detection similar to an Ethernet system. But unlike Ethernet the CAN system can handle high data transmission rates. In essence the node is looking into traffic to see if a higher priority node should be allowed to pass. Each CAN node on the network will have its own network unique identifier code, and nodes may be grouped based on function. The data message is then transmitted with its unique identifying code onto the network. Each node and control module on the network will perform and acceptance test of the transmission to determine if it is relative or not based on its identifier. Relative information is processed and non-relative information is ignored. Then the system segments transmissions based on the priority identifier (Figure 11-10) of the data package. The priority is determined by the unique number of the identifier, with lower number identifiers having higher priority. This guarantees higher priority node identifier messages access to the network and lower priority node messages will be automatically retransmitted in the next available bus cycle based on priority.

Since the CAN protocol technology allows for many nodes on one set of wiring, the overall vehicle wiring harness size is greatly reduced. A twisted pair wired network contains a CAN (+) and a CAN (−) wire. The CAN bus is a differential bus system where the data signal from the CAN (+) wire is a mirror image of the CAN (−) network wire (Figure 11-11). The combination of the twisted pair network wiring combined with the differential bus data eliminates the effect of EMF noise on the data transmission. Multiple networks on the vehicle can be linked together by gateways if necessary (Figure 11-12). Class C high speed data flows on one network, Class B mid speed data flows on a second network, while Class A low speed data flows on a third network. As an example, the intake air temperature (IAT) sensor will place its data on the network data bus allowing any control module on the network direct access to the information without the need for one control module (i.e., BCM) requesting the information from another control module (i.e., PCM). The BCM has direct access to information without having to request it from the PCM.

In 1996 the EPA specified that all vehicles be able to transmit generic scan tool data. However, proprietary data, any data other than P0 codes and data steams, were free to use any other protocol the manufacturer chose. The CAN PCM still transmits data to the DLC in SAE's generic scan tool protocol as specified by the EPA, for generic scan tool data communication, such as generic DTC's. But in order to access all the functions available you will

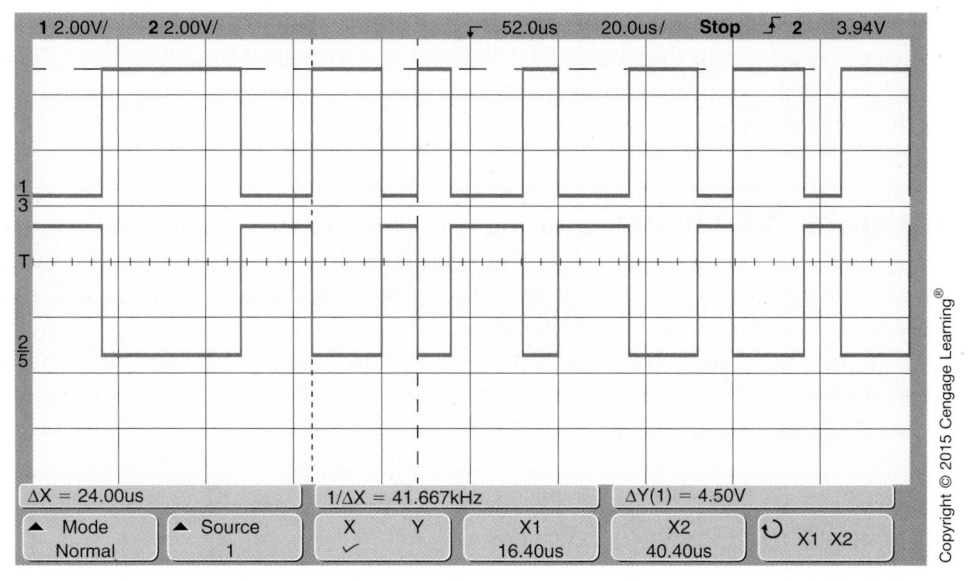

FIGURE 11-11 **Waveform of the CAN+ and the CAN– twisted pair data bus network wires.**

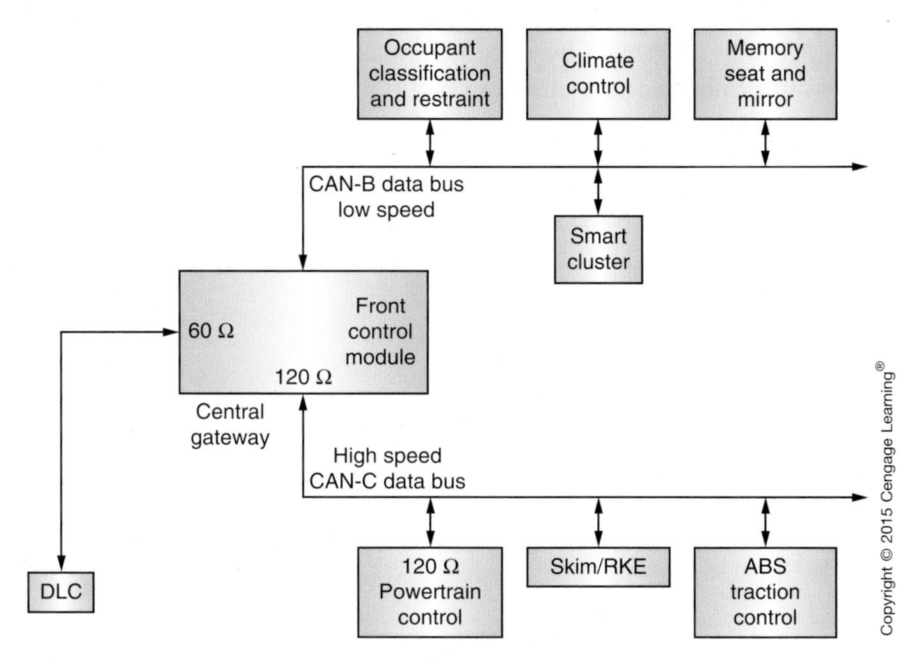

FIGURE 11-12 **An example of a typical CAN bus network.**

need to have a scan tool, which is compatible with CAN if that is the vehicles network operation system. The EPA emission regulations for the 2008 model year have specified CAN as the new scan tool communications protocol for all vehicles sold in the United States, providing the repair technician more data for trouble shooting emission failures. With CAN the industry finally has a single standard for on-board diagnostic communication.

The CAN protocol still allows access to the typical DTC information and data streams, but with enhanced DTC detail. A scan tool is also capable of bi-directional communication directly with a smart sensor or actuator node as well as other control modules on the network. In addition flash calibration for almost all nodes on the network will become common place. A smart sensor is capable of reporting the result of internal voltage drops, opens, grounds, and other self-test features. The network has the ability to take faulty sensors off line and can self diagnose the difference between a faulty device or circuit.

The EPA has required pass-thru flash programming using a standard PC connected to the Internet or using a data disk beginning in 2003 and required by 2008. The CAN is currently the standard communication protocol and scan tools are the interface device between the vehicle and the Internet or PC.

> **AUTHOR'S NOTE: When diagnosing suspension systems and other electronic systems on modern vehicles, the scan tool you are using must be compatible with the networks and computers on the vehicle being diagnosed. The scan tool must be able to receive and transmit data at the same speed as the network system. Scan tools must be updated so they are compatible with the network systems.**

PROGRAMMED RIDE CONTROL SYSTEM

System Design

The **programmed ride control (PRC) system** adjusts shock absorber and strut damping.

The **programmed ride control (PRC) system** is available on some Ford cars. Some import cars have a similar system. The damping action of the front and rear struts and shocks is automatically controlled by the PRC system to provide improved ride and handling characteristics under various driving conditions. The main components in the PRC system are (Figure 11-13):

1. Steering sensor
2. Brake sensor
3. Speed sensor
4. Struts and shocks with electric actuators
5. Control module
6. Powertrain control module (PCM)
7. Firm and plush shock relays
8. Mode select switch
9. Mode indicator light

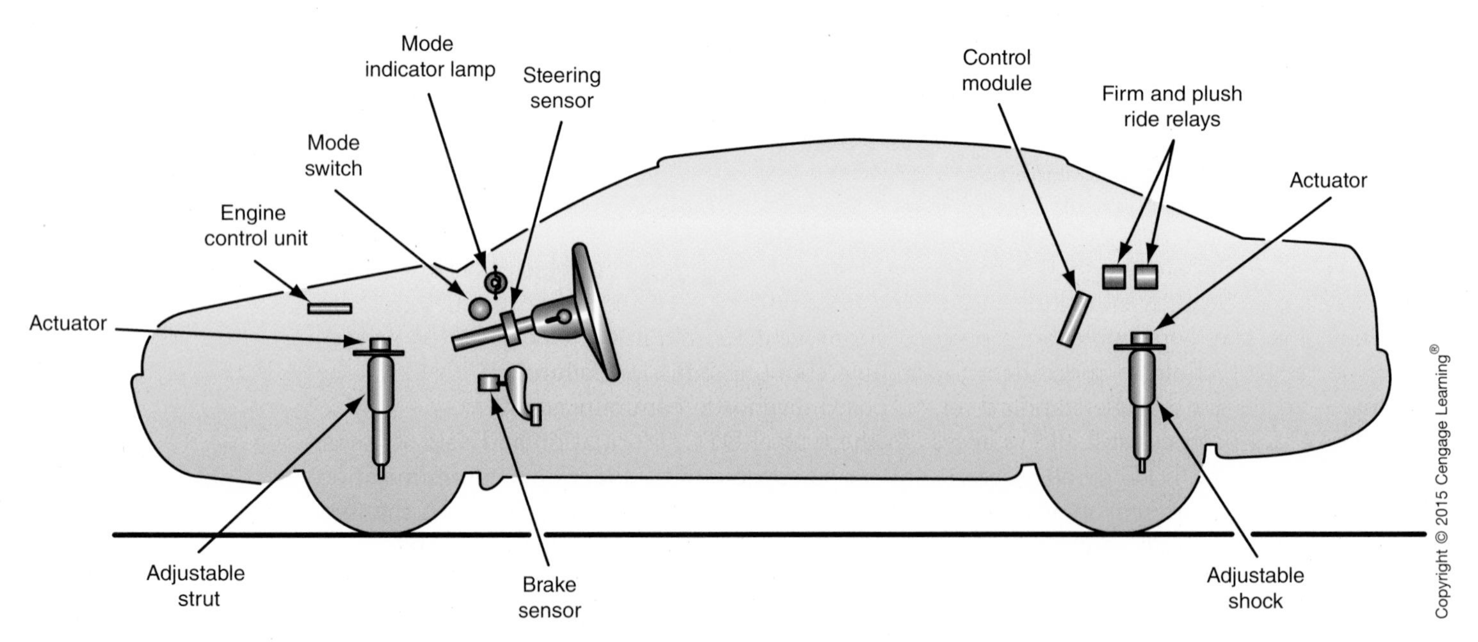

FIGURE 11-13 **Programmed ride control (PRC) system components.**

Steering Sensor

The **steering sensor** is mounted on the steering column. This sensor contains a pair of **light-emitting diodes (LEDs)** and a matching pair of **photo diodes**. A slotted disc attached to the steering shaft rotates between the LEDs and photo diodes when the steering wheel is turned (Figure 11-14). This disc contains 20 slots spaced at 9 degree intervals. A signal is sent from the steering sensor to the control module in relation to the amount and speed of steering wheel rotation.

Shop Manual
Chapter 11, page 442

Brake Sensor

The brake sensor is a normally open (NO) switch mounted in the brake control valve assembly (Figure 11-15). When the brake fluid pressure reaches 400 pounds per square inch (psi) or 2,758 kilopascals (kPa), the **brake pressure switch** closes and sends a signal to the control module.

The electronic engine control IV (EEC IV) or the electronic engine control V (EEC V) system refers to a computer system that controls many outputs, such as fuel injection and spark advance, on most Ford products.

Vehicle Speed Sensor

The vehicle speed sensor is usually mounted in the speedometer cable outlet of the transaxle or transmission (Figure 11-16). This sensor sends a vehicle speed signal to the control module. This signal is also used by the PCM in the EEC IV system.

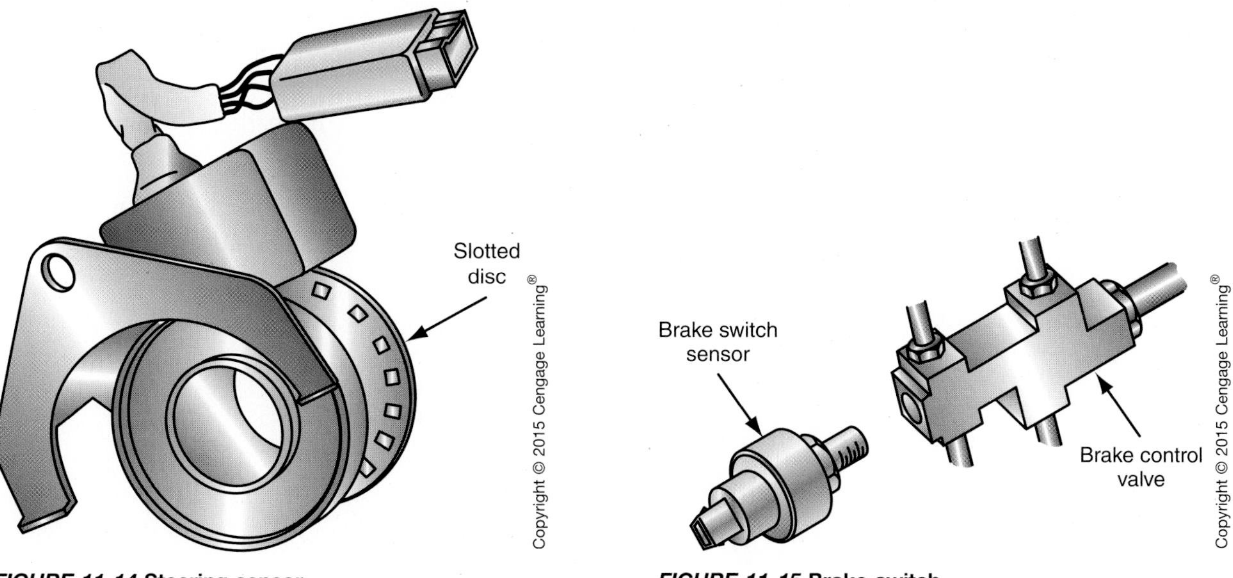

FIGURE 11-14 **Steering sensor.**

FIGURE 11-15 **Brake switch.**

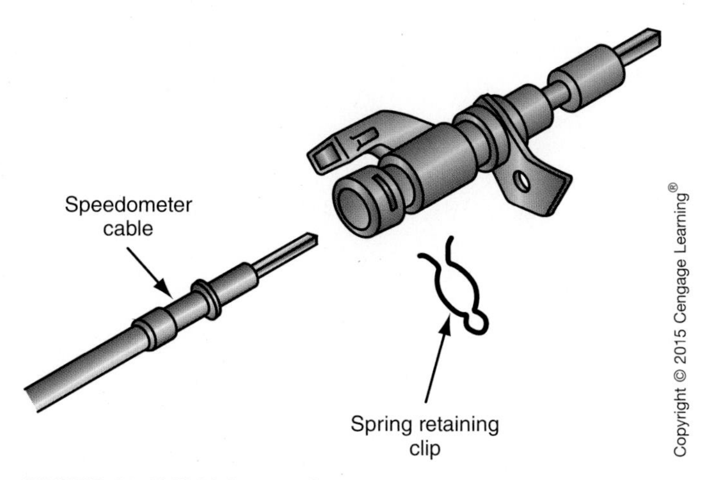

FIGURE 11-16 **Vehicle speed sensor.**

Strut and Shock Actuators

An **actuator** in a strut or shock absorber varies damping action when activated and deactivated.

An **actuator** is positioned in the top of each strut and shock (Figure 11-17). Each actuator contains a single pole armature, a pair of permanent magnets, and a position switch. When current is applied through the plush relay to the armature, the magnetic fields of the armature and the permanent magnets repel each other (Figure 11-18). This action causes clockwise armature rotation until the armature hits the internal stop. Under this condition, the leaf-spring switch is open in the position sensor circuit and no signal is returned to the PRC control module.

If current is applied through the firm relay to the armature, there is an attraction between the magnetic fields of the armature and permanent magnets. This attraction causes counterclockwise armature rotation until the armature contacts the internal stop. The armature rotates an internal strut or shock valve to restrict oil movement and provide increased suspension damping. In the firm position, the leaf-spring switch closes and sends a feedback signal to the control module. The armature movement is 60 degree and armature response time is 30 milliseconds (ms).

Operation

The **firm relay** energizes the strut actuators in the firm mode.

The **soft relay** supplies voltage to the strut actuators in the soft mode.

When the mode select switch is in the Auto position, the **firm relay** is de-energized and the **soft relay** is energized. Under this condition, current flows through the plush ride relay and the shock and strut actuators. This current is then routed to ground through the firm ride relay. If the vehicle is driven under normal speed and relatively straight-ahead conditions, this mode remains in operation.

The following conditions cause the PRC system to switch from the auto mode to the firm mode:

1. Vehicle speed above 83 miles per hour (mph) or 133 kilometers per hour (km/h)
2. Engine acceleration at 90 percent throttle opening or 8 psi (55 kPa) turbo boost pressure

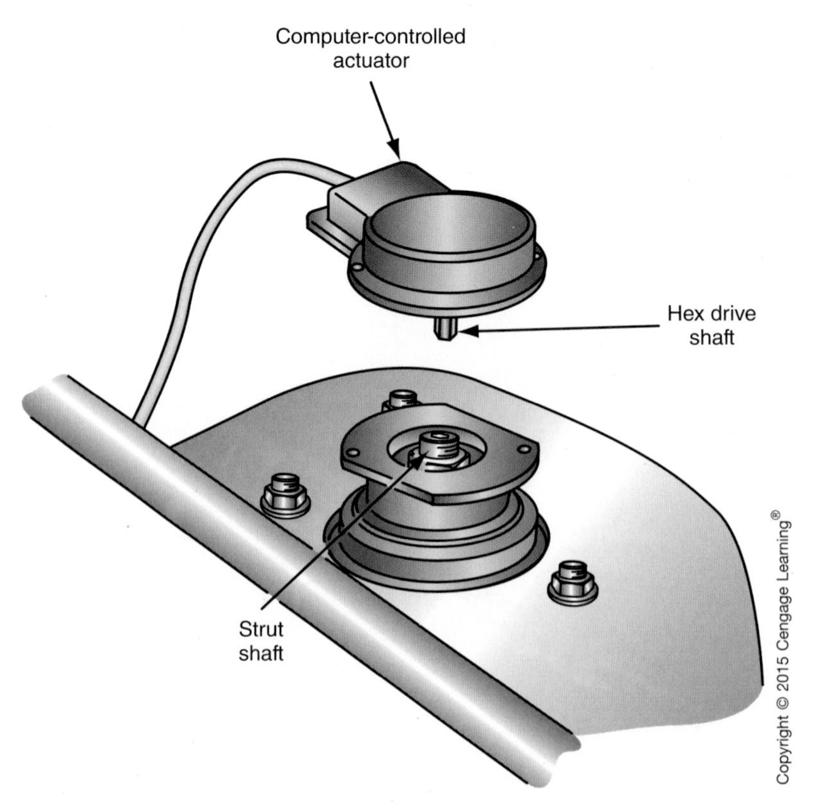

FIGURE 11-17 **Strut actuator.**

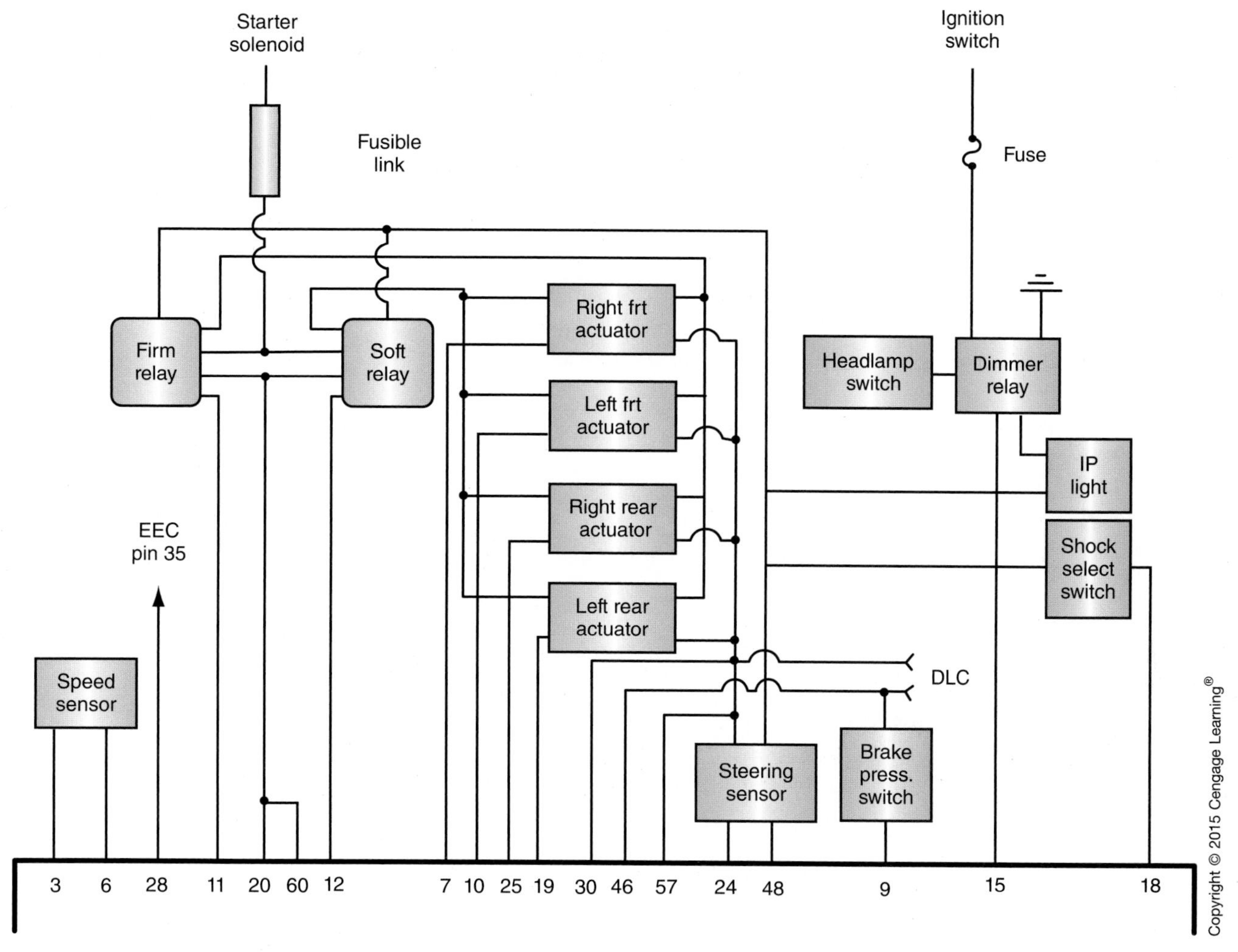

FIGURE 11-18 **Firm relay, strut actuators, and other PRC system components.**

3. Lateral vehicle acceleration above 0.35 g
4. Brake pressure of 400 psi (2,758 kPa) or more

The vehicle acceleration signal is sent from the **throttle position sensor (TPS)** to the PCM in the EEC IV system. This signal is relayed to the PRC control module. Lateral vehicle acceleration is sensed by the steering sensor.

When the PRC control module receives an input signal that requires firm ride control, the control module energizes the firm relay and de-energizes the soft relay. This action results in current flow from the firm ride relay through the shock and strut actuators, and the soft ride relay to ground. Therefore, shock and strut actuator current is reversed and the armature in each actuator moves the shock and strut valves to the Firm position.

The mode indicator light in the tachometer glows when the PRC system is in the Firm mode (Figure 11-19). During the first 80 seconds after the ignition switch is turned on, the PRC system does not respond to changes in vehicle direction. This action allows the PRC control module to calculate the straight-ahead position.

If the mode select switch is placed in the Firm position, the system remains in the firm mode continually. In this mode, the mode indicator light remains on. In the firm mode, the shocks and struts provide approximately three times the damping action on the extension stroke compared to the normal mode.

The **throttle position sensor (TPS)** is usually a potentiometer connected to the throttle shaft. When the throttle is opened, a movable contact slides around a rotary variable resistor, and this contact movement changes the sensor voltage signal in relation to throttle opening. The TPS signal informs the PCM in the EECV system regarding the amount of throttle opening.

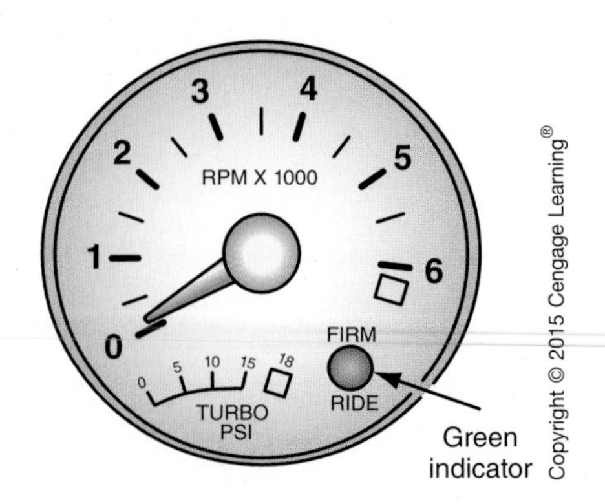

FIGURE 11-19 **Mode indicator light.**

Shop Manual
Chapter 11,
page 424

Air springs support the chassis weight in an air suspension system.

CAUTION:
The ball joint studs in the upper ball joints do not have a press fit in the steering knuckle. When loosening these ball joint nuts, a hex holding feature on the ball joint stud prevents the stud from rotating when loosening the ball joint nuts. If the upper ball joint stud rotates in the aluminum knuckle, the knuckle opening may be damaged.

Electronic Air Suspension System Components

Air Springs

In an air suspension system, the **air springs** replace the coil springs in conventional suspension systems. These air springs have a composite rubber and plastic membrane that is clamped to a piston located in the lower end of the spring. An end cap is clamped to the top of the membrane and an air spring valve is positioned in the end cap. The air springs are inflated or deflated to provide a constant vehicle trim height. Front air springs are mounted between the control arms and the crossmember (Figure 11-20). The lower end of these air springs is retained in the control arm with a clip, and the upper end is positioned in a crossmember spring seat. The front shock absorbers are mounted separately from the air springs.

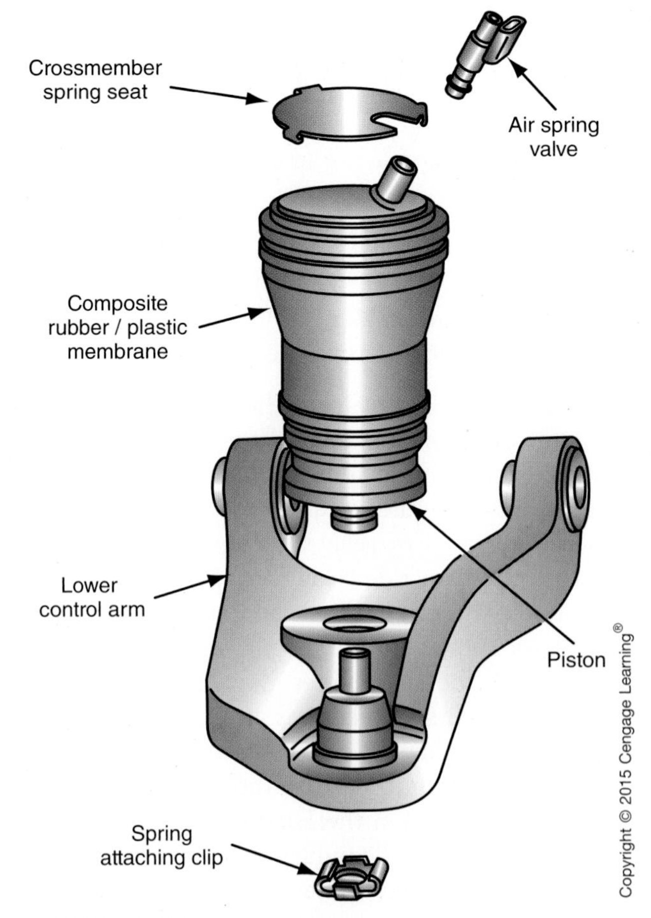

FIGURE 11-20 **Front air spring.**

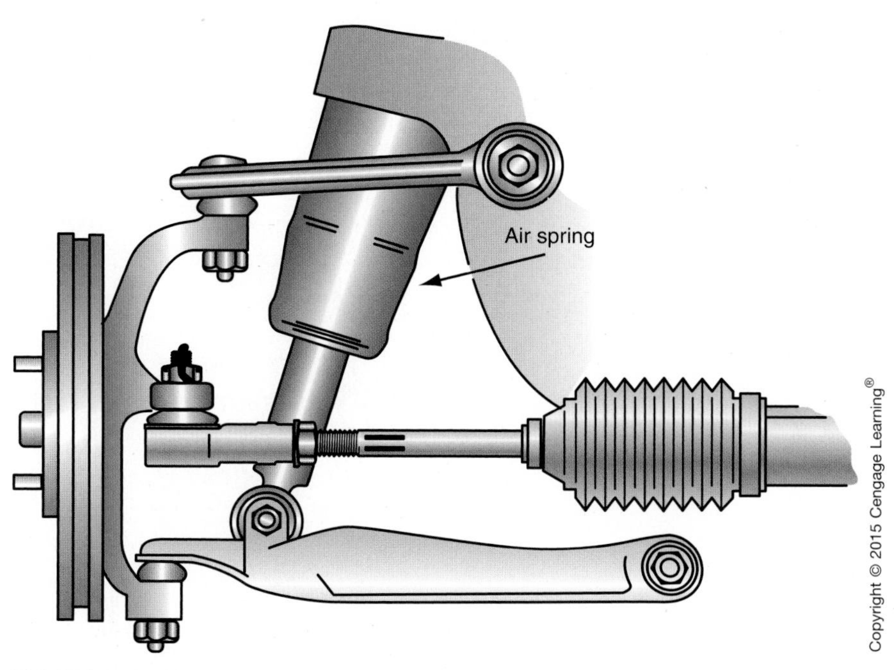

FIGURE 11-21 **Air spring mounted on the shock absorber.**

In some modern air suspension systems, the air springs are mounted and sealed on the shock absorbers (Figure 11-21). The lower end of the shock absorber is attached to the lower control arm through an insulating bushing, and the upper end of the shock absorber is attached to the chassis through an insulating mount. This type of air suspension system has aluminum front lower control arms and spindles, and forged steel upper control arms. Ball joints are mounted in the outer ends of the upper and lower control arms. The aluminum suspension components reduce the unsprung weight and improve ride quality. Reducing vehicle weight also improves fuel economy. The upper and lower ball joints are an integral part of the control arms, and these ball joints cannot be replaced separately.

Other vehicles have the air springs mounted over the front and rear struts, and these struts contain a solenoid actuator that varies the strut valve opening to control ride firmness (Figure 11-22 and Figure 11-23). These strut actuators are similar to the ones explained previously in this chapter on the PRC system. Some of these air suspension systems that control ride firmness are not driver adjustable. The suspension module controls the strut firmness automatically in relation to the module inputs. This type of system may be called an automatic air suspension system. Some vehicles with solenoid actuators in the struts have up to four suspension modes that may be selected by the driver. One premium luxury car has these driver selectable suspension modes:

1. Comfort—provides a smooth luxurious ride.
2. Automatic—the suspension computer provides the best possible combination of comfort and handling based on speed, driver style, and road conditions.
3. Dynamic—stiffest, lowest, sportiest, most aerodynamic suspension mode.
4. Life—for rougher roads, steep approaches, and deep snow. This mode is used only for low speed driving.

When the driver selects a suspension mode, the suspension module positions the strut actuators to provide the desired ride quality, and this module also adjusts the air spring pressure to provide the appropriate ride height.

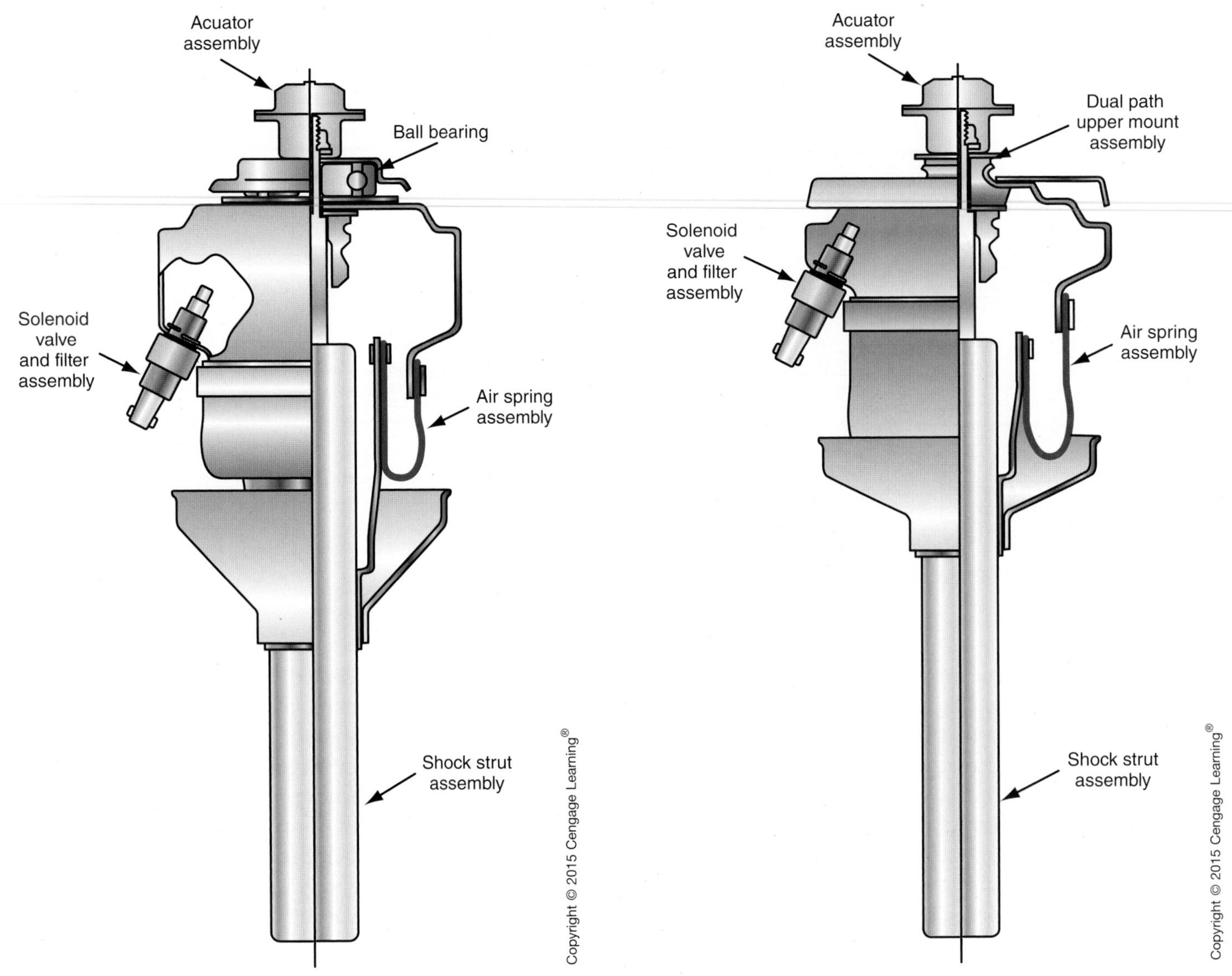

FIGURE 11-22 **Front air spring and strut assembly.**

FIGURE 11-23 **Rear air spring and strut assembly.**

This type of air suspension system dramatically reduce body roll and pitch that occurs during cornering and hard braking. Because this air suspension system lowers the ride height at higher speeds, it improves aerodynamic efficiency.

The rear air springs are similar to the front air springs and also have similar mountings. Some rear air springs are mounted between the rear suspension arms and the frame with the shock absorbers mounted separately from the air springs (Figure 11-24). Other rear air springs are mounted over the rear shock absorbers, and the shock absorbers are mounted between the lower control arms and the frame. Some modern air suspension systems on four-wheel-drive vehicles have rear knuckles with **cross-axis ball joints**. The lower cross-axis ball joint is a round insulating bushing mounted in the lower control arm, and a bolt attaches this bushing to the lower end of the knuckle. The upper cross-axis ball joint is a round insulating bushing mounted in the top of the knuckle, and a bolt attaches this bushing to the upper control arm (Figure 11-25). In this rear suspension system the upper and lower control arms are made from aluminum. An adjustable toe link is connected from the knuckle to the frame to provide a rear toe adjustment. Rear

Cross-axis ball joints contain large insulating bushings in place of typical ball joints.

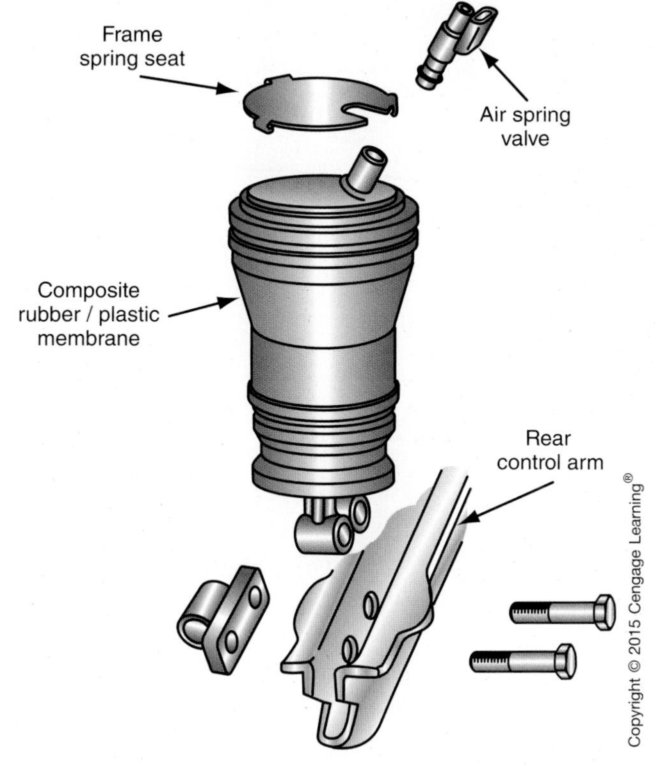

FIGURE 11-24 **Rear air spring mounting.**

wheel camber can be adjusted by installing a camber adjustment kit in place of the upper knuckle-to-control arm retaining bolt.

Air Spring Valves

An **air spring solenoid valve** allows air to flow into and out of an air spring.

An **air spring solenoid valve** is mounted in the top of each air spring (Figure 11-26). These valves are an electric solenoid-type valve that is normally closed. When the valve winding is energized, plunger movement opens the air passage to the air spring. Under this condition, air may enter or be exhausted from the air spring. Two O-ring seals are located on the end of the valves to seal them into the air spring cap. The valves are installed in the air spring cap with a two-stage rotating action similar to a radiator pressure cap.

Air Compressor

A single piston in the air compressor is moved up and down in the cylinder by a crankshaft and connecting rod (Figure 11-27). The armature is connected to the crankshaft, and therefore the rotating action of the armature moves the piston up and down. Armature rotation occurs when 12V are supplied to the compressor input terminal. Intake and discharge valves are located in the cylinder head. An air dryer that contains a silica gel is mounted on the compressor. This silica gel removes moisture from the air as it enters the system.

Nylon air lines are connected from the compressor outlets to the air spring valves. The compressor operates when it is necessary to force air into one or more air springs to restore the vehicle trim height.

An air **vent valve** is located in the compressor cylinder head (Figure 11-28). This normally closed electric solenoid valve allows air to be vented from the system. When it is necessary to exhaust air from an air spring, the air spring valve and vent valve must be energized at the same time with the compressor shut off. Air exhausting is necessary if the vehicle trim height is too high.

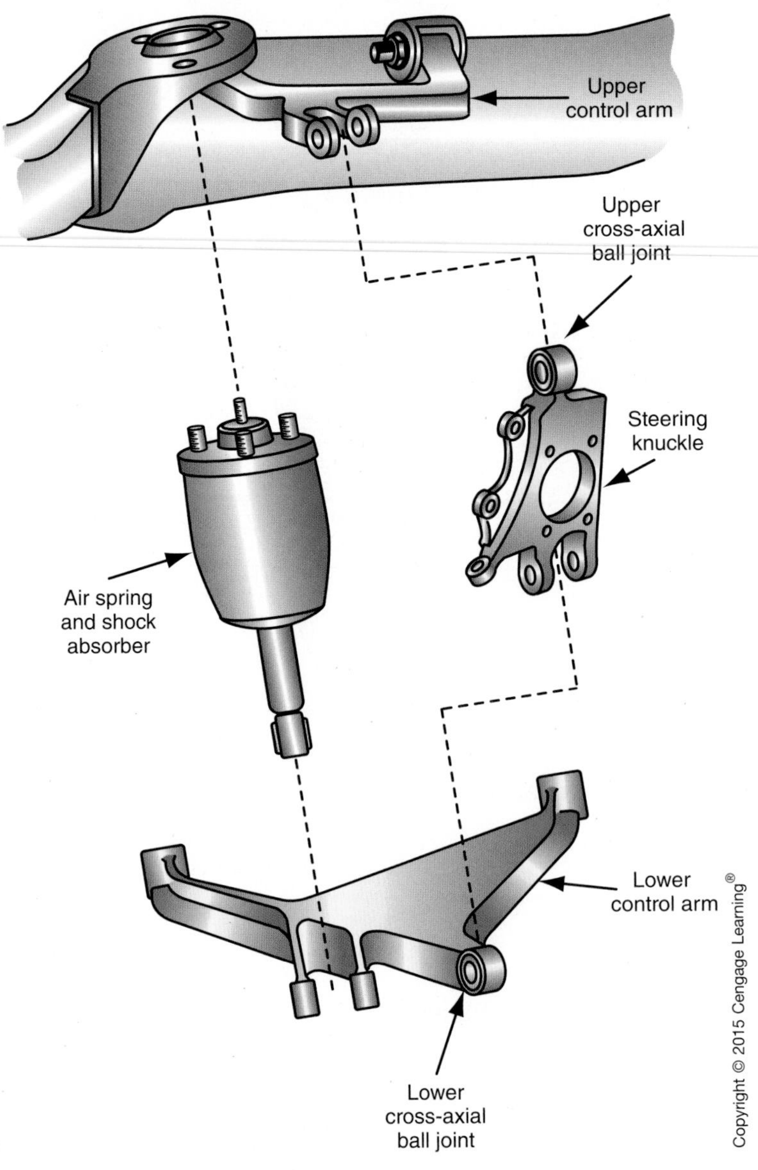

FIGURE 11-25 **Rear air suspension system with cross-axis ball joints and air spring mounted over the shock absorber.**

Compressor Relay

When the compressor relay is energized, it supplies 12V through the relay contacts to the compressor input terminal (Figure 11-29). The relay contacts open the circuit to the compressor if the relay is de-energized. An electronic relay is used in some air suspension systems.

Control Module

The control module is a microprocessor that operates the compressor, vent valve, and air spring valves to control the amount of air in the air springs and maintain the trim height. The control module is located in the trunk (Figure 11-30) on some models. On other models, the module is mounted under the instrument panel above the parking brake.

The control module turns on the suspension service indicator light in the roof panel or instrument panel to alert the driver when a suspension defect occurs. Diagnostic capabilities are designed into the suspension module. On some vehicles the suspension module is called a vehicle dynamics module (VDM). On many air suspension systems the suspension

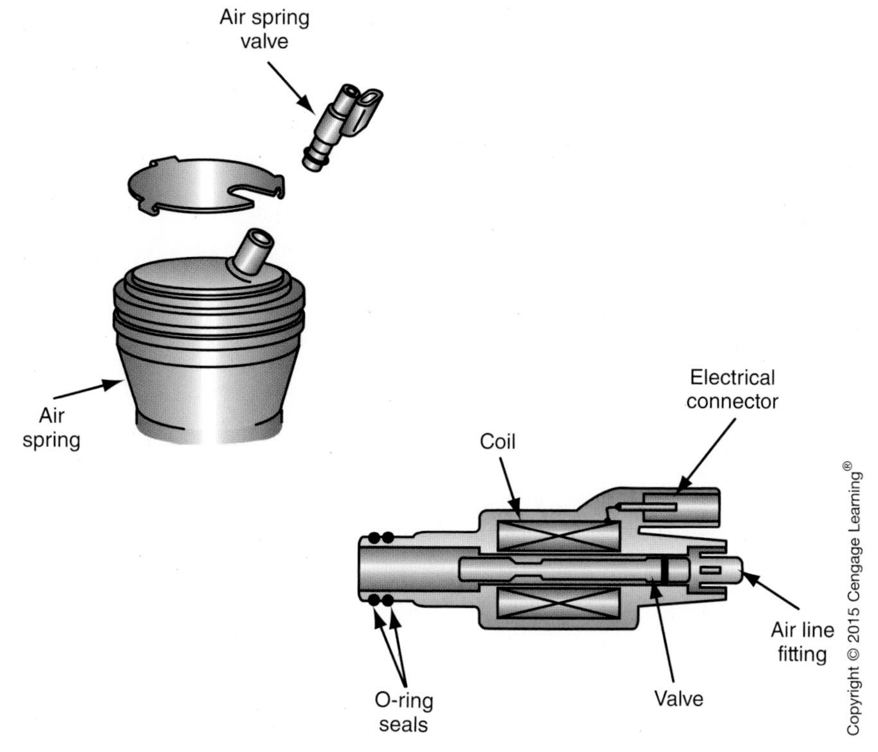

FIGURE 11-26 **Air spring valve.**

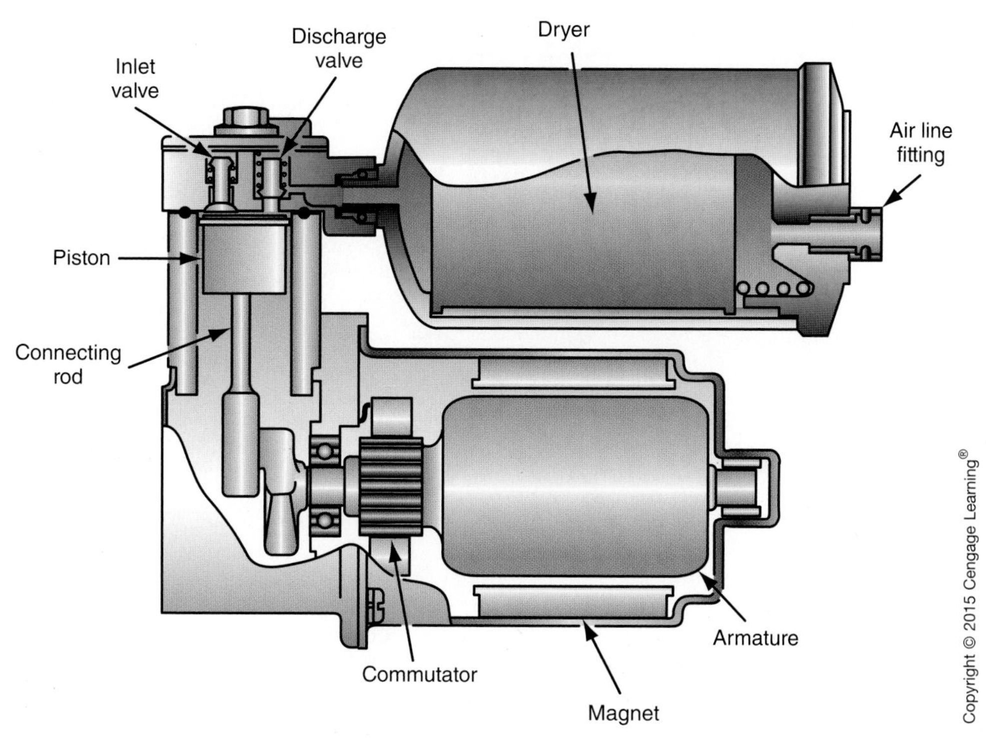

FIGURE 11-27 **Air compressor.**

module is interconnected via data links to some of the other on-board computers and the data link connector (DLC) under the dash. A scan tool may be connected to the DLC to diagnose the air suspension and other electronic systems on the vehicle. The data links allow data transmission between the on-board computers and the DLC. For example, the

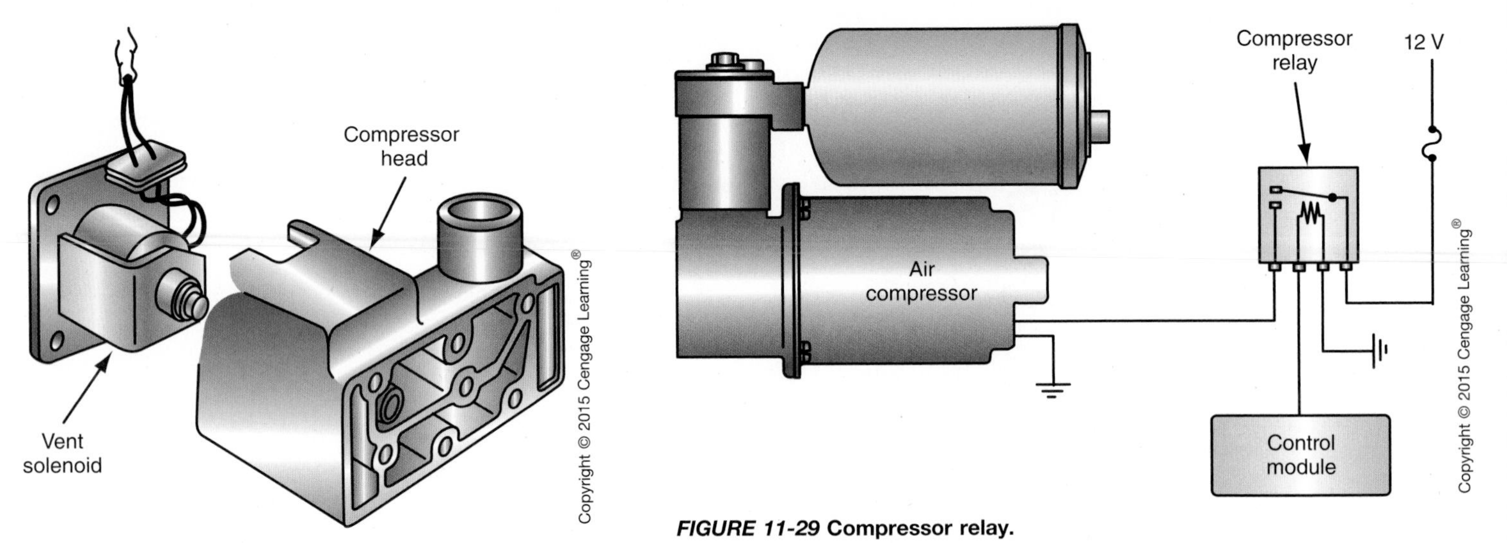

FIGURE 11-28 **Air vent valve.**

FIGURE 11-29 **Compressor relay.**

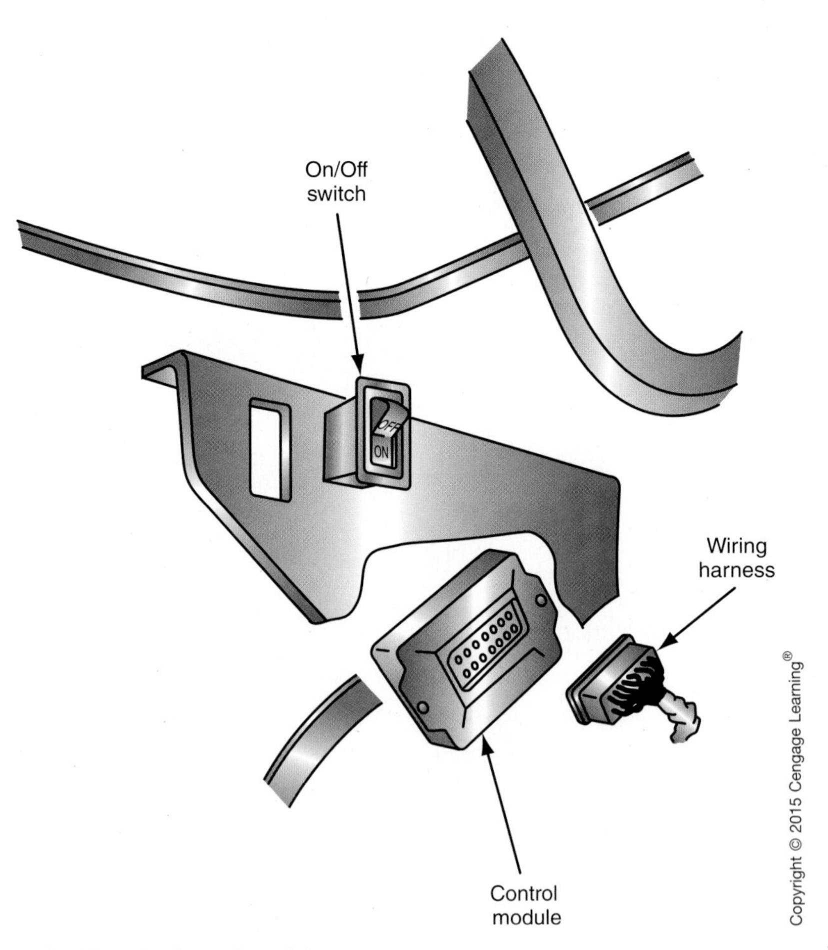

FIGURE 11-30 **Control module.**

vehicle speed sensor (VSS) signal may be sent through connecting wires to the powertrain control module (PCM) that controls engine functions. If the VSS signal is required by the suspension module, the PCM transmits the VSS signal through the data links. If an electronic defect occurs in a modern air suspension system, a diagnostic trouble code (DTC) is set in the suspension module memory. When a scan tool is connected to the DLC, the suspension module transmits the DTC through the data links to the DLC and scan tool.

On/Off Switch

The On/Off switch opens and closes the 12V supply circuit to the suspension module. This switch is located in the trunk (Figure 11-31). Depending on the vehicle make and model year, one or two panels in the trunk may have to be removed to access the On/Off switch. The On/Off switch must be turned off before the vehicle is hoisted, jacked, or towed. Certain air suspension service procedures may require this switch to be placed in the Off position.

WARNING: If the vehicle is hoisted, jacked, or towed with the electronic air suspension switch in the On position, personal injury or vehicle damage may occur.

Height Sensors

In the air suspension system, there are two front **height sensors** located between the lower control arms and the crossmember. A single rear height sensor is positioned between the suspension arm and the frame (Figure 11-32). Each height sensor contains a magnet slide that is attached to the upper end of the sensor. This magnet slide moves up and down in the lower sensor housing as changes in vehicle trim height occur (Figure 11-33). The lower sensor housing contains two electronic switches that are connected through a wiring harness to the control module.

When the vehicle is at **trim height**, the switches remain closed and the control module receives a trim height signal. If the magnet slide moves upward, the above trim switch opens and a **lower vehicle command** is sent from the height sensor to the module. When this signal is received by the module, it opens the appropriate air spring valve and the vent valve. This action exhausts air from the air spring and corrects the above trim height condition. Downward magnet slide movement closes the above trim switch and opens the below trim switch. If this action occurs, the height sensor sends a **raise vehicle command** to the module. When the control module receives this signal, it energizes the compressor relay and starts the compressor. The control module opens the appropriate air spring valve, and this action forces air into the air spring to correct the below trim height condition. The height sensors are serviced as a unit.

Some air suspension systems have **electronic rotary height sensors**. Each rotary height sensor contains a permanent magnet rotor and a **Hall element** (Figure 11-34). An

Height sensors send an electric signal to the control module in relation to curb riding height.

Trim height refers to the distance between the chassis and the road surface measured at a specific location recommended by the vehicle manufacturer.

Electronic rotary height sensors have an internal rotating element and these sensors send voltage signals to the control module in relation to the curb riding height.

A **Hall element** is an electronic device that produces a voltage signal when the magnetic field approaches or moves away from the element.

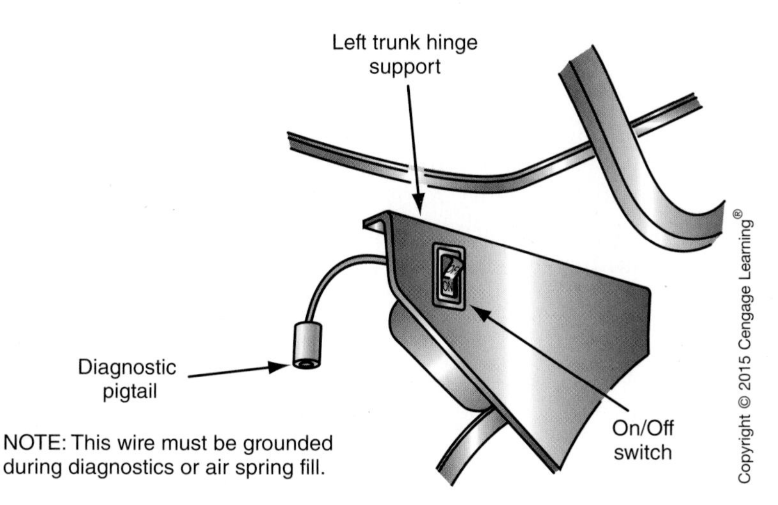

FIGURE 11-31 **On/Off switch.**

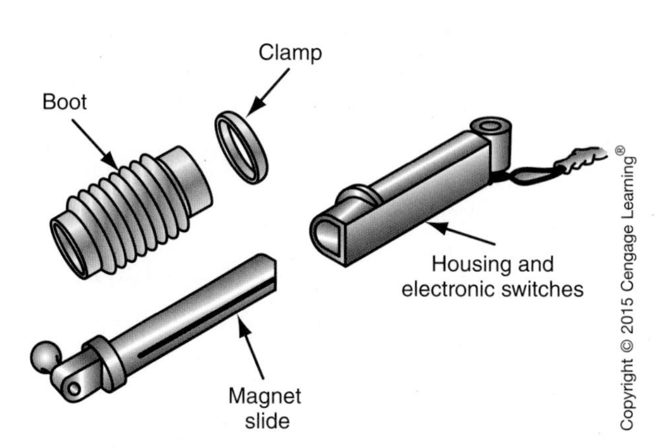

FIGURE 11-32 **Height sensor.**

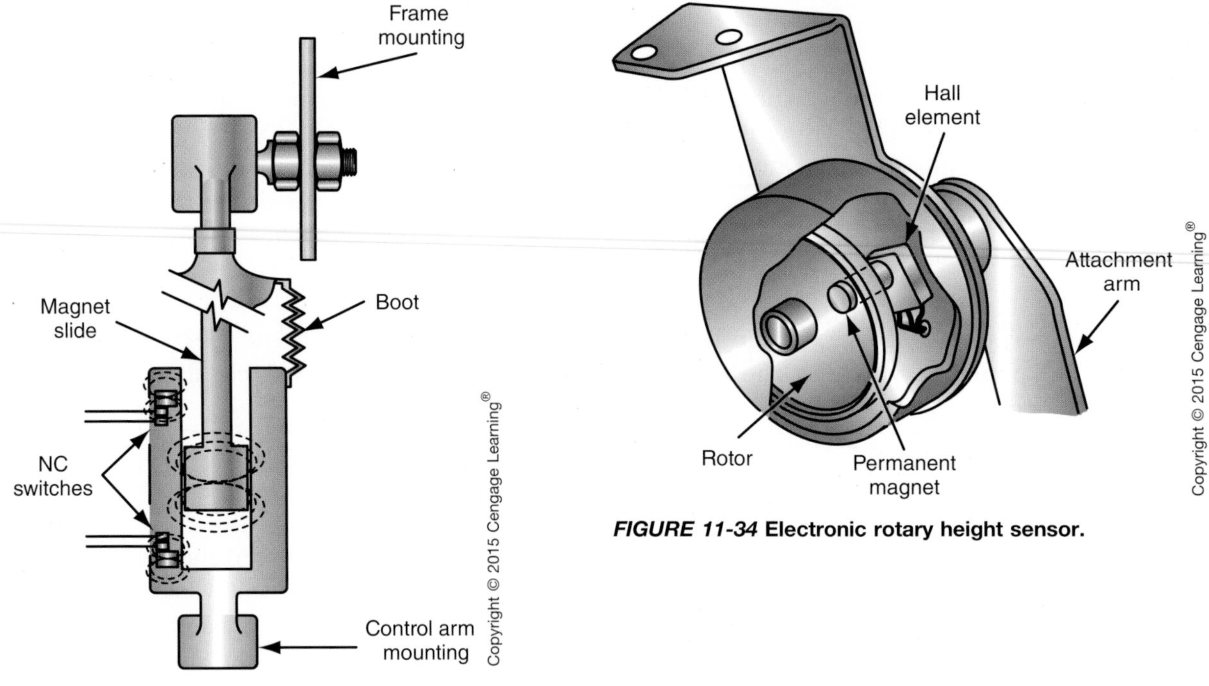

FIGURE 11-33 **Rear height sensor mounting.**

FIGURE 11-34 **Electronic rotary height sensor.**

FIGURE 11-35 **Height sensor mounting.**

CAUTION:
Never attempt to probe the electronic switches in slide-type height sensors. This action may damage the sensor.

arm on the height sensor is attached to the rotor. The height sensor body is mounted on the chassis and a linkage is connected from the sensor arm to the suspension (Figure 11-35). Suspension movement rotates the permanent magnet rotor and changes the voltage signal in the Hall element. The voltage signal from a rotary height sensor is proportional to trim height, above trim height, and below trim height.

Warning Lamp

When the control module senses a system defect, the module turns on the air suspension warning lamp in the roof console or instrument panel to inform the driver that a problem exists (Figure 11-36). If the air suspension system is working normally, the warning lamp will be on for 1 second when the ignition switch is turned from the Off to the Run position. After this time, the warning lamp should remain off. This lamp does not operate with

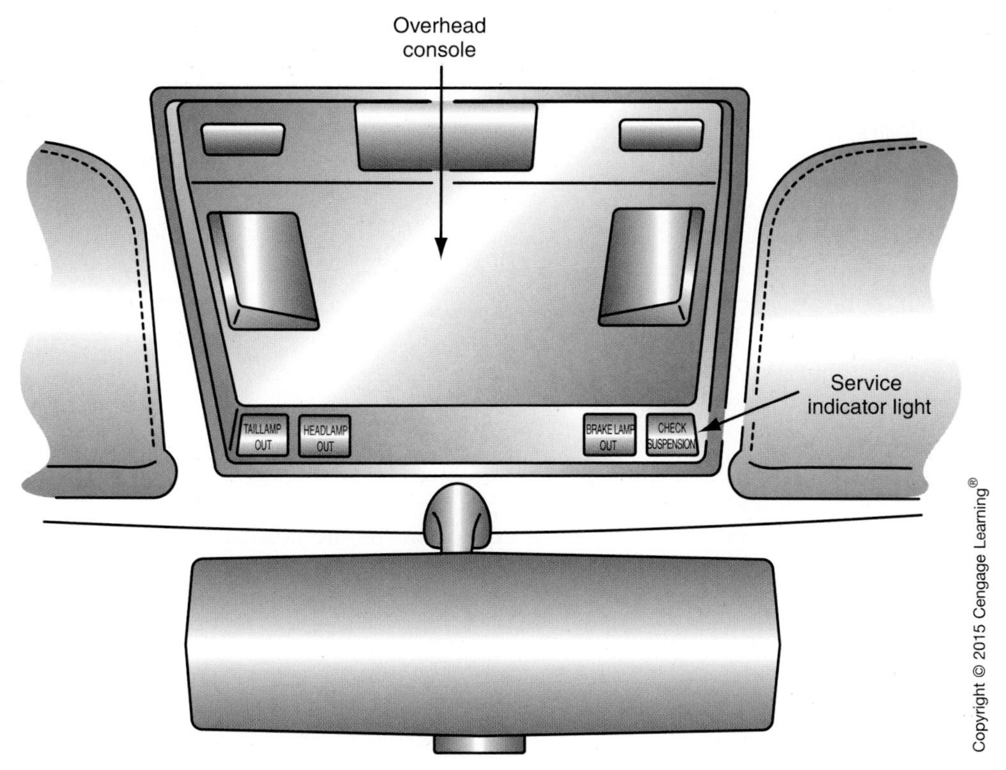

FIGURE 11-36 **Air suspension warning light.**

the ignition switch in the start position. The warning lamp is used during the self-diagnostic procedure and the spring fill sequence.

On some vehicles, the air suspension warning light is replaced with a CHECK SUSPENSION message in the instrument panel message center. The suspension module provides a CHECK SUSPENSION message if an electrical defect occurs in the air suspension system. An AIR SUSPENSION SWITCHED OFF message appears in the message center if the air suspension switch is in the Off position.

Electronic Air Suspension System Operation

The operation of an air suspension system varies depending on the vehicle make and model year. The following section discusses typical air suspension operation.

An electronically controlled suspension system may be called an active suspension system.

Ignition Switch Off

The electronic air suspension system is fully operational for one hour after the ignition switch is turned from the Run to the Off position. During this time, lower vehicle commands are completed unless a height sensor was providing a high signal when the ignition switch was turned off. After a 1-hour period, raise vehicle commands are acted upon and lower vehicle commands are ignored. The air compressor run time is limited to 15 seconds for rear springs and 30 seconds for front springs.

Ignition Switch in Run Position

When the ignition system has been in the Run position for less than 45 seconds, raise vehicle commands are completed immediately, but lower vehicle commands are ignored. If the ignition switch has been in the Run position for more than 45 seconds, the operation is as follows:

1. If a door is opened with the brake pedal released, raise vehicle commands are completed immediately, but lower vehicle commands are serviced after the door is closed. This action prevents an open door from catching on curbs or other objects.
2. If the doors are closed and the brake pedal is released, all commands are serviced by a 45-second averaging method to prevent excessive suspension height corrections on irregular road surfaces.
3. If the brake is applied and a door is open, raise vehicle commands are completed immediately, but lower vehicle commands are ignored.
4. When the doors are closed and the brake pedal is applied, all commands are ignored by the control module. If a command to raise the rear suspension is in progress under these conditions, this command will be completed. This action prevents correction of front end jounce while braking.

General Operation

When a height sensor sends a raise vehicle command to the control module and the other input signals are acceptable, the module grounds the compressor relay winding and starts the compressor. The module also opens the appropriate air spring valve to allow air flow into the air spring (Figure 11-37).

On non-computer-controlled suspension systems, the trim height may be called the curb riding height.

The rear air valve solenoids always operate together, but the front solenoids may be energized independently. When the correct chassis trim height is obtained, the control module opens the circuit from the compressor relay winding to ground and de-energizes the air spring valve. This action shuts off the compressor and traps the air in the air spring.

If a lower vehicle request is sent from a height sensor to the control module and the other input signals are acceptable, the control module opens the air vent valve and appropriate air spring valve. When this action occurs, air is released from the air spring until the correct trim height is obtained. The trim height signal from the height sensor to the module causes the module to close the air vent valve and the air spring valve.

Specific Control Module Operation

Shop Manual
Chapter 11, page 428

Commands are completed by the module in this order: rear up, front up, rear down, front down. When the ignition switch is in the Run position and a command cannot be completed within 3 minutes, the module turns on the air suspension warning lamp. This lamp remains on until the ignition switch is turned off. On some older models, all the control module memory is erased when the ignition switch is turned off. Therefore, the warning lamp may not indicate a defect immediately if the ignition switch is turned from the Off to the Run position. When a system defect causes the module to illuminate the warning lamp with the ignition in the Run position, other commands may be completed by the module. Commands from the front and rear height sensors are never completed simultaneously.

If an electrical defect occurs in a modern air suspension system, a diagnostic trouble code (DTC) is set and retained in the suspension module memory. In these systems, DTCs are transmitted via the data links to the data link connector (DLC) under the instrument panel, and the DTCs may be displayed on a scan tool connected to the DLC.

Air Suspension System Design Variations

Rear Load-Leveling Air Suspension System

Some vehicles have air suspension on only the rear wheels, and these systems may be referred to as rear load-leveling air suspension systems. These air suspension systems have basically the same air springs, compressor and relay, On/Off switch, and rear height sensor as the air suspensions described previously. If additional weight is placed in the trunk, the rear height sensor signals the suspension control module to raise the rear suspension

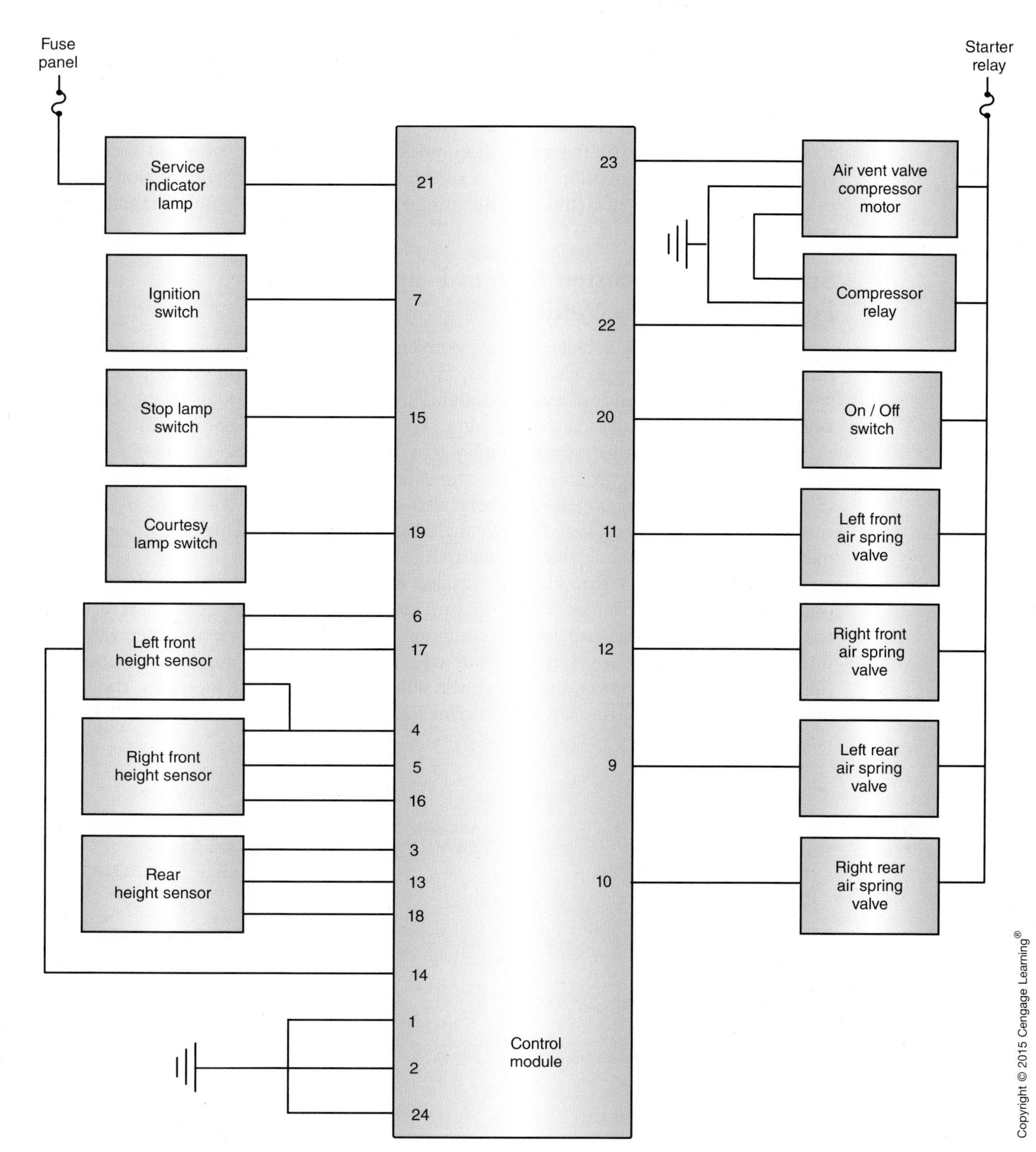

FIGURE 11-37 **Air suspension wiring diagram.**

height to the specified trim height. When the extra weight is removed from the trunk, the suspension height is higher than specified, and the air suspension system lowers the rear trim height to the specified height.

Air Suspension System with Speed Leveling Capabilities

Some vehicles have an air suspension system on the front and rear wheels, which is similar to the air suspension systems described earlier in this chapter. However, these systems have an input signal from the vehicle speed sensor (VSS) to the suspension module. On some vehicles the signal from the VSS is sent to the powertrain control module (PCM) that controls engine functions. The PCM transmits the VSS signal through the data links from the PCM to the suspension module. When the VSS signal indicates the vehicle is traveling above 65 mph (105 km/h), the suspension module opens the vent valve and spring solenoid valves to lower the suspension height a specific amount. Under this condition, the vehicle is more dynamically efficient and this improves high-speed vehicle stability and fuel economy.

Special Air Suspension System Features on Four-Wheel-Drive Vehicles

Some four-wheel-drive vehicles with a conventional front torsion bar suspension and a leaf-spring rear suspension also have an air suspension system on all four wheels. These front and rear suspension systems maintain the specified ride height during two-wheel-drive operation. The air suspension system has the same components as described previously, except this air suspension system has front and rear air shock absorbers in place of air springs. An air solenoid is connected in the air line to each air shock absorber (Figure 11-38). The air shock absorbers also contain electrical solenoids that rotate the shock absorber valves and vary the ride firmness (Figure 11-39). This type of air suspension may be called an automatic ride control (ARC) system.

When driving on a smooth road surface in two-wheel drive, the suspension module positions the shock absorber valves to provide a soft ride. If the vehicle is driven on a rough road surface, the suspension's computer switches the shock absorber valves to the firm mode. When the driver selects four-wheel high mode, the suspension module starts the compressor and opens the air shock absorber valves, and increases the air shock absorber pressure until the suspension trim height is raised 1 in. (25.4 mm). If the vehicle

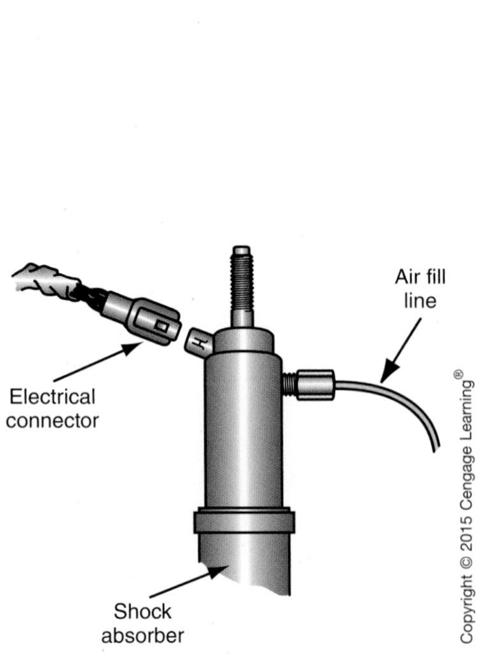

FIGURE 11-38 **Electrical and air fill line connections to the shock absorber.**

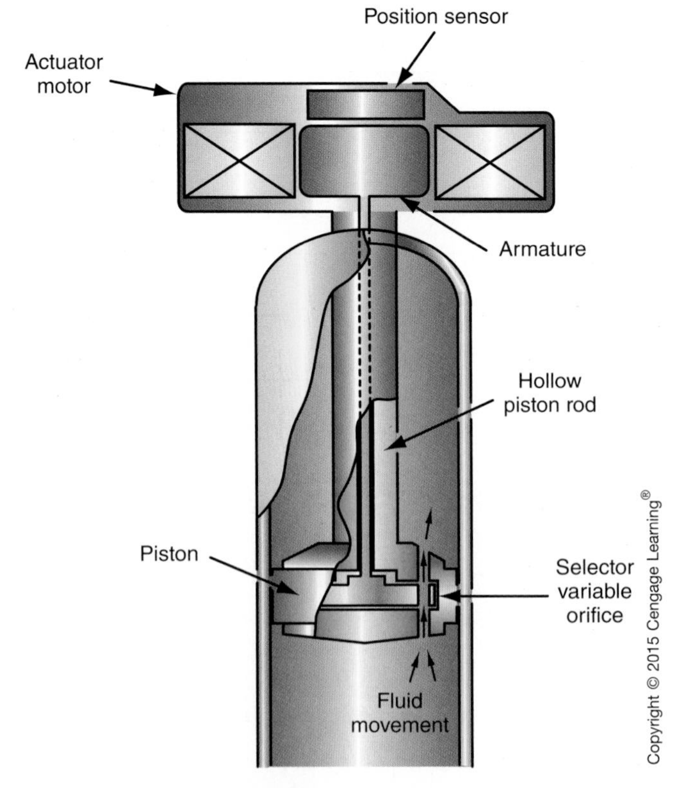

FIGURE 11-39 **Strut actuator.**

speed exceeds 58 mph (93 km/h), the module returns the suspension height to the specified trim height. If the driver selects the four-wheel low mode, the suspension module operates the compressor and air shock absorber valves to increase the trim height 2 in. (50.8 mm). If the vehicle speed exceeds 30 mph (48 km/h in the four-wheel low mode), the module adjusts the suspension height to 1 in. (25.4 mm) above the specified trim height.

Vehicle Dynamic Suspension System

Shop Manual
Chapter 11, page 432

Some SUVs are equipped with a **vehicle dynamic suspension (VDS)** system that is similar to the air suspension systems described previously in this chapter. The VDS system has these components:

1. Off/On service switch
2. Two front height sensors
3. One rear height sensor
4. Compressor with internal vent solenoid and air dryer
5. Control module
6. Air lines
7. Front and rear combined air springs and shock absorbers
8. Four air spring solenoids
9. Compressor relay

When increased air pressure is required in an air spring, the control module closes the compressor relay, starts the compressor, and opens the solenoid on the appropriate air spring. To vent air from an air spring, the control module must energize the air spring solenoid and the vent valve. The VDS system has three operating modes:

1. The kneel mode is provided by the suspension module when the ignition switch is in the Off or Lock position and all doors, liftgate, and liftgate glass are closed. In this mode the module opens the vent valve and air spring valves, and slowly reduces the suspension height to 1 in. (25.4 mm) below the specified trim height. This mode improves the ease of entering and exiting the vehicle.
2. When the ignition switch is On, and the transmission is initially shifted into Drive or Reverse, and all the doors, liftgate, and liftgate glass are closed, the VDS switches from the kneel mode to the trim mode, which provides the normal trim height. The VDS system also switches to the trim height mode if the module detects a vehicle speed above 15 mph (24 km/h). Transitions between modes require approximately 30 to 45 seconds.
3. On four-wheel-drive models the off-road height mode is provided when the driver selects the four-wheel low mode and the vehicle speed is less than 25 mph (40 km/h). In this mode, the module starts the compressor and opens the air spring valve to increase the air spring pressure until the suspension height is 1 in. (25.4 mm) above the specified trim height.

The VDS system also maintains the specified trim height in relation to the weight placed in the vehicle. If a heavy load is placed in the rear of the vehicle, the rear height sensors transmit low trim height signals to the module, and the module opens the rear air spring solenoids and starts the compressor to restore the proper rear suspension trim height. The system stores front and rear trim height when a door or the rear liftgate is opened. The module maintains this suspension height even if weight is added to or removed from the vehicle. When all the doors, liftgate, and liftgate glass are closed, the system returns to normal operation. The VDS system makes limited height adjustments for 40 minutes after the ignition switch is turned Off.

The powertrain control module (PCM) is a computer that controls such output functions as electronic fuel injection, spark advance, emission devices, and trans-axle shifting.

Electronic Suspension Control (ESC) System

General Description

The **electronic suspension control (ESC) system** controls damping forces in the front struts and rear shock absorbers in response to various road and driving conditions. The ESC system changes shock and strut damping forces in 1 to 12 milliseconds, whereas other suspension damping systems require a much longer time interval to change damping forces. It requires about 200 milliseconds to blink your eye. This gives us some idea how quickly the ESC system reacts. On some older models the ESC system may be called a continuously variable road sensing suspension (CVRSS).

The ESC module receives inputs regarding vertical acceleration, wheel-to-body position, speed of wheel movement, vehicle speed, and lift/dive (Figure 11-40). The CVRSS module evaluates these inputs and controls a solenoid in each shock or strut to provide suspension damping control. The solenoids in the shocks and struts can react much faster compared with the strut actuators explained previously in some systems.

The ALC system is similar to the rear load-leveling suspension system explained previously in this chapter.

The ESC module also controls the speed-dependent steering system called MagnaSteer® and the automatic level control (ALC). This MagnaSteer® system is similar to the electronic variable orifice (EVO) steering explained in Chapter 13 under conventional and electronic rack-and-pinion steering gears. The ALC system controls the air pressure in the rear air shock absorbers to maintain the proper rear suspension height.

Inputs

Position Sensors. A **wheel position sensor** is mounted at each corner of the vehicle between a control arm and the chassis (Figure 11-41 and Figure 11-42). These sensor inputs provide analog voltage signals to the ESC module regarding relative wheel-to-body movement and the velocity of wheel movement (Figure 11-43). The rear position sensor inputs also provide rear suspension height information to the ESC module, and this information is used by the module to control the rear suspension trim height. All four position sensors have the same design. The wheel position sensors may be linear-type or rotary-type.

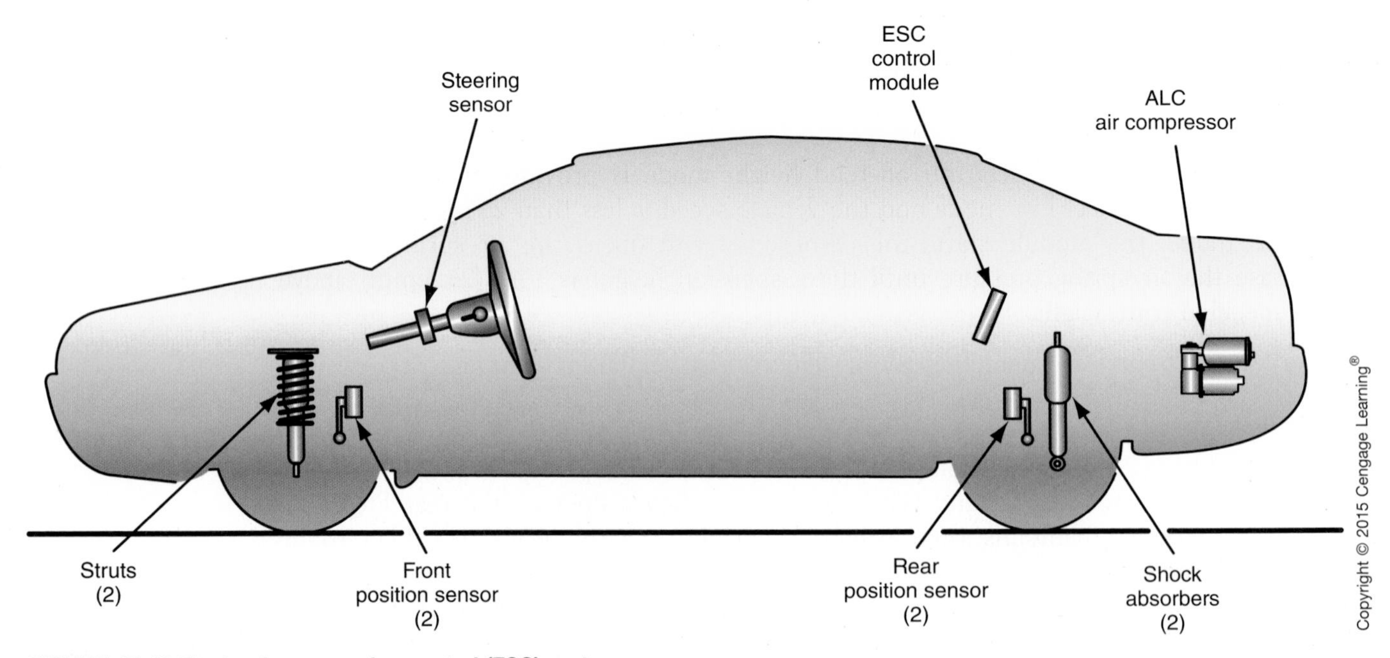

FIGURE 11-40 **Electronic suspension control (ESC) system.**

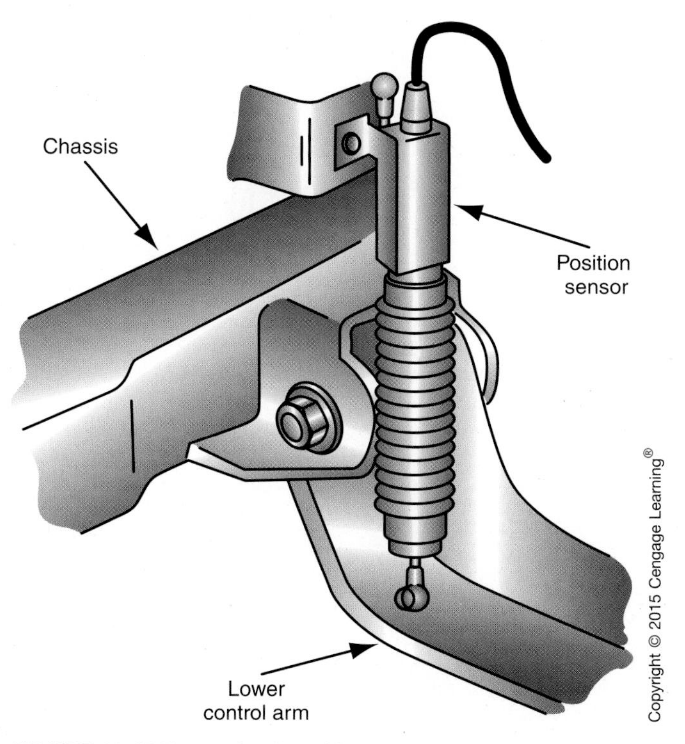

FIGURE 11-41 **Front wheel position sensor.**

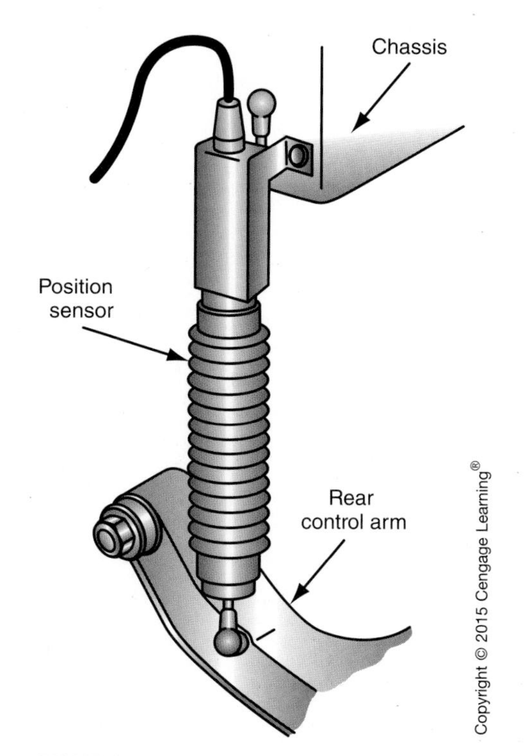

FIGURE 11-42 **Rear wheel position sensor.**

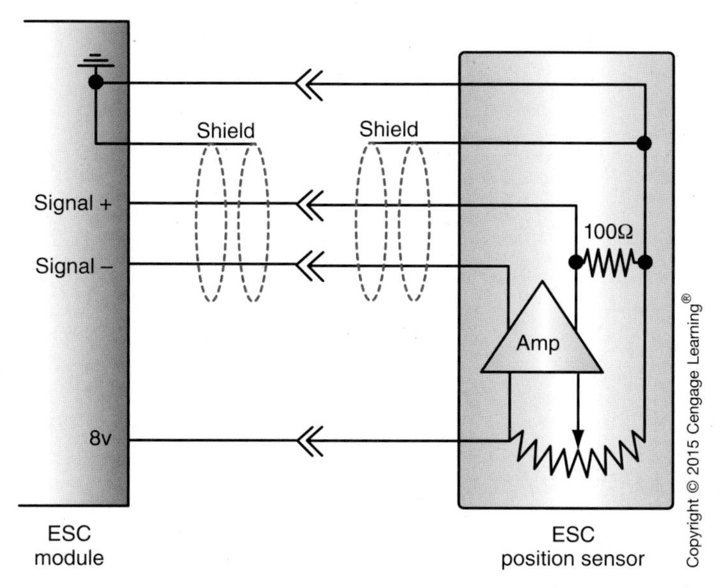

FIGURE 11-43 **Position sensor internal design and wiring diagram.**

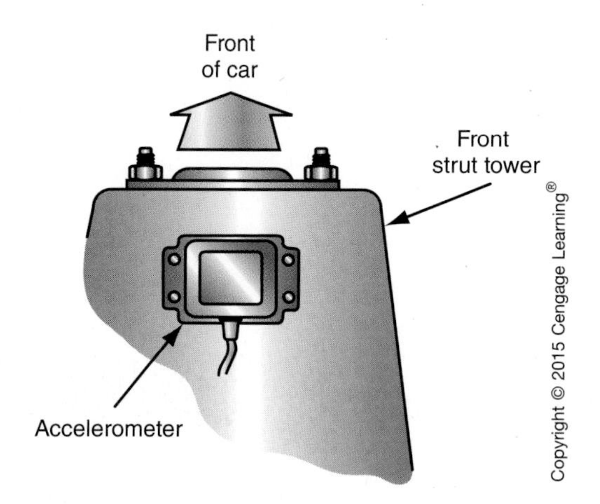

FIGURE 11-44 **Front accelerometer mounting location.**

Accelerometer. An **accelerometer** is mounted on each corner of the vehicle. These inputs send information to the ESC module in relation to vertical acceleration of the body. The front accelerometers are mounted on the strut towers (Figure 11-44), and the rear accelerometers are located on the rear chassis near the rear suspension support (Figure 11-45). All four accelerometers are similar in design, and they send analog voltage signals to the ESC module (Figure 11-46). On some later model vehicles, the four accelerometers are replaced by a single accelerometer under the driver's seat. On other late-model vehicles the accelerometer(s) are eliminated.

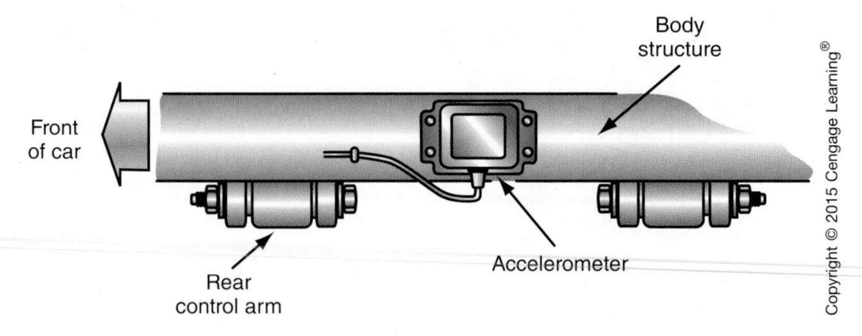

FIGURE 11-45 **Rear accelerometer position.**

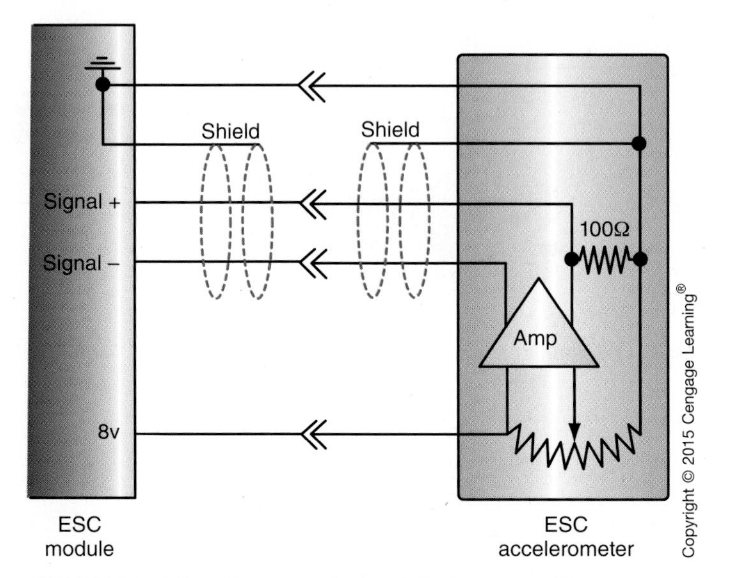

FIGURE 11-46 **Accelerometer internal design and wiring diagram.**

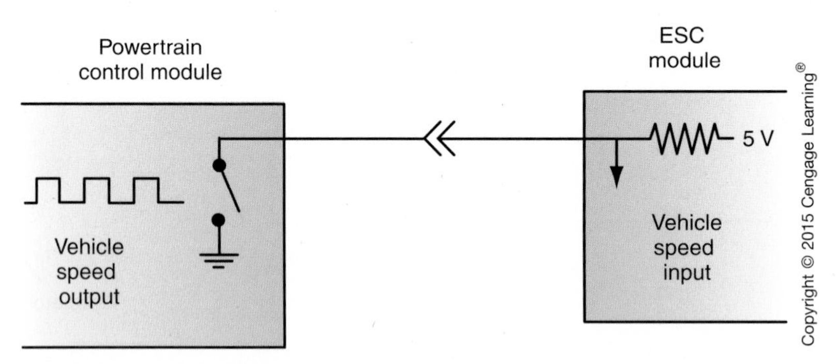

FIGURE 11-47 **The vehicle speed sensor (VSS) signal is sent to the powertrain control module (PCM) and transmitted to the ESC module.**

Vehicle Speed Sensor. The **vehicle speed sensor (VSS)** is mounted in the transaxle. This sensor sends a voltage signal to the powertrain control module (PCM) in relation to vehicle speed (Figure 11-47). The VSS signal is transmitted via data links from the PCM to the ESC module.

Lift/Dive Input. The **lift/dive input** is sent from the PCM to the ESC module (Figure 11-48). Suspension lift information is obtained by the PCM from the throttle position, vehicle speed, and transaxle gear input signals. The PCM calculates suspension dive information from the rate of vehicle speed change when decelerating.

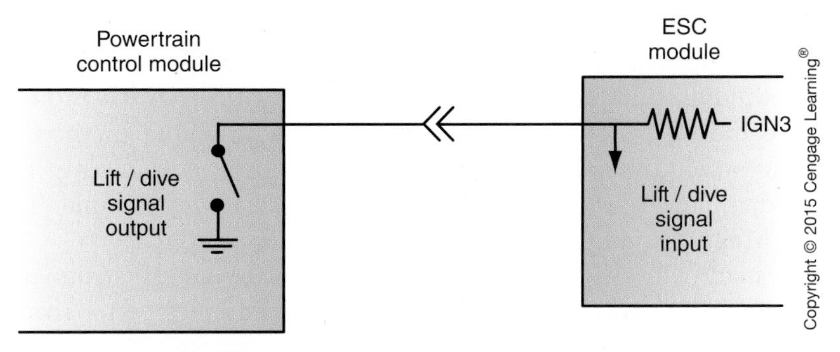

FIGURE 11-48 **The lift-dive signal is sent from the powertrain control module (PCM) to the ESC module.**

Electronic Suspension Control Module

Electronic suspension control module contains three microprocessors that control the ESC, MagnaSteer® system, and automatic level control (ELC). The ESC module is mounted on the right side of the electronics bay in the trunk. Extensive self-diagnostic capabilities are programmed into the ESC module.

Outputs

Shock Absorbers and Struts with Magneto-Rheological (MR) Fluid. The shock absorbers or struts in the ESC system contain magneto-rheological (MR) fluid. This fluid is a synthetic fluid containing suspended iron particles. An electric winding is mounted in each shock absorber housing, and the ends of each shock absorber winding are connected to the ESC module. When the shock absorber windings are not energized by the ESC module, the iron particles in the MR fluid are dispersed randomly. Under this condition the MR fluid has a mineral oil–like consistency, and this fluid flows easily through the shock absorber orifices to provide soft ride quality.

When the ESC module energizes the shock absorber windings, the magnetic field around the winding aligns the iron particles in the MR fluid into thick fibrous structures. In this condition the MR fluid has a jelly-like consistency that does not flow easily through the shock absorber orifices (Figure 11-49). This fluid change provides firm ride quality. When a shock absorber coil is energized, the amount of attraction between the fibrous particles is proportional to the magnetic field strength surrounding the shock absorber winding, and this field strength is controlled by the current flow through the winding. The computer provides very precise control of the current flow through each shock absorber

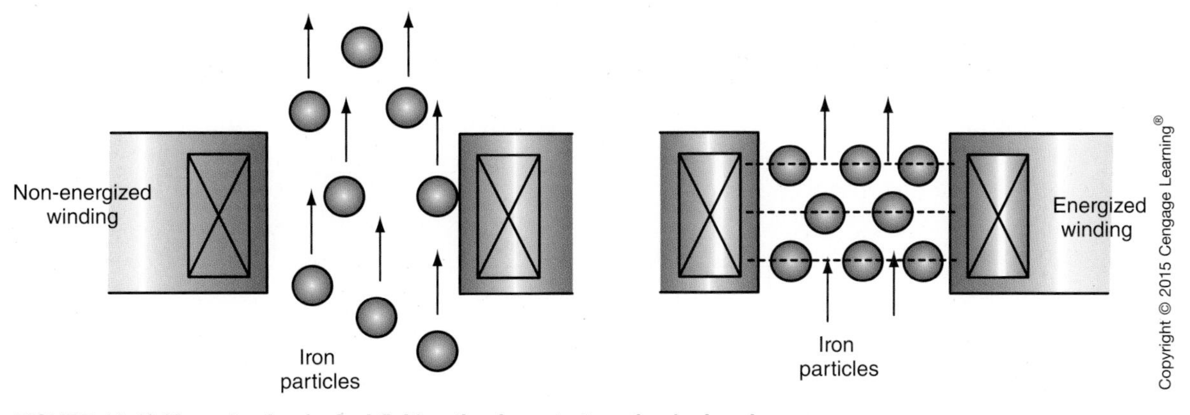

FIGURE 11-49 **Magneto-rheological fluid action in a strut or shock absorber.**

or strut winding to supply a very broad shock absorber and strut damping range. Based on the wheel position sensor and other inputs, the ESC module can energize each individual shock absorber winding many times per second with a **pulse width modulated (PWM)** signal. Therefore the ESC system with magneto-rheological fluid in the shock absorbers or struts provides an almost infinite variation in shock absorber damping. The ESC system can change the damping characteristics of the MR fluid in 1 millisecond. The ESC system may be called a **MagneRide system**.

A **pulse width modulated** signal is a signal with a variable on time that may be used by a computer to control an output.

A **MagneRide** suspension system has magneto-rheological fluid in the shock absorbers.

The ESC system with MR fluid in the shock absorbers or struts provides greatly improved control of pitch and body roll motions, which supplies better road-holding capabilities, steering control, ride quality, and safety. Recent updates to the ESC system with MR fluid in the shock absorbers or struts include the following:

1. Improved computer software to provide a broader damping range and more precise control of the shock actuators.
2. Expanded sensor inputs including brake pressure, vehicle yaw rate, steering angle, and engine torque.
3. Improved computer networks to allow increased and faster data transmission.

These updates provide improved vehicle dynamics and ride quality without any increase in packaging or weight.

Damper Solenoid Valves. On some older models each strut or shock damper contains a solenoid that is controlled by the ESC module. Each **damper solenoid valve** provides a wide range of damping forces between soft and firm levels. Strut or shock absorber damping is controlled by the amount of current supplied to the damper solenoid in each strut or shock absorber. Battery voltage and ignition voltage are supplied through separate fuses to the ESC module (Figure 11-50). If the damper solenoids are not energized, the struts provide minimum damping force. When the damper solenoids are energized, the struts provide increased damping force for a firmer ride. The ESC module switches the voltage supplied to the damper solenoid in each strut on and off very quickly with a 2 kilohertz pulse width modulated (PWM) action. If the ESC module keeps the damper solenoid in a

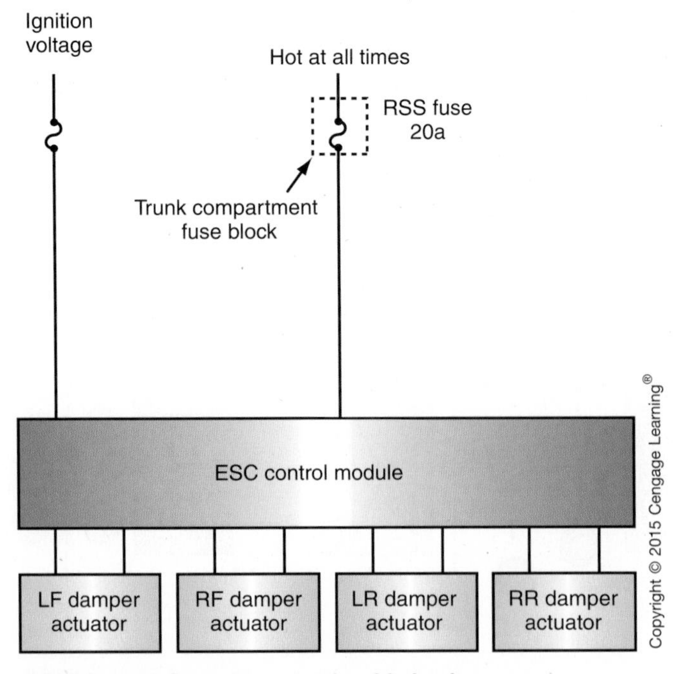

FIGURE 11-50 **Strut damper solenoid circuit.**

strut energized longer on each cycle, current flow is increased through the strut damper solenoid. Under this condition, strut damping force is increased to provide a firmer ride. The ESC module provides precise, variable control of the current flow through each strut or shock damper solenoid to achieve a wide range of damping forces in the struts. The ESC system can change the shock absorber damping forces in 10 to 12 milliseconds.

Each damper solenoid is an integral part of the damper assembly and is not serviced separately. The ESC system operates automatically without any driver-controlled inputs. The fast reaction time of the ESC system provides excellent control over ride quality and body lift or dive, which provides improved vehicle stability and handling. Since the position sensors actually sense the velocity of upward and downward wheel movements and the damper solenoid reaction time is 11 to 12 milliseconds, the ESC module can react to these position sensor inputs very quickly. For example, if a road irregularity drives a wheel upward, the ESC module switches the damper solenoid to the firm mode before that wheel strikes the road again during the downward movement.

Resistor Module. In some older models, the resistor module contains four resistors encapsulated in a ceramic material. This resistor module is mounted in the right rear quarter panel inside the trunk (Figure 11-51).

When the ESC module switches a damper solenoid on, the module provides a direct ground for the solenoid, and full voltage is dropped across the solenoid winding to energize the solenoid very quickly. Under this condition, a higher current flow is supplied through the damper solenoid winding and the ESC module to ground. Since it is undesirable to maintain this higher current flow through the damper solenoid for any longer than necessary, the ESC module switches a resistor in the resistor module into the damper solenoid circuit after this circuit is energized for 15 milliseconds. On later model vehicles, the resistor module is discontinued because the ESC module controls the strut damper solenoids with a PWM signal.

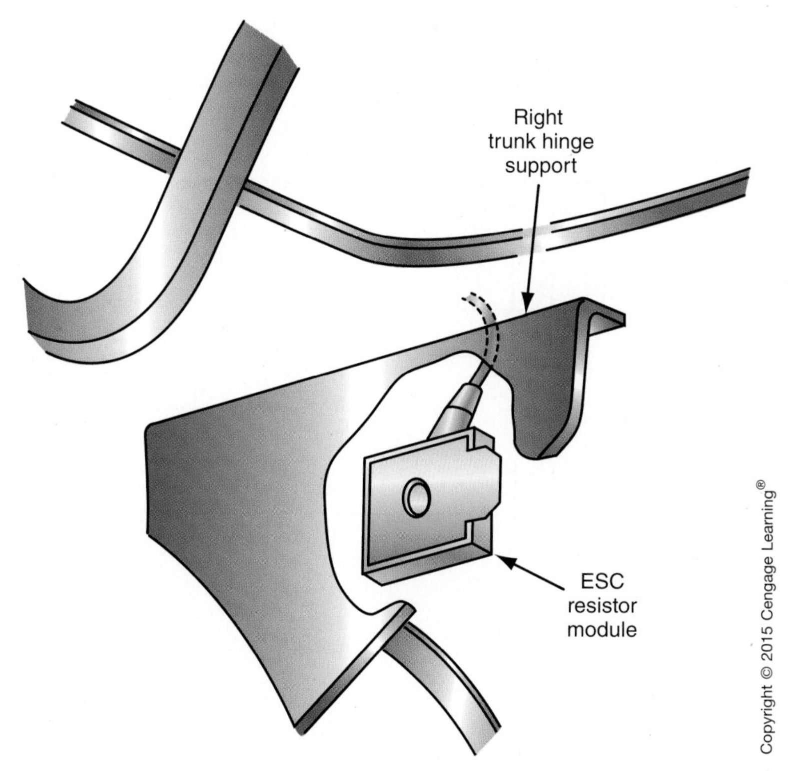

FIGURE 11-51 **Resistor module mounting location.**

This resistor reduces the voltage drop across the damper solenoid, which lowers the current flow. This lower current flow is high enough to hold the damper solenoid in the On mode. Each damper solenoid circuit is basically the same.

Rear Automatic Level Control

The **automatic level control (ALC)** system maintains the rear suspension trim height regardless of the rear suspension load. If a heavy object is placed in the trunk, the rear wheel position sensors send below trim height signals to the ESC module. When this signal is received, the ESC module grounds the ALC relay winding and closes the relay contacts that supply voltage to the compressor motor (Figure 11-52).

Once the compressor starts running, it supplies air through the nylon lines to the rear air shocks and raises the rear suspension height (Figure 11-53). When trim height signals are received from the rear wheel position sensors, the ESC module opens the compressor relay winding circuit and stops the compressor.

If a heavy object is removed from the trunk, the rear wheel position sensors send above trim height signals to the ESC module. Under this condition, the ESC module energizes the exhaust solenoid in the compressor assembly, and this action releases air from the rear air shocks. When the rear wheel position sensors send rear suspension trim height signals to the ESC module, this module shuts off the exhaust solenoid.

An independent ALC system is used on cars without the ESC system. In these systems, the computer is not required and a single suspension height sensor is used. This height sensor contains electronic circuits that control the compressor relay and the exhaust solenoid. This electronic circuit limits the compressor run time and the exhaust solenoid on time to 7 minutes.

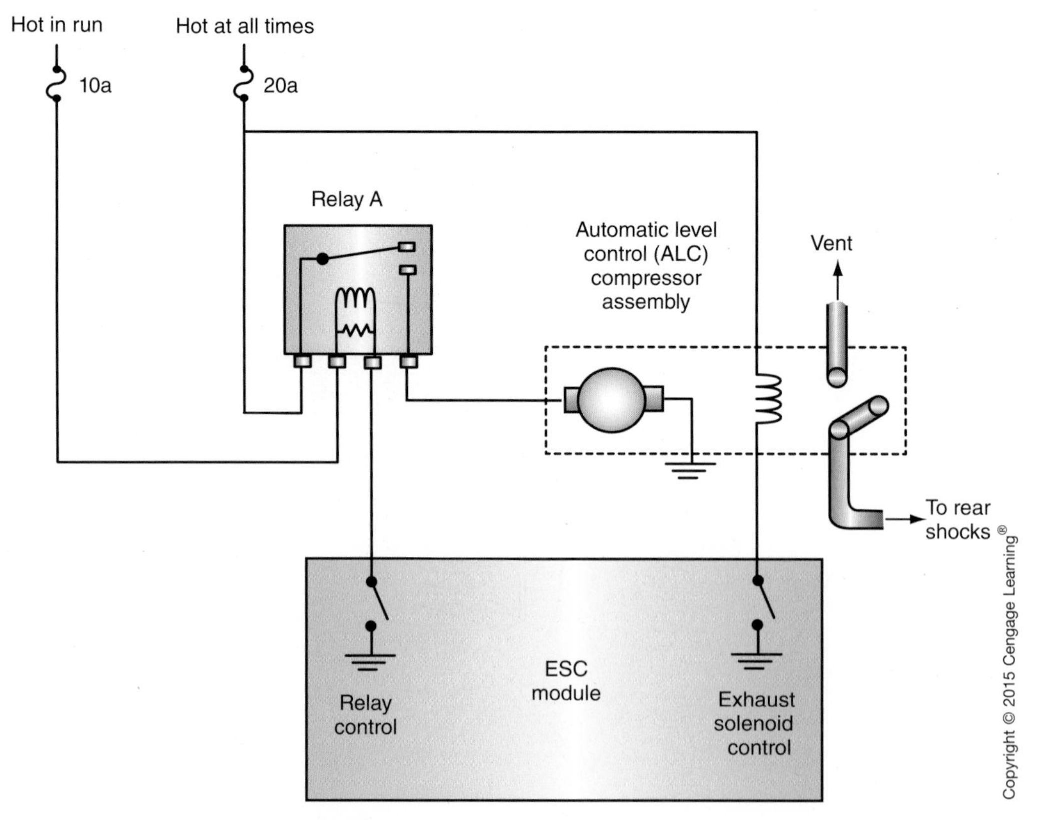

FIGURE 11-52 **Rear automatic level control (ALC).**

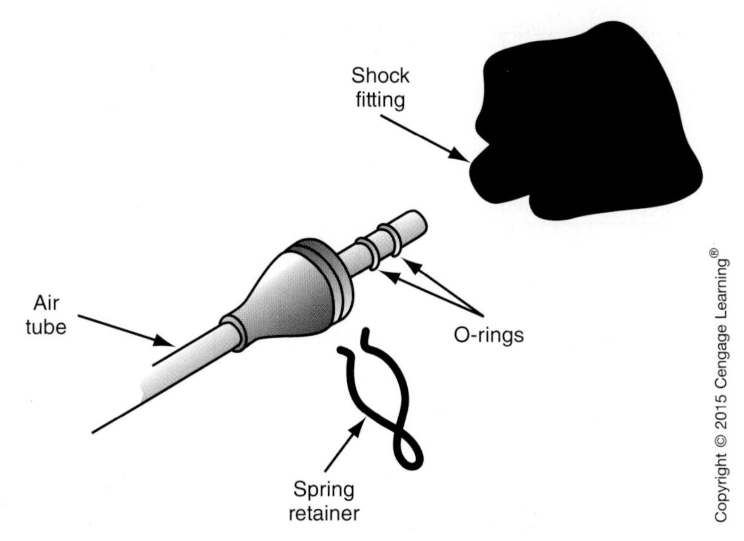

FIGURE 11-53 **Nylon air line and rear shock air line fitting.**

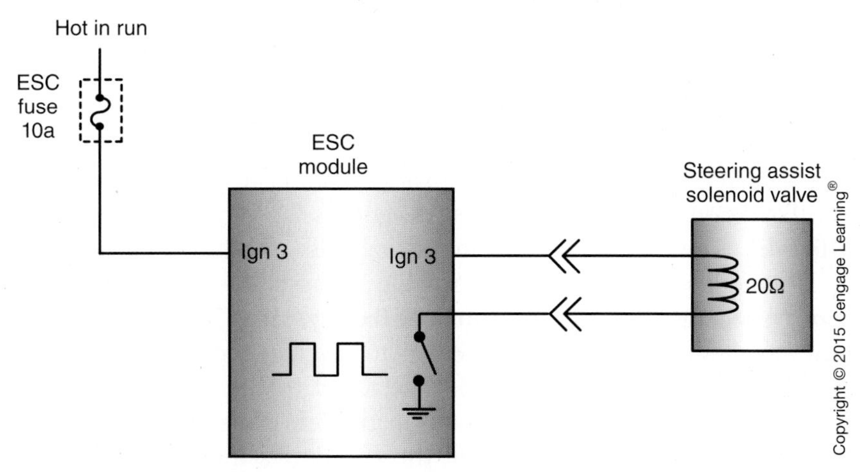

FIGURE 11-54 **Steering solenoid and ESC module wiring diagram.**

Speed Sensitive Steering System

Shop Manual
Chapter 11,
page 441

On some models the ESC module operates a solenoid in the **speed sensitive steering (SSS) system** to control the power steering pump pressure in relation to vehicle speed (Figure 11-54). This action varies the power steering assist levels.

The ESC module varies the on time of the steering solenoid. This action may be referred to as pulse width modulation (PWM). When the solenoid is in the Off mode, the power steering pump supplies normal power assist. Below 10 mph (16 km/h), the computer operates the steering solenoid to provide full power steering assist (Figure 11-55). This action reduces steering effort during low-speed maneuvers and parking.

As the vehicle speed increases, the ESC module operates the steering solenoid so the power steering assist is gradually reduced to provide increased road feel and improved handling.

On later model cars, the speed sensitive steering (SSS) is called speed dependent steering or MagnaSteer®. The module that controls the MagnaSteer® is contained in the electronic brake and traction control module (EBTCM).

Vehicle Stability Control

Many vehicles manufactured in recent years are equipped with a vehicle stability control system. A **vehicle stability control system** provides improved control if the vehicle begins to swerve sideways because of slippery road surfaces, excessive acceleration, or a combination of

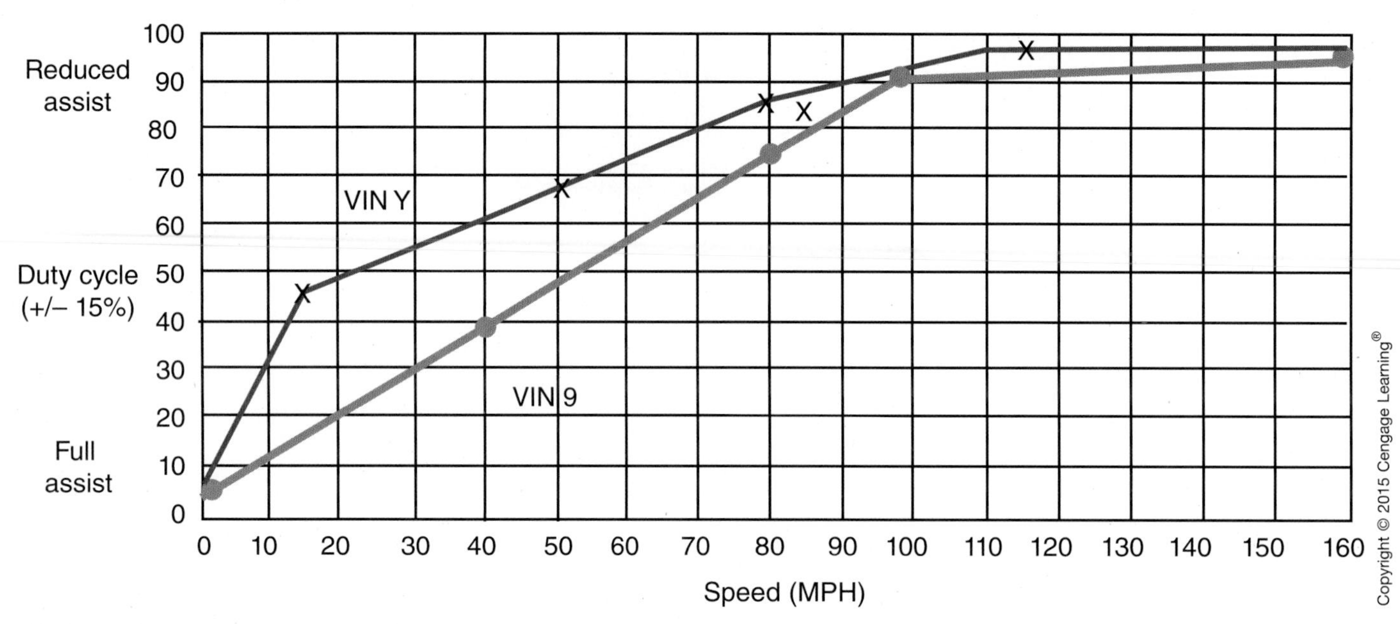

FIGURE 11-55 **Power steering assist in relation to vehicle speed.**

The **antilock brake system (ABS)** prevents wheel lockup during a brake application.

The **traction control system (TCS)** prevents drive wheel slippage.

The **electronic brake and traction control module (EBTCM)** controls ABS, TCS, and Stabilitrak® functions.

The **brake pressure modulator valve (BPMV)** controls brake fluid pressure to the wheel calipers or cylinders.

these two conditions. Therefore, a vehicle stability control system provides increased vehicle safety. Vehicle stability control systems have various brand names depending on the vehicle manufacturer. For example, on General Motors vehicles the vehicle stability control system is called **Stabilitrak®**. The Stabilitrak® system is available on many General Motors cars and some SUVs. The module that controls the Stabilitrak® system is combined with the **antilock brake system (ABS)** module and **traction control system (TCS)** module (Figure 11-56). This three-in-one module assembly is referred to as the **electronic brake and traction control module (EBTCM)**. The EBTCM is attached to the **brake pressure modulator valve (BPMV)** and this assembly is mounted in the left front area in the engine compartment. A data link is connected between all the computers including the EBTCM and the ESC module (Figure 11-57). The combined EBTCM and ESC systems may be referred to as

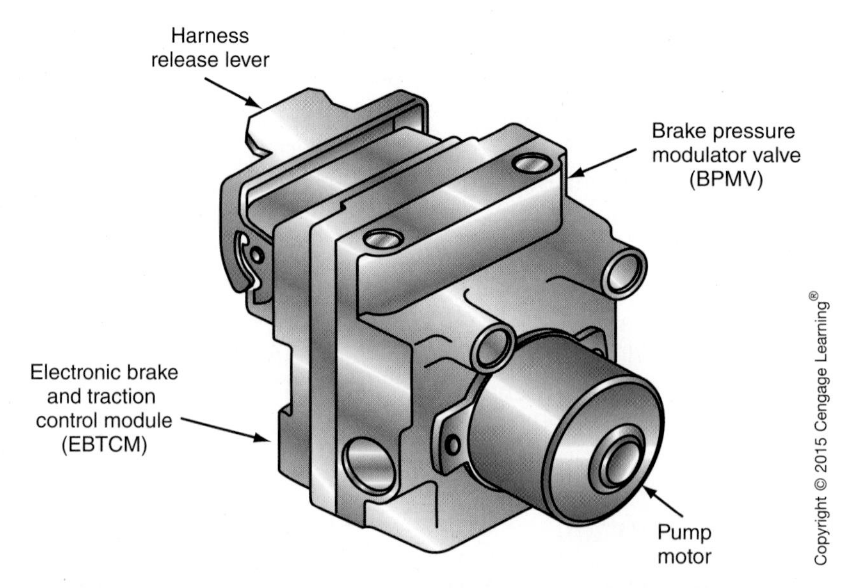

FIGURE 11-56 **The electronic brake and traction control module (EBTCM) contains the antilock brake system (ABS), traction control system, and stability control modules.**

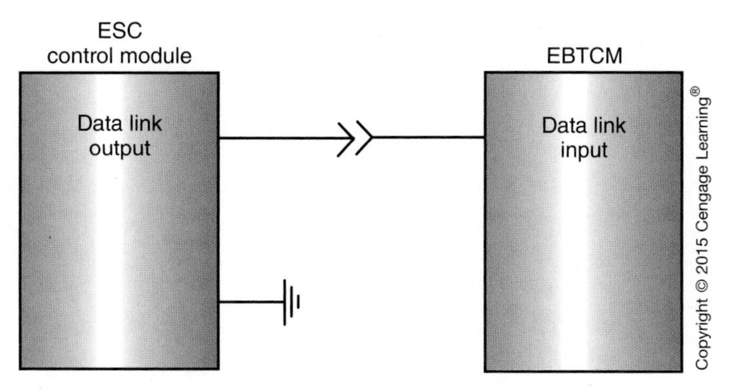

FIGURE 11-57 **Data link between the EBTCM and ESC modules.**

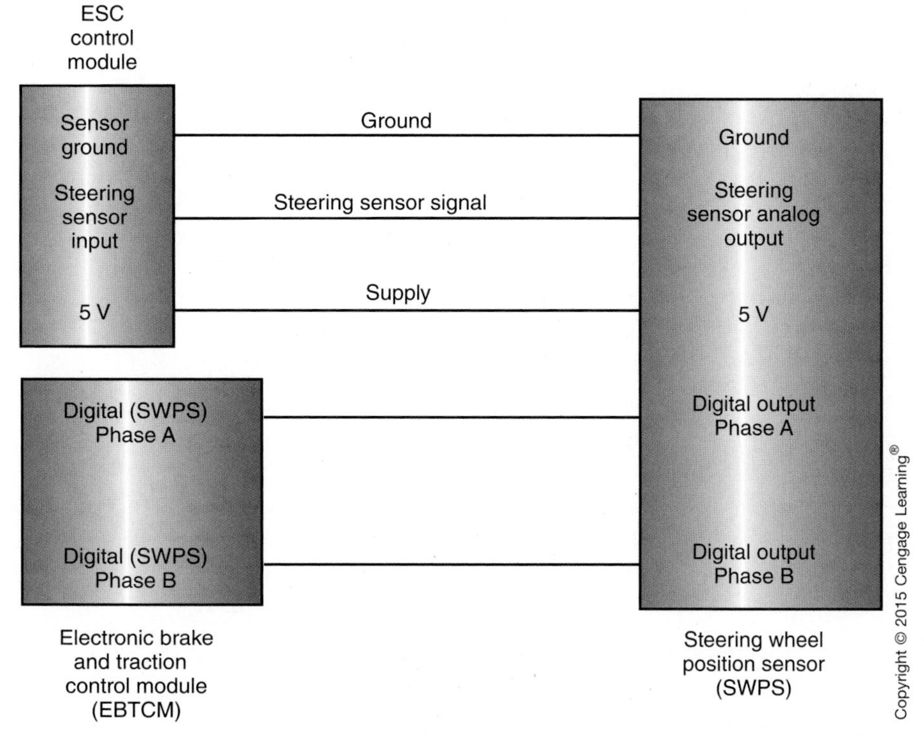

FIGURE 11-58 **The steering wheel position sensor (SWPS) is connected to both the ESC module and the EBTCM.**

the integrated chassis control system 2 (ICCS2). Some sensors such as the **steering wheel position sensor (SWPS)** are hard-wired to both the EBTCM and the ESC module (Figure 11-58). The signals from other sensors may be sent to one of these modules and then transmitted to the other module on the data link. The data link also transmits data from these modules to the instrument panel cluster (IPC) during system diagnosis. This allows the IPC to display diagnostic information.

The **steering wheel position sensor** supplies a voltage signal in relation to the amount and speed of steering wheel rotation.

This book is concerned with suspension and steering systems. Because the stability control system operates in cooperation with the ABS and TCS systems, a brief description of these systems is necessary.

AUTHOR'S NOTE: Statistics compiled by the National Highway Traffic Safety Administration (NHTSA) indicate that stability control systems can reduce the incidence of single-vehicle accidents in SUVs by 63 percent.

Antilock Brake System (ABS) Operation

Wheel speed sensors are mounted at each wheel. In this ABS system, the wheel speed sensors are integral with the front or rear wheel bearing hubs. These wheel bearing hubs with the integral wheel speed sensors are non-serviceable (Figure 11-59). Each wheel speed sensor contains a toothed ring that rotates past a stationary electromagnetic wheel speed sensor. This sensor contains a coil of wire surrounding a permanent magnet. As the toothed ring rotates past the sensor, an alternating current (AC) voltage is produced in the sensor. This voltage signal from each wheel speed sensor is sent to the EBTCM. As wheel speed increases, the frequency of AC voltage produced by the wheel speed sensor increases proportionally. During a brake application, the wheels slow down, and the frequency of AC voltage in the wheel speed sensors also decreases. If a wheel is nearing a lockup condition during a hard brake application, the frequency of the AC voltage from that wheel speed sensor becomes very slow. The EBTCM detects impending wheel lockup from the frequency of AC voltage signals sent from the wheel speed sensors.

A **hold valve** opens and closes the fluid passage to each wheel caliper.

When energized, a **release valve** reduces pressure in a wheel caliper.

The brake pressure modulator valve (BPMV) contains a number of electrohydraulic valves. These valves are operated electrically by the EBTCM. These valves in the BPMV are connected hydraulically in the brake system. A **hold valve** and a **release valve** are connected in the brake line to each wheel (Figure 11-60). If a wheel speed sensor signal indicates an impending wheel lockup condition, the EBTCM energizes the normally open hold solenoid connected to the wheel that is about to lock up. This action closes the solenoid and isolates the wheel caliper from the master cylinder to prevent any further increase in brake pressure. If the wheel speed sensor signal still indicates an impending wheel lockup, the EBTCM keeps the hold solenoid closed and opens the normally closed release solenoid momentarily. This action reduces wheel caliper pressure to reduce brake application force and prevent wheel lockup. The EBTCM pulses the hold and the release solenoids on and off to supply maximum braking force without wheel lockup.

When the hold and the release valves are pulsated during a prolonged antilock brake function, the brake pedal fades downward as brake fluid flows from the release valves into the accumulators. To maintain brake pedal height during an antilock brake function, the EBTCM starts the pump in the BPMV at the beginning of an antilock function. When

FIGURE 11-59 **Wheel speed sensor.**

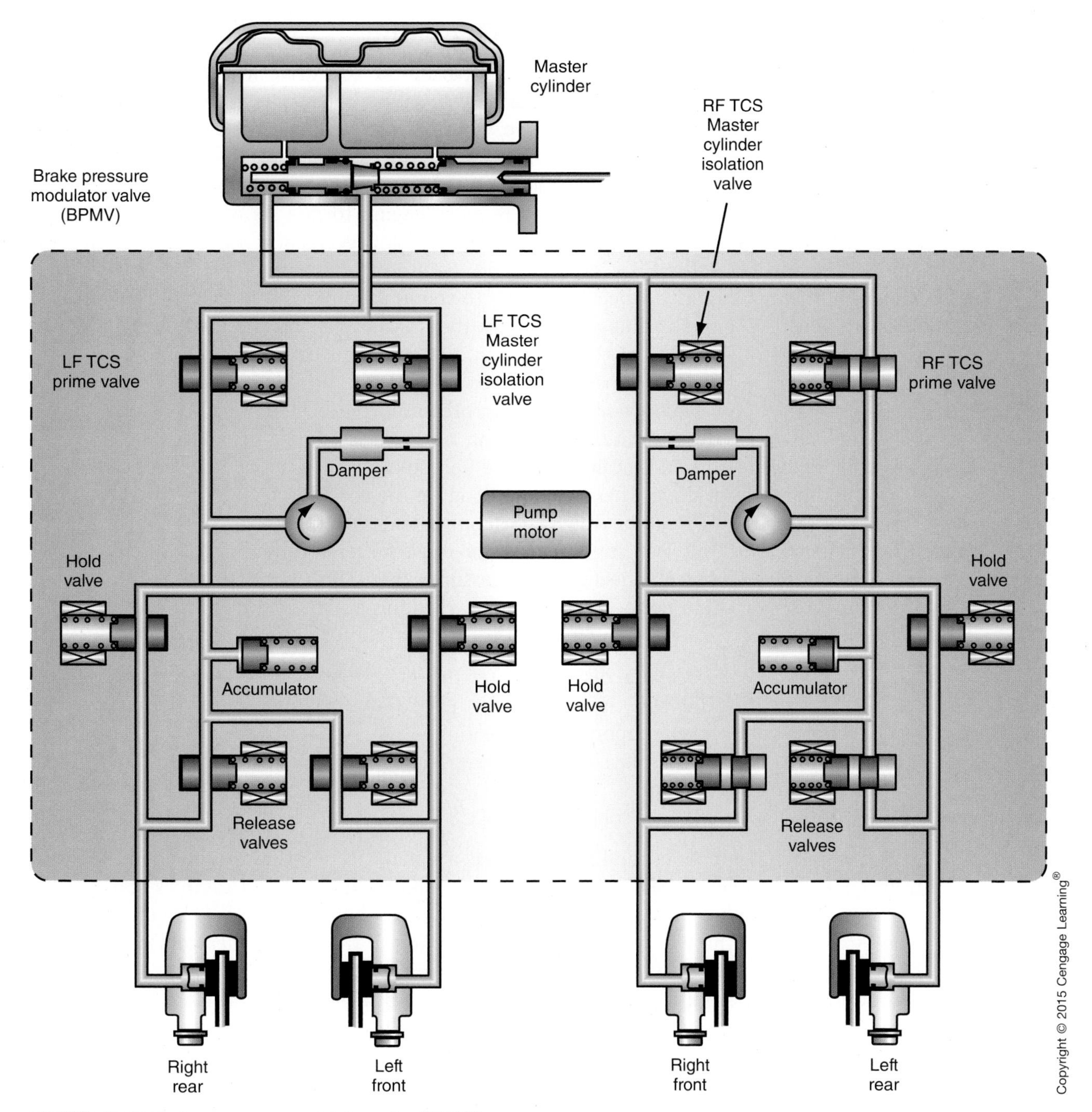

FIGURE 11-60 **Brake pressure modulator valve (BPMV).**

the pump motor is started, the pump supplies brake fluid pressure to the hold valves and wheel calipers. Pump motor pressure is also supplied back to the master cylinder. Under this condition, the driver may feel a firmer brake pedal and pedal pulsations and may hear the clicking action of the hold and the release solenoids.

ANTILOCK and BRAKE warning lights are mounted in the instrument panel. Both of these lights are illuminated for a few seconds after the engine starts. If the amber ANTILOCK light is on with the engine running, the EBTCM has detected an electrical fault in the ABS system. Under this condition, the EBTCM no longer provides an ABS function, but normal power-assisted brake operation is still available. When the red BRAKE warning light is illuminated with the engine running, the parking brake may be on, the master cylinder may be low on brake fluid, or there may be a fault in the ABS system.

Traction Control System (TCS) Operation

The EBTCM detects drive wheel spin by comparing the two drive wheel speed sensor signals. Wheel spin on both drive wheels is detected by comparing the wheel speed sensor signals on the drive wheels and non-drive wheels. If a wheel speed sensor signal informs the EBTCM that one or both drive wheels are spinning, the EBTCM enters the traction control mode. First, the EBTCM requests the PCM to reduce the amount of engine torque supplied to the drive wheels. The PCM reduces engine torque by retarding the ignition spark advance and shutting off the fuel injectors for a very short time. The PCM then sends a signal back to the EBTCM regarding the amount of torque delivered to the drive wheels. If one or both drive wheels continue to spin on the road surface, the EBTCM energizes the normally closed prime valve, closes the normally open isolation valve, and starts the pump in the BPMV. Under this condition, the prime valve opens and the pump begins to move brake fluid from the master cylinder through the prime valve to the pump inlet. The closed isolation valve prevents the pump pressure from being applied back to the master cylinder. Under this condition, the pump pressure is supplied through the normally open hold valve to the brake caliper on the spinning wheel. This action stops the wheel from spinning. If both drive wheels are spinning on the road surface, the EBTCM operates both prime valves and isolation valves to supply brake fluid pressure to both drive wheel brake calipers. During a TCS function, the EBTCM pulses the hold and the release solenoids on and off to control wheel caliper pressure. The EBTCM limits the traction control function to a short time period to prevent overheating brake components. During a TCS function, these messages may be displayed in the driver information center (DIC):

1. TRACTION ENGAGED is displayed after the TCS is in operation for 3 seconds.
2. TRACTION SUSPENDED is displayed if the EBTCM has discontinued the TCS function to prevent brake component overheating.
3. TRACTION OFF is displayed if the driver places the TCS switch on the instrument panel in the Off position.
4. TRACTION READY is displayed if the TCS switch is turned from Off to On.

During a TCS function, the EBTCM sends a signal through the data link to the powertrain control module (PCM). When this signal is received, the PCM disables some of the fuel injectors to reduce engine torque. This action also helps to prevent drive wheel spin. The PCM disables the two injectors at the beginning of the firing order and in the center of the firing order. Depending on the speed of drive wheel spin, the PCM may disable every second injector in the firing order. The injectors are disabled for a very short time period. The TCS system improves drive wheel traction, and this system also prevents the tendency for the vehicle to swerve sideways when one drive wheel is spinning. Therefore, the TCS system increases vehicle safety.

A **lateral accelerometer** supplies a voltage signal in relation to sideways movement of the chassis.

Yaw is erratic, side-to-side deviation from a course. The **yaw rate sensor** supplies a voltage signal in relation to rotational chassis speed during a sideways swerve.

Vehicle Stability Control

To provide stability control, the EBTCM uses two additional input signals from the **lateral accelerometer** and the **yaw rate sensor**. The lateral accelerometer is mounted under the front passenger's seat (Figure 11-61). The EBTCM sends a 5V reference voltage to the lateral accelerometer. If the vehicle is driven straight ahead, the chassis has zero lateral acceleration. Under this condition, the lateral accelerometer provides a 2.5V signal to the EBTCM. If the vehicle begins to swerve sideways because of slippery road conditions, high-speed cornering, or erratic driving, the lateral accelerometer signal to the EBTCM varies from 0.25V to 4.75V, depending on the direction and severity of the swerving action.

The yaw rate sensor is mounted under the rear package shelf (Figure 11-62). Some yaw rate sensors contain a precision metal cylinder whose rim vibrates in elliptical shapes. The vibration and rotation of this metal cylinder is proportional to the rotational speed of

FIGURE 11-61 **Lateral accelerometer.**

FIGURE 11-62 **Yaw rate sensor.**

the vehicle around the center of the cylinder. On some models, the lateral accelerometer and yaw rate sensors are combined in a single sensor. The sensor mounting location varies depending on the vehicle make and model year.

The EBTCM sends a 5V reference voltage to the yaw rate sensor. If the vehicle chassis experiences zero yaw rate, the yaw rate sensor sends a 2.5V signal to the EBTCM module. If the vehicle begins to swerve sideways, the yaw rate sensor provides a 0.25V to 4.75V signal to the EBTCM, depending on the direction and severity of the swerve. The EBTCM also uses the wheel speed sensor signals for stability control. If the vehicle begins to swerve sideways, the EBTCM energizes the normally closed prime valve and closes the normally open isolation valve connected to the appropriate front wheel; then it starts the pump in the BPMV. Under this condition, the prime valve opens and the pump begins to move brake fluid from the master cylinder through the prime valve to the pump inlet. The closed isolation valve prevents the pump pressure from being applied back to the master cylinder. Under this condition, the pump pressure is supplied through the normally open hold valve to the brake caliper on the appropriate front wheel. Applying the brake on the front wheel pulls the vehicle out of the swerve and prevents the complete loss of steering control. If the EBTCM detects an electrical fault in the stability control system, STABILITY REDUCED is displayed in the DIC. If the EBTCM enters the stability control mode, STABILITY ENGAGED is indicated in the DIC.

In Figure 11-63, two vehicles driving side-by-side are negotiating a lane change to the left. The vehicle on the right has vehicle stability control, and the vehicle on the left does not have this system. As the vehicle on the left begins to turn, the rear of the vehicle begins to swing around. When the vehicle on the right starts to turn, the rear of the vehicle swerves slightly and the right front brake is applied by the vehicle stability control system to prevent this swerve. Further into the turn, the driver attempts to steer the car on the left, but this car enters into an uncontrolled swerve with loss of steering control. As the car on the right continues into the turn, the rear of the vehicle swerves slightly, but the vehicle stability control system again applies the right front brake momentarily to prevent this swerve. The car with the stability control system completes the turn while maintaining directional stability, but the vehicle without stability control goes into an uncontrolled swerve with complete loss of directional control. The vehicle stability control system improves vehicle safety! The other charts in Figure 11-64 indicate that yaw rate, vehicle slip angle, and lateral acceleration are greatly reduced on a vehicle with a stability control system.

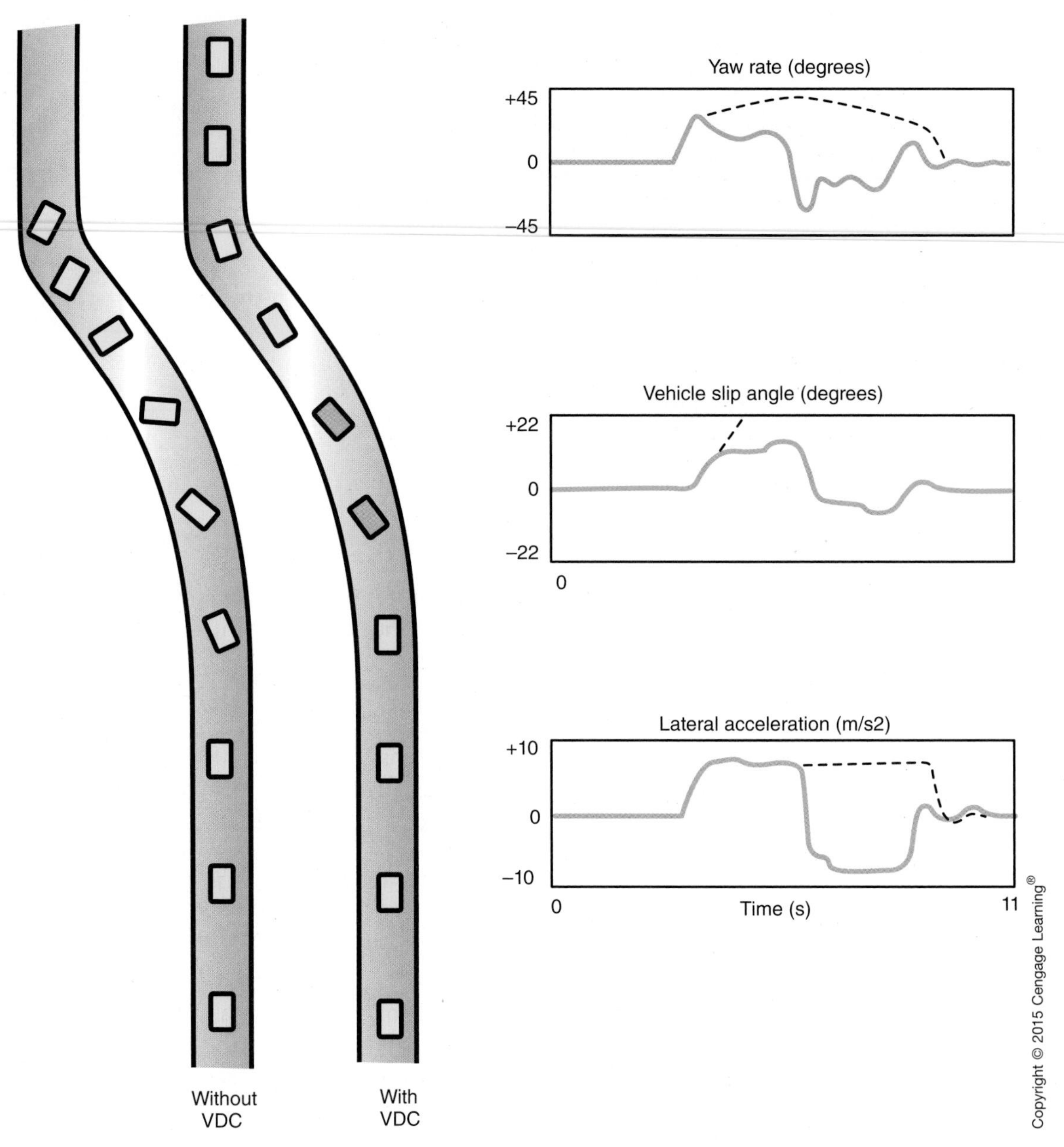

FIGURE 11-63 **Comparison during a turn between a vehicle with a stability control system and a vehicle with no stability control system.**

Active Roll Control Systems

Two independent automotive component manufacturers have developed active roll control systems in response to concerns about sport utility vehicles (SUVs) that roll over more easily compared with cars because of the SUVs' higher center of gravity. The active roll control systems were developed in response to this concern. The active roll control system contains a control module, accelerometer, speed sensor, fluid reservoir, electrohydraulic pump, pressure control valve, directional control valve, and a hydraulic actuator in both the front and rear stabilizer bars (Figure 11-64). The accelerometer and speed sensor may be common to systems other than the active roll control. The electrohydraulic pump may also be used as the power steering pump. The active roll control system may be called a roll stability control (RSC) system or a dynamic handling system (DHS). Some active roll control systems have a gyro sensor that provides a voltage signal to the control module in relation to the vehicle roll speed and angle.

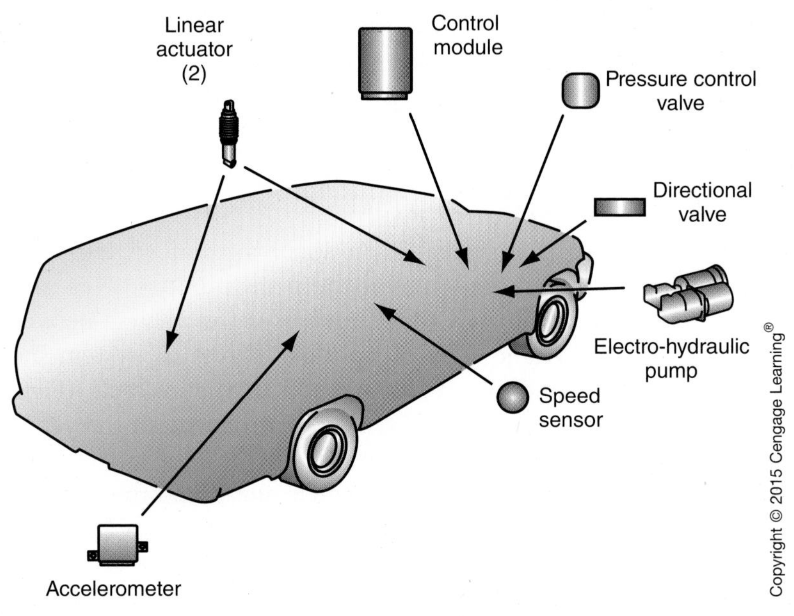

FIGURE 11-64 **Active roll control system components.**

When the vehicle is driven straight ahead, the active roll control system does not supply any hydraulic pressure to the linear actuators in the stabilizer bars. Under this condition, both stabilizer bars move freely until the linear actuators are fully compressed. This action provides improved individual wheel bump performance and better ride quality. If the chassis begins to lean while cornering, the module operates the directional valve so it supplies fluid pressure to the linear actuators in the stabilizer bars. This action stiffens the stabilizer bar and reduces body lean (Figure 11-65). The active roll control system increases safety by reducing body lean, which decreases the possibility of a vehicle rollover.

One vehicle manufacturer has chosen to reduce the yaw force on a new model by reducing the weight at the front and rear of the vehicle, and shifting the weight toward the midpoint on the car. This was accomplished by manufacturing the hood, trunk lid,

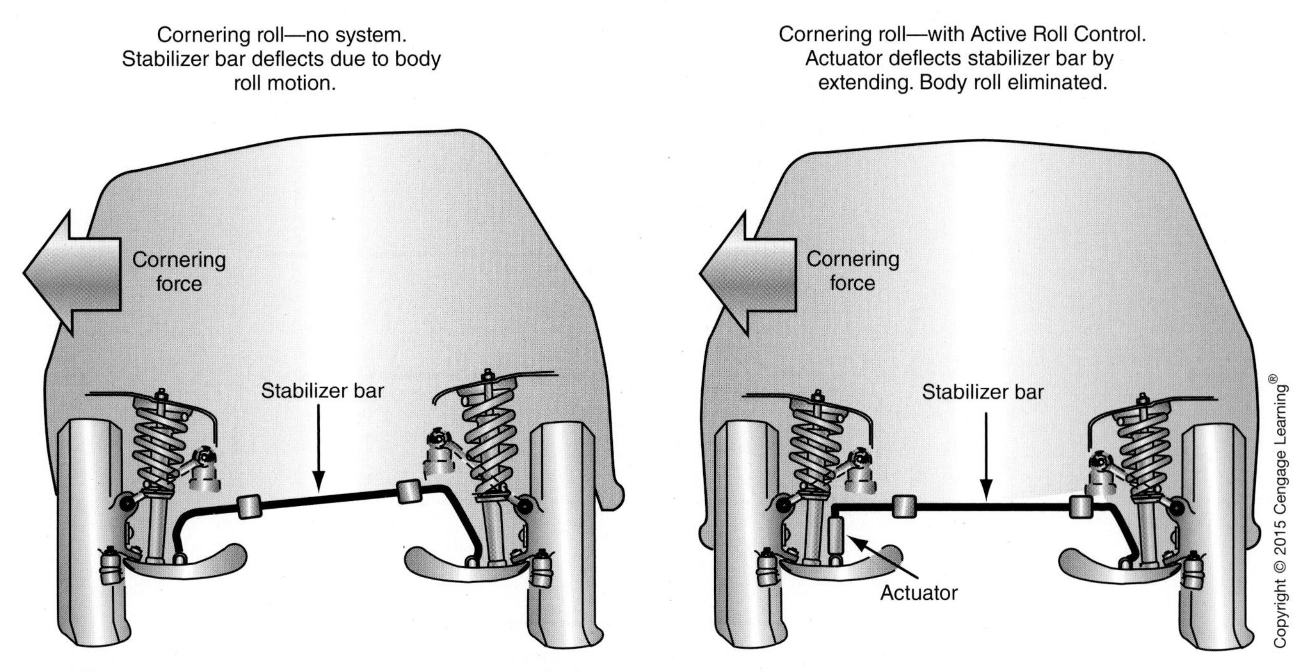

FIGURE 11-65 **Active roll control system operation while cornering.**

and bumper beams from aluminum. Engine weight was reduced by installing hollow camshafts, and using a lighter weight block and cylinder heads. These changes reduced the weight of the vehicle by 180 lbs (80 kg), which reduced the yaw force and also improved fuel economy and performance.

Adaptive Cruise Control (ACC) Systems

Adaptive cruise control (ACC) is available on some current vehicles. The ACC system has a long-range radar sensor mounted behind the front bumper. This radar sensor sends signals to the ACC computer. The ACC computer uses the vehicle network to remain in constant contact with the PCM and ABS computers. The ACC system measures the distance to the vehicle in front and the relative speed of that vehicle (Figure 11-66). The ACC system can detect objects more than 330 ft. (100 m) ahead of the vehicle. A dash control allows the driver to set the distance between the vehicle and the vehicle ahead.

If the ACC system detects a slower moving vehicle ahead and the distance to that vehicle becomes less than the driver-adjusted setting, the ACC system uses the throttle control and limited brake application with a 0.3g braking force to slow down the vehicle. When this action is taken, a small green car icon is illuminated in the head up display (HUD). If the distance to the vehicle in front is still decreasing after this action is taken, the icon in the HUD turns from green to yellow. When the ACC system senses that a collision is imminent, a large red car icon with yellow flashes is illuminated in the HUD and the ACC system activates a beeper.

When the lane ahead of the vehicle is clear, the ACC system maintains the cruising speed set by the driver. The ACC system can detect another vehicle crossing from an adjacent lane into the lane in front of the vehicle and take appropriate action if there is not sufficient distance between the two vehicles. The radar signals in the ACC system provide excellent performance even in adverse weather conditions.

Lane Departure Warning (LDW) Systems

Some current vehicles are equipped with **lane departure warning (LDW) systems**. The LDW system has a video camera mounted behind the rear view mirror (Figure 11-67). This camera uses software to monitor highway lane markings (Figure 11-68). The camera measures the distance from the vehicle to the lane markings and the lateral velocity of the vehicle in relation to the lane markings to determine if the vehicle is moving out of the lane. Signals from the camera are sent to the LDW module. The VSS is also sent to the LDW module. The LDW system does not operate below 45 mph (72 km/h). If the vehicle speed is above 45 mph (72 km/h) and the camera signal indicates the vehicle is moving out of the lane, the

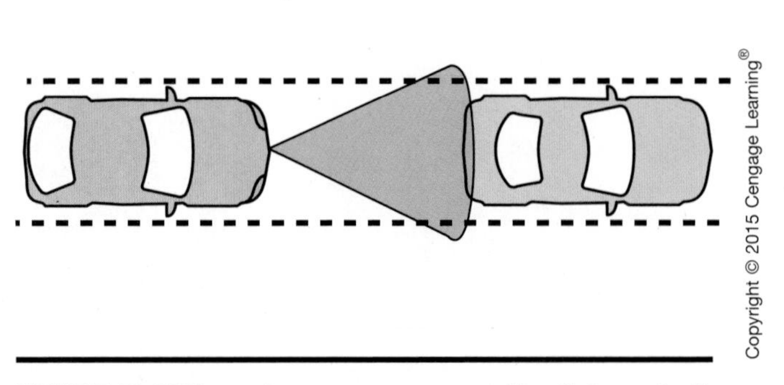

FIGURE 11-66 **The radar sensor measures the distance to the vehicle in front and the relative speed of this vehicle.**

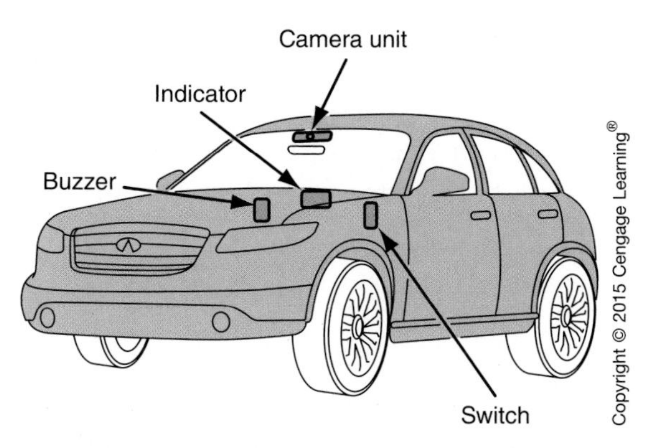

***FIGURE 11-67* Components in the lane departure warning (LDW) system.**

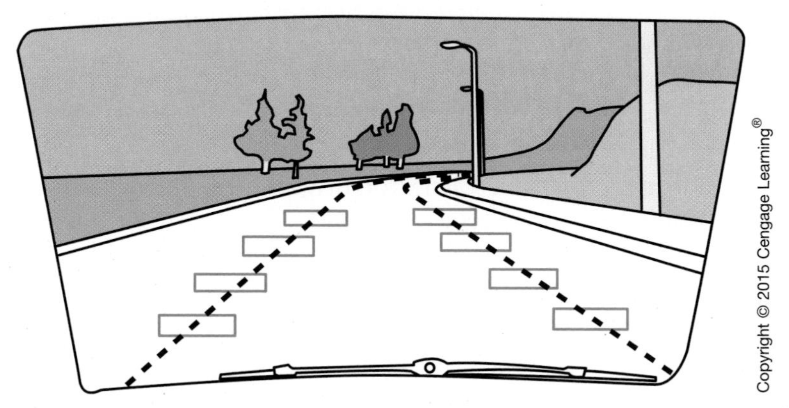

***FIGURE 11-68* The LDW camera measures the distance from vehicle to the highway lane markings.**

LDW module operates a buzzer and an indicator warning in the instrument panel. Activation of the turn signals temporarily disables the LDW system to prevent incorrect LDW warnings.

Some LDW systems have lane-departure prevention capabilities. If the vehicle is moving toward the lane markings on the left, the ABS and stability control computer lightly pulse the brakes on the right side of the vehicle. When the wheels on the right side slow down, the wheels on the left side turn faster than the wheels on the right side, and this action steers the car away from the highway markers on the left. Other LDW systems activate the electronic steering on the vehicle to steer the vehicle away from the lane markers if a lane departure is starting to occur. A dash switch allows the driver to turn the LDW system on or off.

> **AUTHOR'S NOTE: Statistics compiled by one of the major car manufacturers indicate that 30 percent of all vehicle accidents in the United States involve rear-end collisions. In half of these collisions, the driver did not apply the brakes before collision occurred. The statistics also indicate that 75 percent of these rear-end collisions occur below 19 mph (30 km/h).**

Collision Mitigation Systems

Some vehicles are equipped with a collision mitigation system that uses radar and camera information to detect a vehicle in front. The collision mitigation computer is in constant communication with the PCM and ABS computers via the vehicle network. If the radar and camera signals indicate a vehicle in front is too close and a collision is imminent, the collision mitigation computer warns the driver with a visual and audible warning. If the driver does not apply the brakes, the collision mitigation computer applies the brakes aggressively with a 0.5 g force to slow down the vehicle. The vehicle manufacturer's information indicates that the collision mitigation system will avoid rear-end collisions up to 9 mph (19 km/h) and mitigate the effects of a collision up to 19 mph (30 km/h). On some vehicles, the collision mitigation system is called a **city safety system**.

Telematics

The use of telematics is expected to increase significantly in the next few years. Vehicle manufacturers and electronic communications companies have not decided the exact information that will be transmitted to vehicles through telematics, but the following are some possible options.

Vehicle manufacturers could use telematics to inform vehicle owners when their vehicles require specific service. This is being done presently to some extent by General Motors through their OnStar system. Telematics could be used for emission testing. At present, 33 states in the United States have emission test centers, and vehicles must be taken to these centers for emission testing at specific intervals. Using telematics, the emission test center could monitor the vehicle emission levels, and if the emission standards are exceeded, the test center could inform the vehicle owner or driver that the vehicle must be taken to a certified repair center for emission service.

Telematics may also be used to improve vehicle safety. At present, a Vehicle Infrastructure and Integration (VII) system has installed two test projects, one in California and another one in the Detroit area. In these test projects, roadside beacons communicate with vehicle networks. In each project, approximately 25 vehicles supplied by car manufacturers are driven on a 75-mile (120 km) stretch of highway where the communication beacons are installed. At hidden intersections, monitoring stations inform the test vehicles that vehicles are approaching the intersection. Other inputs inform the test vehicles regarding heavy traffic congestion or black ice.

Telematics may be defined as wireless communication between the telephone system and vehicle networks.

INTERESTING FACT

Each year in the United States, approximately 43,000 deaths are caused by traffic accidents. This is approximately the equivalent of a 747 aircraft crashing each week.

AUTHOR'S NOTE: The following are some interesting facts about the OnStar communications system:

1. OnStar is a factory-installed option on General Motors vehicles. One-year OnStar service is included in the new GM vehicle purchase. Today there are approximately 5 million OnStar subscribers, and the subscriber list is expected to double by 2011.
2. A sophisticated module in each vehicle allows communication between the vehicle and an OnStar center at the push of a button in the vehicle. This communication is enabled by the vehicle battery and a special external antenna on each vehicle that combine cell reception, global positioning system (GPS), and XM satellite radio to provide range and performance that far exceed hand-held cell phones and GPS devices.
3. Approximately 3,000 advisors work at three OnStar call centers and receive up to 100,000 calls in three languages on a busy day. At the beginning of an OnStar call, the driver must select the English, Spanish, or French language option, and a special routing system directs the call to an advisor speaking the selected language.
4. Ninety-five percent of emergency calls are answered in 5 seconds. The OnStar module on the vehicle notifies a center if the air bags deploy. An advisor calls the vehicle in an attempt to contact the driver or vehicle occupants. If the advisor cannot contact the driver or occupants, the advisor notifies emergency response

personnel regarding the vehicle location. The OnStar system handles approximately 2,200 vehicle crash responses per month.

5. The OnStar system processes about 600,000 requests for driving directions each month. OnStar subscribers purchase approximately 31 million minutes of OnStar time per month.
6. Current OnStar modules are connected to the CAN data network on the vehicle. This connection allows the OnStar module to monitor the powertrain, ABS, traction control, and air bag systems. On current vehicles, the OnStar module can read about 1,600 DTCs and monitor the engine oil life, emission levels, and tire pressure. About 3 million OnStar subscribers have signed up for a monthly vehicle condition report that informs the subscriber regarding vehicle defects in any of the monitored systems. The monthly vehicle condition report also informs the subscriber regarding any recall repairs that have not been completed on their vehicle. Approximately, 61,000 calls are processed monthly for instant remote diagnosis.
7. If an OnStar subscriber is locked out of one's vehicle, he or she can phone an OnStar advisor and provide his or her OnStar number. The advisor can then unlock the doors on the subscriber's vehicle. If an OnStar advisor is informed that a vehicle is stolen, the advisor can use GPS to provide the vehicle location. Select 2009 General Motors vehicles have OnStar's Stolen Vehicle Slowdown service. If police are following a stolen vehicle, they can inform an OnStar advisor and request that the vehicle be stopped. The advisor will flash the vehicles external lights without the driver's awareness. When police confirm that the vehicle lights have flashed, the OnStar advisor will send a signal to the stolen vehicle that causes the electronic throttle control to remain continually in the idle position. All other vehicle functions such as steering and brakes operate normally. The OnStar subscriber cannot request that the vehicle be stopped. At present, OnStar receives about 700 stolen vehicle calls per month. It is estimated that each year there are 30,000 high-speed chases in the United States each year, and 25 percent of these end in injury, and 300 deaths are a result of these chases.
8. OnStar has expanded beyond the United States and Canada and is now offered in China and Mexico.

SUMMARY

- In a programmed ride control (PRC) system, the steering sensor informs the control module regarding the amount and speed of steering wheel rotation.
- The PRC system switches from the normal to the firm mode during high-speed operation, braking, hard cornering, and fast acceleration.
- The struts and shock absorbers in some PRC systems provide three times as much damping action in the firm mode as in the normal mode.
- The accelerometer in a CCR system contains a mercury switch and this accelerometer sends a vehicle acceleration signal to the control module.
- In a CCR system, the accelerometer signal or the vehicle speed signal may inform the control module to switch from the normal to the firm mode.
- An electronic air suspension system maintains a constant vehicle trim height regardless of passenger or cargo load.

TERMS TO KNOW

Accelerometer
Actuator
Adaptive cruise control system
Air spring solenoid valve
Air springs
Antilock brake system (ABS)
Automatic level control (ALC)
Brake pressure modulator valve (BPMV)
Brake pressure switch
City safety system
Controller Area Network (CAN)
Cross-axis ball joints
Damper solenoid valve
Data bus network
Electronic brake and traction control module (EBTCM)
Electronic rotary height sensors
Electronic suspension control (ESC) system

TERMS TO KNOW
(continued)

Firm relay
Hall element
Height sensors
Hold valve
Lane departure warning (LDW) system
Lateral accelerometer
Lift/dive input
Light-emitting diodes (LEDs)
Lower vehicle command
MagneRide system
Network
Photo diodes
Programmed ride control (PRC) system
Raise vehicle command
Release valve
Soft relay
Speed sensitive steering (SSS) system
Stabilitrak®
Steering sensor
Steering wheel position sensor (SWPS)
Throttle position sensor (TPS)
Traction control system (TCS)
Trim height
Vehicle dynamic suspension (VDS)
Vehicle speed sensor (VSS)
Vehicle stability control system
Vent valve
Wheel position sensor
Wheel speed sensors
Yaw rate sensor

- To exhaust air from an air spring, the air spring solenoid valve and the vent valve in the compressor head must be energized.
- To force air into an air spring, the compressor must be running and the air spring solenoid valve must be energized.
- The air spring valves are retained in the air spring caps with a two-stage rotating action much like a radiator cap.
- An air spring valve must never be loosened until the air is exhausted from the spring.
- Voltage is supplied through the compressor relay points to the compressor motor. This relay winding is grounded by the control module to close the relay points.
- The On/Off switch in an electronic air suspension system supplies 12V to the control module. This switch must be off before the car is hoisted, jacked, towed, or raised off the ground.
- If a car door is open, the control module does not respond to lower vehicle commands in an electronic air suspension system.
- When the brake pedal is applied and the doors are closed in an electronic air suspension system, the control module ignores all requests from the height sensors.
- In an electronic air suspension system, if the doors are closed and the brake pedal is released, all requests to the control module are serviced by a 45-second averaging method.
- If the control module in an electronic air suspension system cannot complete a request from a height sensor in 3 minutes, the control module illuminates the suspension warning lamp.
- In an automatic air suspension system, the control module controls suspension height and strut damping automatically without any driver controlled inputs.
- Rotary height sensors in automatic air suspension systems contain Hall elements. These sensors send voltage signals to the control module in relation to the amount and speed of wheel jounce and rebound.
- Some air suspension systems reduce trim height at speeds above 65 mph (105 km/h) to improve handling and fuel economy.
- The air suspension system on some four-wheel-drive vehicles increase suspension ride height when the driver selects four-wheel-drive high or four-wheel-drive low.
- The air suspension system on some four-wheel-drive vehicles have the capability to provide firmer shock absorber valving in relation to transfer case mode selection, vehicle speed, and operating conditions.
- The electronic suspension control system changes shock and strut damping forces in 10 to 12 milliseconds.
- In the electronic suspension control system, the module controls suspension damping, rear electronic level control, and speed sensitive steering automatically without any driver-operated inputs.
- A vehicle stability control system applies one of the front brakes if the rear of the car begins to swerve out of control. This action maintains vehicle direction control.
- An adaptive cruise control (ACC) system applies the vehicle brakes lightly and warns the driver if the system senses inadequate distance to the vehicle in front.
- A lane departure warning (LDW) system warns the driver if the vehicle begins to leave the current lane. Some LDW systems apply the vehicle brakes on one side or steers the vehicle back into the current lane when lane departure begins to occur.
- A collision mitigation system warns the driver and applies the vehicle brakes aggressively if a rear-end collision is about to occur with the vehicle in front.

REVIEW QUESTIONS

Short Answer Essays

1. Describe the operation of the steering sensor in a programmed ride control (PRC) system.
2. Describe the purpose of the vehicle speed sensor signal in a programmed ride control (PRC) system.
3. Explain how air is forced into an air spring in a rear load-leveling air suspension system.
4. Describe the action taken by the control module if the control module in an electronic air suspension system receives a lower vehicle command from a rear suspension sensor with the doors closed, the brake pedal released, and the vehicle traveling at 60 mph (100 km/h).
5. Describe the action taken by the control module if the engine is running with a door open, and the control module receives a lower vehicle command from the height sensor in a rear load-leveling air suspension system.
6. List the conditions when the On/Off switch in an electronic air suspension system must be turned off.
7. Describe the conditions required for the control module to turn on the suspension warning lamp continually with the engine running in an electronic air suspension system.
8. Explain why the control module in an electronic air suspension system services all commands by a 45-second averaging method when the doors are closed and the brake pedal is released.
9. Explain why the suspension warning lamp in an electronic air suspension system may not indicate a defect immediately when the engine is started.
10. Explain how a vent solenoid is damaged by reversed battery polarity in a rear load-leveling air suspension system.

Fill-in-the-Blanks

1. The armature response time is ________________ milliseconds in a programmed ride control (PRC) system strut.
2. In a programmed ride control (PRC) system, if the car is accelerating with the throttle wide open, the PRC system is in the ________________ mode.
3. When the programmed ride control (PRC) mode switch is in the Auto position, the control module changes to the firm mode if lateral acceleration exceeds ________________.
4. In a programmed ride control (PRC) system, lateral vehicle acceleration is sensed from the ________________ sensor.
5. In a rear load-leveling air suspension system, the control module energizes the compressor relay when a ________________ command is received.
6. Two height sensors are mounted on the ________________ suspension in an electronic air suspension system.
7. In an electronic air suspension system two hours after the ignition switch is turned off, ________________ ________________ commands are completed, but ________________ commands are ignored.
8. In a rear load-leveling air suspension system, if the On/Off switch in the trunk is off, the system is ________________.
9. An electronic rotary height sensor contains a ________________ ________________.
10. In a continuously variable road sensing suspension system, the module senses vehicle lift and dive from some of the ________________ ________________ inputs.

Multiple Choice

1. While discussing a programmed ride control (PRC) system:
   A. The brake system pressure must be 300 psi (2,068 kPa) before this mode change occurs.
   B. A PRC system switches from the auto mode to firm mode if the vehicle accelerates with 90 percent throttle opening.
   C. The PRC system switches to the firm mode if lateral acceleration exceeds 0.25 g.
   D. The mode indicator light in the tachometer is illuminated in the plush ride mode.
2. To increase the rear trim height in an electronic air suspension system, the control module:
   A. Starts the compressor and opens the vent valve.
   B. Starts the compressor and closes the rear air spring valves.
   C. Stops the compressor and opens the rear air spring valves.
   D. Starts the compressor and opens the rear air spring valves.

3. All these statements about air springs, shock absorbers, and struts are true EXCEPT:
   A. Air springs can be mounted separately from the shock absorbers.
   B. Air springs can be mounted over the front and rear struts.
   C. Some struts contain a solenoid actuator that controls strut firmness.
   D. Some air suspension systems have seven driver selectable operating modes.
4. The magneto-rheological fluid in the shock absorbers or struts in an ESC system contains:
   A. Automatic transmission fluid.
   B. Suspended iron particles.
   C. Power steering fluid.
   D. 5W-30 engine oil.
5. To sense the distance to the vehicle in front, an adaptive cruise control system uses:
   A. A video camera.
   B. A short-range radar signal.
   C. A digital camera.
   D. A Long-range radar signal.
6. In an ESC system with magneto-rheological fluid in the shock absorbers or struts, the control module can vary the shock absorber firmness in:
   A. 1 millisecond.
   B. 5 milliseconds.
   C. 10 milliseconds.
   D. 152 milliseconds.
7. All of these statements about network systems are true EXCEPT:
   A. A network system can be a single-wire or dual-wire system.
   B. Some network systems operate between 0V and 12V.
   C. Some vehicles have two network systems.
   D. A fiber-optic network system has a very high data transmission rate.
8. In some traction control systems if the control module senses drive wheel spin, the first action taken by the control module is to:
   A. Retard the spark advance and shut off the fuel injectors.
   B. Apply the brakes on both non-drive wheels.
   C. Apply the brake on the spinning drive wheel.
   D. Apply the brake on the non-spinning drive wheel.
9. On a vehicle with a stability control system, if icy road conditions cause the vehicle to begin swerving sideways, the stability control module:
   A. Applies the brakes on both rear wheels.
   B. Applies the brakes on both front wheels.
   C. Applies the brake on one front wheel.
   D. Applies the brake on one front wheel and the opposite rear wheel.
10. All of these statements about a rear load-leveling suspension system are true EXCEPT:
    A. The On/Off switch is mounted in the vehicle trunk.
    B. The control module operates the compressor relay.
    C. If a door is open, the control module completes the lower suspension height commands.
    D. The rear suspension has one, non-serviceable suspension height sensor.

Chapter 12

Power Steering Pumps

Upon completion and review of this chapter, you should be able to understand and describe:

- The difference between a conventional V-belt and a serpentine belt.
- The advantages of a serpentine belt compared with a conventional V-belt.
- The main components in a power-assisted rack and pinion steering system, and explain the steering gear mounting position.
- Two different types of power steering pump reservoirs.
- The difference between a hydroboost power steering system and an integral power steering system.
- Three different types of power steering pump rotor designs.
- The power steering pump operation while driving with the front wheels straight ahead.
- The power steering pump operation while the vehicle is turning a corner.
- The power steering pump pressure relief operation, and explain when this operation occurs.
- The design and purpose of an electrohydraulic power steering module.
- Electrohydraulic power steering (EHPS) systems.
- Hybrib electric vehicle (HEV) operation.

Introduction

Power steering systems have contributed to reduced driver fatigue and made driving a more pleasant experience. Nearly all power steering systems at present use fluid pressure to assist the driver in turning the front wheels. Since driver effort required to turn the front wheels is reduced, driver fatigue is decreased. The advantages of power steering have been made available on many vehicles, and safety has been maintained in these systems.

There are several different types of power steering systems, including integral, rack and pinion, hydroboost, and linkage type. In any of these systems, the power steering pump is the heart of the system because it supplies the necessary pressure to assist steering.

The power steering pump drive belt is a simple, but very important, component in the power steering system. A power steering pump in perfect condition will not produce the required pressure for steering assist if the drive belt is slipping. Various types of steering systems, drive belts, and pump designs are described in this chapter.

Power Steering Pump Drive Belts

Many power steering pumps are driven by a multi-ribbed belt that surrounds the crankshaft pulley and the power steering pump pulley. If multi-ribbed belt is used to drive other components in addition to the power steering pump, such as the water pump, alternator, air-conditioning compressor it is referred to as a serpentine belt (Figure 12-1). The serpentine belt is much wider than a conventional V-belt, and the underside of the belt has a number of small ribbed grooves. Many serpentine belts have spring-loaded automatic belt tensioners that eliminate periodic belt tension adjustments. Since the serpentine belt may be used to drive all the belt-driven components, these components are placed on the same vertical plane, which saves a considerable amount of underhood space. The smooth back of the serpentine belt may also be used to drive one of the components. Regardless of the type of belt, the belt tension is critical. A power steering pump will never develop full pressure if the belt is slipping.

Serpentine belts may be constructed of either Neoprene or ethylene propylene diene M-class rubber commonly called EPDM. The majority of belts are constructed of EPDM, but vehicles produced prior to 2001 or inexpensive aftermarket replacement belts may still be made of neoprene. Visually it is difficult to tell the two materials apart. Older neoprene belts were designed to last between 50,000-60,000 miles and often showed signs of wear by that time. EPDM belts are designed to last 80,000-100,000 miles and seldom show signs of outward visual wear unless there is a problem. The EPDM belt is more elastic than a standard neoprene belt and resists cracking even at higher mileage. A better indicator of when to replace EPDM belts is rib wear. Belts are designed to have clearance between the rib peaks and the pulley grooves. All belts are exposed to dirt, grit, rocks, road salt, and water. Over time these contaminants cause the EPDM belt to gradually lose material on the ribbing similar to the way a tire wears out causing the belt to ride deeper in the grooves. A 5 to 10 percent loss of material is enough to cause belt slippage, overheating, and hydroplaning. Hydroplaning occurs when the belt ribs sit deeper in the pulley grooves due to wear and there is not enough room for water to escape. Instead the water is trapped between the belt ribs and the pulley grooves lifting the belt away from the pulley and causing slippage. A slipping serpentine belt can cause check engine lights associated with misfire codes, air-conditioning compressor codes, reduced alternator output, reduced engine cooling, and poor air-conditioning performance to name a few. Belts should be checked for wear beginning at 50,000 miles. See Chapter 3 in the Shop

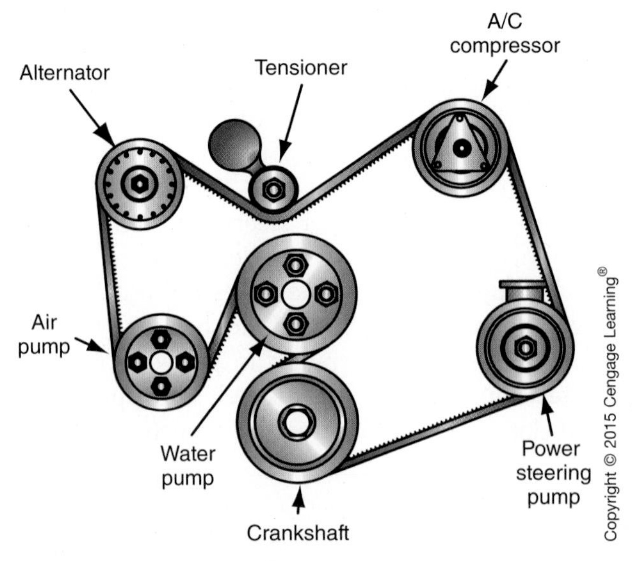

FIGURE 12-1 **Serpentine belt.**

Manual for further detail on diagnosing both Neoprene and EPDM belts, and how to measure for belt grove wear. It is wise to replace the drive belt when any pulley driven component is replaced.

Another change that has occurred in belts is the use of stretch fit belts that self tension on the drive pulleys. They do not require any mechanical adjustment and are used in limited applications, such as a single drive belt for and air-conditioning compressor. As the name implies, they contain an elastomeric material and with the use of a special tool they are stretched on to the pulley system, there is no mechanical means to loosen or tighten the belt and as with the standard EPDM belt are designed to last 80,000-100,000 miles.

Some current vehicles are equipped with a stretchy belt that does not require a belt tensioner or belt adjustment. The stretchy belt has a similar appearance compared to a conventional serpentine belt. However, stretchy belts have tensile cords made from a polyamide material that is three times more elastic compared with the cords in a conventional serpentine belt. In a stretchy belt, the rubber layers around the cord layer have superior durability and improved adhesion to the cord layer to accommodate the stretching action. Eliminating the mechanical belt tensioner saves weight, space, and financial cost. The stretchy belt is usually installed only around two pulleys such as the air-conditioning compressor and crankshaft pulleys, and the remaining belt-driven components have a serpentine-belt drive.

If a stretchy belt requires replacement, the vehicle manufacturer recommends cutting the old belt to remove it. A special tool is used to lift and guide the new stretchy belt onto the pulley.

An earlier drive belt design was the V-belt. The sides of a V-belt are the friction surfaces that drive the power steering pump (Figure 12-2). If the sides of the belt are worn and the lower edge of the belt is contacting the bottom of the pulley, the belt will slip. The power steering pump pulley, crankshaft pulley, and any other pulleys driven by the V-belt *must be properly aligned.* If these pulleys are misaligned, excessive belt wear occurs.

WARNING: Always keep hands, tools, and equipment away from rotating belts and pulleys. If any of these items become entangled in rotating belts, personal injury and equipment damage may result.

WARNING: Always keep long hair tied back while working in the automotive shop! If long hair is entangled in rotating belts and pulleys, personal injury will result.

Shop Manual
Chapter 12,
page 460

AUTHOR'S NOTE: It has been my experience that the most common cause of power steering pump complaints is the pump drive belt. A loose, worn, or dry belt may cause a chirping noise at idle or squealing during acceleration. A loose or worn belt may cause a humming-type vibration noise. Intermittent or continual hard steering may be caused by a loose pump drive belt. A loose power steering pump belt may even cause a complaint of hard steering when driving on wet days, because the belt has more tendency to slip when it is wet. A loose belt may cause low power steering pump pressure when testing the pump. Therefore, when diagnosing power steering pump problems, you should always inspect the pump drive belt and measure the belt tension before any other diagnostic tests are performed.

A **remote reservoir** is mounted externally from the power steering pump.

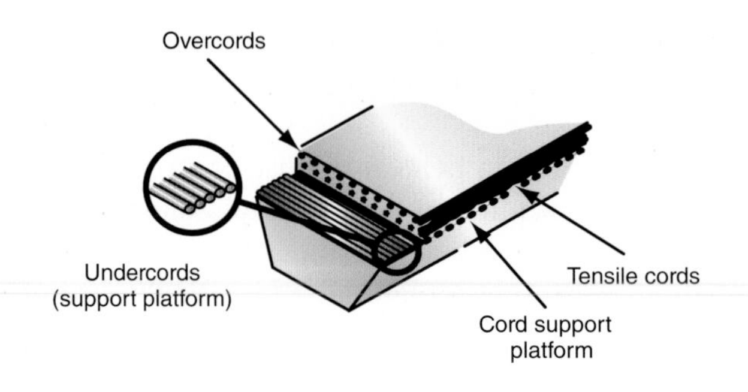

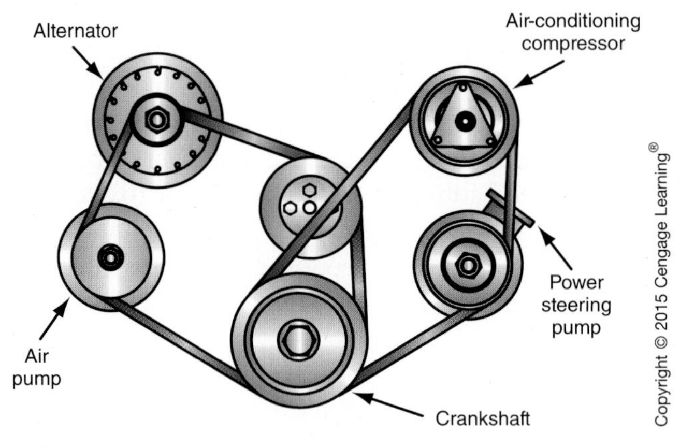

FIGURE 12-2 **Conventional V-belt.**

Types of Power-Assisted Steering Systems

Rack and Pinion Steering System

The **rack and pinion power steering system** is used on most front-wheel-drive cars. In this steering system, the power steering pump is bolted to a bracket on the engine, and the pump is driven by a belt from the crankshaft. In most front-wheel-drive cars, the engine is mounted transversely, and the steering gear is mounted on the cowl behind the engine or on the crossmember below the engine (Figure 12-3).

In many power steering systems, an **integral fluid reservoir** is attached to the power steering pump. A dipstick is mounted in the reservoir for fluid level checking (Figure 12-4).

In an **integral power-assisted steering system**, the control valve and power cylinder are contained in the steering gear.

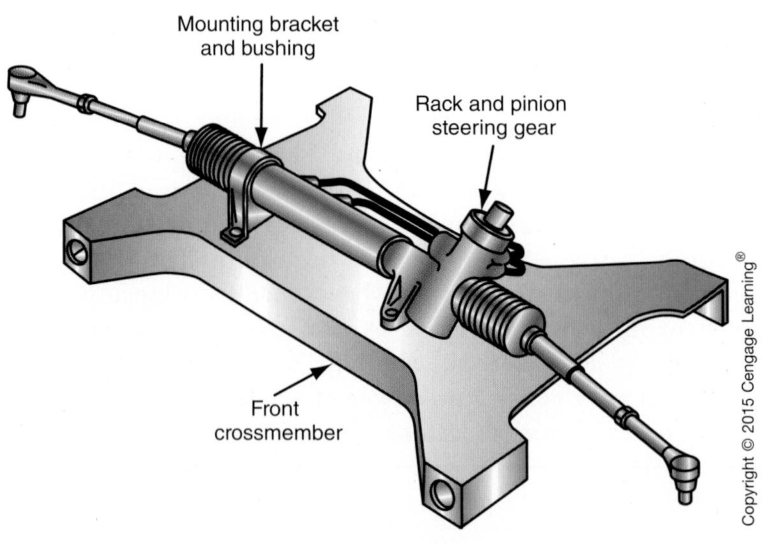

FIGURE 12-3 **Power steering gear mounting.**

FIGURE 12-4 **Power steering reservoir and dipstick.**

High-pressure hose

Rack and pinion steering gear

Low-pressure hose

FIGURE 12-5 **Power steering pump to gear hoses.**

A high-pressure hose and a low-pressure return hose are connected from the pump to the steering gear (Figure 12-5). The high-pressure hose is usually a steel-braided hose with appropriate fittings in each end. The return hose is a rubber-braided hose that is clamped to the fittings on the pump and gear.

Some power steering pumps have a **remote fluid reservoir**. This type of system is commonly used on small cars with limited underhood space where it may be difficult to access an integral reservoir on the power steering pump. The remote reservoir is placed in a convenient position, and the return hose is connected from the steering gear to the reservoir. A second return hose is routed from the reservoir to the pump (Figure 12-6).

A BIT OF HISTORY

The number of independent automotive repair facilities and the number of individuals employed by these shops have increased steadily in the last decade. In 1995, there were 63,844 independent automotive mechanical repair facilities in the United States, and they employed 244,430 workers. In 2010, 723,400 technicians and mechanics were employed in the United States with an expected growth of 17 percent (or 124,800 jobs) over the next 10 years.

Integral Power-Assisted Steering System

In the **integral power-assisted steering system**, the pump is bolted to a bracket on the engine, and the recirculating ball steering gear is mounted on the frame beside the engine. This type of steering system is used on many rear-wheel-drive cars and light-duty trucks. The pump is driven by a belt from the crankshaft and an integral reservoir is mounted on the pump. A high-pressure hose and a return hose are connected from the pump to the steering gear (Figure 12-7).

Hydroboost Power-Assisted Steering System

The **hydroboost power-assisted steering system** is used on many light-duty trucks with diesel engines and on some cars. Since the hydroboost system does not use manifold vacuum for brake assist, this system is suitable for diesel engines, which have reduced intake manifold vacuum compared with gasoline engines. The power steering on these systems is similar to an integral unit, but the power steering pump pressure is also applied to the hydroboost unit in the brake master cylinder. Hydraulic lines are routed from the power steering pump to the steering gear, and another set of lines is connected from the pump to the master cylinder (Figure 12-8). In the hydroboost system, the power steering pump pressure applied to the master cylinder pistons acts as a brake booster to assist the driver in applying the brakes. The hydroboost system does not have a conventional vacuum booster on the master cylinder.

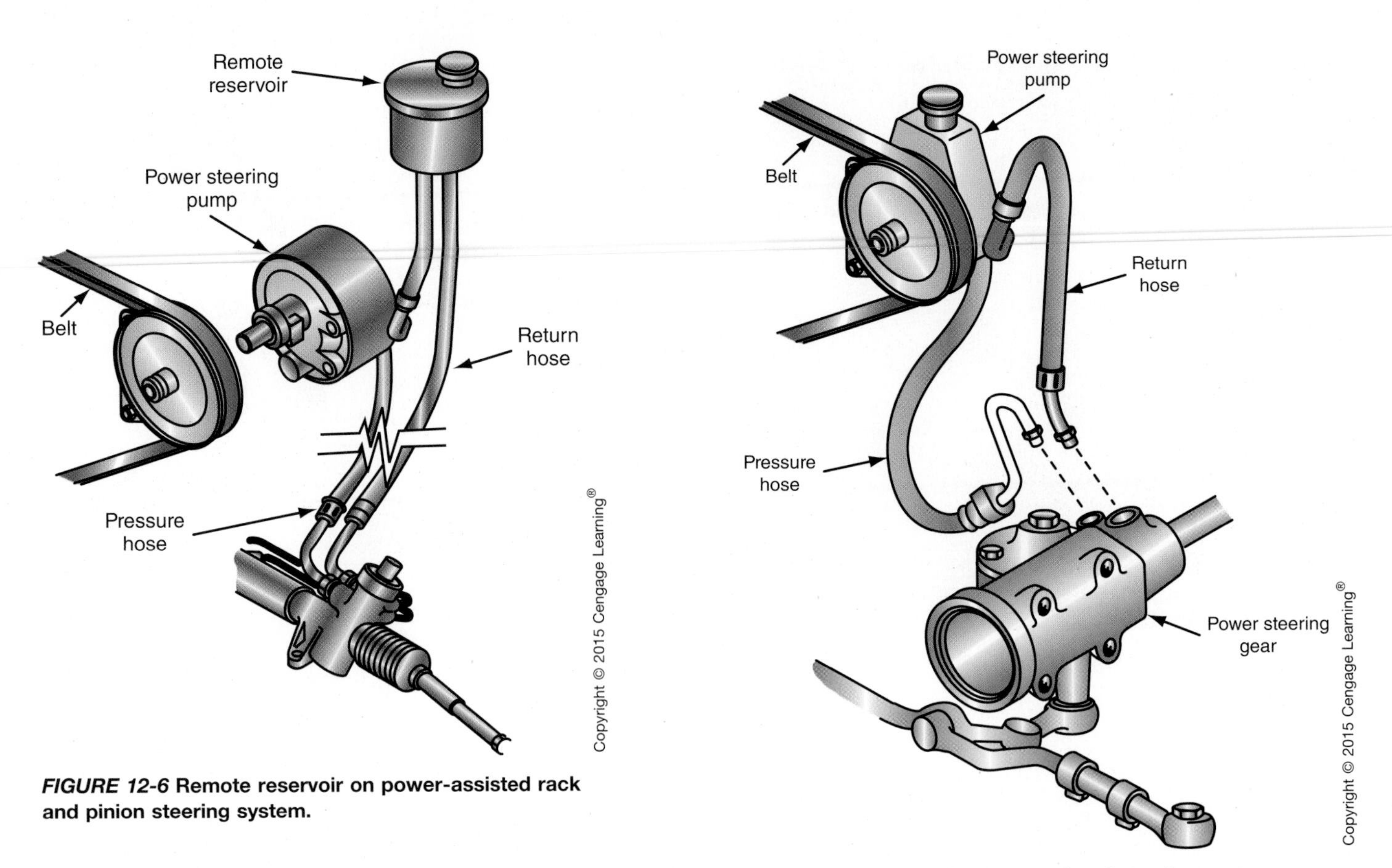

FIGURE 12-6 **Remote reservoir on power-assisted rack and pinion steering system.**

FIGURE 12-7 **Integral power-assisted steering system.**

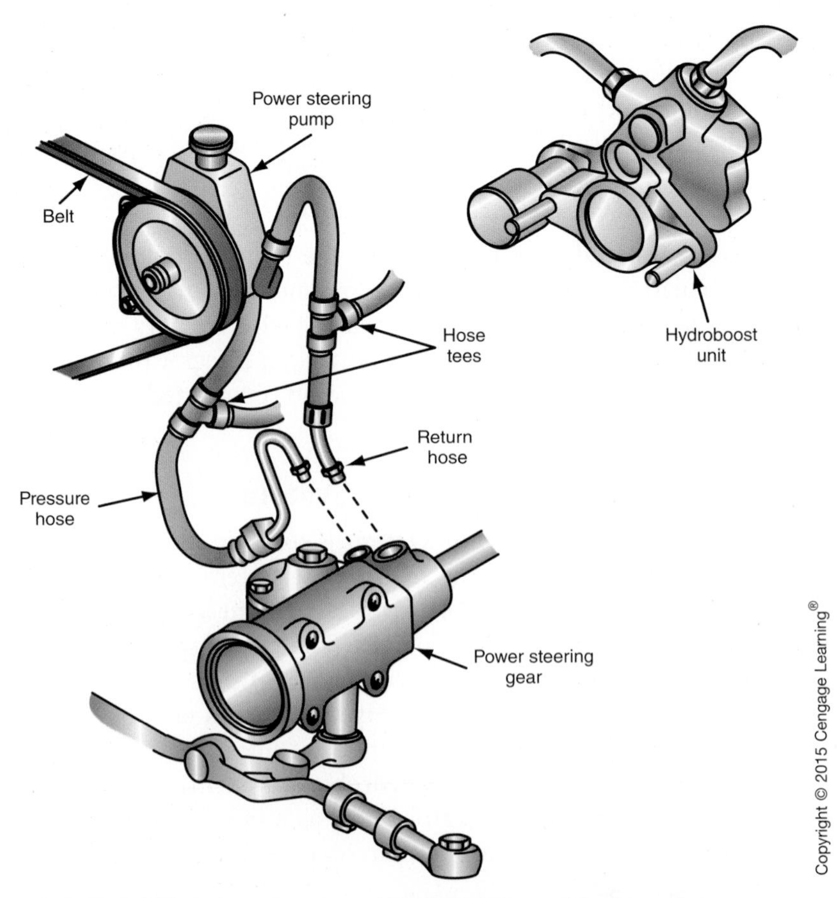

FIGURE 12-8 **Hydroboost power-assisted steering and brake system.**

Power Steering Pump Design

Various types of power steering pumps have been used by car manufacturers. Many **vane-type power steering pumps** have flat vanes that seal the pump rotor to the elliptical pump cam ring (Figure 12-9). Other vane-type power steering pumps have rollers to seal the rotor to the cam ring. In some pumps, inverted, U-shaped slippers are used for this purpose. The major differences in these pumps are in the rotor design and the method used to seal the pump rotor in the elliptical pump ring. The operating principles of all three types of pumps are similar.

A balanced pulley is pressed on the steering pump drive shaft. This pulley and shaft are belt-driven by the engine. A spring-loaded lip seal at the front of the pump housing prevents fluid leaks between the pump shaft and the housing. The oblong pump reservoir is made from steel or plastic. A large O-ring seals the front of the reservoir to the pump housing (Figure 12-10).

Shop Manual Chapter 12, page 474

Smaller O-rings seal the bolt fittings on the back of the reservoir. The combination cap and dipstick keep the fluid reserve in the pump and vent the reservoir to the atmosphere. Some power steering pumps have a variable assist steering actuator in the back of the pump housing. The PCM operates this actuator to provide increased steering assist at low vehicle speeds.

Shop Manual Chapter 12, page 475

The rotating components inside the pump housing include the shaft and rotor with the vanes mounted in the rotor slots. A seal between the output shaft and the housing prevents oil leaks around the shaft. As the pulley drives the pump shaft, the vanes rotate inside an

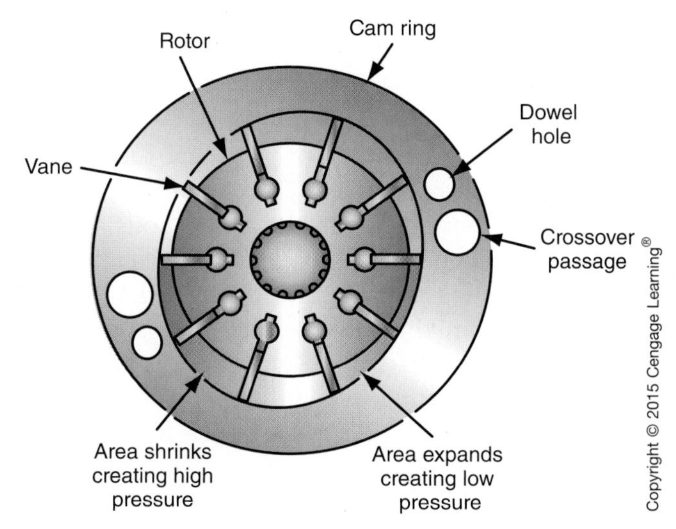

FIGURE 12-9 **Power steering pump rotor and vanes.**

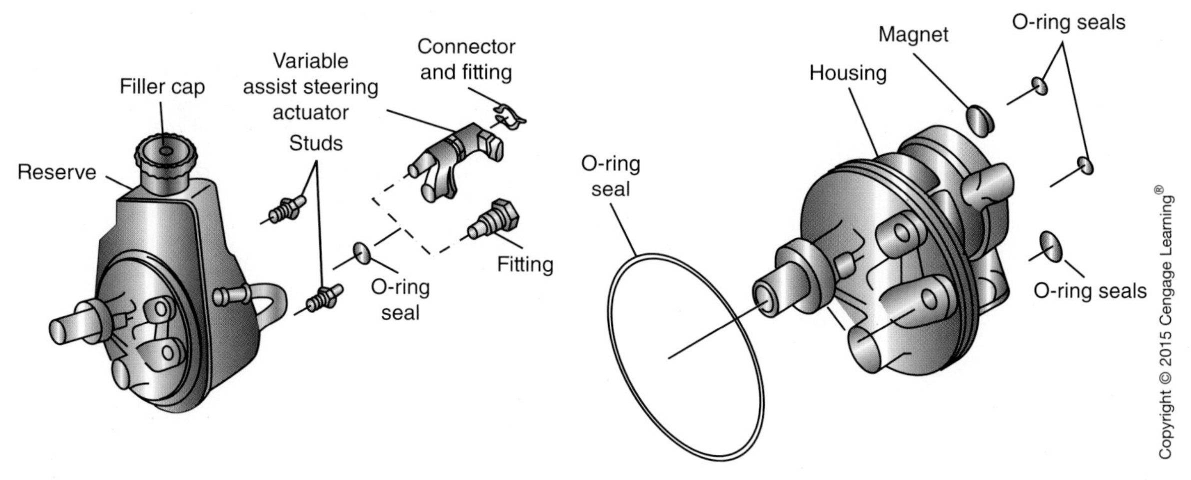

FIGURE 12-10 **Power steering pump housing and reservoir.**

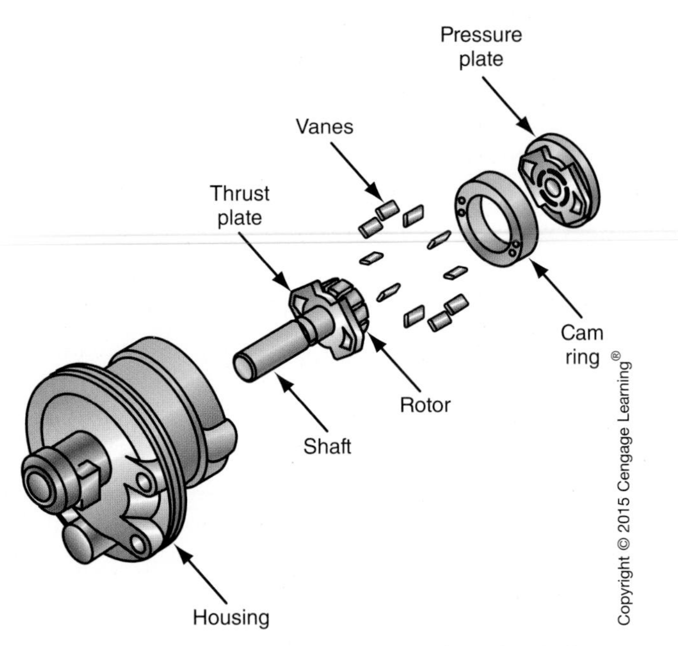

FIGURE 12-11 **Power steering pump housing with shaft and rotor, vanes, cam ring, and pressure plate.**

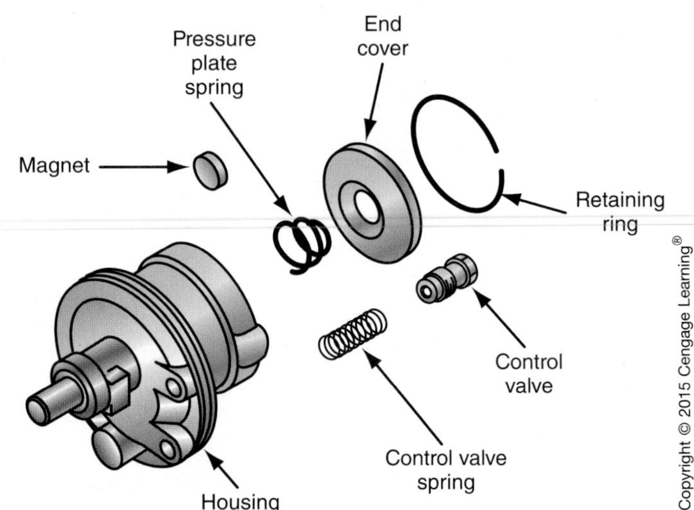

FIGURE 12-12 **Power steering pump housing assembly with end cover, flow control valve, and magnet.**

elliptically shaped opening in the **cam ring**. The cam ring remains in a fixed position inside the pump housing. A pressure plate is installed in the housing behind the cam ring (Figure 12-11).

CAUTION:
Avoid prying on the power steering pump reservoir, because this action can damage or puncture the reservoir.

A spring is positioned between the pressure plate and the end cover, and a retaining ring holds the end cover in the pump housing. The flow control valve is mounted in the pump housing, and a magnet is positioned on the pump housing to pick up metal filings rather than allowing them to circulate through the power steering system (Figure 12-12).

The flow control valve is a precision-fit valve controlled by spring pressure and fluid pressure. Any dirt or roughness on the valve results in erratic pump pressure. The flow control valve contains a pressure relief ball (Figure 12-13). High-pressure fluid is forced past the control valve to the outlet fitting. A high-pressure hose connects the outlet fitting to the inlet fitting on the steering gear. A low-pressure hose returns the fluid from the steering gear to the inlet fitting in the pump reservoir.

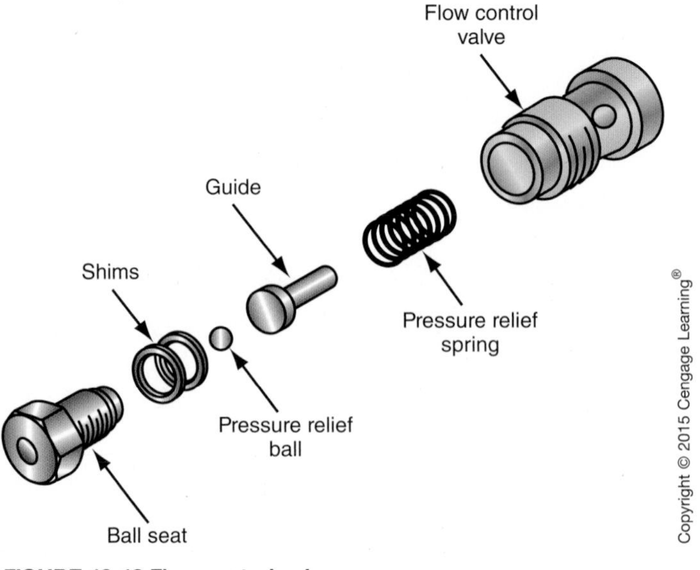

FIGURE 12-13 **Flow control valve.**

POWER STEERING PUMP OPERATION

Vane and Cam Action

As the belt rotates the rotor and vanes inside the cam ring, centrifugal force causes the vanes to slide out of the rotor slots. The vanes follow the elliptical surface of the cam. When the area between the vanes expands, a low-pressure area occurs between the vanes and fluid flows from the reservoir into the space between the vanes. As the vanes approach the higher portion of the cam, the area between the vanes shrinks and the fluid is pressurized.

High-pressure fluid is forced into two passages on the thrust plate. These passages reverse the fluid direction. Fluid is discharged through the cam crossover passages and pressure plate openings to the high-pressure cavity and **flow control valve**. The fluid is discharged from the flow control valve through the outlet fitting (Figure 12-14). Though most of the fluid is discharged from the pump, some fluid returns to the bottom of the vanes past the flow control valve and through the bypass passage.

The **flow control valve** controls the fluid from the pump to the steering gear.

A machined **venturi** orifice is located in the outlet fitting. Fluid passage applies pressure from the venturi area to the spring side of the flow control valve. When fluid flow in the venturi increases, the pressure in this area decreases.

A **venturi** is a narrow area in a pipe through which a liquid or gas is flowing. When liquid or gas flow increases in the venturi, the pressure drops proportionally. If liquid or gas flow decreases, the pressure in the venturi increases.

Power Steering Pump Operation at Idle Speed

At idle speed with the wheels straight ahead, the flow of fluid from the pump to the steering gear is relatively low. Under this condition, the steering gear valve is positioned so the fluid discharged from the pump is directed through the valve and the low-pressure hose to the pump vane inlet. With low fluid flow, the pressure is higher in the venturi and control orifice. This higher pressure is applied to the spring side of the flow control valve and helps the valve spring keep this flow control valve closed (Figure 12-15). Under this condition, pump pressure remains low, and all the fluid is discharged from the pump through the steering gear valve and back to the pump inlet.

Power Steering Pump Operation at Higher Engine Speeds

When engine speed is increased, the power steering pump delivers more fluid than the system requires. This increase in fluid flow creates a pressure decrease in the **outlet fitting venturi**. This pressure reduction is sensed at the spring side of the flow control valve, which allows the pump discharge pressure to force the flow control valve partially open. Under this condition, the excess fluid from the pump is routed past the flow control valve to the pump inlet. If the steering wheel is turned, the fluid discharged from the pump rushes into the pressure chamber in the steering gear.

The **outlet fitting venturi** controls the pressure supplied to the flow control valve.

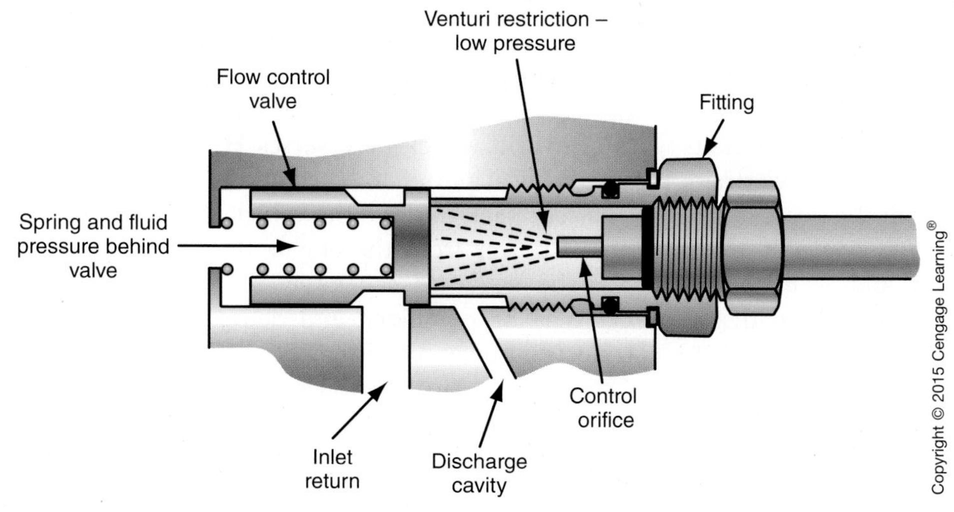

FIGURE 12-14 **Power steering pump fluid flow to control valve and outlet fitting.**

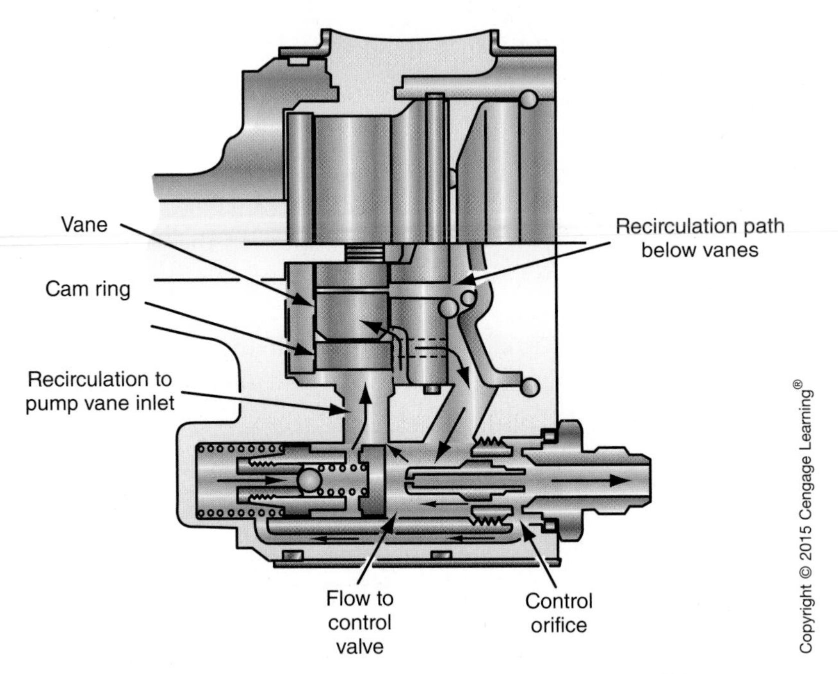

FIGURE 12-15 **Flow control valve operation under various conditions.**

Shop Manual
Chapter 12,
page 473

Power Steering Pump Operation During Pressure Relief

If the steering wheel is turned fully in either direction until the steering linkage contacts the steering stops, the rack piston is stopped. Under this condition, pump pressure could become extremely high and damage hoses or other components. When the rack piston stops, the flow from the pump decreases, but the high pump pressure is still directed through the steering gear valve to the rack piston. Since the flow of fluid through the venturi and control orifice is reduced, a higher pressure is present in this area. This extremely high pressure is also supplied to the spring end of the flow control valve. At a predetermined pressure, the **pressure relief ball** in the center of the flow control valve is unseated and the flow control valve moves to the wide-open position. This action allows some of the high pressure in the steering system to return to the pump inlet, which limits pump pressure to a maximum safe value.

The **pressure relief ball** limits the maximum power steering pump pressure.

Hybrid Vehicles and Power Steering Systems

Advantages and Types of Hybrid Electric Vehicles

Shop Manual
Chapter 12,
page 478

Hybrid electric vehicles (HEVs) have two power sources: typically a small displacement gasoline engine and a high-voltage battery pack that supplies voltage and current to an electric drive motor(s).

The two most common types of hybrid vehicles are **series HEVs** and **parallel HEVs**. In a series HEV powertrain, the mechanical power from engine is combined with electric power from the generator and/or battery pack to drive the vehicle (Figure 12-16). In a parallel HEV powertrain, the mechanical power from the engine or electric power from the generator and/or battery pack may be delivered separately to the drive wheels or the power from both of these sources may be combined and delivered to the drive wheels (Figure 12-17). In a parallel HEV, electric power alone may be supplied to the drive wheels to drive the vehicle, whereas in a series HEV, electric power cannot drive the vehicle.

In a **series HEV**, electric power from the battery pack and/or generator and mechanical power from the engine are supplied together to the drive wheels.

The advantages of an HEV are increased fuel economy and reduced emissions. It is extremely important that all other systems and vehicle design are engineered to help achieve these objectives. Many HEVs have a smaller displacement engine compared with

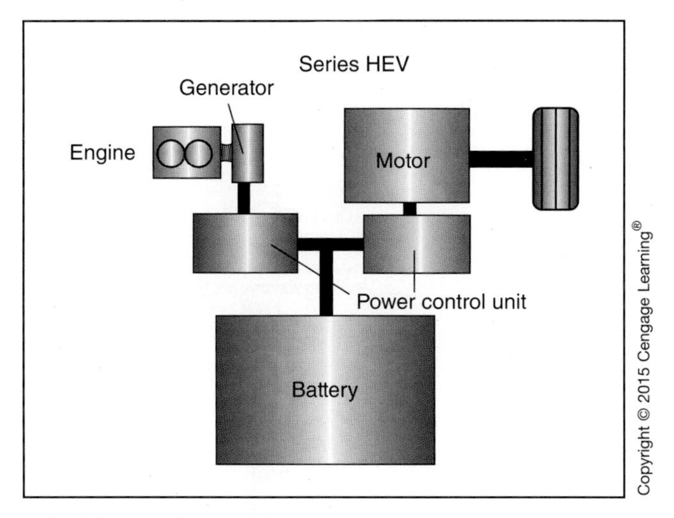

FIGURE 12-16 **Series HEV.**

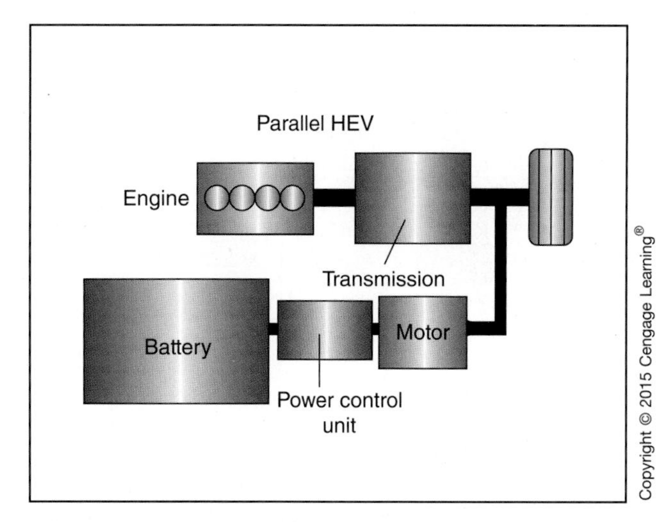

FIGURE 12-17 **Parallel HEV.**

an equivalent-size non-hybrid vehicle. In a typical HEV, the smaller displacement engine coupled with the electric propulsion motor provides equivalent, or nearly equivalent, power compared with that provided in a non-hybrid vehicle. The smaller displacement engine in the HEV supplies improved fuel economy compared with that provided by the larger engine in the non-hybrid vehicle. Improved fuel economy is also achieved because the electric propulsion motor supplies some of the power to the drive wheels. In many HEVs, under specific operating conditions, the engine is shut off and power is supplied to the drive wheels only by the electric motor to provide an additional fuel saving.

In a **parallel HEV**, electric power from the battery pack and/or generator and mechanical power from the engine may be supplied separately to the drive wheels or these two power sources may be combined.

On many HEVs, the engine is stopped to conserve fuel and reduce emissions when the vehicle is standing still and the engine is warmed up and idling. Under this condition, the power steering must be active and ready to provide power steering assist if the driver turns the steering wheel. To meet this requirement, electrohydraulic power steering pumps are presently installed on some HEVs.

Hybrid Powertrain Components

Engine to Transaxle Coupling

In some HEVs, the engine and propulsion motors are coupled through a torque splitting planetary gear set to an electronically controlled continuously variable transaxle (CVT). The engine is coupled to the carrier in the planetary gear set (Figure 12-18). The propulsion motor is connected to the planetary ring gear, and the generator is attached to the planetary sun gear. The drive gear that transfers torque to the transaxle is attached to the electric motor drive shaft. This arrangement allows torque from the engine, the propulsion motor, or both the engine and propulsion motor to be transferred to the transaxle and drive wheels. The sun gear is the smallest gear in the planetary gear set.

Propulsion Motor/Generator

The propulsion motor is an alternate current (AC) synchronous permanent magnet, liquid-cooled type. Some propulsion motors are rated at 40 hp (30 kW) at 940 to 2,000 rpm and 225 ft.-lb (305 Nm) peak torque at 0 to 940 rpm. During deceleration and braking, the propulsion motor acts as a generator to recharge the batteries. This action is called **regenerative braking**. In Figure 12-19, the propulsion motor is identified as MG2 and the generator is identified as MG1. Some propulsion motors have higher torque and horsepower output depending on the motor design. Later model propulsion motors are capable of turning at speeds in excess of 12,000 rpm.

Regenerative braking occurs during vehicle deceleration when the drive motor becomes a generator and supplies current to recharge the batteries.

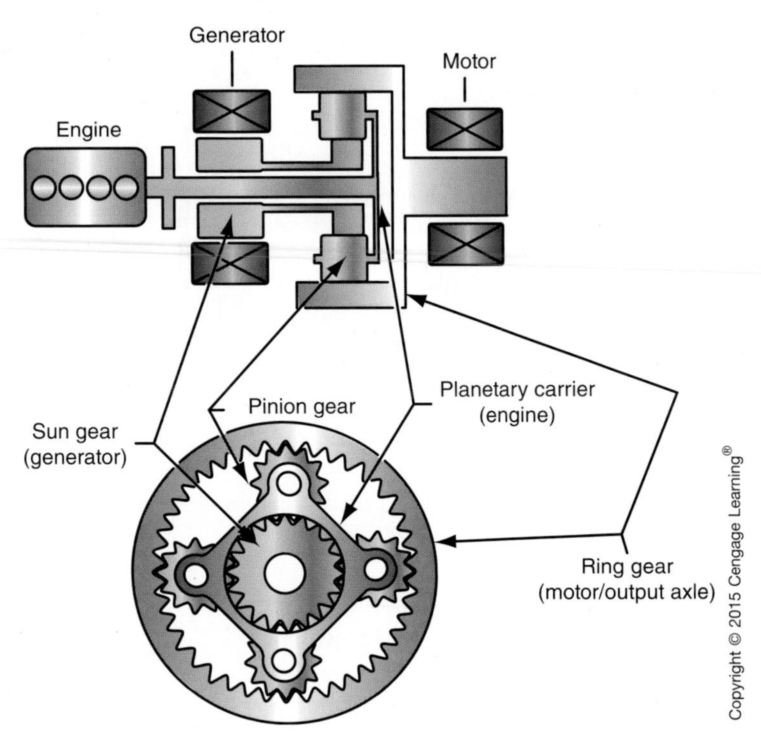

FIGURE 12-18 **Planetary gear coupling between engine, electric motor, and transaxle.**

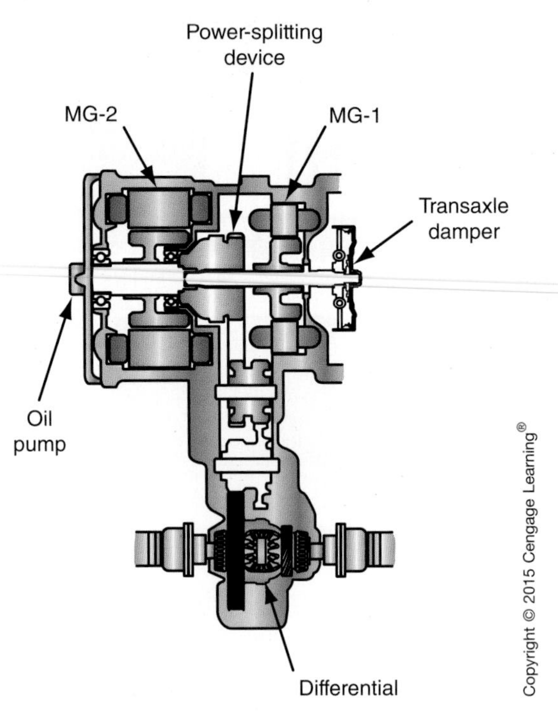

FIGURE 12-19 **Electric drive motor (MG-2) and generator (MG-1).**

Generator/Starter

The generator is an AC synchronous type that produces voltage and current to recharge the batteries and power the electrical system. When starting the engine, current is supplied through the generator stator windings, and the generator cranks the engine. Under normal driving conditions, current from the generator is routed to the electric drive motor to increase the torque supplied to the drive wheels. In later model HEVs, generator design has been improved to allow higher generator rpm and improved generator output. In some current HEVs, the generator is capable of turning at 10,000 rpm, whereas some early model generators turned at 6,500 rpm.

Inverter

The inverter contains the power control unit that controls the voltage and current supplied by the batteries to the propulsion motor. The battery pack supplies high direct current (DC) voltage to the inverter and the inverter converts the DC voltage to three-phase AC voltage that is supplied to the propulsion motor. The inverter also controls the voltage and current supplied from the propulsion motor to the batteries during regenerative braking. Because the voltage and current supplied from the batteries to the propulsion motor may be very high, the inverter contains insulated gate bipolar transistors (IGBT) to control this circuit (Figure 12-20). This high voltage and current controlled by the inverter produces a considerable amount of heat, and the inverter is cooled by a dedicated cooling system to dissipate this heat. Heavy cables are connected between the batteries, inverter, and the propulsion motor. For easy identification, these cables and connectors have orange insulation.

WARNING: The high voltage in circuit between the batteries, inverter, and propulsion motor could electrocute a technician. It is very important to follow the vehicle manufacturer's recommended procedures when diagnosing and servicing HEVs.

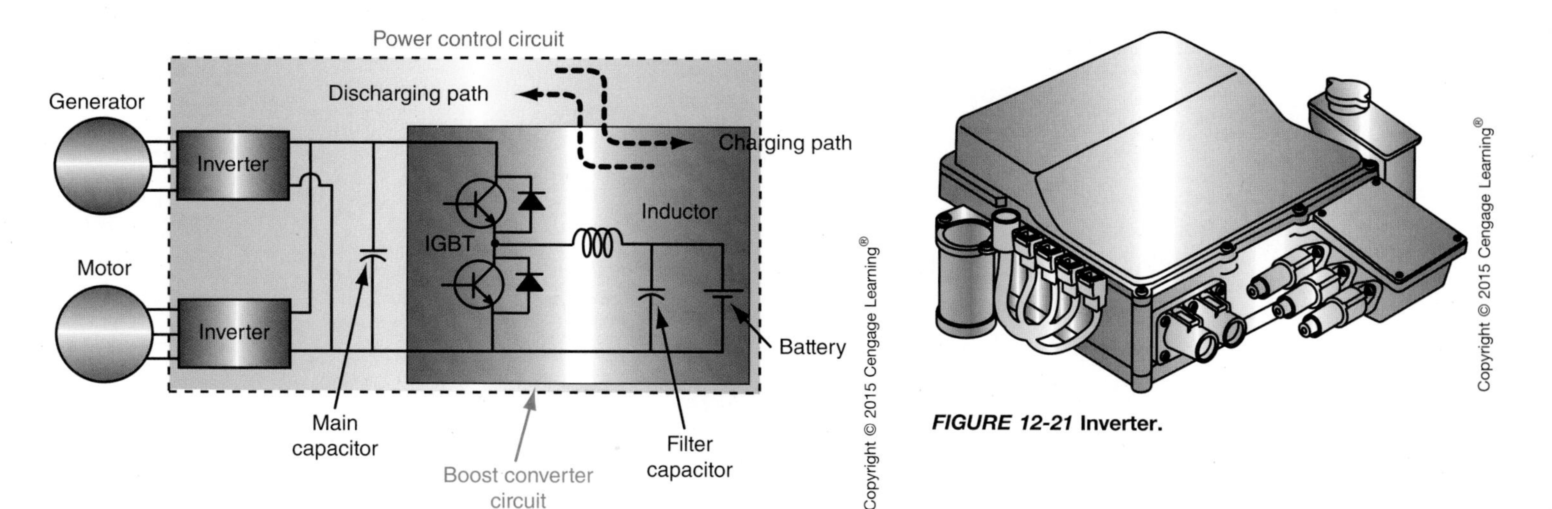

FIGURE 12-20 **High voltage power circuit in the inverter.**

FIGURE 12-21 **Inverter.**

WARNING: Never touch, cut, pierce, or open any orange high voltage cable, or high voltage component. This action may cause severe electrical shock or electrocution.

The AC voltage and current supplied from the generator to the 12V battery and electrical accessories is converted to DC voltage by the inverter and controlled in 14V–15V range. The inverter is illustrated with the cover removed in Figure 12-21.

Battery Pack and Related Cables

Some battery packs contain 240 cylindrical nickel-metal-hydride cells connected in series. Each cell has 1.2V for a total of 288V. Some HEVs have a higher voltage depending on the number of cells in the battery. The battery cells are connected in groups, with six cells in each group. The battery pack is installed behind the rear seat in some HEVs. The electrolyte is a gel in the nickel-metal-hydride battery pack; therefore, leakage is not a great concern. A cooling fan helps cool the battery pack. In the unlikely case of battery overcharging, a battery pack vent hose allows vapors from the battery pack to be vented outside the vehicle trunk.

Many battery packs have a long warranty period such as 10 year, 90,000 mi. (150,000 km). Positive and negative high voltage cables are connected between the battery pack and the inverter. These cables are routed under the vehicle floor pan and they are completely insulated from the chassis to avoid any possibility of electrical shock when touching the chassis.

HEV components increase vehicle weight, which reduces vehicle performance and increases fuel consumption. Therefore, it is very important to reduce the weight of HEV components and also design other vehicle components with less weight.

Some HEVs are currently equipped with lithium-ion batteries, which are 40 percent lighter and take up 24 percent less space compared with a nickel-metal-hydride battery with the same power output.

AUTHOR'S NOTE: One of the major Japanese vehicle manufacturers has just introduced a lithium-ion battery with a new internal design that has only one-half the volume but 1.5 times the power output compared with previous lithium-ion batteries.

HEV Indicators

Some HEVs have normally open battery-pack relays that open and close the circuit between the battery pack and the inverter. The vehicle computer controls these relays.

A READY indicator in the dash informs the driver when the relays are closed and the batteries are connected to the inverter. When the ignition switch is off, the relays in the high voltage circuit are open and the READY indicator is off. If the ignition switch is on and the relays are closed, voltage is supplied from the battery pack to the inverter, and the READY indicator is illuminated (Figure 12-22). The engine may stop and start any

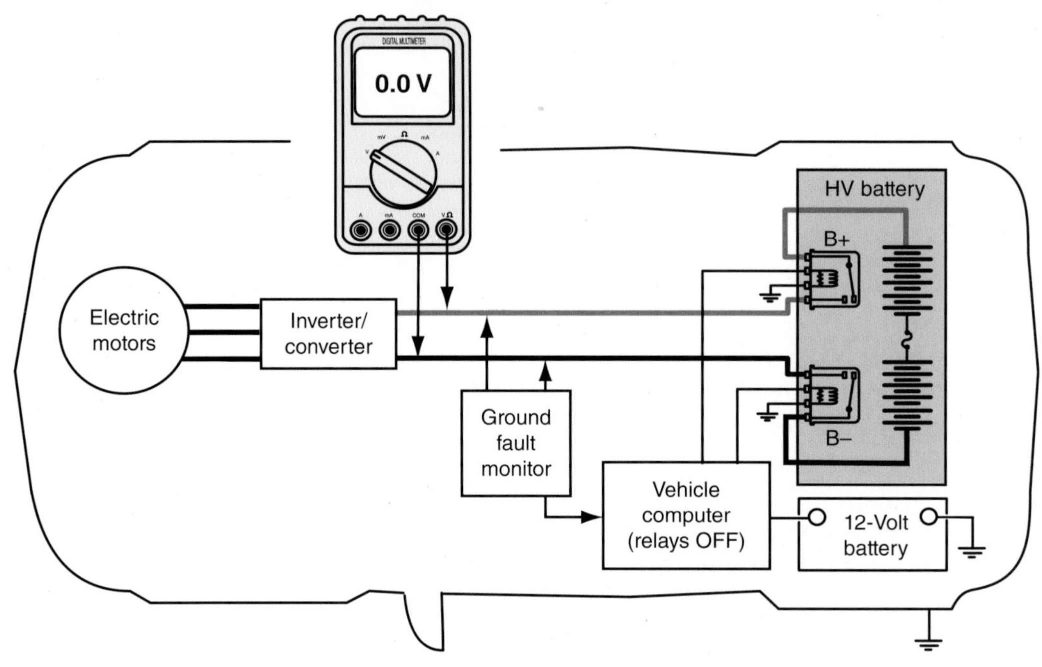

High Voltage System – Vehicle Shut Off (READY -off)

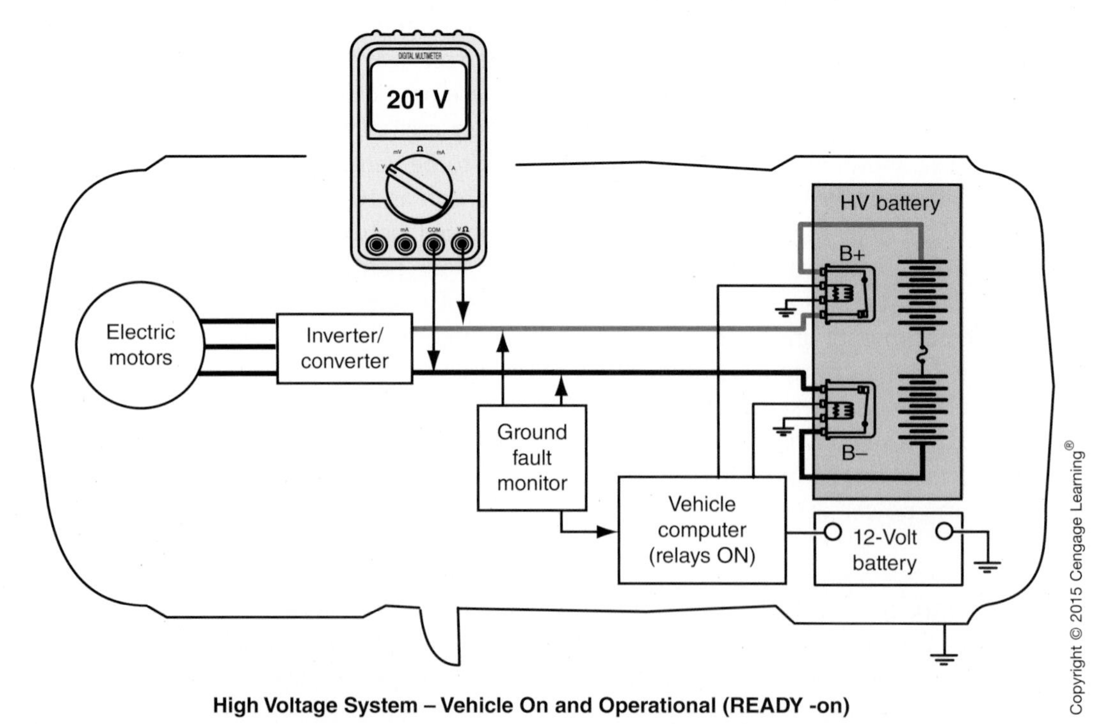

High Voltage System – Vehicle On and Operational (READY -on)

FIGURE 12-22 **High voltage electrical system.**

FIGURE 12-23 **Master warning light.**

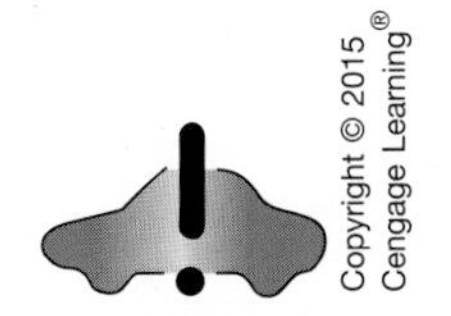

FIGURE 12-24 **Hybrid warning light.**

time while the READY indicator is illuminated. Never assume the electrical system is off just because the engine is stopped. The electrical system is off when the READY indicator is off. If the vehicle is involved in a collision, and the air bags are deployed, the battery pack relays automatically open and disconnect the voltage supplied from the battery pack to the inverter. When the engine is running, a ground fault monitor continuously monitors the high voltage system for high voltage leakage to the metal chassis. If leakage occurs, the vehicle computer illuminates the master warning light in the instrument cluster (Figure 12-23) and the hybrid warning light in the liquid crystal display (LCD) (Figure 12-24). The conventional 12V lead acid battery is located in the truck near the battery pack. The positive cable from the 12V battery is routed to the engine compartment with the high voltage cables from the battery pack. HEV systems and indicators vary depending on the vehicle make and model year.

HEV Operation

When the engine is at normal operating temperature, and the vehicle is starting off from a stop or operating at very low speed and light load, only electrical power from the battery pack is supplied to the propulsion motor to drive the vehicle (Figure 12-25).

During normal driving, mechanical power from the engine and electric power supplied from the generator to the propulsion motor are used to drive the vehicle (Figure 12-26).

If the engine is operating at wide throttle opening, mechanical power from the engine and electric power from the generator and battery pack are supplied to drive the vehicle (Figure 12-27).

During vehicle deceleration, the propulsion motor acts as a generator and supplies current to charge the batteries (Figure 12-28).

If the battery state of charge becomes low, more of the electric power from the generator is transmitted to the battery pack for recharging (Figure 12-29).

Power Steering and Hybrid Vehicles

Many HEVs have a stop/start function that allows the engine to be shut off when the vehicle has come to a complete stop, with the engine idling. When the ignition switch is on and the driver steps on the accelerator pedal, the vehicle starts off, and the engine may restart

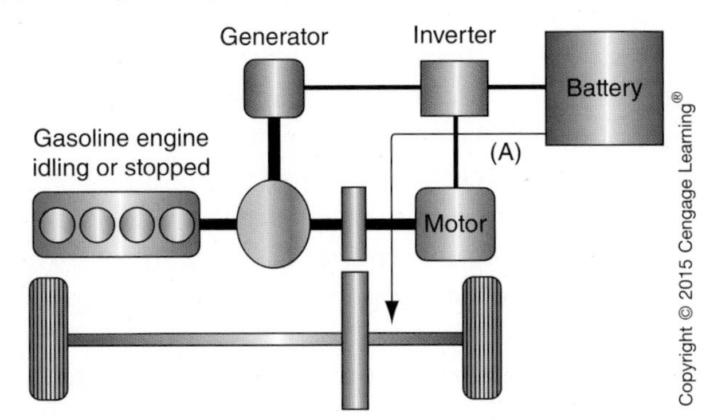

FIGURE 12-25 **HEV starting off or operating at low speed and light load.**

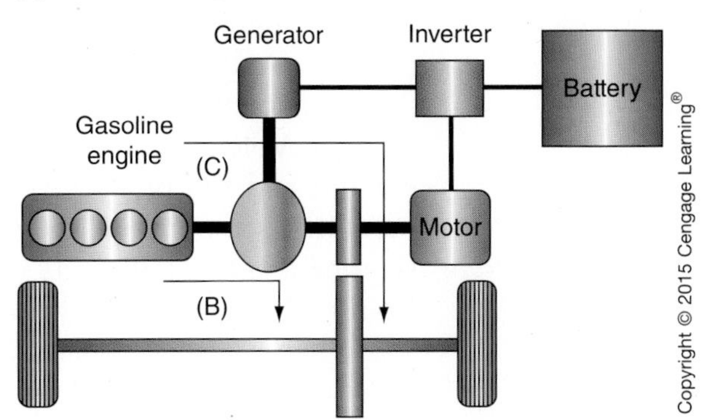

FIGURE 12-26 **HEV operation during normal driving.**

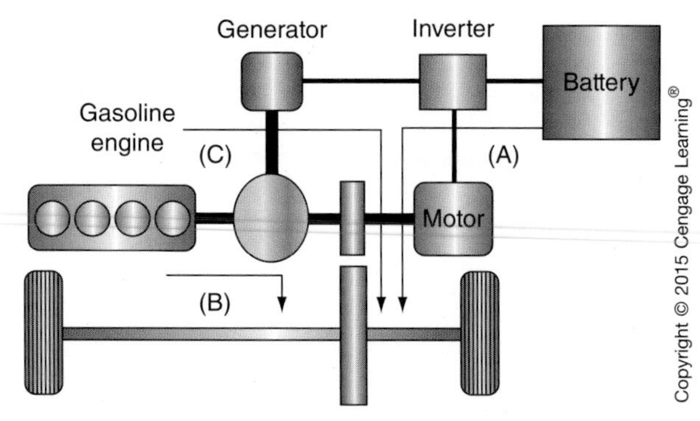

FIGURE 12-27 **HEV operation at wide open throttle.**

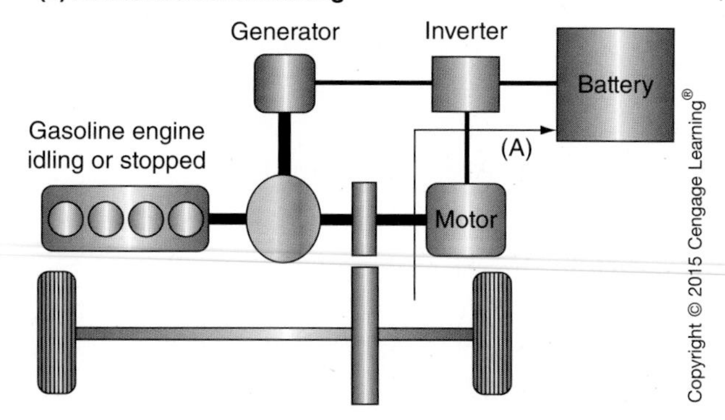

FIGURE 12-28 **HEV operation during deceleration.**

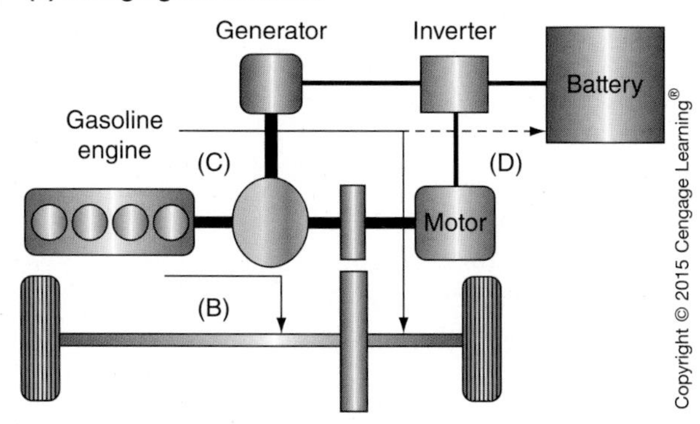

FIGURE 12-29 **HEV operation with low battery state of charge.**

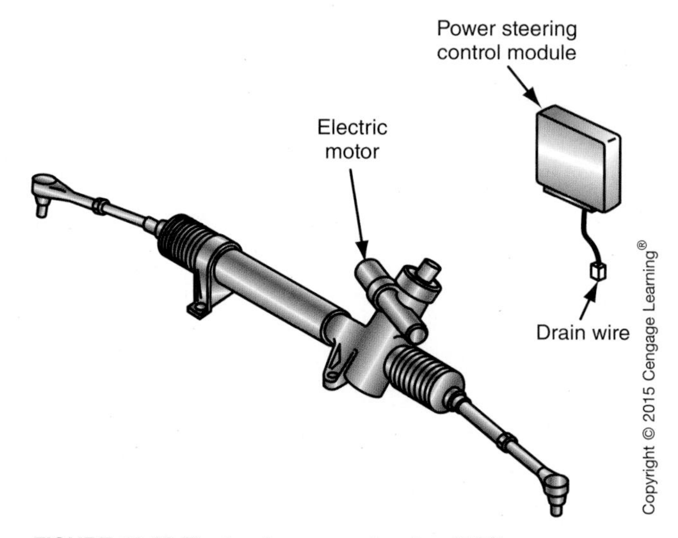

FIGURE 12-30 **Electronic power steering (EPS).**

at any time depending on throttle opening and other factors. The power steering must be operational even during the engine stop part of the stop/start function. To maintain power steering operation during the stop function, some vehicle manufacturers use an electronic power steering (EPS) gear. Some of these EPSs have an electric motor driving the steering gear pinion (Figure 12-30). A power steering control module supplies current to the EPS motor to supply the proper amount of steering assist. The EPS is fully operational when the ignition switch is turned on.

Other HEVs have an electrohydraulic power steering pump to maintain power steering pump operation during the stop function. This type of pump has an electric motor driving the pump rather than a belt drive (Figure 12-31). The electrohydraulic power steering pump contains the pump and electric drive motor, reservoir with filler cap, and the control module. Voltage signals are sent from the steering wheel position sensor in the steering column and the brake pedal position sensor actuated by the brake pedal to the control module. In response to these signals, the control module determines the amount of power steering assist required. In some electrohydraulic power steering pumps, 42V are supplied from the inverter in the HEV to the power steering pump.

Shop Manual
Chapter 12, page 486

When a higher voltage is supplied to the power steering pump motor, less current is required to drive the motor. For example, electrical energy is measured in watts (W) and watts are calculated by multiplying the amperes by the volts. When 12V are supplied to a

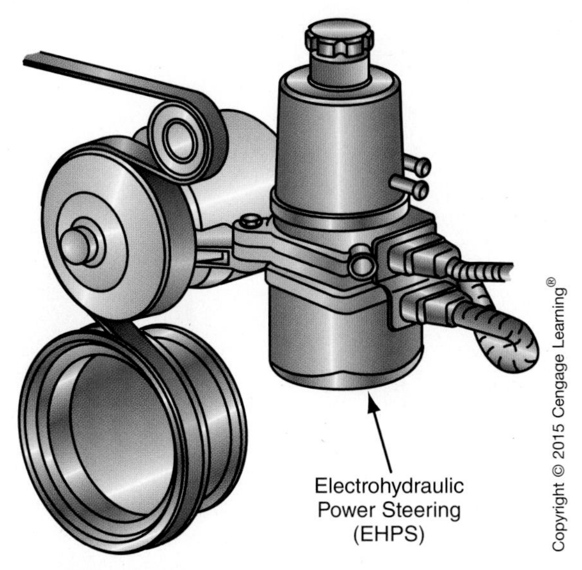

FIGURE 12-31 **Electrohydraulic power steering (EHPS) module.**

headlight, 3 amperes may be required to illuminate the light. The electrical energy required to illuminate the light is $12 \times 3 = 36$ W. If 18V are supplied to the headlight, only 2 amperes are required to illuminate the light. The advantage of increasing the voltage is that the same watts can be obtained with fewer amperes.

AUTHOR'S NOTE: In 2008, one leading manufacturer's worldwide sales of HEVs surpassed 1.5 million units. The average silicon content on a non-hybrid vehicle consumes approximately 30 percent of an 8 in. (3.1 cm) silicon wafer. The silicon content of a hybrid vehicle requires the full 8 in. (3.1 cm) wafer. Approximately, 50 percent of the cost of a hybrid vehicle is in the hardware and software. Toyota has taken 70 percent of the cost of the hybrid technology out of the hybrid Prius, improved its fuel economy, and increased the size of the chassis/body.

Various Types of HEVs

Belt-Driven HEVs

Some HEVs have a belt-driven motor/generator with an improved drive belt and tensioner compared with a conventional generator (Figure 12-32). The latest belt-driven HEV system features a stop/start function, regenerative braking, intelligent battery charging, and 4 kW of energy are supplied from the battery pack to the motor/generator for up to 5 seconds during wide throttle, heavy load driving. Some belt-driven HEVs supply a modest electric-only propulsion below 8 mph (8 km/h). Lithium-ion batteries will soon be available in some belt-driven HEV systems.

Plug-in HEVs

Several major vehicle manufacturers are scheduled to introduce plug-in HEVs (PHEVs) in their showrooms during 2010. A PHEV may be driven only by an electric motor(s), but a small gasoline or diesel engine is installed in the vehicle for extra power or to drive a generator and recharge the batteries. A typical PHEV may be driven for 40 mi. (64 km) using

FIGURE 12-32 **Belt-driven hybrid system components.**

only electric power. When the batteries reach a specific state of discharge, the engine starts and drives the generator to recharge the batteries.

Extended Range Electric Vehicle

An extended range electric vehicle (EREV) is similar to a PHEV. The EREV has a small gasoline or diesel engine that drives a generator to recharge the batteries. The engine only drives the generator; it cannot supply power to the drive wheels. The Chevrolet Volt is an EREV.

Fuel Cell Vehicles

Fuel cell vehicles (FCVs) are not presently available in the automotive showrooms. However, many car manufacturers around the world are working on this technology. Several vehicle manufacturers have designed and produced FCVs, and a significant number of these vehicles have been released to various fleet owners and officials to gain field experience and determine how these vehicles perform in the real world.

An FCV has an electric-drive motor(s), and the fuel cell supplies electricity to this drive motor(s). The fuel cell must be supplied with hydrogen as a fuel, and the fuel cell produces electricity as the hydrogen and air flow through the cell. There are various types of fuel cells, but the most common type is the proton exchange membrane (PEM) fuel cell. The fuel cell actually contains a large group of cells. The only by-product from fuel cell operation is water vapor. A PEM fuel cell is compared to an internal combustion engine in Figure 12-33. The FCV requires a source of high voltage and current such as a battery pack and/or ultra-capacitor to supply extra current to the drive motor during operation at wide throttle openings. The disadvantages of FCVs are:

1. The size, weight, and cost of the FCV components. Fuel cells are very expensive at present.
2. The lack and cost of hydrogen refueling facilities.
3. The cost and availability of FCVs.
4. Difficulty in storing hydrogen on-board vehicles; liquid hydrogen must be stored under pressure in special insulated tanks. It is difficult to store a sufficient amount of gaseous hydrogen on-board a vehicle.

The advantages of FCVs are:

1. Reduced dependency on high-priced hydrocarbon fuels.
2. Hydrogen may be obtained from several different sources.
3. Greatly reduced tailpipe and evaporative emissions.

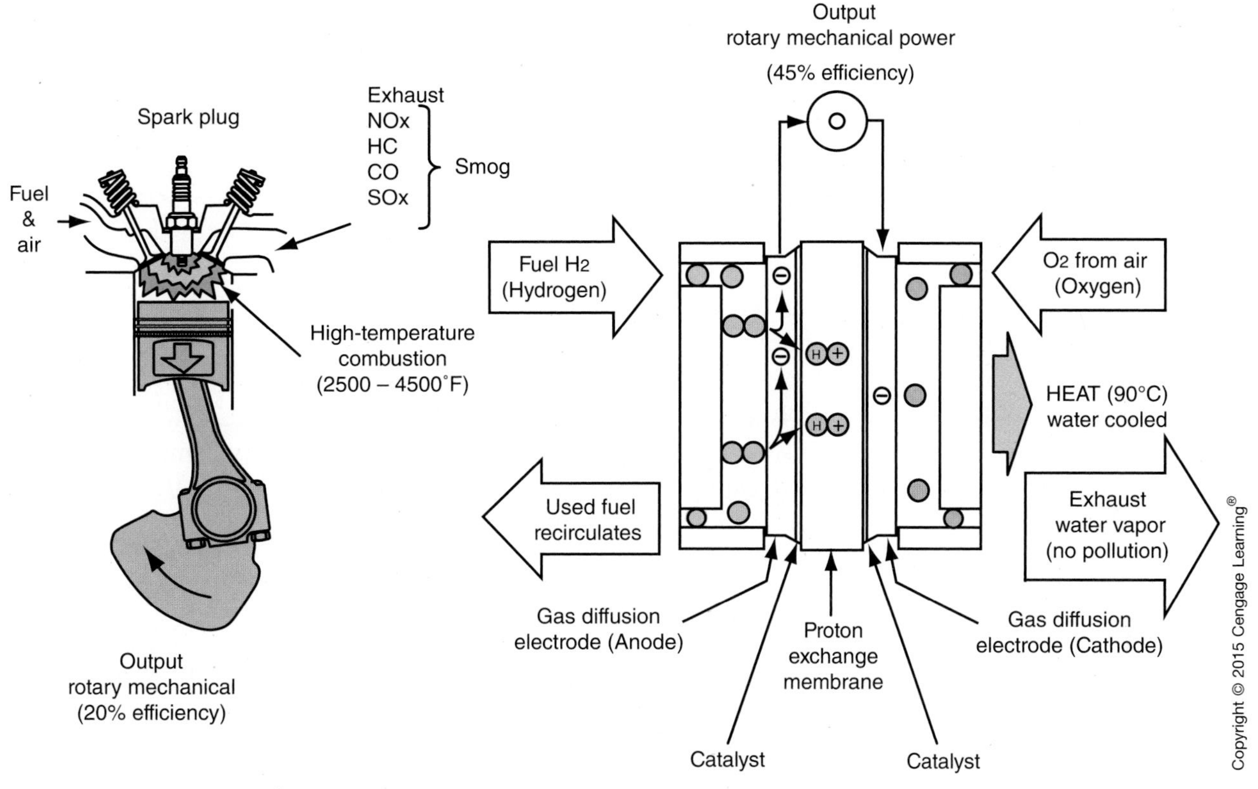

FIGURE 12-33 **Comparison between an internal combustion engine and a fuel cell.**

SUMMARY

- A multi-ribbed belt or a V-belt on older vehicles may be used to drive the power steering pump.
- A serpentine belt is a multi-ribbed belt that is used to drive all the belt-driven components. This allows these components to be on the same vertical plane.
- Proper belt tension is extremely important for adequate power steering pump operation.
- Many belts today are made of EPDM.
- A rack and pinion steering gear is usually mounted on the cowl or the front crossmember.
- Rack and pinion steering gears are used in many front-wheel-drive cars.
- Integral power steering gears are usually mounted on the vehicle frame.
- Integral power steering systems are found on many rear-wheel-drive cars and light-duty trucks.
- A power steering pump may have an integral or a remote fluid reservoir.
- In a hydroboost power-assisted steering system, fluid pressure from the power steering pump is applied to the steering gear and the brake master cylinder. The power steering pump pressure acts as a booster to assist the driver in applying the brakes.
- A power steering pump may have a vane, roller, or slipper-type rotor assembly, but all three types of pumps operate on the same basic principle.
- The flow control valve in a power steering pump is moved toward the closed position by spring pressure and fluid pressure from the venturi in the pump outlet fitting.

TERMS TO KNOW

Cam ring
Flow control valve
Hydroboost power-assisted steering system
Integral power-assisted steering system
Integral fluid reservoir
Outlet fitting venturi
Parallel HEVs
Pressure relief ball
Rack and pinion power steering system
Regenerative braking
Remote fluid reservoir
Series HEVs

TERMS TO KNOW
(continued)

Vane-type power steering pumps
Venturi

- When a vehicle is driven at low speeds with the front wheels straight ahead, the power steering pump fluid flow is lower. Under this condition, the pressure in the outlet fitting venturi is higher. This higher pressure together with the spring tension keeps the flow control valve closed. In this valve position, a small amount of fluid is routed past the flow control valve to the pump inlet.
- If engine speed is increased, the pump delivers more fluid than the system requires. This increase in pump flow reduces pressure in the outlet fitting venturi. This pressure decrease is applied to the spring side of the flow control valve. This action allows the flow control valve to move toward the partially open position, and excessive pump flow is returned past the flow control valve to the pump inlet.
- If the driver turns the steering wheel, fluid rushes from the pump outlet into the steering gear pressure chamber. Under this condition, the flow control valve moves toward the closed position to maintain pump pressure and flow to the steering gear.
- If the front wheels are turned all the way against the stops, the power steering pump pressure could become high enough to damage hoses or other components. Under this condition, the high pump pressure unseats a pressure relief ball in the center of the flow control valve. This action allows some pump flow to move past the pressure relief ball to the pump inlet, which limits pump pressure.
- Most HEVs are powered by a small gasoline engine and a battery pack, generator, and electric motor.
- In a series HEV, mechanical power from the engine is combined with electric power from the battery pack and generator and supplied to the drive wheels.
- In a parallel HEV, electric power only from the battery pack and generator may be supplied to the drive wheels or a combination of electric power and engine power may be sent to the drive wheels.
- The stop/start function in an HEV allows the engine to stop when the vehicle is not moving and the engine is at normal temperature and idling. When the driver steps on the accelerator pedal, vehicle operation is immediately restored.
- A belt-driven HEV system has a belt-driven motor/generator and provides a stop/start function, regenerative braking, and electric power boost to the drive wheels.
- In a PHEV, only electric power is supplied to the drive wheels, and a small gasoline or diesel engine drives a generator to recharge the battery pack.
- An FCV has only electric power supplied to the drive wheels. A fuel cell and battery pack supply the electricity to drive the vehicle, and hydrogen and air are supplied to the fuel cell to create the electricity.

REVIEW QUESTIONS

Short Answer Essays

1. Describe the types of materials a multi-ribbed or serpentine belt may be constructed of.
2. Explain two advantages of a serpentine belt.
3. List two possible locations for a power steering pump reservoir.
4. How is the power steering pump output effected as engine speed is increased but no additional power steering assist is required such as in straight ahead driving?
5. Describe how the brake boost pressure is obtained in a hydroboost power steering system.

6. Explain how the fluid pressure is produced in a vane-type power steering pump.
7. List the forces that move the flow control valve toward the closed position in a power steering pump.
8. Describe the operation of a venturi.
9. Explain the operation of the flow control valve when the steering wheel is turned.
10. Describe the operation of the power steering pump when the steering wheel is rotated all the way to the right or left until the front wheels contact the stops.

Fill-in-the-Blanks

1. On a multi-ribbed belt made of EPDM a ________________ percent loss of material is enough to cause belt ________________, ________________ and ________________.
2. Excessive belt wear occurs if pulleys are ________________.
3. Many serpentine belts have a spring-loaded ________________.
4. The master cylinder in a hydroboost system does not have a conventional ________________.
5. The purpose of the vanes in the power steering pump rotor is to ________________ the rotor in the elliptical cam ring.
6. As the power steering pump shaft and rotor turn, ________________ ________________ causes the vanes to move outward against the cam ring.
7. A machined venturi is located in the power steering pump ________________ ________________.
8. When fluid movement through the venturi increases, the pressure in the venturi ________________.
9. With the engine running and the front wheels straight ahead, the power steering pump pressure is ________________.
10. The pressure relief ball in a power steering pump is forced open when the front wheels are turned against the ________________.

Multiple Choice

1. The most likely result of a power steering drive-belt that is worn is:
   A. Belt slipping.
   B. Belt breaking.
   C. Belt contamination.
   D. Wear on the power steering pump pulley.
2. All of these statements about a serpentine belt are true EXCEPT:
   A. A serpentine belt can be used to drive all the belt-driven components.
   B. The back of a serpentine belt can drive one component.
   C. A serpentine belt made of EPDM is susceptible to cracking.
   D. A serpentine belt can have an automatic belt tensioner.
3. A hydroboost power steering system:
   A. Is used on many small front-wheel-drive vehicles.
   B. Uses pressure from the power steering pump to provide brake boost.
   C. Has a conventional vacuum brake booster.
   D. Has a very high capacity power steering pump.
4. All of the following statements about multi-ribbed belts made of EPDM are true EXCEPT:
   A. EPDM belt is more elastic than a standard neoprene belt and resists cracking even at higher mileage.
   B. EPDM belts are designed to last 50,000-60,000 miles.
   C. Belts are designed to have clearance between the rib peaks and the pulley grooves.
   D. EPDM belt to gradually lose material on the ribbing similar to the way a tire wears.
5. All of these statements about integral-type power steering pump sealing are true EXCEPT:
   A. A large O-ring seal is positioned between the reservoir and the pump housing.
   B. A lip seal on the pump drive shaft prevents oil leaks around the shaft.
   C. Lip seals are mounted on bolt fittings on the back of the reservoir.
   D. The cap and dipstick provide a seal on the top of the reservoir neck.
6. A power steering system experiences repeated belt failures with normal power steering system operation. The most likely cause of this problem is:
   A. A misaligned power steering pump pulley.
   B. A loose power steering pump belt.
   C. Low fluid level in the power steering reservoir.
   D. A worn cam ring in the power steering pump.

7. During normal power steering pump operation:
   A. The flow control valve spring moves this valve toward the open position.
   B. The fluid pressure from the outlet fitting venturi moves the flow control valve toward the open position.
   C. Pump flow increases as the engine speed increases.
   D. The pressure relief valve opens when the engine speed is increased with the front wheels straight ahead.

8. All of these statements about power steering pump design and operation are true EXCEPT:
   A. An increase in fluid movement through the outlet fitting venturi causes a pressure decrease in the venturi.
   B. An increase in the flow control valve opening allows more fluid to return to the pump inlet.
   C. The rotor is attached to the pump shaft and must rotate with this shaft.
   D. Fluid pressure from the pump vanes moves the flow control valve toward the closed position.

9. All of the following statements about hybrid electric vehicles that have an electrohydraulic power steering pump EXCEPT:
   A. The power steering pump has an electric driven motor.
   B. A control module determines the amount of power assist required.
   C. An electric motor drives the steering gear.
   D. Some are sullied 42V from the HEV inverter.

10. A vehicle experiences repeated ruptures of the high-pressure power steering hose. The most likely cause of this problem is:
   A. A pressure relief ball sticking closed.
   B. A flow control valve sticking open.
   C. Worn pump vanes and cam ring.
   D. A partially restricted power steering return hose.

Chapter 13

Rack and Pinion Steering Gears and Four Wheel Steering

Upon completion and review of this chapter, you should be able to understand and describe:

- The advantages of a rack and pinion steering system compared to a recirculating ball steering gear and parallelogram steering linkage.
- How the tie rods are connected to the rack.
- The purpose of the rack bearing and adjuster plug.
- Two possible mounting positions for the rack and pinion steering gear.
- The fluid movement in a rack and pinion steering gear during a right turn.
- The fluid movement in a rack and pinion steering gear during a left turn.
- The operation of the spool valve and rotary valve.
- The operation of the power rack and pinion steering gear when hydraulic pressure is not available.
- The purpose of the breather tube.
- The main differences between a Saginaw and a TRW power rack and pinion steering gear.
- The advantages of an electronic variable orifice (EVO) steering system.
- The input sensors in an EVO steering system.
- The driving conditions when an EVO steering system provides increased power steering assistance.
- The operation of a rack-drive electronic power steering system during right and left turns.
- The operation of the steering shaft torque sensor in a column-drive electronic power steering system.
- The operation and advantages of active front steering (AFS).
- The advantages of four-wheel steering (4WS).
- How the rear steering rack is driven.
- The two inputs used by the control unit to operate the rear steering system.
- Negative phase and positive phase steering.
- The input sensors in an electronically controlled four-wheel steering system and give the location of each sensor.
- The type of signal produced by a wheel speed sensor.
- The steering action of the rear wheels in relation to vehicle speed in an electronically controlled four-wheel steering system.
- The operation and advantages of rear active steering (RAS) and four-wheel active steering (4WAS).

INTRODUCTION

During the late 1970s and 1980s, the domestic automotive industry converted much of its production from larger rear-wheel-drive (RWD) cars to smaller, lightweight, and more fuel efficient front-wheel-drive (FWD) cars. These FWD cars required smaller, lighter components wherever possible. Manual and power rack and pinion steering gears are lighter and more compact than the recirculating ball steering gears and parallelogram steering linkages used on most RWD cars. Therefore, rack and pinion steering gears are ideally suited to these compact FWD cars.

Steering systems have not escaped the electronics revolution. Many cars are presently equipped with electronic variable orifice (EVO) steering, which provides greater power assistance during low-speed cornering and parking for increased driver convenience. Some cars are now equipped with electronic power steering. In these systems, an electric motor in the steering gear provides steering assist.

MANUAL RACK AND PINION STEERING GEAR MAIN COMPONENTS

Rack

Shop Manual
Chapter 13,
page 497

The **rack** is a toothed bar that slides back and forth in a metal housing. The steering gear housing is mounted in a fixed position on the front cross member or on the firewall. The rack takes the place of the idler and pitman arms in a parallelogram steering system and maintains the proper height of the tie rods so they are parallel to the lower control arms. The rack may be compared to the center link in a parallelogram steering linkage. Bushings support the rack in the steering gear housing. Sideways movement of the rack pulls or pushes the tie rods and steers the front wheels (Figure 13-1).

Pinion

Spur gear teeth are cut so they are parallel to the gear rotational axis.

Helical gear teeth are curved to increase the amount of tooth contact between a pair of meshed gears. Helical gears tend to operate more quietly than spur gear teeth and provide increased strength compared with spur gear teeth.

The **pinion** is a toothed shaft mounted in the steering gear housing so the pinion teeth are meshed with the rack teeth. The pinion may contain **spur gear teeth** or **helical gear teeth**. The upper end of the pinion shaft is connected to the steering shaft from the steering column. Therefore, steering wheel rotation moves the rack sideways to steer the front wheels. The pinion is supported on a ball bearing in the steering gear housing.

Tie Rods and Tie Rod Ends

The tie rods are similar to those used on parallelogram steering linkages. A spring-loaded ball socket on the inner end of the tie rod is threaded onto the rack. When these ball sockets are torqued to the vehicle manufacturer's specification, a preload is placed on the ball socket. A bellows boot is clamped to the housing and tie rod on each side of the steering gear, and these boots keep contaminants out of the ball socket and rack.

A tie rod end is threaded onto the outer end of each tie rod. These tie rod ends are similar to those used on parallelogram steering linkages. A jam nut locks the outer tie rod end to the tie rod.

Rack Adjustment

Steering kickback is the movement of the steering wheel caused by a front wheel striking a road irregularity.

A rack bearing is positioned against the smooth side of the rack. A spring is located between the rack bearing and the rack adjuster plug that is threaded into the housing. This adjuster plug is retained with a locknut. The rack bearing adjustment sets the preload between the rack and pinion teeth, which affects **steering kickback**, harshness, and noise.

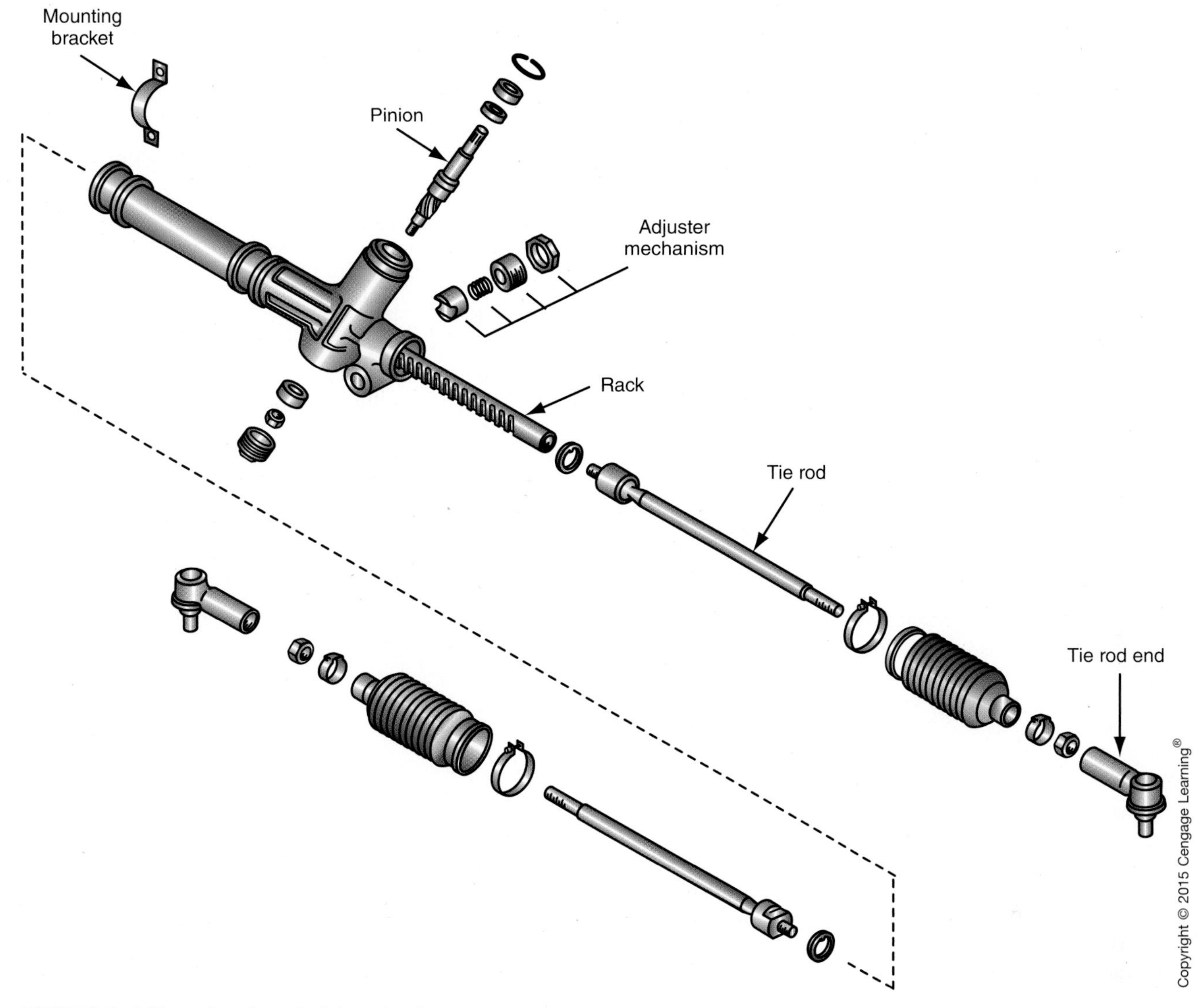

FIGURE 13-1 **Manual rack and pinion steering gear components.**

Steering Gear Ratio

When the steering wheel is rotated from **lock-to-lock** or **stop-to-stop**, the front wheels turn about 30° each in each direction from the straight-ahead position. Therefore, the total front wheel movement from left to right is approximately 60°. With a steering ratio of 1:1, 1° of steering wheel rotation would turn the front wheels 1°, and 30° of steering wheel rotation in either direction would result in lock-to-lock front wheel movement. This steering ratio is much too extreme because the slightest steering wheel movement would cause the vehicle to swerve. The steering gear must have a ratio that allows more steering wheel rotation in relation to front wheel movement.

Rotation of the steering wheel from extreme left to extreme right is called **lock-to-lock** or **stop-to-stop**.

A **steering ratio** of 15:1 is acceptable, and this ratio provides 1° of front wheel movement for every 15° of steering wheel rotation. To calculate the steering ratio, divide the lock-to-lock steering wheel rotation in degrees by the total front wheel movement in degrees. For example, if the lock-to-lock steering wheel rotation is 3.5 turns, or 1,260°, and the total front wheel movement is 60°, the steering ratio is 1,260/60 = 21:1. As a general rule, large, heavy cars have higher numerical steering ratios than small, lightweight cars.

Steering ratio is the relationship between steering wheel rotation in degrees and front wheel movement to the right or left in degrees.

Manual Rack and Pinion Steering Gear Mounting

Large rubber insulating grommets are positioned between the steering gear and the mounting brackets. These bushings help prevent the transfer of road noise and vibration from the steering gear to the chassis and passenger compartment. The rack and pinion steering gear may be attached to the front cross member or to the cowl. Proper steering gear mounting is important to maintain the parallel relationship between the tie rods and the lower control arms. The firewall is reinforced at the steering gear mounting locations to maintain the proper steering gear position.

Advantages and Disadvantages of Rack and Pinion Steering

As mentioned earlier, the rack and pinion steering gear is lighter and more compact than a recirculating ball steering gear and parallelogram steering linkage. Therefore, the rack and pinion steering gear is most suitable for FWD unibody vehicles.

Shop Manual
Chapter 13, page 500

Since there are fewer friction points in the rack and pinion steering than in the recirculating ball steering gear with a parallelogram steering linkage, the driver has a greater feeling of the road with rack and pinion steering gear. However, fewer friction points reduce the steering system's ability to isolate road noise and vibration. Therefore, drivers of a vehicle with a rack and pinion steering system may have more complaints of road noise and vibration transfer to the steering wheel and passenger compartment.

Power Rack and Pinion Steering Gears

Design and Operation

The **rack piston** is attached to the rack, and this piston moves horizontally in the steering gear chamber.

Rack seals are mounted at each end of the rack chamber and on the rack piston. The seals at each end of the rack chamber prevent leaks between the rack and the chamber.

A power-assisted rack and pinion steering gear uses the same basic operating principles as a manual rack and pinion steering gear, but in the power-assisted steering gear, hydraulic fluid pressure from the power steering pump is used to reduce steering effort. A **rack piston** is integral with the rack, and this piston is located in a sealed chamber in the steering gear housing. Hydraulic fluid lines are connected to each end of this chamber, and **rack seals** are positioned in the housing at ends of the chamber. A seal is also located on the rack piston (Figure 13-2).

The following description assumes that the power steering rack is mounted behind the front wheel spindle, such as on the vehicle firewall or cradle assembly. When a driver is completing a left turn, fluid is pumped into the left side of the fluid chamber and exhausted from the right chamber area. This hydraulic pressure on the left side of the rack piston helps the pinion move the rack to the right (Figure 13-3, view A).

When a right turn is made, fluid is pumped into the right side of the fluid chamber, and fluid flows out of the left end of the chamber. Thus, hydraulic pressure is exerted on the right side of the rack piston, which assists the pinion gear in moving the rack to the left (Figure 13-3, view B). Since the steering gear is mounted behind the front wheels, rack movement to the left is necessary for a right turn, whereas rack movement to the right causes a left turn.

Rotary Valve and Spool Valve Operation

Torsion bar twisting during steering wheel rotation moves the spool valve in relation to the rotary valve.

The **rotary valve** is mounted over the spool valve on the steering gear pinion.

Fluid direction in the steering gear is controlled by a rotary valve attached to the pinion assembly (Figure 13-4). A stub shaft on the pinion assembly is connected to the steering shaft and wheel. The pinion is connected to the stub shaft through a **torsion bar** that twists when the steering wheel is rotated and springs back to the center position when the wheel is released. A rotary valve body contains an inner **spool valve** that is mounted over the torsion bar on the pinion assembly.

When the front wheels are in the straight-ahead position, fluid flows from the pump through the high-pressure hose to the center **rotary valve** body passage. Fluid is then routed through the valve body to the low-pressure return hose and the pump reservoir (Figure 13-5).

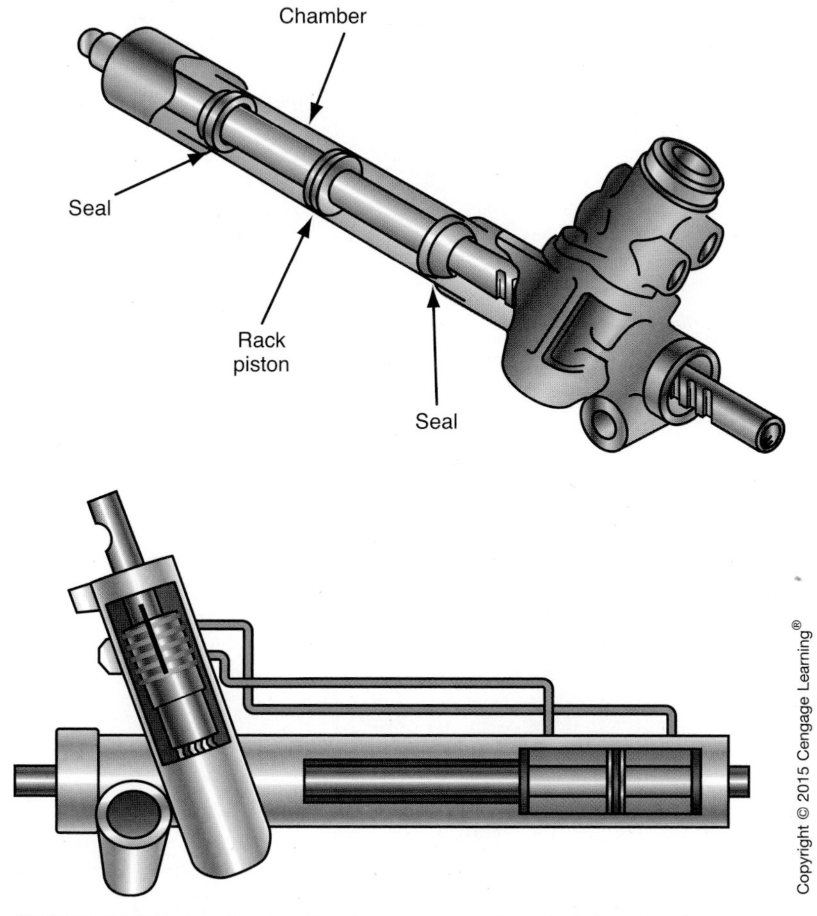

FIGURE 13-2 **Hydraulic chamber in a power rack and pinion steering gear.**

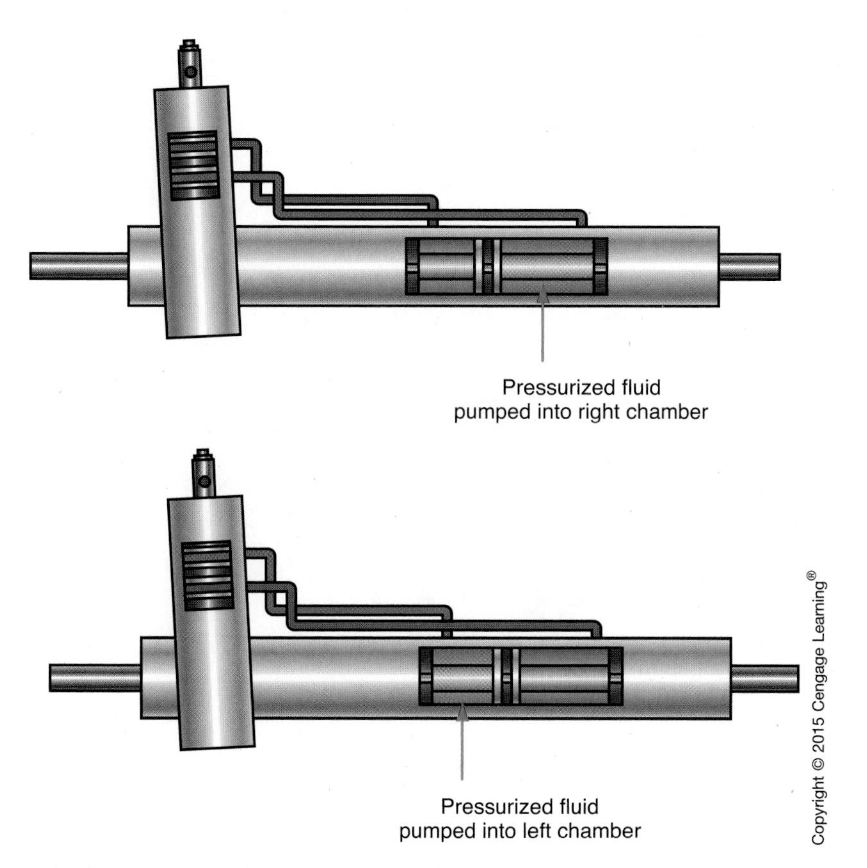

FIGURE 13-3 **Rack movement during left and right turns.**

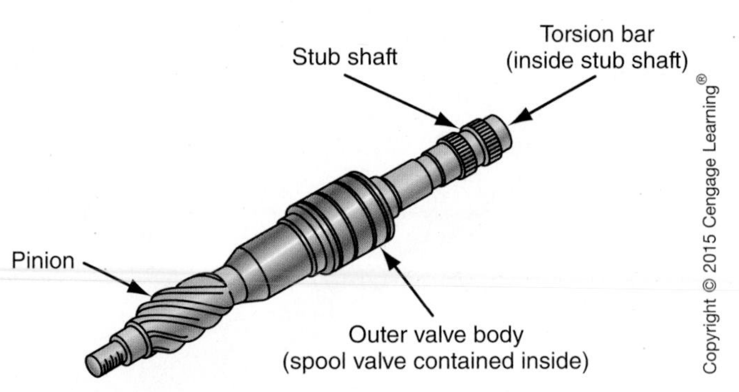

FIGURE 13-4 **Pinion assembly for a power rack and pinion steering gear.**

Item	Part Number	Description
1	3A697	Fluid reservoir (early build vehicles)
2	3A697	Fluid reservoir (late build vehicles)
3	3691	Supply hose (reservoir-to-pump) (early build vehicles)
4	3A713	Return hose (cooler-to-reservoir) (early build vehicles)
5	3691	Supply hose (reservoir-to-pump) (late build vehicles)
6	3A713	Return hose (cooler-to-reservoir) (late build vehicles)
7	3489	Fluid reservoir bracket (early build vehicles)
8	3A733	Pulley
9	3A764	Power steering pump
10	3504	Steering gear
11	3A131	Tie rod end
12	3280	Inner tie rod
13	3332	Inner tie rod boot
14	3A713	Return line (gear-to-cooler)
15	3D746	Fluid cooler
16	3A719	Pressure line (pump-to-gear)
17	9F274	Return hose retainer clip
18	3F886-AA/ 3F886-BA	O-ring (1 each required)

FIGURE 13-5 **Power rack and pinion steering gear with connecting hoses and lines.**

Many power steering systems contain a fluid cooler connected in the high-pressure hose between the pump and the steering gear. The fluid cooler is like a small radiator. Air flows through the fins on the cooler and cools the fluid flowing through the internal cooler passages. Some power steering systems have a remote fluid reservoir connected in the low pressure hose between the steering gear and the pump (Figure 13-5). Many power steering systems have a fluid filter, and this filter is often mounted in the remote reservoir.

Teflon rings or O-rings seal the rotary valve ring lands to the steering gear housing. A lot of force is required to turn the pinion and move the rack because of the vehicle weight on the front wheels. When the driver turns the wheel, the stub shaft is forced to turn. However, the pinion resists turning because it is in mesh with the rack, which is connected to the front wheels. This resistance of the pinion to rotation results in torsion bar twisting. During this twisting action, a pin on the torsion bar moves the spool valve with a circular motion inside the rotary valve. If the driver makes a left turn, the spool valve movement aligns the inlet center rotary valve passage with the outlet passage to the left side of the rack piston. Therefore, hydraulic fluid pressure applied to the left side of the rack piston assists the driver in moving the rack to the right.

When the driver turns the steering wheel, the amount of feeling that he or she senses regarding front wheel turning is called **feel of the road**.

When a right turn is made, twisting of the torsion bar moves the spool valve and aligns the center rotary valve passage with the outlet passage to the right side of the rack piston (Figure 13-6). Under this condition, hydraulic fluid pressure applied to the rack piston helps the driver move the rack to the left. The torsion bar provides a **feel of the road** to the driver.

When the steering wheel is released after a turn, the torsion bar centers the spool valve and power assistance stops. If hydraulic fluid pressure is not available from the pump, the power steering system operates like a manual system, but steering effort is higher. When the torsion bar is twisted to a designed limit, tangs on the stub shaft engage with drive tabs on the pinion. This action mechanically transfers motion from the steering wheel to the rack and front wheels. Since hydraulic pressure is not available on the rack piston, greater steering effort is required. If a front wheel raises going over a bump or drops into a hole, the tie rod pivots along with the wheel. However, the rack and tie rod still maintain the left-to-right wheel direction.

In power rack and pinion steering gears, a condition that causes excessive steering wheel turning effort when the vehicle is first started may be called "morning sickness."

The rack boots are clamped to the housing and the rack. Since the boots are sealed and air cannot be moved through the housing, a **breather tube** is necessary to move air from one boot to the other when the steering wheel is turned (Figure 13-7). This air movement through the vent tube prevents pressure changes in the bellows boots during a turn.

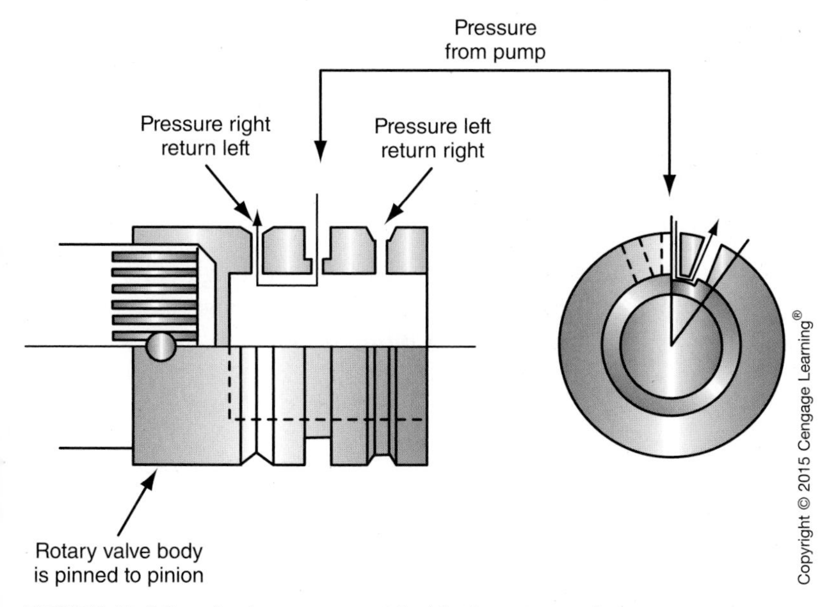

FIGURE 13-6 **Spool valve movement inside the rotary valve.**

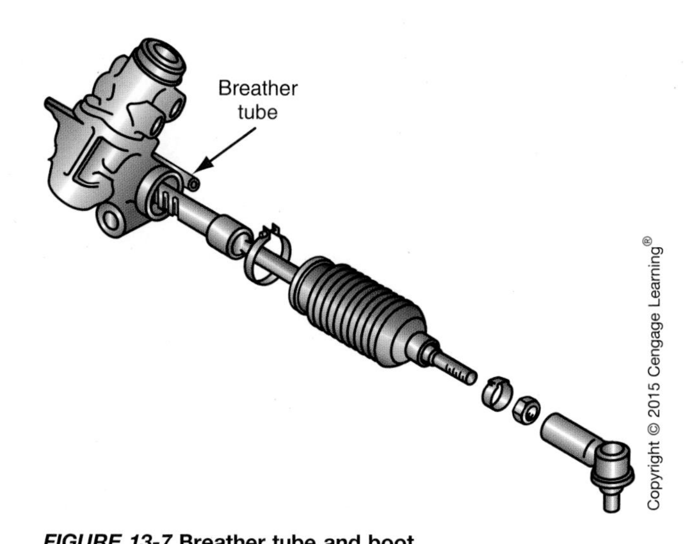

FIGURE 13-7 **Breather tube and boot.**

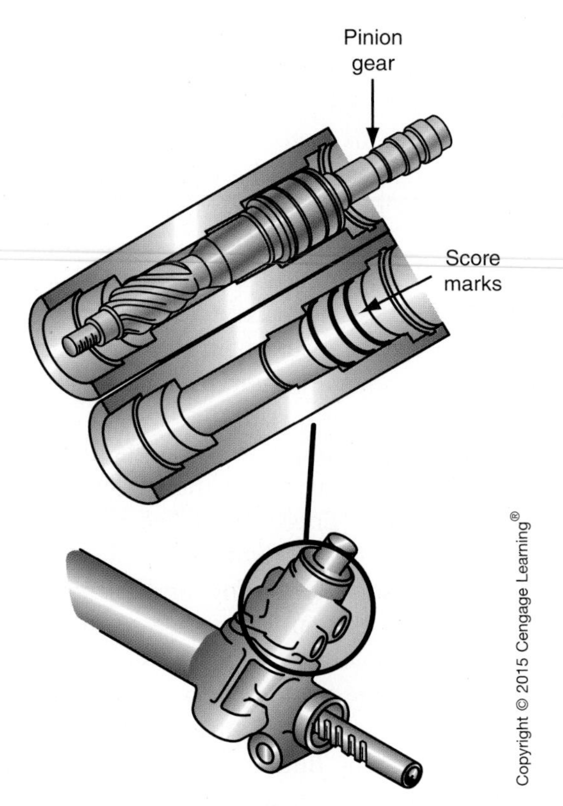

FIGURE 13-8 **Score marks in a rack and pinion steering gear housing.**

AUTHOR'S NOTE: It has been my experience that one of the most common problems with power rack and pinion steering gears is a condition that causes excessive steering wheel turning effort when the vehicle is first started in the morning. After the steering wheel is turned several times the condition disappears. This problem is caused by grooves worn in the aluminum pinion housing by the seals on the control valve (Figure 13-8). When this condition is present, steering gear replacement is usually required. As an automotive technician, you will be required to diagnose and correct this problem.

TYPES OF POWER RACK AND PINION STEERING GEARS

Power Rack and Pinion Steering Gear

Many vehicles have a rack and pinion steering gear manufactured by Saginaw (Figure 13-9). Some vehicles are equipped with a TRW rack and pinion steering gear. This type of steering gear is similar to the Saginaw gear except for the following differences:

1. Method of tie rod attachment
2. Bulkhead oil seal and retainer
3. Pinion upper and lower bearing hardware

On both the Saginaw and TRW power rack and pinion steering gears, the tie rods are connected to the ends of the rack. This type of steering gear may be referred to as **end take-off (ETO)**. On other power rack and pinion steering gears, the rack piston and cylinder are positioned on the right end of the rack and the tie rods are attached to a movable

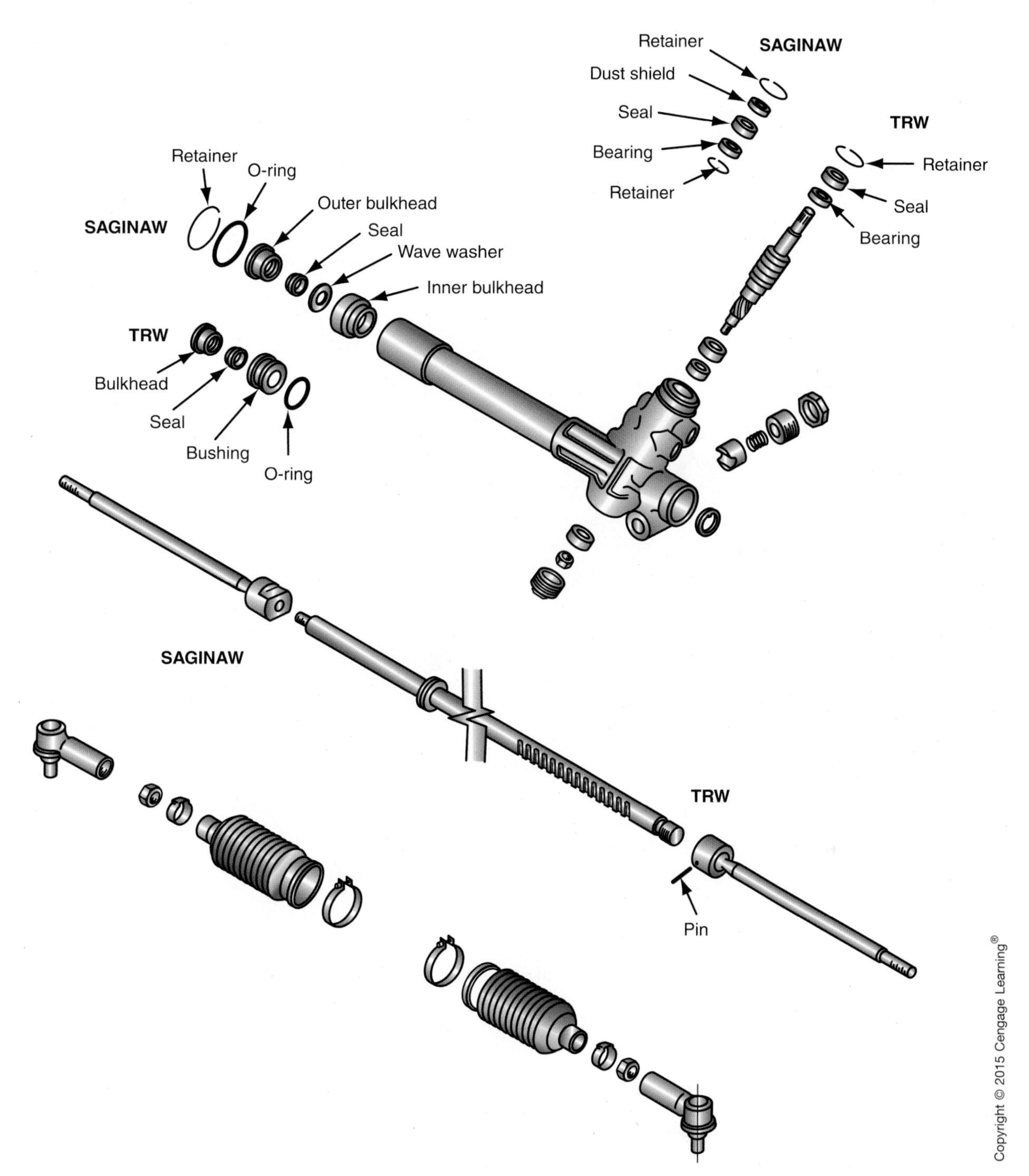

FIGURE 13-9 **Saginaw and TRW power rack and pinion steering gear.**

sleeve in the center of the gear (Figure 13-10). This type of steering gear may be called **center take-off (CTO)**.

The Toyota power rack and pinion steering gear has a removable control valve housing surrounding the control valve and pinion shaft (Figure 13-11). In this steering gear, claw washers are used to lock the inner tie rod ends to the rack. Apart from these minor differences, the Toyota power rack and pinion gear is similar to the Saginaw and TRW gears.

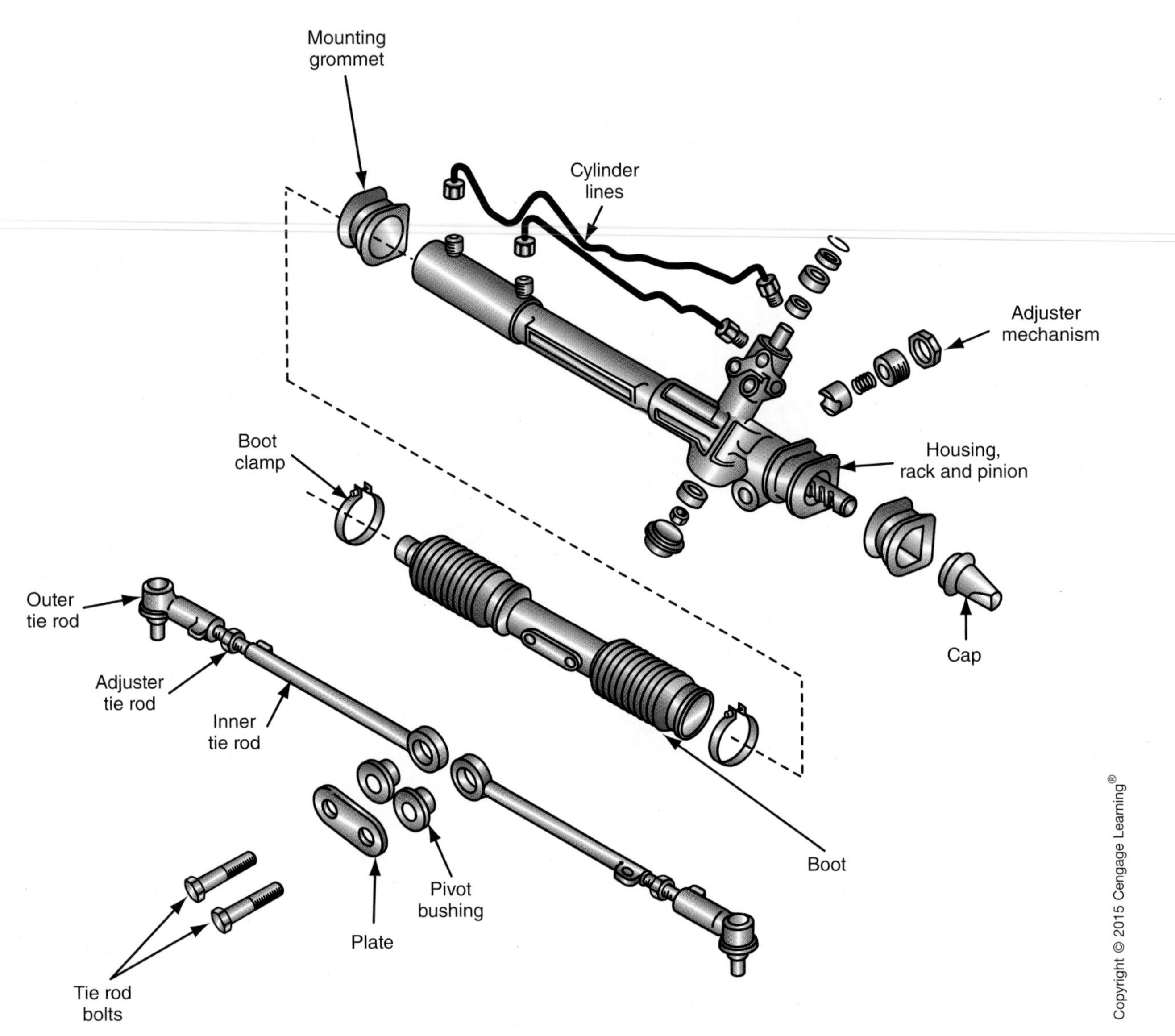

FIGURE 13-10 **Power rack and pinion steering gear with center take-off tie rods.**

WARNING: When working on any power steering system always wear protective gloves and use caution, because the system hoses, components, and fluid can be very hot if the system has been in operation for a period of time.

Electronic Variable Orifice Steering

Input Sensors

The **electronic variable orifice (EVO) steering** system is standard on many late model vehicles. The EVO steering system provides high-power steering assistance during low-speed cornering and parking and normal power steering assistance at higher speeds for proper road feel. High-power steering assistance during low-speed cornering and parking increases driver convenience.

The **steering wheel rotation sensor** is mounted on the steering column, and a shutter disc attached to the steering shaft rotates through the sensor when the steering wheel

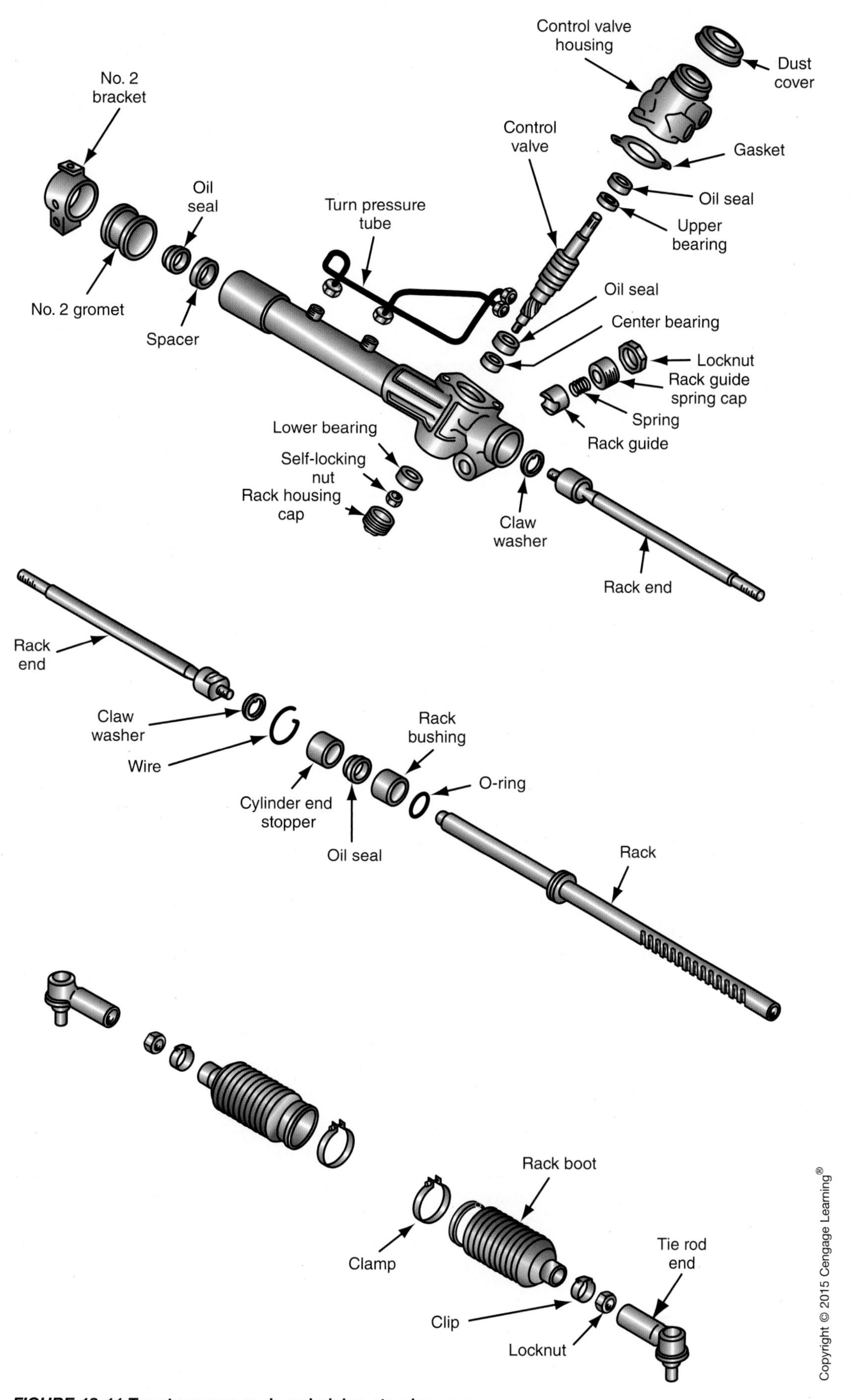

FIGURE 13-11 **Toyota power rack and pinion steering gear.**

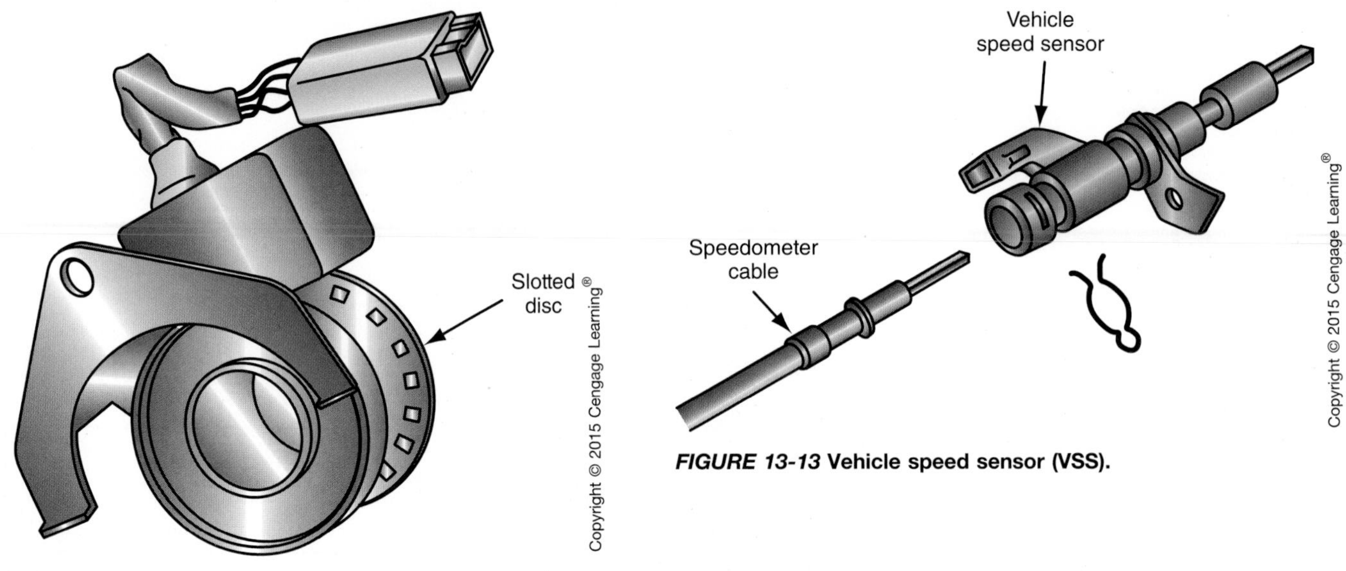

FIGURE 13-12 **Steering wheel rotation sensor.**

FIGURE 13-13 **Vehicle speed sensor (VSS).**

CAUTION:
Never short across, or ground, any terminals in a computer-controlled system unless you are instructed to do so in the vehicle manufacturer's service manual. Such action may damage expensive electronic components.

is rotated. A row of slots is positioned near the outer edge of the shutter disc (Figure 13-12). When these slots rotate through the sensor, a steering wheel rotation speed signal is sent from the sensor to the control module.

The **vehicle speed sensor (VSS)** is mounted in the transaxle or transmission, and this sensor sends a signal to the control module in relation to vehicle speed (Figure 13-13). The VSS signal is also used for other purposes.

Control Module

On some vehicles, the control module is mounted in the trunk (Figure 13-14). This module continually monitors the input signals from the VSS and the steering wheel rotation sensor (Figure 13-15). Some models have a combined EVO steering system and rear air suspension system. On these models, the control module operates the EVO steering system and the rear air suspension system. In the combined EVO steering and rear air suspension system, the inputs and outputs from both of these systems are connected to the same control module. (The rear air suspension system is explained in Chapter 11.) If the EVO system is used alone without the air suspension system, a different control module is required.

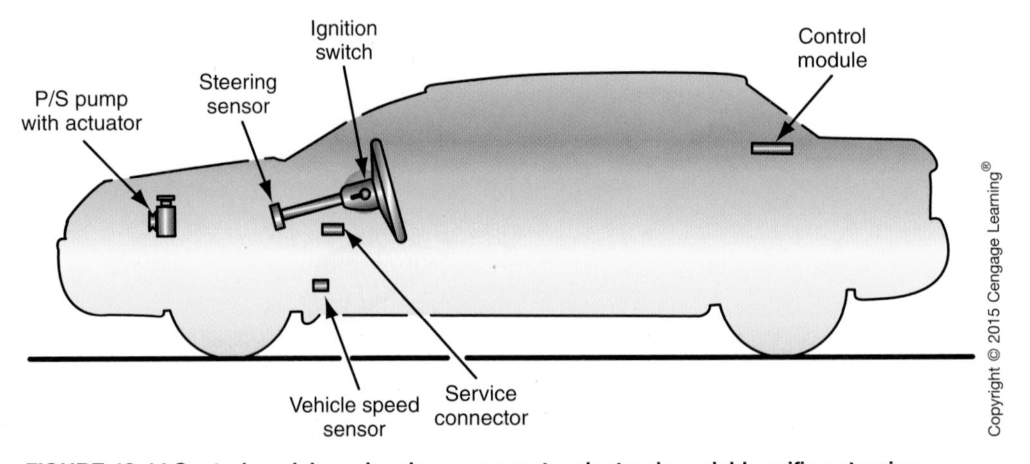

FIGURE 13-14 **Control module and main components, electronic variable orifice steering.**

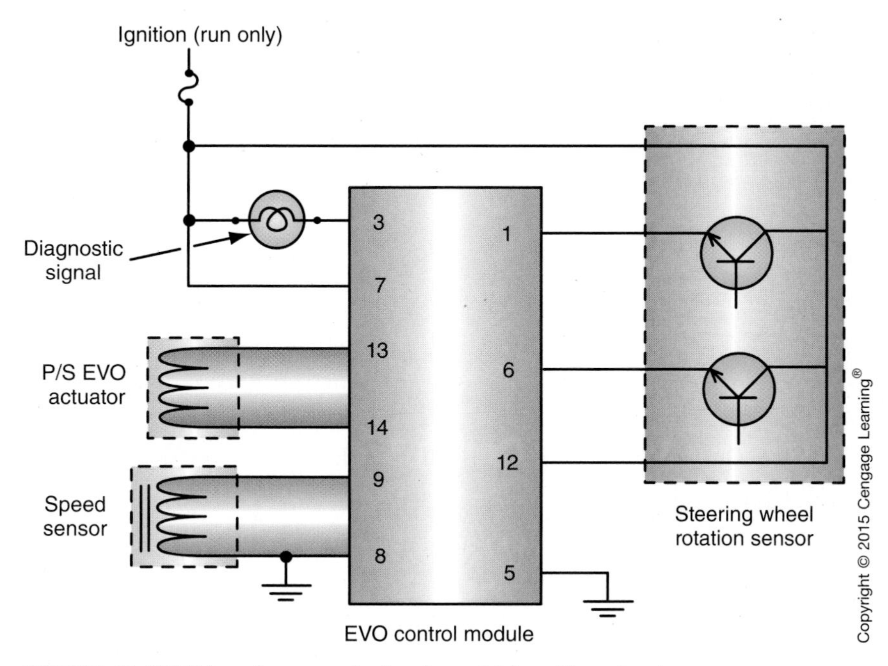

FIGURE 13-15 **Wiring diagram, electronic variable orifice steering.**

Output Control

A varying current flow is sent from the control module through the EVO actuator in the power steering pump (Figure 13-16). The actuator swivels freely when it is installed in the power steering pump. As the control module changes the current flow through the actuator, the actuator supplies a variable pressure to the spool valve in the power steering pump. Two wires are connected from the actuator to the control module. The power steering pump is mounted directly to the engine to reduce noise, vibration, and harshness (NVH).

The control module positions the actuator and spool valve to provide full power steering assistance under these conditions:

1. Vehicle speed less than 10 mph (16 km/h)
2. Steering wheel rotation above 15 rpm

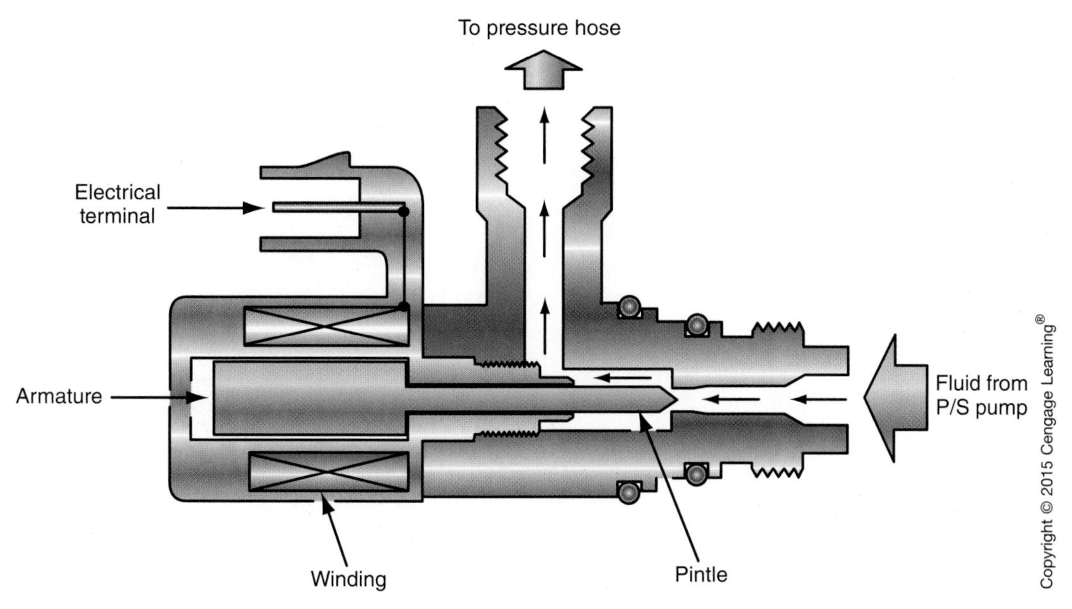

FIGURE 13-16 **Electronic variable orifice (EVO) actuator removed from the power steering pump.**

The full power-assist mode reduces driver effort required to turn the steering wheel during low-speed cornering and parking for increased convenience. In the full power-assist mode, the control module supplies 30 milliamps (mA) to the actuator.

The control module positions the actuator and spool valve to reduce power steering assistance under these conditions:

1. Vehicle speed above 25 mph (40 km/h)
2. Steering wheel rotation below 15 rpm

The reduced power-assist mode provides adequate road feel for the driver. In the reduced power-assist mode, the control module supplies 300 mA to the actuator. Above 88 mph (132 km/h), 590 mA is supplied from the control module to the actuator.

Saginaw Electronic Variable Orifice Steering Design and Operation

The term **variable effort steering (VES)** replaces the previous EVO terminology on some General Motors vehicles. In the EVO system, the vehicle speed sensor input is sent to the EVO controller. This controller supplies a pulse width modulated (PWM) voltage to the actuator solenoid in the power steering pump. The controller also provides a ground connection for the actuator solenoid (Figure 13-17).

A **pulse width modulated (PWM) voltage signal** has a variable on time.

When the vehicle is operating at low speeds, the controller supplies a **pulse width modulated (PWM) voltage signal** to position the actuator solenoid plunger so the power steering pump pressure is higher (Figure 13-18). Under this condition, greater power assistance is provided for cornering or parking. If the vehicle is operating at higher speed, the controller changes the PWM signal to the actuator solenoid, and the solenoid plunger is positioned to reduce power steering pump pressure. This action reduces power steering assistance to provide improved road feel for the driver.

In a Magnasteer II VES system as vehicle speed changes or lateral acceleration occurs, the amount of effort required to steer the vehicle can be varied dependent on the

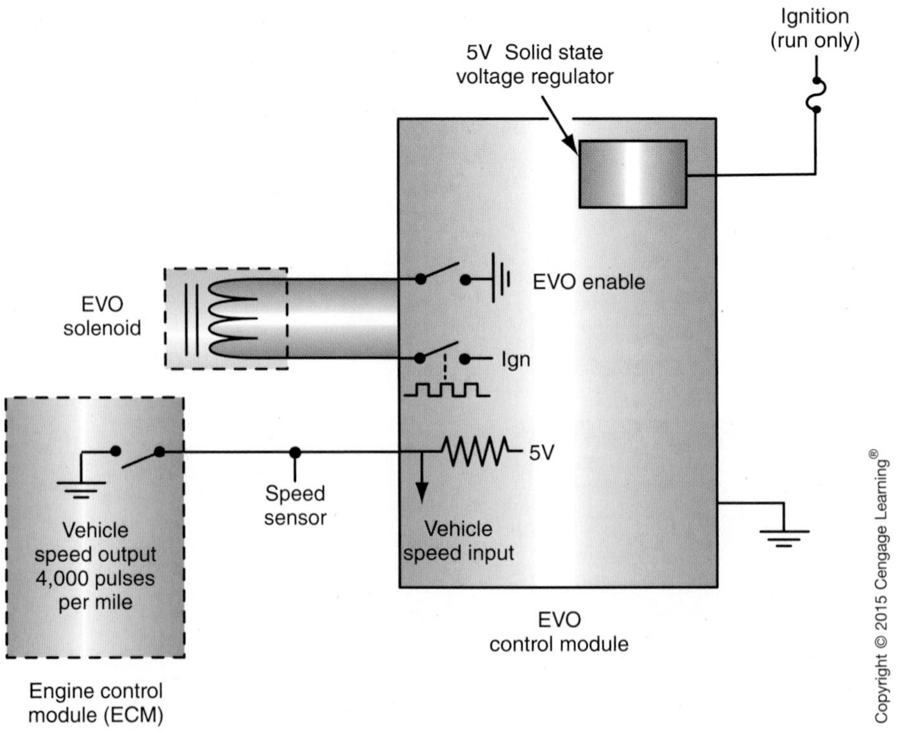

FIGURE 13-17 **Electronic variable orifice (EVO) steering system.**

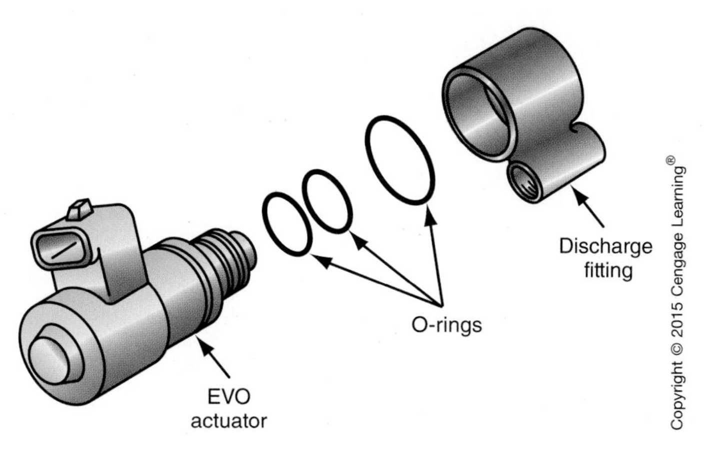

FIGURE 13-18 **Actuator solenoid.**

requirements. During low speed operation increased steering assist is provided for minimal steering effort to facilitate parking maneuvers and easy turning. While at high speeds less assist is provided which provides better road feel, firmer steering, and improved directional stability. If the system senses lateral acceleration the steering effort will become firmer in an effort to reduce oversteering by the driver. The Electronic Brake Control Module (EBCM) controls a bi-directional rotary magnetic actuator solenoid mounted in the steering gear (Figure 13-19). The computer that operates the **Magnasteer system** is combined with the electronic brake and traction control module (EBCM), and this module is usually mounted in the left-front corner of the engine compartment (Figure 13-20). Two wires are connected from the steering actuator to the EBCM (Figure 13-21).

The **Magnasteer system** uses an electromagnetic actuator, a multiple pole ring permanent magnet, a pole piece, and an electromagnetic coil in the steering gear to vary steering effort. The EBCM varies steering assist by regulating the amount of current flow to the electromagnet.

The steering actuator solenoid contains a pole piece with 16 magnetic segments and a coil. A matching 16-segment permanent magnet is attached to the spool valve. As the steering wheel and the spool valve rotate, the 16 segments on the spool valve move into and out of alignment with the segments on the actuator pole piece. The EBCM can reverse the current flow through the steering actuator, and this action reverses the magnetic poles on the actuator segments. At low vehicle speeds, the EBCM supplies a negative current flow, 2–3 amp, and command through the steering actuator so the magnetic poles on the actuator repel the permanent magnet segments on the spool valve. This repelling action assists the driver to turn the steering wheel and reduces steering effort. At medium speeds of about 45 mph no current is supplied and steering is assisted by hydraulic pressure only. At higher speeds, the EBCM provides a positive current flow of up to 2–3 amps through

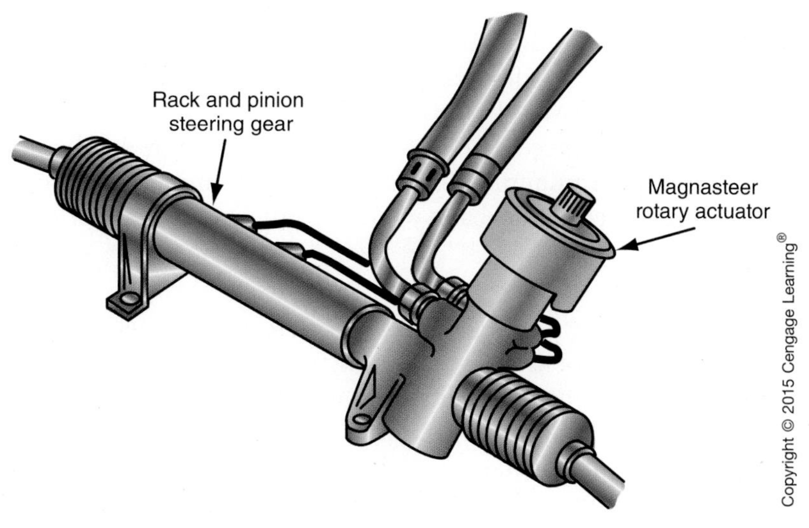

FIGURE 13-19 **Magnasteer, actuator solenoid.**

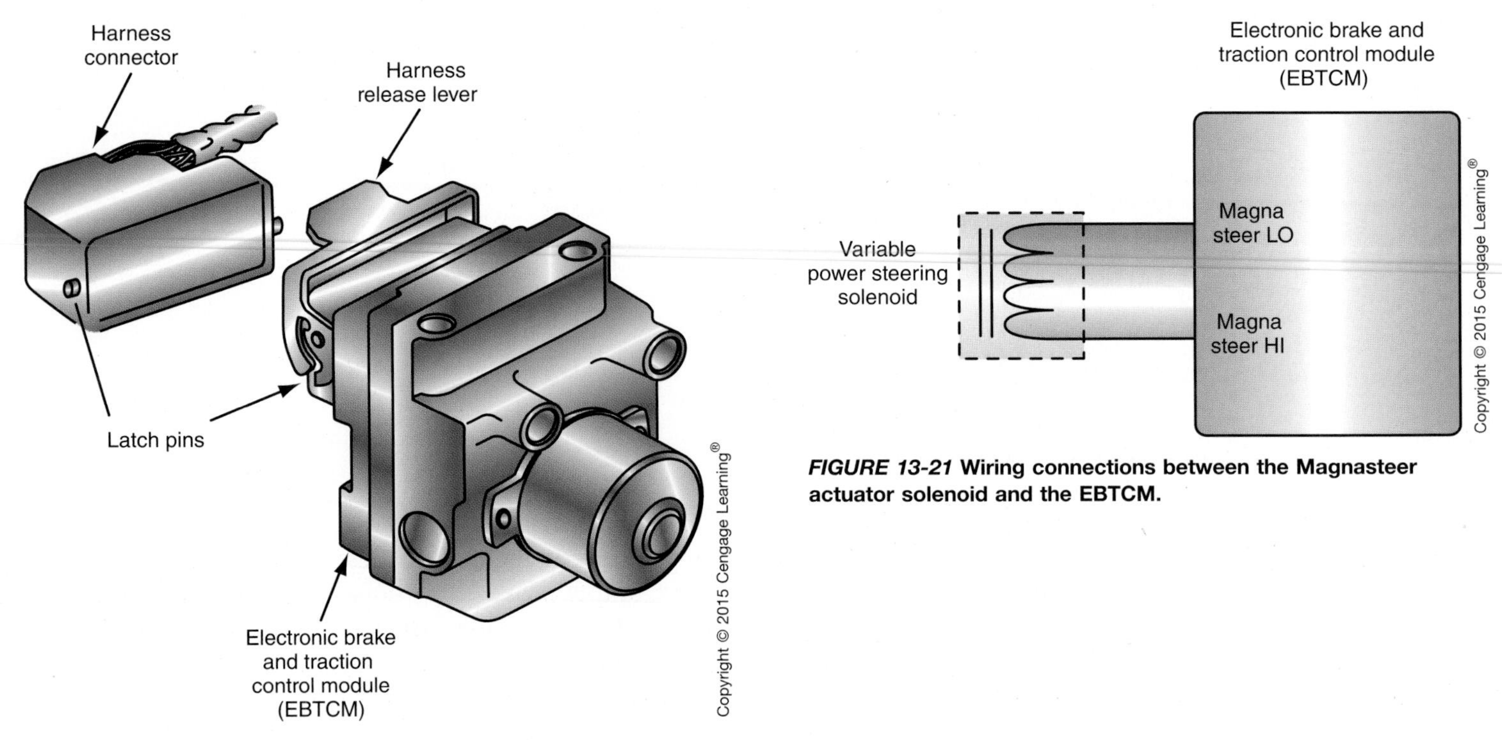

FIGURE 13-20 **Electronic brake and traction control module (EBTCM).**

FIGURE 13-21 **Wiring connections between the Magnasteer actuator solenoid and the EBTCM.**

Shop Manual
Chapter 13,
page 519

the steering actuator, and this action reverses the poles on the actuator segments. Under this condition, the actuator segments are attracted to the permanent magnet segments on the spool valve. This action increases steering effort to improve driver road feel. The EBCM also has the capability to vary the current flow through the steering actuator to provide variable steering effort in relation to vehicle speed. If the EBCM detects a malfunction in the circuitry or actuator the system will supply zero amps to the actuator and the steering will be assisted by hydraulic pressure only.

Rack-Drive Electronic Power Steering

Electronic power steering systems (EPS) may be classified as rack-drive, column-drive, or pinion-drive. All three types of EPS have a rack and pinion steering gear. In a **rack-drive EPS** the electric motor that provides steering assist is coupled to the rack in the steering gear. A **column-drive EPS** has an electric assist motor coupled to the steering shaft in the steering column, and in a **pinion-drive EPS** the electric motor is coupled to the steering gear pinion. The EPS system is light and compact compared to rack and pinion steering gear with hydraulic steering assist, because the power steering pump and hoses are not required on the EPS system. Since the EPS does not require a power steering pump driven by engine power, this system minimizes engine power loss and reduces fuel consumption. The EPS system reduces steering kickback while providing a linear steering feel.

System Components

The rack and pinion steering gear changes rotary steering wheel motion to transverse motion of the rack. The motor that provides electric steering assist is designed into the steering gear (Figure 13-22). The steering sensor is mounted in the pinion shaft. This sensor sends input voltage signals to the EPS control unit in relation to the direction, amount, and torque of the steering wheel rotation.

If the vehicle has an automatic transaxle, a VSS is mounted in the transaxle. This sensor sends voltage input signals to the transmission control module (TCM) in relation to

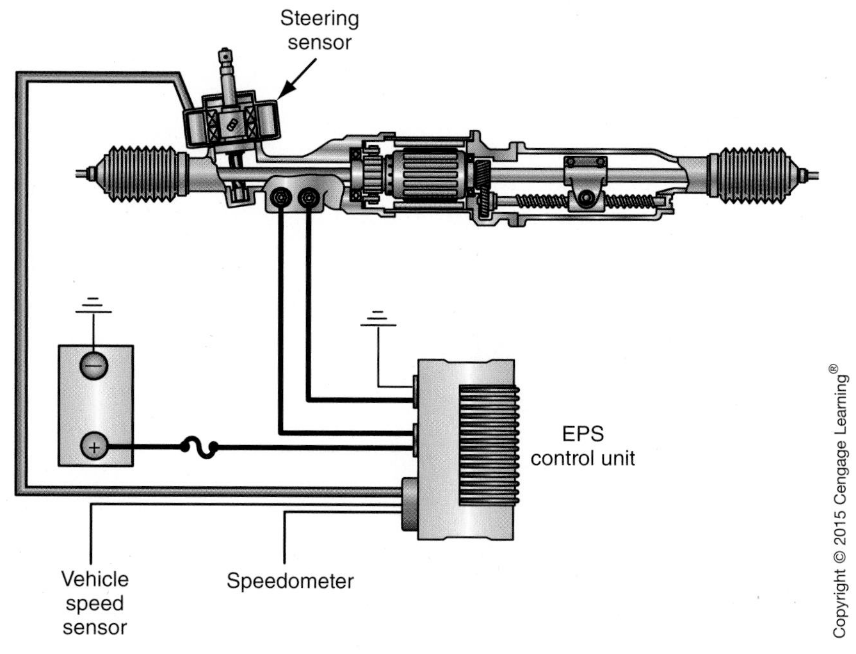

FIGURE 13-22 **Steering gear with electronic assist and related system components.**

transaxle output shaft rotational speed. These voltage signals are transmitted from the TCM to the EPS control unit. When the car is equipped with a manual transaxle, a differential speed sensor sends voltage signals to a pulse unit in relation to differential rotational speed. These voltage signals are transmitted from the pulse unit to the EPS control unit. The TCM and pulse unit is located in the rear of the vehicle (Figure 13-23). The

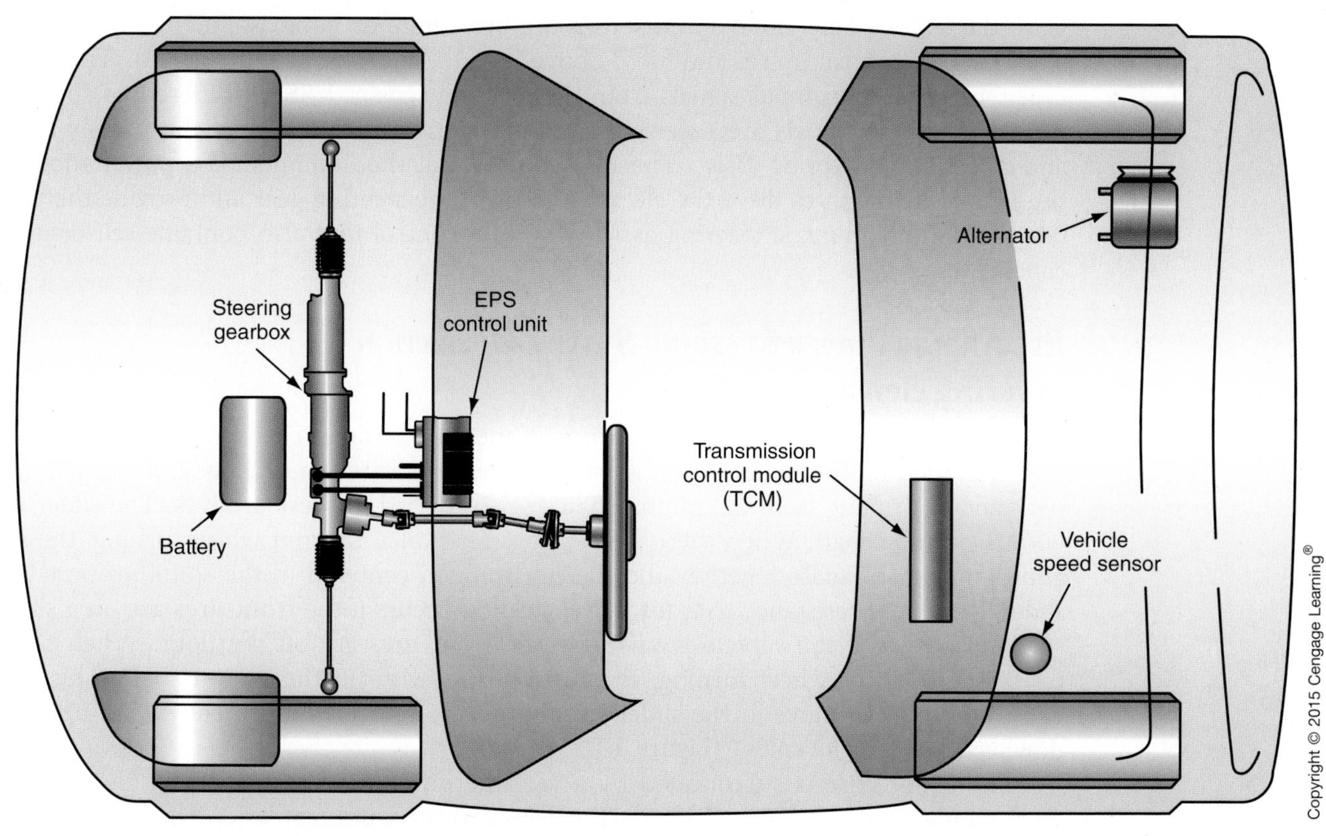

FIGURE 13-23 **Electronic power steering (EPS) system component locations.**

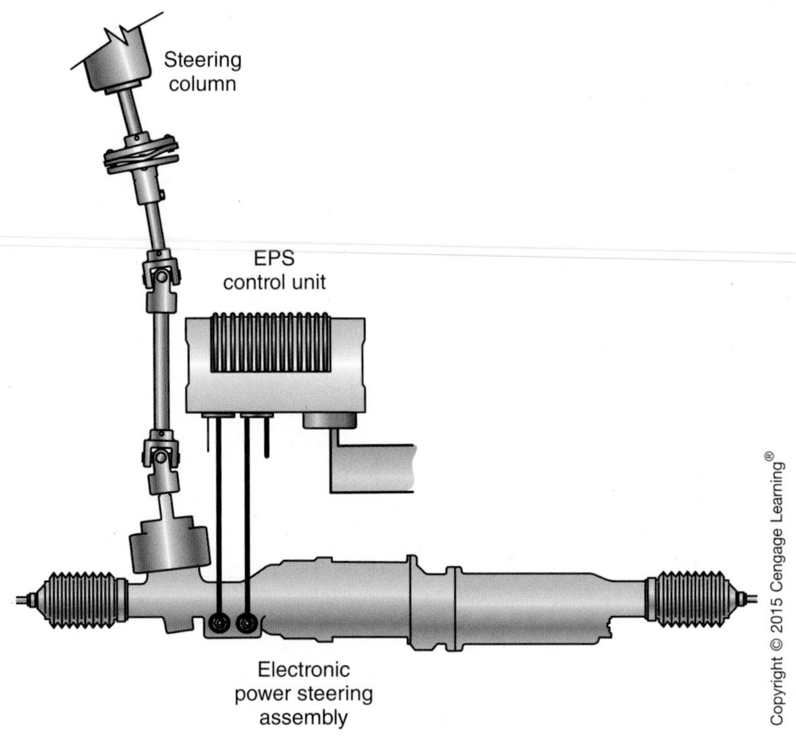

FIGURE 13-24 **Electronic power steering (EPS) control unit and steering gear.**

speedometer also sends voltage signals to the EPS control unit. The voltage signals from the speedometer act as a double check and backup in a situation where the voltage signals from the VSS or differential speed sensor are inaccurate or nonexistent.

The EPS control unit is mounted above the steering gear (Figure 13-24). This control unit receives voltage input signals from the steering sensor, VSS, or differential speed sensor. When these signals are received, the EPS control unit calculates the proper amount and direction of steering assist. The EPS control unit then commands a power module in the EPS control unit to drive the electric motor in the steering gear and provide the proper direction and amount of steering assist. The EPS control unit also contains self-diagnostic capabilities.

Electronic Power Steering System Operation

Steering Sensor

The steering sensor contains a torque sensor and an interface. This sensor contains a slider core mounted on the pinion shaft. A spiral groove is located in each side of the slider, and two pins protruding from the pinion shaft are positioned in these grooves. The slider turns with the pinion shaft. When there is very little resistance to front wheel turning, the slider and pinion shaft rotate together and the pins remain centered in the spiral grooves in the slider. Very little resistance to front wheel turning occurs if the front tires are on a slippery road surface, or if the vehicle is raised so the front tires are off the floor. When there is resistance to front wheel turning, the torsion bar twists in the pinion shaft. This action causes the pins to move in the slider spiral grooves, and this movement causes upward or downward slider movement (Figure 13-25).

The slider core is surrounded by a variable differential transformer, and the slider moves upward or downward inside the transformer windings when the steering wheel

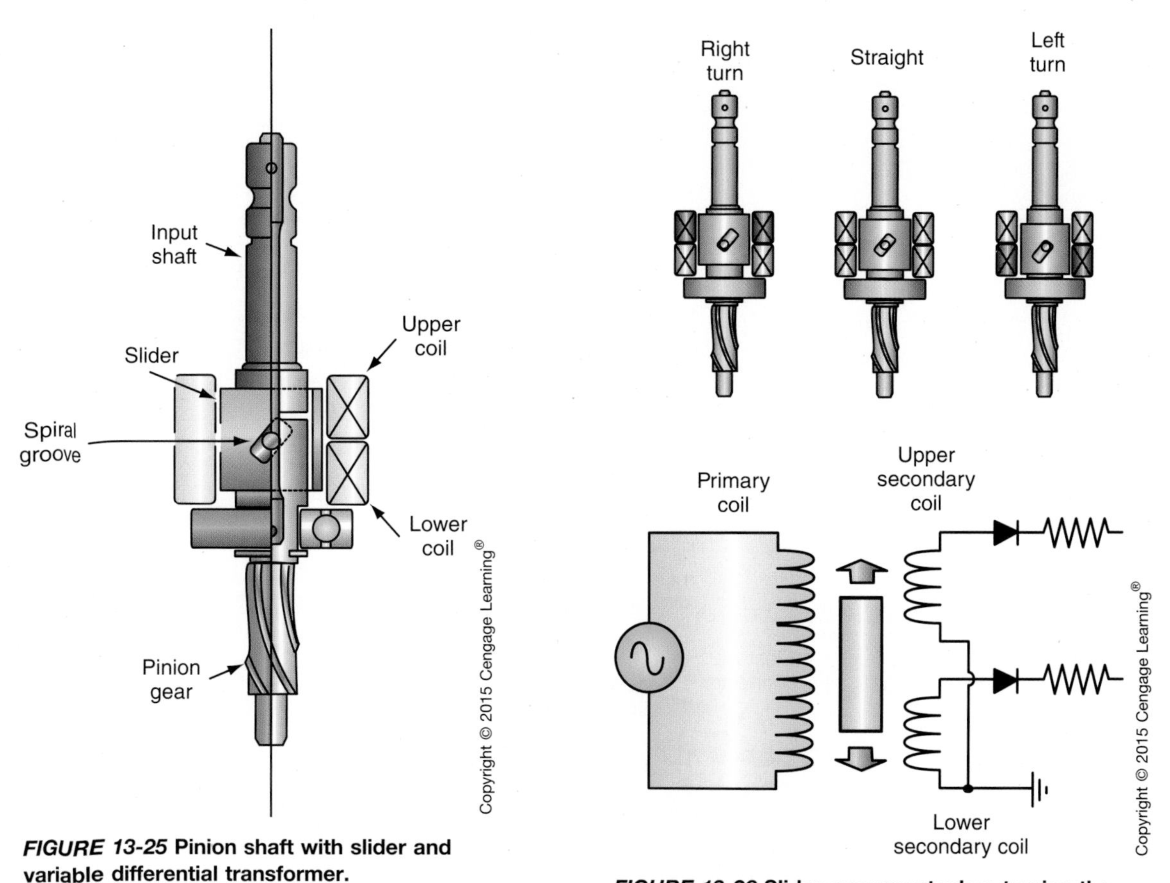

FIGURE 13-25 **Pinion shaft with slider and variable differential transformer.**

FIGURE 13-26 **Slider movement when turning the steering wheel.**

is turned (Figure 13-26). The transformer has a primary coil and upper and lower secondary coils. When the ignition switch is turned on, an oscillation circuit in the steering sensor supplies an alternating current to the primary transformer coil. As the current alternates back and forth through the primary coil, the magnetic field is continually building up and collapsing around this coil. This rapidly expanding and collapsing magnetic field induces voltages in the upper and lower secondary coils. The position of the slider determines whether the voltage is induced in the upper or lower secondary coil.

Shop Manual
Chapter 13, page 520

While driving straight ahead, the slider is centered vertically between the upper and lower secondary coils. Under this condition, the voltage induced in these coils is equal. When voltages are equal in the upper and lower secondary coils, the voltage signals from these coils to the EPS control unit indicate the car is being driven straight ahead, or the steering wheel is being turned with no resistance and no electric power assist is required.

If the steering wheel is turned to the right, the slider moves upward. This slider position causes more induced voltage in the upper secondary coil and less induced voltage in the lower secondary coil. When these voltage signals from the upper secondary coil are sent to the EPS control unit, this control unit supplies current to the electric motor on the rack so the motor rotates in the appropriate direction to provide the proper amount of steering assist to the right (Figure 13-26).

When the steering wheel is turned to the left, the slider moves downward. This slider position causes more induced voltage in the lower secondary coil and less induced voltage

Steering condition	Slider movement	Induction voltage on secondary coil
Steering to right (load steering)	Upward shift	Voltage on upper coil increases, and voltage on the lower decreases
Advancing straight ahead (no load steering)	Neutral	Voltage on the upper and lower coils are equal
Steering to left (load steering)	Downward shift	Voltage on lower coil increases, and voltage on the upper decreases

FIGURE 13-27 **Summary of steering sensor and transformer operation.**

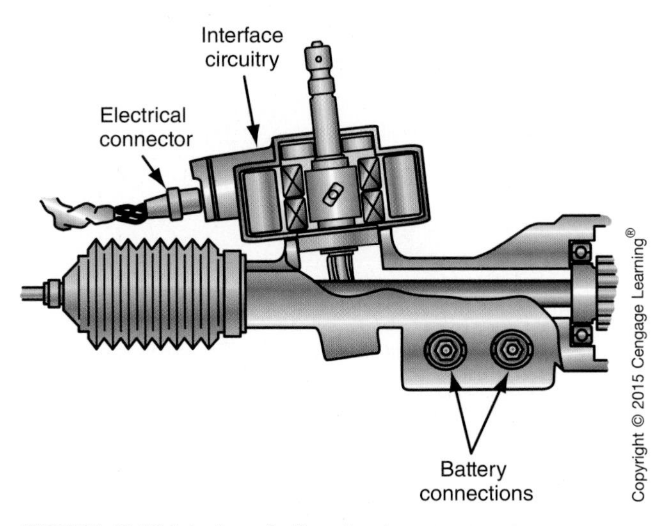

FIGURE 13-28 **Interface in the steering sensor.**

in the upper secondary coil. When these voltage signals from the lower secondary coil are sent to the EPS control unit, the control unit supplies current to the electric motor on the rack so the motor rotates in the appropriate direction to provide the proper amount of steering assist to the left (Figure 13-27).

The voltage signals from the upper and lower secondary transformer coils are sent through the interface in the steering sensor to the EPS control unit (Figure 13-28). The interface rectifies the alternating current (AC) voltage signals from the upper and lower transformer coils to direct current (DC) voltage signals and amplifies or increases the signal strength.

Electric Motor and Steering Gear Operation

The armature in the electric motor is hollow, and the rack extends through the center of this armature. Ball bearings are mounted between the outer diameter of the armature and the steering gear housing to support the armature. Two spring-loaded brushes are mounted on opposite sides of the commutator on one end of the armature. A gear with helical teeth is mounted on the other end of the armature. The teeth on the armature gear are in constant mesh with a matching gear on the ball screw shaft (Figure 13-29).

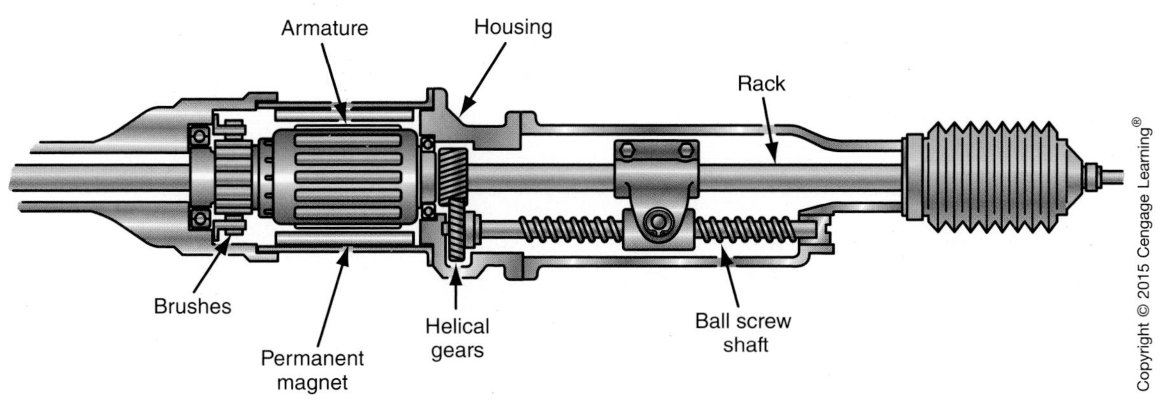

FIGURE 13-29 **Steering gear and electric motor.**

A recirculating ball screw nut is mounted on the ball screw shaft. Ball bearings are mounted in grooves in the ball screw shaft and recirculating ball screw nut. The recirculating ball screw is bolted to the steering gear rack. The ball screw shaft and recirculating ball screw nut are similar in design to the worm shaft and ball nut in a recirculating ball steering gear. A permanent magnet is mounted in the steering gear housing, and this magnet surrounds the armature core. There is a small clearance between the armature core and the permanent magnet.

When the steering wheel is turned, electric current is supplied from the power module in the EPS control unit through the brushes and armature windings to ground. This current flow creates strong magnetic fields around the armature windings. These magnetic fields around the armature windings react with the magnetic field of the permanent magnet and cause armature rotation in the proper direction to supply steering assist. When the armature rotates, the gear on the armature shaft drives the gear on the ball screw shaft. Ball screw shaft rotation moves the recirculating ball screw nut on the shaft. Since the recirculating ball screw nut is bolted to the steering gear rack, movement of this ball screw nut provides steering assist in the proper direction. The power module can reverse the armature rotation to provide steering assist in either direction by reversing the polarity of the brushes on the commutator. When the brush polarity is reversed, the current flow through the armature windings is reversed and this changes the direction of armature rotation.

The recirculating ball screw nut is designed so the ball bearings roll between the grooves in the ball screw shaft and the grooves in the recirculating ball screw nut. Ball bearings coming out of the recirculating ball screw nut move through a tube and re-enter the recirculating ball screw nut at the other end (Figure 13-30). The ball bearings in the grooves in the ball screw shaft and recirculating ball screw nut allow this nut to move on the shaft with very low friction.

Steering Gear Motor Current Control

The power module in the EPC control unit contains a driving circuit, current sensor, field effect transistor (FET) drive circuit, power relay, and fail safe relay. When the ignition switch is turned on and the engine is cranked, the EPS control unit closes the power relay and fail safe relay to make the EPS system operational. These relays actually close when the alternator begins producing voltage while the engine is cranking. Voltage signals are sent from the alternator and the ignition switch to the EPS control unit (Figures 13-31 and 13-32). The EPS system remains operational if the engine stalls

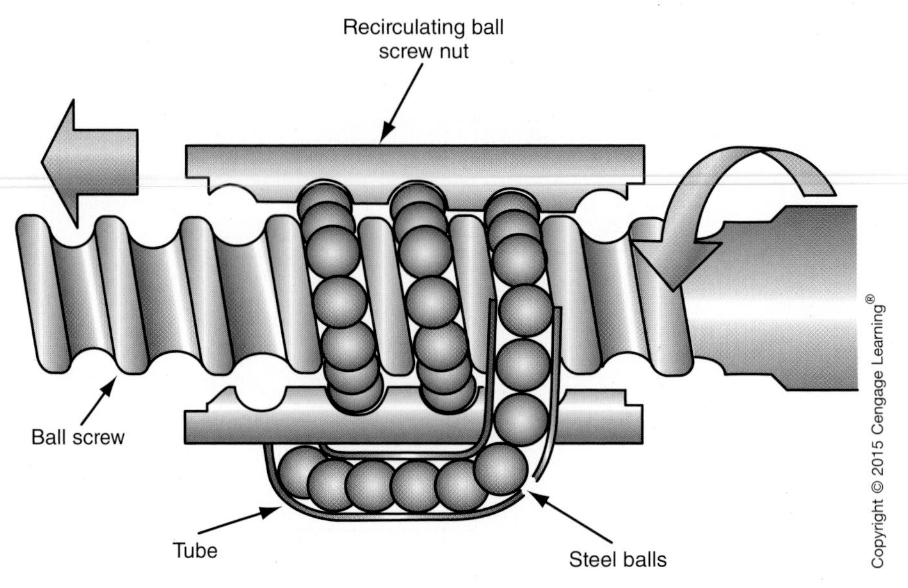

FIGURE 13-30 **Ball screw shaft and recirculating ball screw nut.**

and the ignition switch is on. If the engine stalls on a hydraulically assisted power steering system, power steering assist is lost and the steering wheel becomes very hard to turn. This action may result in a collision if the engine stalls while turning a corner. Since the EPS system is still operational if the engine stalls and the ignition switch is on, the EPS system reduces the possibility of a collision resulting from loss of power steering assist during an engine stall.

When the driver turns the steering wheel, the steering sensor input voltage signals inform the EPS control unit that steering assist is necessary. The steering sensor input voltage signals also inform the EPS control unit regarding the direction and amount of steering assist required. If the driver supplies more torque to the steering wheel, the steering sensor input voltage signal indicates to the EPS control unit that more steering assist is necessary.

When the EPS control unit receives input voltage signals from the steering sensor indicating that steering assist is required, this control unit signals the FET drive circuit in the power module. This drive circuit supplies a PWM voltage to the motor brushes with the proper polarity to provide the required direction and amount of steering assist. A PWM voltage signal is a pulsating voltage signal with a constant frequency but has a variable on time. One cycle of a PWM signal is a specific length of time that includes one Off and one On signal. If the on time lasts for 40 percent of each cycle time and the off time lasts for 60 percent of the cycle time, the motor current remains lower and this reduces motor speed and steering assist. When the on time lasts for 60 percent of each cycle time, and the off time lasts for 40 percent of the cycle time, the motor current is higher and power steering assist is increased (Figure 13-33). The FET drive circuit operates the four FET transistors in the FET bridge to supply the proper direction and amount of voltage and current through the armature windings to provide the necessary direction and amount of steering assist (Figure 13-34).

While the electric assist motor is operating, the motor current flows through a current sensor in the power module regardless of the direction of motor rotation. The current sensor sends a feedback voltage signal to the EPS control unit. If the motor current exceeds a predetermined average motor current for the current operating condition, the EPS control

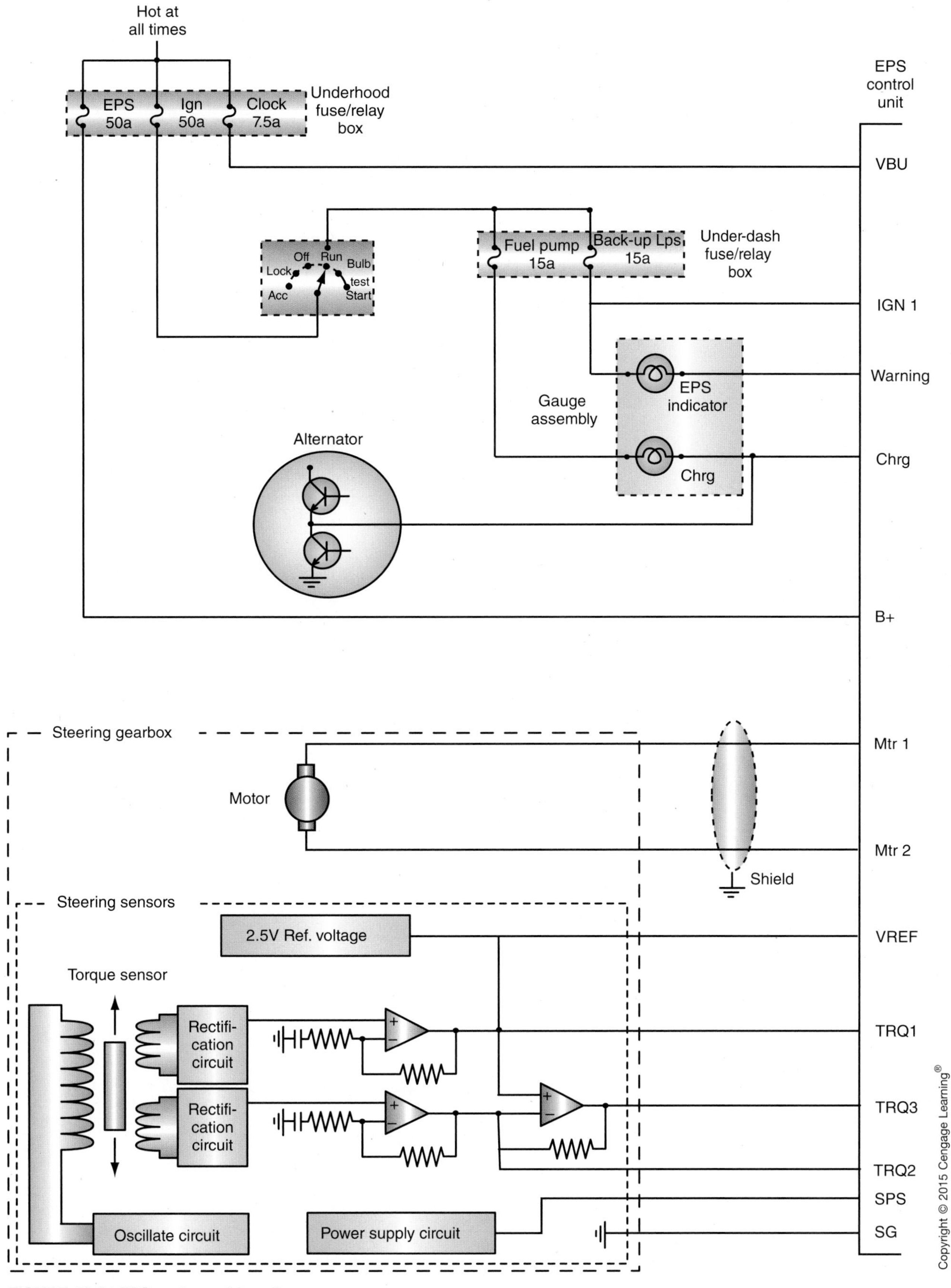

FIGURE 13-31 **EPS system wiring diagram.**

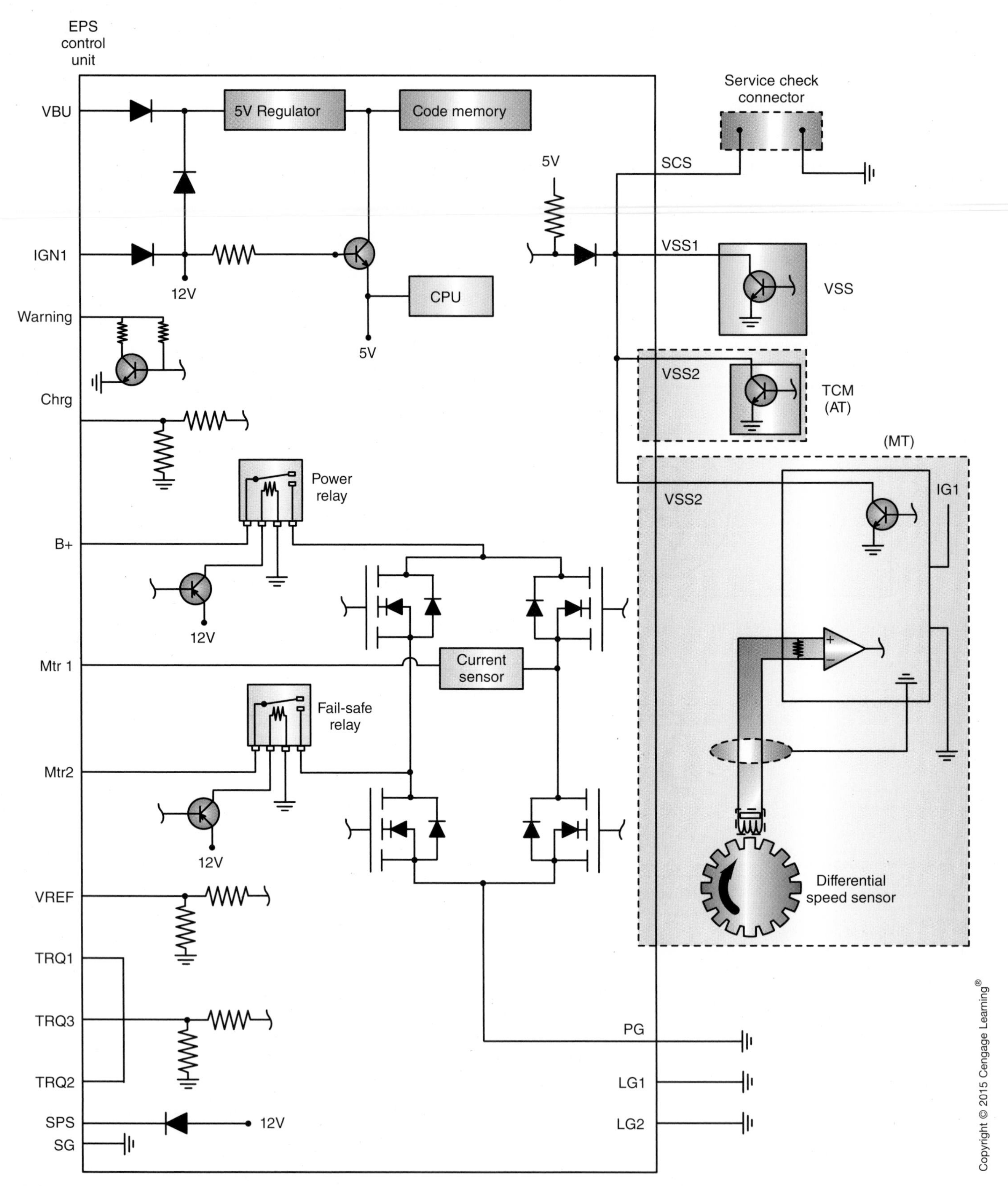

FIGURE 13-32 **EPS system wiring diagram (continued).**

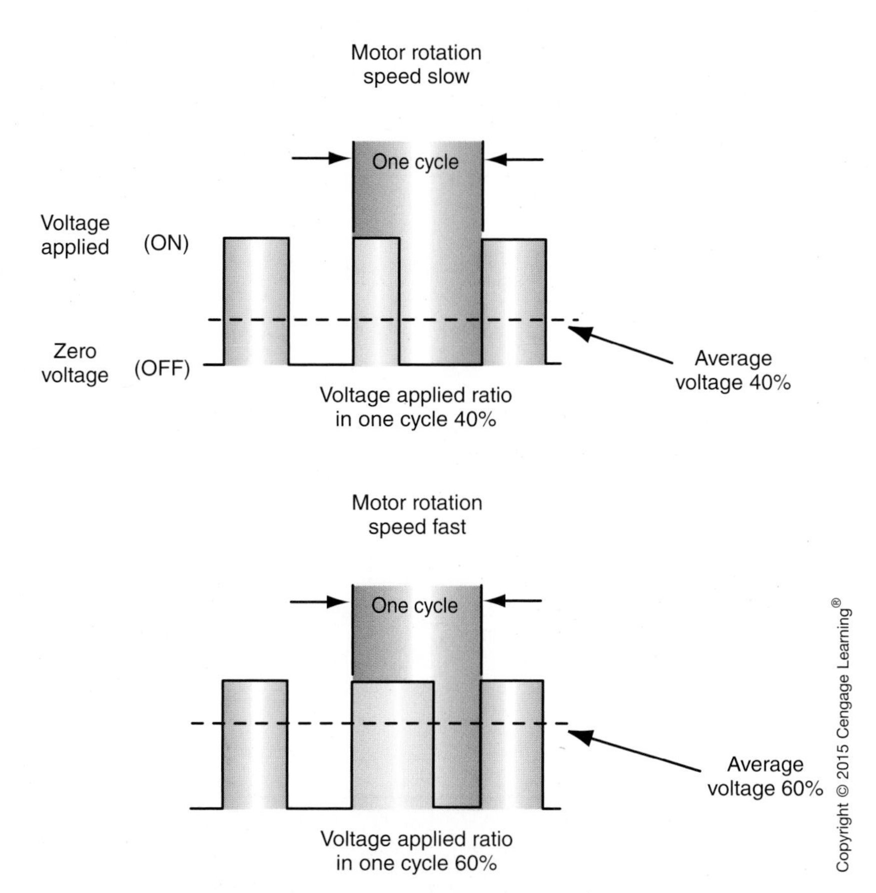

FIGURE 13-33 **Pulse width modulated (PWM) voltage signal.**

unit signals the drive circuit in the power module to reduce the motor current to prevent motor overheating. If the steering wheel is turned fully in one direction and held in this position, the motor current becomes much higher. Under this condition, the current sensor signals the EPS control unit, and this unit signals the drive circuit in the power module to reduce motor current to protect the motor.

If an electrical defect occurs in the EPS system, the EPS control unit illuminates an EPS indicator light in the instrument panel to inform the driver that a fault is present in the EPS system (Figure 13-35). Under this condition, the EPS unit opens the fail safe and power relays to make the EPS system inoperative. When this action occurs, manual steering without any power assist is available.

If one of the front wheels strikes a large road irregularity, force from the front wheel is transferred to the steering gear rack. Rack movement attempts to move the ball screw nut on the ball screw shaft. This action tries to rotate the ball screw shaft and armature. A specific amount of force is required to move the ball screw nut and rotate the ball screw shaft and armature because the armature windings have to rotate through the magnetic field between the permanent magnets. This resistance to movement of the ball screw nut, ball screw shaft, and armature helps prevent road shock from being supplied through the steering gear to the steering wheel. When a very high road shock is transferred to the steering wheel, this road shock moves the pinion shaft and slider in the steering sensor. When this EPS control unit receives this steering sensor input signal, the EPS control unit immediately energizes the armature windings to oppose and cancel the road shock applied to the steering gear. Therefore, the EPS steering gear reduces road shock transferred from the front wheels to the steering wheel.

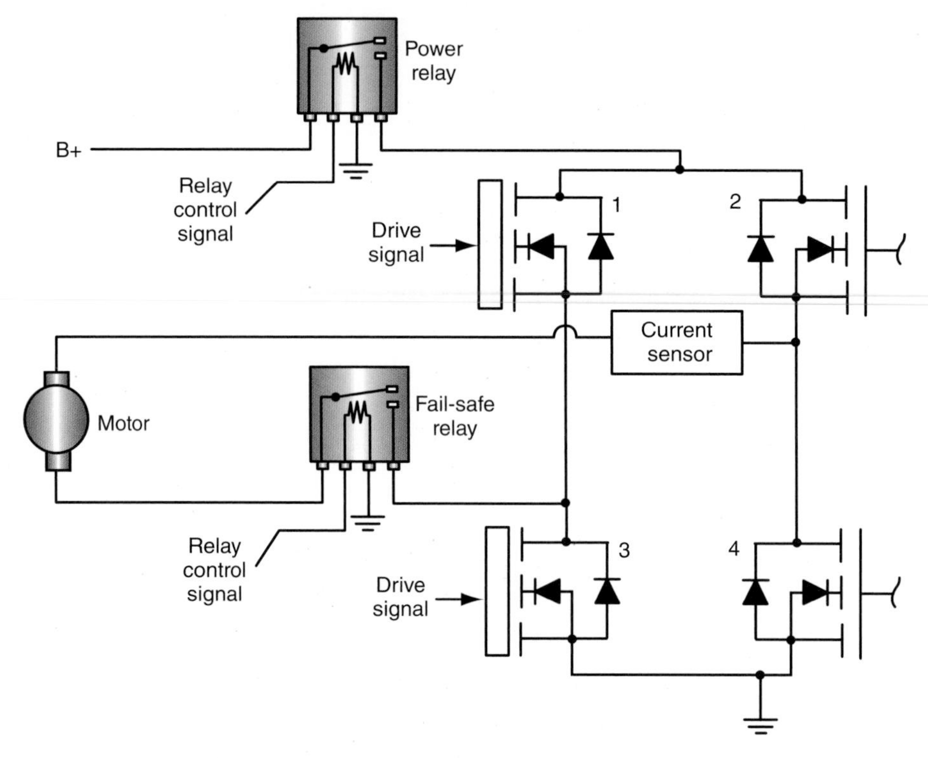

Steering condition	FET (1)	FET (2)	FET (3)	FET (4)	Motor operation
Steering to right	PWM	OFF	OFF	ON	Operates in direction steering to the right
Straight ahead	OFF	OFF	OFF	OFF	Stops
Steering to left	OFF	PWM	ON	OFF	Operates in direction steering to the left

FIGURE 13-34 **Drive circuit and field effect transistor (FET) bridge in the power module.**

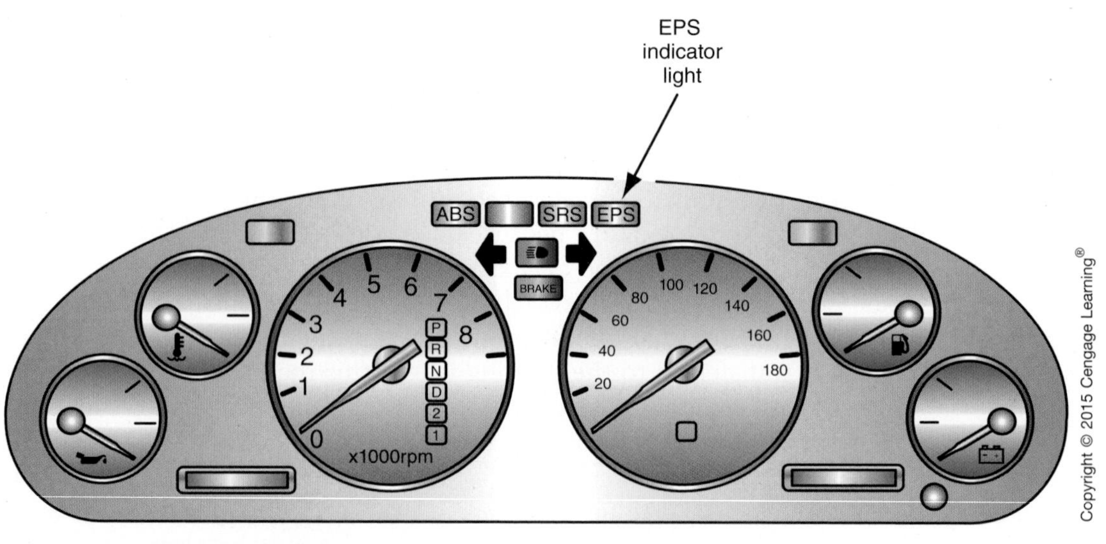

FIGURE 13-35 **EPS indicator light.**

Column-Drive Electronic Power Steering

Some vehicles are presently equipped with a column-drive EPS. In this type of EPS, a motor/module assembly is bolted to the lower end of the steering column (Figure 13-36). A combined steering wheel position sensor and steering shaft torque sensor is mounted in the steering column at the motor/module attachment point. The module in the motor/module assembly is called the power steering control module (PSCM). Mounting the EPS motor/module assembly under the instrument panel provides underhood space for other components, subjects the assembly to less rigorous temperatures, and may provide better protection during a collision.

Input Sensors

The steering shaft torque sensor is the most important input used by the PSCM to supply the proper amount and direction of steering assist. The steering column contains a torque sensor input shaft connected from the steering wheel to the sensor, and an output shaft connected from the sensor to the steering shaft coupler. A torsion bar mounted inside the steering shaft torque sensor separates the input and output shafts. When the steering wheel is turned in either direction, the torque supplied from the steering wheel to the torsion bar causes this bar to twist. Increased torsion bar twisting results in higher voltage signals from the steering shaft torque and position sensors. The steering shaft torque sensor is a dual sensor that provides two different voltage signals (Figure 13-37). During a right turn the voltage from sensor 1 increases and the voltage from sensor 2 decreases. The voltage signal range from each sensor is 0.25 V to 4.75 V. While completing a left turn, the voltage from sensor 1 decreases and the voltage from sensor 2 increases. The steering shaft torque sensor voltage signals inform the PSCM regarding the direction and amount of steering wheel torque.

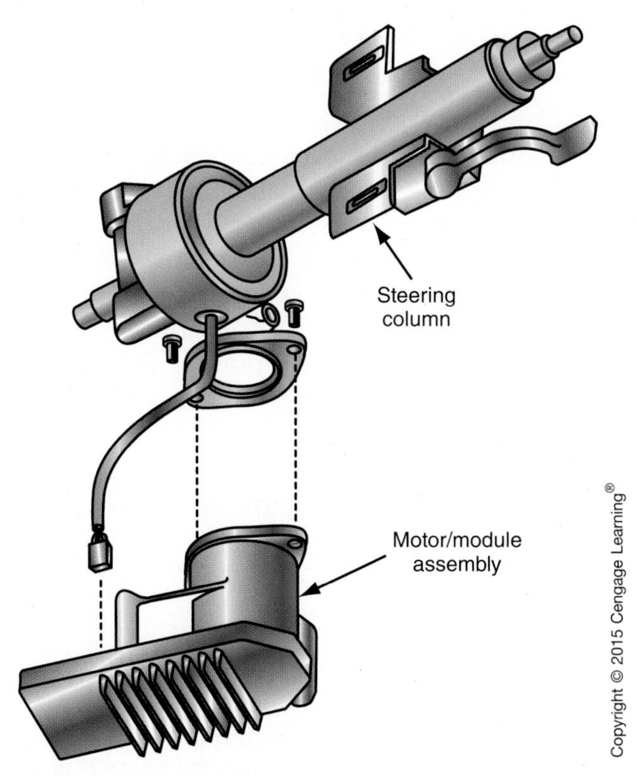

FIGURE 13-36 **Column-drive electronic power steering system.**

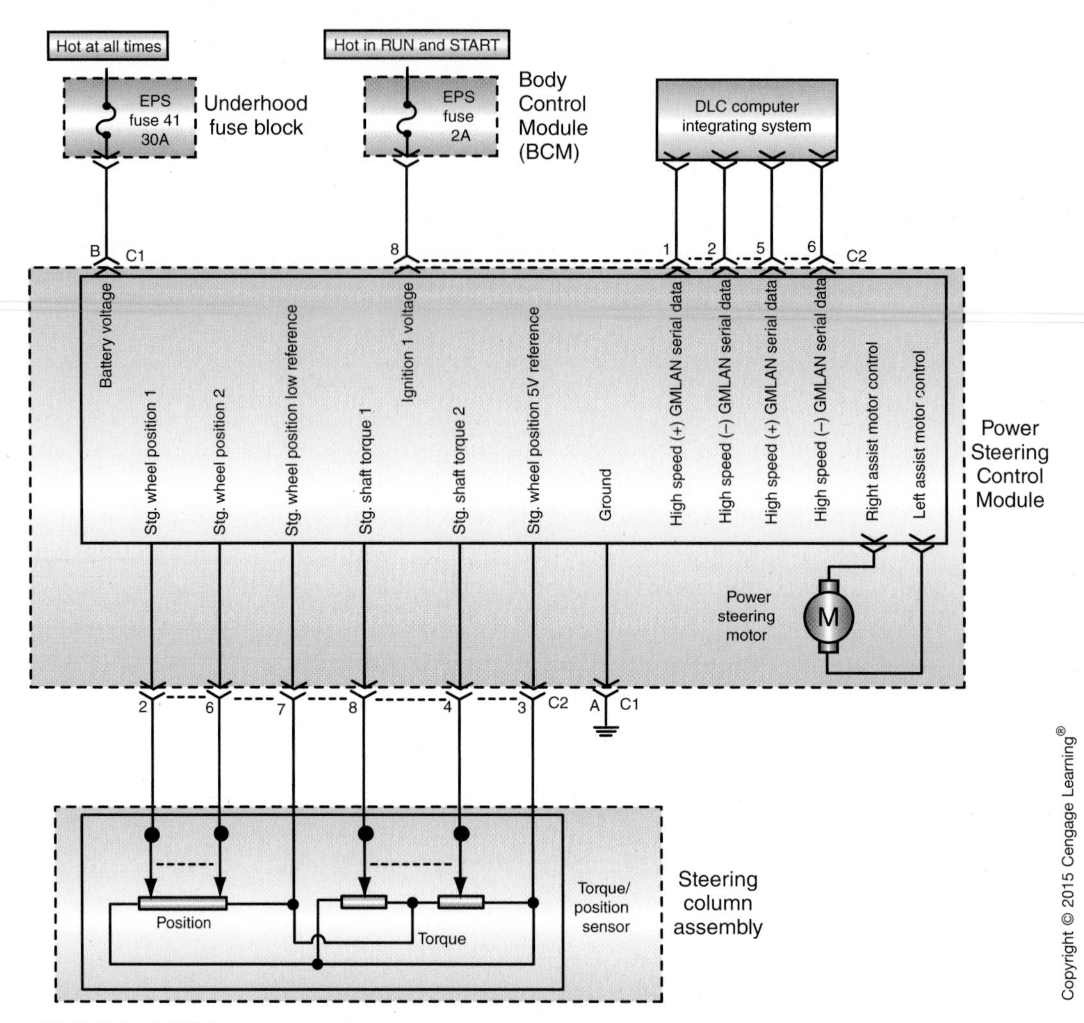

FIGURE 13-37 **Wiring diagram for column-drive electronic power steering.**

The steering wheel position sensor also contains dual voltage sensors that operate in the 0 V–5 V range. During steering wheel rotation the voltage from these two sensors should remain within 2.5 V–2.8 V of each other. The steering wheel position sensor informs the PSCM regarding the position of the steering wheel. The combined steering shaft torque and position sensor assembly is serviced only as a unit.

The PSCM also receives a vehicle speed input signal from the powertrain control module (PCM). This voltage signal is transmitted from the PCM to the PSCM through the interconnecting data links.

Power Steering Motor

The power steering motor is a 12 V brushless DC reversible motor rated at 65 amperes. This motor is coupled to the steering shaft through a worm gear and reduction gear located in the steering column housing.

Column-Drive Electronic Power Steering Operation

When the input signals indicate the driver is beginning to turn the steering wheel, the PSCM supplies the proper amount of current to the power steering motor to provide the necessary power steering assist. At low vehicle speeds the PSCM provides more steering assist for easy steering wheel rotation during parking maneuvers. The PSCM reduces the motor current and provides reduced steering assist when the vehicle is driven at higher speeds to supply improved road feel and directional stability. If the EPS motor becomes overheated, the PSCM enters a protection mode that reduces motor current to avoid thermal damage to system components. The PSCM has the capability to detect electronic system defects in the EPS system. When a defect is detected, a POWER STEERING warning message is displayed in the driver information center (DIC) in the instrument panel and the malfunction indication light (MIL) may also illuminate. The PSCM must be programmed with the correct steering tuning in relation to the vehicle powertrain configuration.

Pinion-Drive Electronic Power Steering

Some hybrid vehicles are equipped with a 12 V EPS, because on these vehicles the engine is automatically shut off during idle operation to conserve fuel and reduce emissions, and the power steering remains operational when the engine is stopped. The EPS also conserves fuel because it is used on demand only when the driver turns the steering wheel. On conventional hydraulic power steering systems the power steering pump operates all the time the engine is running. Some current hybrid vehicles have a pinion-drive EPS. The electric motor on the EPS is coupled to the pinion in the rack and pinion steering gear (Figure 13-38). Except for the drive motor and related drive gears, the other components in the EPS are conventional in design. The EPS is lubricated for life and has no reservoir, fluid, or hoses.

The EPS contains an electromagnetic-type rotational sensor mounted in the steering gear input shaft. This rotational sensor changes a rotational signal to a voltage signal representing the direction and amount of torque supplied from the steering wheel to the input shaft. The PSCM is mounted on the cowl above the steering gear. The PSCM is interconnected to other on-board computers through the data link system. Additional inputs such as vehicle speed are transmitted from the PCM through the data links to the PSCM. When the PSCM receives input signals from the rotational sensor and the PCM, the PSCM supplies the proper amount and direction of current flow to the electric motor in the EPS to provide the required steering assist. A drain wire is connected from the PSCM to ground. This wire prevents the EPS from emitting electromagnetic

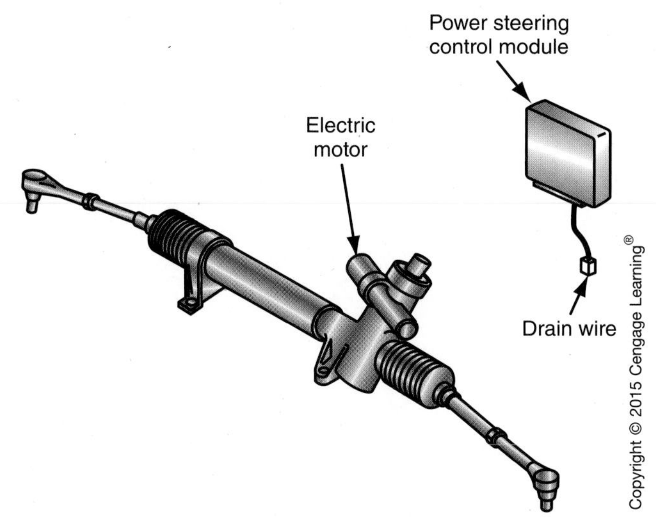

FIGURE 13-38 **Pinion-drive electronic power steering (EPS) gear.**

interference that could affect other electrical/electronic systems in the vehicle. If a defect occurs in the EPS electrical system, a SERVICE PWR STEERING message illuminates in the instrument panel.

Active Steering Systems

Introduction

Shop Manual
Chapter 13, page 525

Active steering is available on a number of luxury vehicles. An active steering system contains many of the same components located in a conventional power rack and pinion steering system. The main difference in the active steering system is that the steering wheel and steering shaft are no longer connected directly to the pinion gear in the steering gear. In an active steering system, a dual planetary gear set is connected between the steering shaft and the pinion gear in the steering gear (Figure 13-39). The steering wheel and shaft are connected to one of the sun gears in a dual planetary gear set. The second sun gear in the planetary gear set is connected to the pinion gear in the steering gear. The two sun gears are connected by a set of planetary pinions. A brushless 3-phase DC electric motor drives the ring gear on the planetary gear sets through a worm drive (Figure 13-40).

The **electronic control unit (ECU)** operates the electric motor to control the steering gear ratio and provide steering angle corrections to improve vehicle stability (Figure 13-41). In the active steering system, there is always a mechanical connection between the steering wheel and the pinion gear in the steering gear. The ECU may drive the electric motor and change the steering gear ratio or steering angle, but the mechanical connection between the steering wheel and the steering gear pinion always leaves the driver in control of the steering. The electric motor in the active steering system never provides turning forces to the front wheels.

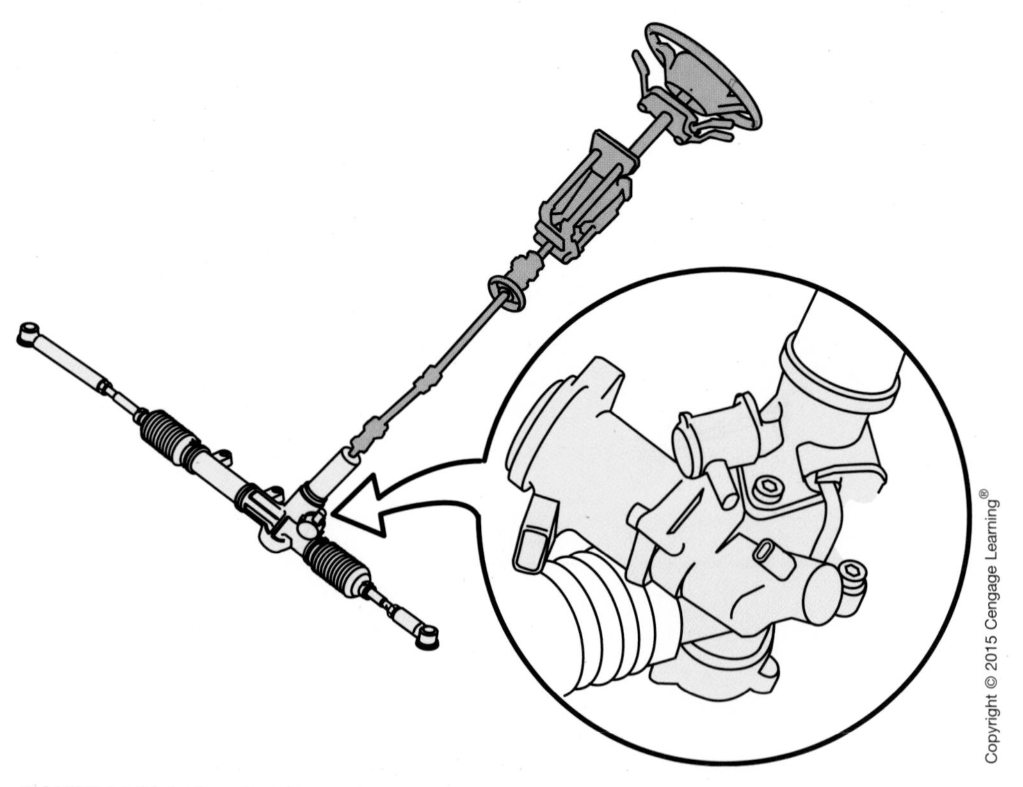

FIGURE 13-39 **Active steering system.**

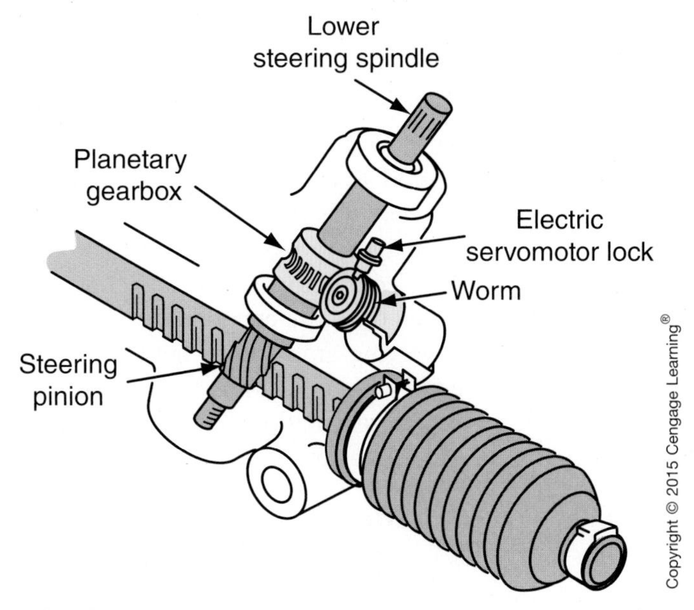

FIGURE 13-40 **Active steering actuator.**

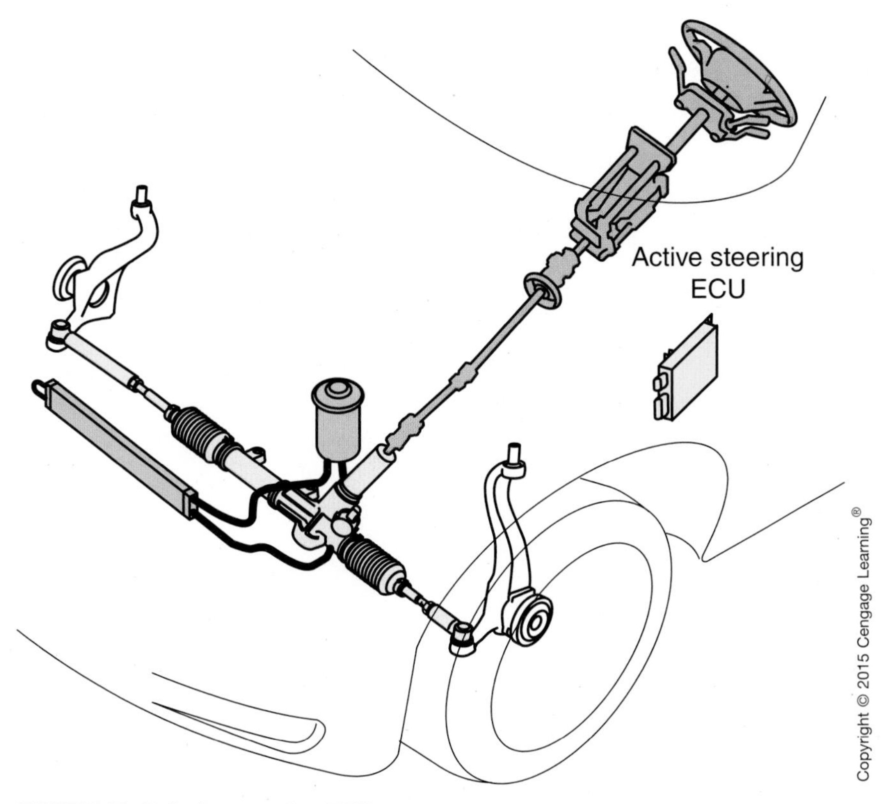

FIGURE 13-41 **Active steering ECU.**

Active Steering System Components

DC Electric Motor and Motor Position Sensor

The electric drive motor has a wound stator and a permanent magnet rotor assembly. The ECU controls drive motor torque by a field-orientated control. The motor position sensor is mounted in the end of the electric motor, and this sensor informs the ECU regarding electric motor position.

Steering Angle Sensor

The **steering angle sensor** is mounted in the steering column switch assembly. The steering angle sensor is an optical sensor with no contacting parts, and this sensor is mounted to the circuit board near the top of the steering column (Figure 13-42). The main components in the steering angle sensor are an **encoded disc** and an optical sensor. The encoded disc is attached to the steering shaft and rotates with this shaft. The encoded disc rotates within the optical sensor (Figure 13-43). The stationary part of the steering angle sensor contains a light-emitting diode (LED), fiber-optical conductor, and a line camera. When the sensor is operating, light from the LED is projected onto the encoded disc through the optical conductor. As the steering wheel is rotated, varying amounts of light from the LED will penetrate the encoded disc and shine on the line camera. The line camera converts these light signals to voltage signals in relation to steering wheel rotation. The type of sensors in an active steering system may vary depending on the model and year of vehicle.

Pinion Angle Sensor and Summation Sensor

The **pinion angle sensor** is mounted in the steering gear at the end of the pinion housing, and this sensor operates on a magneto-resistive principle. The pinion angle sensor sends a signal to the ECU regarding pinion position. The **summation sensor** is mounted in the

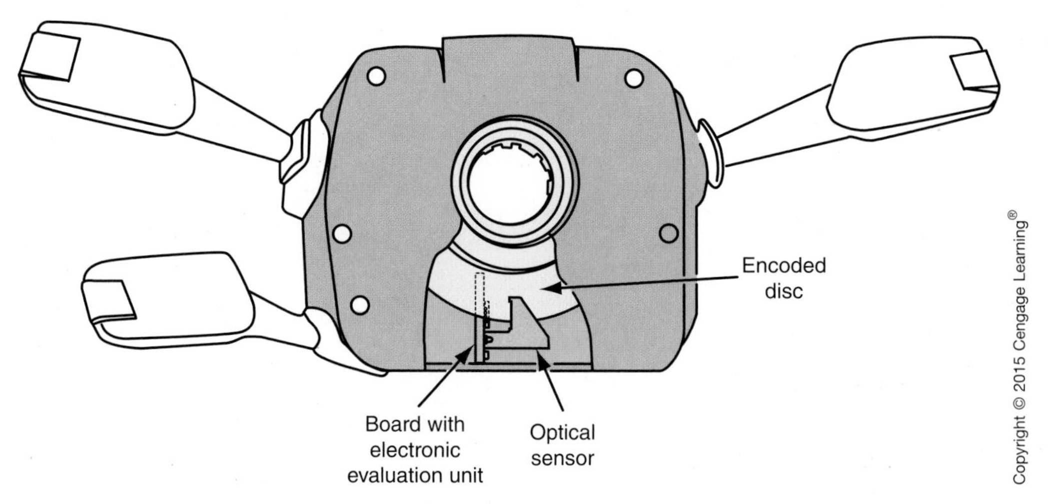

FIGURE 13-42 **Steering angle sensor mounting.**

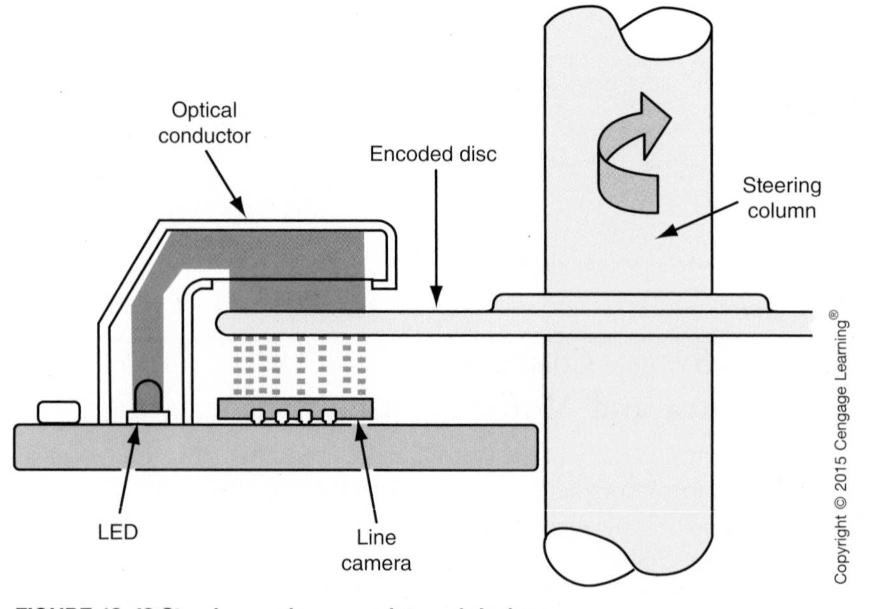

FIGURE 13-43 **Steering angle sensor internal design.**

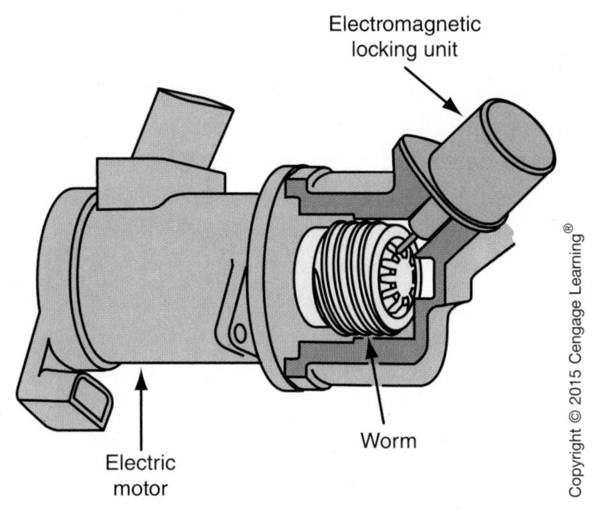

FIGURE 13-44 **Actuator motor locking mechanism.**

steering gear housing, and this sensor sends a signal to the ECU in relation to total steering gear rack movement.

Electronic Control Unit (ECU)

The ECU contains two microprocessors that control the electric motor, steering pump, servotronic valve, and electric locking unit. The ECU communicates with other ECUs and the active steering system input sensors via a CAN network.

Electric Locking Unit (ELU)

The **electric locking unit (ELU)** contains an electric solenoid and a lock pin. When the ELU solenoid is not energized by the ECU, the lock pin drops into one of the slots in the worm drive gear. This action locks the worm drive and drive motor (Figure 13-44). This locking action is taken by the ECU if a safety-related defect occurs in the active steering system. When the worm drive is locked, the driver maintains normal steering control. Under normal conditions, the ECU operates the ELU solenoid and pulls the lock pin away from the notches in the worm drive gear, and this allows normal active steering system operation.

Active Steering Operation

The actuator motor and planetary gear set have the capability to vary the steering gear ratio in relation to vehicle speed. The ECU receives voltage input signals from the system sensors, and in relation to these inputs, the ECU operates the actuator motor to vary the steering gear ratio. While cornering at low speeds, the steering gear ratio approximates 10:1. With this ratio, the driver does not have to rotate the steering wheel as much to obtain more front wheel turning action. When the vehicle is stationary, less than two steering wheel turns are required to turn the front wheels from lock-to-lock. This steering gear ratio allows the driver to apply the least amount of turning action to obtain a large amount of front wheel turning action. While parking and cornering at low speeds, this type of steering system requires less driver hand-to-hand action on the steering wheel, which increases safety and reduces driver fatigue.

To reduce the steering ratio at low speeds, the actuator motor is driven in the same direction as the steering input, which decreases the steering ratio, and this action over-drives the steering input. This steering gear action may be compared to walking on an escalator. If you walk in the same direction as the escalator is moving, you multiply the total walking result.

As the vehicle speed increases, the ECU operates the actuator motor and planetary gear set to increase the steering ratio. At higher vehicle speeds, the steering gear ratio

Yaw motion is an erratic deviation from an intended course such as a sideways swerve when driving a vehicle.

A BIT OF HISTORY

Electronically controlled systems on vehicles have decreased emission levels, increased fuel economy, and improved driver comfort and convenience. In the near future, electronic systems on vehicles may be interconnected with intelligent transportation systems (ITS) to speed up traffic flow and reduce accidents. Some traffic experts indicate that Americans lose 2 billion working hours each year because of traffic congestion and related delays. It is estimated these traffic delays cost American employers $40 billion per year. Therefore, we can appreciate that a system to increase traffic movement could save us billions of dollars each year.

Oakland County, Michigan, claims to be the world's largest operational test of an ITS system. This system uses 400 video cameras at 200 intersections to transmit information to a central computer that switches traffic light operation to match traffic flow. A similar system called Faster And Safer

may be 20:1 and the increase in steering gear ratio dampens any sudden or excessive steering input by the driver. The increase in steering gear ratio under-drives the steering input. This action may be compared to walking against the movement of an escalator. The walking action of the person is cancelled to some extent by the escalator movement, and the person has to walk more to cover the same distance. This steering gear action reduces the possibility of the driver causing the vehicle to go into a sideways swerve (**yaw motion**) by excessive steering wheel rotation. Any excessive steering wheel rotation by the driver is immediately counteracted by an increase in steering ratio by the active steering system. This active steering system action reduces the possibility of yaw vehicle motion. Many vehicles with active steering systems are also equipped with dynamic stability control (DSC), which greatly reduces the possibility of the vehicle swerving sideways out of control. On a vehicle equipped with active steering, the DSC system will not have to operate as frequently. If an electronic defect occurs in the active steering system, the system enters a fail silent mode in which the active steering system is inoperative and the driver has normal control of the steering.

Power Steering System

The electric-drive power steering pump and the active steering system are controlled by the same ECU, which contains two microprocessors. The power steering pump is a high capacity vane-cell pump that is mounted immediately in front of the steering gear (Figure 13-45). The active steering system provides faster front wheel steering angles, and this action requires a higher capacity power steering pump compared with conventional hydraulic-assisted power steering systems. The active steering ECU operates the **electronically controlled orifice (ECO) valve** in the power steering pump to control the fluid flow from the pump (Figure 13-46). The power steering pump supplies fluid pressure from the ECO valve in the pump to the **servotronic valve** in the steering gear. The active steering ECU operates the servotronic valve to increase power steering pump pressure and reduce steering effort and driver fatigue at low speeds (Figure 13-47). At higher vehicle speeds, the ECU operates the servotronic valve to reduce power steering pump pressure and increase steering effort to provide improved road feel and steering control. On some models, the active steering ECU directly operates the servotronic valve.

On other models, the software for the servotronic valve operation is stored in the active steering ECU, and this ECU sends signals through one of the CAN networks to the **safety and gateway module (SGM)**. On the basis of these input signals, the SGM operates the servotronic valve and the ECO valve.

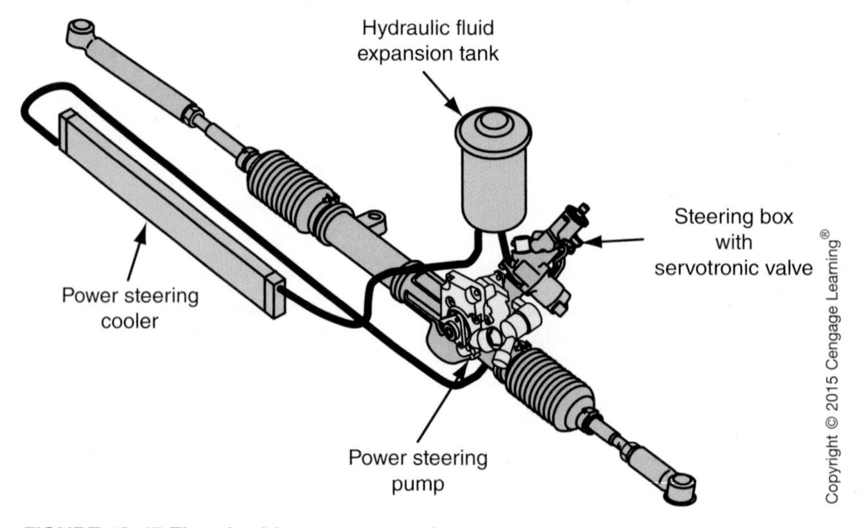

FIGURE 13-45 **Electric-drive power steering system.**

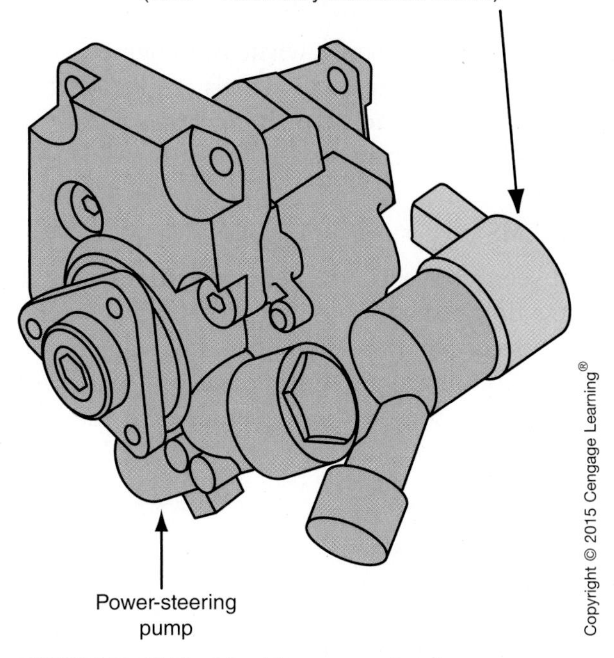

FIGURE 13-46 **Electric drive power steering pump.**

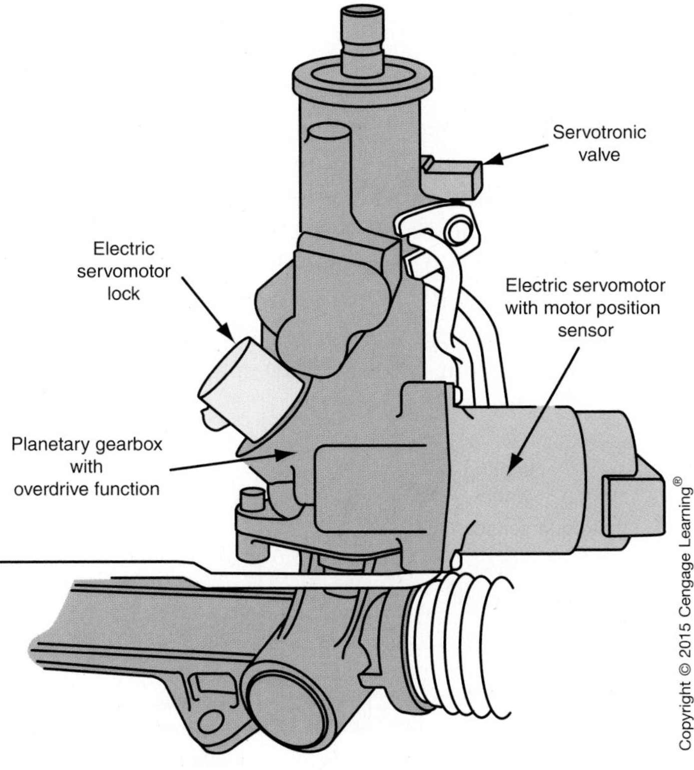

FIGURE 13-47

A BIT OF HISTORY (continued)

Travel Through Traffic Routing and Advanced Controls (FAST-TRAC) is installed in Oakland County, California. The ITS or FAST-TRAC systems may contain Advanced Traveler Information Systems (ATIS). In these systems, on-board computers on vehicles receive information regarding driver information and route guidance from roadside beacons. The FAST-TRAC system in Oakland County, California, has decreased left-turn accidents by 69 percent and increased rush hour vehicle speeds by 19 percent. The U.S. Secretary of Transportation has indicated a commitment to installing ITS systems in 75 metropolitan areas in the United States with a goal of reducing travel time by 15 percent in these areas.

The input signals required for servotronic valve operation are these:

1. Road speed signals from the wheel speed sensors to the DSC unit via one of the CAN networks to the active steering ECU.
2. Engine status signal from the **digital motor electronics (DME)** control unit via one of the CAN networks to the active steering ECU.

3. Terminal status from the **car access system (CAS)** via one of the CAN networks to the active steering ECU.

The servotronic valve is operated only when the engine is running. The servotronic valve enters a default mode if any of the input signals are incorrect, or if there is an electrical defect in the servotronic valve winding or connecting wires. In the default mode, the servotronic valve does not affect steering assist.

Steer-by-Wire Systems

The main difference between conventional steering systems and steer-by-wire systems is that steer-by-wire systems do not have any mechanical linkage between the steering wheel and the front wheels. In a steer-by-wire system, steering wheel input is supplied to torque sensor, gear, DC motor, and motor angle sensor (Figure 13-48). Input signals are sent to a controller from the motor angle sensor, torque sensor, and motor current sensor. On the basis of these input signals, the controller supplies output voltage signals to the motor drivers. The motor driver connected to the DC motor in the steering gear operates this motor to supply the desired steering angle.

Steer-by-wire systems are not available in automotive showrooms at present, but they may be available in the future. Steer-by-wire systems could have many possible advantages such as active steering control, improved vehicle maneuverability and stability, and steering system tuning to specific types of driving conditions. However, before steer-by-wire steering systems are commercialized, concerns regarding reliability and confirmation of advantages must be completely satisfied.

Four-Wheel Steering Systems

If a car with conventional front wheel steering is parallel parked at a curb between two vehicles, this car may be driven from the parking space without hitting the car in front if the front wheels are turned all the way to the left (Figure 13-49, view A). When the same car is equipped with 4WS and the rear wheels steer in the same direction as the front wheels at low speed, the car will not steer out of the parking space without striking the vehicle in front (Figure 13-49, view B).

When the car in the same parking space has a 4WS system that steers the rear wheels in the opposite direction to the front wheels at low speed, the car steers out of the parking

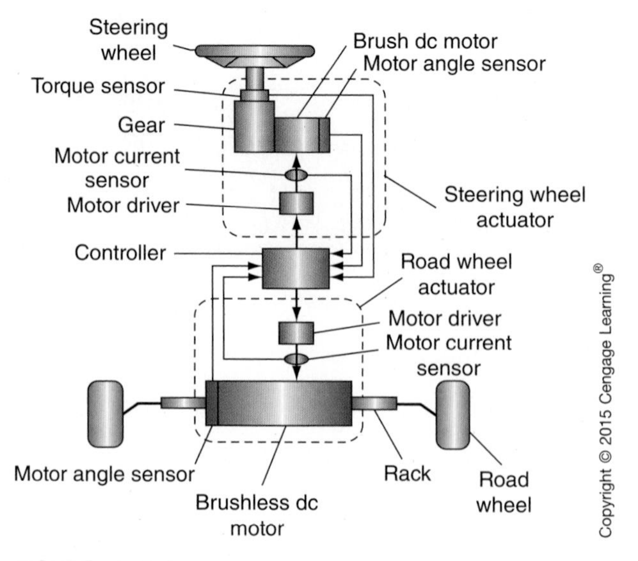

FIGURE 13-48 **Steer-by-wire system.**

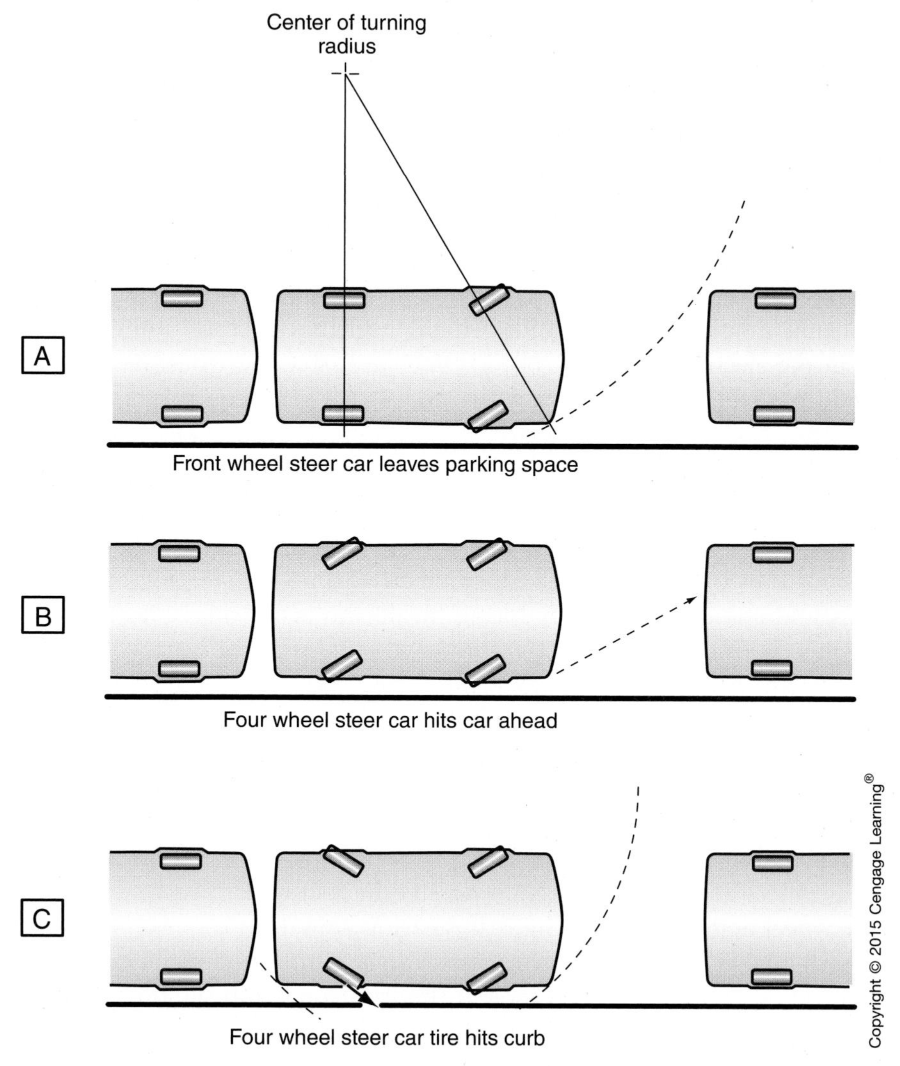

FIGURE 13-49 **Parallel parking with conventional front wheel steering and four-wheel steering.**

space with plenty of distance between the vehicle parked in front (Figure 13-49, view C). When the rear wheels steer in the opposite direction to the front wheels, the rear wheels steer toward the curb. This action causes the right rear tire to strike the curb if the car is parked close to the curb. Therefore, the maximum rear **steering angle** must be considerably less than the maximum front wheel steering angle to help prevent this problem. A car with 4WS has a smaller **turning circle**, or turning radius, compared to a vehicle with conventional front wheel steering. This improves maneuverability while parking.

The rear wheel steering in a 4WS system may be controlled in relation to vehicle speed or the amount of steering wheel rotation. At low vehicle speeds or with a considerable amount of steering wheel rotation, the rear wheels are steered in the opposite direction to the front wheels. When the vehicle is operating at higher speeds or with a small amount of steering wheel rotation, the rear wheels are steered in the same direction as the front wheels.

When a vehicle is cornering at higher speeds, centrifugal force tends to move the rear of the vehicle sideways. This action causes the rear tires to slip sideways on the road surface. This process may be called **sideslip**. The vehicle speed and the severity of the turn determine the amount of sideslip. If sideslip is excessive, the car may spin around, causing the driver to lose control. When the 4WS system steers the rear wheels in the same direction as the front wheels at higher speeds, sideslip is reduced and vehicle stability is

Steering angle is the number of degrees that the wheels turn when steered.

Turning circle is the diameter of a circle completed by a vehicle when the front wheels are turned fully to the right or left.

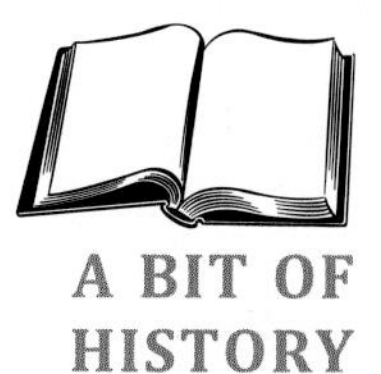

A BIT OF HISTORY

The **four-wheel steering (4WS)** concept has been researched by automotive engineers for several decades. In 1962, an engineer representing one of the Japanese manufacturers stated at the Japanese Automotive Engineering Association's technical meeting, "A major improvement in control and stability may be anticipated by means of an automatic rear wheel steering system." In the late 1970s, Honda and Mazda were actively engaged in 4WS development. However, it was not until the late 1980s that 4WS was introduced on a significant number of Honda and Mazda cars.

In a **four-wheel steering (4WS)** system all four wheels are turned to steer the vehicle.

improved. The higher speed same-direction steering angle is considerably less than the low-speed opposite-direction steering angle.

Shop Manual
Chapter 13,
page 528

ELECTRONICALLY CONTROLLED FOUR-WHEEL STEERING

System Overview

In an **electronically controlled 4WS system**, rear wheel steering is controlled electronically.

Some cars are equipped with an **electronically controlled 4WS system**. On the electronically controlled 4WS system, there is no mechanical connection between the front steering gear and the **rear steering actuator**. This rear steering actuator is now controlled by a 4WS control unit mounted in the trunk behind the left rear seat (Figure 13-50). The 4WS control unit in the electronically controlled system uses steering wheel rotational speed, vehicle speed, and front steering angle information to calculate and control the rear steering angle.

Rear Steering Actuator

Shop Manual
Chapter 13,
page 531

The rear steering actuator may be compared to an electric steering gear. This actuator contains an electric motor that drives a steering rack through a ball screw mechanism (Figure 13-51). Conventional tie rods are connected from the steering actuator to the rear steering arms and spindles. A return spring inside the actuator moves the rear wheels to the straight-ahead position when the ignition switch is turned off or when a defect occurs in the 4WS system. A main rear wheel angle sensor and a sub rear wheel angle sensor are mounted on top of the rear steering actuator.

CAUTION:
Some electronic sensor wires have a special shield surrounding them to prevent electromagnetic interference (EMI) from other voltage sources. If this shield is damaged or removed, computer system operation may be adversely affected. Do not reroute sensor wires close to other voltage sources such as spark plug wires.

INPUT SENSORS

Input sensors in the electronically controlled 4WS include the following:

1. Main rear wheel angle sensor in the rear steering actuator.
2. Sub rear wheel angle sensor in the rear steering actuator.
3. Main steering wheel angle sensor in the steering column under the combination switch.
4. Sub front wheel steering angle sensor in the front rack and pinion steering gear.
5. Conventional rear ABS wheel speed sensors.
6. Conventional VSS.

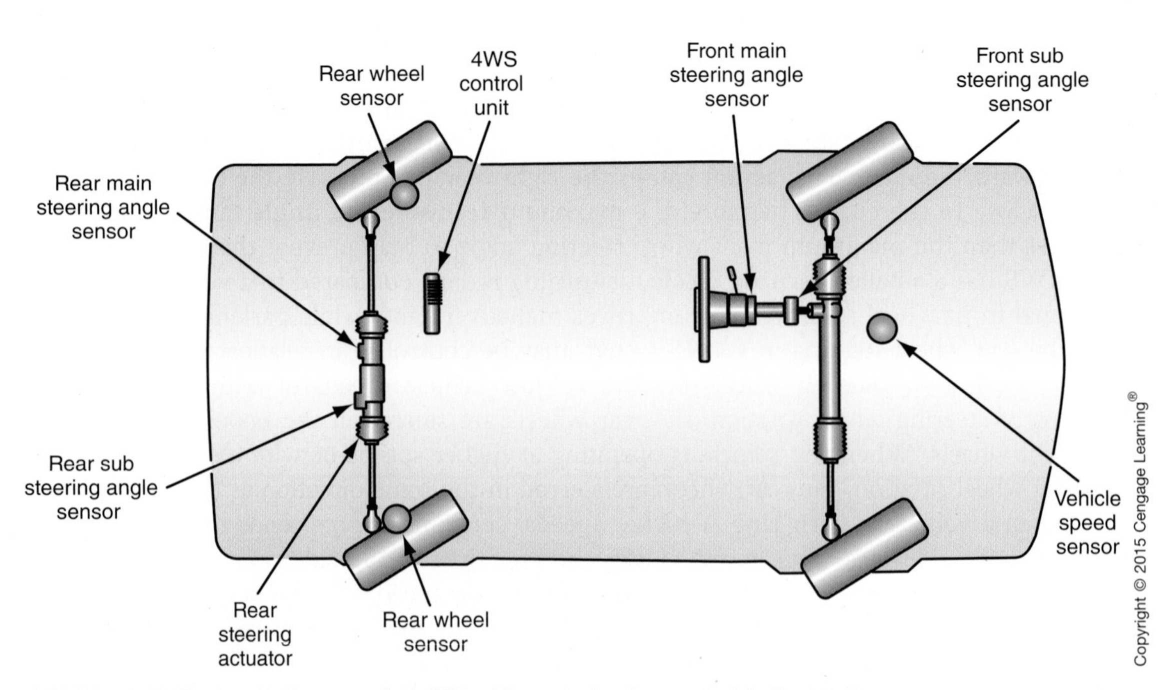

FIGURE 13-50 **Electronically controlled 4WS with control unit mounted in the trunk.**

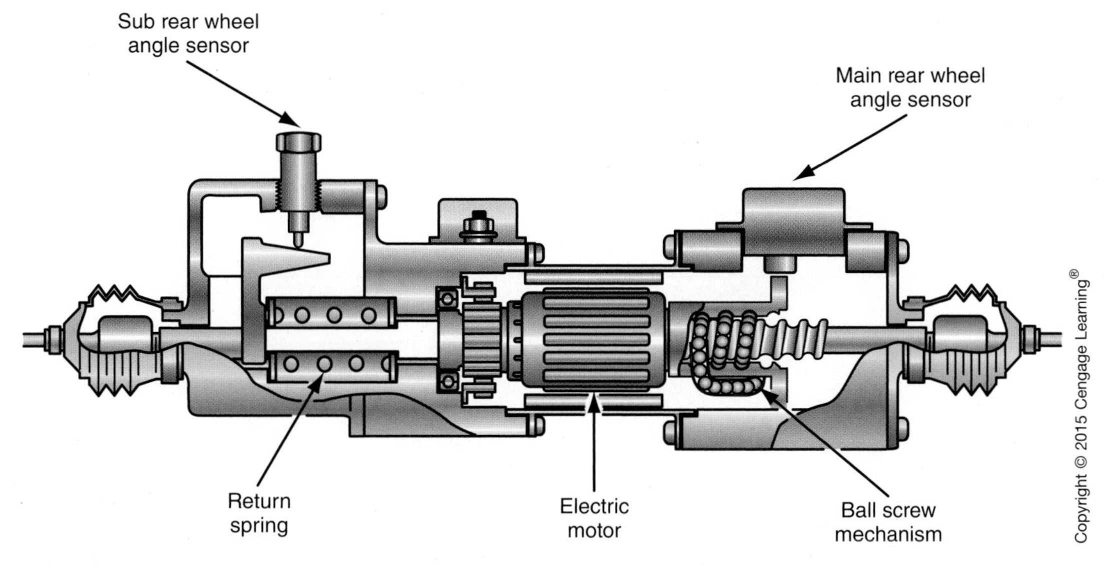

FIGURE 13-51 **Rear steering actuator, electronically controlled 4WS system.**

Four-Wheel Steering System Operation

4WS Control Unit Operation

When the engine is running, the 4WS control unit continually receives information from all the input sensors. If the steering wheel is turned, the 4WS control unit analyzes information from the vehicle speed sensor, main steering wheel angle sensor, sub front wheel angle sensor, main rear wheel angle sensor, sub rear wheel angle sensor, and the rear wheel speed sensors. The 4WS control unit calculates the proper rear wheel steering angle and then sends battery voltage to the rear steering actuator motor to provide this rear steering angle (Figure 13-52).

Transistors may be defined as automatic electronic relays that have no moving parts and are made from semiconductor materials.

Battery voltage is sent to the rear steering actuator motor through two heavy-duty output **transistors**. One of these transistors conducts current during a right turn, whereas the

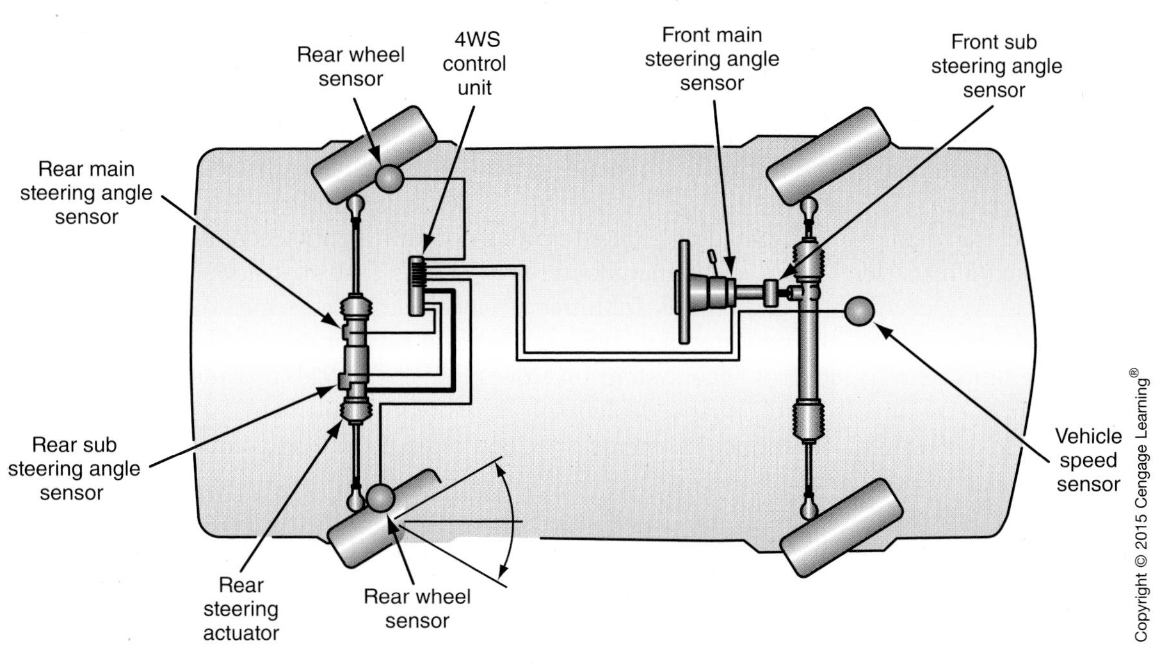

FIGURE 13-52 **The 4WS control unit analyzes input sensor information, calculates the required rear steering angle, and operates the rear steering actuator motor to provide the proper rear steering angle.**

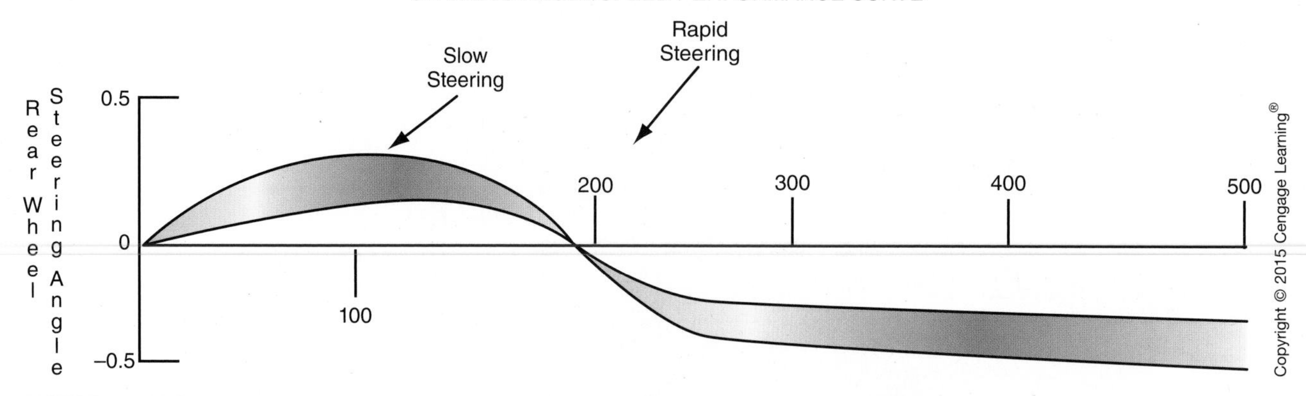

FIGURE 13-53 **Rear steering angle in relation to vehicle speed and steering wheel rotation.**

other transistor is activated during a left turn. The main rear wheel angle sensor and the sub rear wheel angle sensor send feedback signals to the 4WS control unit, indicating the proper rear steering angle has been supplied.

4WS Operating Characteristics

When the vehicle speed is less than 18 mph (29 km/h), the rear wheels immediately begin to steer in the opposite direction to the front wheels if the steering wheel is turned (Figure 13-53). Maximum rear steering angle is 6° at 0 mph. The rate of rear steering angle decreases in relation to vehicle speed, and at 18 mph (29 km/h) the amount of rear steering angle is almost zero.

When the vehicle speed increases above 18 mph (29 km/h), the rear wheels steer in the same direction as the front wheels through the first 200° of steering wheel rotation. The rear steering angle reverts to the opposite phase if the steering wheel is rotated more than 200° in this vehicle speed range. When the vehicle speed is 60 mph (96 km/h) and the steering wheel rotation is 100°, the rear wheels steer about 1° in the same direction as the front wheels. If the steering wheel is rotated 500° slowly at this speed, the rear wheels are steered about 1° in the opposite direction to the front wheels.

Quadrasteer system was available on some GM light-duty trucks and SUVs from 2002–2005. The Quadrasteer system is a 4WS system that improves low speed maneuverability, high speed stability, and towing capability. The Quadrasteer system has an electronically powered rear steering system. Vehicles with a Quadrasteer system require a higher output alternator because of the additional electrical load of the rear wheel steering actuator motor.

While General Motors Quadrasteer system did not gain wide acceptance and was discontinued after several years other manufacturers' systems that are very similar remain in production. One of these systems is Infinities 4 wheel active steering. Infinities 4 wheel active steering adjusts the rear steering angle and ratio in relation to vehicle speed to create a more fluid sense of control. The system improves low-speed feel, provides more responsive mid-speed turn-in, and enhanced high-speed stability. The following description will contain details on the major components that are often found in a four-wheel steering system.

Rear Wheel Steering Control Module and Driver Select Switch

The module controls the entire rear wheel steering functions. A 125A mega fuse in an underhood fuse holder supplies voltage to the control module at all times (Figure 13-54). Battery voltage is also supplied to the control module from the 4WS fuse, and the control module receives ignition voltage through the 10A ignition fuse.

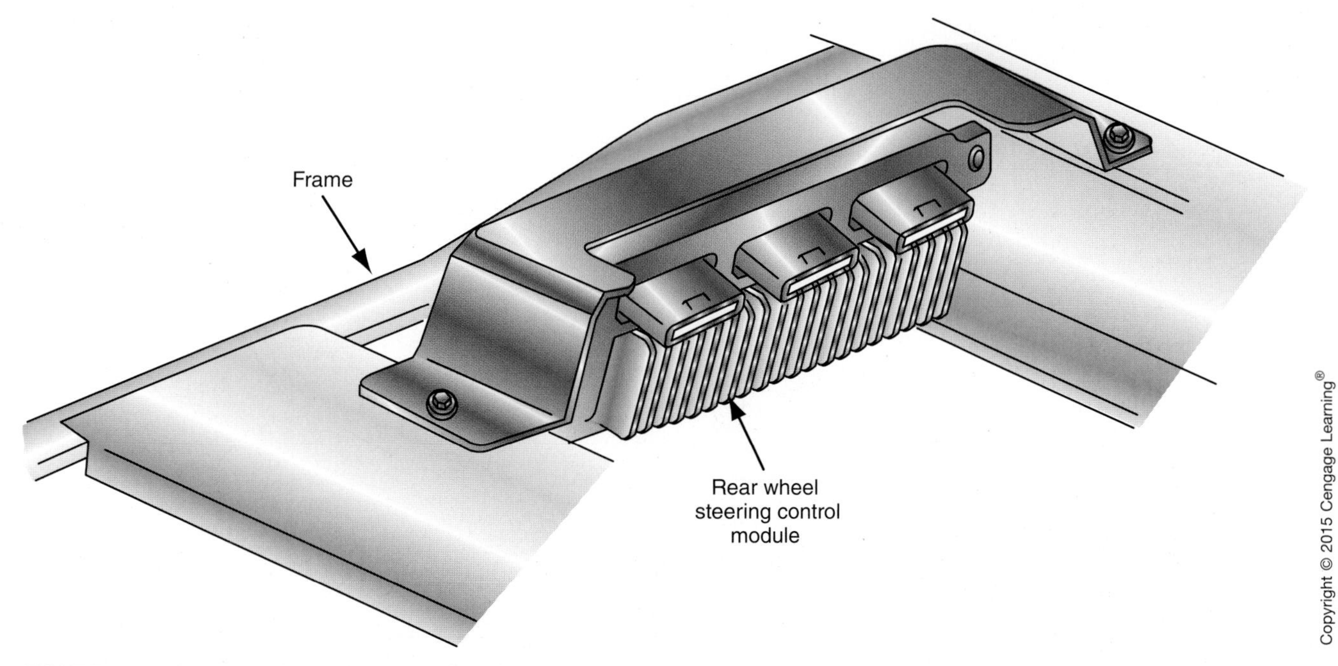

FIGURE 13-54 **Rear wheel steering control module.**

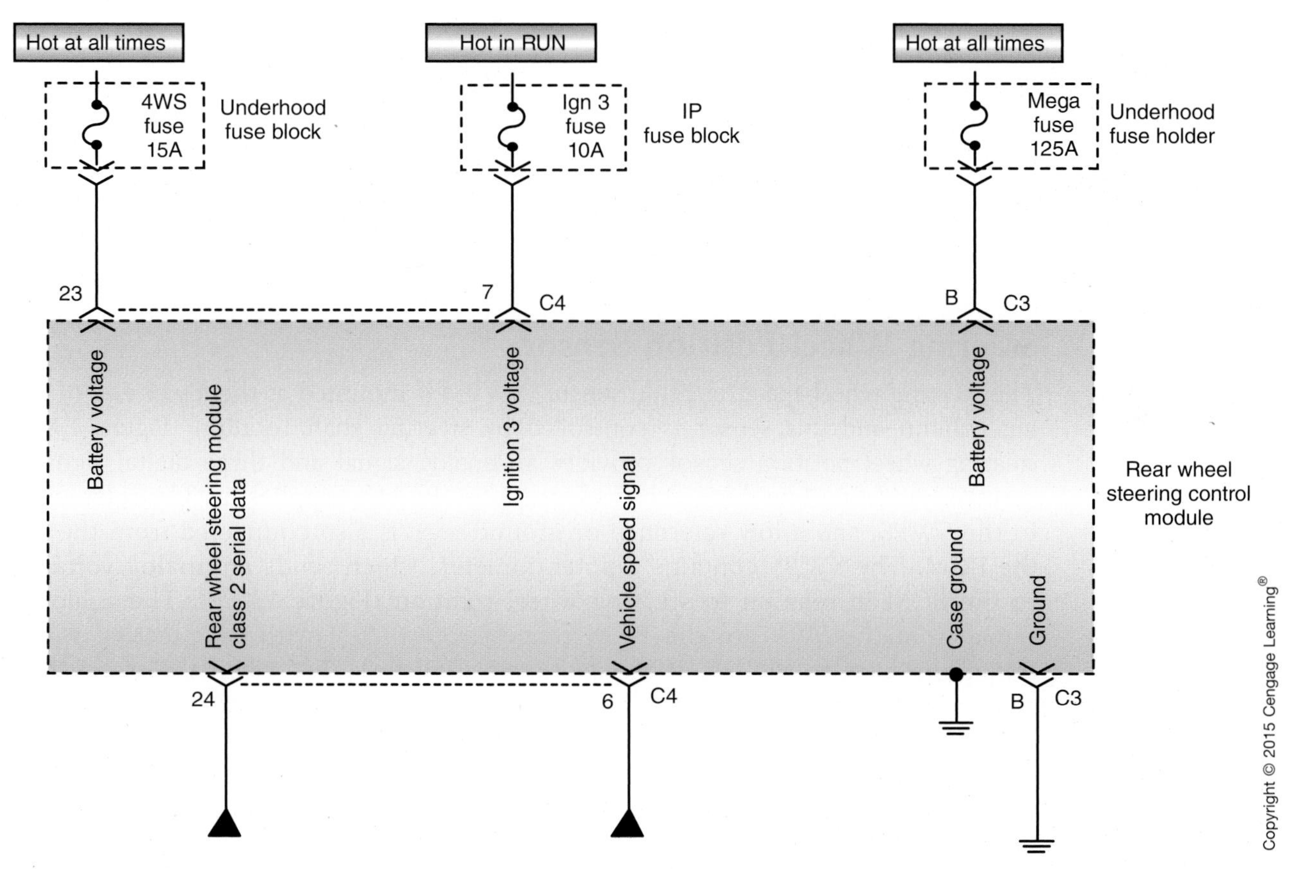

FIGURE 13-55 **Battery and ignition voltage inputs to the rear wheel steering control module.**

The rear wheel steering mode select switch is mounted in the instrument panel (Figure 13-55). The driver uses the rear wheel steering select switch to select 4WS, 2WS, or 4WS trailer mode. Each time the driver selects a rear wheel steering mode, an input signal is sent from the mode select switch to the control module (Figure 13-56). LED indicators in the switch inform the driver regarding the selected rear wheel steering mode.

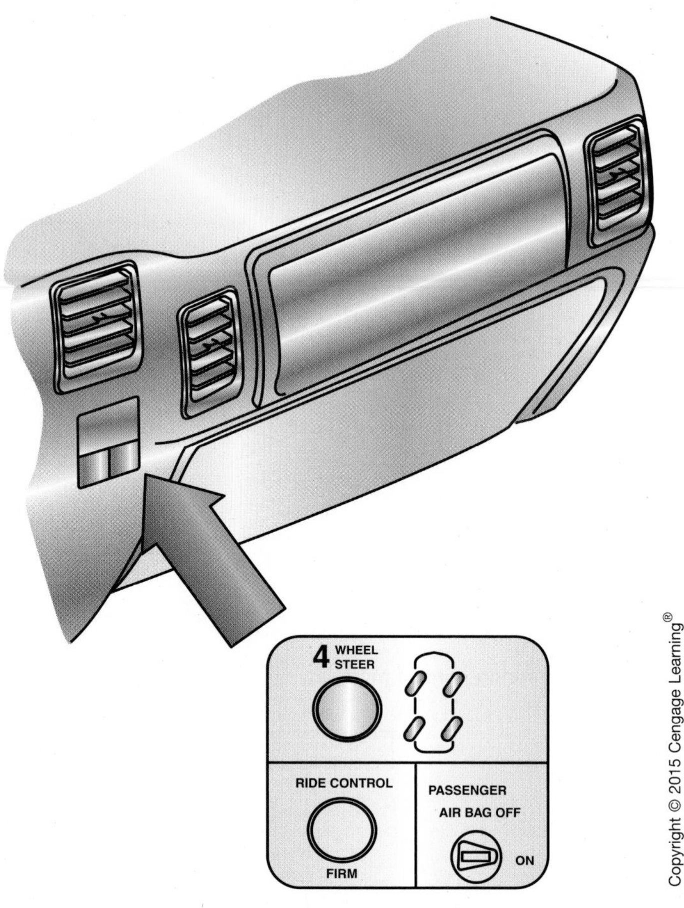

FIGURE 13-56 **Rear wheel steering mode select switch.**

Steering Wheel Position Sensor

The steering wheel speed/position sensor (SWPS) is mounted at the lower end of the steering column and this sensor is controlled by steering shaft rotation (Figure 13-57). The steering wheel position sensor provides an analog signal and three digital signals to the control module. The body control module (BCM) supplies a 5 V reference signal to the SWPS, and a low reference or ground wire is also connected from the SWPS to the BCM. The SWPS contains a potentiometer, which sends an **analog voltage signal** to the BCM in relation to steering wheel rotation (Figure 13-58). The analog voltage signal from the SWPS to the BCM ranges from 0.25 V with the steering wheel positioned one turn to the left of the center position to 4.75 V when the steering wheel is positioned one turn to the right of the center position. With the steering wheel in the center position, the SWPS analog voltage signal is 2.5 V. When the steering wheel is turned more than one turn to the right or left of the center position, the SWPS signal does not change. The BCM relays the SWPS analog voltage signal through class 2 data links to the rear wheel steering control module.

The SWPS sends digital signals through the phase A, phase B, and marker pulse circuits directly to the control module. The marker pulse digital signal is displayed on a scan tool as High if the steering wheel is positioned between 10° to the left or 10° to the right of the center position. If the steering wheel is positioned more than 10° to the right or left of the center position, the pulse marker signal is displayed as Low. The phase A and phase B signals are displayed on a scan tool as High or Low as the steering wheel is rotated. These signals change from High to Low every one degree of steering wheel rotation.

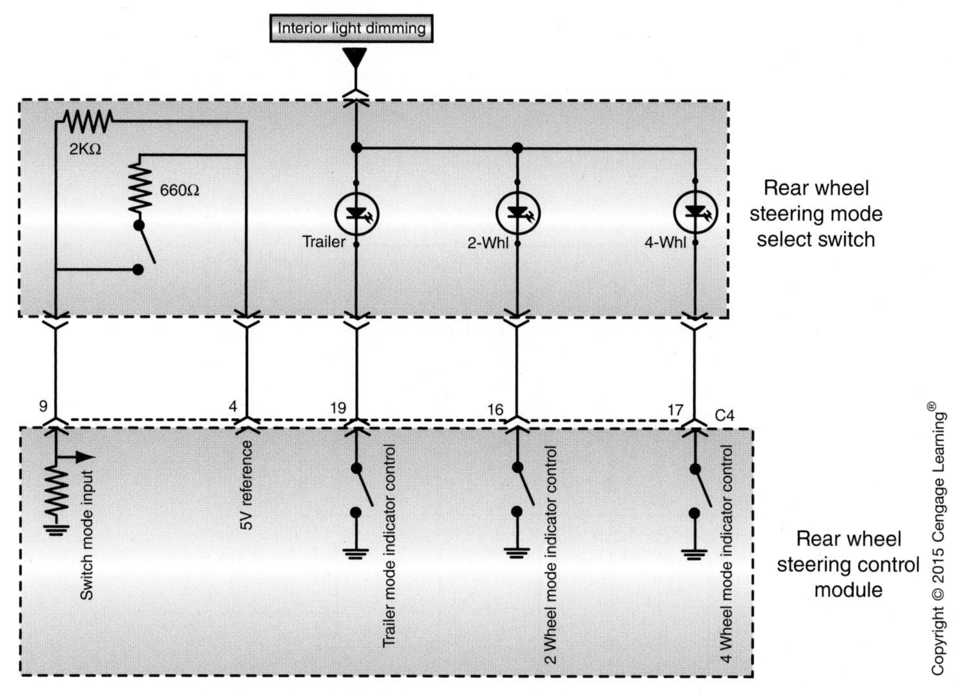

FIGURE 13-57 **Rear wheel steering mode select switch inputs to the control module.**

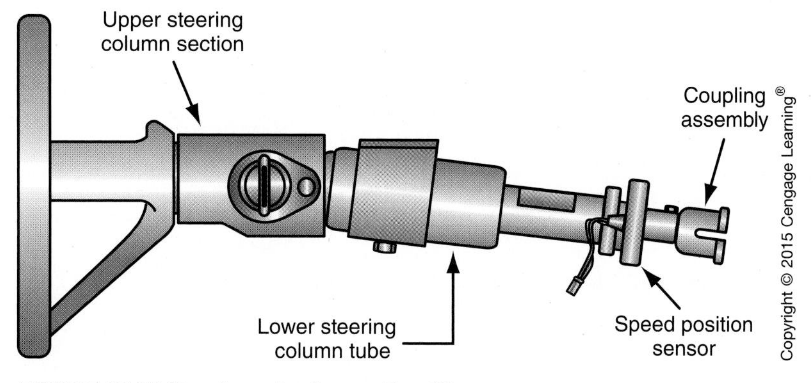

FIGURE 13-58 **Steering wheel speed/position sensor.**

Rear Wheel Position Sensor

The rear wheel position sensor is mounted on the lower side of the rear wheel steering gear (Figure 13-59). The rear wheel position sensor has two signal circuits connected to the rear wheel steering control module (Figure 13-60). The position 1 signal is a linear measurement of voltage per degree of rear steering position sensor rotation. For the position 1 input the measurement in degrees is from −620° to the left to +620° to the right. The voltage signal from the position 1 input is 0.25 V to 4.75 V. If the signal voltage from position 1 is 0.25 V, the steering wheel has been rotated −600° past center. When the signal voltage from position 1 is 4.75 V, the steering wheel has been rotated +600° past center. The voltage signal from position 2 increases or decreases from 0.25 V to 4.75 V every 180° of steering wheel rotation.

Rear Steering Gear

The rear steering gear is a rack and pinion–type gear mounted on the differential cover. The rack is connected through tie rods and outer tie rod ends to the steering knuckles. The outer tie rod ends are threaded onto the tie rods and retained with a jam nut. The inner ends of

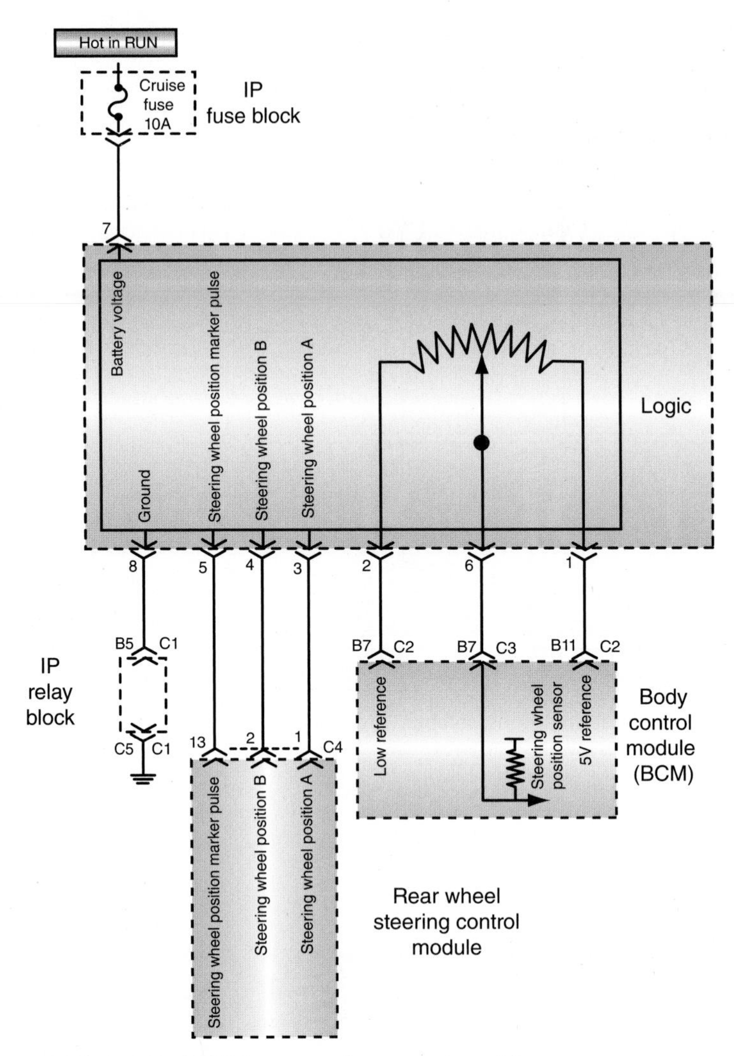

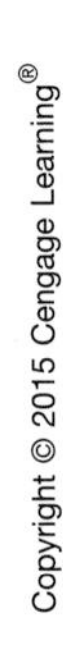

FIGURE 13-59 **SWPS inputs to the control module.**

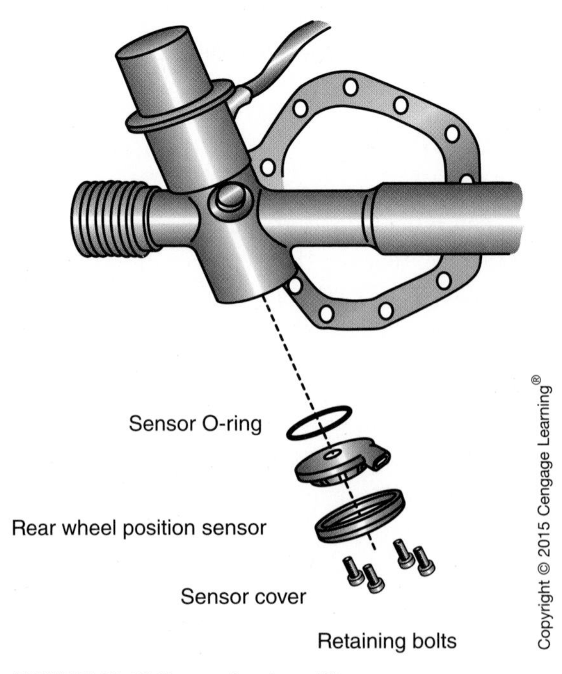

FIGURE 13-60 **Rear wheel position sensor.**

the tie rods are clamped to the ends of the steering gear rack. Special tools are required to remove and install the inner ends of the tie rods on the rack. The steering knuckles pivot on upper and lower ball joints that are bolted into extensions on the differential housing (Figure 13-61). Constant velocity (CV) joints are mounted near the outer end of each rear axle shaft. Splined shafts extend from the outer side of the CV joints into the rear hub and bearing assemblies that are bolted to the steering knuckles (Figure 13-62). A nut, lock, and

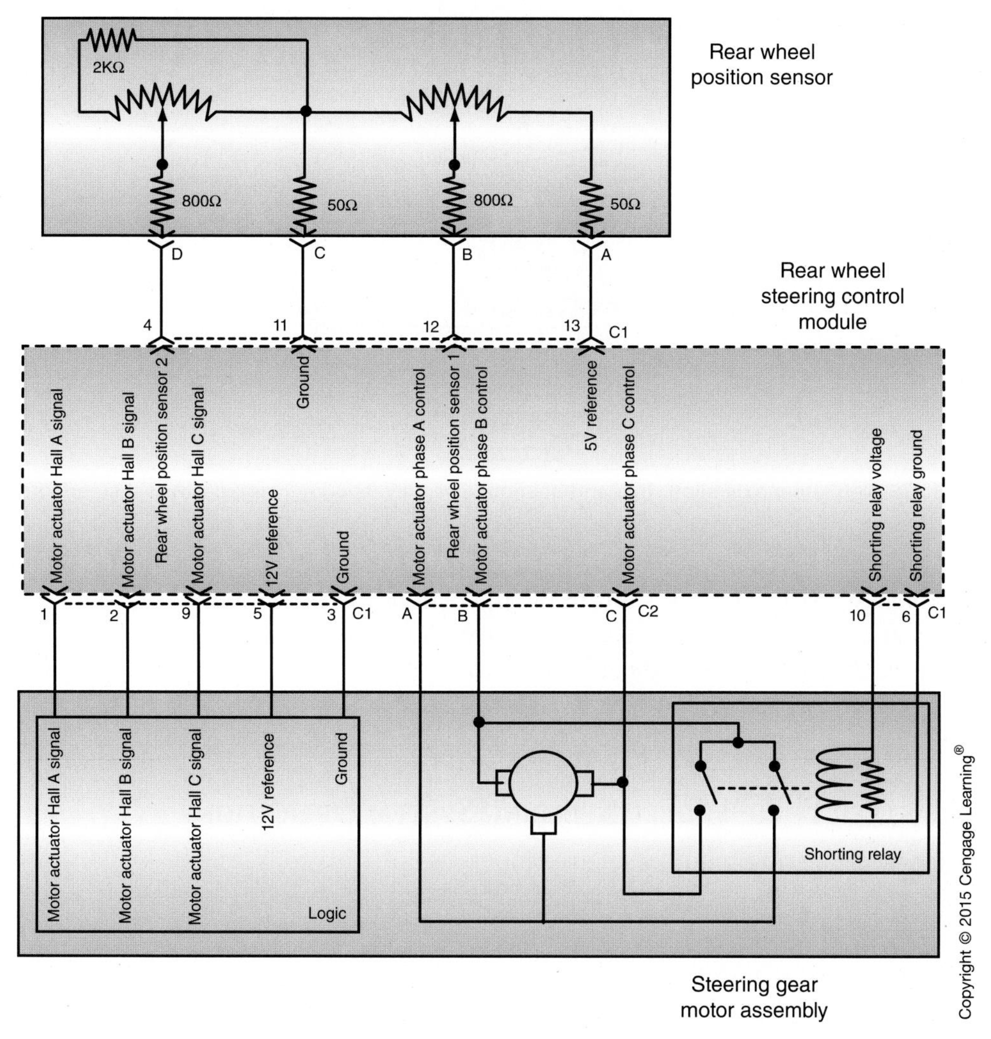

FIGURE 13-61 **Rear wheel position sensor inputs to the rear wheel steering control module.**

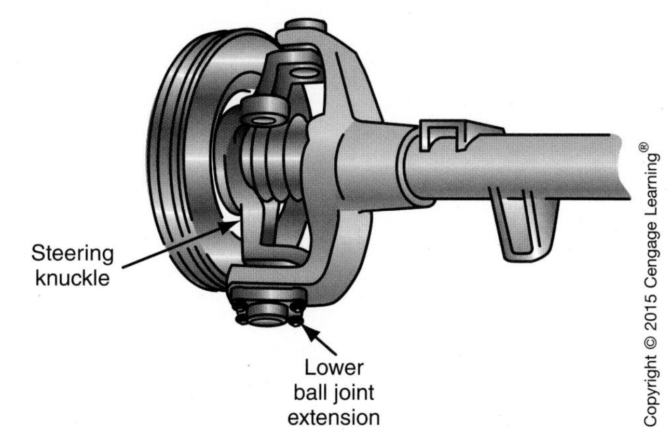

FIGURE 13-62 **Lower ball joint, rear wheel steering.**

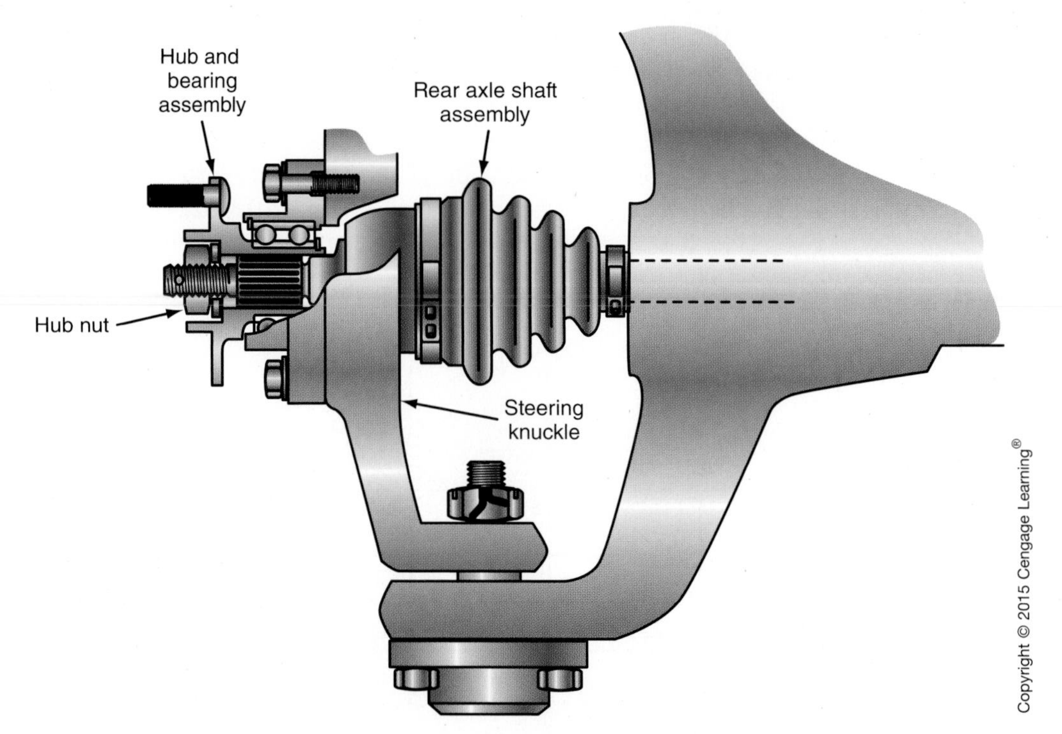

FIGURE 13-63 **Rear axle shaft, hub and bearing assembly, and knuckle, rear wheel steering.**

cotter pin retain the outer ends of the splined shafts in the hub and bearing assemblies. The brake rotors are retained on the studs in the hub and bearing assemblies. The hub and bearing assemblies are not serviceable. The electric drive motor for the rear steering gear is mounted on the top of the steering gear, and this motor is protected by a shield and skid plate attached to the differential (Figure 13-63). The electric motor drives the steering gear rack to provide rear wheel steering. The maximum rear wheel steering angle is 12° in either direction.

Quadrasteer Four-Wheel Steering System Operation

The rear wheel steering module receives inputs from these sources:

1. Battery voltage
2. Ignition voltage
3. Class 2 serial data links
4. Steering wheel position sensor analog, marker pulse, and phase signals
5. Rear wheel position sensor signals
6. Vehicle speed sensor signal from the instrument panel cluster (IPC)
7. Rear wheel steering mode switch

In response to these input signals, the module commands the rear wheel steering motor to operate the rear wheel steering gear and provide the proper rear wheel steering angle.

In the 2-wheel steer mode the rear wheels are held in a centered position, and the rear wheel steering is disabled.

In the 4-wheel steer mode the rear wheel steering can provide negative phase steering, **neutral phase steering**, or **positive phase steering**. In the negative phase the rear wheels are steered in the opposite direction to the front wheels, and this phase occurs at low vehicle speeds. When the rear wheels are steered in the opposite direction to the front wheels, the vehicle turning radius is reduced and parking maneuverability is improved (Figure 13-64). In the neutral phase the rear wheels remain centered. If an electronic defect occurs in the rear wheel steering system, the rear wheels remain in the neutral phase. When the rear wheel

Neutral phase steering occurs when the rear wheels are centered in the straight-ahead position.

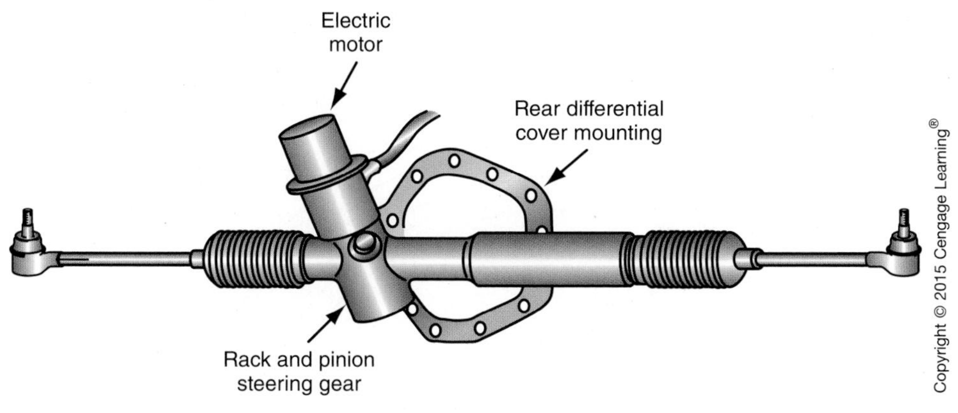

FIGURE 13-64 **Four-wheel steering gear and motor.**

steering control module commands positive phase steering, the module drives the electric motor and steers the rear wheels in the same direction as the front wheels. This action reduces **yaw forces** on the rear of the vehicle and improves steering control at higher speeds. The transition from negative phase to positive phase steering occurs at 40 mph (65 km/h).

Yaw forces tend to cause the rear of the vehicle to swerve sideways.

In the 4-wheel steer tow mode, the degrees of positive phase steering is increased compared to the 4-wheel steer mode, and the transition speed from negative phase steering to positive phase steering occurs at 25 mph (40 km/h). Below this speed the negative phase steering is similar to the negative phase steering in the 4-wheel steer mode.

WARNING: When diagnosing, servicing, or adjusting a 4WS system, it is very important to follow the diagnostic, service, and adjustment procedures in the vehicle manufacturer's service manual. Failure to follow these procedures may cause improper rear wheel steering operation and reduced vehicle stability that could result in a vehicle collision.

General Motors engineers decided to install the Quadrasteer system on an SUV because of the advantages of this system when hauling a trailer. When the rear vehicle wheels are steered in the opposite direction to the front wheels during low-speed turning, the trailer follows the true vehicle path more closely than it does with a two-wheel steering system. When backing up a trailer, steering the rear wheels in the opposite direction to the front wheels provides better trailer response to vehicle steering inputs, and this action makes it easier to back the trailer into the desired position. When steering the vehicle at higher speeds such as lane changing, the positive steering action of the rear wheels reduces the articulation angle between the tow vehicle and the trailer. This action reduces the lateral forces applied to the rear of the tow vehicle by the trailer, which in turn reduces yaw velocity gain and improves trailer stability. On a full-size SUV, the Quadrasteer system reduces turning circle diameter from 45 ft. to 33.9 ft. (Figure 13-65).

AUTHOR'S NOTE: Previous to 2002, four-wheel steering was available on a few imported cars; however, these systems were never sold in large numbers. The Quadrasteer system is the first four-wheel steering system to be offered on an SUV and marketed on the basis of improved steering control when trailer hauling. The question resulting from this application of four-wheel steering systems is: Will these systems be widely accepted by the motoring public or will they be considered too expensive for the advantages they provide?

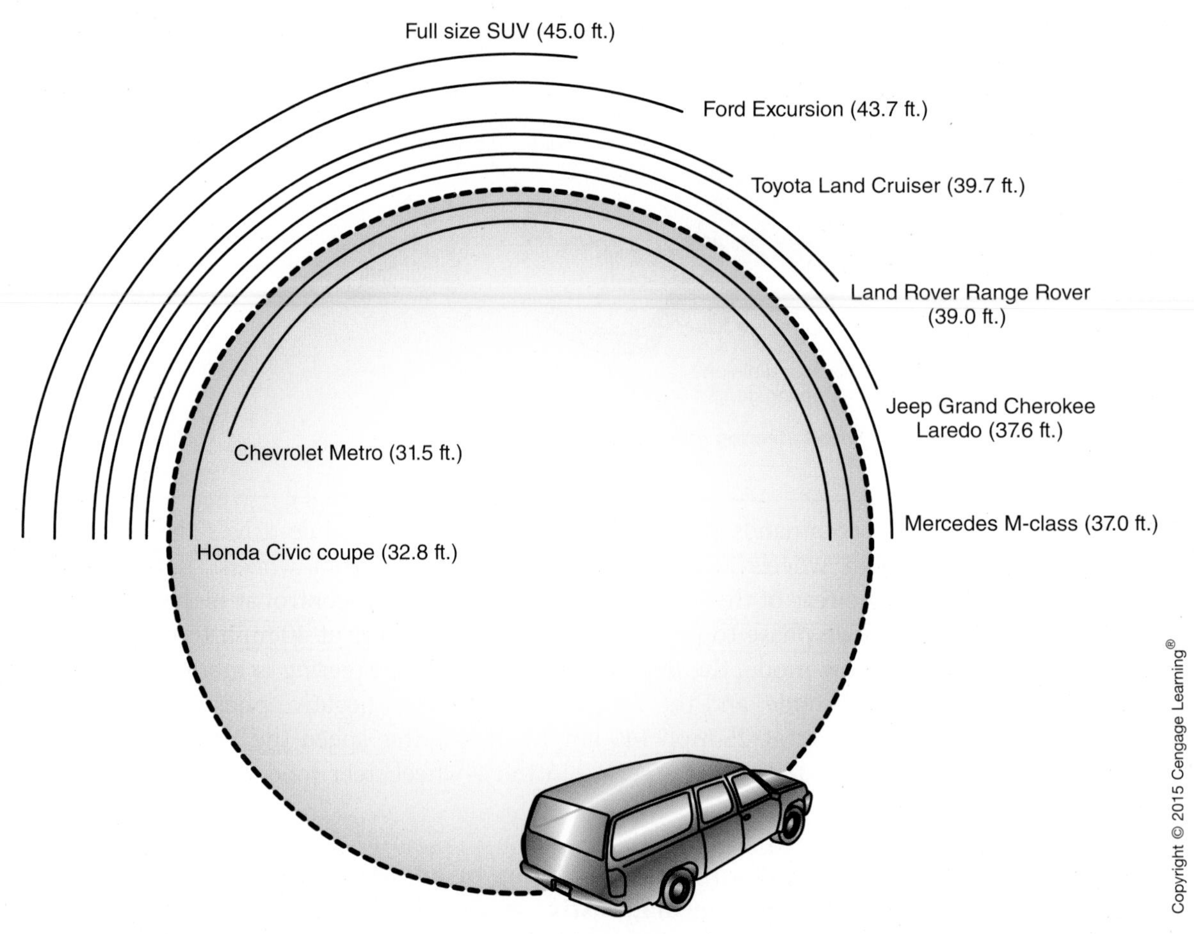

FIGURE 13-65 **Turning circle diameter, four-wheel steering system.**

As an automotive technician, you must keep up to date on the changes in automotive electronics, and these changes are occurring at a very rapid pace. It has been my experience that one of the best ways to keep up to date is to join professional technicians' organizations such as the Service Technicians Society (STS), or the International Automotive Technicians' Network (IATN). As a member of these organizations, you will be able to obtain information on the latest automotive electronics technology and the solutions to diagnostic problems related to automotive electronics.

Rear Active Steering System

Rear active steering (RAS) is featured as an available option on some Infiniti luxury car model packages. The manufacturer claims these cars are the most technologically advanced Infiniti models ever developed, and also claims the RAS is probably the most significant new technology on the vehicle. This technology may or may not stay part of the luxury car segment. Often it depends more on whether the feature sells cars or not than it does on the viability of the technology. The RAS makes the steering and handling characteristics on this long-wheelbase car more nimble and agile compared to models without this technology.

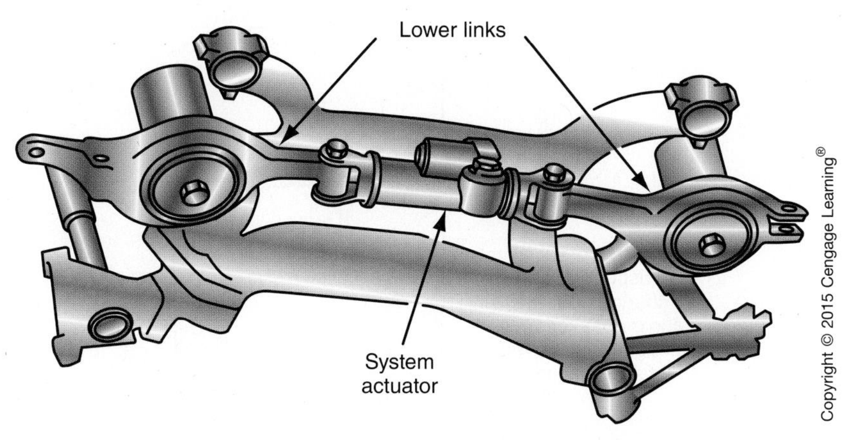

FIGURE 13-66 **Rear steering actuator and adjustable lower links.**

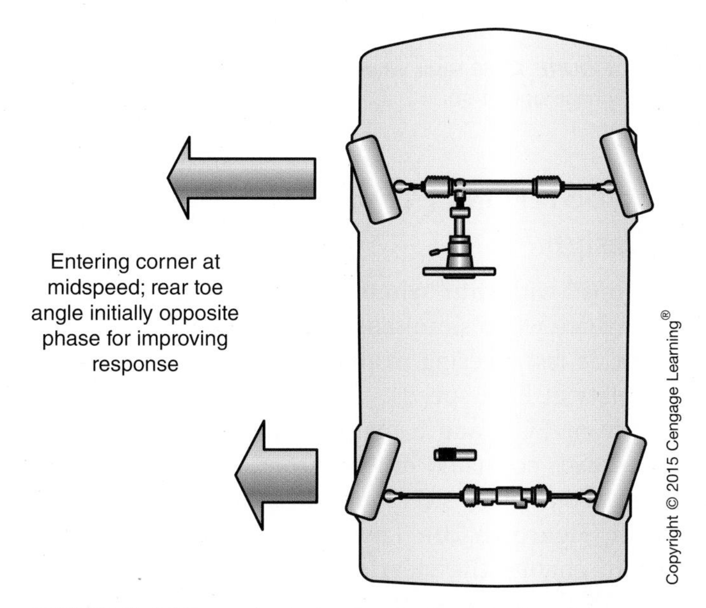

FIGURE 13-67 **Rear wheel steering during negative phase operation.**

The RAS improves steering and handling by adjusting the geometry of the rear suspension in relation to steering input and vehicle speed. The RAS control unit calculates the ultimate vehicle dynamics from the input signals from a group of sensors. In response to these inputs the control unit operates the electric actuator to lengthen or shorten each lower rear suspension link (Figure 13-66). This action provides a possible 1° turn on each rear wheel.

When the vehicle begins turning a corner, driving around a curve, or making a lane change at mid-speed, the control unit operates the RAS actuator to initially turn the rear wheels in the negative phase mode in relation to the front wheels (Figure 13-67). This action provides improved turn-in response.

When the steering wheel rotation indicates the vehicle is exiting a corner or making a lane change at mid-speed, the control unit operates the RAS actuator to turn the rear wheels in the positive phase mode in relation to the front wheels (Figure 13-68). Under this condition yaw forces are reduced on the rear of the vehicle to provide improved vehicle stability.

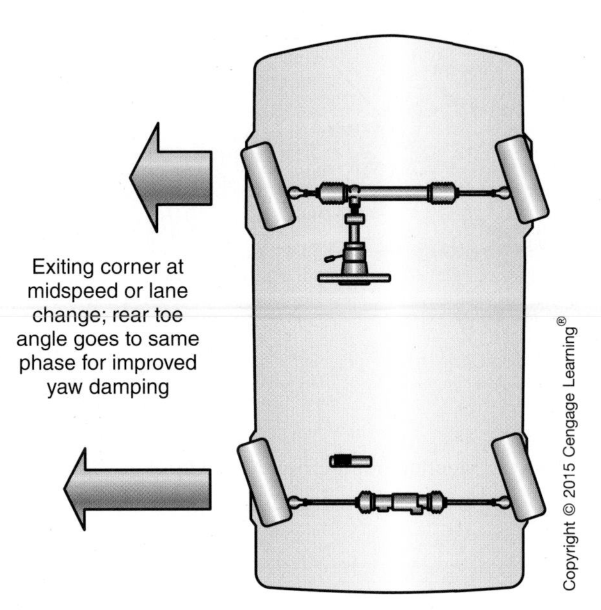

Exiting corner at midspeed or lane change; rear toe angle goes to same phase for improved yaw damping
Copyright © 2015 Cengage Learning®

FIGURE 13-68 **Rear wheel steering during positive phase operation.**

Four-Wheel Active Steering (4WAS)

4WAS System Design

Four-wheel active steering (4WAS) may be defined as front and rear wheel steering in which a computer(s) can electronically change the front and rear steering angles.

Some vehicles are equipped with **four-wheel active steering (4WAS)** to provide improved overall steering performance while simultaneously reducing steering effort. The 4WAS system is designed to provide fast steering responses in the low to medium speed range combined with vehicle stability at high speeds.

The 4WAS has a front ECU and a main (rear) ECU. The 4WAS system has a front steering actuator mounted coaxially in the front steering shaft (Figure 13-69). The front actuator contains a front wheel steering angle sensor, a front lock solenoid valve, a front motor, and a gear shaft (Figure 13-70). The front ECU operates the motor and gear shaft in the front actuator to change the steering ratio. The lock solenoid valve in the front steering actuator locks this actuator so the steering ratio cannot change if a defect occurs in the system.

The rear steering actuator body is attached to a chassis member, and the outer ends of the actuator shaft are linked to the rear wheels. The rear steering actuator contains a rod attached to the rear steering arms (item 1), a motor (item 4), motor shaft and drive gear (item 5), a driven gear (item 7), a housing assembly (item 3), and a rear wheel steering angle sensor (item 6) (Figure 13-71).

Shop Manual
Chapter 13, page 533

The main ECU is mounted in the trunk, and controls the rear steering actuator, and the front ECU is mounted under the dash and controls the front steering actuator. The software in the ECUs contains a reference model that contains the desired dynamic steering characteristics. The ECUs operate the front and rear steering actuators to conform to the reference model in the software.

The main control unit calculates the front and rear steering angles that will provide the best steering performance and vehicle stability based on the front and rear steering angle sensor inputs and the vehicle speed input. The vehicle speed input is sent from the antilock brake system (ABS) control unit through the CAN communication network to the main control unit. Engine speed signals are transmitted from the engine electronic control unit (ECU) via the CAN network to the main control unit. The 4WAS warning light is mounted in the unified meter in the instrument panel, and operated by the main control unit.

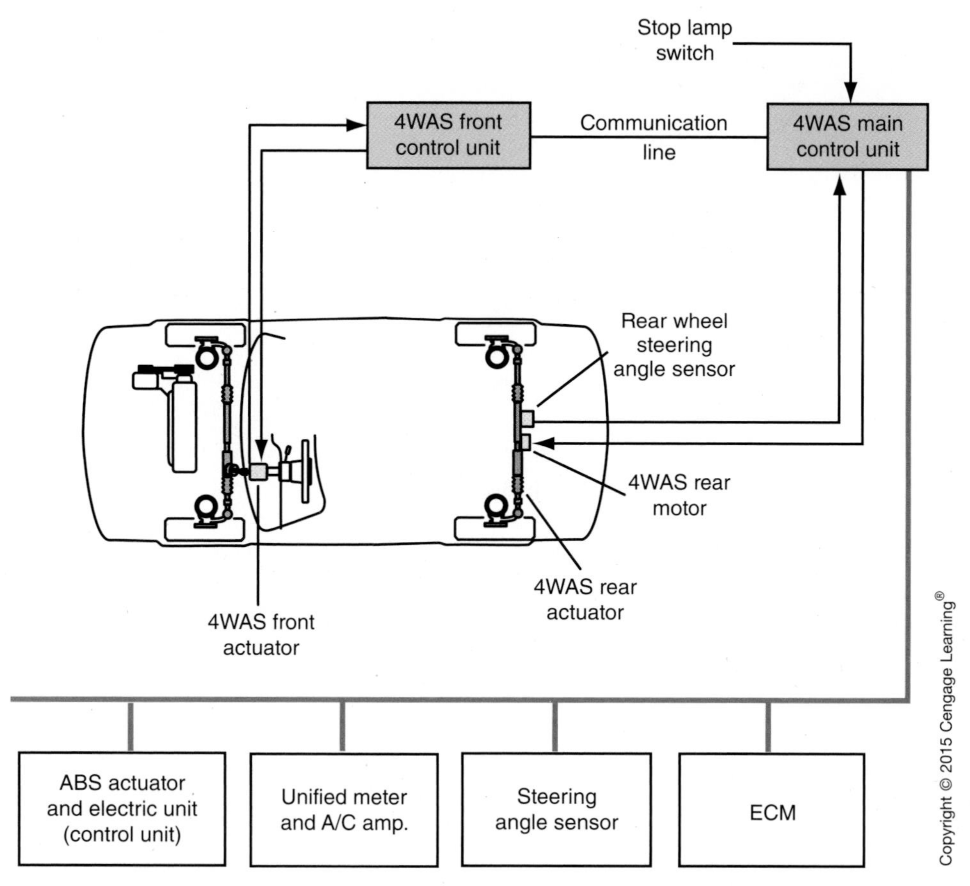

FIGURE 13-69 **4WAS system.**

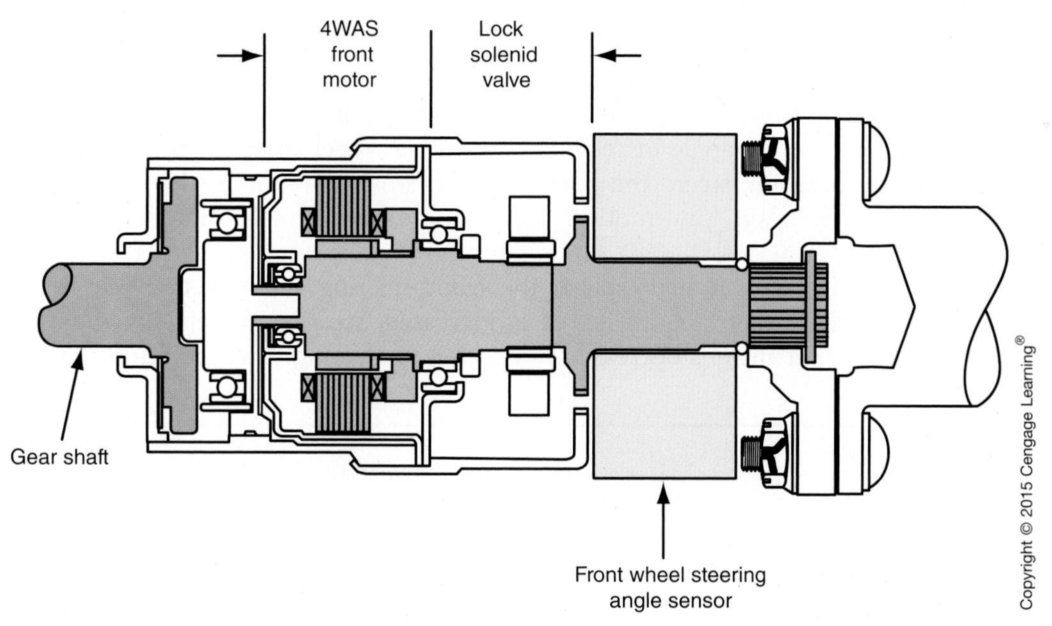

FIGURE 13-70 **4WAS front actuator.**

4WAS System Operation

When cornering at low speed the front control unit operates the front steering actuator to reduce the steering gear ratio. This action increases the front steering angle with less steering wheel movement and reduced driver steering effort (Figure 13-72).

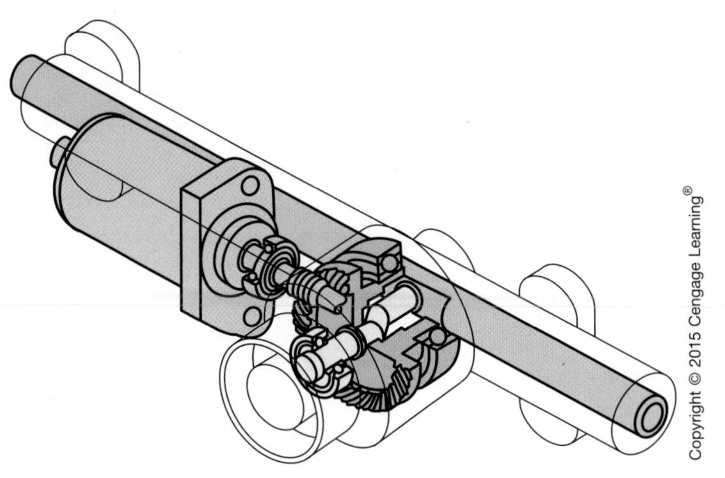

FIGURE 13-71 **4WAS rear actuator.**

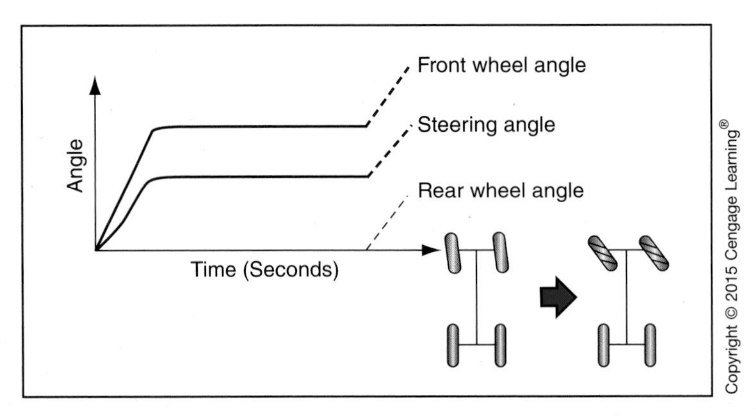

FIGURE 13-72 **Steering angles while cornering at low speed.**

When changing lanes in the medium-speed range, the front steering actuator can increase the front steering angle in relation to steering wheel rotation, and simultaneously the rear steering actuator steers the rear wheels a small amount in the same direction. When this action is taken, yaw motion, lateral force, and vehicle slide slip are reduced (Figure 13-73).

When changing lanes at high speed, the front steering actuator increases the front steering ratio so the driver steering effort is increased, steering wheel movement is also increased, and steering angle is decreased in relation to steering wheel rotation. This action

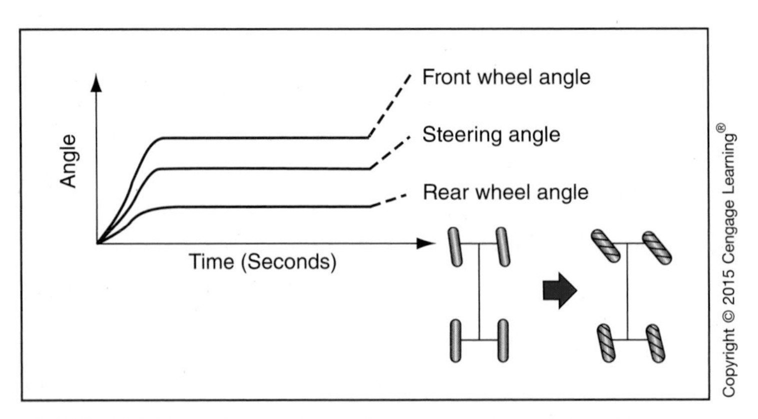

FIGURE 13-73 **Steering angles while changing lanes at medium speed.**

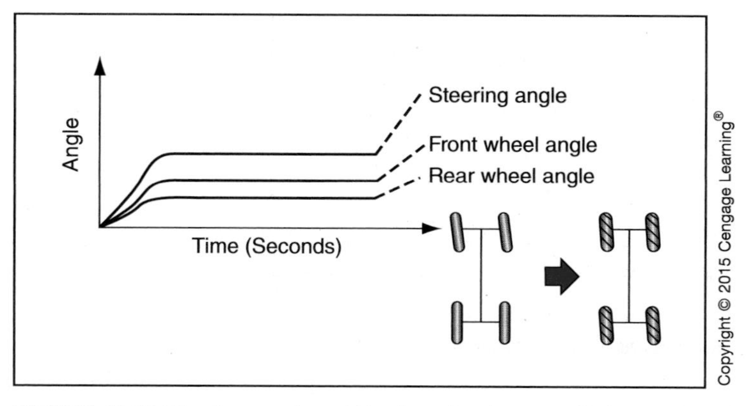

FIGURE 13-74 **Steering angles while changing lanes at high speed.**

provides more road feel to the driver. Simultaneously, the rear steering actuator steers the rear wheels a few degrees in the same direction (Figure 13-74). This action provides improved steering response and vehicle stability.

The 4WAS system has an EPS system. The 4WAS main control unit operates a solenoid valve in the EPS system (Figure 13-75). If this solenoid valve is open, some of the power steering pump pressure flows through the solenoid and returns through the steering gear passages to the power steering reservoir. Under this condition, power steering pump pressure is reduced. When the solenoid valve is closed, the solenoid blocks the flow of fluid through the valve, and this action increases power steering pump pressure. When the engine is idling or the vehicle is operating at low speeds, the main control unit energizes and closes the 4WAS solenoid valve. This action prevents the power steering pump pressure from flowing through the solenoid valve, and power steering pump pressure, and power steering assist are increased, resulting in a decrease in steering effort and reduced driver fatigue.

If the vehicle is driven at higher speeds, the 4WAS main control unit de-energizes the solenoid valve, allowing it to open. Under this condition, some of the power steering fluid pressure flows through the solenoid valve to the reservoir. This action decreases power steering pump pressure and increases steering effort for improved road feel.

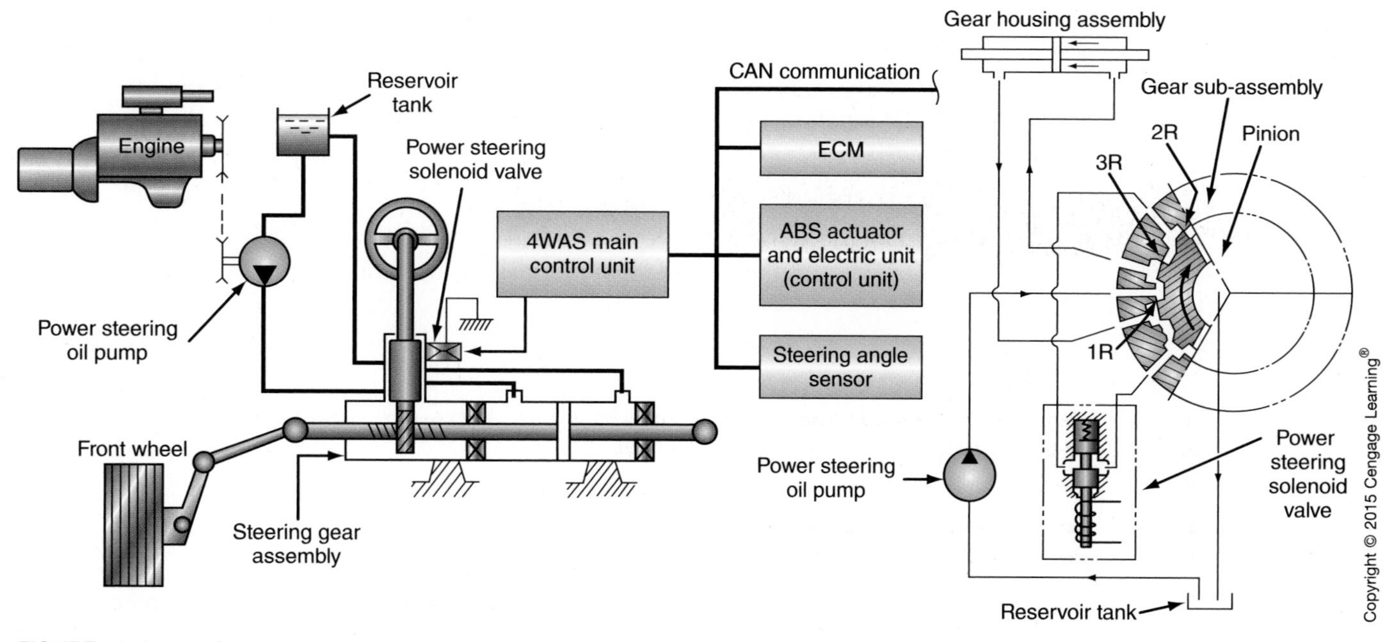

FIGURE 13-75 **4WAS EPS system.**

SUMMARY

TERMS TO KNOW

Analog voltage signal
Breather tube
Car access system (CAS)
Center take-off (CTO)
Column-drive EPS
Digital motor electronics (DME)
Electric locking unit (ELU)
Electronic control unit (ECU)
Electronically controlled orifice (ECO) valve
Electronically controlled 4WS system
Electronic variable orifice (EVO) steering
Encoded disc
End take-off (ETO)
Feel of the road
Four-wheel steering (4WS)
Four-wheel active steering (4WAS)
Helical gear teeth
Lock-to-lock
Magnasteer system
Neutral phase steering
Pinion
Pinion angle sensor
Pinion-drive EPS
Positive phase steering
Pulse width modulated (PWM) voltage signal
Quadrasteer
Rack
Rack-drive EPS
Rack piston
Rack seals
Rear active steering (RAS)

- Manual or power rack and pinion steering systems are lighter and more compact than recirculating ball steering gears and parallelogram steering linkages.
- The rack and pinion steering gear must be mounted so the rack maintains the tie rods in a parallel position in relation to the lower control arms.
- The rack takes the place of the idler arm and pitman arm in a parallelogram steering linkage.
- The pinion teeth may be spur or helical.
- The inner tie rods are connected to the rack through spring-loaded inner ball sockets.
- The rack bearing and adjuster plug maintains proper preload between the rack and pinion teeth.
- A rack and pinion steering gear may be mounted on the firewall or front crossmember.
- A rack and pinion steering system has fewer friction points than a recirculating ball steering gear and parallelogram steering linkage.
- Reducing the number of friction points in a steering system provides more road feel and reduces the steering system's ability to isolate road vibration and noise.
- Steering ratio is the relationship between steering wheel movement and front wheel movement to the right or left.
- During a left turn with a power rack and pinion steering gear, fluid pressure is directed to the left side of the rack piston, and fluid is released from the right side of this piston. This action moves the rack to the right to complete the left turn.
- During a right turn with a power rack and pinion steering gear, fluid pressure is directed to the right side of the rack piston, and fluid is released from the left side of this piston. Under this condition, the rack is forced to the left to help the driver complete the right turn.
- The fluid pressure in a power rack and pinion steering gear is directed to the appropriate side of the rack piston by the spool valve and rotary valve position.
- The spool valve position is controlled by torsion bar twisting during a right or left turn.
- If hydraulic pressure is not available in a power rack and pinion steering gear, stop tangs on the stub shaft and pinion provide steering action, but steering effort is much higher.
- The differences between Saginaw and TRW power rack and pinion steering gears are mainly in the method of tie rod attachment, bulkhead oil seal and retainer, and upper and lower pinion bearing hardware.
- The Toyota power rack and pinion steering gear has removable control valve housing and claw washers to lock the inner tie rods to the rack.
- The electronic variable orifice (EVO) steering system provides high power steering assistance during low-speed cornering and parking and normal power assistance at higher speeds.
- The two inputs in the EVO system are the steering wheel rotation sensor and the vehicle speed sensor.
- The rear wheel steering system in a four-wheel steering system is usually electronically controlled.
- Steering the rear wheels in the opposite direction to the front wheels at low speed provides a shorter turning circle, or radius, for easier maneuvering.
- Steering the rear wheels in the same direction as the front wheels at higher speeds reduces rear sideslip and improves vehicle stability while cornering or changing lanes.

- In an electronically controlled 4WS system, there is no mechanical connection between the front steering gear and the rear steering gear.
- In an electronically controlled 4WS system, the rear steering actuator contains an electric motor that is controlled by the 4WS control unit in response to various input sensor signals.
- In a Quadrasteer system, the control unit uses inputs from the vehicle speed sensor (VSS) and the front steering position sensor to control the rear steering actuator.
- A higher output alternator is required on a Quadrasteer system because of the higher current draw of the rear actuator.
- Negative phase steering occurs at lower speeds when the rear wheels are steered in the opposite direction as the front wheels.
- Positive phase steering occurs at higher speeds when the rear wheels are steered in the same direction as the front wheels.
- Steering the rear wheels in the same direction as the front wheels at higher speeds reduces the lateral forces applied to the rear of a vehicle, and this action reduces yaw velocity.
- When towing a trailer, steering the rear wheels in the same direction as the front wheels at higher speeds improves trailer stability.

TERMS TO KNOW
(continued)

Rear steering actuator
Rotary valve
Safety and gateway module (SGM)
Servotronic valve
Sideslip
Spool valve
Spur gear teeth
Steering angle
Steering angle sensor
Steering kickback
Steering ratio
Steering wheel rotation sensor
Stop-to-stop
Summation sensor
Torsion bar
Transistor
Turning circle
Variable effort steering (VES)
Vehicle speed sensor (VSS)
Yaw forces
Yaw motion

REVIEW QUESTIONS

Short Answer Essays

1. Explain why the rack and pinion steering gear is ideally suited for unibody front-wheel-drive vehicles.
2. Explain the advantage of a 4WS system while parking a vehicle.
3. Explain the purpose of the front main steering wheel angle sensor in an electronically controlled 4WS system.
4. Explain the purpose of the bellows boots in a rack and pinion steering gear.
5. Describe two possible mounting positions for rack and pinion steering gears.
6. Explain why a rack and pinion steering gear provides improved feel of the road compared with a recirculating ball steering gear and parallelogram steering linkage.
7. Describe the fluid movement in the rack piston chamber during a right turn.
8. Explain how the spool valve is moved inside the rotary valve in a power rack and pinion steering gear.
9. Explain the difference between spur and helical pinion gear teeth and explain the advantage of the helical design.
10. Describe the driving conditions required for an electronic variable orifice (EVO) steering system to provide high-power steering assistance.

Fill-in-the-Blanks

1. The preload on the rack and pinion teeth affects steering harshness, noise, and ________________.
2. Compared to a recirculating ball steering gear and parallelogram steering linkage, a rack and pinion steering system has a reduced ability to isolate road noise and vibration because the rack and pinion system has fewer ________________.
3. During a left turn, the power steering pump forces fluid into the ________________ side of the rack chamber, and fluid is removed from the ________________ side of this chamber if the rack is mounted behind the front wheel spindle.
4. During a turn, fluid is directed to the appropriate side of the rack piston chamber by the ________________ valve and ________________ valve position.
5. When the torsion bar twists, it changes the position of the ________________ valve.
6. In an electronically controlled 4WS system, the electronic module operates a(n) ________________ ________________ to steer the rear wheels.
7. The front main steering angle sensor in an electronically controlled 4WS system is mounted in the ________________ ________________.
8. In an EVO steering system, the power steering assistance is decreased at higher speeds to provide improved ________________ ________________.
9. The input sensors in an EVO steering system are the vehicle speed sensor and the ________________ ________________ ________________ ________________.
10. During negative phase steering, the rear wheels are steered in the ________________ ________________ to the front wheels.

Multiple Choice

1. Compared to a recirculating ball steering gear, a rack and pinion steering gear provides:
   A. Improved capability to isolate road noise.
   B. Better capability to reduce steering harshness on road irregularities.
   C. Increased capability to absorb steering kickback.
   D. Increased road feel.
2. All of these statements about a 4WS system are true EXCEPT:
   A. A 4WS system reduces yaw forces on the vehicle at higher speeds.
   B. A 4WS system provides a smaller turning circle.
   C. A 4WS system reduces tire wear.
   D. A 4WS system improves vehicle maneuverability at low speeds.
3. All of these statements about a power rack and pinion steering gear mounted behind *the front wheel spindle are* true EXCEPT:
   A. During a left turn fluid pressure is directed to the left side of the rack piston.
   B. During a right turn fluid pressure is directed to the right side of the rack piston.
   C. Fluid is returned through the vent tube positioned in the bellows boot.
   D. The spool valve is moved by torsion bar twisting during a turn.
4. In a column-driven electric power steering system all of the following are true EXCEPT:
   A. A motor module assembly is bolted to the lower end of the steering column.
   B. The steering shaft torque sensor is the most important input used by the power steering control module.
   C. At low speeds more steering assist is provided.
   D. The steering system may have a 20:1 steering ratio during low-speed cornering.
5. In a power rack and pinion steering gear the breather tube:
   A. Vents the right rack piston chamber during a left turn.
   B. Moves air from one boot to the other during a turn.
   C. Allows air into the rotary valve area while driving straight ahead.
   D. Prevents pressure buildup in both rack piston chambers when driving straight ahead.
6. The power steering pump belt breaks on a rack and pinion power steering system. Under this condition:
   A. The torsion bar may be broken when the front wheels are turned to the right or left.
   B. The steering operates like a manual steering system, and steering effort is higher.
   C. The spool valve no longer moves inside the rotary valve when the steering wheel is turned.
   D. Power steering fluid may be forced from the rack seals when the steering wheel is turned.

7. In electronic variable orifice (EVO) steering systems:
   A. Steering assistance is increased if the vehicle is cornering at high speeds.
   B. Steering assistance is increased if the vehicle is driven straight ahead at highway speeds
   C. Full power steering assistance is provided if the vehicle speed is below 10 mph (16 km/h) and steering wheel rotation is above 15 rpm.
   D. The control unit increases steering assistance at higher speed to provide improved road feel.
8. If the lock-to-lock steering wheel rotation is 3.25 turns and the total front wheel movement is 78 degrees the steering ratio is:
   A. 15.5:1
   B. 15:1
   C. 16:1
   D. 19.5:1
9. If an electrical defect occurs in a 4WS system the:
   A. Rear wheels remain in the centered position.
   B. Positive phase rear wheel steering is cancelled.
   C. Negative phase rear wheel steering is increased.
   D. Positive phase rear wheel steering angle is increased, and negative phase rear wheel steering is decreased.
10. All of the following are true about a Magnasteer power steering system EXCEPT:
   A. As vehicle speed changes or lateral acceleration occurs the amount of effort required to steer the vehicle can be varied depending on the requirements.
   B. At high speeds more assist is provided, which provides better road feel, firmer steering, and improved directional stability.
   C. If the system senses lateral acceleration the steering effort will become firmer in an effort to reduce oversteering by the driver.
   D. During low speed operation increased steering assist is provided for minimal steering effort to facilitate parking maneuvers and easy turning.

Chapter 14

Recirculating Ball Steering Gears

Upon completion and review of this chapter, you should be able to understand and describe:

- The purpose of a steering gear.
- The advantage of a recirculating ball steering gear compared with earlier worm and roller or cam and lever steering gears.
- The purpose of the worm shaft preload adjustment.
- How the sector shaft is rotated in a manual recirculating ball steering gear.
- The purpose of the interference fit between the sector shaft teeth and the recirculating ball teeth, and explain how this fit is obtained.
- The difference between constant ratio sector teeth and variable ratio sector teeth in a recirculating ball steering gear, and explain the advantage of the variable ratio sector teeth.
- Gear ratio.
- The term "faster" steering as it relates to steering gears.
- The power steering fluid movement in a power recirculating ball steering gear with the engine running and the front wheels straight ahead.
- The power steering fluid movement in a power recirculating ball steering gear during a left turn.
- The power steering fluid movement during a right turn.
- How kickback action is prevented in a power recirculating ball steering gear.

A BIT OF HISTORY

Steering gear design progressed from the crude, high-friction worm and gear of the early 1900s to the Ross cam and lever gear and Saginaw worm and roller gear of the 1920s. The Saginaw worm and roller steering gear was the forerunner of the modern, low-friction recirculating ball steering gear.

Introduction

The purpose of the steering gear box is to provide a mechanical advantage that allows the driver to turn the front wheels with a reasonable amount of effort. In the early 1900s, steering gears were a **worm and gear** or **worm and sector design**. These steering gears gave the driver a mechanical advantage to turn the front wheels, but they created a lot of friction.

The **Ross cam and lever steering gear** was introduced in 1923. The cam in this gear was a spiral groove machined into the end of the steering shaft. A pin on the pitman shaft was mounted in the spiral groove in the steering shaft. When the steering wheel and shaft were turned, the pin was forced to move, and this action rotated the pitman shaft. When a front wheel struck a road irregularity, this steering gear design prevented serious **kickback** on the steering wheel. However, this steering gear design still created a considerable amount of friction and required higher steering effort.

In the mid-1920s, Saginaw Steering Division of General Motors Corporation developed the **worm and roller steering gear**. In this steering gear, the **sector** became a roller, which greatly reduced friction and steering effort. The Saginaw worm and roller steering gear was the forerunner of the recirculating ball steering gear that has been widely used on rear-wheel-drive cars for many years.

In a **worm and gear** or **worm and sector design**, the sector teeth are in direct contact with the worm.

Manual Recirculating Ball Steering Gears

Design and Operation

In a recirculating ball steering gear, the steering wheel and shaft are connected to the **worm shaft**. Ball bearings support both ends of the worm shaft in the steering gear housing. A seal above the upper worm shaft bearing prevents oil leaks, and an adjusting plug is provided on the upper worm shaft bearing to adjust **worm shaft bearing preload**. Proper preloading of the worm shaft bearing is necessary to eliminate **worm shaft endplay** and to prevent **steering gear free play** and **vehicle wander**. A ball nut is mounted over the worm shaft, and internal threads or grooves on the ball nut match the grooves on the worm shaft. Ball bearings run in ball nut and worm shaft grooves (Figure 14-1).

Steering **kickback** refers to a strong and sudden movement of the steering wheel in the opposite direction to which the steering wheel is turned. This kickback action tends to occur if a front wheel strikes a road irregularity during a turn.

When the worm shaft is rotated by the steering wheel, the ball nut is moved up or down on the worm shaft. The gear teeth on the ball nut are meshed with matching gear teeth on the **pitman shaft sector**. Therefore, ball nut movement causes pitman shaft sector rotation. Since the pitman shaft sector is connected through the pitman arm and steering linkage to the front wheels, the front wheels are turned by the pitman shaft

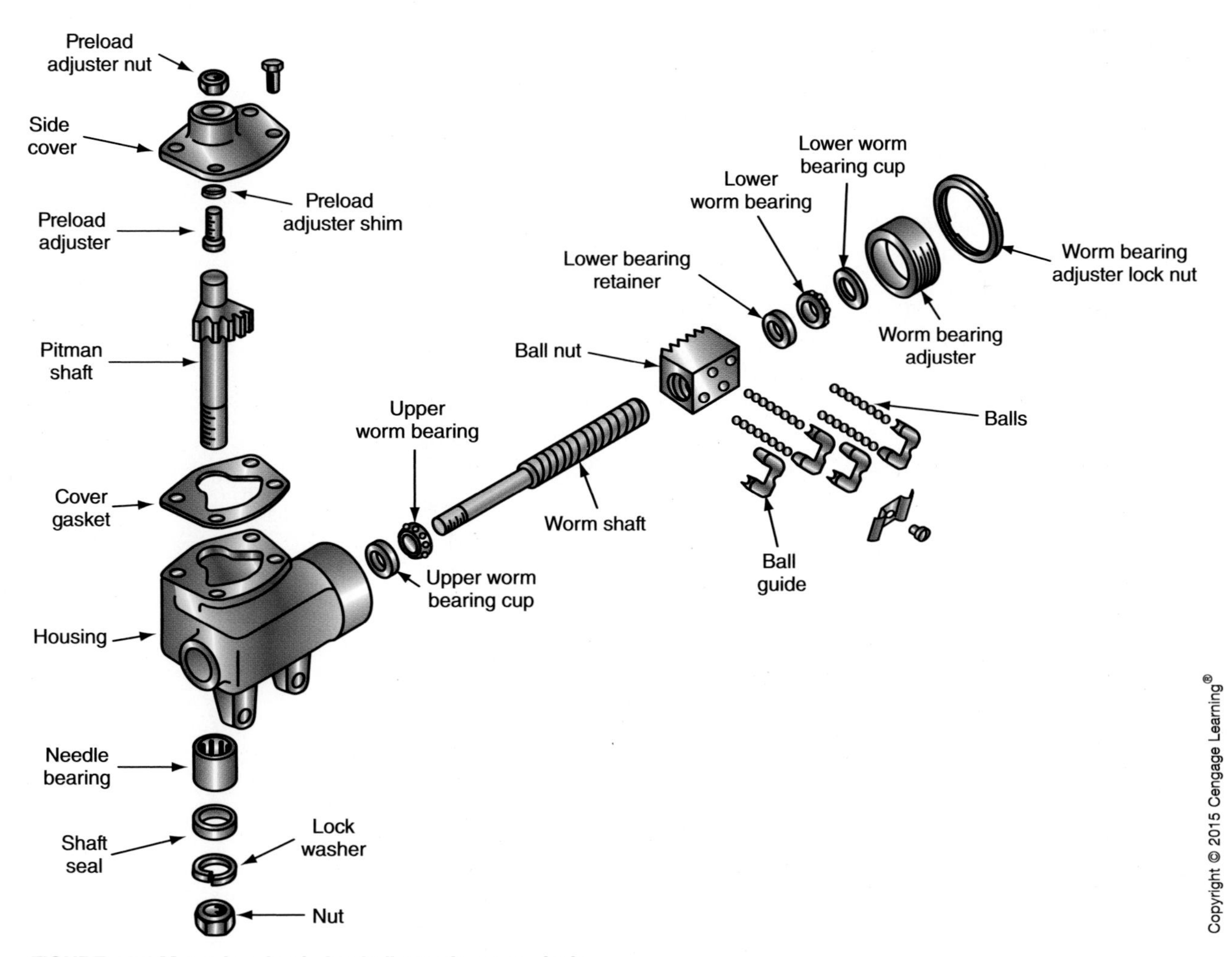

FIGURE 14-1 **Manual recirculating ball steering gear design.**

A **sector** may be defined as a part of a gear.

The **worm shaft** is a spiral gear connected through the steering shaft to the steering wheel.

Worm shaft bearing preload is a condition where all endplay is removed and there is a slight tension placed on the bearing.

Worm shaft endplay refers to movement of this shaft between the bearings on which the shaft is mounted.

Steering gear free play refers to the amount of steering wheel rotation before the front wheels begin to turn right or left.

Vehicle wander is the tendency of a vehicle to steer to the right or left as it is driven straight ahead.

The **pitman shaft sector** is meshed with the ball nut teeth.

sector. The lower end of the pitman shaft sector is usually supported by a bushing or a needle bearing in the steering gear housing. A bushing in the side cover supports the upper end of this shaft.

When the front wheels are straight ahead, an **interference fit** exists between the sector shaft teeth and ball nut teeth. This interference fit eliminates **gear tooth backlash** when the front wheels are straight ahead and provides the driver with a positive feel of the road. Proper axial adjustment of the sector shaft is necessary to obtain the necessary interference fit between the sector shaft and worm shaft teeth. A sector shaft adjuster screw is threaded into the side cover to provide axial sector shaft adjustment (Figure 14-2).

Manual recirculating ball steering gears have sector gear teeth designed to provide a constant ratio, whereas power recirculating ball steering gears usually have sector gear teeth with a variable ratio (Figure 14-3). The sector gear teeth have equal lengths in a constant ratio steering gear, but the center sector gear tooth is longer compared with the other teeth in a variable ratio gear. The variable ratio steering gear varies the amount of mechanical advantage provided by the steering gear in relation to steering wheel position. This variable ratio provides **faster steering**. The steering **gear ratio** in a constant ratio manual steering gear is usually 15:1 or 16:1, whereas the average variable ratio steering gear ratio may be 13:1. When the same types of steering gears are compared, a higher numerical ratio provides reduced steering effort and increased steering wheel movement in relation to the amount of front wheel movement.

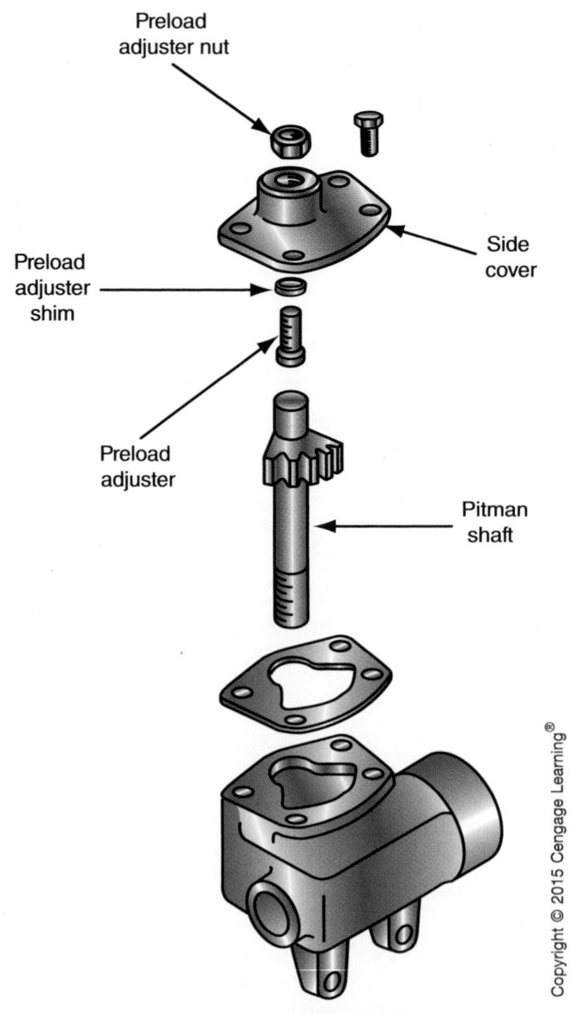

FIGURE 14-2 **Sector shaft adjusting nut.**

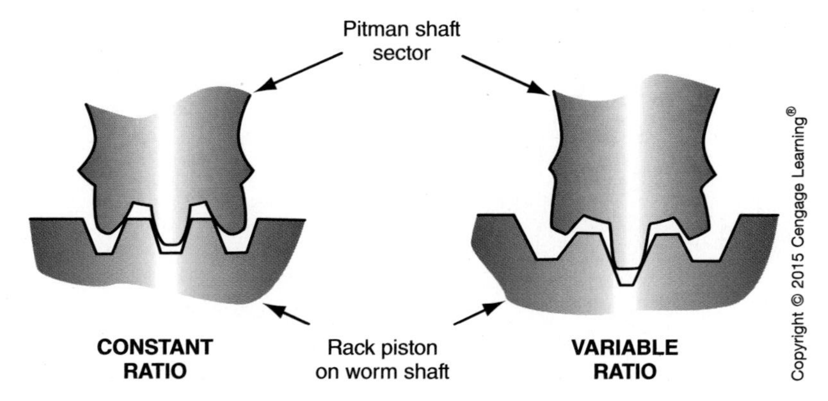

FIGURE 14-3 **Constant and variable ratio steering gears.**

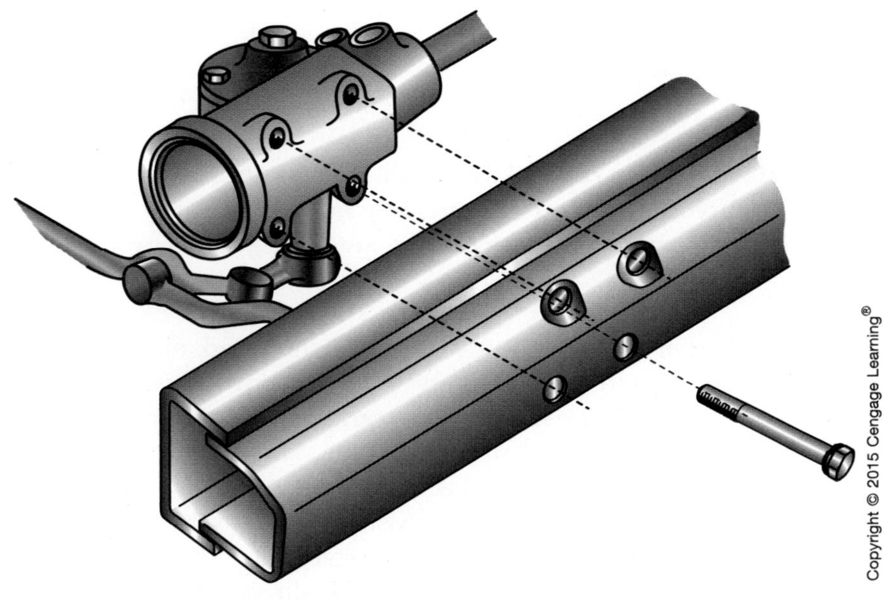

FIGURE 14-4 **Steering gear mounting on the vehicle frame.**

Many recirculating ball steering gears are bolted to the frame with hard steel bolts (Figure 14-4). These bolts must be tightened to the vehicle manufacturer's specified torque.

Power Recirculating Ball Steering Gears

Design and Operation

The ball nut and pitman shaft sector are similar in manual and power recirculating ball steering gears. In the power steering gear, a **torsion bar** is connected between the steering shaft and the worm shaft. Since the front wheels are resting on the road surface, they resist turning, and the parts attached to the worm shaft also resist turning. This turning resistance causes torsion bar twist when the wheels are turned, and this twist is limited to a predetermined amount. The worm shaft is connected to the **rotary valve** body, and the torsion bar pin also connects the torsion bar to the worm shaft. The upper end of the torsion bar is attached to the steering shaft and wheel. A stub shaft is mounted inside the rotary valve and a pin connects the outer end of this shaft to the torsion bar. The pin on the inner end of the stub shaft is connected to the **spool valve** in the center of the rotary valve (Figure 14-5).

Gear tooth backlash refers to movement between gear teeth that are meshed with each other.

A steering gear with a lower numerical ratio may be called a **faster steering** gear compared with a steering gear with a higher numerical ratio.

Gear ratio refers to the relationship between the rotation of the drive and driven gears. If 13 turns of the drive gear are necessary to obtain one turn of the driven gear, the gear ratio is 13:1.

CAUTION:
Hard steel bolts can be used for steering gear mounting. Bolt hardness is indicated by the number of ribs on the bolt head. Harder bolts have five, six, or seven ribs on the bolt heads. Never substitute softer steel bolts in place of the original harder bolts, because these softer bolts may break, allowing the steering gear box to detach from the frame. This action results in a loss of steering control.

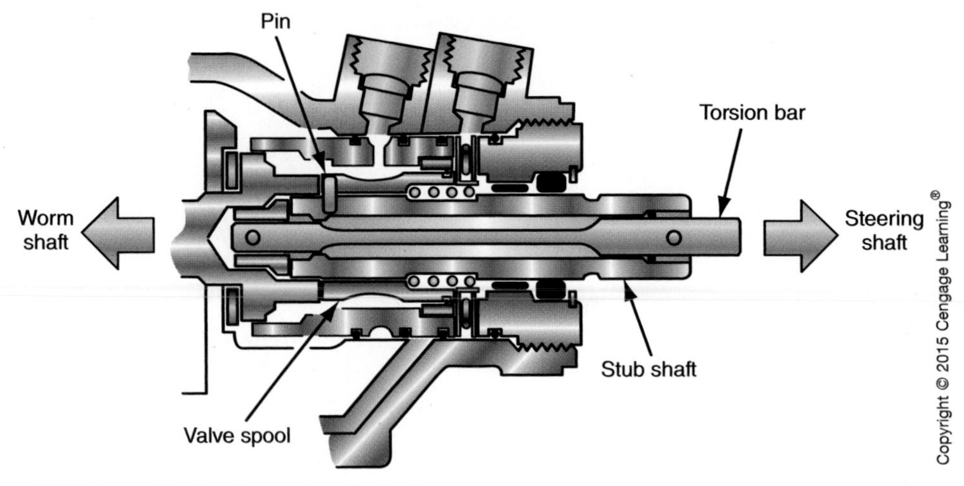

FIGURE 14-5 **Torsion bar and stub shaft.**

Shop Manual
Chapter 14,
page 552

WARNING: Many recirculating ball steering gears are mounted near the exhaust manifold, which can be extremely hot. Use caution and wear protective gloves when inspecting or servicing the steering gear to avoid burns to hands and arms.

The **rotary valve** is mounted over the top of the spool valve in a steering gear.

When the car is driven with the front wheels straight ahead, oil flows from the power steering pump through the spool valve, rotary valve, and low-pressure return line to the pump inlet (Figure 14-6). In the straight-ahead steering gear position, oil pressure is equal on both sides of the recirculating ball piston, and the oil acts as a cushion that prevents road shocks from reaching the steering wheel.

The **spool valve** is mounted inside the rotary valve. The twisting action of the torsion bar moves the spool valve in relation to the rotary valve.

If the driver makes a left turn, torsion bar twist moves the valve spool inside the rotary valve body so that oil flow is directed through the rotary valve to the left-turn holes in the spool valve (Figure 14-7). Since power steering fluid is directed from these left-turn holes

Shop Manual
Chapter 14,
page 553

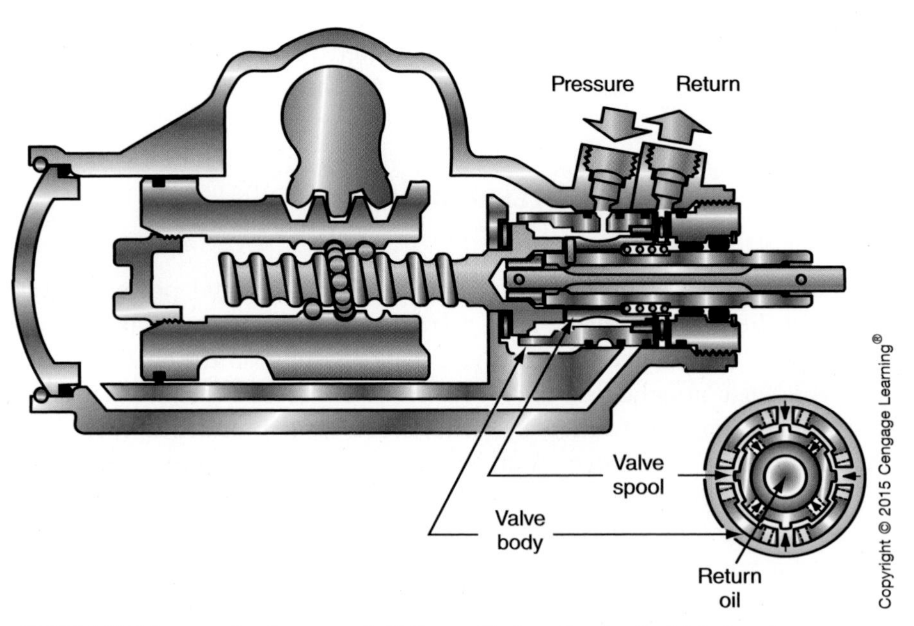

FIGURE 14-6 **Power steering gear fluid flow with the wheels straight ahead.**

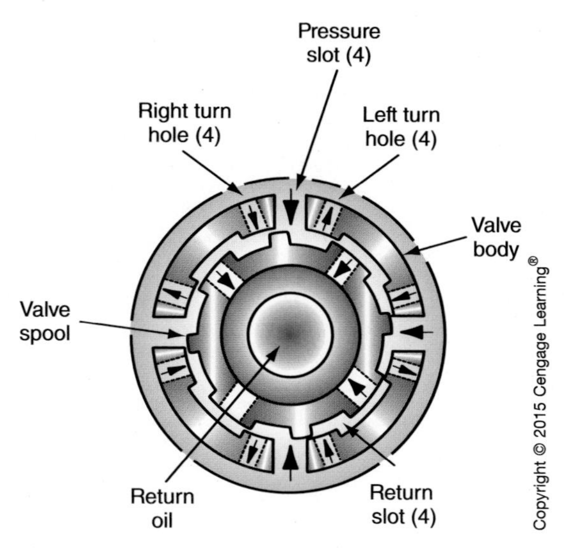

FIGURE 14-7 **Spool valve position during a left turn.**

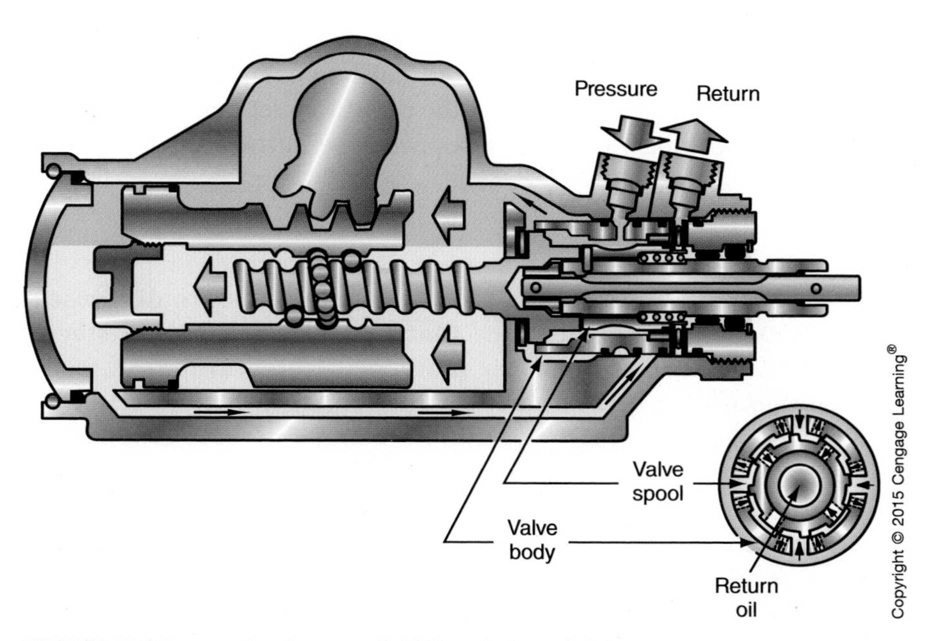

FIGURE 14-8 **Power steering gear fluid flow during a left turn.**

to the upper side of the recirculating ball piston (Figure 14-8), this hydraulic pressure on the piston assists the driver in turning the wheels to the left.

When the driver makes a right turn, torsion bar twist moves the spool valve so that oil flows through the spool valve, rotary valve, and a passage in the housing to the pressure chamber at the lower end of the ball nut piston (Figure 14-9). During a right turn, hydraulic pressure applied to the lower end of the recirculating ball piston helps the driver turn the wheels.

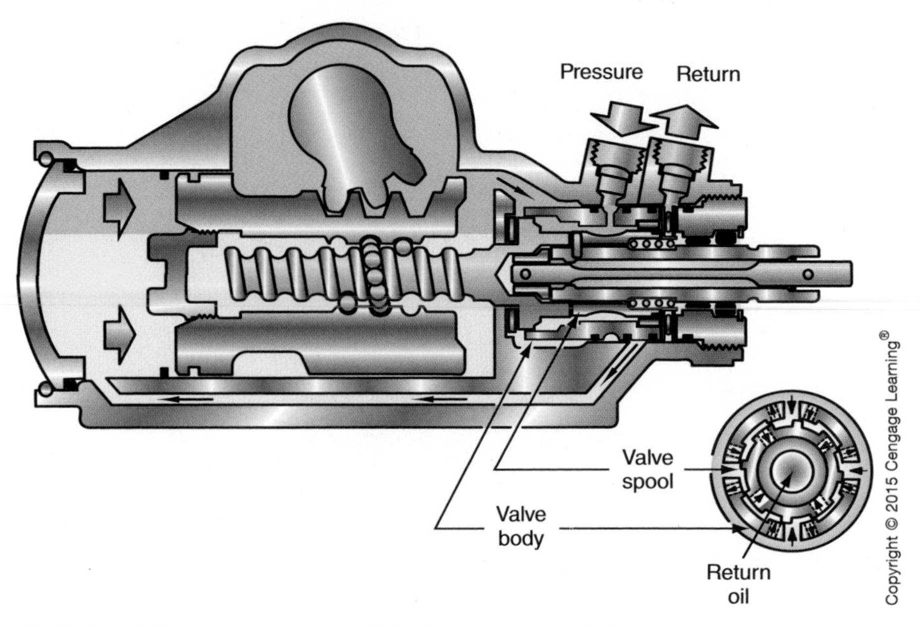

FIGURE 14-9 **Power steering gear fluid flow during a right turn.**

During a turn, if a front wheel strikes a bump and the front wheels are driven in the direction opposite the turning direction, the recirculating ball piston tends to move against the hydraulic pressure and force oil back out the pressure inlet port. This action would create a kickback on the steering wheel, but a poppet valve in the pressure inlet fitting closes and prevents kickback action.

AUTHOR'S NOTE: After road testing many vehicles to diagnose steering problems, it has been my experience that the rack and pinion steering gear and related linkage tends to transfer more road shock from the front wheels to the steering wheel compared with a recirculating ball steering gear. The design of the recirculating ball steering gear and the related steering linkage resists the transfer of road shock from the front wheels to the steering wheel. The reason for this resistance is the linkage design in which the steering arms are connected through the tie rods to the center link, and this link is connected through the pitman arm to the steering gear. In a rack and pinion steering gear, the tie rods are connected directly between the steering arms and the rack, and the rack is connected through the pinion to the steering wheel. Compared with a recirculating ball steering gear, a rack and pinion steering gear with its related linkage has fewer friction points and a more direct connection between the front wheels and the steering wheel. However, the rack and pinion steering gear does provide the driver with a greater feel of the road.

You need to be aware of the characteristics of these steering systems when diagnosing steering problems. When you understand the basic characteristics of these steering systems, you are able to immediately recognize normal and abnormal conditions, and this allows you to provide a fast, accurate diagnosis.

SUMMARY

- Early model steering gears such as the worm and gear or Ross cam and lever created a lot of friction and required higher steering effort.
- The modern recirculating ball steering gear reduces friction and steering effort.
- In a manual recirculating ball steering gear, a worm shaft adjusting plug provides a worm shaft preload adjustment.
- In a recirculating ball steering gear, rotation of the steering wheel causes the ball nut to move up and down on the worm shaft.
- With the front wheels straight ahead in a recirculating ball steering gear, an interference fit exists between the ball nut teeth and the sector shaft teeth.
- Axial sector shaft adjustment provides the proper interference fit between the ball nut teeth and the sector shaft teeth.
- Sector shaft teeth may have a constant ratio design or a variable ratio design.
- In a power recirculating ball steering gear with the front wheels straight ahead, equal pressure is applied to both sides of the recirculating ball piston, and the fluid from the pump is directed through the spool valve and rotary valve to the return hose and pump inlet.
- In a power recirculating ball steering gear, the spool valve movement inside the rotary valve is controlled by torsion bar twist.
- When the front wheels are turned in a power recirculating ball steering gear, torsion bar twist moves the spool valve inside the rotary valve, and this valve movement directs the power steering fluid to the appropriate side of the recirculating ball piston to assist steering.

TERMS TO KNOW

Faster steering
Gear ratio
Gear tooth backlash
Interference fit
Kickback
Pitman shaft sector
Ross cam and lever steering gear
Rotary valve
Sector
Spool valve
Steering gear free play
Torsion bar
Vehicle wander
Worm and gear
Worm and roller steering gear
Worm and sector design
Worm shaft
Worm shaft bearing preload
Worm shaft endplay

REVIEW QUESTIONS

Short Answer Essays

1. Describe the advantage of the manual recirculating ball steering gear compared with the previous cam and lever steering gear, and explain how this advantage is obtained.
2. Explain three purposes of the worm shaft preload adjustment.
3. Explain the purpose of the interference fit between the ball nut teeth and the sector shaft teeth in a manual recirculating ball steering gear, and describe how this interference fit is obtained.
4. Explain the difference in design between constant ratio and variable ratio sector shaft teeth.
5. Define the term faster steering.
6. Define steering gear ratio, and explain the effect of this ratio on steering effort with a manual recirculating ball steering gear.
7. Describe the flow of power steering fluid with the engine running and the front wheels straight ahead in a power recirculating ball steering gear.
8. Describe how torsion bar twist occurs in a recirculating ball steering gear.
9. Explain the purpose of torsion bar twist in a power recirculating ball steering gear.
10. Explain how kickbacks are prevented in a power recirculating ball steering gear.

Fill-in-the-Blanks

1. A manual recirculating ball steering gear provides reduced ________________ and ________________ compared with the earlier cam and lever manual steering gear.
2. If a bearing has preload, there is a slight ________________ on the bearing.
3. When the steering has free play, there is some ________________ movement before the front wheels start to turn.
4. The worm shaft end plug provides a worm shaft bearing ________________ adjustment.
5. In a manual recirculating ball steering gear, the interference fit between the ball nut and sector shaft teeth is obtained by proper ________________ sector shaft adjustment.
6. Compared with a constant ratio steering gear with a ratio of 13:1, a constant ratio steering gear with a ratio of 16:1 requires ________________ steering effort.
7. When a faster steering system is compared with a slower steering system, the faster system provides more front wheel movement with ________________ steering wheel rotation.
8. A variable ratio steering gear varies the amount of ________________ in relation to steering wheel position.
9. In a power recirculating ball steering gear with the front wheels straight ahead and the engine running, the fluid on each side of the recirculating ball piston helps prevent ________________ from reaching the steering wheel.
10. In a power recirculating ball steering gear during a turn, the ________________ movement directs the power steering fluid pressure to the appropriate side of the recirculating ball piston.

Multiple Choice

1. In a power recirculating ball steering gear, kickback is prevented by:
   A. The preload between the ball nut teeth and sector gear teeth.
   B. The bending of the torsion bar.
   C. A poppet valve in the pressure inlet fitting.
   D. The flexible coupling in the steering shaft.
2. When the worm shaft and bearings are properly installed and adjusted in a power recirculating ball steering gear:
   A. The worm shaft bearings should be preloaded.
   B. The worm shaft should have the specified end play.
   C. The worm shaft end play should be eliminated without any preload on the bearings.
   D. There must be an interference fit between the ball nut teeth and the worm shaft teeth.
3. All of these statements about recirculating ball steering gears are true EXCEPT:
   A. The steering gear provides a mechanical advantage.
   B. The steering gear reduces steering effort.
   C. Some early model steering gears had a worm and sector design.
   D. Ross cam and lever steering gears provided a very low internal friction.
4. A typical steering gear ratio in a manual steering gear is:
   A. 16:1.
   B. 18:1.
   C. 19:1.
   D. 22:1.
5. When comparing manual recirculating ball steering gears a higher numerical ratio provides:
   A. Increased steering effort.
   B. Increased steering wheel movement in relation to front wheel movement.
   C. Reduced kickback force on the steering wheel.
   D. Improved steering control.
6. In a manual recirculating ball steering gear:
   A. The interference fit between the sector shaft teeth and ball nut teeth is present when the front wheels are straight ahead.
   B. The proper interference fit between the sector shaft teeth and ball nut teeth is obtained by radial sector shaft movement.
   C. The interference fit between the sector shaft teeth and ball nut teeth becomes tighter when the front wheels are turned to the right or left.
   D. Tightening the worm shaft adjusting plug tightens the interference fit between the sector shaft teeth and ball nut teeth.

7. All of these statements about steering gear ratios are true EXCEPT:
   A. When a steering gear with a 15:1 ratio is compared with a steering gear with a 13:1 ratio, the gear with the 15:1 ratio requires more steering wheel rotation to turn the front wheels.
   B. When a steering gear with a 15:1 ratio is compared with a steering gear with a 13:1 ratio, the gear with the 15:1 ratio provides reduced steering effort.
   C. When a steering gear with a 15:1 ratio is compared with a steering gear with a 13:1 ratio, the gear with the 15:1 ratio may be called a faster steering gear.
   D. The 13:1 steering gear ratio may be used on a smaller, lighter car and the 15:1 steering gear ratio may be used on a larger, heavier vehicle.

8. In power recirculating ball steering gears:
   A. A variable ratio steering gear has sector shaft teeth with equal lengths.
   B. A variable ratio steering gear has a constant mechanical advantage regardless of steering wheel position.
   C. The torsion bar is connected between the steering shaft and the sector shaft.
   D. During a turn, power steering fluid is directed to the proper end of the ball nut piston by spool valve movement.

9. While driving with the front wheels straight ahead with a power recirculating ball steering gear:
   A. The torsion bar is twisted to move the spool valve inside the valve body.
   B. The fluid pressure is higher on the upper sideof the ball nut piston compared with the lower side.
   C. The fluid pressure on each side of the ball nut piston cushions road shocks from reaching the steering wheel.
   D. The return passage from the steering gear valve body to the power steering pump is nearly closed.

10. When a vehicle with a power recirculating ball steering gear is making a right turn:
    A. Torsion bar twisting and spool valve movement has moved the spool valve and aligned the pressure ports with the right-turn ports.
    B. Fluid pressure is supplied to the upper side of the ball nut piston.
    C. Fluid pressure is exhausted from the lower side of the ball nut piston.
    D. Fluid pressure gradually leaks past the ball nut piston to reduce steering assist during long, gradual turns.

GLOSSARY
GLOSARIO

Note: **Terms are highlighted in color,** followed by **Spanish translation in bold.**

Accelerometer An input sensor that senses vehicle acceleration in a computer-controlled suspension system.

Acelerómetro Sensor de entrada que advierte la aceleración del vehículo en un sistema de suspensión controlado por computadora.

Active coils Located in the center area of a coil spring.

Muelles activos Ubicado en el centro de un muelle espiral.

Active suspension system A computer-controlled suspension system with double-acting solenoids at each wheel.

Sistema activo de suspensión Sistema de suspensión controlado por computadora con solenoides de doble acción montados en cada una de las ruedas.

Actuator An electronically operated solenoid in a strut or shock absorber that controls firmness.

Actuador Un solenoide operado electrónicamente en un tirante o un amortiguador que controla la firmeza.

Adaptive cruise control system Maintains a specific distance between the vehicle being driven and the vehicle being followed.

Sistema adaptable de marcha a velocidad de crucero Mantiene una distancia específica entre el vehículo que se conduce y el vehículo que se sigue.

Adjustable strut A strut with a manual-operated adjustment for strut firmness.

Tirante ajustable Un tirante de ajuste operado por mano que determina la firmeza del tirante.

Air bag deployment module The air bag and deployment canister assembly that is mounted in the steering wheel for the driver's side air bag or in the dash panel for the passenger's side air bag.

Unidad de despliegue del Airbag El conjunto del Airbag y elemento de despliegue montado en el volante de dirección para proteger al conductor, o en el tablero de instrumentos para proteger al pasajero.

Air spring An air-filled membrane that replaces the conventional coil springs in an air suspension system.

Muelle de aire Membrana llena de aire que reemplaza los muelles helicoidales convencionales en un sistema de suspensión de aire.

Air spring solenoid valve An electrically operated solenoid that allows air to flow in and out of an air spring.

Muelle de aire con válvula activada por solenoide Solenoide activado eléctricamente que permite el flujo libre de aire en un muelle de aire.

All-season tires Tires with special tread designed to improve traction on snow or ice, while providing acceptable noise levels on smooth road surfaces.

Neumáticos para toda época Neumáticos con una huella especial diseñada para mejorar la tracción en la nieve o en el hielo, a la misma vez que provee niveles aceptables de ruido cuando se conduce el vehículo en un camino cuya superficie es lisa.

American Petroleum Institute (API) An organization in charge of engine oil classifications and many other areas in the petroleum industry.

Instituto Americano del Petróleo (API) Organización que tiene a su cargo la clasificación del aceite de motor y varias otras áreas en la industria del petróleo.

Ampere (A) Ampere is a unit of measurement of electrical current flow.

Amperio (A) El amperio es una unidad de medida del flujo de la corriente eléctrica.

Amplitude The extent of a vibratory movement.

Amplitud El magnitud de un movimiento de vibración.

Analog voltage signal A voltage signal that varies continuously within a specific range.

Señal de voltaje análogo Un señal que varía continuamente dentro de una banda específica.

Angular bearing load A load applied to a bearing in a direction between horizontal and vertical.

Carga del cojinete angular Una carga aplicada a un cojinete en una dirección entre horizontal y vertical.

Antilock brake system (ABS) A computer-controlled brake system that prevents wheel lockup during brake applications.

Sistema antiblocante de frenos Un sistema controlado por computadora que previene el bloqueo en las ruedas durante la aplicación de los frenos.

Asymmetrical An arrangement marked by irregularity such as different shapes, sizes, and positions.

Asimétrico Un arreglo caracterizado por la irregularidad tal como las diferencias de forma, tamaño y posición.

Asymmetrical leaf spring Has the same distance from the spring center bolt to each end of the spring.

Ballesta de hojas asimétricas Tiene la misma distancia desde el perno del centro del resorte a cada extremo del resorte.

Axle windup The tendency of the rear-axle housing to rotate in the opposite direction to the wheel and tire rotation during hard vehicle acceleration.

Bobinado del eje La tendencia de la carcasa del eje trasero de rotar en la dirección opuesta a la rotación de la rueda y el neumático durante la aceleración fuerte del vehículo.

Atmospheric pressure The pressure exerted on the earth by the atmosphere, which is 14.7 psi at sea level.

Presión atmosférica Presión que la atmósfera ejerce sobre la tierra. Dicha presión es 14.7 psi [peso sobre unidad de superficie] al nivel del mar.

Automatic level control (ALC) System maintains the rear suspension trim height regardless of the rear suspension load. If a heavy object is placed in the trunk, the rear wheel position sensors send below trim height signals to the ESC module.

Sistema de Control de nivel automático (CNA) Mantiene la altura del centrado de la suspensión trasera sin tomar en cuenta la carga de la suspensión trasera. Si un objeto se coloca en la cajuela, los sensores de posición de la rueda trasera envían señales de altura por debajo del centrado al módulo de CSE.

Axle offset A condition in which the complete rear axle assembly has turned so one rear wheel has moved forward and the opposite rear wheel has moved rearward.

Descentralización del eje Condición que ocurre cuando todo el conjunto del eje trasero ha girado de manera que una de las ruedas traseras se ha movido hacia adelante y la opuesta se ha movido hacia atrás.

Axle sideset A condition in which the rear axle assembly has moved sideways from its original position.

Resbalamiento lateral del eje Condición que ocurre cuando el conjunto del eje trasero se ha movido lateralmente desde su posición original.

Axle thrustline Is a line extending forward from the center of the rear axle at a 90° angle.

Eje de directriz de presiones Es una línea que se extiende hacia enfrente desde el centro del eje trasero a un ángulo de 90°.

Axle tramp The repeated lifting of the differential housing during extremely hard acceleration.

Barreta del eje Levantamiento repetido del alojamiento del diferencial durante una aceleración sumamente rápida.

Ball bearings Have round steel balls between the inner and outer races.

Cojinete de bolas Tiene bolas de acero entre las carreras interiores y exteriores.

Bead filler A piece of rubber positioned above the bead that reinforces the sidewall and acts as a rim extender.

Relleno de la pestaña de la llanta Pieza de caucho ubicada sobre la pestaña que sirve para reforzar la pared lateral y actúa como una extensión de la llanta.

Bead wire A group of circular wire strands molded into the inner circumference of the tire that anchors the tire on the rim.

Cable de pestaña Grupo de cordones circulares de cables moldeados en la circunferencia interior del neumático que sujetan el neumático a la llanta.

Bearing hub unit A complete wheel bearing and hub unit to which the wheel rim is bolted.

Conjunto de cubo y cojinete Conjunto total de cojinete y cubo de rueda al que se emperna la llanta de la rueda.

Bearing seals Circular metal rings with a sealing lip on the inner edge. Bearing seals are usually mounted on both sides of a bearing to prevent dirt and moisture from entering the bearing while sealing the lubricant in the bearing.

Juntas de estanqueidad del cojinete Anillos metálicos circulares con un reborde de estanqueidad en el borde interior. Normalmente las juntas de estanqueidad del cojinete se montan en ambos lados de un cojinete para evitar la entrada de suciedad y humedad mientras sellan el lubricante dentro del mismo.

Bearing shields Circular metal rings attached to the outer race that prevent dirt from entering the bearing but allow excess lubricant to flow out of the bearing.

Protectores para cojinetes Anillos metálicos circulares fijados a la arandela exterior que evitan la entrada de suciedad al cojinete, pero permiten la salida de exceso de lubricante del mismo.

Belt cover A nylon cover positioned over the belts in a tire that helps to hold the tire together at high speed, and provides longer tire life.

Cubierta de correa Cubierta de nilón ubicada sobre las correas de un neumático. Esta cubierta ayuda a conservar la solidez del neumático cuando se conduce el vehículo a gran velocidad, además de proporcionarle mayor durabilidad.

Belted bias-ply tire A type of tire construction with the cord plies wound at an angle to the center of the tire, and fiberglass or steel belts mounted under the tread area of the tire.

Neumático de correas con estrías diagonales Tipo de neumático fabricado con estrias arregladas a un ángulo con respecto al centro del neumático, y correas de fibra de vidrio o de acero montadas debajo de la huella del mismo.

Bias-ply tire A type of tire construction with the cord plies wound at an angle to the center of the tire and no belts surrounding the cords.

Neumático de estrías diagonales Un tipo de construcción de neumático con estrías arregladas en un ángulo con respecto al centro del neumático y sín correas envolviendo las estrías.

Brake-by-wire A computer-controlled brake system with no direct mechanical connection between the brake pedal and the master cylinder.

Frenado alámbrico Un sistema de frenos controlado por computadora sin una conexión mecánica directa entre el pedal de freno y el cilíndro maestro.

Brake pressure modulator valve (BPMV) A group of solenoid valves mounted in a single valve body that controls fluid pressure in the brake system during antilock brake, traction control, and vehicle stability control functions.

Válvula moderador de presión de frenado (BPMV) Un grupo de válvulas de solenoide montado en una sóla caja de válvula que controla la presión de los fluidos en el sistema de frenos durante las operaciones de freno antibloqueante, control de tracción, y control de la estabilidad del vehículo.

Brake pressure switch An electrical switch operated by brake fluid pressure.

Interruptor de presión del freno Interruptor eléctrico activado por la presión del disco de freno.

Braking and deceleration torque The torque applied to the rear axle assembly during deceleration and braking in a rear wheel drive vehicle.

Torsión de deceleración y frenaje La torsión impuesta en el ensamblado del eje trasero durante la deceleración y el frenado en un vehículo de tracción trasera.

Breather tube A small metal tube that allows air to flow between the bellows boots in a rack and pinion steering gear during a turn.

Tubo respiradero Pequeño tubo metálico que permite el flujo libre de aire entre las botas de fuelles en un mecanismo de dirección de cremallera y piñón durante un viraje.

Bump steer The tendency of the steering to veer suddenly in one direction when one or both front wheels strike a bump.

Cambio de dirección ocacionado por promontorios en el terreno Tendencia de la dirección a cambiar repentinamente de sentido cuando una o ambas ruedas delanteras golpea un promontorio.

Cam ring A metal ring with an elliptical-shaped inner surface on which the vanes make contact in a power steering pump.

Anillo de levas Un anillo metálico con una superficie interior de forma elíptica en la cual las aletas hacen contacto en una bomba de dirección hidraulica.

Car access system (CAS) A system that controls specific vehicle functions and operates in conjunction with the safety and gateway module on a vehicle with an active steering system.

Sistema de acceso al automóvil (CAS, en inglés) Un sistema que controla las funciones específicas del vehículo y que opera en conjunción con el módulo de la puerta de enlace y de seguridad de un vehículo con un sistema de dirección activo.

Carbon dioxide (CO_2) A greenhouse gas that is a byproduct of gasoline or diesel fuel combustion.

Dióxido de carbono (CO_2) Un gas de efecto invernadero que es un subproducto de la combustión de la gasolina o del combustible diesel.

Caster angle The angle between an imaginary line through the center of the tire and wheel and another imaginary line through the centers of the lower ball joint and upper strut mount viewed from the side.

ángulo de inclinación El ángulo entre una línea imaginaria que pasa por el centro de un neumático y la rueda y otra línea que pasa por los centros de la articulación esférica inferior y la montadura superior del apoyadero visto de un lado.

Center link A long rod connected from the pitman arm to the idler arm in a steering linkage.

Biela motriz central Varilla larga conectada del brazo pitman al brazo auxiliar en un cuadrilátero de la dirección.

Center take-off (CTO) The tie rods are attached to a movable sleeve in the center of the gear.

Conexión central (Center take-off, CTO) Los brazos de dirección se conectan a una manga móvil en el centro del engranaje.

Channel frame A type of frame construction in which a steel channel is positioned on the top and bottom of the vertical frame web.

Armazón fabricado con ranuras Tipo de fabricación del armazón en la que una ranura de acero es colocada en las partes superior e inferior de la malla del armazón vertical.

City safety system On some vehicles, the collision mitigation system is called a city safety system.

Sistema de seguridad de la ciudad En algunos vehículos, el sistema de mitigación de choques se conoce como sistema de seguridad de la ciudad.

Clock spring electrical connector A conductive ribbon in a plastic case mounted on top of the steering column that maintains electrical contact between the air bag inflator module and the air bag electrical system.

Conector eléctrico de cuerda de reloj Cinta conductiva envuelta en una cubierta plástica montada en la parte superior de la columna de dirección, que mantiene contacto eléctrico entre la unidad infladora y el sistema eléctrico del Airbag.

Collision mitigation system Operates specific vehicle functions to help prevent a collision.

Sistema de alivio de colisiones Opera funciones especificas dei vehiculo para ayudar a evitar una colisión.

Column-drive EPS Has an electric assist motor coupled to the steering shaft in the steering column.

Transmisión de columna de GPE (guiado de propulsión electrónica) Tiene un motor de corriente eléctrica que acompaña al eje de dirección en la columna de la dirección.

Compact spare tire A spare tire and wheel that is much smaller than the other tires on the vehicle, and is designed for short-distance driving at low speed.

Neumático compacto de repuesto Conjunto de neumático y rueda de repuesto mucho más pequeño que los demás neumáticos del vehículo, diseñado para viajes de corta distancia a poca velocidad.

Complete box frame A type of vehicle frame shaped like a rectangular steel box.

Fabricación completa Tipo de armazón en forma de caja rectangular de acero.

Compressible A substance is compressible when an increase in pressure causes a decrease in volume.

Compresibile Una sustancia es compresible cuando un aumento en la presión produce una disminución en el volumen.

Compression loaded A suspension ball joint mounted so the vehicle weight is forcing the ball into the joint.

Cargada de presión Junta esférica de la suspensión montada de manera que el peso del vehículo presiona la bola hacia el interior de la junta.

Conicity Tire conicity refers to a condition where the plies and/or belts are not level across the tire tread. When a tire has conicity, the plies and/or belts are somewhat cone shaped and may cause the vehicle to pull in one direction when driven.

Conicidad La conicidad de un neumático se refiere al estado en el que las capas y/o las fajas no están niveladas en la trocha del neumático. Cuando un neumático presenta conicidad, las capas y/o las fajas tienen de alguna manera forma cónica y pueden hacer que el vehículo tire hacia una dirección cuando se conduce.

Controller Area Network (CAN) A high speed serial bus communication network. The CAN protocol has been standardized by the International Standards Organization (ISO) as ISO 11898 standard for high speed and ISO 11519 for low speed data transfer.

Red de área de controlador (CAN) Una red de comunicaciones de bus serial de alta velocidad. El protocolo CAN ha sido estandarizado por la Organización Internacional de Estandarización (ISO) como la norma ISO 11898 para la transferencia de datos a alta velocidad y la norma ISO 11519 para la transferencia de datos a baja velocidad.

Cord plies Surround both tire beads and extend around the inner surface of the tire to enable the tire to carry its load.

Pliegues de cuerda Rodean ambas cejas de los neumáticos y se extienden alrededor de la superficie interor del neumático para permitir que el nuemático soporta su carga.

Cross-axis ball joints Contain large insulating bushings in place of typical ball joints.

Articulaciones de rótula a través del eje Contienen pasa tapas grandes en lugar de las rótulas esféricas comunes.

Cycle A series of events that repeat themselves regularly.

Ciclo Una serie de eventos que se repitan regularmente.

Damper solenoid valve An electronically operated hydraulic valve that controls shock absorber or strut damping.

Válvula amortiguador del solenoide Una válvula hidráulica operada electronicamente que controla el amortiguador o el amortiguamiento del apoyadero.

Data bus network Circuit over which serial data are transmitted.

Red de datos de bus Circuito por el cual se transmiten datos seriales.

Diamond-frame condition A frame that is diamond-shaped from collision damage.

Condición forma de diamante del armazón Armazón que ha adquirido la forma de un diamante a causa del impacto recibido durante una colisión.

Digital motor electronics (DME) A computer that controls specific vehicle functions and operates in conjunction with the safety and gateway module on a vehicle with an active steering system.

Sistema electrónico digital del motor (DME, en inglés) Una computadora que controla funciones específicas del vehículo y opera en conjunción con el módulo de la puerta de enlace y de seguridad de un vehículo con un sistema de dirección activo.

Digital voltage signal A voltage signal that is either high or low.

Señal digital de tensión Señal de tensión que es alta o baja.

Directional stability The tendency of the vehicle steering to remain in the straight-ahead position when driven straight ahead on a smooth level road surface.

Estabilidad direccional Tendencia de la dirección del vehículo a permanecer en línea recta al ser así conducido en un camino cuya superficie es lisa y nivelada.

Double-row ball bearing A bearing with two adjacent rows of circular steel balls.

Cojinete de bolas de doble hilera Cojinete que contiene dos hileras adyacentes de bolas circulares de acero.

Double wishbone rear suspension system A multilink suspension system in which the links may be shaped like a wishbone.

Suspensión trasera de doble horquilla Sistema de suspensión de empalme múltiple en el que a las bielas motrices se les puede dar forma de espoleta.

Dynamic balance The balance of a wheel in motion.

Equilibrio dinámico Equilibrio de una rueda en movi-miento.

Electronically controlled orifice (ECO) valve A computer-controlled solenoid valve in the power steering pump that controls fluid flow from the pump to the servotronic valve.

Válvula de orificio controlado electrónicamente (ECO, en inglés) Una válvula solenoide controlada por computadora que se encuentra en la bomba de dirección de potencia que controla el flujo de fluidos desde la bomba hasta la válvula servotrónica.

Electronically controlled 4WS system A four-wheel steering system in which the rear wheel steering is controlled electronically.

Dirección en las cuatro ruedas controlada electrónicamente Sistema de dirección en las cuatro ruedas en el que la dirección de la rueda trasera se controla electrónicamente.

Electronic brake and traction control module (EBTCM) A module that controls antilock brake, traction control, and vehicle stability control functions.

GLOSSARY
GLOSARIO

Módulo electrónico de control de freno y tracción (EBTCM) Un módulo que controla las funciones de control del freno antibloqueo, control de tracción, y la estabilidad del vehículo.

Electronic control unit (ECU) A computer that receives input signals and controls output functions.

Unidad de control electrónico (ECU, en inglés) Una computadora que recibe señales de entrada y controla las funciones de salida.

Electric locking unit (ELU) Locks the worm drive and drive motor in an active steering system if a safety-related defect occurs.

Unidad de bloqueo eléctrico (ELU, en inglés) Bloquea la transmisión por tornillo sinfín y el motor de impulsión de un sistema de dirección activo si se produce un defecto relacionado con la seguridad.

Electronic rotary height sensor Has an internal rotating element and this sensor sends voltage signals to the control module in relation to the curb riding height.

Sensores electrónicos rotativos de altura Tienen un elemento rotativo interno, y estos sensores envían señales de voltaje al módulo de control con relación a la curva de la altura de rodaje.

Electronic suspension control (ESC) system Controls damping forces in the front struts and rear shock absorbers in response to various road and driving conditions.

Sistema de control de suspensión electrónica (CSE) Controla las fuerzas de amortiguamiento en las barras transversales frontales y los amortiguadores neumáticos traseros como respuesta a las variadas condiciones de manejo y del camino.

Electronic variable orifice (EVO) steering A computer-controlled power steering system in which the computer operates a solenoid to control power steering pump pressure in relation to vehicle speed.

Dirección de orificio variable electrónico Sistema de dirección hidráulica controlado por computadora en el que la computadora activa un solenoide para controlar la presión de la bomba de la dirección hidráulica de acuerdo a la velocidad del vehículo.

Electronic vibration analyzer (EVA) A tester that measures vibration amplitude.

Analizador electrónico de vibraciones (EVA) Un comprobador que mide el amplitud de las vibraciones.

Encoded disc Works with an optical steering angle sensor to supply a voltage signal in relation to steering wheel rotation.

Disco codificado Funciona con un sensor de ángulo de la dirección óptico para proporcionar una señal de voltaje en relación con la rotación del volante.

End take-off (ETO) The tie rods are connected to the ends of the rack.

Conexión en los extremos (End take-off, ETO) Los brazos de dirección se conectan a los extremos de la cremallera.

Energy The ability to do work.

Energía Capacidad para realizar un trabajo.

Energy-absorbing lower bracket A steering column bracket that protects the driver by allowing the column to move away from the driver when impacted by the driver in a collision.

Soporte absorbente de energía Un soporte de la columna de dirección que proteja al conductor permitiendo que la columna se aleja del conductor cuando éste la golpea en una colisión.

Faster steering A steering gear that turns the front wheels from lock-to-lock with less steering wheel rotation.

Dirección más rápida Mecanismo de dirección que gira las ruedas delanteras de un extremo al otro con menos movimiento del volante de dirección.

Feel of the road A feeling experienced by a driver during a turn when the driver has a positive feeling that the front wheels are turning in the intended direction.

Sensación del camino Sensación experimentada por un conductor durante un viraje cuando está completamente seguro de que las ruedas delanteras están girando en la dirección correcta.

Firm relay A relay in a computer-controlled suspension system that supplies voltage to the strut actuators and moves these actuators to the firm position.

Relé de fijación Relé en un sistema de suspensión controlado por computadora que le provee tensión a los accionadores de los montantes y los lleva a una posición fija.

Five-link suspension A five-link suspension has five attachment locations between each side of the suspension and the frame.

Suspensión de cinco enlaces Una suspensión de cinco enlaces tiene cinco lugares de conexión entre cada lado de la suspensión y el chasis.

Flow control valve A special valve that controls fluid movement in relation to system demands.

Válvula de control de flujo Válvula especial que controla el movimiento del fluido de acuerdo a las exigencias del sistema.

Fluted lip seal A seal lip with fine grooves that direct the lubricant back into the reservoir rather than leaking past the seal lip.

Junta de estanqueidad de reborde estriado Reborde de una junta de estanqueidad con ranuras finas que dirigen el lubricante hacia el tanque en vez de permitir que el mismo se escape del reborde.

Force Implies the exertion of strength, which may be physical or mechanical.

Fuerza Implica el esfuerzo excesivo físico o mecánico.

Four-wheel alignment Measurement and adjustment of wheel alignment angles at the front and rear wheels.

Alineación de cuatro ruedas La medida y el ajuste de los ángulos de alineación de las ruedas en las ruedas delanteras y traseras.

Four-wheel active steering (4WAS) A steering system in which the rear wheels are steered electronically in relation to the front wheel steering angle and vehicle speed.

Dirección activa con tracción en las cuatro ruedas (4WAS, en inglés) Un sistema de dirección en el que las ruedas traseras se direccionan electrónicamente en relación con el ángulo de dirección de las ruedas delanteras y la velocidad del vehículo.

Four-wheel steering (4WS) A steering system in which both the front and rear wheels turn right or left to steer the vehicle.

Dirección en las cuatro ruedas Sistema de dirección en el que tanto las ruedas traseras como las delanteras giran hacia la derecha o la izquierda para dirigir el vehículo.

Frame buckle A frame that is accordion-shaped to some extent from collision damage.

Encorvamiento del armazón Armazón que ha adquirido la forma de un acordeón a causa del impacto recibido durante una colisión.

Frame sag A frame that is bent downward in the center from collision damage.

Hundimiento del armazón Armazón torcido hacia abajo en el centro a causa del impacto recibido durante una colisión.

Frame twist A frame that is bent from collision damage so one corner is higher in relation to the opposite corner.

Torcedura del armazón Armazón torcido a causa del impacto recibido durante una colisión; como resultado de esta torcedura un ángulo será más alto con relación al ángulo opuesto.

Friction The resistance to motion when the surface of one object is moved over the surface of another object.

Fricción Resistencia al movimiento cuando la superficie de un objeto se mueve sobre la superficie de otro.

Full-wire open-end springs Coil spring ends that are cut straight off and sometimes flattened, squared, or ground to a D-shape.

Muelles de extremo abierto con cable cerrado Extremos de muelles helicoidales cortados en línea que a veces son aplanados, cuadrados o afilados en forma de D.

Garter spring A circular spring behind a seal lip.

Muelle jarretera Muelle circular ubicado detrás de un reborde de una junta de estanqueidad.

Gas-filled shock absorber Contains a gas charge and hydraulic fluid.

Amortiguadore lleno de gas Contienens una carga de gas y fluido hidráulico.

Gear ratio The relationship between the drive gear and driven gear in a gear set.

Relación de engranaje La relación entre el engranaje de impulso y el engranaje mandado en un juego de engranajes.

Gear tooth backlash Movement between gear teeth that are meshed together.

Juego entre los dientes del engranaje Movimiento entre los dientes del engranaje que están endentados entre sí.

Geometric vehicle centerline An imaginary line through the exact center of the front and rear wheels.

Linea de eje central geométrica Línea imaginaria a través del centro exacto de las ruedas delanteras y traseras.

Geometric centerline The exact centerline of the vehicle chassis.

Línea central geométrica La línea central exacta del chasis del vehículo.

Geometric centerline alignment A wheel alignment procedure in which the front wheel toe is adjusted to specifications using the geometric centerline as a reference.

Alineación de la línea central geométrica Procedimiento de alineación de las ruedas en el que se ajusta el tope de la rueda delantera según las especificaciones utilizando la línea central geométrica como punto de referencia.

Greenhouse gas A gas that helps to form a blanket above the earth's atmosphere.

Gas de efecto invernadero Un gas que ayuda a formar un manto por encima de la atmósfera de la tierra.

Gussets Pieces of metal welded into a corner where two pieces of metal meet to increase component strength.

Esquinero Las piezas de metal soldadas en una esquina en donde se juntan dos piezas de metal para aumenter la fuerza del componente.

Hall element Is an electronic device that produces a voltage signal when the magnetic field approaches or moves away from the element.

Elemento Hall Es un dispositivo electrónico que produce una señal de voltaje cuando el campo magnético se acerca o se aleja del elemento.

Heavy-duty coil spring Compared with conventional coil springs, heavy-duty springs have larger wire diameter and 3% to 5% greater load-carrying capability.

Muelle helicoidal para servicio pesado Comparado con los muelles helicoidales convencionales, los muelles para servicio pesado tienen cables de diámetro más grande y una capacidad de carga de un 3% a un 5% mayor.

Heavy-duty shock absorbers Compared with conventional shock absorbers, heavy-duty shock absorbers have improved seals, a single tube to reduce heat, and a rising rate valve for precise spring control.

Amortiguadores para servicio pesado Comparados con los amortiguadores convencionales, los amortiguadores para servicio pesado tienen juntas de estanqueidad mejoradas, un tubo único para reducir el calor, y una válvula de vástago ascendente para un control preciso del muelle.

Height sensor Sends a signal to the suspension computer in relation to chassis height.

Sensor de altura Envía una señal a la computadora de suspensión referente a la altura del chasis.

Helical gear teeth Are positioned at an angle in relation to the gear centerline.

Dientes de engranaje helicoidales Ubicados a un ángulo con relación a la línea central del engranaje.

Hertz (Hz) A measurement for the speed at which an electronic signal or vibration cycles from high to low.

Hercio Una medición de la velocidad a la cual una señal electrónica o una vibración pasa de alta a baja.

High-frequency transmitter An electronic device that sends high-frequency voltage signals to a receiver. Some wheel sensors on computer-controlled wheel aligners send high-frequency signals to a receiver in the wheel aligner.

Transmisor de alta frecuencia Un dispositivo electróncio que manda las señales de alta frecuencia a un receptor. Algunos sensores de ruedas en los alineadores de ruedas controlados por computadora mandan los señales de alta frecuencia a un receptor en el alineador de ruedas.

High-strength steels (HSS) Steels that have above-average strength compared with other body components. High-strength steels are used in some unitized body parts.

Aceros de alta resistencia (HSS) Los aceros que tienen una resistencia superior a lo normal comparado a los otros componentes de la carrocería. Los aceros de alta resistencia se usan en algunos partes de un monocasco.

Hold valve A valve that prevents fluid return to the master cylinder during traction control or vehicle stability control functions.

Válvula de retención Una válvula que previene que los fluidos regresan al cilíndro maestro duante las funciones de control de tracción o control de estabilidad del vehículo.

Horsepower A measurement for the amount of power delivered by an internal combustion engine or an electric motor.

Caballo de fuerza Una medida de la cantidad de potencia entregada por un motor de combustión interna o por un motor eléctrico.

Hybrid vehicles Have two power sources that can supply torque to the drive wheels. In most applications the two power sources are a gasoline engine and an electric motor(s).

Vehículos híbridos Tienen dos fuentes de potencia que pueden proporcionar par motor a las ruedas de arranque. En la mayoría de las aplicaciones las dos fuentes de potencia son los motores de gasolina y eléctricos.

Hydroboost power-assisted steering system Uses fluid pressure from the power steering pump to supply brake power assistance.

Sistema de dirección asistido hidráulicamente Utiliza la presión del fluido de la bomba de la dirección hidráulica para reforzar la potencia del freno.

Hydro-forming Is the process of using extreme fluid pressure to shape metal.

Hidroformación Es el proceso de usar presión extrema de líquidos para darle forma a un metal.

Hydroplaning Occurs when water on the pavement is allowed to remain between the pavement and the tire tread contact area. This action reduces friction between the tire tread and the road surface, and can contribute to a loss of steering control.

Hidrodeslizamiento Sucede cuando se deja agua entre el pavimento y la superficie de rodadura. Esta acción reduce la fricción entre la superficie de rodadura y la superficie del camino, y puede contribuir a una pérdida de control en la dirección.

Idler arm A short, pivoted steering arm bolted to the vehicle frame and connected to the steering center link.

Brazo auxiliar Brazo de dirección corto y articulado, empernado al armazón del vehículo y conectado a la biela motriz central de dirección.

Inactive coils Inactive coils located at the top and bottom ends of a coil spring introduce force into the spring when a wheel strikes a road irregularity.

Bobinas inactivas Las bobinas inactivas localizadas en las extremidades superiores e inferiores de un resorte helicoidal introducen la fuerza en el resorte cuando una rueda choque contra una irregularidad en el camino.

Included angle The sum of the camber and steering axis inclination (SAI) angles.

Ángulo incluído La suma de los ángulos de la inclinación y la inclinación del eje de dirección (SAI).

Independent rear suspension A rear suspension system in which one rear wheel moves upward or downward without affecting the opposite rear wheel.

Suspensión trasera independiente Sistema de suspensión trasera en el que una rueda trasera se mueve de manera ascendente o descendente sin afectar la rueda trasera opuesta.

Inertia The tendency of an object to remain at rest, or the tendency of an object to stay in motion.

Inercia La tendencia de un objeto a permanecer inmóvil, o la tendencia de un objeto a continuar en movimiento.

Inner race The race in the center of a ball or tapered roller bearing that supports the balls or rollers.

Anillo interior El anillo en el centro de una bola o de un cojinete de rodillos cónicos que apoya las bolas o los rodillos.

Integral power-assisted steering system A power steering system in which the power-assisted components are integral with the steering gear.

Sistema de dirección de astencia hidráulica integral Sistema de dirección hidráulica en el que los componentes asistidos forman parte integral del mecanismo de dirección.

Integral reservoir A reservoir that is part of the power steering pump.

Tanque de la bomba Tanque que forma parte de la bomba de la dirección hidráulica.

Integral fluid reservoir A reservoir that is part of the power steering pump.

Tanque de líquido integral Tanque que forma parte de la bomba de la dirección hidráulica.

Interference fit A precision fit between two components that provides a specific amount of friction between the components.

Ajuste a interferencia Ajuste de precisión entre dos componentes que provee una cantidad específica de fricción entre los mismos.

Jounce Jounce is the action of the spring deflecting or compressing, storing energy.

Rebote Rebote es la acción del resorte al desviar o comprimir la energía de almacenamiento.

Jounce travel Upward wheel and suspension movement.

Sacudida Movimiento ascendente de la rueda y de la suspensión.

Kickback A force supplied to the steering wheel when one of the front wheels strikes a road irregularity.

Contragolpe Fuerza que se le suministra al volante de dirección cuando una de las ruedas delanteras golpea una irregularidad en el camino.

King pin inclination (KPI) The angle of a line through the center of the king pin in relation to the true vertical centerline of the tire viewed from the front of the vehicle.

Inclinación de la clavija maestra Ángulo de una línea a través del centro de la clavija maestra con relación a la línea central vertical real del neumático vista desde la parte frontal del vehículo.

Ladder frame design A frame with crossmembers between the side rails. These crossmembers resemble steps on a ladder.

Diseño de armazón de tipo escalera Armazón con travesaños entre las vigas laterales. Dichos travesaños se parecen a los peldaños de una escalera.

Lane departure warning (LDW) system Warns the driver if the vehicle drifts out of the lane in which it is driven.

Sistema de advertencia de salida del carril (LDW, en inglés) Advierte al conductor si el vehículo se aparta del carril en el que se está desplazando.

Lateral accelerometer Sends a voltage signal to the vehicle stability control computer in relation to lateral rear chassis movement.

Acelómetro lateral Manda una señal de voltaje a la computador de control de estabilidad del vehículo en relación con el movimiento lateral del trasero del chasis.

Lift/dive input A computer-controlled suspension system input signal regarding front end lift during acceleration and front end dive while braking.

Señal de entrada de ascenso y descenso Señal de entrada del sistema de suspensión controlado por computadora referente al ascenso del extremo delantero durante la aceleración y al descenso del extremo delantero durante el frenado.

Light-emitting diode (LED) A diode that emits light when current flows through the diode.

Diodo emisor de luz (LED) Diodo que emite luz cuando una corriente fluye a través del mismo.

Linear-rate coil spring A coil spring with equal spacing between the coils, one basic shape, and constant wire diameter. These springs have a constant deflection rate regardless of load. If 200 pounds compress the spring 1 inch, 400 pounds compress the spring 2 inches.

Muelle helicoidal de capacidad lineal Muelle helicoidal con separación igual entre los espirales, una forma básica, y un diámetro de cable constante. Estos muelles tienen una capacidad de desviación constante sin importar la carga. Si 200 libras comprimen el muelle una pulgada, 400 libras comprimen el muelle 2 pulgadas.

Liner A synthetic gum rubber layer molded to the inner surface of a tire for sealing purposes.

Revestimiento Capa de caucho sintético moldeada en la superficie interior de un neumático para sellarlo herméticamente.

Lithium-based grease A special lubricant that is used on the rack bearing in manual rack and pinion steering gears.

Grasa con base de litio Lubricante especial utilizado en el cojinete de la cremallera en mecanismos de dirección de cremallera y piñón manuales.

Live axle rear suspension system A live axle rear suspension system is one in which the differential housing, wheel bearings, and brakes act as a unit.

Sistemas de suspensión trasera del eje motriz La suspensión trasera del eje motriz es aquella en la cual la caja del diferencial, los rodamientos y los frenos actúan como una unidad.

Load-carrying ball joint A ball joint that supports the weight of the chassis.

Junta esférica con capacidad de carga Junta esférica que apoya el peso del chasis.

Load-leveling shock absorbers Shock absorbers to which air pressure is supplied to increase their load-carrying capability.

GLOSSARY
GLOSARIO

Amortiguadores con nivelación de carga Amortiguadores a los que se suministra presión de aire para aumentar su capacidad de carga.

Load rating A rating that indicates the load-carrying capability of a tire.

Clasificación de carga Clasificación que indica la capacidad de carga de un neumático.

Lock-to-lock A complete turn of the front wheels from full right to full left, or vice versa.

Vuelta completa de las ruedas Viraje completo de las ruedas delanteras desde la extrema derecha hasta la extrema izquierda, o viceversa.

Lower vehicle command A command, or signal, sent from a height sensor to the suspension computer that indicates the suspension height must be lowered.

Orden para bajar el vehículo Orden o señal enviada desde un sensor de altura a la computadora del sistema de suspensión que indica que debe bajarse la altura de la suspensión.

MacPherson strut front suspension system A suspension system in which the strut is connected from the steering knuckle to an upper strut mount, and the strut replaces the shock absorber.

Sistema de suspensión delantera de montante MacPherson Sistema de suspensión en el que el montante se conecta del muñón de dirección a un montaje del montante superior; el montante reemplaza el amortiguador.

Magnasteer system A power steering system that varies the steering effort in relation to vehicle speed.

Sistema Magnasteer Un sistema de dirección de potencia que varía el esfuerzo de dirección en relación con la velocidad del vehículo.

MagneRide system A suspension system with computer-controlled shock absorbers containing magneto-rheological fluid.

Sistema MagneRide Un sistema de suspensión con amortiguadores controlados por computadora que contienen fluido magneto-reológico.

Magnesium alloy wheels Wheels manufactured from magnesium mixed with other metals.

Ruedas de aleación de magnesio Ruedas fabricadas con magnesio mezclado con otros metales.

Magneto-rheological fluid (MR fluid) A synthetic fluid that contains numerous small, suspended metal particles. It can be found in some electronically controlled shock absorbers and struts.

Líquido magnetorreológico Líquido sintético que contiene numerosas partículas metálicas suspendidas y pequeñas. Puede encontrarse en algunos amortiguadores y barras transversales electrónicamente controlados.

Mass The measurement of an object's inertia.

Masa La medida de la inercia de un objeto.

Memory steer Occurs when the steering does not return to the straight-ahead position after a turn, and the steering attempts to continue turning in the original turn direction.

Dirección de memoria Condición que ocurre cuando la dirección no regresa a la posición de línea recta después de un viraje, y la dirección intenta continuar girando en el sentido original.

Momentum An object gains momentum when a force overcomes static inertia and moves the object.

Impulso Un objeto cobra impulso cuando una fuerza supera la inercia estática y mueve el objeto.

Mono leaf spring A leaf spring with a single leaf.

Muelle de lámina singular Muelle de lámina con una sola hoja.

Motorist Assurance Program (MAP) A program that establishes uniform parts inspection guidelines to improve customer satisfaction with the automotive industry.

Programa de aseguranza del conductor (MAP) Una programa que establece los requerimientos uniformes de la inspección de partes para mejorar la satisfacción del cliente con la industria automotríz.

Mud and snow tires Tires with a special tread that provides improved traction when driving in mud or snow.

Neumáticos para condiciones de lodo y nieve Los neumáticos que tienen una huella especial que provee tracción mejorada para conducir en el lodo o la nieve.

Multilink front suspension A suspension system in which the top of the knuckle is supported by two links connected to the chassis.

Suspensión delantera de bielas múltiples Un sistema de suspensión en el cual la parte superior del muñon se soporta por las dos bielas conectadas al chasis.

Multilink independent rear suspension system It has upper and lower links, and a third link connects the upper link to the top of the knuckle through a bearing.

Sistema de enlaces múltiples de suspensión trasera independiente Posee un enlace superior e inferior, y un tercer enlace que conecta el enlace superior a la parte superior del nudillo a través de un cojinete.

Multiple-leaf spring A leaf spring with more than one leaf.

Muelle de láminas múltiples Muelle de lámina con más de una hoja.

Natural vibration frequencies The frequency at which an object tends to vibrate.

Frecuencias de vibración naturales La frecuencia a la cual tiende a vibrar un objeto.

Needle roller bearing A bearing containing small circular rollers.

Cojinete de rodillos con agujas Cojinete que contiene rodillos circulares pequeños.

Negative camber Occurs when the camber line through the center of the tire is tilted inward in relation to the true vertical centerline of the tire viewed from the front of the vehicle.

Combadura negativa Condición que ocurre cuando la combadura a través del centro de un neumático se inclina hacia adentro con relación a la línea central vertical real del neumático vista desde la parte frontal del vehículo.

Negative caster Is the angle of a line through the center of the tire that is tilted forward in relation to the true vertical centerline of the tire viewed from the side.

Avance negativo del pivote de la rueda Es el ángulo de una línea a través del centro de la llanta que está inclinado hacia enfrente en relación con la verdadera línea central vertical de la rueda vista por un lado.

Negative offset A rim with negative offset has the centerline outboard of the mounting face.

Desviación negativa La llanta con desviación negativa tiene la línea central fuera de la borda de la superficie de montaje.

Negative-phase steering Occurs when the rear wheels are steered in the opposite direction to the front wheels.

Dirección de fase negativa Se aplica al modo cuando las ruedas traseras se mueven en dirección opuesta a las ruedas frontales.

Negative scrub radius Is present when the steering axis inclination line meets the road surface outside the true vertical centerline of the tire at the road surface.

Radio matorral negativo Condición que ocurre cuando la inclinación del pivote de dirección entra en contacto con la superficie del camino fuera de la línea central vertical real del neumático en la superficie del camino.

Negative offset Occurs when the wheel centerline is located outboard from the vertical wheel mounting surface.

Desfasaje negativo Se produce cuando la línea central está ubicada fuera de la superficie vertical de montaje de la rueda.

Network Data-transmitting wires that interconnect various computers on a vehicle.

Red Cables que transmiten datos que interconectan diversas computadoras de un vehículo.

Neutral phase steering Occurs when the rear wheels are centered in the straight-ahead position on a four-wheel steering system.

Direccional de fase neutral Sucede cuando las ruedas traseras se centran en una posición hacia enfrente.

Nitrogen tire inflation Tires inflated with nitrogen rather than air.

Inflado de neumáticos con nitrógeno Neumáticos inflados con nitrógeno en lugar de aire.

Non-load-carrying ball joint A ball joint that maintains suspension component location, but does not support the chassis weight.

Junta esférica sin capacidad de carga Junta esférica que mantiene la ubicación del componente de la suspensión, pero no apoya el peso del chasis.

Occupational Safety and Health Administration (OSHA) Regulates working conditions in the United States.

Dirección para Seguridad y Salud Industrial Rige las condiciones de trabajo en los Estados Unidos.

Ohms (Ω) The ohm is a unit of measurement of electrical resistance.

Ohmios (Ω) Ohmio es una unidad de medida para la resistencia eléctrica.

Ohm's law Ohm's law defines the relationship between voltage, resistance, and current.

Ley de Ohmio La Ley de Ohmio define la relación entre la tensión (o voltaje), la resistencia y la corriente.

Outer race The outer part of a bearing that supports the balls or tapered rollers.

Anillo exterior La parte exterior de un cojinete que apoya las bolas o los cojinetes de rodillo cónico.

Outlet fitting venturi A narrowing of the passage through the outlet fitting in a power steering pump.

Conexiones de salida venturi Estrechamiento del pasaje a través de la conexión de salida en la bomba de la dirección hidráulica.

Parallel HEVs A hybrid electric vehicle in which the drive wheels may be driven by the internal combustion engine, electric motor, or both.

HEV paralelos Un vehículo híbrido eléctrico en el cual las ruedas motrices se pueden impulsar mediante el motor de combustión interno, el motor eléctrico o ambos.

Parallelogram steering linkage A steering linkage in which the tie rods are parallel to the lower control arms.

Cuadrilátero de la dirección en paralelograma Cuadrilátero de la dirección en el que las barras de acoplamiento son paralelas a los brazos de mando inferiores.

Perimeter-type frame A vehicle frame in which the side rails are bent outward so these rails are near the side of the body.

Armazón de tipo perímetro Armazón del vehículo en el que las vigas laterales están dobladas hacia afuera para que queden cerca del lado de la carrocería.

Photo diode A diode that provides a voltage signal when light shines on the diode. The light source may be a light-emitting diode (LED).

Fotodiodo Diodo que provee una señal de tensión cuando es iluminado por la luz. La fuente de luz puede ser un diodo emisor de luz (LED).

Pigtail spring ends An end of a coil spring that is wound to a smaller diameter.

GLOSSARY
GLOSARIO

Extremos de muelle enrollados en forma de espiral Extremo de un muelle helicoidal devanado a un diámetro más pequeño.

Pinion The drive gear in rack and pinion steering gear.

Piñón Engranaje de mando en el mecanismo de dirección de cremallera y piñón.

Pinion angle sensor Provides a voltage signal in relation to steering pinion position.

Sensor de ángulo del piñón Proporciona una señal de voltaje de acuerdo con la posición del piñón de dirección.

Pinion-drive EPS The electric motor is coupled to the steering gear pinion.

GPE(guiado de propulsión electrónica) por tracción del piñón El motor eléctrico acompaña al piñón de la dirección.

Pitman arm A short steel arm connected from the steering gear sector shaft to the steering center link.

Brazo pitman Brazo corto de acero conectado del árbol del mecanismo de dirección a la biela motriz central de dirección.

Pitman shaft sector A shaft in a recirculating ball steering gear that is connected to the pitman shaft, and the gear teeth on the sector shaft are meshed with the worm gear.

Sector de eje pitman Un eje de un engranaje de bola recirculatoria que se conecta al eje pitman, y los dientes del engranaje del eje del sector se endentan con el engranaje sinfín.

Positive camber Occurs when the camber line through the centerline of the tire is tilted outward in relation to the true vertical tire centerline viewed from the front.

Combadura positiva Condición que ocurre cuando la combadura a través de la línea central del neumático se inclina hacia afuera con relación a la línea central vertical real vista desde la parte frontal.Is the angle of a line through the center of the tire that is tilted rearward in relation to the true vertical centerline of the tire viewed from the side.

Positive caster Is the angle of a line through the center of the tire that is tilted rearward in relation to the true vertical centerline of the tire viewed from the side.

Avance positivo del pivote de la rueda Es el ángulo de una línea a través del centro de la llanta que está inclinada hacia atrás en relación con la verdadera línea central vertical de la llanta vista por el lado.

Positive-phase steering Is applied to the mode when the rear wheels are steered in the same direction as the front wheels.

Guiado de fase positiva Se aplica cuando las ruedas traseras se guían en la misma dirección de las delanteras.

Positive offset Occurs when the wheel centerline is located inboard from the vertical wheel-mounting surface.

Desfasaje positivo Se produce cuando la línea central de las ruedas se encuentra dentro de la superficie vertical de montaje de las ruedas.

Positive scrub radius Occurs when the SAI line through the center of the lower ball joint and upper strut mount meets the road surface inside the true vertical centerline of the tire at the road surface as viewed from the front.

Radio matorral positivo Condición que ocurre cuando la línea SAI a través del centro de la junta esférica inferior y el montaje del montante superior entra en contacto con la superficie del camino dentro de la línea central vertical real del neumático en la superficie del camino vista desde la parte frontal.

Pre-safe systems React during the few milliseconds before a collision occurs to increase driver and passenger safety.

Presistema de protección Reacción durante los pocos milisegundos antes de que ocurra un choque para aumentar la seguridad del conductor y del pasajero.

Pressure relief ball A spring-loaded ball that opens at a specific pressure and limits pressure in a hydraulic system.

Bola de alivio de presión Bola con cierre automático que se abre a una presión específica y limita la presión en un sistema hidráulico.

Programmed ride control (PRC) system A computer-controlled suspension system in which the computer operates an actuator in each strut to control strut firmness.

Sistema de control programado del viaje Sistema de suspensión controlado por computadora en el que la computadora opera un accionador en cada uno de los montantes para controlar la firmeza de los mismos.

Pulse width modulation (PWM) A method of computer control in which the computer cycles an output on and off with a variable on time.

Modulación de duración de impulsos Método de control de computadoras en el que la computadora produce un ciclo de rendimiento a intérvalos; lo que produce un trabajo efectivo variable en cada ciclo.

Pulse width modulated (PWM) voltage signal Is a signal with a variable on time that may be used by a computer to control an output.

Señal modulada de la anchura entre impulsos Es una señal con una variable de tiempo que puede usar una computadora para controlar una potencia de salida.

Puncture sealing tires A tire with a special compound on the inner tire surface that seals punctures when the puncturing object is removed from the tire.

Neumáticos autoselladores Neumático con un compuesto especial en la superficie de la parte interior que sella pinchazos cuando se le remueve el objecto punzante al neumático.

Pyrotechnic Device that contains an explosive and an ignition source.

Pirotecnia Dispositivo que contiene un explosivo y una fuente de ignición.

Quadrasteer® A type of 4WS system used on some SUVs.

Quadrasteer® Un tipo de suspensión de cuatro ruedas utilizado en algunos modelos de SUV.

Rack A horizontal shaft in a rack and pinion steering gear containing teeth that are meshed with the pinion teeth.

Cremallera Eje horizontal en un mecanismo de dirección de cremallera y piñón que contiene los dientes que se engranan con los del piñón.

Rack and pinion power steering system A type of power steering in which the steering gear contains a pinion gear that is meshed with teeth on the horizontal rack.

Sistema de dirección hidráulica de cremallera y piñón Tipo de dirección hidráulica en la que el mecanismo de dirección contiene un piñón que se engrana con los dientes de la cremallera horizontal.

Rack and pinion steering linkage A rack and pinion steering gear with the tie rods connected from the rack ends or rack center to the steering arms.

Cuadrilátero de la dirección de cremallera y piñón Mecanismo de dirección de cremallera y piñón con las barras de acoplamiento conectadas de los extremos de la cremallera o del centro de la misma a los brazos de dirección.

Rack-drive EPS The electric motor that provides steering assist is coupled to the rack in the steering gear.

Cremallera de dirección GPE(guiado de propulsión electrónica) El motor eléctrico que proporciona ayuda de dirección acompaña a la cremallera en el mecanismo de dirección.

Rack piston A piston mounted on the rack in a power rack and pinion steering gear. Hydraulic pressure is supplied to this piston for power steering assistance.

Pistón de la cremallera Pistón montado en la cremallera en un mecanismo de dirección hidráulica de cremallera y piñón. A este pistón se le suministra presión hidráulica para reforzar la dirección hidráulica.

Rack seals Are positioned between the rack and the housing in a rack and pinion steering gear.

Juntas de estanqueidad de la cremallera Ubicadas entre la cremallera y el alojamiento en un mecanismo de dirección de cremallera y piñón.

Radial bearing load A load applied in a vertical direction.

Carga de marcación radial Carga aplicada en dirección vertical.

Radial-ply tires A tire in which the carcass plies are positioned at 90° in relation to the center of the tire, and steel or fiberglass belts are mounted under the tread.

Neumáticos de cordón radial Neumático fabricado con las estrias de armazón colocadas a un ángulo de 90° con relación al centro del neumático, y correas de acero o de fibra de vidrio montadas debajo de la huella.

Raise vehicle command A command, or signal, sent from a height sensor to the computer in a computer-controlled suspension system indicating the chassis height must be raised.

Orden para levantar el vehículo Orden o señal enviada desde un sensor de altura a la computadora en un sistema de suspensión controlado por computadora que indica que debe levantarse la altura del chasis.

Reactive Having the capability to cause a chemical reaction with another substance.

Reactivo Que tiene la capacidad de producir una reacción química con otra sustancia.

Real-time damping (RTD) system A term used to designate the road-sensing suspension system in onboard diagnostics.

Sistema de amortiguamiento en tiempo real Término utilizado para designar el sistema de suspensión con equipo sensor en pruebas de diagnóstico realizadas en el vehículo mismo.

Rear active steering (RAS) Has recently been introduced as standard equipment on the 2006 Infinity M series luxury car. The RAS makes the steering and handling characteristics on this long-wheelbase car more nimble and agile compared with models without this technology.

Dirección activa trasera (DAT) Últimamente se presentó como un equipo estándar en la serie del automóvil de lujo de la serie Infinity M del 2006. La DAT hace que la dirección y las características de manipulación de este automóvil de paso largo sean más ligeras y ágiles comparadas con los modelos que no poseen esta tecnología.

Rear-axle bearing A bearing that supports the rear axle in the housing on a rear wheel drive vehicle.

Cojinete del eje trasero Cojinete que apoya el eje trasero en el alojamiento en un vehículo de tracción trasera.

Rear axle offset A condition in which the complete rear axle assembly has turned so one rear wheel has moved forward, and the opposite rear wheel has moved rearward.

Desviación del eje trasero Condición que ocurre cuando todo el conjunto del eje trasero ha girado de manera que una de las ruedas traseras se ha movido hacia adelante y la opuesta se ha movido hacia atrás.

Rear axle sideset A condition in which the rear axle assembly has moved sideways from its original position.

Resbalamiento lateral del eje trasero Condición que ocurre cuando el conjunto del eje trasero se ha movido lateralmente desde su posición original.

Rear steering actuator An assembly that controls rear wheel steering in some four-wheel steering systems.

Accionador de la dirección trasera Conjunto que controla la dirección de la rueda trasera en algunos sistemas de dirección en las cuatro ruedas.

GLOSSARY
GLOSARIO

Rear suspension system The rear suspension system plays a very important part in ride quality and in the control of suspension and differential noise, vibration, and shock.

Sistema de suspensión trasera El sistema de suspensión trasera desempeña un papel muy importante en la calidad de la conducción y en el control de la suspensión y el ruido, la vibración y el impacto en el diferencial.

Rear wheel camber The tilt of a line through the rear tire and wheel centerline in relation to the true vertical centerline of the rear tire and wheel.

Comba de las ruedas traseras El ángulo de una línea que atraviesa el eje mediano del neumático trasero y de la rueda en relación al eje mediano verdadero del neumático trasero y la rueda.

Rear wheel toe The distance between the front edges of the rear wheels in relation to the distance between the rear edges of the rear wheels.

Tope de la rueda trasera Distancia entre los bordes frontales de las ruedas traseras con relación a la distancia entre los bordes traseros de las ruedas traseras.

Rebound Rebound is the action of the spring expanding and releasing stored energy.

Rebote Rebote es la acción del resorte al expandir y liberar la energía de almacenamiento.

Rebound travel The downward spring and wheel action resulting from rebound. This downward spring and wheel action is called rebound travel.

Viaje del rebote Rebote es la acción del resorte al expandir y liberar la energía de almacenamiento. Esta acción descendente del resorte y la rueda se conoce como viaje del rebote.

Receiver An electronic device in some computer-controlled wheel aligners that receives high-frequency voltage signals from the wheel sensors.

Receptor Un dispositivo electrónico en algunos alineadores controlados por computadoras que recibe los señales de alta frequencia de los sensores de las ruedas.

Regenerative braking A system on a hybrid or electric vehicle that allows the drive wheels to turn the electric drive motor so it recharges the battery during vehicle deceleration.

Frenado regenerativo Un sistema en un vehículo híbrido o eléctrico que permite que las ruedas motrices giren el motor de impulsión eléctrica para que recargue la batería durante la desaceleración del vehículo.

Regular-duty coil spring A coil spring supplied to handle average loads to which the vehicle is subjected. This type of spring has smaller wire diameter compared with a heavy-duty spring.

Muelle helicoidal para servicio normal Muelle helicoidal provisto para sostener la carga normal a la que el vehículo está expuesto. Este tipo de muelle tiene un cable de un diámetro más pequeño comparado con un muelle para servicio pesado.

Release valve A valve that releases fluid pressure from a wheel caliper during antilock brake operation.

Válvula descargador Una válvula que descarga la presión de fluido del calíbre de la rueda durante la operación de frenado antibloqueante.

Remote reservoir A reservoir mounted separately from the power steering pump.

Depósito remoto Un depósito montado separadamente de la bomba de dirección hidráulica.

Remote fluid reservoir A reservoir mounted separately from the power steering pump.

Tanque de líquido remoto Un tanque montado por separado de la bomba de dirección hidráulica.

Replacement tires Tires purchased to replace the original tires that were supplied by the vehicle manufacturer.

Neumáticos de repuesto Neumáticos que se compran para reemplazar los neumáticos originales provistos por el fabricante del vehículo.

Resonance A reinforcement of sound in a vibrating body caused by sound waves from another body.

Resonancia Una fortificación del sonido en un cuerpo vibrante causada por las ondas sónicas de otro cuerpo.

Resource Conservation and Recovery Act (RCRA) States that hazardous material users are responsible for hazardous materials from the time they become a waste until the proper waste disposal is completed.

Ley de Conservación y Recuperación de Recursos Establece que los usuarios de materiales peligrosos se encarguen de estos materiales desde el momento en que se convierten en desperdicios hasta que se lleve a cabo la eliminación adecuada de los mismos.

Road crown The higher center of a road surface in relation to the edges of this surface.

Corona de camino Centro más alto de la superficie de un camino con relación a los bordes de la misma.

Road force variation Is a condition that occurs when tire sidewalls do not have equal stiffness around the complete area of the sidewalls.

Variación de la fuerza en carretera Es una condición que sucede cuando las paredes laterales de las llantas no tienen la rigidez equitativa alrededor del área completa de las paredes laterales.

Road variables Variables such as weight in the vehicle or road surface that affect wheel alignment while driving.

Variables del camino Condiciones variables, como por ejemplo el peso en el vehículo o la superficie del camino que afectan la alineación de la rueda durante un viaje.

Roller bearing A roller bearing contains precision machined rollers that have the same diameter at both ends. Designed primarily to carry radial loads, but they can withstand some thrust load.

Cilindro del cojinete El cilindro del cojinete contiene cilindros con torneado de precisión con el mismo diámetro en ambos extremos. Se diseñaron inicialmente para transportar cargas radiales, pero pueden soportar algunas cargas de empuje.

Rolling elements The balls or rollers and the separator in a bearing.

Elementos rodantes Las bolas o rodillos y el separador en un cojinete.

Ross cam and lever steering gear A type of steering gear used in the early 1900s that had a spiral groove in the lower end of the steering shaft meshed with a pin on the pitman shaft.

Mecanismo de dirección de leva y palanca Ross Tipo de mecanismo de dirección utilizado a principios de siglo que tenía una ranura espiral en el extremo inferior del eje de dirección engranado con un pasador en el árbol pitman.

Rotary valve A valve mounted with the spool valve in a power steering gear. The position of these two valves directs power steering fluid to the appropriate side of the rack or power piston.

Válvula rotativa Válvula montada con la válvula de carrete en un mecansimo de dirección hidráulica. La posición de estas dos válvulas dirige el fluido de la dirección hidráulica al lado apropiado de la cremallera o del pistón impulsor.

Run-flat tires Tires designed to operate safely without any air pressure for a specific distance.

Neumáticos de no presión Los neumáticos diseñados a operar sin peligro sín presión por distancias específicas.

Safety and gateway module (SGM) Operates the servotronic valve and the electronically controlled orifice valve in some power steering systems.

Módulo de puerta de enlace y seguridad (SGM, en inglés) Opera la válvula servotrónica y la válvula de orifi cio controlado electrónicamente en algunos sistemas de dirección de potencia.

Scrub radius The distance between the SAI line and the true vertical centerline of the tire at the road surface.

Radio matorral Distancia entre la línea SAI y la línea central vertical real del neumático en la superficie del camino.

Sector A shaft in a recirculating ball steering gear that is connected to the pitman shaft. The gear teeth on the sector shaft are meshed with the worm gear.

Sector Árbol en un mecanismo de dirección con bola recirculante que se conecta al árbol pitman. Los dientes del engranaje en el eje sector se endentan con el engranaje sinfín.

Semiellliptical springs Semielliptical springs have individual leaves stacked with the shortest leaf at the bottom and the longest leaf at the top.

Resortes semielípticos Los resortes semielípticos poseen láminas individuales apiladas con la lámina más corta en el fondo y la lámina más larga en la parte superior.

Semi-independent rear suspension A rear suspension system in which one rear wheel has a limited amount of movement without affecting the opposite rear wheel.

Suspensión trasera semi-independiente Sistema de suspensión trasera en el que una de las ruedas traseras tiene una cantidad limitada de movimiento sin afectar la rueda trasera opuesta.

Separator A component that prevents contact between two other parts.

Separador Componente que evita que otras dos piezas entren en contacto.

Series HEVs A hybrid electric vehicle in which the internal combustion engine and the electric drive motor both supply torque to the drive wheels.

HEV de serie Un vehículo híbrido eléctrico en el cual tanto el motor de combustión interna como el motor de impulsión eléctrica proporcionan potencia a las ruedas motrices.

Serpentine belt A ribbed V-belt drive system in which all the belt-driven components are on the same vertical plane.

Correa serpentina Sistema de transmisión con correa nervada en V en el que todos los componentes accionados por una correa se encuentran sobre el mismo plano vertical.

Servotronic valve A computer-controlled valve in an active steering system that controls power steering pump pressure in relation to vehicle speed.

Válvula servotrónica Una válvula controlada por computadora en un sistema de dirección activo que controla la presión de la bomba de dirección de potencia de acuerdo con la velocidad del vehículo.

Shock absorber A hydraulic mechanism connected between the chassis and the suspension to control spring action.

Amortiguador Un mecanismo hidráulico conectado entre el chasis y la suspensión para controlar la acción del muelle.

Shock absorber ratio The amount of extension control compared with the amount of compression control.

Relación del amortiguador Cantidad de control de extensión comparado con la cantidad de control de compresión.

Short-and-long arm front suspension systems Suspension systems in which the upper control arm is shorter than the lower control arm.

Sistemas de suspensión de brazos largos y cortos Un sistema suspensión en el cual el brazo de control superior es más corto que el brazo de control inferior.

Sideslip The tendency of the rear wheels to slip sideways because of centrifugal force while a vehicle is cornering at high speed.

GLOSSARY
GLOSARIO

Patinaje Tendencia de las ruedas traseras a deslizarse lateralmente a causa de la fuerza centrífuga mientras un vehículo hace un viraje a gran velocidad.

Side sway Occurs when one side of a vehicle frame is bent inward.

Desviación Ocurre cuando un lado del bastidor del vehículo esta torcido hacia adentro.

Sidewall The area between the tread and the bead of a tire.

Pared lateral Área entre la huella y la pestaña de un neumático.

Silencer A component designed to reduce noise.

Silenciador Componente diseñado para disminuir el ruido.

Single-row ball bearing A bearing with a single row of balls.

Cojinete de bola de una sola fila Cojinete que contiene una sola fila de bolas.

Slip angle The actual angle of the front wheels during a turn compared with the turning angle of the front wheels with the vehicle at rest.

Ángulo de patinaje Ángulo real de las ruedas delanteras durante un viraje comparado con el ángulo de giro de las ruedas delanteras cuando se ha detenido la marcha del vehículo.

Snapring A circular steel ring with some tension designed to snap into a groove and retain a component.

Anillo de resorte Anillo circular de acero con un poco de tensión diseñado para ajustarse dentro de una ranura y sujetar un componente en su posición.

Society of Automotive Engineers (SAE) A society of professional engineers that provides many member and industry services, such as the development of industry standards and the communication of engineering information through publications and conferences.

Sociedad de Ingenieros de Automóviles (SAE) Sociedad de ingenieros profesionales que les provee muchos servicios a sus miembros y a la industria, como por ejemplo, el desarrollo de normas para la industria, y la comunicación de información sobre ingeniería mediante publicaciones y conferencias.

Sodium-based grease A special lubricant with a sodium base that may be required on some steering components.

Grasa a base de sodio Lubricante especial con una base de sodio que podrían necesitar algunos componentes de la dirección.

Soft relay A relay in a computer-controlled suspension system that supplies voltage to the strut actuators in the soft mode.

Relé blando Un relé en un sistema de suspensión controlado por computadora que suministra el voltaje a los accionadores de los montantes en un modo blando.

Solid axle beam A solid axle beam is usually a transverse inverted U-section channel connected between the rear wheels in a semi-independent rear suspension system.

Viga del eje sólido Una viga del eje sólido suele ser un canal transversal de sección U invertida, conectado entre las ruedas traseras en un sistema de suspensión trasera semi-independiente.

Speed rating A tire rating that indicates the maximum safe vehicle speed that a tire will withstand.

Clasificación de la velocidad Clasificación de un neumático que indica la velocidad máxima que podrá resistir un neumático.

Speed sensitive steering (SSS) system A computer-controlled steering system that varies the steering effort in relation to vehicle speed.

Dirección sensible a la velocidad Sistema de dirección controlado por computadora que varía el esfuerzo necesario de la dirección de acuerdo a la velocidad del vehículo.

Spherical bearing A bearing shaped like a sphere and used in some tilt steering wheels.

Cojinete esférico Cojinete en forma de esfera utilizado en algunos volantes de dirección inclinables.

Spiral cable A conductive ribbon that is mounted in a plastic container on top of the steering column and maintains electrical contact between the air bag inflator module and the air bag electrical system. A spiral cable may be called a clock spring electrical connector.

Cable espiral Cinta conductiva montada en un recipiente plástico sobre la columna de dirección que mantiene el contacto eléctrico entre la unidad infladora y el sistema eléctrico del Airbag. El cable espiral se conoce también como conector eléctrico de cuerda de reloj.

Spool valve Positioned with the rotary valve in a power steering gear. These two valves direct power steering fluid to the appropriate side of the rack or power piston to provide steering assistance.

Válvula de carrete Ubicada con la válvula rotativa en un mecanismo de dirección hidráulica. Estas dos válvulas dirigen el fluido de la dirección hidráulica al lado apropiado de la cremallera o del pistón impulsor para reforzar la dirección.

Spring insulator Positioned between the ends of a coil spring and the spring mounting surfaces to reduce the transfer of noise and vibration from the spring to the chassis.

Aisladore de muelle Ubicados entre los extremos de un muelle helicoidal y las superficies para el montaje de muelles con el propósito de reducir la transferencia de ruido y la vibración del muelle al chasis.

Springless seal A seal lip with no spring behind the lip.

Junta de estanqueidad sin muelle Reborde de una junta de estanqueidad que no tiene un muelle detrás del mismo.

Spring-loaded seal A seal lip with a garter spring behind the lip to increase lip tension.

Junta de estanqueidad con cierre automático Reborde de una junta de estanqueidad con un muelle jarretera detrás del reborde para aumentar la tensión del mismo.

Sprung weight Is the vehicle weight that is supported by the coil springs.

Peso suspendido Es el peso del vehículo que soportan los muelles en espiral cilíndrica.

Spur gear teeth Gear teeth that are parallel to the centerline of the gear.

Dientes de engranaje rectos Dientes del engranaje que son paralelos a la línea central del engranaje.

Stabilitrak® A computer-controlled system that provides vehicle stability control by reducing sideways swerving.

Stabilitrak® Un sistema controlado por computadora que provee el control de estabilidad del vehículo asi disminuyendo viraje lateral.

Stabilizer bar A round steel bar, connected between the front or rear lower control arms, that reduces body sway.

Barra estabilizadora Una barra de acero circular conectada entre los brazos de control inferiores delanteros o traseros, que reducen el balanceo del cuerpo.

Static balance Refers to the balance of a tire and wheel at rest.

Equilibrio estático Se refiere al equilirio de un neumático y una rueda cuando se ha detenido la marcha del vehículo.

Steering angle The angle of the front wheel on the inside of a turn compared to the angle of the front wheel on the outside of the turn.

Ángulo de la dirección Ángulo de la rueda delantera en el interior de un viraje comparado con el ángulo de la rueda delantera en el exterior del viraje.

Steering angle sensor An optical-type sensor that supplies a voltage signal in relation to steering wheel rotation.

Sensor del ángulo de la dirección Un sensor de tipo óptico que proporciona una señal de voltaje de acuerdo con la rotación del volante

Steering axis inclination (SAI) The angle of a line through the center of the upper strut mount and lower ball joint in relation to the true vertical centerline of the tire viewed from the front of the vehicle.

Inclinación del pivote de dirección Ángulo de una línea a través del centro del montaje del montante superior y la junta esférica inferior con relación a la línea central vertical real del neumático vista desde la parte frontal del vehículo.

Steer-by-wire A computer-controlled steering system with no direct mechanical connection between the steering wheel and the steering gear.

Dirección alámbrica Un sistema de dirección controlado por computadora sín una conexión mecánica directa entre el volante de dirección y el aparato de dirección.

Steering gear free play The amount of steering wheel movement before the front wheels begin to turn.

Juego en el aparato de dirección Cantidad de movimiento del volante de dirección antes de que las ruedas delanteras comiencen a girar.

Steering drift The tendency of the steering to drift slowly to one side when the vehicle is driven straight ahead on a smooth, straight road surface.

Desviación de la dirección La tendencia de la dirección a desviarse poco a poco hacia un lado mientras que el vehículo se conduce en línea recta sobre un camino liso y nivelado.

Steering kickback Road force supplied back to the steering wheel.

Tensión de retroceso de la dirección Fuerza del camino que se vuelve a suministrar al volante. Sensor de posició.

Steering pull The tendency of the steering to pull constantly to one side or the other when driving the vehicle straight ahead on a smooth, level road surface.

Tiro en la dirección La tendencia de la dirección a tirar en una manera constante hacia un lado u otro mientras que el vehículo se conduce en línea recta sobre un camino liso y nivelado.

Steering ratio The number of degrees of steering wheel rotation in relation to the number of degrees of front wheel movement.

Relación de la dirección Número de grados de rotación del volante de dirección con relación al número de grados de movimiento de la rueda delantera.

Steering sensor A sensor that sends a voltage signal to the computer in relation to the amount and speed of steering wheel rotation in a programmed ride control (PRC) system.

Sensor de dirección Un sensor que manda una señal de voltaje a la computadora en relación a la cantidad y la velocidad de la rotación del volante de dirección en un sistema de viaje controlado (PRC).

Steering wander The tendency of the steering to pull to the right or left when the vehicle is driven straight ahead on a smooth road surface.

Desviación de la dirección Tendencia de la dirección a desviarse hacia la derecha o hacia la izquierda cuando se conduce el vehículo en línea recta en un camino cuya superficie es lisa.

Steering wheel angle sensor (SWAS) Provides a voltage signal on relation to steering wheel rotation.

Sensor del ángulo del volante (SWAS, en inglés) Proporciona una señal de voltaje de acuerdo con la rotación del volante.

Steering wheel position sensor (SWPS) A sensor that produces a voltage signal in relation to the amount and velocity of steering wheel rotation.

Sensor de posición de la dirección (SWPS) Un sensor que produce una señal de voltaje en relación a la cantidad y la velocidad de rotación del volante de dirección.

Steering wheel rotation sensor An input sensor in a computer-controlled suspension or steering system that sends a signal to the computer in relation to the amount and speed of steering wheel rotation.

Sensor de la rotación del volante de dirección Sensor de entrada en un sistema de suspensión controlado por computadora o en un sistema de dirección que le envía una señal a la computadora referente a la cantidad y a la velocidad de la rotación del volante de dirección.

Stop-to-stop Steering wheel rotation from extreme left to extreme right.

Parada a parada Rotación del volante de dirección desde la extrema izquierda hasta la extrema derecha.

Struts Components connected from the top of the steering knuckle to the upper strut mount that maintain the knuckle position and act as shock absorbers to control spring action.

Montantes Componentes conectados de la parte superior del muñón de dirección al montaje del montante superior que mantienen la posición del muñón y actúan como amortiguadores para controlar el movimiento de ascenso y descenso del muelle.

Summation sensor Provides a voltage signal in relation to steering rack position.

Sensor de suma Proporciona una señal de voltaje de acuerdo con la posición del soporte de la dirección.

Symmetrical leaf spring A symmetrical leaf spring has the same distance from the center bolt to the front and rear of the spring.

Hoja de muelle simétrica La hoja de muelle simétrica posee la misma distancia desde el centro del perno hasta la parte frontal y trasera del resorte.

Synthetic rubber A type of rubber developed in a laboratory.

Caucho sintético Tipo de caucho fabricado en un laboratorio.

Tapered roller bearing A bearing containing tapered roller bearings mounted between the inner and outer races.

Cojinete de rodillos cónicos Cojinete que contiene cojinetes de rodillos cónicos montados entre los anillos interiores y exteriores.

Taper-wire closed-end springs Coil spring ends that are tapered to a smaller diameter.

Muelles helicoidales de extremos cónicos Extremos de muelles helicoidales a los que se les da forma de cono para reducir su diámetro.

Telematics Using an in-vehicle phone for communication, with the vehicle driver regarding vehicle diagnostics, emissions, or traffic- and safety-related concerns.

Telemática Uso de un teléfono incorporado al vehículo para comunicarse con el conductor del vehículo acerca de problemas de diagnóstico, emisiones, tránsito y seguridad.

Temperature rating A tire rating that indicates the ability of the tire to withstand heat.

Clasificación de la temperatura Clasificación de un neumático que indica la capacidad del neumático de resistir el calor.

Tension loaded A ball joint mounted so the vehicle weight tends to pull the ball out of the joint.

Cargada de tensión Junta esférica montada de manera que el peso del vehículo tiende a remover la bola de la junta.

Throttle position sensor (TPS) Sends an analog voltage signal to the engine computer in relation to throttle opening.

Sensor de posición del acelerador (TPS, en inglés) Envía una señal de voltaje analógica a la computadora del motor en relación con la apertura del acelerador.

Thrust angle The angle between the thrust line and the vertical centerline of the vehicle.

Ángulo de empuje El ángulo entre la línea de empuje y el eje mediano vertical del vehículo.

Thrust bearing load This type of load may be called an axial load.

Carga del cojinete de empuje Este tipo de carga puede llamarse una carga.

Thrust line An imaginary line positioned at a 90° angle to the center of the rear axle and extending forward.

Directriz de presiones Una línea imaginaria en posición de un ángulo de 90° con relación al centro del eje trasero y que se extiende hacia enfrente.

Thrust line alignment A wheel alignment in which the thrust line is used as a reference for front wheel toe adjustment.

Alineación de la línea de empuje Alineación de la rueda en la que se utiliza la línea de empuje como punto de referencia para el ajuste del tope de la rueda delantera.

Tie rod A rod connected from the steering arm to the rack or center link, depending on the type of steering linkage.

Barra de acoplamiento Varilla conectada del brazo de dirección a la cremallera o a la biela motriz central, dependiendo del tipo de cuadrilátero de la dirección.

Tire belts Belts that are placed under the tire tread to provide longer tread wear. Belts are usually made from steel or polyester.

Banda de refuerzo del neumático Las bandas que se colocan debajo de la banda de rodamiento para aumentar la durabilidad de la banda de rodamiento. Las bandas suelen ser fabricadas del acero o del poliester.

Tire chains May be placed over the tires to improve traction when driving on ice or snow.

Cadenas antideslizantes Pueden colocarse sobre los neumáticos para mejorar la tracción cuando se conduce el vehículo en el hielo o la nieve.

Tire contact area The part of the tire in contact with the road surface when the tire is supporting the vehicle weight.

Área de contacto del neumático Parte del neumático en contacto con la superficie del camino cuando el neumático apoya el peso del vehículo.

Tire free diameter The distance between the outer edges of the tread measured on a horizontal line through the center of the wheel.

Diámetro libre del neumático Distancia entre los bordes exteriores de la huella que se mide sobre una línea horizontal a través del centro de la rueda.

Tire performance criteria (TPC) Information molded on the tire sidewall indicating that the tire meets the manufacturer's performance standards for traction, endurance, dimensions, noise, handling, and rolling resistance.

Criterio sobre el rendimiento del neumático Información moldeada en la pared lateral del neumático que indica que el neumático cumple con las normas de rendimiento establecidas por el fabricante sobre la tracción, acción, dimensiones, ruido, movilización, y resistencia al rodaje.

Tire placard Often attached to the rear face of the driver's door, the tire placard provides information regarding maximum vehicle load, tire size, and cold inflation pressure.

Cartel del neumático Comúnmente fijado a la cara posterior de la puerta del conductor, el cartel del neumático provee información referente a la carga máxima del vehículo, el tamaño del neumático, y la presión fría de inflación.

Tire pressure The amount of air pressure contained in the tire to allow the tire to carry a load.

Presión del neumático La cantidad del presión de aire contenido en el neumático para permitir que el neumático soporta una carga.

Tire rolling diameter The distance between the outer edges of the tread measured on a vertical line through the center of the wheel when the tire is supporting the vehicle weight.

Diámetro del rodaje del neumático Distancia entre los bordes exteriores de la huella que se mide sobre una línea vertical a través del centro de la rueda cuando el neumático apoya el peso del vehículo.

Tire treads The part of the tire in contact with road surface.

Bandas de rodamiento del neumático La parte del neumático que se pone en contacto con la superficie del camino.

Tire valves Mounted in the wheel rim, the tire valves allow air to enter or be exhausted from the tire.

Válvulas del neumático Estas válvulas están montadas en la llanta de la rueda y permiten la entrada o la salida de aire desde el neumático.

Toe-in A condition that is present when the distance between the front edges of the front or rear wheels is less than the distance between the rear edges of the wheels.

Convergencia Condición que ocurre cuando la distancia entre los bordes frontales de las ruedas delanteras o traseras es menor que la distancia entre los bordes traseros de las mismas.

Toe-out A condition that is present when the distance between the front edges of the front or rear wheels is more than the distance between the rear edges of the wheels.

Divergencia Condición que ocurre cuando la distancia entre los bordes delanteros de las ruedas delanteras o traseras es mayor que la distancia entre los bordes traseros de las mismas.

Toe-out on turns The steering angle of the wheel on the inside of a turn compared with the steering angle of the wheel on the outside of the turn.

Divergencia durante un viraje El ángulo de la dirección de la rueda en el interior de un viraje comparado con el ángulo de la dirección de la rueda en el exterior del viraje.

Toe plate A metal plate surrounding the steering column and attached to the vehicle floor.

Placa de pie Una placa de metal que rodea la columna de dirección y conectada al piso del vehículo.

Torque A twisting force.

Par de torsión Fuerza de torcimiento.

Torque steer The tendency of the steering on a front-wheel-drive vehicle with unequal drive axles to pull to one side during hard acceleration.

Dirección de torsión Tendencia de la dirección en un vehículo de tracción delantera con ejes de mando desiguales a desviarse hacia un lado durante una aceleración rápida.

Torsion bar A steel bar connected from the chassis to the lower control arm. As the vehicle weight pushes the chassis downward, the torsion bar twists to support this weight. Torsion bars are used in place of coil springs.

Barra de torsión Barra de acero conectada del chasis al brazo de mando inferior. Mientras el peso del vehículo presiona el chasis hacia abajo, la barra de torsión se tuerce para apoyar este peso. Las barras de torsión se utilizan en lugar de los muelles helicoidales.

Track bar Some rear suspension systems have a track bar connected from one side of the differential housing to the chassis to prevent lateral chassis movement.

Barra transversal Algunos sistemas de suspensión trasera poseen una barra transversal conectada de un lado de la caja del diferencial al chasis, para evitar el movimiento lateral del chasis.

Tracking Refers to the position of the rear wheels in relation to the front wheels.

Encarrilamiento Se refiere a la posición de las ruedas traseras con relación a las ruedas delanteras.

Traction control system (TCS) A computer-controlled system that prevents drive wheel spinning.

Sistema de control de tracción (TCS) Un sistema controlado por computador que previene que giran las ruedas de propulsión.

Traction rating A tire rating indicating the traction capabilities of the tire.

Clasificación de tracción Clasificación de un neumático que indica las capacidades de tracción del mismo.

Transitional coils Coils located between the inactive and active coils in variable rate coil springs.

Bobinas de transición Las bobinas localizadas entre las bobinas inactivas y activas en los resortes helicoidales de relación variable.

Transistors Fast-acting electronic switches with no moving parts.

Transistores Interuptores electrónicas de acción rápida sín partes móviles.

Travel-sensitive strut A strut with the capability to adjust its firmness in relation to the amount of piston travel inside the strut.

Montante sensible al movimiento Montante con la capacidad de ajustar su firmeza de acuerdo a la cantidad de movimiento del pistón dentro del montante.

Tread wear rating A tire rating indicating the wear capabilities of the tread that allow customers to compare tire life expectancy.

Clasificación del desgaste de un neumático Clasificación de un neumático que indica las capacidades de desgaste de la huella y le permite a los clientes comparar el índice de durabilidad del neumático.

Trim height The specified, or normal, suspension height in a computer-controlled suspension system.

Altura de la suspensión Altura especificada, o normal, de la suspensión en un sistema de suspensión controlado por computadora.

Tubular frame A frame member with a circular, or tubular, design.

Fabricación del armazón en forma tubular Pieza del armazón diseñada en forma circular o tubular.

Turning circle The turning angle of one front wheel in relation to the opposite front wheel during a turn.

Círculo de giro Ángulo de giro de una de las ruedas delanteras con relación a la rueda delantera opuesta durante un viraje.

Turning radius The turning angles of the front wheels around a common center point.

Diámetro del giro Los ángulos de giro de las ruedas frontales alrededor de un punto central común.

Uniform Tire Quality Grading (UTQG) System Information including tread wear, traction, and temperature ratings that is molded into the tire sidewall, and is required by the Department of Transportation (DOT).

Sistema de sobre la clasificación de la calidad uniforme de neumáticos Información moldeada en la pared lateral del neumático que incluye el desgaste de la huella, la tracción, y las clasificaciones de temperatura y que es requerida por el Ministerio de Transporte.

Unitized body A body design that does not have a frame because each body component is a supporting member.

Carrocería unificada Diseño de la carrocería que no tiene armazón porque cada uno de sus componentes es una pieza suplementaria.

Unsprung weight Is the vehicle weight that is not supported by the coil springs.

Peso no suspendido Es el peso del vehículo que no soportan los muelles en espiral cilíndrica.

Upper strut mount A mount connected between the strut and the strut tower. Front upper strut mounts contain a bearing that allows strut rotation.

Montaje de los montantes superiores Montaje conectado entre el montante y la torre del montante. Los montajes frontales y superiores contienen un cojinete que permite la rotación del montante.

Vacuum A pressure that is less than atmospheric pressure.

Vacío Presión menor que la de la atmósfera.

Vane-type power steering pumps Pumps that have rotors with metal vanes that provide a seal between the rotor and the pump cam.

Bombas de dirección hidráulica tipo aletas Las bombas que tienen rotores con aletas de metal que proveen un sello entre el rotor y la leva de la bomba.

Variable effort steering (VES) A computer-controlled steering system that provides reduced steering effort at low vehicle speeds and increased steering effort at higher vehicle speeds.

Dirección de esfuerzo variable Sistema de dirección controlado por computadora que provee un menor esfuerzo de dirección cuando el vehículo viaja a velocidad baja y un mayor esfuerzo de dirección cuando el vehículo viaja a gran velocidad.

Variable-rate coil springs Rather than having a standard spring deflection rate, these springs have an average spring rate based on load at a predetermined deflection.

Muelles helicodal de capacidad variable En vez de tener una capacidad de desviación de muelle estándar, estos muelles tienen un valor promedio de elasticidad basado en la carga a una desviación predeterminada.

V-belt A drive belt with a V-shape.

Correa en V Correa de transmisión en forma de V.

Vehicle dynamic suspension (VDS) System found in some SUVs that is similar to air suspension systems.

Suspensión dinámica del vehículo (SDV) Sistema que se encuentra en algunas camionetas SUV que es similar a los sistemas de suspensión de aire.

Vehicle speed sensor (VSS) An input sensor that sends a voltage signal to the engine computer in relation to vehicle speed.

Sensor de la velocidad del vehículo Sensor de entrada que le envía una señal de tensión a la computadora del motor referente a la velocidad del vehículo.

Vehicle stability control system A computer-controlled system that prevents vehicle swerving, especially during hard acceleration on slippery road surfaces.

Sistema de control de estabilidad del vehículo Un sistema controlado por computadora que previene que se desvía el vehículo, especialmente durante una aceleración fuerte sobre un camino cuyo superficie es resbalosa.

Vehicle wander The tendency of the steering to pull to the left or right when the vehicle is driven straight ahead on a smooth road surface.

Desviación de la marcha del vehículo Tendencia de la dirección a desviarse hacia la izquierda o hacia la derecha cuando se conduce el vehículo en línea recta en un camino cuya superficie es lisa.

Vent valve An electrically operated valve that vents air from an air spring in a computer-controlled suspension system.

Válvula de respiración Válvula activada eléctricamente que da salida al aire desde un muelle de aire en un sistema de suspensión controlado por computadora.

Venturi A narrow portion of an air passage.

Venturi Una porción estrecha de un pasaje de aire.

Vibration A rapid motion of particles, or a component that produces sound.

Vibración Un movimiento rápido de los partículos, o de un componente que produce un sonido.

Voltage (V) Voltage is a unit of measurement of electrical pressure.

Voltaje (V) El voltaje es una unidad de medida de la presión eléctrica.

Volume Volume is the length, width, and height of the space occupied by an object.

Volumen El volumen es la longitud, la anchura y la altura del espacio ocupado por un objeto.

Watts rod A rod connected from the chassis to the rear suspension to reduce body side sway, usually referred to as a tracking bar.

Barra wats Una barra conectada del chasis a la suspensión trasera para disminuir la oscilación lateral de la carrocería, suele referirse como una barra de tracción.

Wear indicators Rubber bars located near the bottom of tire treads. When the tread is worn to a specific depth, these bars become visible.

Indicadores del desgaste Las barras del caucho localizadas cerca la parte inferior de las bandas de rodamiento de los neumáticos. Cuando las bandas de rodamiento se han gastado a una profundidad específica, estas barras son visibles.

Weight The measurement of the earth's gravitational pull on an object.

Peso La medida de la atracción gravitacional de la tierra en un objeto.

Wheel alignment May be defined as an adjustment and refitting of suspension parts to original specifications that ensures design performance.

Alineación de una rueda Puede definirse como un ajuste y una reparación de las piezas de la suspensión según las especificaciones originales, lo que asegura el rendimiento del diseño.

Wheelbase The distance between the center of the front and rear wheels.

Distancia entre ejes Distancia entre el centro de las ruedas delanteras y traseras.

Wheel offset The distance between the vertical wheel-mounting surface and the centerline of the wheel.

Desfasaje de rueda La distancia entre la superficie vertical de montaje de ruedas y la línea central de la rueda.

Wheel position sensor Is connected between each front and rear lower control arm and the chassis in a road-sensing suspension system. This sensor provides a computer input signal in relation to the amount and velocity of wheel movement.

Sensor para la posición de las ruedas Se conecta entre cada uno de los brazos de mando delantero y trasero y el chasis en un sistema de suspensión con equipo sensor. Este sensor envía una señal de entrada a la computadora referente a la cantidad y la velocidad del movimiento de la rueda.

Wheel rims Circular steel, aluminum, or magnesium components on which the tires are mounted. The wheel rim and tire assembly is bolted to the wheel hub.

Llantas de la rueda Componentes circulares de acero, aluminio o magnesio sobre los que se montan los neumáticos. El conjunto de la llanta de la rueda y el neumático se emperna al cubo de la rueda.

Wheel sensor An electronic unit attached to each wheel and connected to a four-wheel aligner.

Sensor de la rueda Una unedad electrónica prendida a cada rueda y conectada a un alineador de cuatro ruedas.

Wheel setback Occurs when one front or rear wheel is moved rearward in relation to the opposite front or rear wheel.

Retroceso de la rueda Ocurre cuando una rueda delantera o trasera se mueve hacia atrás con relación a la rueda delantera o trasera opuesta.

Wheel shimmy Rapid inward and outward oscillations of the front wheels.

GLOSSARY
GLOSARIO

Bailoteo de la rueda Oscilaciones rápidas hacia adentro y hacia afuera de las ruedas delanteras.

Wheel speed sensor Sends an AC voltage signal to the antilock brake system computer in relation to wheel speed.

Sensor de velocidad de la rueda Manda una señal de voltaje de corriente alterna a la computadora del sistema de freno antibloqueo en relación a la velocidad de la rueda.

Wheel tramp Rapid upward and downward wheel and tire movement.

Recorrido de la rueda Movimiento rápido ascendente y descendente de la rueda y el neumático.

Work The result of applying a force.

Trabajo Resultado de la aplicación de una fuerza.

Worm and gear or worm and sector design A steering gear developed in the early 1900s that required high steering effort because of internal friction.

Diseño del sinfín y engranaje o del sinfín y sector Un engranaje de dirección desarrollado en los principios de los años 1900 que requerían un gran esfuerzo en maniobro de dirección debido a la fricción interna.

Worm and roller steering gear A gear that contains a wormshaft and a roller-type sector.

Engranaje de dirección con sinfín y rodillo Un engranaje que contiene un eje de sinfín y un sector tipo rodillo.

Worm shaft The gear meshed with the pitman shaft sector in a recirculating ball steering gear.

Árbol del sinfín Engranaje endentado con el sector del árbol pitman en un mecanismo de dirección de bola recirculante.

Worm shaft bearing preload The amount of tension placed on the bearing by the adjustment procedure.

Carga previa del cojinete del árbol del sinfin Cantidad de tensión aplicada al cojinete por el procedimiento de ajuste.

Worm shaft endplay The distance between the fully upward and fully downward worm shaft positions in a recirculating ball steering gear.

Holgadura del árbol del sinfín Distancia entre las posiciones completamente ascendente y descendente del árbol del sinfín en un mecanismo de dirección de bola recirculante.

Yaw forces Tend to cause the rear of the vehicle to swerve sideways.

Fuerzas de dirección tienden a causar que la parte posterior del vehículo viren bruscamente hacia los lados.

Yaw motion The tendency of the rear of a vehicle to swerve sideways during a turn.

Movimiento de derrape La tendencia de la parte posterior de un vehículo de desplazarse lateralmente durante un giro.

Yaw rate sensor Sends a voltage signal to the vehicle stability control computer in relation to sideways chassis movement.

Sensor de cantidad de desviación Manda una señal de voltaje a la computadora de control de estabilidad del vehículo en relación del movimiento lateral del chasis.

A

B

C

INDEX

D

E

INDEX

INDEX

INDEX

T

IT WAS A DARK AND STORMY NIGHT, SNOOPY

BY CHARLES M. SCHULZ

BALLANTINE BOOKS • NEW YORK

A Ballantine Book

Published by The Random House Publishing Group

www.ballantinebooks.com

www.snoopy.com

Library of Congress Control Number: 2003110572

ISBN 0-345-44272-5

Design by Diane Hobbing of Snap-Haus Graphics

Cover design by United Media

Manufactured in the United States of America

First Edition: March 2004

10 9 8 7 6 5 4 3 2

IT WAS A DARK AND STORMY NIGHT, SNOOPY

PEANUTS
by SCHULZ
C'MON, I'VE GOT THE BOARD..

WHAT ARE WE PLAYING? DO I GET FIRST MOVE?

BEFORE WE BEGIN, I'D LIKE TO SAY SOMETHING..

I HAVE A FEELING THAT I MAY BE A POOR LOSER..
1-1-95

I MEAN, IF I LOSE, DO I HAVE TO BE NICE ABOUT IT?

I MEAN, WHAT IF I THINK I'M GONNA WIN, BUT I DON'T, CAN I KICK THE BOARD OVER?!

IF YOU KICK THE BOARD OVER, I'M GOING TO PICK YOU UP, AND THROW YOU RIGHT OUT THE FRONT DOOR INTO A SNOWBANK!

NICE GAME!
SCHULZ
© 1994 United Feature Syndicate, Inc.

VACATION'S OVER, HUH, MA'AM? WE'RE ALL BACK.. IT'S GOOD TO SEE YOU AGAIN..
1/2

THAT ISN'T OUR REGULAR TEACHER, SIR.. IT'S A SUBSTITUTE...

SORRY, MA'AM..FROM WHERE I'M SITTING, ALL TEACHERS LOOK ALIKE..
SCHULZ
© 1995 United Feature Syndicate, Inc.

YOU'RE SURE SHE'S A SUBSTITUTE TEACHER, MARCIE?

OUR REGULAR TEACHER IS TALLER AND WEARS GLASSES
1-3

I APOLOGIZE AGAIN, MA'AM.. I DIDN'T REALIZE YOU WERE A SUBSTITUTE..
© 1995 United Feature Syndicate, Inc.

CAME RIGHT IN OFF THE BENCH, HUH?
SCHULZ

YES, MA'AM..WE'VE MISSED YOU AT SCHOOL..WE THOUGHT MAYBE YOU'VE BEEN SICK..WE BROUGHT YOU SOME FLOWERS..

WHO ARE WE? WE'RE A COUPLE OF YOUR STAR PUPILS..

CHECK THE ADDRESS AGAIN, MARCIE..
1-4

I KNEW IT.. WE HAVE THE WRONG HOUSE!
SCHULZ
© 1995 United Feature Syndicate, Inc.

SHE KEPT THE FLOWERS..

I KNEW WE WERE AT THE WRONG HOUSE..I KNEW THAT WASN'T OUR TEACHER..
THAT LADY KEPT THE FLOWERS..
1-5

OUR TEACHER IS TALLER AND WEARS GLASSES..
SHE KEPT THE FLOWERS..

IF YOU'RE SO WORRIED ABOUT THE FLOWERS, GO BACK AND GET THEM..
WHICH HOUSE WAS IT?
SCHULZ

OUR REGULAR TEACHER IS BACK, SIR..
Z

WELCOME BACK, MA'AM! IT'S GOOD TO SEE YOU AGAIN..
1-6

SCHULZ
SHE STEPPED OUT OF THE ROOM, SIR..THAT'S THE CUSTODIAN..

IT'S TIME TO WAKE UP MY BIG BROTHER..

HE TOLD ME NOT TO LET HIM OVERSLEEP..

IT'S HARD TO KNOW JUST HOW TO WAKE UP SOMEBODY
1-7

I NEVER KNOW QUITE HOW TO SAY IT...

WELCOME TO THE REAL WORLD!
SCHULZ

PEANUTS
by SCHULZ
ALMOST DONE?
NO, MA'AM.. MY DOG DIDN'T EAT MY HOMEWORK..
HE WROTE IT!
©1995 United Feature Syndicate, Inc.
1-8
SCHULZ

IT'S GOING TO BE COLD TONIGHT SO MAYBE YOU'D BETTER COME AND SLEEP INSIDE..

1-9

WHO ARE YOU TALKING TO?

WELL, WHAT SHALL WE READ TONIGHT?

"GONE WITH THE BEAGLE," "THE MALTESE BEAGLE," OR "SNOW WHITE AND THE SEVEN BEAGLES"?
1-10

WE NEVER GET PAST THE THIRD BEAGLE..
Z

1-11

I LEAD AN EXCITING LIFE..

ASK YOUR DOG IF HE WANTS TO COME OUT AND PLAY..

HE'S THINKING ABOUT IT..

HEADS, I GO OUT..TAILS I STAY IN...

TWO OUT OF THREE..HEADS, GO OUT..TAILS, STAY IN... ONE MORE TIME...OKAY, ONE MORE TIME...

ALL RIGHT, I'LL ASK HIM..

THAT LITTLE KID WANTS YOU TO COME OUT AND PLAY HOCKEY..

OKAY, WE'LL PLAY THREE TWENTY-MINUTE PERIODS, AND I GET TO DRIVE THE ZAMBONI!

HERE WE GO!

THE WORLD FAMOUS STOMACH SLIDE..

ALL RIGHT, WHO WANTED THE HOT WATER BOTTLE?

PEANUTS
by SCHULZ
TOYS

© 1995 United Feature Syndicate, Inc.

1-15

B
A
R
K

WOOF!
SCHULZ

1-16
DON'T PAY ANY ATTENTION TO THEM.. THEY'RE JUST BEING RUDE!
© 1995 United Feature Syndicate, Inc.
SCHULZ

WHY DO WE HAVE TO GO ON A FIELD TRIP?
1-17

SO WE CAN LEARN MORE ABOUT NATURE..

JUST WHAT I NEED.. A CHANCE TO STUDY RAIN UP CLOSE..
© 1995 United Feature Syndicate, Inc.
SCHULZ

WHEN I WRITE A PAPER, I TRY TO SHOW THE TEACHER THAT SHE HAS SOMEHOW ENRICHED MY EXPERIENCE..

This is my report on the dumb field trip we took yesterday.
1-18

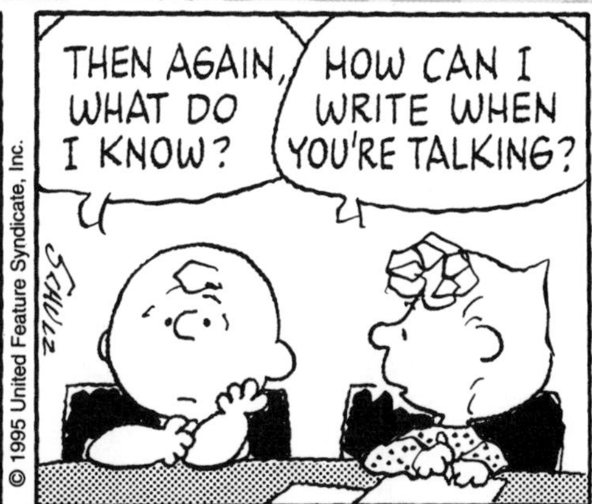
THEN AGAIN, WHAT DO I KNOW?
HOW CAN I WRITE WHEN YOU'RE TALKING?
© 1995 United Feature Syndicate, Inc.
SCHULZ

1-19
MY TEACHER'S MAD AT ME..

SHE'S MAD BECAUSE I CALLED OUR FIELD TRIP DUMB
WHAT ARE YOU GOING TO DO?
© 1995 United Feature Syndicate, Inc.

I'M TRYING TO FIGURE OUT HOW I CAN PUT THE BLAME ON YOU..

WE'RE GOING TO BE GONE FOR A WHILE SO WE WANT YOU TO GUARD THE HOUSE
© 1995 United Feature Syndicate, Inc.

1-20
NO, THIS HOUSE..

1-21 © 1995 United Feature Syndicate, Inc.
DON'T WORRY.. NO ONE WILL DARE TO GO NEAR OUR HOUSE..

PEANUTS
by SCHULZ
YOU KNOW WHAT MY GRAMPA SAYS?
HE SAYS THAT KIDS TODAY DON'T KNOW HOW TO HAVE FUN..
© 1995 United Feature Syndicate, Inc.
HE SAYS WE HAVE TOO MANY ORGANIZED SPORTS..
HE BELIEVES IN THE SIMPLE THINGS..
HE SAYS WE'RE MISSING OUT ON ALL THE SIMPLE THINGS..
DO YOU THINK MAYBE HE'S RIGHT?
I CAN'T HEAR YOU..YOU'VE GOT A SIMPLE BOX OVER YOUR HEAD..
1-22
SCHULZ

YOU SHOULD WRITE A NOVEL BASED ON MY LIFE..
1-23

THE MATERIAL IS THERE.. ALL YOU'D NEED IS A GOOD TITLE..
© 1995 United Feature Syndicate, Inc.

Dull City
SCHULZ

1-24
YOU COULD WRITE A WHOLE NOVEL ABOUT MY LIFE..

IT COULD START RIGHT FROM THE DAY I WAS BORN
© 1995 United Feature Syndicate, Inc.

Born Crabby
SCHULZ

YOU KNOW, THE MAIN CHARACTER IN YOUR NOVEL DOESN'T HAVE TO BE EXACTLY LIKE ME..

YOU CAN MAKE A FEW MINOR CHANGES IF YOU HAVE TO..
1-25
© 1995 United Feature Syndicate, Inc.

GOOD.. IN MY NOVEL YOU'RE A GOLDEN RETRIEVER..
SCHULZ

WHAT KIND OF A NOVEL DO YOU CALL THIS?

IT DOESN'T HAVE ANY PEOPLE IN IT! ALL IT HAS IS SIX DOGS!
© 1995 United Feature Syndicate, Inc.

I'LL ADD ANOTHER DOG..
SCHULZ
1-26

RING!

YOUR DOG IS A TERRIBLE WRITER!

© 1995 United Feature Syndicate, Inc.

WHY WOULD ANYONE COME TO A PERSON'S FRONT DOOR, AND TELL HIM HIS DOG IS A TERRIBLE WRITER?
THE NEIGHBORHOOD IS GOING DOWNHILL
SCHULZ
1-27

1-28
REGRETS.. I WONDER IF OTHER PEOPLE HAVE AS MANY REGRETS AS I HAVE..
I REGRET THE WAY I ACTED THAT NIGHT AT THE OBEDIENCE SCHOOL PROM.. SORT OF..
SCHULZ
© 1995 United Feature Syndicate, Inc.

PEANUTS
by SCHULZ

OVER HERE..

YES, SIR..WE'RE LOOKING FOR A BAG OF DOG FOOD..
TRASH
TV listin

"FOR PUPPIES".. NO, YOU'RE NOT A PUPPY ANYMORE..
CHOW
CHOW
VITAMIN PLUS
VITAMIN PLUS
ACTIVE
PUPPY
1-29

"LITE," "ACTIVE," "ADULT," "YOUNG ADULT."
CHOW

"MATURE DOGS," "OLD DOGS," "GOLDEN YEARS DOGS."
HERE WE ARE..
GRAVY

YES, SIR..THIS ONE..

I DIDN'T KNOW THEY HAD A BRAND FOR "DOGS WHO UNDERSTAND A LITTLE FRENCH AND FLEW DURING WORLD WAR I"
I'VE LEARNED TO BECOME A SMART SHOPPER..
SCHULZ

NO, WE WEREN'T REQUIRED TO READ A BOOK THIS MONTH..
1-30

WHAT?! YOU MEAN I READ A BOOK WHEN I DIDN'T HAVE TO?!!

YOU MEAN I READ IT FOR NOTHING?!

MARCIE, WHAT AM I GOING TO DO?
IS THERE ANYTHING I CAN TAKE?

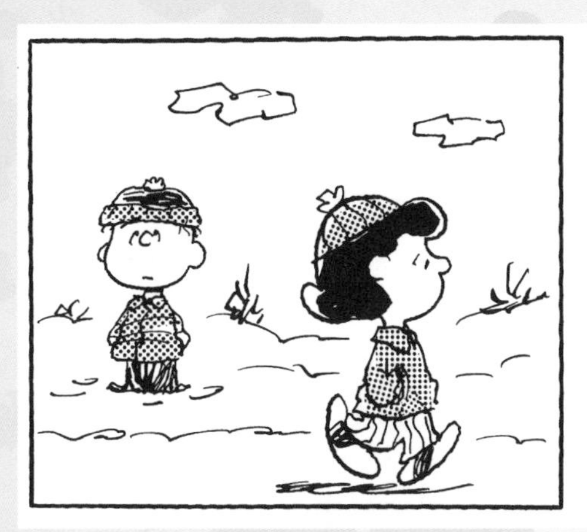

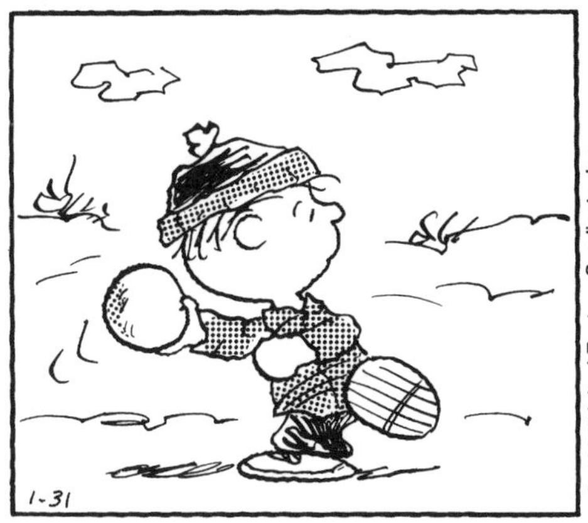
1-31

THROW THAT SNOWBALL AT ME, YOU BLOCKHEAD, AND YOU'LL REGRET IT FOR THE REST OF YOUR LIFE!

YOU'LL LIE AWAKE AT NIGHT, AND YOU'LL ASK YOURSELF OVER AND OVER, "WHY DID I DO IT?"

BUT MAYBE I'LL ASK MYSELF, "WHY DIDN'T YOU DO IT?"

BECAUSE SHE'D PROBABLY TURN AROUND, AND KICK ME INTO A SNOWBANK!
YOU WERE LUCKY!!
2-1

DON'T MELT MY SNOWMAN!

DON'T! DON'T! PLEASE, DON'T!
© 1995 United Feature Syndicate, Inc.

STOP! PLEASE, STOP!! STOP!!

RATS! YOU CAN TALK TO THE MOON, BUT THE SUN WON'T LISTEN..
SCHULZ
2-2

MY SNOWMAN MELTED.. I'LL NEVER SEE HIM AGAIN..
2-3

YOU MUST HAVE BEEN VERY CLOSE..
© 1995 United Feature Syndicate, Inc.

I WAS WITH HIM WHEN HE MELTED..
SCHULZ

YOU'RE STANDING RIGHT WHERE MY SNOWMAN WAS STANDING WHEN HE MELTED..
SPOOKY..
2-4 © 1995 United Feature Syndicate, Inc.
SCHULZ

PEANUTS
by SCHULZ
FREE
GO
BACK

ARE WE PLAYING WINTER RULES?
2-5

WINTER RULES? WHY WOULD WE BE PLAYING WINTER RULES?

BECAUSE IT'S SNOWING OUTSIDE..WHEN THE WEATHER IS BAD, GOLFERS PLAY WINTER RULES..

WE'RE INSIDE THE HOUSE, YOU BLOCKHEAD!
HOW ABOUT A HANDICAP? **DO** I GET A HANDICAP?
© 1995 United Feature Syndicate, Inc.

NO, YOU DON'T GET A HANDICAP..
IN HORSE RACING AND GOLF THEY HAVE HANDICAPS..

FORGET IT!
IF I LOSE, CAN I HIT YOU OVER THE HEAD WITH THE EMPTY BOX?

TRY THAT JUST ONCE, AND I'LL THROW YOU OUT THE FRONT DOOR!

SEE? WE SHOULD HAVE PLAYED WINTER RULES!
SCHULZ

NO ONE EVER ASKS ME WHAT I THINK ABOUT SOMETHING..

WELL, WHAT DO YOU THINK ABOUT SOMETHING?
© 1995 United Feature Syndicate, Inc.

WHO KNOWS?
2-6

PSYCHIATRIC HELP 5¢
I DON'T KNOW.. I JUST FEEL LONELY A LOT OF THE TIME..
THE DOCTOR IS IN

WHY DON'T YOU TRY TAKING DANCE LESSONS?
DANCE LESSONS?
THE DOCTOR

WHAT IF NO ONE WILL DANCE WITH ME?
2-7
© 1995 United Feature Syndicate, Inc.

THEN YOU'LL BE A LONELY PERSON WHO KNOWS HOW TO DANCE
THE DOCTOR IS IN

YES MA'AM.. I WAS TOLD TO SIGN UP FOR DANCE LESSONS BECAUSE I NEED TO BE MORE OUTGOING..
Dance Studio

2-8
MAYBE IF I'M MORE OUTGOING, I WON'T BE SO LONELY..
© 1995 United Feature Syndicate, Inc.

YES, THAT'S MY DOG... HE'S VERY OUTGOING..

WHAT ARE YOU DOING HERE, KID?
I'M TAKING DANCE LESSONS SO I CAN LEARN TO BE MORE OUTGOING

MAYBE IF I'M MORE OUTGOING, I WON'T BE SO LONELY..
2-9

ASK ME TO DANCE, AND I'LL OUTGO YOU RIGHT OVER YOUR HEAD!
© 1995 United Feature Syndicate, Inc.

ACTUALLY, BEING LONELY ISN'T ALL THAT BAD..
SCHULZ

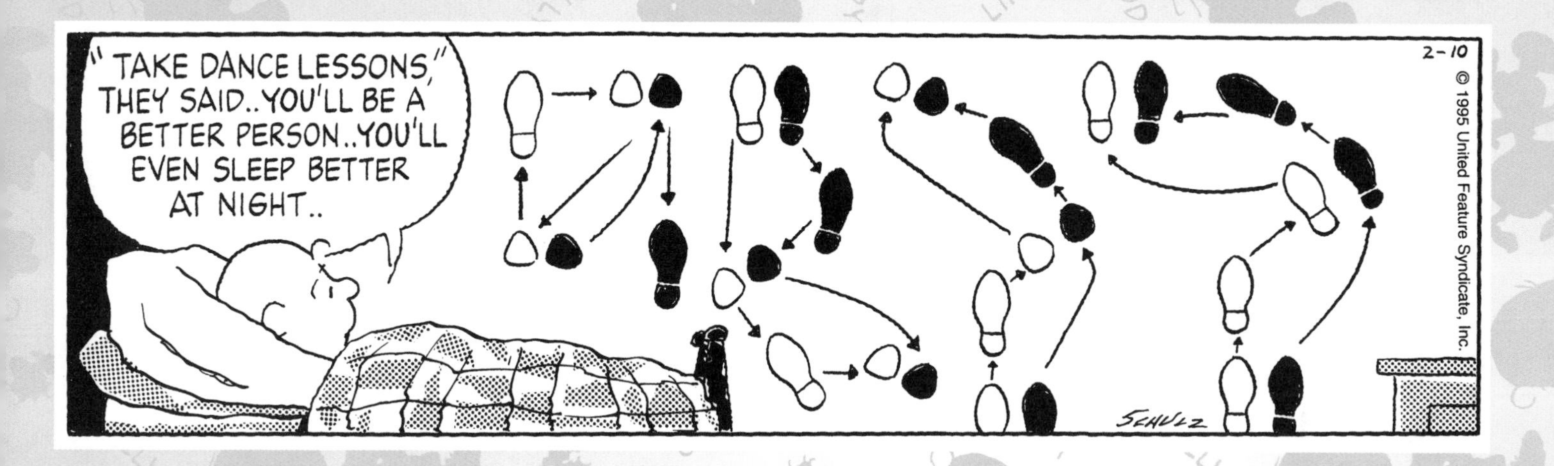

"TAKE DANCE LESSONS," THEY SAID..YOU'LL BE A BETTER PERSON..YOU'LL EVEN SLEEP BETTER AT NIGHT..
2-10
© 1995 United Feature Syndicate, Inc.
SCHULZ

HI! MY NAME IS EMILY..WOULD YOU LIKE TO BE MY DANCE PARTNER?

WELL, SURE.. BUT I'M NOT VERY GOOD..
2-11

IT'LL BE EASY.. "SLOW, SLOW, QUICK, QUICK"
© 1995 United Feature Syndicate, Inc.

HOW ABOUT, "SLOW, SLOW, PANIC, PANIC"?
SCHULZ

PEANUTS

by SCHULZ

love, Sally

I'M GOING OUT TO DELIVER MY VALENTINES..

HI, GIVE THIS TO YOUR BROTHER, WILL YOU?
"TO MY SWEET BABOON"

NOT "BABOON"! "BABBOO"!
WHAT'S A "BABBOO"?

A TERM OF ENDEARMENT.. GIVE IT TO HIM..

I'M NOT HER "SWEET BABBOO"!
© 1995 United Feature Syndicate, Inc

HE SAYS HE'S NOT YOUR "SWEET BABBOO"!

OF COURSE, HE IS!! WHAT'S THE MATTER WITH HIM?!!
2-12

GIMMEE THAT!

WELL, DID YOU DELIVER ALL YOUR VALENTINES?
I ONLY HAD ONE, AND I GAVE IT TO YOUR DOG!

WHAT'S A "BABBOO"?
SCHULZ

I'M TAKING THIS BALLROOM DANCE CLASS, SEE, AND THIS GIRL COMES UP TO ME, AND ASKS ME TO DANCE..

SHE ASKED ME! CAN YOU BELIEVE IT? ME! SHE ASKED ME!
2-13

THE TEACHER IS TALKING TO YOU, CHARLIE BROWN..
ME?
© 1995 United Feature Syndicate, Inc.
SCHULZ

SOMETIMES IT'S NICE TO LIE IN BED, AND THINK OF SOMETHING WONDERFUL THAT'S HAPPENED TO YOU..
HI! MY NAME IS EMILY..WOULD YOU LIKE TO BE MY DANCE PARTNER?
HI! MY NAME IS EMILY..WOULD YOU LIKE TO BE MY DANCE PARTNER?
HI! MY NAME IS EMILY..WOULD YOU LIKE TO BE MY DANCE PARTNER?
HI! MY NAME IS EMILY..WOULD YOU LIKE TO BE MY DANCE PARTNER?
2-14
© 1995 United Feature Syndicate, Inc.

WELL, I'M OFF TO MY DANCE CLASS..EMILY IS PROBABLY WAITING TO DANCE WITH ME AGAIN..

HERE I COME, SWEET EMILY! HERE COMES YOUR DANCE PARTNER!

WHIRL AROUND THE FLOOR WITH ME, EMILY..

"SOME ENCHANTED EVENING.."
© 1995 United Feature Syndicate, Inc.
2/15

BONK!
SCHULZ

2-16
EMILY?
MA'AM, WHERE IS EMILY?
Dance Studio

WHAT DO YOU MEAN THERE'S NO EMILY? SHE WAS MY DANCE PARTNER!

SHE HAD LONG HAIR AND SOFT HANDS..

WE DANCED, AND DANCED AND DANCED..

IT WAS AN ENCHANTED EARLY AFTERNOON
SCHULZ

EMILY! IT'S SO NICE TO BE DANCING WITH YOU AGAIN!
2-17

JUST TO SEE YOU, AND TO HOLD YOU, AND..

MA'AM? WHO AM I DANCING WITH?

WHO AM I TALKING TO? WHO...

OH, GOOD GRIEF!
SCHULZ

I CAN'T BELIEVE IT! I'M POSITIVE THERE WAS A GIRL NAMED EMILY, AND THAT SHE ASKED ME TO DANCE..

YOU'VE FOOLED YOURSELF BEFORE, CHARLIE BROWN, BUT YOU'RE NOT THE ONLY ONE, YOU KNOW..LOTS OF PEOPLE FOOL THEMSELVES..
2-18

I SHOULD HAVE DANCED WITH THEM..
SCHULZ

PEANUTS
by SCHULZ
COULD IT BE?

COULD IT REALLY BE?!

THE MALTESE FALCON!

I'VE FOUND THE MALTESE FALCON!

2-19
A MILLION DOLLARS!

"THE STUFF THAT DREAMS ARE MADE OF.."

THE BLACK BIRD! THE MALTESE FALCON..

OR DAFFY DUCK!
SCHULZ

PSYCHIATRIC HEL
5¢
YOU SAID IF I TOOK DANCE LESSONS, I WOULDN'T BE LONELY ANYMORE
THE DOCTOR IS IN
2-20

YOU WERE SUPPOSED TO DANCE WITH REAL GIRLS!

I STILL THINK EMILY WAS REAL... I DON'T KNOW WHAT HAPPENED..
THE DOCTOR IS IN

I'D TELL YOU, BUT YOU'VE USED UP YOUR FIVE CENTS..
SCHULZ

ASK YOUR DOG TO COME OUT AND PLAY..
SORRY.. HE'S BUSY..

2-21
BUSY?

GUARDING THE BEANBAG IS BUSY..

MY GRAMMA SAYS SHE ALWAYS BROUGHT AN APPLE TO SCHOOL FOR HER TEACHER..

I'M WONDERING IF MAYBE I SHOULD DO THAT..

BRING HER SIX OR SEVEN, MARCIE, AND SHE CAN MAKE A PIE!

HA HA HA HA HA HA!
2-22

NOTHING, MA'AM.. JUST IDLE CHATTER..

WHEN I FIXED YOUR SUPPER TONIGHT, I TRIED SOMETHING NEW

INSTEAD OF STIRRING IT CLOCKWISE, I STIRRED IT COUNTERCLOCKWISE..
© 1995 United Feature Syndicate, Inc.

I COULD TELL THE DIFFERENCE RIGHT AWAY..
SCHULZ
2-23

2-24

© 1995 United Feature Syndicate, Inc.

YOU LIKE THE NEW TRAMPOLINE?
SCHULZ

2-25
DO YOU THINK DOGS CHANGE AS THEY GET OLDER LIKE PEOPLE?
I'M SURE THEY DO..
I CAN'T READ THE FINE PRINT AS WELL ANYMORE
© 1995 United Feature Syndicate, Inc.
SCHULZ

PEANUTS
by SCHULZ

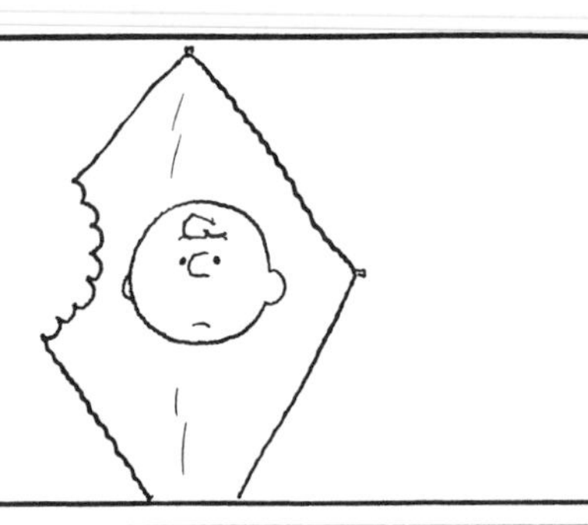

THANK YOU, CHARLIE BROWN, FOR LETTING ME WATCH YOU FLY YOUR KITE..

I'M GLAD TO HAVE YOU ALONG, RERUN.. IT'S GOING TO BE FUN..

2-26

I JUST SAW THE WORST THING I'VE EVER SEEN IN MY LIFE!

CHARLIE BROWN'S KITE GOT CAUGHT IN A TREE, AND THE TREE ATE HIS KITE! IT WAS TERRIBLE!!

MAYBE IT'S WRONG TO BRING A KITE INTO THE WORLD THE WAY THINGS ARE TODAY..
SCHULZ

THIS IS CALLED A LEASH..
2-27

WHEN THE MASTER AND HIS DOG GO FOR A WALK, THE LEASH IS ATTACHED TO THE DOG'S COLLAR..

OKAY, LET'S REVIEW WHAT WE HAVE JUST LEARNED..

WHAT IS THIS CALLED AND HOW IS IT USED?
I DON'T HAVE THE SLIGHTEST IDEA..
SCHULZ

THE TEACHER SAYS WE'RE SUPPOSED TO READ A BIOGRAPHY NEXT WEEK..

WHOSE BIOGRAPHY ARE YOU GOING TO READ, SIR?
2/28

I DON'T KNOW..SOMEBODY WHO DIDN'T LIVE VERY LONG
SCHULZ

THOSE YOUNG TUMBLEWEEDS ARE HARD TO BREAK..
3-1
SCHULZ

© 1995 United Feature Syndicate, Inc.
SCHULZ 3-2
"PLEASE, SIR," REPLIED OLIVER, "I WANT SOME MORE."
HE COULD SAY THAT.. HE DIDN'T HAVE A DISH IN HIS MOUTH..

NOW, LOOK HERE..JUST BECAUSE YOU'RE MY OLDER SISTER, DOESN'T MEAN THAT...

3-3
YES, IT DOES!
© 1995 United Feature Syndicate, Inc.

I GUESS IT DOES..
SCHULZ

"LITTLE WOMEN"

3-4
THIS IS THE KIND OF BOOK YOU SHOULD TRY TO WRITE..
© 1995 United Feature Syndicate, Inc.

"Little Beagles"
SCHULZ

PEANUTS
by SCHULZ

ASK YOUR DOG IF HE WANTS TO COME OUT AND PLAY MARBLES..

FETCH!
© 1995 United Feature Syndicate, Inc.

3-5

DOGS DON'T PLAY MARBLES..
SCHULZ

ASK YOUR DOG IF HE WANTS TO COME OUT AND PLAY CARDS..

3-6

ONLY IF HE CAN BRING HIS OWN DECK..
SCHULZ

THIS IS MY MARBLE COLLECTION.. I HAVE AGGIES, SHOOTERS, IMMIES, MILKIES, BUMBOOZERS, DOBIES AND GLIMMERS..
3-7

WHAT KIND DO YOU HAVE?
ROUND ONES..
SCHULZ

THIS IS HOW WE PLAY MARBLES, RERUN.. FIRST, WE DRAW A BIG CIRCLE OR RING...

THEN WE EACH PUT SOME MARBLES IN THE RING..

NOW, BECAUSE YOU'RE A BEGINNER, WE WON'T PLAY FOR KEEPS.. WE'LL JUST PLAY FOR FUN..

DO YOU UNDERSTAND WHAT THAT MEANS?
IF I WIN, IT'S FUN.. IF I DON'T WIN, IT'S NO FUN..
SCHULZ
3-8

PUT THE MARBLE IN YOUR FINGERS LIKE THIS, AND THEN FLIP IT WITH YOUR THUMB..

3-9
RATS!

WHY IS EVERYTHING IN LIFE SO HARD?
SCHULZ

I'VE BECOME A MARBLE PLAYING FANATIC!
3-10

THIS SUMMER I'M GOING TO ENTER THE "ROLLEY HOLE" CHAMPIONSHIP IN "STANDING STONE STATE PARK" IN TENNESSEE..
© 1995 United Feature Syndicate, Inc.

WHEREVER THAT IS..
IT'S NOT ACROSS THE STREET..
SCHULZ

WHY DOES A PERSON HAVE TO LIE AWAKE ALL NIGHT WORRYING ABOUT EVERYTHING?
3-11
© 1995 United Feature Syndicate, Inc.

WHY SHOULD A PERSON BE BURDENED WITH ALL THE CARES OF THE WORLD?
TRY SLEEPING
SCHULZ

PEANUTS
by SCHULZ

WHAT'S THIS?
IT'S CALLED "SARCASTIC REMARKS FOR MANAGERS"

BONK!
© 1995 United Feature Syndicate, Inc.

3-12
AH! HERE'S A GOOD ONE..

"MAYBE YOU CAN SUE THE BALL FOR HEAD HARASSMENT!"

HA HA HA HA HA HA!

HERE'S ANOTHER GOOD ONE..
EVERYBODY HATES OUTFIELDERS
SCHULZ

SEE THAT HOUSE WE'RE COMING TO? THEY SAY IT'S A HAUNTED HOUSE..

3-13
I DON'T BELIEVE IN HAUNTED HOUSES..
© 1995 United Feature Syndicate, Inc.

BUT I'LL WALK ON THIS SIDE OF YOU..
SCHULZ

© 1995 United Feature Syndicate, Inc.
3-14
SO I HAD TO TELL THE TEACHER THAT I JUST DIDN'T KNOW..
MAYBE SOME QUESTIONS DON'T HAVE AN ANSWER
LIKE WHAT?
LIKE DID JESUS EVER OWN A DOG?
SCHULZ

CHIRP?
3-15

WOOF!
© 1995 United Feature Syndicate, Inc.

CHIRP!
SCHULZ

3-16
© 1995 United Feature Syndicate, Inc.
IF THAT DOG IS A BOTHER, WHY DON'T YOU SEND HIM HOME?
IF I JUMP DURING A BAD DREAM, HE KEEPS ME FROM JUMPING TOO HIGH..
SCHULZ

THIS CONCLUDES MY REPORT ON THE SUN..

YOU'RE NEXT, SIR..

© 1995 United Feature Syndicate, Inc.
THIS IS MY REPORT ON THE WIND..
SCHULZ
3-17

3-18

IT'S RAINING HERE, TOO, HUH?
© 1995 United Feature Syndicate, Inc.

SCHULZ

PEANUTS
by SCHULZ

© 1995 United Feature Syndicate, Inc.
3-19

I SUPPOSE THIS IS WHY TUMBLEWEED BALL NEVER BECAME VERY POPULAR..
SCHULZ

ASK YOUR DOG IF HE WANTS TO COME OUT AND PLAY..

HE SAID, "NO"

HOW COULD HE SAY, "NO"? DOGS CAN'T TALK..
HE WHISPERED
© 1995 United Feature Syndicate, Inc.

DOGS CAN WHISPER?
SCHULZ
3-20

YOU HAVE THIS ALL WRONG, SIR..

THE TEACHER SAID THIS WAS A "TRUE OR FALSE" TEST..
© 1995 United Feature Syndicate, Inc.

NO WONDER.. I THOUGHT SHE SAID, "TRICK OR TREAT"
3-21
SCHULZ

"ALL IS WELL".. THAT'S MY NEW PHILOSOPHY

I LIKE IT.. GOOD FOR YOU

3-22
© 1995 United Feature Syndicate, Inc.

I DON'T KNOW WHAT IT MEANS
SCHULZ

"ALL IS WELL".. THAT'S MY NEW PHILOSOPHY..
3-23

BEING A DOG, I DON'T SUPPOSE YOU HAVE A PHILOSOPHY, DO YOU?

SCHULZ

DON'T TELL ME YOU'RE GOING OUT TO PLAY BASEBALL AGAIN..

YOU'RE GONNA LOSE! YOU ALWAYS LOSE!

I'M YOUR SISTER, AND I KNOW!

FAMILY VALUES
SCHULZ
3-24

HEY, PITCHER..TELL YOUR CATCHER I THINK HE'S KIND OF CUTE..

TELL YOUR RIGHT FIELDER I THINK SHE'S INSANE..
3-25

SHE SAYS SHE THINKS YOU'RE KIND OF CUTE, AND HE SAYS HE THINKS YOU'RE INSANE..

I'M NOT A PITCHER, I'M A MESSAGE CENTER
SCHULZ

PEANUTS
by SCHULZ
HI! IT'S ME AGAIN!
WHAT CAN I DO FOR YOU, RERUN?
ASK YOUR DOG IF HE WANTS TO COME OUT AND PLAY..
TELL HIM WE'LL RUN AROUND IN CIRCLES, AND ROLL ON THE GROUND AND ACT SILLY..
AND THEN, AFTERWARDS, I'LL GIVE HIM A COOKIE..
HA HA HA HA!
HEE HEE HEE!
HA HA HA HA!!
© 1995 United Feature Syndicate, Inc.
3-26
WHAT HAPPENED? I THOUGHT YOU WERE HAVING FUN..
THE COOKIE HAD COCONUT ON IT!

EIGHT ROCK IN THE CORNER!
3-27
© 1995 United Feature Syndicate, Inc.

YOU KNOW WHAT I'D DO IF I WERE YOU?

WELL, I'LL TELL YOU.. BUT FIRST, TELL ME WHAT YOU WERE GOING TO DO BEFORE I ASKED YOU IF YOU KNEW WHAT I'D DO IF I WERE YOU..
3-28
© 1995 United Feature Syndicate, Inc.

WHY ARE YOU SO HARD TO TALK TO?

THIS IS SUPPOSED TO BE A GOOD MOVIE..
IT'S ABOUT A DOG WHO GETS LOST
AND HE HAS TO CROSS A MOUNTAIN TO GET HOME..
AND HE ALSO RESCUES A CAT..
HE WHAT?
3/29
© 1995 United Feature Syndicate, Inc.

IT'S TOO BAD YOU DIDN'T LIVE DURING WORLD WAR I ..
3-30

YOU COULD HAVE BEEN A CARRIER PIGEON..YOU COULD HAVE DELIVERED IMPORTANT MESSAGES THROUGH ENEMY FIRE!

© 1995 United Feature Syndicate, Inc.

?
OH, I DON'T KNOW.. LIKE, "WHEN IS LUNCH?"
SCHULZ

YES, MA'AM..REQUEST PERMISSION TO USE THE PENCIL SHARPENER...

REQUEST PERMISSION TO BORROW A PENCIL..

REQUEST PERMISSION TO BORROW SOME NOTEBOOK PAPER..
© 1995 United Feature Syndicate, Inc.

REQUEST PERMISSION TO GO BACK TO KINDERGARTEN AND START OVER..
3-31
SCHULZ

Gentlemen, I have decided to become a shepherd.
4-1

Please send me a dozen sheep.

And a book of directions.
SCHULZ
© 1995 United Feature Syndicate, Inc.

PEANUTS
by SCHULZ
DO YOU KNOW ANYTHING ABOUT GHOSTS?
I'VE BEEN WORRIED LATELY THAT MY HOUSE IS HAUNTED..
DO YOU THINK MAYBE YOU COULD GO IN AND FIND OUT?
YOU CHASED THEM ALL OUT? GOOD FOR YOU
© 1995 United Feature Syndicate, Inc.
I'D NEVER BE ABLE TO SLEEP AGAIN IF I KNEW MY HOUSE WAS HAUNTED
SCHULZ
4-2

RERUN, WHAT'S THE MATTER?
SOME KID WON ALL MY MARBLES..

YOU'RE JUST A BEGINNER..YOU SHOULDN'T HAVE BEEN PLAYING FOR "KEEPS"
I KNOW.. YOU TOLD ME THAT

HE EVEN WON MY SHOOTER.. ALL I HAVE LEFT IS THIS EMPTY SACK..

I'M SURPRISED HE DIDN'T TAKE THAT, TOO..
HE COULDN'T.. I RAN ALL THE WAY HOME..
SCHULZ
4-3

SO THIS KID WON ALL MY MARBLES..I DIDN'T KNOW WE WERE PLAYING FOR "KEEPS"..

AM I TOO TRUSTING, CHARLIE BROWN?

NO, THERE'S NOTHING WRONG WITH BEING TRUSTING.. YOU'RE JUST YOUNG..

I'M WORKING ON THAT
SCHULZ
4-4

THERE HE IS! THAT'S THE KID WHO WON ALL MY MARBLES..

WELL, I'LL GO OVER AND TALK TO HIM..
WHY DON'T YOU HIT HIM WITH A STICK?
4-5

I JUST WANT TO TELL HIM I THINK HE TOOK ADVANTAGE OF YOU..
IF YOU SNEAKED UP BEHIND HIM, YOU COULD HIT HIM WITH A STICK..
SCHULZ

I'VE NEVER BEEN SO MAD IN ALL MY LIFE!
WHAT HAPPENED?

THAT KID OVER THERE! HE WON ALL MY MARBLES!! HE NEVER TOLD ME WE WERE PLAYING FOR "KEEPS"!
4/6

WHY DON'T YOU SNEAK UP BEHIND HIM, AND HIT HIM WITH A STICK?
SCHULZ

HI, KID.. YOU WANNA PLAY MARBLES? C'MON, I'LL SHOW YOU HOW..

IT'S EASY.. WE'LL JUST PLAY FOR FUN..
NO, WE'LL PLAY FOR "KEEPS"
4-7

FOR "KEEPS"? YOU MEAN YOU'VE PLAYED BEFORE?

LET'S SAY I DIDN'T JUST ARRIVE ON THE STAGECOACH..
SCHULZ

YOU SURE YOU WANNA PLAY ME, KID? I'M "JOE AGATE"! I NEVER LOSE!
YOU NEVER LOSE BECAUSE YOU ONLY PLAY BEGINNERS, AND YOU TELL THEM YOU'RE PLAYING FOR FUN UNTIL YOU WIN, AND THEN YOU SAY, "KEEPS!" AND YOU TAKE ALL THEIR MARBLES!!
4-8

YOU TRASH TALKIN' ME, KID?
"KNUCKLE DOWN," JOE! THIS IS FOR "KEEPS"!
SCHULZ

PEANUTS
I COULD BE PITCHER..I COULD BE CATCHER..I COULD BE ANYTHING..

I WANNA PLAY ON YOUR TEAM, CHARLIE BROWN

I DON'T KNOW, RERUN.. YOU'RE KIND OF SMALL

WHAT ABOUT YOUR SHORTSTOP? HE'S NOT ONLY SMALL, HE'S ALSO A DOG!!

WHY SHOULD A DOG BE PLAYING INSTEAD OF A HUMAN BEING?

WELL, GO ASK HIM..
I WILL! I'LL ASK HIM, AND I'LL TELL HIM!
4-9

DOGS CAN'T PLAY BASEBALL! I'M A HUMAN BEING, AND I'M THE ONE WHO SHOULD BE PLAYING!

© 1995 United Feature Syndicate, Inc.
SO, WHAT DO YOU THINK?

ANY OTHER POSITIONS OPEN?

WHAT'S GOING ON?
CHARLIE BROWN IS PLAYING "JOE AGATE" A GAME OF MARBLES..

THEY SAY "JOE AGATE" IS THE BEST PLAYER THIS SIDE OF THE MISSISSIPPI..
4-10

WHICH SIDE OF THE MISSISSIPPI ARE WE ON?
I DON'T HAVE THE SLIGHTEST IDEA..
SCHULZ

4-11
PLINK!

PLINK

PLINK!

PLINK!

PLINK!

WHO ARE YOU, KID? WHERE'D YOU LEARN TO SHOOT LIKE THAT?
"COOL THUMB" BROWN..JUST A STRANGER PASSING THROUGH..
SCHULZ

HERE YOU GO, RERUN..I WON BACK ALL YOUR MARBLES..
WOW! I CAN'T BELIEVE IT! YOU BEAT "JOE AGATE"!

WILL YOU TEACH ME TO SHOOT LIKE THAT?

NO, RERUN.. I HAVE TO MOVE ON..
4/12

SHANE! SHANE! COME BACK!
I CAN'T STAND IT..
SCHULZ

WHAT HAPPENED TO ALL MY GOLF TEES?
SCHULZ
4-13 © 1995 United Feature Syndicate, Inc.

IT SAYS HERE THAT BIRDS HAVE LARGER BRAINS THAN FISH..
4-14

THAT SHOULD MAKE YOU FEEL GOOD..

NOW YOU KNOW YOU'RE SMARTER THAN A FISH..

© 1995 United Feature Syndicate, Inc.

HOW ARE YOU AT THINKING UNDER WATER?
SCHULZ

SOMETIMES I LIE AWAKE AT NIGHT, AND I THINK, "MAYBE I CAN CHANGE MY LIFE AROUND"
4-15
© 1995 United Feature Syndicate, Inc.

THEN A VOICE COMES TO ME OUT OF THE DARK, "SURE, MAKE A LOT OF PAPER-WORK FOR THE REST OF US.."
SCHULZ

PEANUTS
by SCHULZ
"CHAPTER TWO"

"WHICH WAY SHOULD HE RUN?"

"THE FOREST WAS DARK"

"DARKER THAN ANYTHING HE HAD EVER SEEN"

© 1995 United Feature Syndicate, Inc.
"THEN HE HEARD A STRANGE NOISE"

"SUDDENLY.."
4-16

SUDDENLY WHAT?!

SUDDENLY THE PERSON READING THE BOOK COULDN'T SEE WHAT HE WAS READING!

SORRY.. I WAS WORRIED ABOUT WHAT WAS GOING TO HAPPEN TO THE BUNNY..
SCHULZ

ASK MY SWEET BABBOO IF HE'D LIKE TO DANCE WITH ME WHILE WE'RE WAITING FOR THE BUS..
I'M NOT HER SWEET BABBOO, AND I WOULDN'T DANCE WITH HER FOR A MILLION DOLLARS!
SOME GUYS JUST HATE TO DANCE..
SOME GUYS JUST LOVE IT..
PUT A RECORD ON, SWEETIE..I'M READY!
U.S. MAIL
4-17
SCHULZ
© 1995 United Feature Syndicate, Inc.

I WAS GOING TO RUN FOR CLASS PRESIDENT, BUT I CHANGED MY MIND..

MAYBE I'LL RUN FOR HALF PRESIDENT
VICE PRESIDENT
© 1995 United Feature Syndicate, Inc.
4-18

WHATEVER
SCHULZ

SLIGHT PROBLEM HERE, MARCIE..

I'VE BEEN WORKING ON THIS GREAT REPORT... WRITING PAGE AFTER PAGE AFTER PAGE..
4-19

SUDDENLY IT HAPPENED..
© 1995 United Feature Syndicate, Inc.

I FORGOT WHO I WAS WRITING ABOUT
SCHULZ

GUESS WHAT.. OUR CLASS HAD A MEETING, AND I'VE BEEN MADE PROGRAM CHAIRMAN..

I GET TO SPEND ALL DAY WATCHING PROGRAMS..

THE PROGRAM CHAIRMAN HAS TO ORGANIZE ALL THE ACTIVITIES AND ENTERTAINMENT FOR EVERY MEETING
4/20
© 1995 United Feature Syndicate, Inc.

I RESIGN!
SCHULZ

She had always been kind.

Sometimes, however, she wondered if she was appreciated.

"Even so," she thought, "I shall always smile and be kind."
4-21
© 1995 United Feature Syndicate, Inc.

Once a Golden Retriever, always a Golden Retriever.
SCHULZ

LIVING IN THE DESERT HAS MANY ADVANTAGES..
© 1995 United Feature Syndicate, Inc.

FOR ONE THING, THERE'S ALWAYS A PLACE TO HANG YOUR TEA BAG..
4-22
SCHULZ

PEANUTS
by SCHULZ

JACKS!

JACK'S WHAT?

THESE ARE JACKS, SEE?

LET'S SAY I'VE TOSSED THE JACKS OUT IN FRONT OF ME..

NOW, I BOUNCE THE BALL, AND TRY TO PICK UP THE JACKS..

4-23
OKAY, YOU TRY IT..

BONK!

THAT WAS FUN, AND WE DIDN'T EVEN NEED THE JACKS!
SCHULZ

ASK YOUR DOG TO COME OUT AND PLAY.. TELL HIM I HAVE A NEW BALLOON..

THE BALLOON MIGHT BREAK, AND EVER SINCE HE RETURNED FROM WORLD WAR I, SUDDEN NOISES FRIGHTEN HIM..
© 1995 United Feature Syndicate, Inc.

HOW WAS THAT FOR AN EXCUSE?
WHEN YOU CLOSED THE DOOR, THE SUDDEN NOISE FRIGHTENED ME..

© 1995 United Feature Syndicate, Inc.

WHAT WAS THAT?
THAT WAS MY FAMOUS JUMP SHOT..

DON'T WAIT UP FOR ME.. I'M GOING OUT TO PLAY A LITTLE MIDNIGHT BASKETBALL..
© 1995 United Feature Syndicate, Inc.

WELL, MAYBE A LITTLE NOON BASKETBALL..

QUICK, MARCIE.. IS THE FIRST QUESTION "TRUE" OR "FALSE"?

"TRUE"
HOW DO YOU KNOW?

BECAUSE I DID MY HOMEWORK, AND STUDIED FOR THE TEST..

4-27
© 1995 United Feature Syndicate, Inc.

WOW!
SCHULZ

LET HIM HIT IT, PITCHER! WE'RE BEHIND YOU A THOUSAND PERCENT!!
4-28

BONK!
© 1995 United Feature Syndicate, Inc.

HOW ABOUT .001?
SCHULZ

4-29

© 1995 United Feature Syndicate, Inc.

THE SOUND OF THE WAVES CRASHING AGAINST THE SIDES OF MY WATER DISH ALWAYS PUTS ME TO SLEEP..
SCHULZ

PEANUTS
SCHULZ

CURSE YOU, RED BARON!

HERE'S THE WORLD WAR I FLYING ACE RETURNING TO THE AERODROME.. HE IS EXHAUSTED..

AH! HE SEES A LIGHT IN THE SMALL FRENCH CAFE..

4-30
© 1995 United Feature Syndicate, Inc.

HELLO, CHARLES? YOUR DOG JUST CAME INTO OUR HOUSE..

YES, HE HAD ONE ROOT BEER, AND FELL ASLEEP AT THE TABLE..

"IT'S A LONG WAY TO TIPPERARY.."

WELL! I'M GLAD TO SEE YOU'RE BACK..

YOU LOOK SAD.. SOMEDAY THE WAR WILL BE OVER..BUT YOU'LL HAVE TERRIBLE MEMORIES, WON'T YOU? AND MAYBE EVEN SOME REGRETS

I REGRET I ONLY DRANK ONE ROOT BEER..
SCHULZ

THE VETERINARIAN SAYS IT'S TIME FOR YOUR ANNUAL CHECKUP..

HE SAYS WE...
5-1

NOW YOU SEE HIM, NOW YOU DON'T..
SCHULZ

ASK YOUR DOG IF HE WANTS TO COME OUT AND CHASE RABBITS..

I THINK HE'LL DO IT IF YOU DO IT HIS WAY..

HIS WAY?

5-2
OKAY, DOWN TO THE CORNER AND ACROSS THE FIELD..
SCHULZ

ASK YOUR DOG IF HE WANTS TO GO RABBIT CHASING AGAIN..

BUT TELL HIM I'M NOT GONNA PUSH HIM AROUND IN A STROLLER!

5-3

LEAD THE WAY, OL' CHAP..

RABBITS RUN TOO FAST..

AS SOON AS I SEE ONE, HE'S GONE..

© 1995 United Feature Syndicate, Inc.

YOU'RE GOING TO WEAR YOURSELF OUT, LAD.. HAVE A CUP OF TEA..
5-4
SCHULZ

ASK YOUR DOG TO COME OUT AND PLAY
WHY DON'T YOU GET YOUR OWN DOG?

MOM WON'T LET ME HAVE A DOG
WELL, YOU CAN'T KEEP BORROWING MINE..
5-5

I TOLD HIM THAT DOGS AREN'T SOMETHING YOU BORROW LIKE A LAWN MOWER OR A CUP OF SUGAR..
© 1995 United Feature Syndicate, Inc.

NEXT TIME TELL HIM TO CALL "LEASE A DOG"
SCHULZ

5-6
© 1995 United Feature Syndicate, Inc.

I DON'T KNOW.. MAYBE SOME FLOWERS DON'T LIKE TO BE PICKED..
!
SCHULZ

PEANUTS®
by
SCHULZ

AND THE FIRST RULE, OF COURSE, IS ALWAYS FOLLOW YOUR LEADER..

AND STAY TOGETHER..REMEMBER, WE'RE A TEAM..
© 1995 United Feature Syndicate, Inc.

ALL RIGHT, TROOPS..HERE'S WHERE WE'LL SPEND THE NIGHT..

OH, YES, THERE'S SOMETHING ELSE I SHOULD HAVE MENTIONED..
5-7

NEVER BUILD A CAMPFIRE WITH YOUR TENT PEGS..
SCHULZ

AND YOU DO IT ALL THE TIME! IT DRIVES ME CRAZY!

WHY CAN'T YOU SEE THAT? WHY?
5-8

WHY DO YOU INSIST ON..
ARE YOU TWO FIGHTING?
© 1995 United Feature Syndicate, Inc.

SHE'S FIGHTING.. I'M JUST SITTING HERE..
SCHULZ

© 1995 United Feature Syndicate, Inc.
MAYBE SOMEDAY YOU CAN EXPLAIN HER TO ME, OKAY?
SCHULZ
5-9

Z

WAKE UP, SIR..IT'S LUNCH TIME..HERE, HAVE AN APPLE...

MBPHPHM BPHMP

NICE GOING, SIR..SHE SAID THAT WAS THE BEST ANSWER YOU'VE GIVEN TODAY..
© 1995 United Feature Syndicate, Inc.

I WAS TRYING TO SPELL "MISSISSIPPI"... WHERE'D THIS APPLE COME FROM?
SCHULZ
5-10

I'VE BEEN GOING TO SCHOOL ALL YEAR, AND I JUST FOUND OUT WHAT OUR ROOM NUMBER IS..

AND WHERE THE DRINKING FOUNTAIN IS, AND THE PENCIL SHARPENER..

AND THAT THE SCHOOL BUS IS YELLOW..
© 1995 United Feature Syndicate, Inc.

THE SECRET OF LIFE IS TO BE OBSERVANT..
SCHULZ
5-11

I SEE YOU'VE STARTED WEARING A BACKPACK TO SCHOOL..
© 1995 United Feature Syndicate, Inc.

I LIKE TO GIVE THE IMPRESSION THAT I'M GOING SOMEPLACE IMPORTANT
5-12
SCHULZ

WOOF!
5-13

© 1995 United Feature Syndicate, Inc.

GOOD COOKIES COME WHEN THEY'RE CALLED
SCHULZ

PEANUTS
by SCHULZ
"E♭"

THAT WAS IT.. "E♭"

DO YOU KNOW WHO ROBERT SCHUMANN WAS?
A COMPOSER

HE HAD A MENTAL BREAKDOWN

AT ONE POINT HE THOUGHT HE HEARD ANGELS SINGING TO HIM

I HAVE THE SAME THING.. I KEEP HEARING SOMETHING, TOO.. IT COMES AND IT GOES...
5-14

LIKE MAYBE A MUSICAL NOTE?
NO, IT'S A DIFFERENT SORT OF SOUND..
E♭?

TRY TO GET THE BALL OVER THE PLATE, YOU BLOCKHEAD!

THAT'S IT!
SCHULZ

IF JESUS HAD OWNED A DOG, WHAT KIND DO YOU SUPPOSE IT WOULD HAVE BEEN?
I DON'T KNOW.. MAYBE A BORDER COLLIE..
I'LL BET THE DOG WOULD HAVE HAD A GOOD LIFE..
UNLESS HE GOT TIRED OF WALKING AROUND IN ALL THAT SAND..
5-15
SCHULZ
© 1995 United Feature Syndicate, Inc.

AND I DREAMED I WAS LYING IN THIS BEAUTIFUL MEADOW..

SUDDENLY, A FLOCK FLEW OVERHEAD..
© 1995 United Feature Syndicate, Inc.

BIRDS?
NO... D-MINUSES!
5/16
SCHULZ

SO LET'S ALL KEEP CONCENTRATING..

IT'S JUST A MATTER OF KEEPING OUR MINDS ON THE GAME..

WE CAN'T LET OUR MINDS WANDER..
© 1995 United Feature Syndicate, Inc.
5/17

SOME PANCAKES WOULD TASTE GOOD RIGHT NOW..
SCHULZ

HERE'S THE WORLD FAMOUS SERGEANT OF THE FOREIGN LEGION LEADING HIS TROOPS ACROSS THE DESERT..

AS THEY MARCH UNDER A MOONLIT SKY, THEY SING A STIRRING FIGHT SONG..
© 1995 United Feature Syndicate, Inc.

"SOME ENCHANTED EVENING" IS NOT A STIRRING FIGHT SONG!
SCHULZ
5-18

Dear Sweetheart,

Dearest Sweetheart,

5-19
© 1995 United Feature Syndicate, Inc.
SCHULZ

ARE YOU ASLEEP, OR ARE YOU RESTING YOUR EYES?
5-20

I WAS ASLEEP, BUT NOW I'M RESTING MY EYES

TELL YOUR EYES TO GET BACK IN THE GAME!
© 1995 United Feature Syndicate, Inc.

EYES ARE UP ALL DAY..THEY NEED LOTS OF REST..
SCHULZ

PEANUTS
by SCHULZ
GOOD GRIEF!

MAY I ASK HOW YOU INTEND TO PLAY SHORTSTOP WITH A SUPPER DISH IN YOUR MOUTH?

5-21

I SHOULD NEVER ASK..
SCHULZ

YES, I KNOW YOU DON'T LIKE TO BE CALLED "SHORTY"
5-22

AND I'VE NEVER CALLED YOU "SHORTY," HAVE I ? NO, I HAVEN'T..

OF COURSE, YOU DON'T EXACTLY HAVE TO DUCK WHEN YOU GO THROUGH A DOOR, DO YOU ?

boot!
SCHULZ

NO, MA'AM..I CAN'T ANSWER ANY QUESTIONS ABOUT KINGS..

OR RIVERS, OR NUMBERS, OR COUNTRIES OR STATES..
5-23

DO YOU HAVE ANY FOOD QUESTIONS?
SCHULZ

SURE, I DON'T CARE WHAT YOU DO..

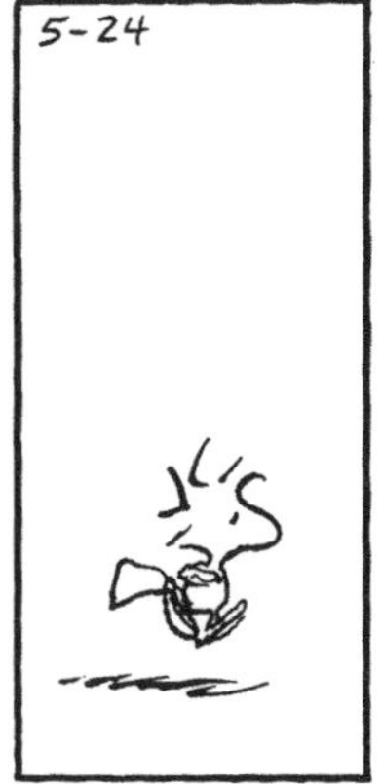
5-24

BUT A REAL VULTURE NEVER SITS IN A CHAIR..
SCHULZ

MARCIE, WHAT BOOK DID OUR TEACHER ASK US TO READ BY TOMORROW?

SHE GAVE US A CHOICE, SIR..
5-25

WE CAN READ "WAR AND PEACE" OR "IF YOU GIVE A MOUSE A COOKIE"

I'LL BE UP ALL NIGHT TRYING TO DECIDE..
SCHULZ

5-26
ASK YOUR BROTHER IF HE WANTS TO SELL HIS DOG..

THERE'S A STUPID KID OUTSIDE WHO WANTS TO KNOW IF YOU'LL SELL YOUR DOG..

AND THEN THEY THREW A BEANBAG AT ME!
SCHULZ

SURE, TAKE IT HOME..

IT'S A GOOD BOOK.. YOU'LL LIKE IT..

BUT HOW ARE YOU GOING TO READ IT IN THE DARK?
SCHULZ
5/27

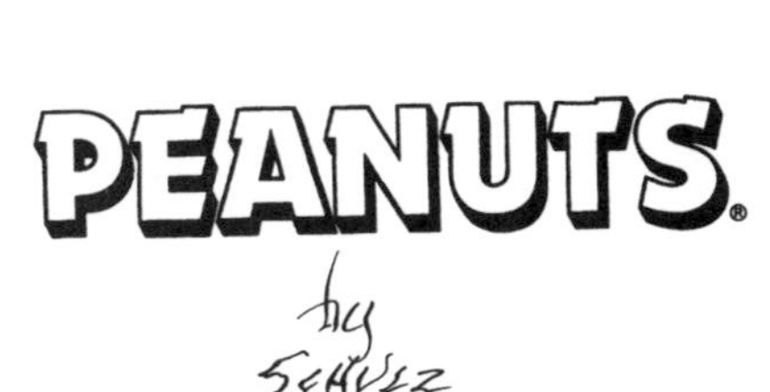
PEANUTS
by
SCHULZ

5-28

SCHULZ

© 1995 United Feature Syndicate, Inc.
5-29
SO YOU DON'T THINK JESUS EVER OWNED A DOG?
NO, I DOUBT IT
BUT WHY?
IF HE HAD A DOG, ALL OF THE APOSTLES WOULD HAVE WANTED ONE..

WILL YOU HELP ME DRAW A MAP?
WHAT DO YOU NEED A MAP FOR?

I'M THINKING OF RUNNING AWAY FROM HOME SO I'LL PROBABLY NEED A MAP..
5-30

OKAY, WHICH WAY ARE YOU PLANNING TO GO, NORTH OR SOUTH?
HOW SHOULD I KNOW?
© 1995 United Feature Syndicate, Inc.

JUST PUT DOWN "THIS WAY" AND "THAT WAY"

THERE.. I'VE DRAWN YOU A MAP SO YOU'LL KNOW WHERE YOU'RE GOING WHEN YOU RUN AWAY FROM HOME..

IT DOESN'T HAVE ENOUGH ARROWS..PUT IN SOME MORE ARROWS...

MAYBE I SHOULD COLOR IN THE RIVERS AND OCEANS..
© 1995 United Feature Syndicate, Inc.
5/31

IF IT'LL HELP..

WHAT YOU WANT TO DO IS GO DOWN TO THE CORNER, AND TURN RIGHT..

KEEP TURNING RIGHT, AND BEFORE YOU KNOW IT, YOU'LL BE BACK HERE WHERE YOU STARTED..

HOW CAN I RUN AWAY FROM HOME IF I COME BACK TO WHERE I STARTED?

AS THEY SAY IN BOARD MEETINGS, "TRUST ME ON THIS"
I'VE NEVER BEEN TO A BOARD MEETING
6-1
SCHULZ

DON'T FOLLOW ME! I'M RUNNING AWAY FROM HOME!

I MAY NEVER COME BACK..

I SAID, DON'T FOLLOW ME!

SCHULZ
I THOUGHT YOU MIGHT NEED AN ATTORNEY..
6-2

WHERE ARE YOU GOING?

I'M RUNNING AWAY FROM HOME..
YOU'RE WHAT?

6-3

I THINK MAYBE YOU'RE RIGHT..
SCHULZ

PEANUTS
by SCHULZ

I CAN'T HEAR IT..
CAN'T HEAR WHAT?

I CAN'T HEAR THE HARP..
SHH..

WHEN WE GET HOME, I'M GOING TO HAVE SOME COLD CEREAL..

OR MAYBE SOME TOAST WITH GRAPE JELLY
LISTEN TO THE MUSIC, SIR..

HERE, HAVE A MINT..
THANK YOU

OH, SURE! TAKE THE BEST ONE!
I DIDN'T TAKE THE BEST ONE..

YOU TOOK AN ORANGE ONE! THE ORANGE ONES ARE MY FAVORITE!
I JUST TOOK WHATEVER WAS ON TOP!

6-4
BOOT!

THIS IS MY REPORT ON THE CONCERT THEY TOOK US TO YESTERDAY.. I NEVER HEARD THE HARP..
SCHULZ

I SEE YOU DECIDED NOT TO LEAVE HOME..

I HAD TO... I WALKED ALL THE WAY AROUND THE BLOCK...
6-5

SCHULZ
MY LIFE DIDN'T GET ANY BETTER SO I CAME HOME..

6-6

SCHULZ

WELL, GO AHEAD AND SHOOT.. WHAT ARE YOU WAITING FOR?

GROWING..
SCHULZ
6-7

6-8
OKAY, YOU LOOK THERE, AND I'LL LOOK OVER HERE..
© 1995 United Feature Syndicate, Inc.

NO, I NEVER EVEN KNEW YOU WORE CONTACT LENSES..

WHY AM I STANDING OUT HERE IN THE RAIN?

BECAUSE YOU LOVE THE GAME..

I HATE THE GAME!
6-9

BECAUSE YOU LOVE THE CATCHER..
© 1995 United Feature Syndicate, Inc.

YOU GOT IT!

TELL YOUR RIGHT FIELDER TO STOP WINKING AT ME..

MY CATCHER SAYS FOR YOU TO STOP WINKING AT HIM!

I CAN'T HELP IT.. IT'S A SPONTANEOUS REACTION..

6-10

BONK!
© 1995 United Feature Syndicate, Inc.

I LOVE THOSE SPONTANEOUS REACTIONS!

PEANUTS
by Schulz
They never called me Ishmael.
I THOUGHT I HEARD TYPING..
YOU'RE GOING TO WRITE MY BIOGRAPHY?
WHAT A GREAT IDEA!
© 1995 United Feature Syndicate, Inc.
6-11
MAKE SURE YOU TELL ABOUT MY HUMBLE BEGINNINGS
YOU PROBABLY KNOW ME BETTER THAN ANYONE..
I HOPE YOU HAVE A GOOD TITLE..
"We Used to Call Him 'Banana Nose'"
SCHULZ

DO YOU KNOW WHAT "REDUNDANT" MEANS?
I'M NOT SURE..
REDUNDANT? THERE'S REDUNDANT..
SCHULZ
© 1995 United Feature Syndicate, Inc.
6-12

ISN'T IT GREAT, MARCIE? NO SCHOOLWORK TO WORRY ABOUT..NOTHING TO DO!

© 1995 United Feature Syndicate, Inc.
I JUST FINISHED ADDRESSING ALL MY CHRISTMAS CARDS, SIR..

6-13
I'LL CALL YOU AGAIN IN SEPTEMBER, MARCIE..

6-14

© 1995 United Feature Syndicate, Inc.

SCHULZ

DID YOU REALLY THINK I WAS GOING TO SHARE THIS ICE CREAM CONE WITH YOU?

YES, I THOUGHT THAT
6-15

DID YOU REALLY THINK WE WERE BOTH GOING TO LICK THE SAME ICE CREAM CONE?
YES, I THOUGHT THAT

ALL RIGHT! HERE, TAKE IT!
NOT AFTER YOU'VE LICKED IT!
SCHULZ

THIS IS A GREAT PLACE FOR ICE CREAM CONES..THEY HAVE ALL KINDS OF FLAVORS..

SCHULZ
WHAT FLAVOR DO YOU THINK YOU WANT?
CAT!
6-16

SOMETHING STRANGE IS HAPPENING TO ME, CHARLIE BROWN..

I KEEP HEARING AN "E FLAT" IN MY HEAD OVER AND OVER..

6-17

HE'S LUCKY.. I KEEP HEARING "BALL FOUR!"
SCHULZ

PEANUTS
by SCHULZ
YAWN
CLOMP!
6-18
AUGHH!
BONK!
© 1995 United Feature Syndicate, Inc.
EXCESSIVELY WEIRD!
SCHULZ

WHAT HAPPENS IF A DOG DOESN'T LIKE THE FAMILY HE'S LIVING WITH?
HE SNEAKS AWAY AT NIGHT, HOPS A FREIGHT, AND HEADS OUT WEST
THEN, AFTER HE'S RICH AND FAMOUS, HE RETURNS TO HIS HOME TOWN AND BECOMES GRAND MARSHAL OF THEIR ANNUAL PARADE
ON THE OTHER HAND, HE'D BETTER STAY FAIRLY CLOSE TO THAT SUPPER DISH..
SCHULZ
6-19
© 1995 United Feature Syndicate, Inc.

SOME OF THE GUYS OVER AT THE PLAYGROUND WERE DISCUSSING CRABBY SISTERS

GUESS WHAT.. I WON!

THEY ALL AGREED THAT I HAVE THE CRABBIEST SISTER IN THE NEIGHBORHOOD
© 1995 United Feature Syndicate, Inc.

I'M A CELEBRITY!
SCHULZ
6-20

YOU THINK I'M CRABBY NOW..WAIT UNTIL I'M FORTY OR FIFTY..

AND WHEN I'M SIXTY, I'LL BE EVEN MORE CRABBY!

BUT JUST WAIT 'TIL I'M EIGHTY.. THEN YOU'LL SEE CRABBY LIKE NOTHING EVER BEFORE!
6/21
© 1995 United Feature Syndicate, Inc.

WHAT ABOUT WHEN YOU'RE NINETY?
THEN I'LL BE REAL NICE..
SCHULZ

I'VE DECIDED I'M NOT GOING TO SUMMER CAMP THIS YEAR..

I THOUGHT ABOUT IT, AND THOUGHT ABOUT IT..
IT SEEMS LIKE YOU GAVE IT A LOT OF THOUGHT..
6-22
© 1995 United Feature Syndicate, Inc.

ABOUT TEN SECONDS
SCHULZ

ASK YOUR DOG IF HE WANTS TO COME OUT AND FROLIC..
© 1995 United Feature Syndicate, Inc.

YESTERDAY YOU ASKED HIM IF HE WANTED TO COME OUT AND PLAY..
6-23

WE'RE UPGRADING THE NEIGHBORHOOD
SCHULZ

ARE YOU SURE? ARE YOU POSITIVE?
6-24

WELL, WE CAN'T TELL UNTIL WE TRY IT, CAN WE?
© 1995 United Feature Syndicate, Inc.

BONK!

OKAY, YOU WERE RIGHT.. I WAS FACING THE WRONG WAY..
SCHULZ

PEANUTS
by SCHULZ
SEE? YOU TWIRL THE ROPE AROUND, AND YOU JUMP UP AND DOWN
WHY WOULD I WANT TO DO THAT?
BECAUSE IT'S FUN! IT'S DOING SOMETHING
WHY DO I HAVE TO DO SOMETHING?
I DON'T LIKE TO DO ANYTHING.. I JUST LIKE TO HANG AROUND..
YOU CAN'T JUST HANG AROUND...
6-25
THAT'S NO WAY TO LIVE! YOU'RE GOING TO WASTE YOUR WHOLE LIFE!
© 1995 United Feature Syndicate, Inc.
DON'T LOOK AT ME... I'M JUST HANGING AROUND..
SCHULZ

6-26
YOU'RE EMOTIONALLY BANKRUPT... SCOTT FITZGERALD WAS EMOTIONALLY BANKRUPT...WE'RE ALL EMOTIONALLY BANKRUPT...
SCHULZ
© 1995 United Feature Syndicate, Inc.

I'VE DECIDED TO TRY TO BE A BETTER PERSON..

BUT NOT RIGHT AWAY, OF COURSE..
© 1995 United Feature Syndicate, Inc.

MAYBE A FEW DAYS FROM NOW..
SCHULZ
6-27

SOMETIMES I LIE AWAKE AT NIGHT, AND I ASK, "WHY AM I HERE?"
© 1995 United Feature Syndicate, Inc.
6/28

THEN A VOICE COMES TO ME THAT SAYS, "WE CAN'T TAKE YOUR QUESTION NOW..WE'RE ALL OUT ROLLERBLADING.."
SCHULZ

6-29
SO WE'LL PUT ALL THE SAND RIGHT ABOUT HERE..
BUT REMEMBER, WHEELBARROWS TIP OVER EASILY SO DON'T PUSH IT TOO FAST..
SCHULZ
© 1995 United Feature Syndicate, Inc.

6-30
MAYBE FIVE OR SIX MORE LOADS..THEN YOU CAN RAKE IT AROUND A BIT..
SCHULZ
© 1995 United Feature Syndicate, Inc.

7-1
..AND THE FEELING YOU'LL HAVE KNOWING YOU HAULED THE SAND FOR YOUR VERY OWN SANDBOX..
SCHULZ
© 1995 United Feature Syndicate, Inc.

PEANUTS
by SCHULZ
!
BAM BAM BAM
IS IT SUPPERTIME ALREADY?
BUT THAT'S NOT YOUR SUPPER DISH.. THAT'S YOUR WATER DISH...
YOU BROUGHT THE WRONG DISH
© 1995 United Feature Syndicate, Inc.
7-2
PRETTY EMBARRASSING, HUH?
EVEN THE BEST OF US MAKE MISTAKES, HUH?
SIGH
I THINK I'LL STAND BY THE FRONT DOOR TONIGHT, AND LET BURGLARS IN..
SCHULZ

I THOUGHT A SANDBOX WAS SUPPOSED TO HAVE A SHOVEL AND A PAIL, AND A HOE AND A RAKE, AND A TRACTOR AND A DUMP TRUCK..

WHERE'S THE SHOVEL AND THE PAIL, AND THE HOE AND THE RAKE, AND THE TRACTOR AND THE DUMP TRUCK?!

7-3
SCHULZ

THE LITTLE RED-HAIRED GIRL IS AT THE DOOR ASKING FOR YOU..

7-4
APRIL FOOL!

THIS IS JULY FOURTH!
WHATEVER..
SCHULZ

ASK YOUR DOG TO COME OUT AND PLAY "CHASE THE STICK"

"THANK YOU FOR YOUR OFFER TO COME OUT AND PLAY..WE ARE BUSY AT THIS TIME, HOWEVER, AND CANNOT ACCEPT YOUR OFFER..WE HOPE YOU WILL BE SUCCESSFUL ELSEWHERE"
7-5
DOGS HAVE REJECTION SLIPS?
SCHULZ

OKAY, HERE WE GO...UP THROUGH THE OL' HOOP!

7-6
© 1995 United Feature Syndicate, Inc.

WE DON'T AGREE ABOUT ANYTHING, DO WE?
SCHULZ

7-7

© 1995 United Feature Syndicate, Inc.
SCHULZ

DID YOU NOTICE THAT I DIDN'T GO TO CAMP THIS YEAR?
7-8

DON'T TALK TOO SOON..SUMMER ISN'T OVER..

THEY COULD STILL DRAG YOU AWAY
© 1995 United Feature Syndicate, Inc.

I'M GOING TO GO CHAIN MYSELF TO MY BED..
SCHULZ

It was a dark and stormy night.

I mean it was really dark and it was really stormy!

WHY ARE YOU SITTING HERE WHEN YOU COULD BE UP THERE FLYING AROUND WITH ALL THOSE OTHER BIRDS?
7-10

I KNOW WHAT YOU MEAN..
© 1995 United Feature Syndicate, Inc.

I'M NOT A JOINER, EITHER
SCHULZ

OKAY, BEFORE WE BEGIN, I'LL READ THE RULES..

GOOD! I LOVE THE RULES.. ONCE YOU KNOW THE RULES, YOU CAN CHEAT..
7-11

WHAT I ALWAYS SAY IS YOU CAN'T REALLY CHEAT UNLESS YOU KNOW THE RULES..
© 1995 United Feature Syndicate, Inc.

THAT'S WHAT I ALWAYS SAY..
SCHULZ

7-12
MOLES HAVE VERY WEAK EYES.. THEY DIG TUNNELS JUST UNDER THE GROUND, AND HUNT WORMS AND INSECTS
© 1995 United Feature Syndicate, Inc.

YOU'RE RIGHT..IT'S A TOUGH WAY TO MAKE A LIVING..
SCHULZ

RIGHT HERE IS WHERE I USED TO STAND WAITING FOR THE SCHOOL BUS! RIGHT HERE ON THIS VERY SPOT!

BUT THAT WAS A LONG TIME AGO..
7-13
© 1995 United Feature Syndicate, Inc.

I'M SURPRISED YOU CAN REMEMBER..
SOME THINGS YOU NEVER FORGET..
SCHULZ

I THINK IT'S TIME YOU AND I HAD A SERIOUS CONVERSATION..

YOU MEAN A LITTLE BACK AND FORTH DIALOGUE?
© 1995 United Feature Syndicate, Inc.

7-14
NO, SOMETHING A LITTLE MORE ONE-SIDED..
SCHULZ

GAME, SET, MATCH..LIFE!
7-15
© 1995 United Feature Syndicate, Inc.
SCHULZ

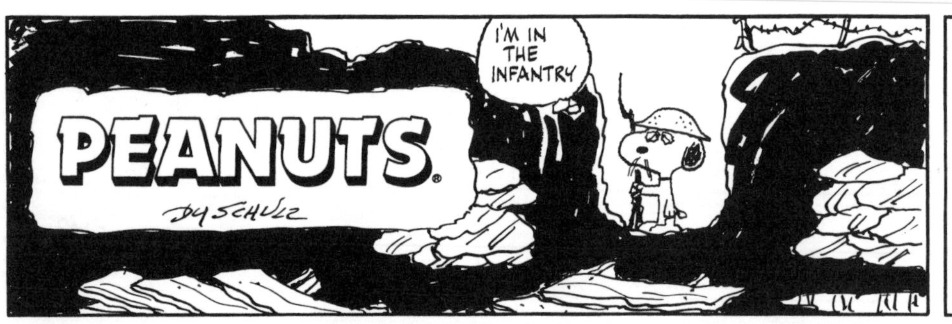
PEANUTS
BY SCHULZ
I'M IN THE INFANTRY

THE FLAK IS HEAVY TODAY..

GOOD GRIEF! THERE'S MY BROTHER SPIKE!

HEY, SPIKE! I WAS FLYING OVER THE TRENCHES WHEN I SAW YOU..
I'M IN THE INFANTRY

ARE YOU GETTING READY TO LEAD A CHARGE OVER THE TOP?

THE LAST TIME I LOOKED OVER THE TOP SOMEONE THREW A ROCK AT ME!

WELL, I HAVE TO GET BACK TO THE AERODROME..I'LL WRITE TO MOM, AND TELL HER I SAW YOU..

TELL HER I'M IN THE INFANTRY, AND I MISS HER TAPIOCA PUDDING..

THAT SPIKE..HE'S A REAL SOLDIER..ALWAYS THINKING OF DUTY...
7-16

WHEN YOU'RE IN THE INFANTRY, YOU NEVER GET TAPIOCA PUDDING..
SCHULZ

"WHAT IS SO RARE AS A DAY IN JUNE?"
© 1995 United Feature Syndicate, Inc.

7-17
THIS IS JULY

WHAT IS SO RARE AS A DAY IN JUNE, AND YOU MISSED IT AGAIN?
SCHULZ

I MEAN, IT'S THE SORT OF THING THAT MAKES YOU WONDER, DOESN'T IT?
© 1995 United Feature Syndicate, Inc.

AND WHAT ARE YOU DOING?

WONDERING
7-18
SCHULZ

HOW COULD A KITE GO DOWN THROUGH THE BACK OF YOUR SHIRT AND COME OUT YOUR SLEEVE?
I SUPPOSE YOU DON'T BELIEVE IN THE LOST ISLAND OF ATLANTIS, EITHER..
SCHULZ
7-19
© 1995 United Feature Syndicate, Inc.

7-20
I SAID, DO YOU SEE A KITE OVER THERE?
YES, BUT I THINK IT'S LOST INTEREST
SCHULZ

7-21
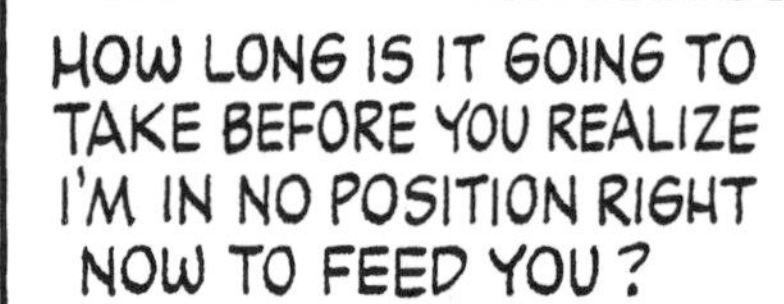
HOW LONG IS IT GOING TO TAKE BEFORE YOU REALIZE I'M IN NO POSITION RIGHT NOW TO FEED YOU?

I'M A QUICK LEARNER
SCHULZ

7-22

THEY DON'T MAKE TREES LIKE THEY USED TO..
SCHULZ

PEANUTS
by
SCHULZ

OH, NO!

AAUGH!

I CAN'T STAND IT!
© 1995 United Feature Syndicate, Inc.

7-23
I JUST CAN'T STAND IT!

WHY ARE YOU WALKING LIKE THAT, BIG BROTHER?

BECAUSE WE LOST AGAIN!

THIS IS CALLED "THE LOSER'S WALK".. THIS IS THE WAY YOU'RE SUPPOSED TO WALK WHEN YOU'VE LOST AGAIN..

ALSO CALLED "THE WHIPPED PUPPY"
SCHULZ

STOP SINGING "SHALL WE GATHER AT THE RIVER"
7-24

HOW DO I LOOK, MARCIE? I'M GOING FISHING!

YOU LOOK GREAT, SIR... HAVE A GOOD TIME..

I WAS HOPING I LOOKED LIKE HUCKLEBERRY SAWYER..
7-25

YES, SIR..I'M GOING FISHING SO I NEED SOME BAIT..

NO, I COULDN'T AFFORD THAT..COULD I BUY JUST ONE WORM?
7-26

HOW ABOUT IF I RENT ONE?

7-27

WOOF!

HOW'S THE FISHING?
© 1995 United Feature Syndicate, Inc.

© 1995 United Feature Syndicate, Inc.
WHAT HAPPENED TO YOU?
I WAS FISHING OFF THE DOCK WHEN YOUR STUPID DOG STARTLED ME, AND I FELL IN THE LAKE..
I TALKED TO SOME OF THE FISH AFTERWARD, AND THEY THOUGHT IT WAS VERY FUNNY..
7-28

I CAME OVER TO SEE ALL THE FISH YOU CAUGHT, SIR..
I DIDN'T CATCH ANY FISH, MARCIE.. I FELL IN THE LAKE..

DID YOU GET WET?
7-29

HA HA HA HA HA HA!
© 1995 United Feature Syndicate, Inc.

YOU ARE INORDINATELY WEIRD, MARCIE..

PEANUTS
by SCHULZ

I'M AWAKE!

YES, MA'AM.. WE WENT TO THE CONCERT, AND HEARD "ADAGIO FOR STRINGS" BY SAMUEL THE BARBER..
7-30

NOT SAMUEL THE BARBER! SAMUEL BARBER! GOOD GRIEF!

CAN'T YOU EVER GET ANYTHING RIGHT?!

WHAT'S WRONG WITH YOU? DON'T YOU EVER LISTEN? DON'T YOU EVER READ?!

SAMUEL THE BARBER! THAT'S NOT EVEN FUNNY! WHEN ARE YOU EVER GOING TO LEARN SOMETHING?!

DON'T YOU EVER THINK?
?

YES, MA'AM.. AND THEN THEY PLAYED "PETER AND THE FOX"
"WOLF," FRANKLIN.. "HAROLD AND THE WOLF"
SCHULZ

WHEN I LOOK AT A CACTUS SHADOW, I CAN TELL WHAT TIME IT IS, WHAT THE TEMPERATURE IS, AND WHAT THE WEATHER WILL BE TOMORROW..

7-31
WHEN YOU DON'T KNOW ANYTHING, YOU HAVE TO PRETEND YOU KNOW EVERYTHING..
SCHULZ

IS THE WORLD FAMOUS ATTORNEY ON HIS WAY TO COURT ?
A VERY IMPORTANT CASE..
8-1

YES, YOUR HONOR..WE SHALL PROVE THAT MY CLIENT NEVER INTENDED TO GO INTO THE GARDEN OF MR. McGREGOR..
SCHULZ

WHAT I'M TRYING TO SAY, YOUR HONOR, IS THAT MY CLIENT WAS SEVERELY WRONGED BY MR. McGREGOR

WHEN MR. McGREGOR CHASED MY CLIENT, THIS INNOCENT LITTLE BUNNY, WITH A RAKE, HE CAUSED HIM GREAT EMOTIONAL DISTRESS..
8-2

SNIF
SNIF

YOUR HONOR, MAY WE HAVE A TEN MINUTE RECESS?
SCHULZ

HAPPY BIRTHDAY, AMY-

PEANUTS
by SCHULZ
!
?
SHE DID IT!
MY OWN SISTER THREW MY BLANKET UP IN A KITE-EATING TREE!
WHY AM I DOING THIS? DOGS CAN'T CLIMB TREES
YOU GOT IT! YOU GOT IT!
8-6
© 1995 United Feature Syndicate, Inc.
WOW! I THOUGHT I'D NEVER SEE THIS BLANKET AGAIN..
CLOMP!
I CAN'T STAND IT!
SCHULZ

I WAS WONDERING IF YOU COULD FLY ME TO MONTANA..

I KNOW YOU GO BY THE COLORS ON THE MAP..
8-7

ON SOME MAPS IT'S ORANGE, AND ON OTHERS IT'S GREEN..

THEN WE CAN'T GO..
SCHULZ

HEY, PILOT..

I NOTICE WE DON'T HAVE PARACHUTES..SHOULDN'T WE HAVE PARACHUTES?

8-8
TRY TO LAND IN THE WATER..
SCHULZ

HOW DO YOU KEEP YOUR PLANE SO CLEAN?

I HAVE A DEDICATED MECHANIC..HE'S ALWAYS ON THE JOB..

8-9
SCHULZ

WHAT IF WE RUN INTO THE RED BARON WHILE WE'RE UP HERE ?
8-10

DON'T WORRY.. HE ALWAYS SHOOTS LOW..
© 1995 United Feature Syndicate, Inc.

SCHULZ

THIS IS RIDICULOUS!

I THOUGHT WE'D BE FLYING OVER MOUNTAINS AND RIVERS! I THOUGHT I'D SEE THINGS!
8-11

ALL I'VE SEEN IS A BACK YARD..
© 1995 United Feature Syndicate, Inc.

IT'S MY FAVORITE BACK YARD..
SCHULZ

SO THIS IS A SOPWITH CAMEL, IS IT ?

I THOUGHT MAYBE IT WAS A ZAMBONI ..

HA HA HA HA!

© 1995 United Feature Syndicate, Inc.
8-12
SCHULZ

PEANUTS
by SCHULZ
C'MON. LET 'IM HIT IT!

BUT NOT TO ME..

OH, NO!

I CAN'T FIND THE BALL!

WELL, KEEP LOOKING!
8-13

HERE, THIS IS ALL I COULD FIND..
© 1995 United Feature Syndicate, Inc.

!

DON'T YELL AT ME.. I'M ONLY THE PITCHER..

HE'S LUCKY I DIDN'T TRY TO THROW A KNUCKLE BIRD!
SCHULZ

8-14
© 1995 United Feature Syndicate, Inc.

SOMETIMES I LIE AWAKE AT NIGHT, AND I ASK, "WHEN WILL IT ALL END?"
8-15
© 1995 United Feature Syndicate, Inc.

THEN A VOICE COMES TO ME THAT SAYS, "RIGHT AFTER THE CREDITS!"

I HOPE THIS ISN'T ONE OF THOSE MOVIES WHERE A KID GOES TO BOARDING SCHOOL, AND EVERYONE IS MEAN TO HIM..
OR WHERE EVERYONE TEASES A GIRL BECAUSE SHE HAS FUNNY HAIR..
I LIKE A MOVIE THAT SHOWS A DOG SLEEPING IN FRONT OF A FIREPLACE FOR TWO HOURS
8/16
© 1995 United Feature Syndicate, Inc.

ALL RIGHT, HOW MUCH IS 5 X 10?
HOW SHOULD I KNOW?

HOW ABOUT 6 X 20?
WHO CARES?

8/17

HOW ABOUT 2 X 11?
ARE YOU KIDDING?

I HOPE THESE AREN'T TOO HARD FOR YOU..
DO I LOOK DISCOURAGED?
SCHULZ

HEY, MANAGER! IT'S TOO HOT TO PLAY TODAY!

YOU'RE ALWAYS COMPLAINING! IT'S EITHER TOO HOT OR IT'S TOO COLD!

HEY, MANAGER! IT'S TOO "JUST RIGHT" TO PLAY TODAY!
SCHULZ
8-18

Dear Pen Pal,
In my last letter I asked if you have a dog.

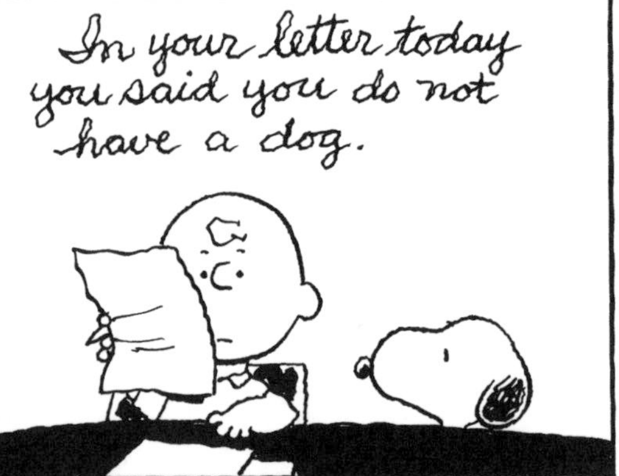
In your letter today you said you do not have a dog.

8-19
WHY WRITE TO SOMEBODY WHO DOESN'T HAVE A DOG?
SCHULZ

PEANUTS

SUPPERTIME

8-20
JUST WAIT 'TIL I FINISH THIS CHAPTER, OKAY?

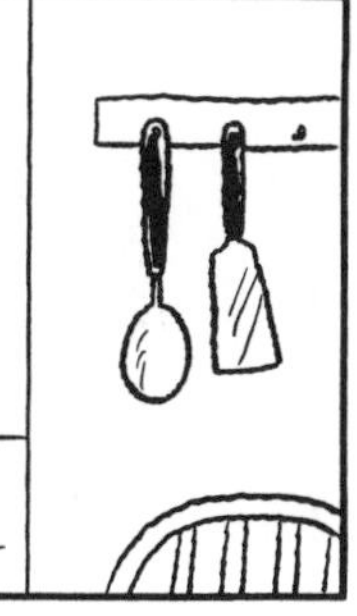

© 1995 United Feature Syndicate, Inc.

I'VE BECOME OBSOLETE!
SCHULZ

THAT LITTLE RED HAIRED GIRL JUST CALLED..

SHE WANTS YOU TO MEET HER AT THE PLAYGROUND, AND PUSH HER ON THE SWINGS..
8-21

AUGUST FOOL!
© 1995 United Feature Syndicate, Inc.
SCHULZ

8-22
SCHULZ
WHEN YOU'RE PLAYING IN A TOURNAMENT, DO THE CROWDS BOTHER YOU?
YES, THEY'RE ALWAYS STANDING IN FRONT OF ME IN THE LINE AT THE HOT DOG STAND..
© 1995 United Feature Syndicate, Inc.

IF THAT LITTLE RED HAIRED GIRL WAS HERE, I COULD PUSH HER ON THE SWING..
© 1995 United Feature Syndicate, Inc.

I'D PUSH HER REAL HIGH, AND SHE'D LAUGH..
8-23

SHE'D SAY, "PUSH ME HIGHER, CHARLIE BROWN!" AND SHE'D LIKE ME BETTER THAN ANYONE..

MAYBE
SCHULZ

WHY ARE YOU ALWAYS YELLING AT ME?
© 1995 United Feature Syndicate, Inc.

WELL, I'M NOT SURE I CAN HELP IT.. IT'S ME..
8-24

IT'S WHAT I DO..
SCHULZ

© 1995 United Feature Syndicate, Inc.
I HEARD COYOTES HOWLING LAST NIGHT.. I WONDER IF THEY'RE DANGEROUS..
I HAVE NO IDEA
MY GRAMPA ALWAYS SAID, "NEVER TRUST A COYOTE WHO'S WEARING A TUXEDO"
8-25
SCHULZ

I HEARD THE COYOTES HOWLING AGAIN LAST NIGHT, CHARLIE BROWN..

IT WAS EERIE
I NEVER HEARD A THING..

HOW ABOUT YOUR DOG?
I WAS UNDER THE COVERS..
8-26
© 1995 United Feature Syndicate, Inc.
SCHULZ

PEANUTS
by SCHULZ
DOG WANTED TO CHASE STICKS

ASK YOUR BROTHER'S DOG TO COME OUT AND PLAY..
WHY DON'T YOU GET YOUR OWN DOG?

MOM WON'T LET ME HAVE A DOG..
WELL, STOP BOTHERING US.. GET OFF OUR PORCH!
8-27

THIS ISN'T A PORCH.. THIS IS A STOOP!
© 1995 United Feature Syndicate, Inc.

A PORCH HAS A RAILING AROUND IT, AND A ROOF OVER IT, AND IT'S ALL WHITE, AND IT HAS A SWING AND SOME ROCKING CHAIRS..

AND A LITTLE TABLE WITH LEMONADE GLASSES, AND WARM NIGHTS, AND FIREFLIES, AND CRICKETS, AND SOFT MUSIC, AND A MOON IN THE SKY..

DO YOU WANT TO GO OUT AND PLAY WITH THAT KID AGAIN?

HE'S WAITING ON OUR STOOP!
SCHULZ

I'VE ALWAYS WONDERED HOW YOU DECIDED TO BECOME A DOG..
THAT'S A GOOD QUESTION

I REMEMBER GOING DOWN THE LIST..
© 1995 United Feature Syndicate, Inc.

8-28
EVERYTHING ELSE WAS TAKEN..
SCHULZ

"Your eyes are beautiful," he said.

"Shall I compare them to a summer day? No, even more."
8-29
© 1995 United Feature Syndicate, Inc.

"Your eyes are like two supper dishes."
SCHULZ

8-30
© 1995 United Feature Syndicate, Inc.
WHO'S THE FELLOW IN THE BLACK HAT?
THAT'S JOE BLACKJACK, THE WORLD FAMOUS RIVER BOAT GAMBLER..
ARE WE GONNA TALK, OR ARE WE GONNA PLAY "OLD MAID"?
SCHULZ

SOMETIMES I WISH THAT JOE BLACKJACK, THE WORLD FAMOUS RIVER BOAT GAMBLER, WOULD LEARN TO SHUFFLE..
SCHULZ
© 1995 United Feature Syndicate, Inc.
8-31

HEY, MARCIE..LET'S GO DO SOME SHOPPING FOR SCHOOL SUPPLIES..

I DID THAT A MONTH AGO, SIR
9-1

SURE, MARCIE..AND I SUPPOSE YOU ALREADY KNOW WHICH COLLEGE YOU'RE GOING TO!
© 1995 United Feature Syndicate, Inc.
SCHULZ

AND I'VE ENROLLED MY THREE KIDS IN PRE-SCHOOL!

YES, SIR.. I NEED SOME SCHOOL SUPPLIES...

A NEW BINDER..A COUPLE OF PENCILS.. AND MAYBE A PEN..
9-2
© 1995 United Feature Syndicate, Inc.
SCHULZ

DO YOU HAVE ANY D-MINUS SPRAY?

PEANUTS®
by SCHULZ
AGAIN!!

EVERY YEAR!

THEY DID IT AGAIN!
© 1995 United Feature Syndicate, Inc.

THEY'RE STARTING SCHOOL AGAIN BEFORE I'M READY!

I DON'T KNOW HOW HIGH MOUNT EVEREST IS..

9-3
YOU DON'T HAVE TO KNOW.. THAT'S WHY YOU GO TO SCHOOL

WHEN YOU GET TO SCHOOL, THEY'LL TELL YOU HOW HIGH MOUNT EVEREST IS..

SCHULZ

WHY WOULD I WANT TO KNOW THAT?

HOW COME WE NEVER WIN? DO YOU THINK HE HAS CARDS UP HIS SLEEVE?
I DON'T THINK HE NEEDS A SLEEVE..
9-4
© 1995 United Feature Syndicate, Inc.

I CAN SEE THE LIGHTS OF THE SCHOOL BUS COMING THROUGH THE FOG

IT'S LIKE A HUGE MONSTER DEVOURING EVERYTHING IN ITS PATH..
9-5

WHERE ARE YOU GOING?
© 1995 United Feature Syndicate, Inc.

WALKING TO SCHOOL

I JUST HEARD SOME PEOPLE OVER AT THE CHURCH SINGING "BRINGING IN THE SHEAVES"

I DON'T EVEN KNOW WHAT A SHEAVE IS..
9-6
© 1995 United Feature Syndicate, Inc.

ME NEITHER, BUT IF I EVER SEE ONE, I'LL BRING IT IN..

YES, MA'AM..I WAS WONDERING IF I MIGHT HAVE A DESK IN THE BACK ROW..
9-7

NO, MA'AM..I UNDERSTAND.. THAT'S LIFE...

LAST ROW, UPPER DECK AT THE BALLPARK.. FIRST ROW IN THE CLASSROOM
SCHULZ

YOUR DOG IS SITTING ON THE BACK PORCH..I THINK HE'S WAITING FOR YOU..

IS HE ALONE?
NO, HE HAS A FRIEND
9/8

SCHULZ

9-9

?
IT HAPPENED YESTERDAY.. THE WHOLE WORLD SUDDENLY TURNED UPSIDE DOWN! YOU'RE THE ONLY ONE WHO IS RIGHT-SIDE UP!

THANK YOU FOR WORRYING ABOUT THE REST OF US..
SCHULZ

PEANUTS
by SCHULZ

HERE'S THE FIERCE JUNGLE ANIMAL PERCHED HIGH IN A TREE WAITING FOR A VICTIM TO COME BY..

AH!

WATCH IT, DOG!

BEFORE YOU THINK ABOUT LEAPING ON ME, REMEMBER THESE THINGS..

REMEMBER MOUNT VESUVIUS! REMEMBER THE JOHNSTOWN FLOOD! REMEMBER THE TITANIC!

DO YOU WANT ME TO RUN THROUGH THEM AGAIN ?

REMEMBER MOUNT VESUVIUS! REMEM..

DOGS ONLY REMEMBER FOOD THINGS..
SCHULZ
9-10

I CAN'T BELIEVE IT.. I GOT EVERY SINGLE ANSWER WRONG..
9/11

THAT'S TOO BAD, SIR..
IT PROVES MY THEORY..
© 1995 United Feature Syndicate, Inc.

I'M A PARCHESI PLAYER IN A CHESS WORLD..
SCHULZ

© 1995 United Feature Syndicate, Inc.
AS I ROUNDED THE CORNER, SANDI PATTI HIT A HIGH NOTE..
9-12
SCHULZ

YES, MA'AM..I UNDERSTAND.. YOU WANT MORE THAN JUST THE MATH ANSWER..
9-13

YOU WANT ME TO EXPLAIN **HOW** I GOT THE ANSWER..
© 1995 United Feature Syndicate, Inc.

I COPIED IT FROM THE KID BEHIND ME..
SCHULZ

I CAN'T GO TO SCHOOL TODAY, MARCIE..
HOW COME?

TELL THE TEACHER I HAVE A SORE THROAT, AND A HEADACHE, AND MY ARM HURTS..
I'LL WRITE THAT DOWN, SIR..
© 1995 United Feature Syndicate, Inc.

"SORE THROAT" IS TWO WORDS, "HEADACHE" IS ONE WORD, AND "ARM HURTS" IS TWO WORDS..
9-14

AND TELL HER I SAID TO HAVE A NICE DAY..
"NICE DAY.." THAT'S TWO WORDS

I MAY BE SICK, BUT YOU'RE WEIRD, MARCIE!
SCHULZ

© 1995 United Feature Syndicate, Inc.
HUMANS HAVE A MORAL SENSE THAT ANIMALS DON'T HAVE, AND WE CAN FORESEE THE CONSEQUENCES OF OUR ACTIONS..
AND YOU KNOW WHO SAID THAT? CARL SAGAN!
CARL WHO?
SCHULZ
9-15

I SHOULD BE OUR TEAM CHEERLEADER AGAIN, SIR..

LISTEN TO THIS ONE..

"NYAHH NYAHH! WE'RE NICER THAN YOU!"
MARCIE, THAT'S THE WORST CHEER I'VE EVER HEARD!
9-16
© 1995 United Feature Syndicate, Inc.

I'LL ADD ANOTHER "NYAHH"
SCHULZ

PEANUTS®
by SCHULZ

DUNCE CAPS Third floor

WHAT DID YOU PUT DOWN FOR THE FIRST QUESTIONS, MARCIE?

"TRUE, FALSE, TRUE, TRUE, AND FALSE"

HA! SHE TRICKED YOU, MARCIE! YOU FELL FOR IT! THOSE AREN'T THE ANSWERS!

THEY'RE NOT?
NO! "PERHAPS, FOR SURE, MAYBE, WHO KNOWS, AND DOWN HOME"
9-17

HOW ABOUT YOU, FRANKLIN?
"TRUE, FALSE, TRUE, TRUE, AND FALSE"

I CAN'T BELIEVE IT! SHE FOOLED YOU, TOO!

BOY, YOU'RE SHARP, MA'AM! YOU FOOLED 'EM ALL!

SCHULZ
EIGHT MORE MONTHS 'TIL SUMMER VACATION..
PRINCIPAL'S OFFICE
© 1995 United Feature Syndicate, Inc.

HERE WE GO! SLAM DUNK!

LOTS OF SLAM, BUT NO DUNK..
9-18

THE BASES ARE LOADED AGAIN, AND THERE'S STILL NOBODY OUT..
9-19

SO WHAT DO YOU THINK?

WE LIVE IN DIFFICULT TIMES..

A TOOTH! I'VE LOST ANOTHER TOOTH!

I MUST BE GETTING OLD..
9-20

PRETTY SOON I'LL BE LOSING ALL MY HAIR..

HI, SPIKE.. HOW ARE THINGS IN THE TRENCHES?

WE NEVER GET TAPIOCA PUDDING..
9-21

WHEN YOU SEE GENERAL PERSHING, ASK HIM WHY YOU NEVER GET TAPIOCA PUDDING WHEN YOU'RE IN THE INFANTRY..

I'LL ASK HIM, BUT HE'LL PROBABLY THROW ME OFF THE EIFFEL TOWER..
SCHULZ

YES, SIR.. I'D LIKE PERMISSION TO SPEAK TO GENERAL PERSHING..

YES, IT'S A VERY DELICATE MATTER..

WELL, MY BROTHER, WHO IS IN THE INFANTRY, WANTS TO KNOW WHY THEY NEVER GET ANY TAPIOCA PUDDING

9-22
SCHULZ

LOOK, SPIKE, I HAD OUR COOK MAKE YOU SOME TAPIOCA PUDDING!

GOOD, HUH? YOU LIKE IT, HUH?

9-23
SPIKE.. SPIKE..

SPIKE.. I HAVE TO TAKE THE BOWL BACK..
SCHULZ

PEANUTS
by SCHULZ
OH, NO!

© 1995 United Feature Syndicate, Inc.

AH-CHOO!

BONK!
9-24

SORRY, MANAGER.. I WAS UNDER THAT ONE, AND THEN I SNEEZED..

"BLESS YOU"

WHEN THE CATCHER COMES OUT TO THE MOUND FOR A CONFERENCE, IT'S USUALLY A DRAMATIC MOMENT..

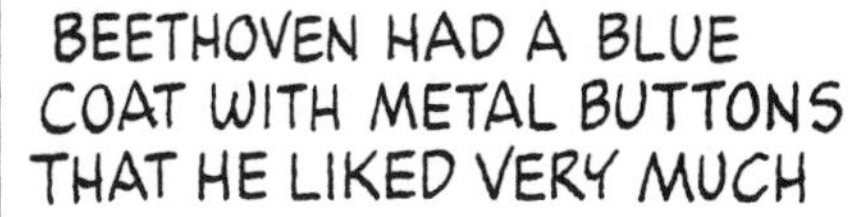
BEETHOVEN HAD A BLUE COAT WITH METAL BUTTONS THAT HE LIKED VERY MUCH

9-25
BUT NOT ALWAYS
SCHULZ

DID YOU KNOW THERE'S A STOREFRONT CHURCH ABOUT FOUR BLOCKS FROM HERE?

9-26
I JUST CAME FROM THERE..

I BOUGHT TWO POUNDS OF FAITH, HOPE AND CHARITY..
SCHULZ

© 1995 United Feature Syndicate, Inc.
9-27
TELL THE SCHOOL BUS TO WAIT!
WE HAVE TO SEE IF HE MAKES THIS SMALL SLAM..
SCHULZ

"AND THEN THIS ONE TEAM BEAT THE OTHER TEAM IN SOME KIND OF GAME.."

"AND THIS TALL GIRL WON OVER THIS SHORT GIRL, AND THE TEAM FROM AROUND HERE SOMEPLACE LOST AGAIN.."
9/28

"AND THIS ONE GUY RAN FASTER THAN THE REST OF THEM, AND THAT'S SPORTS FOR TONIGHT, AND WHO CARES?"

SO! WHAT DO YOU HEAR FROM BEETHOVEN LATELY?

KLUNK!

I NEED A BETTER CONVERSATION OPENER..
9-29
SCHULZ

WHAT WOULD BE YOUR REACTION IF I TOLD YOU SUPPER MIGHT BE ELEVEN SECONDS LATE TONIGHT?

9-30

ALL RIGHT, I'LL GO BACK IN THE KITCHEN, AND SEE WHAT I CAN DO..

HAS HE NO IDEA HOW MUCH CAN GO WRONG IN ELEVEN SECONDS?
SCHULZ

PEANUTS
by SCHULZ
PAWPET SHOW TODAY
"I GIVE IT TEN STARS," SAYS LARRY KING
PAWPET SHOW TODAY "THE FLYING MAGIC CARPET"
CUTE
BUT WHAT ABOUT THE FLYING MAGIC CARPET?
WHERE IS IT?
I CAME TO SEE A FLYING MAGIC CARPET..
© 1995 United Feature Syndicate, Inc.
"TWENTY PERCENT OFF ON ALL CARPETS DURING MONTH OF OCTOBER"
10-1
SCHULZ

YOU'RE GOING TO THE MOON?
10-2

BUT YOU HAVE TO WEAR ONE OF THOSE THINGS ON YOUR HEAD..

© 1995 United Feature Syndicate, Inc.

WELL, YES... SORT OF LIKE THAT..
SCHULZ

SO YOU'RE REALLY GOING TO THE MOON..

I'M SURPRISED

I HOPE YOU REALIZE THAT WHEN YOU GET THERE, YOU'RE GOING TO BE ALL ALONE..
10-3
© 1995 United Feature Syndicate, Inc.

SCHULZ

WHY DO I HAVE A FEELING THAT SOMEONE IS WATCHING ME?
© 1995 United Feature Syndicate, Inc.
10-4 SCHULZ

IS THIS THE MOON?
10-5

SOMETIMES I WONDER..

HOW DID I GET HERE?
© 1995 United Feature Syndicate, Inc.

WELL, I LIKE TO THINK I CAME UP THE HARD WAY..
SCHULZ

YES, WE'RE STANDING ON THE MOON, AND THAT'S THE EARTH WE'RE LOOKING AT..
10-6

BEFORE YOU GO BACK THERE'S SOMETHING OVER HERE YOU SHOULDN'T MISS..
© 1995 United Feature Syndicate, Inc.

MOON ROCKS FOR SALE
SCHULZ

YOU'RE BACK FROM THE MOON? WOW! WHAT AN ADVENTURE! TELL ME WHAT YOU SAW

A WEIRD CREATURE? LET'S SEE THE PICTURE..

THAT'S MY BROTHER SPIKE! MY FAVORITE BROTHER! GOOD OL' WONDERFUL SPIKE
10-7
© 1995 United Feature Syndicate, Inc.
SCHULZ

HE SOLD YOU A MOON ROCK?

PEANUTS
by SCHULZ

WHAT ARE YOU WATCHING?
"CITIZEN KANE"... I'VE NEVER SEEN IT..

"ROSEBUD" WAS HIS..
10-8
© 1995 United Feature Syndicate, Inc.

DON'T TELL HIM! WHAT ARE YOU TRYING TO DO, RUIN THE WHOLE MOVIE FOR HIM ?!

ARE YOU OUT OF YOUR MIND? WHAT'S WRONG WITH YOU?

BOY, THAT WAS CLOSE! I SAVED YOUR LIFE!

WHAT'S THIS? I THOUGHT YOU WERE WATCHING "CITIZEN KANE".
PLEDGE BREAK
SCHULZ

PEOPLE ALWAYS WONDER WHAT THE CATCHER SAYS TO THE PITCHER WHEN HE GOES OUT TO THE MOUND..

ANOTHER SEASON DOWN THE DRAIN..
10-9

© 1995 United Feature Syndicate, Inc.

IT'S NOT WORTH WONDERING ABOUT..
SCHULZ

"AND THE KING LOVED THE PEOPLE, AND THE PEOPLE KIND OF LOVED THE KING.."

"AND THEY ALL PRETTY MUCH LOVED ONE ANOTHER AND EVERY-ONE WAS SORT OF HAPPY.."
10-10
© 1995 United Feature Syndicate, Inc.

JUST TIPTOEING AROUND THE TRUTH, MA'AM..
SCHULZ

WHAT'S THAT?
THIS IS THE CHANGE THE CLERK GAVE ME

GRAMPA SAYS IN THE OLD DAYS THEY USED TO COUNT IT OUT FOR YOU..
COUNT IT OUT? WHY WOULD THEY DO THAT?
10-11

TO BE POLITE
© 1995 United Feature Syndicate, Inc.

POLITE?
SCHULZ

HEY, MARCIE, DO YOU HAVE A ROUGH OR EVEN VAGUE IDEA WHAT OUR HOMEWORK IS FOR TOMORROW?

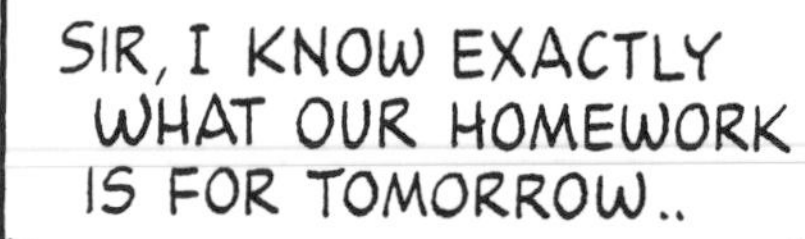

SIR, I KNOW EXACTLY WHAT OUR HOMEWORK IS FOR TOMORROW..
10-12

MARCIE, DO YOU HAVE A ROUGH OR EVEN VAGUE IDEA HOW ANNOYING YOU ARE?
SCHULZ

BONK!
10-13

HOW ABOUT THAT, SIR? WHEN I KICKED THE BALL, MY GLASSES, MY SOCK, AND MY SHOE ALL FLEW OFF, AND THE BALL CAME DOWN, AND HIT ME ON THE HEAD!

THIS IS A HUMOROUS GAME, ISN'T IT, SIR?
ONLY THE WAY YOU PLAY IT, MARCIE
SCHULZ

I STILL THINK WE SHOULD HAVE PARACHUTES
10-14

HERE..TAKE A HANDKERCHIEF, TIE A STRING AT EACH CORNER, AND YOU HAVE A PARACHUTE...TRY IT..

KLUNK

PROBABLY SHOULD HAVE GONE TO JUMP SCHOOL

SCHULZ

PEANUTS
!

WOW!

WHAT'S THIS?
A TOOTH.. I LOST ANOTHER TOOTH..

THAT'S GREAT! YOU CAN PUT IT UNDER YOUR PILLOW TONIGHT, AND THE TOOTH FAIRY WILL BRING YOU A WHOLE BUNCH OF MONEY!

WHY WOULD SHE DO THAT?
THEY NEED TEETH.. THEY USE THEM TO MAKE PIANO KEYS..

PIANO KEYS?
10-15
© 1995 United Feature Syndicate, Inc.

I'M FIVE YEARS OLD.. I'VE BEEN AROUND FOR FIVE YEARS, AND I'VE NEVER BELIEVED ANYTHING ANYONE HAS EVER TOLD ME..

IF I COULD TALK, KID, I'D TELL YOU ABOUT THE "TOOTH BEAGLE"!
SCHULZ

ALL RIGHT, TROOPS, BEFORE WE START, ARE THERE ANY QUESTIONS?

?
WHY DON'T WE HAVE HORSES?
10-16

BECAUSE WE DON'T, THAT'S WHY!
© 1995 United Feature Syndicate, Inc.

OKAY, I'LL ADMIT IT.. IF WE HAD HORSES, WE COULD GALLOP..
SCHULZ

NO, MY DOG ISN'T HOME.. HE WENT ON A HIKE WITH HIS TROOPS..

10-17
© 1995 United Feature Syndicate, Inc.

DO DOGS HAVE TROOPS?
SCHULZ

WHAT A VIEW!

IF YOU MEN HAD THE EYES OF AN EAGLE, YOU COULD SEE THE OCEAN..
© 1995 United Feature Syndicate, Inc.

YES, IF YOU HAD TAMBOURINES, YOU COULD DANCE A JIG..
SCHULZ
10-18

NO, CONRAD, I DON'T THINK YOU SHOULD HAVE YOUR BEAK PIERCED..
© 1995 United Feature Syndicate, Inc.
10-19
SCHULZ

ASK YOUR DOG TO COME OUT AND PLAY "CHASE THE STICK"
HE'S NOT BACK YET FROM HIS HIKE..

10-20

© 1995 United Feature Syndicate, Inc.

TELL HIM TO HURRY BACK..
SCHULZ

KEEP GOING, TROOPS... WE'RE ALMOST HOME..

I KNOW YOUR FEET HURT.. MINE DO, TOO..
© 1995 United Feature Syndicate, Inc.

YES, I KNOW MICKEY MOUSE HAS SHOES..
10-21
SCHULZ

PEANUTS
by SCHULZ
NOT FOR SALE
RING
I WAS WONDERING IF YOU'D BE WILLING TO SELL YOUR DOG..
HOW MUCH MONEY DO YOU HAVE?
HERE, YOU TELL ME..I DON'T KNOW ANYTHING ABOUT MONEY..
WELL, YOU HAVE A DIME, A NICKEL AND A PENNY..
10-22
YOU HAVE SIXTEEN CENTS
IS THAT ENOUGH?
NO, YOU'D NEED TEN MILLION DOLLARS!
© 1995 United Feature Syndicate, Inc.
HOW WOULD IT BE IF I GAVE YOU SIXTEEN CENTS NOW, AND THE REST LATER?
SCHULZ

SOMETIMES I LIE AWAKE AT NIGHT, AND I ASK, CAN MY GENERATION LOOK TO THE FUTURE WITH HOPE?
10-23

THEN, OUT OF THE DARK, A VOICE COMES TO ME THAT SAYS, "WHY, SURE..WELL, I MEAN...THAT IS...IT SORT OF DEPENDS.. I MEAN..IF...WHEN..WHO..WE..AND..."
SCHULZ

WHEN ARE YOU GOING TO GIVE ME ANOTHER MUSIC BOX?

GIRLS LOVE MUSIC BOXES..

LOOK AT ME..I'M A MUSIC BOX! LA DUM TE DA DUM TE DA TE DUM

CLOSE YOUR LID!
10-24
SCHULZ

10-25
I THINK I'LL ASK THE DRIVER TO DROP ME OFF AT THE MALL..
I WONDER IF HE'D STOP SOMEPLACE WHERE WE COULD GET SOME DOUGHNUTS..
I WONDER IF HE'D DROP ME OFF AT THE COURTHOUSE..
SCHULZ

RERUN, YOU'VE GOT YOUR SHOES ON THE WRONG FEET..

I'M TEACHING THEM A LESSON..

I'M TRYING TO SHOW THEM THEY CAN'T ALWAYS HAVE THEIR OWN WAY..
SCHULZ
10-26

Dear Author,
Congratulations! We have decided to publish your novel.

First printing will be one copy.
10-27

If we sell it, we'll print another.
SCHULZ

IT'S FROM YOUR PUBLISHER.. THEY PRINTED ONE COPY OF YOUR NOVEL..
10-28

IT SAYS THEY HAVEN'T BEEN ABLE TO SELL IT..

THEY SAY THEY'RE SORRY.. YOUR BOOK IS NOW OUT OF PRINT..
SCHULZ

PEANUTS
by SCHULZ
CHARLIE BROWNNN..

AGAIN?

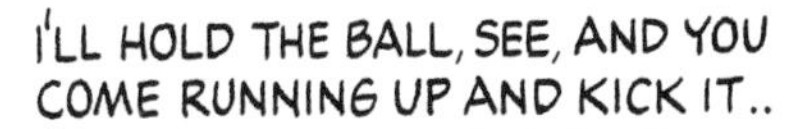
I'LL HOLD THE BALL, SEE, AND YOU COME RUNNING UP AND KICK IT..

SURE, AND WHAT HAPPENS IF YOU PULL THE BALL AWAY?

I FLY THROUGH THE AIR, AND LAND ON MY BACK, AND KILL MYSELF..THAT'S WHAT HAPPENS!

YOU COULD ALWAYS SUE..

SHE'S RIGHT.. IF SHE PULLS THE BALL AWAY, I'LL SUE!
10-29

SCHULZ

10-30
THAT SIGN SAID GOLFERS DRIVING A CART SHOULD OBEY THE 90° RULE..
I FAILED GEOMETRY
© 1995 United Feature Syndicate, Inc.
SCHULZ

I CAN'T HELP YOU WITH YOUR HOMEWORK TONIGHT.. I WANT TO WATCH THIS PROGRAM...

THEN I SHALL SEEK ELSEWHERE FOR A SPIRITUAL CLOAK TO SHIELD ME FROM MY TEACHER'S ICY WRATH..
10-31
© 1995 United Feature Syndicate, Inc.

HERE.. WE'RE SUPPOSED TO DO ALL OF THESE PROBLEMS ON PAGE NINE..
SCHULZ

SO WHAT ARE YOU DOING, MARCIE?
11-1

JUST LYING IN BED READING

DO YOU WEAR YOUR GLASSES WHEN YOU'RE SLEEPING?

SURE, SO I CAN SEE WHAT I'M DREAMING
© 1995 United Feature Syndicate, Inc.

THERE'S NO DOUBT IN MY MIND, MARCIE, THAT YOU ARE EXCESSIVELY WEIRD..
SCHULZ

YES, MA'AM..THAT'S WHAT THEY SAID..

" BRIGHT AND CLEAR.."
11-2

THE WEATHER... NOT ME..
SCHULZ

DO ME A FAVOR..

TELL THE BUS DRIVER TO GO SLOW BECAUSE I HAVE TO FINISH READING THIS BOOK BEFORE WE GET THERE

11-3

YESTERDAY HE WENT AROUND A CORNER, AND I FELL INTO THE AISLE..
HAVE YOU TRIED READING AT HOME?
MAYBE I CAN GET A WINDOW SEAT..
SCHULZ

I CAN DO IT! I'VE MASTERED THE TECHNIQUE!
11-4

NOTHING CAN STOP ME FROM PUTTING THIS BALL THROUGH THE HOOP!

NOTHING!

SCHULZ

PEANUTS
by
SCHULZ
SIGH

PSYCHIATRIC
HELP 5¢
THE DOCTOR
IS IN
SO YOU SAY YOU'VE BEEN DEPRESSED..
11-5

YES, WE HAD A BAD SEASON..

IS BASEBALL THE ONLY THING YOU EVER THINK ABOUT?
ACTUALLY, WE DIDN'T HAVE A REAL BAD TEAM..IT WAS OUR RIGHT FIELDER WHO WAS SO BAD!

WHAT WAS WRONG WITH HER?
SHE'S THE WORST PLAYER IN THE HISTORY OF THE GAME!
THE DOCTOR

BUT SHE WAS CUTE, WASN'T SHE?
SHE COULDN'T CATCH A BALL IF IT WAS HANDED TO HER!

WELL, MAYBE IF YOU'D GET THE BALL OVER THE PLATE NOW AND THEN, WE'D WIN A GAME, HUH?

FIVE CENTS, PLEASE..
SCHULZ

PSYCHIATRY HAS CHANGED A LOT LATELY..

I HEARD THE COYOTES HOWLING AGAIN LAST NIGHT, CHARLIE BROWN..

THERE ARE NO COYOTES AROUND HERE, LINUS..

IF THEY TRY TO GET INTO OUR HOUSE, WHERE SHOULD I HIDE?
11-6

COYOTES CAN'T READ SO I HIDE IN THE BOOKCASE..
SCHULZ

SEE, IF YOU STAY OVERNIGHT WITH ME, CHARLIE BROWN, MAYBE YOU'LL HEAR THE COYOTES HOWLING..
THERE ARE NO COYOTES AROUND HERE, LINUS..
11-7
© 1995 United Feature Syndicate, Inc.

IF THE PHONE RINGS, IT'LL BE FOR ME... I TOLD GENERAL PERSHING I'D BE HERE..
SCHULZ

SOMETIMES I'LL BE LYING IN BED LIKE THIS..SORT OF HALF ASLEEP..
11-8

THEN, SUDDENLY, I HEAR COYOTES HOWLING.. THEY SOUND SAD AND LONELY... THEN I GET DEPRESSED..

© 1995 United Feature Syndicate, Inc.

I THOUGHT I HEARD COYOTES HOWLING ONCE, BUT IT WAS A DOUGHNUT CALLING ME..
SCHULZ

11-9

HERE THEY COME! DIVING OUT OF THE SUN!! BULLETS WHINING ALL AROUND!! ENEMY PLANES ABOVE, BELOW, AND ON ALL SIDES!!!
SCHULZ

WAKE UP! A SURPRISE ATTACK BY THE ENEMY! ALL NURSES REPORT TO THEIR UNITS!

HERE'S THE WORLD WAR I FLYING ACE RACING BACK TO THE AERODROME..

11-10

WHY IS YOUR STUPID DOG RUNNING THROUGH THE HOUSE?! CAN'T ANYONE SLEEP AROUND HERE?!!

I WONDER IF HE HEARD THE COYOTES HOWLING..

SCHULZ

YOU'RE LEAVING, CHARLIE BROWN? BUT YOU DIDN'T HEAR THE COYOTES HOWL..
11-11

TELL HIM TO TAKE HIS STUPID DOG HOME, AND DON'T COME BACK!

I JUST HEARD ONE..
SCHULZ

PEANUTS
by SCHULZ

GET READY, MARCIE..IT'S THE KICKOFF!
I HATE THIS GAME..

© 1995 United Feature Syndicate, Inc.
11-12

WHAT HAPPENED? WHERE AM I?
YOU'RE STILL ALIVE..

I HAVE TO FIND A NEWSPAPER..
A WHAT?

THIS IS A ROUGH GAME..

I DECIDED I SHOULD PROBABLY WEAR A HELMET..
SCHULZ

11-13
I DIDN'T KNOW THAT TROOPS FROM THE FOREIGN LEGION RODE THE SCHOOL BUS..
OUR CAMEL BROKE DOWN
SCHULZ
© 1995 United Feature Syndicate, Inc.

SCHULZ 11-14
WHY ARE TROOPS FROM THE FOREIGN LEGION GOING INTO OUR SCHOOL, CHARLIE BROWN?
ALL RIGHT, MEN.. ACCORDING TO MY MAP THIS SHOULD BE FORT ZINDERNEUF..
© 1995 United Feature Syndicate, Inc.

11-15
YES, MA'AM, HE'S MY DOG.. WELL, SORT OF.. NO, I DON'T KNOW WHAT HE'S DOING HERE..
ASK HER HOW LONG I HAVE TO KEEP THESE TROOPS STANDING AT ATTENTION..
© 1995 United Feature Syndicate, Inc.
SCHULZ

YES, MA'AM.. I'LL TAKE THEM DOWN..

OUR TEACHER WANTS ME TO TAKE YOU TO THE PRINCIPAL..
I NOTICED SHE HAD DIFFICULTY PRONOUNCING "LÉGION ÉTRANGÈRE"... PROBABLY KNOWS VERY LITTLE FRENCH..
© 1995 United Feature Syndicate, Inc.
SCHULZ 11-16

© 1995 United Feature Syndicate, Inc.
11-17
YES, SIR.. WELL, YOU MAY NOT KNOW IT, BUT FORT ZINDERNEUF IS UNDER ATTACK..
WE ARE VERY GENEROUS IN PEACE..IF YOU SURRENDER IMMEDIATELY, YOU GET A FREE BALLOON..
SCHULZ

© 1995 United Feature Syndicate, Inc. 11-18
I'M SURPRISED TO SEE YOU AND YOUR TROOPS AT HOME..
WE DIDN'T REALIZE THAT FORT ZINDERNEUF IS CLOSED ON SATURDAY..
SCHULZ

PEANUTS
by SCHULZ
I THOUGHT I HEARD SOMEONE SINGING "BRINGING IN THE SHEAVES"
RING!
HI! I'M FROM THAT NEW STOREFRONT CHURCH A FEW BLOCKS UP THE STREET
WOULD YOU LIKE TO COME TO OUR CHURCH?
I ALREADY HAVE A CHURCH..
WE HAVE GOOD SINGING
11-19
SO DO WE..DO YOU SING "BLESSED ASSURANCE"?
DO YOU SING "SWEET HOUR OF PRAYER"?
WE GIVE OUR MONEY IN ENVELOPES..
WHO CARES?
© 1995 United Feature Syndicate, Inc.
MY DAD IS THE PREACHER
GOOD FOR HIM
HE SAYS THE WORLD IS COMING TO AN END..
I FEEL THAT WAY EVERY TIME I GET ON THE SCHOOL BUS
ANYWAY, YOU KNOW WHERE WE ARE..
I'M ALWAYS HERE
THOSE DEEP THEOLOGICAL DISCUSSIONS WEAR ME OUT..
SCHULZ

ALL RIGHT, HERE'S THE WAY IT'S GONNA WORK..

I TOSS THE BALL INTO THE AIR, AND IT GOES THROUGH THE HOOP..
11-20

I WAS TALKING TO **YOU**!
SCHULZ

SO WE'RE IN THIS COFFEE SHOP, SEE, TRYING TO DECIDE ABOUT DESSERT..

"HOW ABOUT ICE CREAM?" SAYS MY DAD.. "GREAT," I SAID.. "I'LL HAVE ZAMBONI"

THEN MY DAD SAYS, "AT THE HOCKEY GAME TONIGHT, DID YOU ENJOY WATCHING THE SPUMONI CLEAN THE ICE?"

HA HA HA HA!
YOU AND YOUR DAD ARE VERY WEIRD, MARCIE..
SCHULZ
11/21

SO WHO ASKED YOU?
11-22

WHO ASKED ME WHAT?

WHO ASKED YOU WHAT YOU WOULD HAVE SAID IF SOMEONE HAD ASKED YOU!

AND WHO ASKED YOU?
DOGS CAN'T TALK
SCHULZ

11-23
SCHULZ
© 1995 United Feature Syndicate, Inc.

OKAY, MARCIE, YOU GO STRAIGHT OUT, THEN CUT LEFT, THEN GO STRAIGHT, AND I'LL THROW YOU A PASS..
HOW CAN I GO STRAIGHT OUT, CUT LEFT, AND GO STRAIGHT WHEN MY GLASSES ARE ALL WET AND I CAN'T SEE A THING?
I THINK IT'S GOING TO BE A TRICK PLAY..
ANYONE TRIES TO GET BY ME GETS HIT WITH THE UMBRELLA!
SCHULZ
11-24
© 1995 United Feature Syndicate, Inc.

HEY! WHERE'D EVERYBODY GO? IT'S NOT RAINING! IT'S ONLY SPRINKLING! WHAT'S A LITTLE MUD?!
WE CAN'T QUIT JUST BECAUSE OF A LITTLE MUD!
HOW COME YOU'RE STILL HERE?
WE FORGOT TO SING THE NATIONAL ANTHEM..
11-25
SCHULZ
© 1995 United Feature Syndicate, Inc.

PEANUTS®
by SCHULZ
SORT OF
MAYBE
COULD BE
WHO KNOWS?
WHY ?

?

TRUE OR FALSE.. "IF THE TRUTH BE KNOWN, IT'S FALSE"

"HE WHO WOULD DISTINGUISH THE TRUE FROM THE FALSE MUST HAVE AN ADEQUATE IDEA OF WHAT IS TRUE AND FALSE"

"RING OUT THE OLD, RING IN THE NEW.. RING OUT THE FALSE, RING IN THE TRUE"

"ALL WAS FALSE AND HOLLOW!"

"TRUE AS THE STARS ABOVE!" "LIVE PURE, SPEAK TRUE!"

"BE SO TRUE TO THYSELF, AS THOU BE NOT FALSE TO OTHERS"

"THAT HE IS MAD, 'TIS TRUE!"

THERE YOU ARE, MA'AM.. FINISHED WITH A FLOURISH!

THIS WAS A MULTIPLE CHOICE QUIZ, SIR..

FAILED WITH A FLOURISH!
SCHULZ
11-26

LOOK AT ALL THE PEOPLE I HAVE TO SEND CHRISTMAS CARDS TO!
IF IT BOTHERS YOU, WHY DO YOU DO IT?

BECAUSE IF I DON'T, THEY'LL HATE ME..

LOOK AT LORETTA.. IF I DON'T SEND HER A CARD, SHE'LL HATE ME..
NO, SHE WON'T

SHE WON'T?
11/27

SO LONG, LORETTA!
SCHULZ

IF I'M GOING TO SEND YOU A CHRISTMAS CARD, I NEED TO KNOW YOUR ADDRESS
I LIVE HERE

I'LL PUT DOWN, "HE LIVES THERE"

11-28
NO, HERE
SCHULZ

PSST! BIG BROTHER! ARE YOU AWAKE?
11-29

I AM NOW
I NEED FIVE DOLLARS

FIVE DOLLARS?!
FOR STAMPS.. FOR MY CHRISTMAS CARDS..

WHERE WOULD I GET FIVE DOLLARS?
I'LL COME BACK IN THE MORNING WHEN YOU'RE NOT SO CRABBY..
SCHULZ

THERE! I'VE MAILED ALL MY CHRISTMAS CARDS!

I THOUGHT YOU DIDN'T HAVE ANY STAMPS..
11-30
© 1995 United Feature Syndicate, Inc.

I DREW MY OWN.. I DREW DICK TRACY, AND POPEYE, AND THE YELLOW KID..
THAT'S ILLEGAL
SOMEONE NEED AN ATTORNEY?

ALL MY CHRISTMAS CARDS CAME BACK!
12-1

THAT'S BECAUSE YOU DREW YOUR OWN STAMPS
© 1995 United Feature Syndicate, Inc.

I COPIED THEM FROM THE NEW CARTOON STAMPS
SHE DREW A BETTER POPEYE THAN THEY DID..

12-2
© 1995 United Feature Syndicate, Inc.
DO YOU THINK I'LL GET ARRESTED FOR DRAWING MY OWN STAMPS?
WELL, YOU MIGHT START LOOKING FOR A GOOD ATTORNEY..
BEFORE WE BEGIN, YOUR HONOR, MAY WE ASK IF YOU RECEIVED THE CHRISTMAS CARD WE SENT YOU?

PEANUTS
by SCHULZ
?
© 1995 United Feature Syndicate, Inc.
SCHULZ
12-3

HOW CAN BIRDS FLY WHEN IT'S SO COLD?

I THINK MY EARS WOULD FREEZE..

12-4
SCHULZ

Z
12-5

SORRY, MA'AM.. I DIDN'T HEAR THE QUESTION...

MY MIND WAS A THOUSAND MILES FROM HERE..

OR LIKE MAYBE ON THE MOON!
PUNISH HER SEVERELY, MA'AM..
SCHULZ

"Dumb"

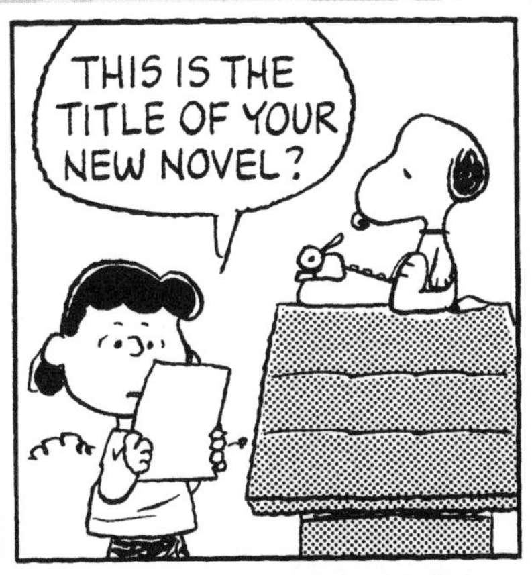
THIS IS THE TITLE OF YOUR NEW NOVEL?

12-6
I THINK YOU CAN DO BETTER THAN THAT

"Beyond Dumb"
SCHULZ

IT'S LUNCH TIME, MA'AM! WE'VE ACCOMPLISHED A LOT THIS MORNING..
12-7

YOU'VE TAUGHT US WELL.. WE'RE ALL BETTER FOR HAVING BEEN IN YOUR PRESENCE..YOU'VE BROUGHT HONOR TO YOUR PROFESSION

THAT'S NOT OUR TEACHER, SIR.. THAT'S THE CUSTODIAN..

SWEEP THOSE FLOORS, JOE..
SCHULZ

12-8

SOMEHOW I HAVE THE FEELING I'VE MISSED THE SCHOOL BUS..
Z
SCHULZ

"Secrets of Life"
Always look ahead.
12-9

Also, always look back over your shoulder.

Make sure you can still see your supper dish.
SCHULZ

PEANUTS
by SCHULZ
© 1995 United Feature Syndicate, Inc.
12-10
DO YOU LIKE WATCHING BASKETBALL GAMES?
I DON'T KNOW..I'VE NEVER SEEN ONE..
SCHULZ

I DIDN'T UNDERSTAND OUR HOMEWORK
SHE SAID TO WRITE A SHORT ESSAY ON THE MOST IMPORTANT WORDS EVER SPOKEN..
..IN THE HISTORY OF THE WORLD..
" DON'T FORGET TO FEED THE DOG "
© 1995 United Feature Syndicate, Inc.
SCHULZ
12-11

YES, MA'AM..I SORT OF NEED YOUR ADVICE..

DO YOU THINK I SHOULD SPEND THE ONLY DOLLAR I HAVE ON A CHRISTMAS PRESENT FOR A GIRL WHO DOESN'T KNOW I EVEN EXIST?
12-12

THANK YOU..
© 1995 United Feature Syndicate, Inc.

I JUST SAVED A DOLLAR..
SCHULZ

IF YOU WORK REAL HARD, AND YOU GET EVERYTHING YOU'VE ALWAYS WANTED, IS IT WORTH IT?
NOT IF YOUR DOG DOESN'T LIKE YOU..
© 1995 United Feature Syndicate, Inc.
12-13
SCHULZ

WHY DON'T YOU GET RID OF THAT BELL SO YOU CAN HEAR WHAT I WANT FOR CHRISTMAS?

12-14
BONK!

ASK YOUR DOG IF HE WANTS TO COME OUT AND PLAY IN THE SNOW..

12-15
FIVE MINUTES..THEN I GO BACK IN BY THE FIRE..
SCHULZ

12-16
WOW!
HAPPY BEETHOVEN'S BIRTHDAY!
?
BIRD MUSIC? NO, BEETHOVEN NEVER WROTE ANY BIRD MUSIC..
SCHULZ
© 1995 United Feature Syndicate, Inc.

PEANUTS
by SCHULZ

IF YOU'RE A REAL SANTA CLAUS, WHERE ARE YOUR REINDEER?

HOW ARE YOU GONNA LAND ON ALL OF THOSE ROOFTOPS AND GO DOWN ALL THOSE CHIMNEYS?

AND AFTER YOU GO DOWN A CHIMNEY, HOW ARE YOU GONNA GET BACK UP?
12-17

AND EVEN IF YOU DO, WHAT MAKES YOU THINK YOUR REINDEER WILL BE WAITING FOR YOU?
© 1995 United Feature Syndicate, Inc.

I'LL GIVE YOU ABOUT THREE HOUSES, AND YOU'LL BE COMPLETELY EXHAUSTED..

I THOUGHT YOU WERE PROBABLY DOWN HERE, BUT THEN I DIDN'T HEAR YOUR BELL ANYMORE..
SCHULZ

IS THERE ANYTHING SPECIAL YOU THINK ABOUT BEFORE EACH SHOT?
JUST BEFORE I SWING, I THINK TO MYSELF, "YOU HAVEN'T HIT A GOOD SHOT ALL DAY..WHAT MAKES YOU THINK YOU'RE GOING TO HIT ONE NOW?"
12-18
SCHULZ
© 1995 United Feature Syndicate, Inc.

RATS! ANOTHER D-MINUS!

IF YOU EVER OWN A YACHT, OR A SAILBOAT, OR A RACEHORSE, SIR, YOU CAN CALL IT "D-MINUS"..

HA HA HA HA!
© 1995 United Feature Syndicate, Inc.

YOU ARE DISPROPORTIONATELY WEIRD, MARCIE..
SCHULZ
12-19

WHO'S THAT MAN, FRANKLIN?
THAT'S OUR PRINCIPAL..

HE LOOKS KIND OF OLD..
THEY SAY HE'S BEEN HERE A LONG TIME..
12/20
© 1995 United Feature Syndicate, Inc.

ASK HIM IF HE EVER KNEW ANY PILGRIMS..
SCHULZ

THIS IS THE TIME OF YEAR WHEN I MISS THE DAISY HILL PUPPY FARM..

WE ALWAYS HAD A CHRISTMAS TREE..
© 1995 United Feature Syndicate, Inc.

12-21
IT'S HARD TO DECORATE A ROCK..
SCHULZ

© 1995 United Feature Syndicate, Inc.
12-22
AND THEN, RIGHT AFTER SCHOOL TODAY, OUR CLASS IS GOING TO ROAST WIENERS AND MARSHMALLOWS..
I KNOW, MARCIE.. I KNOW..
SCHULZ

YOU'RE WRITING A LETTER TO SANTA CLAUS? FORGET IT, RERUN!

HE DOESN'T HAVE TIME TO READ ALL THOSE LETTERS.. I MEAN, HE STANDS ON THE CORNER ALL DAY RINGING THAT BELL..WHEN HE GETS HOME, HE'S TIRED..HE DOESN'T WANT TO READ A BUNCH OF WHINY LETTERS!
© 1995 United Feature Syndicate, Inc.

12-23
TELL HIM LUCY SAYS, "HI!"
SCHULZ

PEANUTS®
by
SCHULZ

NO, JUST BE CAREFUL WHEN YOU DRIVE OVER MY EXTENSION CORDS..MERRY CHRISTMAS!
SCHULZ 12-24
© 1995 United Feature Syndicate, Inc.

12-25
DID SCROOGE EVER HAVE A DOG? YES, HE HAD ONE WHEN HE WAS YOUNGER, BUT THE DOG LEFT HIM..
© 1995 United Feature Syndicate, Inc.
SCHULZ

Dear Gramma,

WHICH GRAMMA GAVE ME THE BOOK FOR CHRISTMAS..THE FAT ONE OR THE SKINNY ONE?
12-26

THE SKINNY ONE

GRAMMAS SHOULD HAVE NAMES LIKE PEOPLE..
© 1995 United Feature Syndicate, Inc.
SCHULZ

WHAT I DON'T UNDERSTAND IS HOW YOU CAN FLY AROUND UP THERE WITHOUT BUMPING INTO ANOTHER BIRD..

12-27
© 1995 United Feature Syndicate, Inc.

NO, I REALIZE YOU'RE NOT STUPID..
SCHULZ

LISTEN TO THOSE GEESE FLYING OVER.."HONK, HONK, HONK, HONK"

THEY KEEP SAYING THE SAME THING OVER AND OVER.. "HONK, HONK, HONK, HONK.."
12-28
© 1995 United Feature Syndicate, Inc.

NO, I DON'T EXPECT THE GETTYSBURG ADDRESS..
SCHULZ

12-29
DO I HEAR SOMEBODY TALKING?
I THINK THE WORM AND THE FISH USED TO KNOW EACH OTHER..
SCHULZ
© 1995 United Feature Syndicate, Inc.

12-30
NEW YEAR'S? DON'T ASK ME..

DOGS CAN'T TELL ONE YEAR FROM ANOTHER..

I KNOW I HAVE MY SUPPER IN THE RED DISH AND MY DRINKING WATER IN THE YELLOW DISH..
© 1995 United Feature Syndicate, Inc.

BUT I DON'T KNOW WHAT YEAR THIS IS..
SCHULZ

PEANUTS
by SCHULZ

WHAT'S THIS?

TICKETS TO A HOCKEY GAME? I LOVE HOCKEY GAMES!

ROW "A" AND ROW "B"? I'LL TAKE ROW "A"..

12-31

RATS! I SHOULD HAVE TAKEN ROW "B"
SCHULZ